LACTIC ACID BACTERIA: GENETICS, METABOLISM AND APPLICATIONS

Lactic Acid Bacteria: Genetics, Metabolism and Applications

Proceedings of the Sixth Symposium on lactic acid bacteria: genetics, metabolism and applications, 19–23 September 1999, Veldhoven, The Netherlands

Edited by

W.N. KONINGS
University of Groningen,
The Netherlands

O.P. KUIPERS
University of Groningen,
The Netherlands

and

J.H.J. HUIS IN 'T VELD
Utrecht University,
The Netherlands

Reprinted from *Antonie van Leeuwenhoek*, Volume 76, Nos. 1–4, 1999

Springer-Science+Business Media, B.V.

A C.I.P. Catalogue record for this book is available from the Library of Congress.

ISBN 978-90-481-5312-1 ISBN 978-94-017-2027-4 (eBook)
DOI 10.1007/978-94-017-2027-4

Printed on acid-free paper

Table of Contents

SHORT LECTURES

Antonie van Leeuwenhoek **76:** 1, 1999.

Editorial

The Sixth Symposium on Lactic Acid Bacteria

In 1981 the initiative was taken to organize a symposium on Lactic Acid Bacteria. This symposium would offer the international scientific community involved in research on Lactic Acid Bacteria the opportunity to present progress reports to the experts in the field and to offer a platform for scientific interactions. The first meeting in Wageningen in 1983 was a great success. Now in hindsight we may conclude that in 1983 the research on LAB was still in a lag-phase and has developed since exponentially. Right now in 1999 we might actually be in mid-log phase, judging from the excellent and exciting contributions presented in this book, by the lecturers of the 6th LAB symposium, from the progress reports described in the poster abstracts and from the large number of participants. Many important aspects of LABs have been analyzed especially from a number of *Lactococcus and Lactobacillus* strains. It will take several more years to study other LABs in similar detail.

In this book the first complete genome of a Lactic Acid Bacterium is presented. The sequencing of the genomes of several other LABs has started already or will begin soon. These genome sequences will open and stimulate new research area on LAB. Functional genomics of the unknown genes will be one research line, a detailed analysis of the metabolic pathways another obvious offspring of this achievement. The relative simplicity of LABs and the enormous amount of information already available will be very helpful in the construction of 'in silico' cells.

But this will be by no means the only challenge for future research. Among others the role of LABs in probiotics and in food manufacturing will remain important topics for future research. It is a pleasure to work in such a dynamic field of microbiology. The excitement that this research brings to the scientists involved will be transparent for those who read the different chapters of this book.

W.N. Konings
University of Groningen, the Netherlands

J.H.J. Huis in 't Veld
Utrecht University, the Netherlands

O.P. Kuipers
University of Groningen, the Netherlands

Antonie van Leeuwenhoek **76:** 3–23, 1999.

From DNA sequence to application: possibilities and complications

Gerard Venema[1,*], Jan Kok[1] & Douwe van Sinderen[2]
[1]*Department of Genetics, Groningen Biomolecular Sciences and Biotechnology Institute, University of Groningen, Kerklaan 30, 9751 NN, Haren, The Netherlands;* [2]*Department of Microbiology and the National Food Biotechnology Centre, University College Cork, Ireland. (*Author for correspondence)*

Key words: lactic acid bacteria, bacteriophage, inducible gene expression, regelatory DNA elements

Abstract

The development of sophisticated genetic tools during the past 15 years have facilitated a tremendous increase of fundamental and application-oriented knowledge of lactic acid bacteria (LAB) and their bacteriophages. This knowledge relates both to the assignments of open reading frames (ORF's) and the function of non-coding DNA sequences. Comparison of the complete nucleotide sequences of several LAB bacteriophages has revealed that their chromosomes have a fixed, modular structure, each module having a set of genes involved in a specific phase of the bacteriophage life cycle. LAB bacteriophage genes and DNA sequences have been used for the construction of temperature-inducible gene expression systems, gene-integration systems, and bacteriophage defence systems.

The function of several LAB open reading frames and transcriptional units have been identified and characterized in detail. Many of these could find practical applications, such as induced lysis of LAB to enhance cheese ripening and re-routing of carbon fluxes for the production of a specific amino acid enantiomer. More knowledge has also become available concerning the function and structure of non-coding DNA positioned at or in the vicinity of promoters. In several cases the mRNA produced from this DNA contains a transcriptional terminator-antiterminator pair, in which the antiterminator can be stabilized either by uncharged tRNA or by interaction with a regulatory protein, thus preventing formation of the terminator so that mRNA elongation can proceed. Evidence has accumulated showing that also in LAB carbon catabolite repression in LAB is mediated by specific DNA elements in the vicinity of promoters governing the transcription of catabolic operons.

Although some biological barriers have yet to be solved, the vast body of scientific information presently available allows the construction of tailor-made genetically modified LAB. Today, it appears that societal constraints rather than biological hurdles impede the use of genetically modified LAB.

Introduction

The central molecule of a living unit is a polymerized string of either ribonucleotides (RNA) or deoxyribonucleotides (DNA). These often huge linear or circular polymers serve two functions: (1) they encode all information needed to sustain and reproduce an individual cell and, (2) they are transmitted during cell division with high fidelity. Alterations are kept to a minimum due to the presence of powerful repair mechanisms and daughter cells are essentially identical to the parent cell, at least during asexual reproduction. When sexual or parasexual systems interfere, the genetic information can be changed more drastically. However, in the natural context, never beyond certain limits set by the degree to which the genetic exchange mechanisms operating in general recombination can tolerate.

The natural constraints on the formation of grossly changed DNA at high frequency in a short period of time were definitely raised with the advent of recombinant DNA technology which permitted to intentionally change the genetic meaning of genomes almost overnight in a way unlikely to occur in Nature. This development not only accelerated the progress of fundamental research to an impressive speed, but also opened avenues for the production of cells and organisms for the pharmaceutical and fermentation industries. In addition, automated DNA sequencing facilities and advanced computer programmes which convert nucleotide sequences to ORF's and link together nucleotide sequences into contigs has permitted entire microbial genomes to be sequenced as well as that of the eukaryotic *Saccharomyces cerevisiae* and, very recently, that of the multicellular *Caenorhabditis elegans*. As more genomes are being sequenced, we expect to understand in a much more detailed way than before, how evolution worked in shaping the genomes in present day organisms. For the first time the genome of a LAB has been sequenced to completion. The sequence of *Lactococcus lactis* strain IL1403 is a major step toward making comparisons with the genome sequences of other LAB, such as that of *Lactobacillus acidophilus*, which is well under way in the framework of the U.S. National Dairy Foods Research Center Programme (T.R. Klaenhammer, pers. comm.). As more LAB genomes are completed, comparing their genetic organization and informational content will undoubtedly shed light on the evolutionary relationships among the diverse members of this group.

After the genomes are sequenced, determining the function of each and every gene contained in the genome of an organism is a tremendous challenge. The general strategy to determine the function of a gene is to produce an interrupted copy by a Campbell-type integration and to analyze the phenotypic effect(s). However, if the gene to be interrupted is essential, this strategy does not work and should be replaced by one in which, in addition to the interrupted copy, an intact copy remains present, which can be turned on at will. For the functional analysis of the *Bacillus subtilis* genome, the vector pMUTin2 meeting these requirements has been developed (Vagner et al. 1998). Upon integration of this vector, transcription of the non-interrupted copy of the essential gene is controlled by the Pspac promoter in an IPTG-dependent fashion. As far as we are aware, such a vector for LAB has not yet been constructed. However, since both suitable integration vectors for LAB (Leenhouts et al. 1991; Maguin et al. 1992; Biswas et al. 1993; Law et al. 1996) and tightly regulated inducible promoters are currently available, a LAB vector with properties similar to pMUTin2 could be easily constructed. Nevertheless, the shear number of all the genes of a genome makes it a formidable task to uncover their functions. Undoubtedly, this task will carry us through the next few decades, even if several groups would join forces in this enterprise for just one species of LAB. Yet, pursuing this goal is worthwhile, as it will provide a wealth of information for both fundamental and applications-oriented research.

Although the phenotype of an organism ultimately depends on its genetic potential, knowledge of the function of all its ORF's is only part of the complete understanding of the organism in its biological context. Ideally, one would like to be able to predict the behaviour of that organism from its genetic composition. This requires knowledge about non-coding DNA sequences, such as those to which regulatory molecules attach to modulate gene expression. Even then, full understanding of the organism would be incomplete without knowledge of how proteins interact to produce a particular phenotype. Rapid progress is being made on this level, owing to the computer-assisted exploitation of the large body of information available on proteins and protein domains. In those fortunate cases where the crystal structure of a protein is known, examples exist in which computer-assisted modelling of related proteins can occur and result in genetic engineering of proteins with desirable properties.

Cells are continuously adapting to their surroundings: nutrients can become limiting and, in the case of LAB, cells are exposed to increasingly lower pH values through their own metabolic activity. Moreover, chemical additions to industrial fermentations are routinely made (e.g., salt during cheese making). How these changes affect the cell's biochemistry, depends on the modulation of specific gene sets, which can be examined by Northern hybridisation at the transcriptional level or by Western blotting at the translational level. These are, of course, very time consuming approaches and in a in a majority of cases Western analysis is not feasible because suitable antibodies are lacking. An alternative to assess the protein-encoding potential of the cell is to use highly standardized 2D gel electrophoresis to construct a 2D protein index as has been done for *B. subtilis* (Bernhardt et al. 1997; Schmid et al. 1997;). However, to link particular protein spots to the corresponding genes is time-consuming and often requires microsequencing and/or mutant production. Moreover, genes which are weakly expressed will remain hidden in this type of analysis.

The recent introduction of DNA array technology to monitor RNA expression of target genes, represents a major step forward in studying genome-wide gene expression. These techniques have now advanced to a state that the transcriptional potential of complete genomes can be monitored on just one, or a very limited number of oligonucleotide chips (for review: see

Schena et al. 1998; Ramsay 1998, and elsewhere in this volume), with a resolving power of only a few mRNA molecules per cell (de Saizieu et al. 1998). In summary, a whole spectrum of sophisticated analytical tools is presently available for quickly assessing gene function, its modification in a predictable way and tracing differential gene expression under a variety of external conditions.

Due to the relatively small size of their chromosomes, the genomic analysis of LAB bacteriophages advanced much more rapidly than that of their hosts. In the following text, some aspects of LAB bacteriophage genomics will be reviewed briefly, emphasizing those bacteriophage elements with potential interest for the industry, This will be followed by a few selected examples in which the (probable) function of a lactococcal gene could be derived from sequence comparisons with bacteria distantly related to LAB. Finally, some attention will be given to non-coding DNA, as well as to those constraints which impede applications of genetically modified LAB.

LAB bacteriophage genomics

The first LAB bacteriophage chromosome sequenced to completion was that of the lytic lactococcal bacteriophage bIL67 (Schouler et al. 1994). Additional (complete) sequences of bacteriophage chromosomes, from both virulent and temperate phages infecting various LAB species, have now become available (Table 1). We are now in the position to compare their overall genetic organization and, by homology comparisons, assign probable functions to a number of their ORF's. Figure 1 displays a schematic representation of such a functional assignment for a specific group of small isometric-headed phages containing both lytic and temperate representatives which infect a variety of LAB species. Comparison of the partial functional maps has allowed a number of generalisations. Figure 2 shows that: (1) genes involved in lysogeny are transcribed divergently from those required for the lytic pathway; (2) the element governing the switch from the lysogenic to the lytic cycle of temperate bacteriophages is located between the lytic and lysogenic genes; and, (3) adjacent to the 3′-end of the genetic switch is a cluster of genes for the initiation and sustenance of bacteriophage chromosome replication. These genes are followed by gene clusters involved in packaging of the chromosomes into the phage heads, and directing bacteriophage morphogenesis and genes required for lysis of the host. Thus, these bacteriophage chromosomes are highly modular with functionally-related genes clustered together in a fixed order. Exchange of modules between interbreeding bacteriophages by recombination has been invoked to explain the variability between and evolution of bacteriophage chromosomes (Botstein 1980). However, in an extensive review on the variability of a number of *Streptococcus thermophilus* bacteriophages, this evolutionary mechanism has been questioned on the basis of the observation that exchanged segments were neither functional units (modules) nor even complete genes (Brüssow et al. 1998). A number of ORF's of bacteriophage ØSfi21 of *S. thermophilus* showed significant identities with ORF's from various LAB-infecting bacteriophages and even from a *Lb. delbrueckii* plasmid. These observations favour the idea that *S. thermophilus* bacteriophages have evolved through horizontal gene transfer between various LAB as their hosts share the same ecological environment.

It is interesting to note that the mechanism by which the chromosomes are incorporated in the bacteriophage heads may differ profoundly, even among closely-related bacteriophages. For example bacteriophage r1t uses a *cos* site and a terminase to produce chromosomes with single- strand complementary termini, whereas Tuc2009 employs the headful mechanism to produce circularly permuted chromosomes. Apparently, lytic phage varieties can be derived from temperate bacteriophages, as illustrated by inspection of the genome organization of bacteriophage LL-H, which, although having retained *attP* and part of the integrase gene, appears to lack the repressor gene and the switch governing the life cycle (Figure 1).

Apart from these general deductions, the question should be raised as to which set of genetic information could be used for (future) applications. Obvious examples are: (1) repressor-operator systems for the construction of LAB strains in which gene expression can be turned on or off at will; (2) genetic elements which could be used to provoke lysis of LAB in the context that lysis of LAB may facilitate cheese ripening; (3) elements that could be used for the stable integration of (foreign) genes in LAB; and, (4) elements which could be used to combat phage infections during fermentations. An example of each of these possibilities will be described in the following pages.

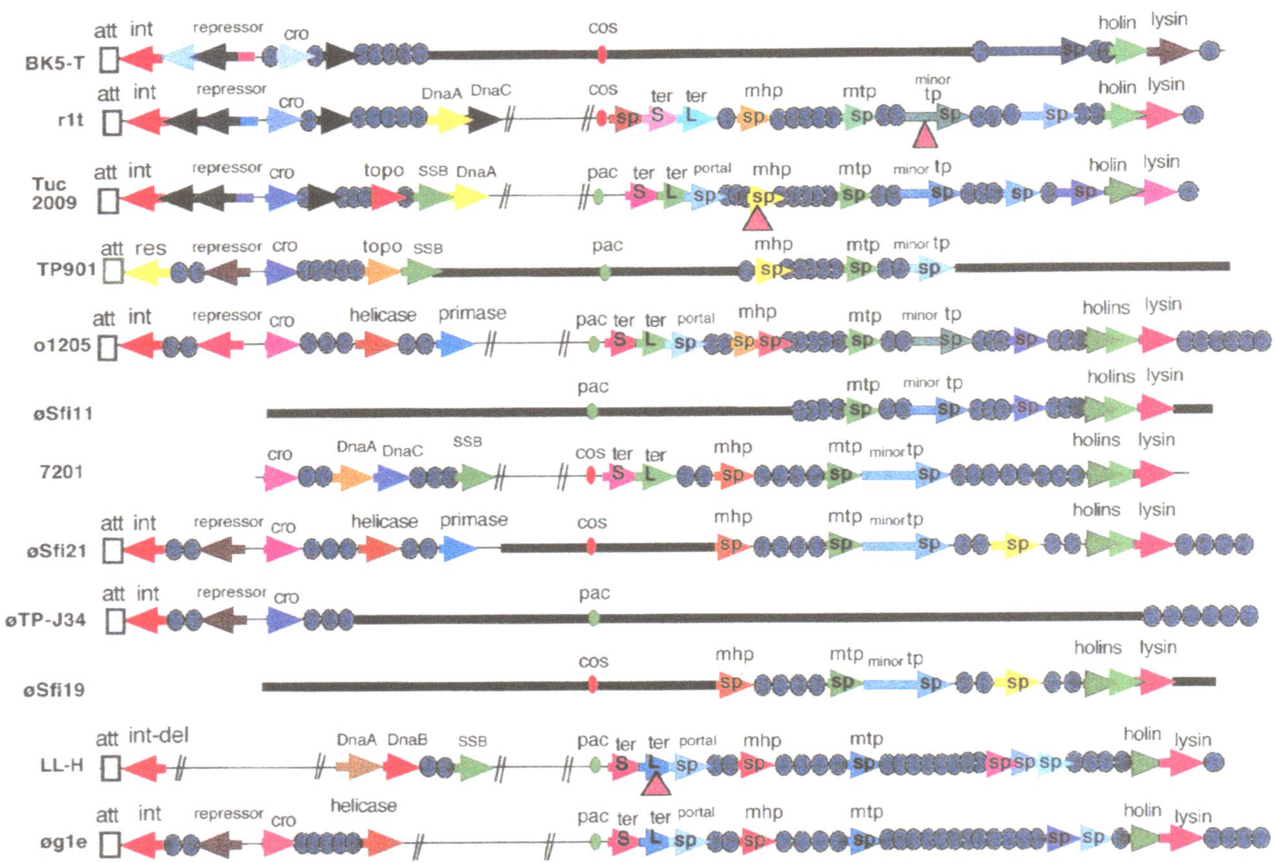

Figure 1. Schematic representation of the genomes of a number of temperate and lytic small isometric-headed bacteriophages infecting various LAB (names in the left-hand margin refer to phages; see also Table 1). Arrows and filled circles (●) depict identified open reading frames (ORF's). The direction of the arrows corresponds to the presumed direction of transcription. Triangles (▲) indicate the position at which an intron interrupts the coding coding sequence of an ORF. A thick black line (▬) indicates sections of phage genomes where no sequence information is available. Genomic sections which were not considered because of lack of known function have been placed between two sets of perpendicular lines (–//—//–). Deduced or determined positions of initiation of packaging, pac or cos, are indicated by green and red filled circles, respectively. Bacteriophage attachment sites (att) are indicated by squares (□). Other abbreviations refer to to the presumed function of the ORF and are as follows: int, integrase; cro, small DNA-binding protein probably involved in the genetic switch; DnaA, DnaB and DnaC, components of the replisome, referring to the *E. coli* replisome; sp, structural protein; terS and terL, small and large subunits of the terminase, respectively; mhp, major head protein; mtp, major tail protein; minor tp, minor tail protein; portal, portal protein; SSB, single-strand DNA binding protein; topo, topoisomerase I; int-del, deleted integrase-encoding ORF.

Table 1. List of LAB bacteriophages of which the genome has been partially or completely sequenced. This table only includes phages of which at least six kb of contiguous sequence has been determined

Phage	cos/pac[a]	Type	Complete sequence available; base pairs	ORF's present on genome	LAB host	Reference
bIL67	cos	prolate, lytic	yes; 22,195	37	*L. lactis* subsp *lactis*	Schouler et al. 1994
c2	cos	prolate, lytic	yes; 22,163	39	*L. lactis* subsp *lactis*	Lubbers et al. 1995
sk1	cos	small isometric, lytic	yes; 28,451	54	*L. lactis* subsp. *cremoris*	Chandry et al. 1997
bIL41	cos	small isometric, lytic	no	NA	*L. lactis* subsp. *lactis*	Parreira et al. 1996
BK5-T	cos	small isometric, temperate	no	NA	*L. lactis* subsp. *cremoris*	Boyce et al. 1995
r1t	cos	small isometric, temperate	yes; 33,350	50	*L. lactis* subsp *cremoris*	van Sinderen et al. 1996
Tuc2009	pac	small isometric, temperate	yes; 38,347	56	*L. lactis* subsp *cremoris*	van Sinderen et al.[1]
TP901-1	pac	small isometric, temperate	no	NA	*L. lactis* subsp *cremoris*	Christiansen et al. 1996 Johnsen et al. 1996 Madsen et al. 1998
LL-H	pac	small isometric, lytic	yes; 34,659	52	*Lb. delbrueckii* subsp. *lactis*	Mikkonen et al. 1996
øg1e	(pac)	small isometric, temperate	yes; 42,259	62	*Lactobacillus* species	Kodaira et al. 1997
øO1205	pac	small isometric, temperate	yes; 43,075	57	*S. thermophilus*	Stanley et al. 1997
ø7201	cos	small isometric, lytic	yes; 35,466	44	*S. thermophilus*	van Sinderen et al.[2]
TP-J34	pac	small isometric, temperate	no	NA	*S. thermophilus*	Neve et al. 1998
øSfi11	pac	small isometric, lytic	no	NA	*S. thermophilus*	Lucchini et al. 1998
øSfi19	cos	small isometric, lytic	no	NA	*S. thermophilus*	Lucchini et al. 1998 Desiere et al. 1998
øSfi21	cos	small isometric, temperate	no	NA	*S.thermophilus*	Desiere et al. 1998 Bruttin et al. 1997 Desiere et al. 1997

[a] cos/pac: phages using cos or pac type of DNA packaging, respectively; NA: not applicable.

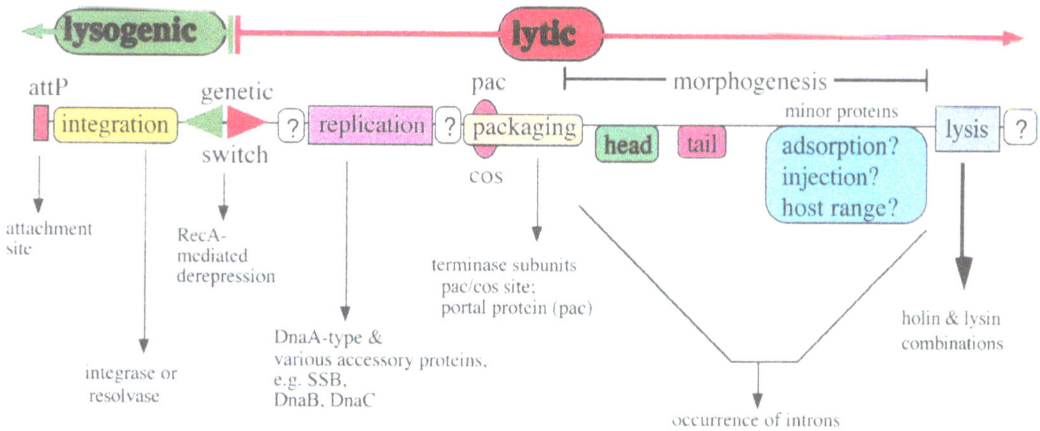

Figure 2. Modular genomic organization found in a large group of small isometric-headed bacteriophages (see Figure 1 and Table 1). Arrows positioned at the top of the figure and coloured in green and red indicate the transcription direction during the lysogenic and lytic life cycle, respectively (note that in lytic phages the DNA region transcribed from lysogenic phages is not, or only partially present). The presumed function of specific DNA modules is indicated in the middle part of the figure and regions with no known function are indicated by question marks. Vertical arrows pointing towards the bottom part of the figure point at additional information on functionality or composition of the specific DNA module.

Inducible gene expression systems derived from bacteriophage

As noted above, all genes involved in the lytic cycle of temperate phages are transcribed oppositely from those governing the lysogenic cycle. In analogy with temperate phages of other bacterial species, this suggests that important regulatory functions could be located in the intergenic region. The two divergently transcribed ORF's flanking this region are ORF 3 and ORF 4 (Figure 1, see r1t, in which ORF3 and ORF4 are indicated as repressor and cro). Examination of the nucleotide sequence of this region of the *L. lactis* phage r1t revealed that it contained two oppositely oriented promoters, P1 and P2, identical to the consensus promoter sequence used by *L. lactis* RNA polymerase. As the -35 sequence of these promoters are part of two direct repeats with dyad symmetry, the intergenic region has the characteristic features of control regions which are instrumental in the decision between the lytic and lysogenic life cycles of temperate phages. Alignments of the deduced *rro* product (Rro) with a number of known and supposed repressors showed that it shared a number of identical amino acids at specific positions with these repressors, including two adjacent amino acids, A and G, where the *E. coli* repressor LexA and the λ CI repressor cleaves during autodigestion. In addition, the protein contained a putative helix-turn-helix motif. Together these features strongly suggested that the Rro was the repressor of phage r1t. This was underscored by experimental evidence: a strain carrying a plasmid in which the transcription of *lacZ* was placed under the control of the lytic promoter P2 produced β-galactosidase when exposed to mitomycin C, whereas a frameshift mutation in *rro* led to constitutive synthesis of β-galactosidase (Nauta et al. 1996). The conclusion that Rro was indeed the repressor controlling the lytic/lysogenic switch in phage r1t provided the basis for the construction of a temperature-inducible gene expression system for lactococci.

A suitable repressor-based temperature-inducible system should meet a number of requirements: the cells should have the wild-type phenotype at low temperatures, whereas at higher temperatures the repressor should be inactive. In addition, the critical temperature interval should be small, as the cells should stay metabolically active at the elevated temperature. The successful development of such a system was facilitated by the availability of the crystal structure of the repressors of the two temperate *E. coli* bacteriophages λ and 434 (Aggarwal et al. 1988; Beamer & Pabo 1992). Characteristically, these repressors have a two domain structure with an N-terminal portion binding to the cognate operators and a C-terminal domain important for dimer formation, dimer-dimer interaction, and repressor self-cleavage (Pabo et al. 1979). The binding specificity of repressors to the cognate operators of temperate bacteriophages resides in the second helix of the helix-turn-helix motif present in the N-terminal repressor domain, whereas the amino acid sequence constituting the turn is important for the correct positioning of the second helix relative to the DNA backbone (see Ptashne 1992). The requirements to be met by a suitable temperature-labile repressor, noted above, are perfectly met by the well-known temperature-sensitive λ repressor CI857, carrying an A66Thr substitution in the N-terminal domain (Lieb 1981). Molecular modelling, using the programme WHAT IF (Vriend 1990) showed that Ala66, contained in helix 4 of the N-terminal domain, is buried in a densely packed hydrophobic pocket. Replacement of Ala66 by Thr would result in distortion of local hydrophobic interactions. Modelling studies using the 3D-structures of the N-terminal domains of the λ and 434 repressors revealed the presence of a structurally similar pocket in the N- terminal domain of Rro, despite the fact that the N-terminal domains of the three repressors shared only 5 conserved amino acids (Figure 3). By randomizing three residues predicted to destabilize the hydrophobic pocket, several temperature-sensitive repressor mutants were obtained. In cells producing one of the mutant repressors, carrying an F50S in addition to a P61A replacement, the induction factor upon raising the temperature from 24 °C to 37 °C and 40 °C was approximately 200 and 600, respectively (Nauta et al. 1997).

Using a random mutagenesis technique, a thermosensitive mutant of the repressor (Rad) of the *Lb. gasseri* bacteriophage Øadh was recently obtained, which contained three amino acid substitutions. Based on tertiary structure analysis, one of the mutations, V48A, corresponded to V54 in the hydrophobic pocket of Rro. However, induction at 42 °C appeared to be quite low. Interestingly, one of the hydrophobic pocket variants (L52D) showed a slightly cryosensitive phenotype. From this observation we may speculate that by proper engineering of the hydrophobic pocket, repressors could be obtained which effectively lose their repressing ability at low temperature, which

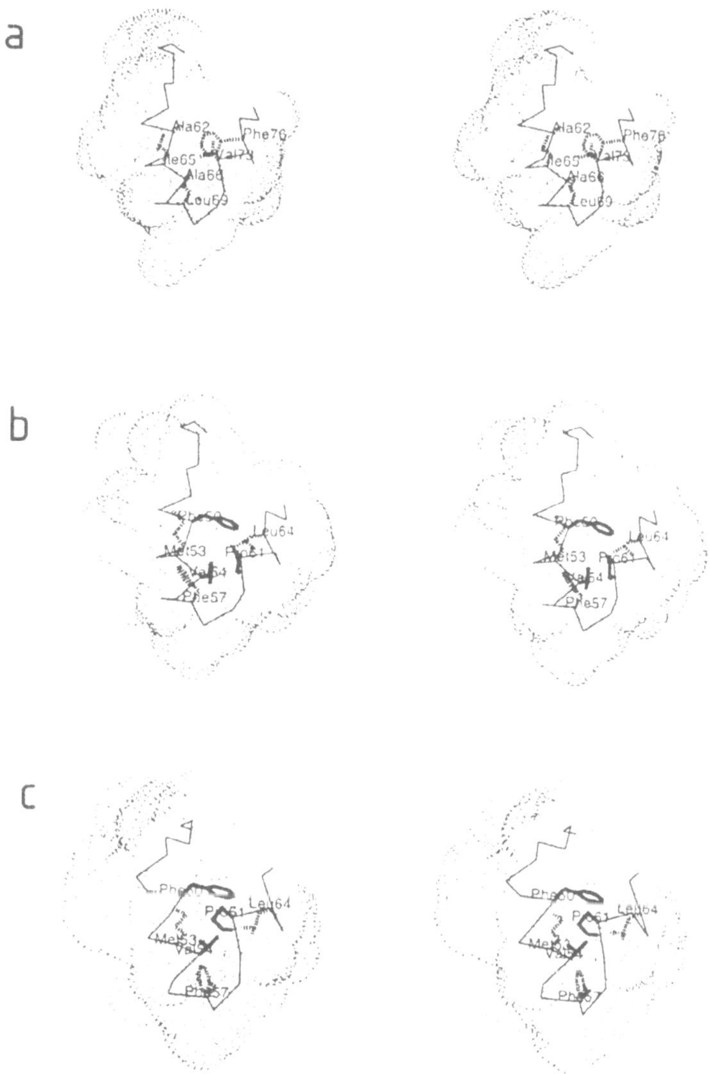

Figure 3. The hydrophobic pockets within the N-terminal domain of λ CI (a) and the Rro models that were obtained using λ CI (b) and 434 CI (c) as templates. The hydrophobic core residues supposed to be important for stability of the pockets are indicated. The side chain of the λ CI residue A66 that has been shown to be critical for stability at elevated temperatures and the side chains of Rro residues that were randimized are depicted in bold. Solvent accessible surfaces are dotted.

might be profitable for *in situ* expression of genes during actual cheese making (G. Engel, pers. comm.).

These systems, in which expression is induced by a simple physical parameter, hold great promise for the induced synthesis of both homologous and heterologous proteins in LAB, important both for fundamental research as well as possible applications.

Lytic genes

The first LAB lysis gene to be successfully cloned, sequenced and characterized was that of the lactococcal prolate-headed bacteriophage ØvML3 (Shearman et al. 1989). Several lysis genes have subsequently been identified in a number of other bacteriophages from both *Lactococcus* and *Lactobacillus* origin (for review: see Gasson 1996). Homology comparisons strongly suggested that ORF 23 of bacteriophage r1t (denoted lysin, Figure 1) could encode the lysin. Proof was obtained by cloning the gene, *lytR*, in the IPTG-inducible plasmid pAG58 in *E. coli*. Subsequent assay of cell wall lytic activity by subjecting cell-free *E. coli* extracts to SDS-polyacrylamide gel electrophoresis showed a prominent clearing zone in the gel containing *M. leisodeikticus* cell walls, corresponding to the expected molecular size of LytR (Nauta 1997).

Phage lysins are generally synthesized without an N-terminal secretory signal peptide needed for *sec*-dependent transport across the cytoplasmic membrane. To gain access to the cell wall substrate, the phage lytic genes are accompanied by an ORF specifying a small hydrophobic protein, thought to form pores in the cytoplasmic membrane, through which the phage lysin gains access to the cell wall substrate (Young 1992; Young and Bläsi 1995). Support for the supposed function of these so-called holins is the presence of two membrane spanning domains, separated by a short sequence likely to adopt a β turn configuration. Supporting evidence that holins act by forming pores in the cytoplasmic membrane came from experiments in which the genes for the supposed holin and that for the lysin of phage r1t were separately and jointly expressed in *E. coli*. Whereas expression of just the lysin gene had little effect on cell viability (due to the absence of a signal peptide the enzyme remains cytoplasmic), expression of just the holin gene reduced the viability by almost 4 log units. When both genes were expressed, viability was reduced by two additional log units. The dramatic effect on viability of expression of the holin gene is in accordance with its presumed pore forming function in the cytoplasmic membrane, which is, apparently, not reconcilable with the cell's life (Nauta 1997). Essentially similar results were obtained in *L. lactis*; induced expression of the lysin gene of bacteriophage øUS3 did not cause significant lysis, partial lysis was observed upon expression of the holin gene, whereas simultaneous expression of both genes resulted in complete lysis of the host (de Ruyter et al. 1997). In summary, two genes appear to be required for the phage progeny to escape from their host, the phage lysin gene and the holin gene. These genes are usually, but not always, closely linked to form the lysis cassette.

Two major (potential) applications of the lysis genes or lysis cassettes have been explored: (1) provoked lysis to accelerate cheese ripening and (2) elimination of undesirable bacteria during fermentation. The former application was implemented in a cheese making experiment where the effect of nisin-induced lysis of an adjunct starter culture on lysis of cells was studied, using cytoplasmic L-lactate dehydrogenase as a marker for lysis. After 4 days of ripening the activity of the marker enzyme had increased 4-fold as compared to the controls, indicating that the system of induced lysis also functioned under actual cheese making conditions (de Ruyter et al. 1997). The latter application has been demonstrated by expression of the lysin gene of a *Listeria monocytogenes* bacteriophage in *L. lactis*. Upon induction, a range of *Listeria* strains were found vulnerable to lysis (Payne et al. 1995).

Integration of genes using bacteriophage DNA-derived sequences

The capacity of temperate bacteriophages to lysogenize their hosts by site-specific integration essentially depends on their attachment site (*attP*), the presence of an identical sequence on the host chromosome (*attB*), and a bacteriophage-encoded integrase (Int). Site-specific recombination has been used to integrate plasmids into the chromosomes of lactococci, lactobacilli and *S. thermophilus*. To make the bacteriophage-derived systems operational, first the critical genetic elements for integration had to be determined, which was established for the *Lactobacillus gasseri* bacteriophage Øadh (Raya et al. 1992), *Lb. delbrueckii* subsp. *bulgaricus* bacteriophage mv4 (Dupont et al. 1995), and the *L. lactis* subsp. *cremoris* bacteriophages Tuc2009 (van de Guchte et al. 1994), TP901-1 (Christiansen et al. 1994), r1t (van Sinderen et al. 1996) and ØLC3 (Lillehaug & Birkeland 1993) as well as for ØSfi21 of *S. thermophilus* (Bruttin et al. 1997b). By incorporating the *int* gene and *attP* into (conditionally) non-replicating plasmids, the plasmids became integrated into the target site, *attB*, of the chromosomes of transformants. Integration at the correct site was even observed when a single gene was ligated to the *int* gene-*attP* cassette to form a circular molecule (Lillehaug et al. 1997). The main advantages of using the *int-attP* cassette over systems in which integration is governed by general recombination are: (1) the integration takes place at the same site used by the bacteriophage chromosome during lysogenisation, thus at a site which disruption is likely to be harmless to the cell, (2) the genetic stability of the integrants under non-selective conditions is usually much higher than those generated by general recombination, (3) the orientation of the integrated structure appears to be fixed and (4) the integrated structure is generally not subject to amplification. With respect to the versatility of use, there is little doubt that the *int-attP* derived system can be used for integration in LAB lysogenizable by bacteriophage from which the casette was obtained. As the integrase should recognize the cognate *attP-attB* pair, conservation of the nucleotide sequence of *attB* might allow integration in a wider range of strains or (sub)species. In fact, this has been demonstrated in *L. lactis* (van de Guchte et al. 1994) and the genus *Lactobacillus* (Dupont et al. 1995)

Combatting bacteriophage propagation with phage DNA

Bacteriophage pose a serious threat to dairy fermentations. Consequently, a considerable body of research continues to be dedicated to bacteriophage-host interrelationships with the aim to inhibit bacteriophage propagation. Various phage defence mechanisms in LAB have been uncovered (see elsewhere in this volume). The discussion below will be limited to non-native systems entirely derived from bacteriophage chromosomes.

The basis for combatting bacteriophage multiplication using a piece of bacteriophage DNA was laid down by Hill et al. (1990). A 4.5-kb DNA fragment from the virulent, small, isometric-headed *L. lactis* bacteriophage Ø50 incorporated into the shuttle vector pSA3, reduced the efficiency of plating of Ø50 by a factor of approximately 3. The plaques, however, were pinpoint in size. The bacteriophage-resistance phenotype was labelled Per (*p*hage-*e*ncoded *r*esistance). Subsequent structural and functional analyses of the *per* region revealed that Per was dependent on the presence in a 1.4-kb fragment of a series of non-coding direct and inverted repeats, hallmarks of phage replicative origins. Consistent with the idea that the *per* region was, in fact, the replication origin of Ø50, was the observation that a plasmid carrying the *per* region was subject to amplification as a concatamer upon phage infection. Also, the level of protection against phage attack correlated with the copy number of the plasmids on which *per* resided (O'Sullivan et al. 1993). Insertion of *per* from the lactococcal virulent isometric-headed bacteriophage Ø31 into a low and a high copy plasmid, reduced the EOP of Ø31 to a mere 0.3 and 7.2×10^{-7}, respectively, demonstrating the potential power of this non-native resistance mechanism. Characteristically, *per*-based systems incorporate two features: titrating out factors required for bacteriophage replication and enabling amplification of *per*-containing plasmids, especially those of the low-copy category.

By combining the latter feature with a regulated middle promoter of bacteriophage Ø31, a system was developed resulting in explosive plasmid amplification and gene expression triggered by infection with a lytic bacteriophage (O'Sullivan et al. 1996). Analysis of the phage-inducible promoter revealed the presence of two sets of phage-inducible tandem transcription start sites in addition to a complete ORF (ORF2) on a 888 nt fragment. When fused to the promoterless *lacZ* gene from *S. thermophilus* in a low copy number plasmid, more than 2000 units of β-galactosidase were produced in *L. lactis* upon infection with Ø31. ORF2 appeared to be a transcriptional activator and a large part of the 5′ end of the 888 bp promoter fragment could be deleted without loss of inducibility. A shortened promoter fragment, carrying only one set of transcriptional start sites, prevented basal expression of β-galactosidase and, after a few further changes, was used in a vector which directed the synthesis of 11,000 units of β-galactosidase upon infection with Ø31. As no synthesis of β-galactosidase was observed in uninfected cells, the system would potentially allow the synthesis of toxic gene products in *L. lactis* (Walker & Klaenhammer 1998). To further exploit this system in phage defence, a 239 bp promoter fragment of the middle promoter of Ø31, designated Ø31*p*, was transcriptionally fused to the restriction cassette *Lla*IR^{+}, consisting of the genes *llaI.1*, *llaI.2* and *llaI.3*. These genes are all essential for restriction activity and when expressed, are lethal for *L. lactis* (O'Sullivan et al., 1995). The efficiency of plating of Ø31 on *L. lactis* expressing the suicide cassette on a high copy-number plasmid was lowered to 10^{-4}, whereas the burst size decreased fourfold (Djordjevic et al. 1997). Bacteriophages that escaped this suicide system appeared to have lost their capacity to effectively fire the suicide cassette. This data prompted the use of a fourfold stronger version of Ø31*p* which reduced the EOP by another factor of 2.2, but somewhat retarded the growth of the host. Combining the suicide cassette with the restriction enhancer *llaC* on the same replicon, which elevates the restriction activity of the suicide cassette, resulted in a further 5-fold reduction of the EOP. In hosts carrying this system and plasmids specifying the abortive phage defences *per31* and *abiA*, the EOP was shown to be less than 10^{-10} (Djordjevic & Klaenhammer 1997).

An essentially different method to combat phage infections concerns the production of antisense RNA directed against bacteriophage mRNA. The potential of this approach was demonstrated in a series of experiments in which anti-sense RNA of a cryptic ORF encoding a protein of 449 amino acids was produced from a constitutive lactococcal promoter. This ORF appeared to be conserved among a number of lactococcal bacteriophages. The complementary RNA produced from the ORF of one of these reduced the EOP by 99.6%, but was also effective against two other bacteriophages in which the ORF was conserved (Kim & Batt 1991). However, subsequent study (Chung et al. 1992) showed that the target RNA largely determined the efficiency of the system. Antisense RNA directed against the activator of the middle genes of Ø31, referred to above, was ineffective against Ø31 propagation (Walker & Klaenhammer 1998). Notwithstanding the failure in some systems to inhibit bacteriophage propagation by means of the antisense methodology, this approach would seem potentially useful, provided the antisense could be targeted to sufficiently conserved pivotal (regulatory) bacteriophage genes which are not abundantly transcribed.

Potentially useful applications of ORF's from LAB

Recently the complete chromosome of *L. lactis* IL1403 has been sequenced (for details see elsewhere in this volume). Undoubtedly, the genetic information derived from the sequence will be helpful in advancing our knowledge of this organism, other lactococcal strains and, likely, LAB species in general. In the text that follows, we will discuss a number of lactococcal ORF's, identified by means of standard genetic analyses, which may be useful for practical applications.

Salt and pH-inducible ORF's

During cheese making procedures NaCl is often added to the curd. This, in addition to the fact that LAB produce lactic acid during growth, resulting in a decreasing pH of the medium, has prompted considerable interest in LAB gene expression in response to varying salt concentration and pH.

In their search for NaCl-inducible ORF's, Sanders et al (1998a) employed a chromosome integration strategy in which a conditionally non-replicating integration vector, pORI13, was used to insert a promoterless *E. coli lacZ* gene randomly in the genome of *L. lactis*. The integrants were assayed for a blue colour on plates containing X-gal and salt, while remaining white in the absence of salt. One integrant showed this phenotype, which depended on the presence of Cl^- rather than of Na^+ ions. Genetic analysis of the integrant showed that pORI13 had integrated in an ORF whose putative product shared homology with GadC of *Shigella flexneri* and *E. coli*. Both Gad's are homologous to a number of amino acid antiporters. Similar to GadC of *S. flexneri*, the lactococcal GadC contains a cluster of 12 highly hydrophobic domains in addition to a conserved domain present in glutamate transporting proteins. GadC probably represents an integral membrane protein involved in glutamate transport. Further analysis of the sequences around the integrated pORI13 revealed the presence of two ORF's flanking *gadC*, labelled *gadR* and *gadB*. GadB is homologous to glutamate decarboxylase from *Synechocystis* sp. (48% identity) and two glutamate decarboxylases from *E. coli* (43 and 44% identity). Northern blot analysis revealed that the *L. lactis gadC* and *gadB* genes are in one operon. Immediately upstream of *gadC*, *gadR* could encode a protein with similarity to positive regulator proteins from *Streptococcus gordonii* (Rgg, 25% identity) and that from *Bacillus thuringiensis* (PlcR, 19% identity). GadR was shown to be the activator of the *gadCB* operon as a *gadR* insertional mutant failed to transcribe the operon. Transcription of *gadCB* was not induced upon lowering of the pH of the medium. However, in the presence of salt and glutamate, the activity of the operon was considerably higher at pH 4.5 than at pH 5.5, suggesting that it might be involved in protection against acid stress. This idea proved to be correct. When wild-type cells were challenged with lactic acid, such that the pH dropped to 3.5, the percent survival was only 5×10^{-4}. Under the same conditions in the presence of NaCl and glutamate, survival increased by a factor of 1000. Increased survival was not observed in a *gadR* mutant which does not produce *gadCB* mRNA (Sanders et al. 1998b). The most plausible explanation for the function of the *gad* system is depicted in Figure 4, in which it is assumed that glutamate is taken up by the amino acid antiporter GadC. After conversion to γ-aminobutyrate by glutamate decarboxylase, it is subsequently externalized by GadC in antiport with a molecule of glutamate. Since the conversion of glutamate to γ-amino butyrate consumes protons, this helps to stabilize the internal pH under acidic external conditions.

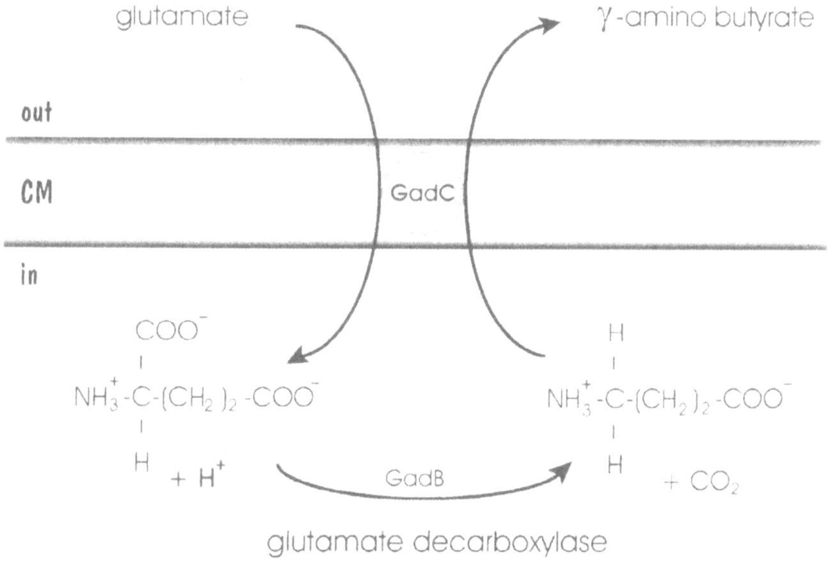

Figure 4. Schematic representation of the stabilization of the internal pH through the activity of the *L. lactis gadBC* operon. The antiporter GadC in the cytoplasmic membrane (CM) internalizes glutamate, which, after conversion to γ-aminobutyrate by GadB, is externalized by GadC in antiport with the optake of glutamate. The conversion of glutamate to γ-aminobutyrate consumes a proton, thus helping in stabilization of the internal pH.

The results of the research summarized above may have at least two practical applications. First, salt and glutamate might be added to protect commercial starter cultures for dairying purposes from adverse effects of low pH. Second, the *gad* cassette, consisting of *gadR* and the *gadCB* promoter could be used for NaCl-induced production of heterologous and homologous proteins. In this case, the induced expression is effected by a simple chemical often used in cheese making. In fact, this cassette is quite effective in inducing lysis *L. lactis* when transcriptionally fused to the lysis cassette of bacteriophage r1t and the *acmA* gene of *L. lactis*, which specifies the autolysin (Sanders et al., 1997).

Based on homology comparisons the function of genes of the *gad* system could be defined reasonably easy. However, when no homology exists with ORF's of other organisms, or because no function could as yet be assigned to homologous ORF's, the situation becomes less clear. An example of the latter of these two possibilities concerns an ORF which is expressed under conditions of low pH at the start of the stationary growth phase of *L. lactis* (Madsen et al. 1999). The putative ORFX product shows a high degree of sequence identity (57.3%) with the hypothetical *B. subtilis* Ybh protein, to which, unfortunately, no function has as yet been ascribed. The P170 promoter of OrfX, like that of *gadCB,* lacks the -35 canonical promoter sequence TTGACA, present in the majority of the promoters so far identified in *L. lactis* (van de Guchte et al. 1992; de Vos & Simons 1994). Interestingly, the inducibility of the promoter of OrfX could be conferred on a constitutive lactococcal promoter (CP25) by substituting the -35 sequence together with 22 bp upstream of this sequence for the corresponding 27 bp of the inducible promoter. Inducible expression of OrfX was increased 150–200 fold upon deletion of 72 bp of the 270 bp untranslated leader of the OrfX mRNA. The modulating activity of this fragment could be conferred on two other constitutive lactococcal promoters. Potentially, the 27 bp element could be used for those situations in which it would be desirable to turn a constitutively expressed gene into an inducible one, whereas the untranslated 72 bp OrfX leader fragment might be exploited for the inhibition of late genes that are not targeted for expression.

LAB protein modules

Analysis of derived amino acid sequences has revealed that proteins are often composed of functional domains and/or motifs such as involved in catalysis, substrate recognition and co-factor binding. A specific example of this modularity with potentially interesting possibilities for application concerns the composition of the major autolysin AcmA of *L. lactis*. The amino acid sequence of the N-terminal moiety of this enzyme (amino acids 65–220) shows approximately 56% identity with muramidase-2 of *Enterococcus hirae* and the autolysin of *Streptococcus faecalis*, whereas the identity in the C-terminal part (amino acid 222–437) was only approximately 24%. Inspection of the amino acid sequence of this part of the lactococcal enzyme showed that it contained three interspersed repeated sequences of 45 amino acids, each containing 16 identical amino acids and 18 conservative replacements (Buist et al., 1995). These repeated motifs are also present in the C-terminal moieties of the corresponding enzymes of *S. faecalis* and *E. hirae*, although in different numbers – 4 in the former and 5 in the latter of these species. Fusing the C-terminal *L. lactis* AcmA domain to TEM β-lactamase, provided with a *Lactococcus*-specific signal peptide, resulted in association of the hybrid protein with the cell wall. Since β-lactamase without the C-terminal repeats was secreted into the medium, the C- terminal part of AcmA apparently functions as a cell wall anchor (Buist 1997), which agrees with data obtained in other lytic systems (López et al. 1995; Sheehan et al. 1997). As discussed in more detail elsewhere in this volume, such anchoring motifs may find use for applications.

Non-coding sequences

Obviously, the identification, characterization and functional analysis of genes is crucial for successful genetic modification. Equally important is detailed knowledge as to how genes targeted for use in genetic modification are regulated. In many cases, non-coding DNA or mRNA is the target for regulatory molecules. The function of such nucleotide sequences will be briefly reviewed because they could be used in many conceivable ways for modulation of gene expression.

Analysis of non-coding sequences is inherently more difficult than that of coding sequences because one needs to consider the significance of the ordering of individual bases rather than of base triplets.

Presently, functional genome analysis is almost completely limited to efforts to determine the function of ORF's, usually by disrupting the ORF's by the integration of specially designed plasmids into the respective genes and subsequent analyses of the phenotype under a variety of different conditions. If the vectors designed for gene interruption are also provided with a reporter gene, differential gene expression may be monitored under different conditions or during growth. Thus ORF's can be identified that are either positively regulated by transcription activators or negatively regulated by repressors binding near to the start of transcription. Fortunately, regulatory proteins often bind as dimers or tetramers to their recognition sites. At the DNA level the target sequences are reflected by the presence of inverted or direct repeats. However, the absence of such repeats does not guarantee that the promoter is *not* regulated, as examplified by the pH-inducible promoter P170, referred to above, in which no secondary structural elements could be detected. Both the pH-inducible P170 as well as the salt inducible P*gad* lack the -35 promoter sequence. Whether or not the absence of -35 sequences are characteristic features of regulated host promoters remains to be established.

The presence of long stretches of untranslated, or partially translated mRNA 5′ to the encoded gene may also be indicative of regulation, especially when these leaders contain (potential) secondary structure. Examples in *L. lactis* illustrating this notion concern a number of operons involved in amino acid biosynthesis. In the 439 nt leader of *trp* operon mRNA several stem-loops could be formed. Two of these, the terminator and antiterminator, mutually exclude each others formation, because they use the same nucleotide stretch in the formation of their stems. Under Trp limitation, the 3′-end of uncharged $tRNA^{trp}$, the acceptor (GCCA), is believed to base pair with the complementary sequence (anti-acceptor), embedded in the so-called T-box, residing in the stem of the antiterminator. This interaction, together with basepairing of the anticodon of $tRNA^{trp}$ with the specifier codon UGG in a stem-loop structure (numbered I) close to the 5'-end of the leader, is believed to stabilize the antiterminator. This prevents the formation of the terminator, and allows transcription to continue over the entire length of the structural genes of *trp* operon (van de Guchte et al., 1998). Apparently, the *his* operon of *L. lactis* is subject to a similar mode of attenuation regulation, as the specifier codon (CAC) is positioned at the same location in stem-loop I as the UGG codon for tryptophan biosynthesis and a T-box is present containing three nucleotides complementary to the 3′-end of $tRNA^{his}$. By comparing the nucleotide sequences of the leaders of *his* mRNA of various distantly related lactococci, it was observed that nucleotide variations generally did not affect the proposed secondary structure elements, thus lending support to the folding model of the leader (Delorme et al. 1999).

The leader of the *leu* operon appears to be partially translated into a 16 amino acid-containing peptide with 3 consecutive leucines. The leader also contains a rho-independent transcription terminator which, under conditions of leucine starvation, will not be formed due to stalling of the ribosomes. This form of regulation clearly represents attenuation regulation of the operon on the basis of the specific codons present in the translated part of the leader (P. Renault, pers. comm.)

Although little is known about the regulation of the other amino acid biosynthetic pathways in LAB, characterization of the 5′ portions of the mRNA involved in these pathways will probably prove valuable in defining whether or not they are subject to similar modes of regulation.

In the examples referred to above, the formation of the terminator is prevented by antiterminator stabilization either by interaction of uncharged tRNA, or by ribosome stalling at the leader. The formation of terminator structures residing in certain mRNA leaders can also be prevented by interaction with Bgl-type regulator proteins, which recognize a conserved nucleotide sequence, labelled RAT (*r*ibonucleic *a*nti*t*erminator; consensus: GGATTGTTACTGCA AAGCAGGCAAAACC, Schnetz et al., 1996) that is part of the 5′-portion of the terminator. This mechanism is well documented in *E. coli* as well as a number of Gram-positive bacteria (for references see Bardowski et al. 1994), and also appears to operate in LAB. The regulatory protein BglR of *L. lactis* classifies as a member of the Bgl family of antiterminator proteins, where its expression is positively regulated by β-glucosides (Bardowski et al. 1994). Antitermination by interaction of Bgl-type proteins with mRNA leaders also appears to operate in lactobacilli. For example, in *Lb. casei*, *lacT* is the first gene of the Lac-PTS operon and encodes a Bgl-type antiterminator protein which destabilizes a transcription terminator on the mRNA preceding *lacT* in a way similar to that in *L. lactis* (Alpert & Siebers 1997). These antiterminator proteins belong to a large class of regulators, which contain a duplicated, conserved domain (PRD), and are controlled by PTS-dependent phosphorylation (for review see Stülke et al. 1998). A model for PTS-mediated control of PRD-containing regulators postulates that in the absence of both glucose and the PTS-specific substrate, phosphorylation of both PRD's inactivates the regulator. In the absence of glucose, but in the presence of the substrate, phosphorylation of one of the PDR's would render the regulator active, while in the presence of both glucose and the substrate, neither of the PDR's would be phoshorylated, resulting in inactivation (Stülke et al. 1998).

A different signature of regulated genes in Gram-positive bacteria is the *cis*-acting *c*atabolite *r*esponsive *e*lement (*cre*), a sequence potentially able to form a hairpin (consensus sequence TGT/ANANCGNTNA/TCA, Weickert & Chambliss 1990; for review, see Hueck & Hillen 1995). As implied in its name, this element plays an important role in carbon catabolite repression (CCR), the phenomenon where expression of many catabolic genes is repressed when a readily metabolizable sugar is available. In contrast to the regulatory secondary structure elements described above, which seem to be invariably positioned in mRNA leaders, *cre*'s are either present upstream or overlap with the promoters, or are found in the 5′-coding moieties of the genes or operons they control. The salient model for the mechanics of CCR in Gram-positive bacteria is that the PTS component, HPr, becomes phosphorylated at Ser-46 by an ATP-dependent kinase in the presence of a readily metabolizable carbon source. Phosphorylated HPr interacts with the key regulatory protein of CCR, CcpA, such that the latter binds to the target *cre*, thus affecting expression of sugar catabolic operons (Hueck & Hillen 1995). Using a variety of experimental approaches including DNase I footprinting and gel retardation assays, convincing support for this model has been obtained (Fujita et al. 1995; Ramseier et al. 1995). Migration of a 26-mer containing *cre* was effectively retarded in the presence of wild-type CcpA and Hpr-Ser46-P, but not in the presence of unphosphorylated HPr. Also, a number of specific CcpA mutants completely failed to bind to the probe in the presence of phoshorylated HPr (Kraus et al. 1998). In further support of the model, when HPr-Ser-46P with a radioactively labeled phosphoryl group was used together with CcpA, the radioactivity was co-retarded with the *cre*-containing probe (W. Hillen, pers. commun.).

The data on the mechanisms underlying CCR in Gram-positive bacteria were obtained with a number of bacilli. Recently *cre* sequences have also been examined in LAB and the data strongly suggest that they are also the targets for CcpA-type proteins. Polyclonal antibodies raised against CcpA from *B. megaterium* cross-reacted with specific proteins of many

Gram-positives, including *L. lactis*, *Lb. casei* and *Lb. plantarum* (Küster et al. 1996). The gene encoding CcpA of *L. lactis* was cloned and sequenced and the protein was shown to repress transcription of the *gal* operon, which contains a *cre* site in the untranslated leader. CcpA activated the *las* operon, encoding the glycolytic enzymes phosphofructokinase, pyruvate kinase and L-lactate dehydrogenase (Luesink et al. 1998). This operon also contains a *cre* located upstream of the -35 promoter sequence. A *cre* sequence is also present in the leader of *sacB* specifying EII^{Suc} of the sucrose gene cluster, but whether this site is required for CCR remains to be established (Luesink et al. 1999a). The gene encoding the protein modulating the *cre* binding activity of CcpA, HPr, was recently also isolated from *L. lactis*. Functional analysis of the protein using site-directed mutagenesis of the serine residue at position 46 demonstrated the role of HPr(Ser-P) as the co-repressor in CCR of the *gal* operon and as co-activator in catabolite activation of the *las* operon in *L. lactis* (Luesink et al. 1999b).

The main players in catabolite repression, *cre*, HPr and CcpA, are also present in lactobacilli. In *Lb. pentosus ccpA* as well as a *cre* sequence overlapping with the -35 sequence of the *xylA* promoter were identified (Lokman et al. 1994; Lokman et al. 1997). Two *cre*'s were found in *Lb. amylovorus*, one overlapping with, the other close to the 3′-end of the -10 sequence of the α-amylase gene (Lokman et al. 1997). In addition, evidence was obtained that both EII^{Man} and EII^{Fru} are important components of CcpA-mediated CCR in *Lb. pentosus* (Chaillou 1999). The *ccpA* gene has been isolated from *Lb. casei* and a mutational analysis indicated that disruption of *ccpA*, or deletion of *cre*, resulted in relief of catabolite repression of the *lacTEGF* operon (Monedero et al. 1997). The overlap of *cre* with various promoters suggests that their interaction with CcpA inhibits the initiation of transcription by RNA polymerase. It might be speculated that binding of CcpA to *cre*'s located upstream of promoters facilitates RNA polymerase binding and thus transcription initiation, whereas association of CppA with *cre*'s located within coding sequences might inhibit mRNA elongation through ribosome stalling.

Another example of a protein that can act as a repressor or as an activator is the global regulator protein, FNR (*f*umarate and *n*itrate as *r*educible substrates), which in *E. coli* is activated to a functional repressor or functional activator under conditions of anoxia (for review see Spiro & Guest 1990; Spiro 1994). FNR is homologous to the cyclic AMP receptor protein CRP, which mediates catabolite repression in Gram-negative bacteria. FNR differs from CRP in being an iron containing monomeric molecule, rather than a metal-free dimer, and in having an N-terminal extension containing four cysteine residues (Melville & Gunsalus 1990; Sharrocks et al. 1990; Green et al. 1991). Recently, the gene encoding an FNR-like protein (FLP) has been isolated from *Lb. casei*. The protein, when expressed in *E. coli*, failed to activate or to interfere with transcription from FNR-dependent promoters. This could be related to differences in the binding sites for FNR and FLP (TTGAT-N4-ATCAA and CCTGA-N4-TCAGG, respectively), in which the symmetrical TGA half-site motifs (underlined) in the former are separated by 6, and in the latter by 4 base pairs. Gel-retardation experiments showed that the reduced (dithiol-containing) form, rather than the disulphide containing form of FLP retarded target DNA, and it was proposed that the interconversion of the two forms, in which cupric ions may have a role, provides a redox-mediated transcriptional switch in *Lb. casei* (Gostic et al. 1998). *L. lactis* contains at least two *flp* genes, *flpA* and *flpB* (J. Kok, unpublished; Gostic et al. 1999). The derived amino acid sequences of FlpA and FpB show a high degree of similarity with that of *Lb. casei*. The *flpB* gene is located in an operon preceded by an FNR-binding site which also contains the transcriptional start site. As the TGA motifs in this site are separated by 6 bp, it resembles a 'true' FNR binding site rather than a FLP binding site. FlpB contains 3 His doublets in the first 14 N-terminal amino acids, providing a 'natural' His tag that allowed easy purification of the protein by means of Ni^{2+} column chromatography. Gel retardation experiments showed that the reduced form of the lacococcal FlpB bound to its FNR box at low pH. Transition metals enhanced FlpB binding and a chelator of transition metals (phenanthrolin) completely abolished binding (J. Kok and A. de Jong, unpublished). *FlpA* is located in an operon analogous to that containing *flpB*. The promoter regions of these two operons are associated with potential FNR sites at positions +4.5 for the *flpA* operon and -42.5 for the *flpB* operon, possibly suggesting that the respective operons are subject to positive and negative regulation, respectively. A *flp* double mutant led to a lower zinc content and increased hydrogen peroxide sensitivity, suggesting that FlpA and FlpB respond to metal-ion or oxidative stress and include systems concerned with the transport and storage of zinc (Gostic et al. 1999).

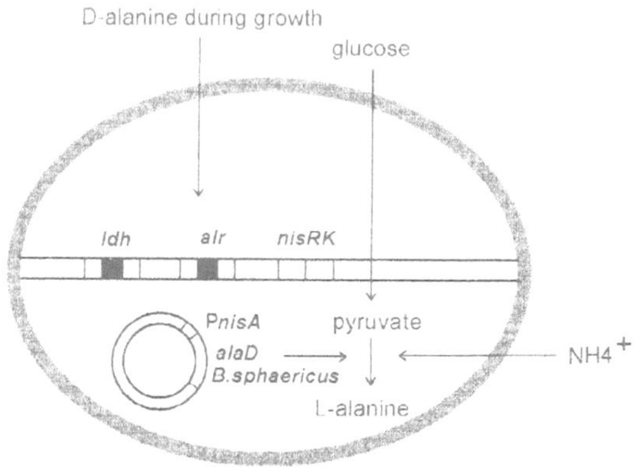

Figure 5. Schematic representation of the genetic structure of *L. lactis* for the production of enantiopure L-alanine. By disrupting the gene encoding lactate dehydrogenase (*ldh*), resting cells at elevated pH in the presence of excess ammonium convert pyruvate almost exclusively to L-alanine, catalysed by *alaD*-encoded alanine dehydrogenase from *B. sphaericus*. The conversion of L- to D-alanine was prevented by disruption of *alr*, specifying the synthesis of alanine racemase. High AlaD production was obtained by nisin activation of P*nisA* (for further details: see text). ◻■◻: interrupted gene

ORF's, regulatory elements and metabolic engineering

The rapidly expanding knowledge of the function of ORF's from LAB and the way they are controlled, offers many opportunities for rational and efficient metabolic engineering. An elegant example, in which packages of initially unrelated knowledge were successfully integrated, concerns the conversion of *L. lactis* from homolactic to homoalanine fermentation. An interesting perspective of alanine-producing dairy strains would be their use as *in situ* producers of the natural sweetener alanine in cheese, buttermilk and yoghurt. Pyruvate, which plays a pivotal role in the sugar metabolism of LAB (for review see de Vos 1996) is the substrate for a number of enzymes which convert this intermediate to various compounds. Conversion of pyruvate to L-lactate by L-lactate dehydrogenase (L-LDH) with NADH as the cofactor was prevented by interrupting of the *ldh* gene through a Campbell-type recombination of an integration plasmid. To re-route the carbon flux towards alanine, the gene encoding alanine dehydrogenase (L-AlaDH) from *Bacillus sphaericus* was expressed in this mutant using the highly efficient nisin- controlled expression system (de Ruyter et al. 1996; Kuipers et al. 1997). The choice for this enzyme was prompted by the fact that it uses pyruvate as substrate, NADH as cofactor, and the km of the *B. sphaericus* enzyme and NADH are very close to that of *L. lactis* L-LDH. Alanine production was measured in a suspension of cells (first pre-loaded with L-AlaDH through nisin induction) in a phosphate buffer containing glucose with an excess of ammonium at alkaline pH, as the pH optimum for amination is between 8.5 and 9. Under these conditions glucose was completely and exclusively converted to alanine. However, the alanine produced still contained 15% of the D-isomer, which is required for peptidoglycan biosynthesis and is produced from the L-form by alanine racemase. To prevent the racemization, the alanine racemase gene in the *L. lactis* strain was disrupted by Campbell-type recombination. To ensure growth of the production cultures, D-alanine was supplied externally, which, after growth, was washed away before the cells were resuspended in the supplemented buffer for production purposes. This procedure resulted in the exclusive production of the L-enantiomer (over 99.5 purity) and represents, for the first time, an example of re-routing pathways towards a metabolite which is not a natural end product of the host (Hols et al. 1999). This example, schematically represented in Figure 5, also demonstrates that the knowledge acquired on ORF's and regulatory elements has created new possibilities for metabolic engineering which only a few years ago would have been considered quite unrealistic.

Complications

Present-day organisms are equipped with genomes that are optimally adapted to their environments. The question arises, why would we then like to improve them. The answer is, of course, that man uses organisms for his own benefit, and consequently, tries to modify them by breeding programmes, selection and, with the advent of the recombinant DNA technology, by genetic modification. With sufficient knowledge about genes and their regulation, which will undoubtedly be accelerated now that the complete sequence of genome of *L. lactis* has been determined, and that of *Lb. acidophilus* will be soon available, almost all modifications to suit the fermentation and pharmaceutical industry are within reach. However two constraints exist, one arising from biological considerations, the other from societal perception.

Biological constraints

Many biological constrains exist, or are conceivable, which preclude the production of GMO's with properties designed for industrial use. Some obvious obstacles will be mentioned briefly. First, procedures should exist for the introduction of (sufficient) DNA into the host to be modified. If not available, well-transformable intermediary hosts can sometimes be used as conjugal donors for the DNA to be transferred. The search for proper transfer systems can be tedious, as examplified by finding the conditions and the proper vector for making *Lb. delbrueckii* subsp. *bulgaricus* transformable (Sasaki et al. 1993). Such searches, in which all possible variables are examined, do not always meet with success as is, for instance, the case for *Oenococcus oeni*, which remains as yet untransformable (M. Zuñiga, pers. comm.).

Second, the expression of heterologous genes in LAB may not be efficient when the codon usages of the host and the donor of the heterologous DNA are seriously mismatched. Conceivably, attempted over-expression of the heterologous material might even titrate away amino acids required for the synthesis of the host's own proteins, thus interfering in a negative way with cellular metabolism. In principle, such complications can be circumvented by redesigning the gene to be expressed, such that the codon content is more comparable to that used by the host, but even for medium-sized genes this is a laborious and costly affair.

Third, the gene product to be synthesized could form insoluble aggregates and, when required to be exported from the cell, fails to do so, or when transported, is degraded in the medium. Whether such problems relate to LAB, has not yet been thoroughly investigated. However, *B. subtilis*, which efficiently secretes various homologous exo-enzymes, often fails to produce exported proteins from heterologous sources (Simonen & Palva 1993).

Fourth, the presence of paralogous genes may impede functional analysis, as the mutant phenotype may be only produced when all paralogues have been inactivated.

In principle, given time and understanding, several of these and other barriers may be surmounted. Fantastic progress has been made during the past 15 years, where our almost complete ignorance of the genetics and molecular biology of *L. lactis* has been converted into a vast body of scientific information. This progress makes one optimistic about overcoming these barriers for the production of genetically modified LAB for improving dairy processing, food products, and making these organisms useful for human and animal health care purposes.

Societal constraints

In the present space of time a substantial proportion of the conscious consumers prefer 'natural' foods. Manipulation of food induces an attitude of caution by the consumer towards these foods. This concern immediately touches on the use of genetically modified LAB for food production. What scientists and industries could do to convince consumers and regulatory public bodies that this concern is unnecessary and unfounded was excellently reviewed by Verrips & van den Berg (1996). In our opinion, to address the consumer's concern and guide modified LAB's through regulatory procedures, modified LAB should retain their original properties as closely as possible, in that: (1) the improvement or new property should preferably be caused by mutation(s) also occurring in Nature; (2) the new property should preferably be caused by rearrangements of the existing genetic information (self-cloning); (3) when the incorporation of heterologous genes is required to produce the desired LAB, these genes should preferentially be obtained from other food-grade or at least harmless organisms; (4) the presence of antibiotic markers should be avoided; and (5) heterologous genes should be stably integrated into the chromosome, rather than carried on plasmids to reduce the possibility of dissimination of the property to other bacteria present in food, the digestive tract, or the environment. Fortunately, a collection of well-defined integration vectors and genetic procedures exists which do not leave behind a trace of vector DNA and meet the points raised in (2) and (3) (de Vos et al. 1989; Leenhouts 1995; Leenhouts et al. 1996; Leenhouts et al. 1998). In addition, when induced gene expression is required, several completely harmless means are at our disposal, such as induction by nisin, temperature, salt and pH shifts.

The following example illustrates how recombinant DNA technology was used by a starter culture company attempting to obtain an improved strain of *L. lactis* in which the improvement was due to a spontaneous mutation. The aim was to isolate a spontaneous mutant of the *L. lactis* biovar. *diacetylactis ald* gene, specifying acetolactate decarboxylase, which converts acetolactate (ALA) into acetoin. A mutation in *ald* would lead to an increased pool of ALA, the immediate precursor of diacetyl, and result in increased diacetyl production. As ALA is also a precursor for the synthesis of the branched amino acids isoleucine, leucine and valine, and the presence of leucine allosterically activates ALD, the pool of ALA is depleted and, consequently, will not be available for the synthesis of diacetyl and the branched amino acids. Thus the cells will cease to grow. However, mutants that produce no ALD accumulate ALA and, therefore, continue to synthesize the branched amino acids even in the presence of leucine. Such mutants will continue to grow in the absence of valine (Swindell et al. 1996; Goupil-Feuillerat et al. 1997). Unfortunately, this selection procedure can not be applied to many dairy lactococci, because they have irreversibly lost their capacity to synthesize branched amino acids. To overcome this problem, several strains of the *diacetylactis* biovar were first transformed with a plasmid carrying the genes for isoleucine and valine production. By subsequent selection for transformants capable of growth in the presence of leucine but in the absence of valine, *aldB* mutants were obtained. Finally, curing of the strain of the plasmid used to make the selection possible, yielded a strain with a plasmid complement that was indistinghuisable from the wild-type and produced approximately 27 times more diacetyl (Curic et al. 1999). Although the modified phenotype of the strain is caused by just one spontaneous mutation, nevertheless its application in the manufacture of dairy products must, according to current EU regulations, still be approved, because it descended from a progenitor bearing a recombinant plasmid.

In our opinion, the need for an approval process is unnecessary. The product rather than the way in which the modification was produced should be taken as the sole criterion for approval, as is currently practiced by US regulatory agencies.

Concluding remarks

Since the development of the first plasmid cloning vector from a small cryptic lactococcal plasmid, which was capable of replication in both Gram-positive and Gram-negative bacteria (Kok et al. 1984) and the development of efficient electro-transformation protocols (Chassy & Flickinger 1987; Luchansky et al. 1988; Holo & Nes 1989), the field of LAB genetics and molecular biology has developed explosively. In Europe, this development was further catalysed by the generous and constant support from the EU, which, through its conditions for proposal granting, united individual research groups into tightly cooperating consortia. With the advent of automated DNA sequencing facilities and the development of advanced computer software for the analysis of ORF's, operons, regulatory elements and even complete genomes, the means has become available to quickly obtain a view of the functionalities of these genetic entities. Extrapolating the past developments to the future, one should anticipate that knowledge of LAB and their interactions with the environment will increase at even a faster rate than before. The recent developments in DNA chip technology represents an extremely valuable asset to analyse ways in which bacteria adapt their metabolism when transcending to stationary growth or after exposure to stress conditions.

Although a tremendous arsenal of techniques, biological tools and equipment is presently available to increase our global and detailed knowledge of LAB, an overwhelmingly large body of work remains to be done. For instance, if hybridization-based expression studies using DNA micro-arrays would show that certain groups of genes are switched on under particular conditions, one would like to know the function of such genes and how they are controlled. If no clues are available as to their possible functions, only a mutation or expression approach will yield that information. To facilitate this, it would be extremely advantageous to structure a programme for gene function analysis, similar to that presently employed for yeast and *B. subtilis*. Now that the complete genome of *L. lactis* has been sequenced and the sequencing of the *Lb. acidophilus* genome is well under way, excellent conditions exist for pursuing this goal.

With our present knowledge, the construction of tailor-made LAB strains for use in the dairy industry is already possible. Ingenious inducible gene expression tools and integration systems, that do not leave behind a trace of vector DNA upon insertion into the LAB chromosomes, have been developed. Yet, these are applied much less than desired, the obvious barrier being the hesitation to use genetically modified LAB, which, in turn derives from public fear for new technologies. The main obstacle for application is, in many cases, not the present state of our scientific knowledge and technical capabilities, but rather public perception. Much needs to be done in terms of frank and open information to modify this perception. Yet, it would seem that this process is already under way. Permission was granted recently to allow a tasting panel to judge the quality of experimental cheeses produced with a LAB which contained and expressed a non-LAB gene. When this encouraging trend continues, we anticipate that the fundamental knowledge on LAB acquired at universities, research institutions and industries will be channelled much more quickly to provide for improved strains, not only for the manufacturing of safer, more wholesome and tastier products, but also for applications in the human and veterinary health care sector.

Acknowledgements

We gratefully acknowledge the permission of several colleagues to quote from preprints or unpublished results and helpful suggestions otherwise, notably M.J. Gasson, M. van de Guchte, W. Hillen, H. Israelsen, E. Johansen, T.R. Klaenhammer, O.P. Kuipers, P.H. Pouwels and P. Renault.

References

Aggarwal AK, Rodgers DW, Drottar M, Ptashne M & Harrison SC (1988) Recognition of a DNA operator by the repressor of phage 434: a view at high resolution. Science 242: 899–907

Alpert CA & Siebers U (1997) The *lac* operon of *Lactobacillus casei* contains *lacT*, a gene coding for a protein of the BglG family of transcriptional antiterminators. J. Bacteriol. 179: 1555–1562

Bardowski J, Ehrlich SD & Chopin A (1994) BglR protein, which belongs to the BglG family of transcriptional antiterminators, is involved in β-glucoside utilization in *Lactococcus lactis*. J. Bacteriol. 176: 5681–5685

Beamer LJ & Pabo CO (1992) Refined 1.8 Å crystal structure of the lambda repressor-operator complex. J. Mol. Biol. 227: 177–196

Bernhardt J, Volker U, Volker A, Antelmann H, Schmidt R, Mach H & Hecker M (1997) Specific and general stress proteins in *Bacillus subtilis* – A two dimensional protein electrophoretic study. Microbiology 143: 999–1017

Biswas I, Gruss A, Ehrlich SD & Maguin E (1993) High efficiency gene inactivation and replacement system for gram-positive bacteria. J. Bacteriol. 175: 3628–3635

Botstein D (1980) A theory of modular evolution for bacteriophages. Ann. N.Y. Acad. Sci. 354: 484–491

Boyce JD, Davidson BE & Hilier AJ (1995) Sequence analysis of the *Lactococcus lactis* temperate bacteriophage BK5-T and demonstration that the phage DNA has cohesive ends. Appl. Environ. Microbiol. 61: 4089–4098

Brüssow H, Bruttin A, Desiere F, Lucchini S & Foley S (1998) Molecular ecology and evolution of *Streptococcus thermophilus* bacteriophages – a review. Virus Genes 16: 95–109

Bruttin A, Desiere F, Lucchini S, Foley S & Brüssow H (1997a) Characterization of the lysogeny DNA module from the temperate *Streptococcus thermophilus* bacteriophage øSfi21. Virology 233: 136–148

Bruttin A, Foley S & Brussow H (1997b) The site-specific integration system of the temperate *Streptococcus thermophilus* bacteriophage ØSfi21. Virology 237: 148–158

Buist G (1997) AcmA of *Lactococcus lactis*, a cell-binding major autolysin. Doctoral thesis, University of Groningen

Buist G, Kok J, Leenhouts KJ, Drabowska M, Venema G & Haandrikman AJ (1995) Molecular cloning and nucleotide sequence of the gene encoding the major peptidoglycan hydrolase of *Lactococcus lactis*, a muramidase needed for cell separation. J. Bacteriol. 177: 1554–1563

Challou S (1999) Regulation of xylose and α-xyloside transport and metabolism in lactobacilli. Doctoral thesis, University of Amsterdam

Chandry PS, Moore SC, Boyce JD, Davidson BE & Hillier AJ (1997) Analysis of the DNA sequence, gene expression, origin of replication and modular structure of the *Lactococcus lactis* lytic bacteriophage sk1. Mol. Microbiol. 26: 49–64

Chassy BM & Flickinger JL (1987) Transformation of *Lactobacillus casei* by electroporation. FEMS Microbiol. Lett. 44: 173–177

Christiansen B, Johnsen MG, Stenby E, Fogensen FK & Hammer K (1994) Characterization of the lactococcal phage TP901-1 and its site-specific integration. J. Bacteriol, 176: 1069–1076

Christiansen B, Brønsted L, Vogensen FK & Hammer K (1996) A resolvase-like protein is required for the site-specific integration of the temperate lactococcal bacteriophage TP901-1. J. Bacteriol. 178: 5164–5173

Chung K, Chung SK & Batt CA (1992) Antisense RNA directed against the major capsid protein of *Lactococcus lactis* subsp. *cremoris* 4-1 confers partial resistance to the host. Appl. Microbiol. Biotechnol. 37: 79–83

Curic M, Stuer-Lauridsen B, Renault P & Nilsson D (1999) A general method for selection of α-acetolactate decarboxylase deficient mutants of *Lactococcus lactis* to improve diacetyl formation. Appl. Environ. Microbiol. 65: 1202–1206

Delorme C, Ehrlich SD & Renault P (1999) Regulation of the expression of the *Lactococcus lactis* histidine operon. J. Bacteriol. 181: 2026–2037

de Ruyter PGGA, Kuipers OP & de Vos WM (1996) Controlled gene expression systems for *Lactococcus lactis* with the food-grade inducer nisin. Appl. Environ. Microbiol. 62: 3662–3667

de Ruyter PGGA, Kuipers OP, Meyer WC & de Vos WM (1997) Food- grade controlled lysis of *Lactococcus lactis* for accelerated cheese ripening. Nature Biotechn. 15: 976–979

de Sazieu A, Certa U, Warrington J, Gray C, Keck W & Mous J (1998) Bacterial transcript imaging by hybridization of total RNA to oligonucleotide arrays. Nature Biotech. 16: 45–48

Desiere F, Lucchini S, Bruttin A, Zwahlen M-C & Brüssow H (1997) A highly conserved DNA replication module from *Streptococcus thermophilus* phages is similar in sequence and topology to a module from *Lactococcus lactis* phages. Virology 234: 372–382

Desiere F, Lucchini S & Brüssow H (1998) Evolution of *Streptococcus thermophilus* bacteriophage genomes by modular exchanges followed by point mutations and small deletions and insertions. Virology 241: 345–356

de Vos WM, Vos P, Simons G & David S (1989) Gene organization and expression in mesophilic lactic acid bacteria. J. Dairy Sci. 72: 3398–3405

de Vos WM & Simons G (1994) Gene cloning and expression systems in lactococci. In: Gasson MJ & de Vos WM, (Eds) (pp. 52–105). Genetics and biotechnology in lactic acid bacteria. Blackie Academic and Professional, Glasgow

de Vos WM (1996) Metabolic engineering of sugar metabolism in lactic acid bacteria. Antonie van Leeuwenhoek 70: 232–242

Djordjevic GM, O'Sullivan DJ, Walker A, Concling MA & Klaenhammer TR (1997) A triggered suicide system designed as a defence against bacteriophages. J. Bacteriol. 179: 6741–6748

Djordjevic GM & Klaenhammer TR (1997) Bacteriophage-triggered defence systems: phage adaptations and design improvements. Appl. Environ. Microbiol. 63: 4370–4376

Dupont L, Boizet-Bonhoure B, Coddeville M, Auvray F & Ritzenthaler P (1995) Characterization of genetic elements required for site-specific integration of *Lactobacillus delbrueckii* subsp. *bulgaricus* bacteriophage mv4 and construction of an integration-proficient vector for*Lactobacilus plantarum*. J. Bacteriol. 177: 586–595

Fujita Y, Miwa Y, Galinier A & Deutcher J (1995) Specific recognition of the *Bacillus subtilis gnt cis*-acting catabolite- responsive element by a protein complex formed between CcpA and seryl-phosphorylated HPr. Mol. Microbiol. 17: 953–960

Gasson MJ (1996) Lytic systems in lactic acid bacteria and their bacteriophages. Antonie van Leeuwenhoek 70: 147–159

Gostick DO, Green J, Irvine AS, Gasson MJ & Guest JR (1998) A novel regulatory switch mediates the FNR-like protein of *Lactobacillus casei*. Microbiology 144: 715–717

Gostick DO, Griffin HG, Shearman CA, Scott C, Green J, Gasson MJ & Guest JR (1999) Two operons that encode FNR-like proteins in *Lactococcus lactis*. Mol. Microbiol. 31: 1523–1535

Goupil-Feuillerat N, Cocaign-Bousquet M, Godon JJ, Ehrlich SD & Renault P (1997) Dual role of α-acetolactate decarboxylase in *Lactococcus lactis* subsp. *lactis*. J. Bacteriol. 179: 6285–6293

Green J, Trageser M, Six S, Unden G & Guest JR (1991) Characterization of the FNR protein of *Escherichia coli*, an iron binding transcriptional regulator. Proc. R. Soc. Lond. Ser B 244: 137–144

Hill C, Miller LA & Klaenhammer TR (1990) Cloning, expression and sequence determination of a bacteriophage fragment encoding resistance in *Lactococcus lactis*. J. Bacteriol. 172: 6419–6426

Holo H & Nes IF (1989) High frequency transformation, by electroporation of *Lactococcus lactis* subsp. *lactis* with glycine in osmoticaly stabilized media. Appl. Environ. Microbiol. 55: 3119–3123

Hols P, Kleerebezem M, Schanck AN, Ferain T, Hugenholtz J, Delcour J & de Vos WM (1999) Conversion of *Lactococcus lactis* from homolactic to homoalanine fermentation through metabolic engineering. Nature Biotech. 7: 588–592

Hueck CJ & Hillen W (1995) Catabolite repression in *Bacillus subtilis*: a global regulatory mechanism for the Gram-positive bacteria? Mol. Microbiol. 15: 395–401

Johnsen MG, Appel KF, Madsen PL, Vogensen FK, Hammer K & Arnau J (1996) A genomic region of lactococcal temperate phage TP901-1 encoding major virion proteins. Virology 218: 306–315

Kim SG & Batt CA (1991) Antisense mRNA-directed bacteriophage resistance in *Lactococcus lactis* subsp. *lactis*. Appl. Environ. Microbiol. 57: 1109–1113

Kodaira K-I, Oki M, Kakikawa M, Watanabe N, Hirakawa M, Yamada K & Taketo A (1997) Genome structure of the *Lactobacillus* temparate phage øg1e: the whole genome sequence and the putative promoter/repressor system. Gene 187: 45–53

Kok J, van der Vossen JMBM & Venema G (1984) Construction of plasmid cloning vectors which also replicate in *Bacillus subtilis* and *Escherichia coli*. Appl. Environ. Microbiol. 48: 726–731

Kraus A, Küster E, Wagner A, Hoffmann K & Hillen W (1998) Identification of a co-repressor binding site in catabolite repressor protein CcpA. Mol. Microbiol. 30: 955–963

Kuipers OP, de Ruyter PGGA, Kleerebezem M & de Vos WM (1997) Controlled overexpression of proteins by lactic acid bacteria. Trends Biotechn. 15: 135–140

Küster E, Luesink EJ, de Vos WM & Hillen W (1996) Immunological crossreactivity to the catabolite control protein CcpA from *Bacilus megaterium* is found in many Gram-positive bacteria. FEMS Microbiol. Lett. 139: 109–115

Law J, Buist G, Haandrikman A, Kok J, Venema G & Leenhouts K (1995) A system to generate chromosomal mutations in *Lactococcus lactis* which allow fast analysis of targetted genes. J. Bacteriol. 177: 7011–7018

Leenhouts K (1995) Integration strategies and vectors. In: Ferretti JJ, Gilmore MS, Kleinhammer TR & Brown F, (Eds) Genetics of Streptococci, Enterococci and Lactococci. (pp 523–530). Dev. Biol. Stand. Basel, Karger

Leenhouts KJ, Kok J & Venema G (1991) Lactococcal plasmid pWVO1 as an integration vector for lactococci. Appl. Environ. Microbiol. 57: 2562–2567

Leenhouts K, Buist G, Bolhuis A, ten Berge A, Kiel J, Mierau I, Dabrowska M, Venema G & Kok J (1996) A general system for generating unlabelled gene replacements for lactococci. Mol. Gen. Genet. 353: 217–224

Leenhouts K, Bolhuis A, Venema G & Kok J (1998) Construction of a food-grade multiple-copy integration system for *Lactococcus lactis*. Appl. Microbiol. Biotechnol. 49: 417–423

Lieb M (1981) A fine structure map of spontaneous and induced mutations in the lambda repressor gene, including insertions of IS elements. Mol. Gen. Genet. 184: 364–371

Lillehaug D & Birkeland NK (1993) Characterization of genetic elements required for site-specific integration of the temperate lactococcal bacteriophage ØLC3 and construction of integration-negative ØLC3 mutants. J. Bacteriol. 175: 1745–1755

Lillehaug D, Nes IF & Birkeland N-K (1997) A highly efficient and stable system for site-specific integration of genes and plasmids into ØLC3 attachment site (*attB*) of the *Lactococcus lactis* chromosome. Gene 172: 6419–6426

Lokman BC, Leer RJ, van Sorge R & Pouwels PH (1994) Promoter analysis and transcriptional regulation of *Lactobacillus pentosus* genes involved in xylose catabolism. Mol. Gen. Genet. 245: 117–125

Lokman BC, Heerikhuisen M, Borsboom Y & Pouwels PH (1997) Characterization of the *Lactobacillus pentosus ccpA* gene and its involvement in catabolite repression. J. Bacteriol. 179: 5391–5397

López R, García E, García P & García JL (1995) Architecture and domain interchange of the pneumococcal cell wall lytic enzymes. In: Ferretti JJ, Gilmore MS, Klaenhammer TR & Brown F, (Eds) Genetics of Streptococci, Enterococci and Lactococci. (pp. 273–281). Dev. Biol. Stand., Basel, Karger.

Lubbers MW, Waterfield NR, Beresford TPJ, Le Page RWF & Jarvis AW (1995) Sequencing and analysis of the prolate-headed lactococcal bacteriophage c2 genome and identification of the structural genes. Appl. Environ. Microbiol. 61: 4348–4356

Lucchini S, Desiere F & Brüssow H (1998) The structural gene module in *Streptococcus thermophilus* bacteriophage øSfi11 shows a hierarchy of relatedness to Siphoviridae from a wide range of bacterial hosts. Virology 246: 63–73

Luchansky JB, Muriana PM & Kaenhammer TR (1988) Application of electroporation for transfer of plasmid DNA into *Lactobacillus, Leuconostoc, Listeria, Pediococcus, Bacillus, Staphylococcus, Enterococcus* and *Propionibacterium*. Mol. Microbiol. 2: 637–646

Luesink EJ, van Herpen REMA, Grossiord BP, Kuipers OP and de Vos WM (1998) Transcriptional activation of the glycolytic *las* operon and catabolite repression of the *gal* operon in *Lactococcus lactis* are mediated by the catabolite control protein CcpA. Mol. Microbiol. 30: 789–798

Luesink EJ, Marugg JD, Kuipers OP & de Vos WM (1999a) Characterization of the divergent *sacBK* and *sacAR* operons involved

in sucrose utilization by *Lactococcus lactis*. J. Bacteriol. 181: 1924–1926
Luesink EJ, Beumer CMA, Kuipers OP & de Vos WM (1999b) Molecular characterization of the *Lactococcus* lactis *ptsH1* operon and analysis of the regulatory role of HPr. J. Bacteriol. 181: 764–771
Madsen PL & Hammer K (1998) Temporal transcription of the lactococcal temperate phage TP901-1 and DNA sequence of the early promoter region. Microbiology 144: 2203–2215
Madsen SM, Arnau J, Vrang A, Givskov M & Israelsen H (1999) Molecular analysis and characterization of a pH and stationary phase inducible promoter P170 from *Lactococcus lactis*. Mol. Microbiol. 32: 75–87
Maguin E, Duwat P, Hege T, Ehrlich D & Gruss A (1992) New thermosensitive plasmid for Gram-positive bacteria. J. Bacteriol. 174: 5633–5638
Melvile SB & Gunsalus RB (1990) Mutations in *fnr* that alter anaerobic regulation of electron- associated genes in *Escherichia coli*. J. Biol. Chem. 265: 18733–18736
Mikkonen M (1996) Doctoral thesis, University of Oulu, Finland
Monodero V, Gosalbes MJ & Pérez-Martínez G (1997) Catabolite repression in *Lactobacillus casei* is mediated by CcpA. J. Bacteriol. 179: 6657–6664
Nauta A, van Sinderen D, Karsens H, Smit E, Venema G & Kok J (1996) Inducible gene expression mediated by a repressor-operator system isolated from *Lactococcus lactis* bacteriophage r1t. Mol. Microbiol. 19: 1331–1341
Nauta A, van den Burg B, Karsens H, Venema G & Kok J (1997) Design of thermolabile bacteriophage repressor mutants by comparative molecular modelling. Nature Biotech. 15: 980–983
Nauta A (1997) Molecular characterization and exploitation of the temperate *Lactococcus lactis* bacteriophage r1t. Doctoral thesis, University of Groningen
Neve H, Zenz KI, Desiere F, Koch A, Heller KJ & Brüssow H (1998) Comparison of the lysogeny modules from the temperate *Streptococcus thermophilus* bacteriophages TP-J34 and Sfi21: implications for the modular theory of phage evolution. Virology 241: 61–72
O'Sullivan DJ, Hill C & Klaenhammer TR (1993) Effect of increasing the copy number of bacteriophage origins of replication, in *trans*, on incoming-phage proliferation. Appl. Environ. Microbiol. 5449–2456
O'Sullivan DJ, Zagula K & Klaenhammer TR (1995) *In vivo* restriction by *LlaI* is encoded by three genes, arranged in an operon with *llIM*, on the conjugative *Lactococcus* plasmid pTR2030. J. Bacteriol. 177: 134–143
O'Sullivan DJ, Walker SA, West SG & Klaenhammer TR (1996) Development of an expression strategy using a lytic phage to trigger explosive plasmid amplification and gene expression. Biotechnology 10: 82–87
Pabo CO, Sauer RT, Sturtevant JM & Ptashne M (1979) The λ repressor contains two domains. Proc. Natl. Acad. Sci. USA 76: 1608–1612
Parreira R, Valyasev R, Lerayer ALS, Ehrlich SD & Chopin M-C (1996) Gene organisation and transcription of a late-expressed region of a *Lactococcus lactis* phage. J. Bacteriol. 178: 6158–6165
Payne J, MacCormic CA, Griffin HG & Gasson MJ (1996) Exploitation of a chromosomally integrated lactose operon for controlled gene expression in *Lactococcus lactis*. FEMS Microbiol. Lett. 136: 19–24
Ptashne M (1992) A genetic switch. Phage λ and higher organisms. Blackwell Scientific Publications and Cell Press, Cambridge, Mass.
Ramseier TM, Reizer J, Küster E, Hillen W. & Saier MH (1995) In vitro binding of the CcpA protein of *Bacillus megaterium* to *cis*-acting catabolite responsive elements (CRE's) of Gram- positve bacteria. FEMS Microbiol. Lett. 129: 207–214
Ramsay G (1998) DNA chips: state-of-the-art. Nature Biotech. 16: 40–48
Raya RR, Fremaux C, De Antoni GL & Klaenhammer TR (1992) Site- specific integration of the temperate bacteriophage Øadh into the *Lactobacillus gasseri* chromosome and molecular characterization of the phage (*attP*) and bacterial (*attB*) attachment sites. J. Bacteriol. 174: 5584–5592
Sanders JW, G Venema & Kok J (1997) A chloride-inducible gene expression cassette and its use in induced lysis of *Lactococcus lactis*. Appl. Environ. Microbiol. 63: 4877–4882
Sanders JW, Venema G, Kok J & Leenhouts K (1998a) Identification of sodium chloride-regulated promoters in *Lactococcus lactis* by single-copy chromosomal fusions with a reporter gene. Mol. Gen. Genet. 257: 681–685
Sanders JW, Leenhouts K, Burghoorn J, Brands JR, Venema G & Kok J (1998b) A chloride-inducible acid resistance mechanism in*Lactococcus lactis* and its regulation. Mol. Microbiol. 27: 299–310
Sasaki T, Ito Y & Sasaki Y (1993) Proc. 4th Symp. on lactic acid bacteria; genetics, metabolism and applications; de Vos WM, JMJ Huis in 'tVeld, & Poolman B., (Eds). Special issue FEMS Microbiol. Rev., poster abstract P8
Schena M, Heller RA, Theriault TP, Konrad K, Lachenmeier E & Davis RW (1998) Microarrays: biotechnology's discovery platform for functional genomics. Trends in Biotechn. 16: 301–306
Schmid R, Bernhardt J, Antelmann H, Volker A, Volker U & Hecker M (1997) Identification of vegetative proteins for a two-dimensional protein index of *Bacillus subtilis*. Microbiology 143: 991–998
Schnetz K, Stülke J, Gertz S, Krüger S, Krieg M, Hecker M & Rak B (1996) LicT, a *Bacillus subtilis* transcriptional antiterminator protein of the BglG family. J. Bacteriol. 178: 1971–1979
Schouler C, Ehrlich SD & Chopin M-C (1994) Sequence and organization of the lactoccal prolate-headed bIL67 phage genome. Microbiology 140: 3061–3069
Sharrocks AD, Green J & Guest JR (1990) *In vivo* and *in vitro* mutants of FNR, the anaerobic transcriptional regulator of *E. coli*. FEBS Lett. 270: 119–132
Shearman CA, Underwood HM, Jury K & Gasson M (1989) Cloning and DNA sequence analysis of a *Lactococcus* bacteriophage lysis gene. Mol. Gen. Genet. 218: 214–221
Sheehan MM, García JL, López R & García P (1997) The lytic enzyme of the pneumococcal phage DP-1: a chimeric lysin of intergeneric origin. Mol. Microbiol. 25: 717–725
Simonen S & Palva I (1993) Protein secretion in *Bacillus* species. Microbiol. Rev. 57: 109–137
Spiro S & Guest JR (1990) FNR and its role in oxygen-regulated gene expression in *Escherichia coli*. FEMS Microbiol. Rev. 75: 399–428
Spiro S (1994) The FNR family of transcriptional regulators. Antonie van Leeuwenhoek 66: 23–36
Stanley E, Fitzgerald GF, Le Marrec C, Fayard B & van Sinderen D (1997) Sequence analysis and characterization of øO1205, a temperate bacteriophage infecting *Streptococcus thermophilus* CNRZ1205. Microbiology 143: 3417–3429
Stülke J, Arnaud M, Rapoport G & Martin-Verstraete I (1998) PRD-a protein domain involved in PTS-dependent induction and carbon catabolite repression of catabolic operons in bacteria. Mol. Microbiol. 28: 865–874

Swindell SR, Benson KH, Griffin HG, Renault P, Ehrlich SD & Gasson MJ (1996) Genetic manipulation of the pathway for diacetyl metabolism in *Lactococcus lactis*. Appl. Environ. Microbiol. 62: 2641–2643

Vagner V, Dervyn E & Ehrlich SD (1998) A vector for systematic gene inactivation. Microbiology 144: 3097–3104

van de Guchte M, Kok J & Venema G (1992) Gene expression in*Lactococcus lactis*. FEMS Microbiol. Rev. 88: 73–92

van de Guchte M, Daly C, Fitzgerald GF & Arendt EK (1994) Identification of *int* and *attP* on the genome of lactococcal bacteriophage Tuc2009 and their use for site-specific plasmid integration in the chromosome of Tuc2009-resistant *Lactococcus lactis* mg1363. Appl. Environ. Microbiol. 60: 2324–2329

van de Guchte M, Ehrlich SD & Chopin A (1998) $tRNA^{TRP}$ as a key element of antitermination in the *Lactococcus lactis trp* operon. Mol. Microbiol. 29: 61–74

van Sinderen D, Karsens H, Kok J, Terpstra P, Ruiters MHJ, Venema G & Nauta A (1996) Sequence analysis and molecular organization of the temperate lactococcal bacteriophage r1t. Mol. Microbiol. 19: 1343–1355

van Sinderen D, Seegers J, McGrath S, Arend E, Leach S, van de Guchte M, Daly C & Fitzgerald[1] GF, (unpublished)

van Sinderen D, Walsh L, Stanley E & Fitzgerald[2] GF, (unpublished)

Verrips CT & van den Berg DJC (1996) Barriers to application of genetically modified lactic acid bacteria. Antonie van Leeuwenhoek 70: 299–316

Vriend G (1990) WHAT IF, a molecular modelling and drug design program. J. Molec. Graph. 8: 52–56

Walker SA & Klaenhammer TR (1998) Molecular characterization of a phage-inducible middle promoter and its transcriptional activation from the lactococcal bacteriophage Ø31. J. Bacteriol. 180: 921–931

Weickert MJ & Chambliss GH (1990) Site directed mutagenesis of a catabolite repression operator sequence in *Bacillus subtilis*. Proc. Natl. Acad. Sci. USA 87: 6238–6242

Young R (1992) Bacteriophage lysis: mechanisms and regulation. Microbiol. Rev. 56: 430–481

Young R & Bläsi (1995) Holins: form and function in bacteriophage lysis. FEMS Microbiol. Rev. 17: 191–205

GENETICS

Antonie van Leeuwenhoek **76:** 27–76, 1999.

Low-redundancy sequencing of the entire *Lactococcus lactis* IL1403 genome

Alexander Bolotin, Stéphane Mauger, Karine Malarme, S. Dusko Ehrlich & Alexei Sorokin*
*Génétique Microbienne, INRA, Domaine de Vilvert, 78352 Jouy en Josas cedex, France. (*Author for correspondence; E-mail: sorokine@biotec.jouy.inra.fr)*

Abstract

Lactococcus lactis is an AT-rich gram positive bacterium phylogenetically close to the genus Streptococcus. Various strains of *L. lactis* are used in dairy industry as starters for cheese making. *L. lactis* is also one of the well characterized laboratory microorganisms, widely used for studies on physiology of lactic acid bacteria. We describe here a low redundancy sequence of the genome of the strain *L. lactis* IL1403. The strategy which we followed to determine the sequence consists of two main steps. First, a limited number of plasmids and λ-phages that carry random segments of the genome were sequenced. Second, sequences of the inserts were used for production of novel sequencing templates by applying Multiplex Long Accurate PCR protocols. Using of these PCR products allowed to determine the sequence of the entire 2.35 Mb genome with a very low redundancy, close to 2. The error rate of the sequence is estimated to be below 1%. The correctness of the sequence assembly was confirmed by PCR amplification of the entire *L. lactis* IL1403 genome, using a set of 266 oligonucleotides. Anotation of the sequence was undertaken by using automatic gene prediction computer tools. This allowed to identify 1495 protein-encoding genes, to locate them on the genome map and to classify their functions on the basis of homology to known proteins. The function of about 700 genes expected to encode proteins that lack homologs in data bases cannot be reliably predicted in this way. The approach which we used eliminates high redundancy sequencing and mapping efforts, needed to obtain detailed and comprehensive genetic and physical maps of a bacterium.

Availability of detailed genetic and physical maps of the *L. lactis* IL1403 genome provides many entries to study metabolism and physiology of bacteria from this group. The presence of 42 copies of five different IS elements in the IL1403 genome confirms the importance of these elements for genetic exchange in Lactococci. These include two previously unknown elements, present at seven and fifteen copies and designated IS1077 and IS983, respectively. Five potential or rudimentary prophages were identified in the genome by detecting clusters of phage-related genes. The metabolic and regulatory potential of *L. lactis* was evaluated by inspecting gene sets classified into different functional categories. *L. lactis* has the genetic potential to synthesise 20 standard amino acids, purine and pyrimidine nucleotides and at least four cofactors. Some of these metabolites, which are usually present in chemically defined media, can probably be omitted. About twenty compounds can be used by *L. lactis* as a sole carbon source. Some 83 regulators were revealed, indicating a regulatory potential close to that of *Haemophilus influenzae*, a bacterium with a similar genome size. Unexpectedly, *L. lactis* has a complete set of late competence genes, which may have concerted transcriptional regulation and unleadered polycistronic mRNAs. These findings open new possibilities for developing genetic tools, useful for studies of gene regulation in AT-rich gram positive bacteria and for engineering of new strains for the diary industry.

Introduction

Lactococcus lactis is probably the best characterised lactic acid bacterium, with highly developed knowledge on physiology and genetics. It is therefore an obvious target for genome sequencing, since the availability of the full sequence would greatly improve the efficiency of numerous studies that are currently being conducted with this organism, or are planned in the near future. We decided to undertake the *L. lactis* genome sequencing project in an academic setting, with resources of a single laboratory, which obviates the need to raise a rather large amount of funding which would be required for sequence determination by a private company, or by a consortium of academic laboratories. For this purpose, we developed an aproach that allowed us to determine the genome sequence with a redundancy much lower than generally used, and at a commensurately lower cost. Due to the low redundancy, the sequence is of a so-called diagnostic quality, allowing to detect essentially all genes that have homologs in other organisms, but not the genes which may be present in the lactococcus genome only.

The approach most often used for sequencing of an entire bacterial genome is based on a high-throughput random clone sequencing (Fleischmann et al. 1995). This method is most efficient if the facilities to produce raw reading data are not limiting. However, the speed of accumulation of new sequencing data decreases exponentially with the linear increase of the number of randomly sequenced templates (Lander & Waterman 1988). This becomes a limiting factor, in particular for relatively large bacterial genomes. Furthermore, the assembly becomes exceedingly difficult if the genome contains numerous repeated elements. The chromosome of *L. lactis* IL1403 was estimated to have a size of about 2.4 Mb (Le Bourgeois et al. 1992). Application of the whole-genome random sequencing and assembly method (Fleischmann et al. 1995) would require producing of 30,000 random gel readings, corresponding to six-fold coverage, and the sequence would be distributed in about 80 contigs. We considered an alternative strategy that would consist in a much less massive random sequencing followed by primer walking on templates generated by Long Range PCR (LR PCR, Cheng et al. 1994). If sequence acquisition and assembly were carried out in parallel, template synthesis can be conducted in such a way that the re-sequencing of the regions that were sequenced already is kept at a minimum. We estimated that such approach may allow to sequence the whole *L. lactis* genome with a coverage of only about two. Some 10,000 sequencing reactions instead of 30,000 would therefore be required. We report here the *L. lactis* genome sequence determined by this approach.

Low-redundancy genome sequencing

The strategy of low-redundancy *L. lactis* genome sequencing is presented in Table 1. This strategy is a compromise between completely random and directed approaches. Its objective was to reduce time and effort for obtaining complete information about gene content and genome organisation. In the first stage, a limited number of randomly chosen clones is sequenced, so that the speed of new sequence accumulation keeps approximately constant. We estimated that this condition is fulfilled up to the genome coverage of about one. A large number of small contigs, separated by sequencing gaps, is present at the end of the first stage. In the second stage randomly chosen clones carrying large inserts are sequenced by primer walking. This allows to close some sequencing gaps and extend the contigs, as verified by sequence assembly, which is carried out in parallel with the acquisition of new sequence. The redundancy is kept low by selecting the primers that correspond to the ends of the extended contigs for the next round of primer walking. This stage is carried out until the rate of acquisition of the new sequence is above about 1 new base for 3 sequenced bases. Finishing of the sequence is achieved by applying another directed procedure, which was designated Multiplex Long Accurate PCR (MLA PCR) and tested in the course of our participation in the *Bacillus subtilis* sequencing project (Sorokin et al. 1996; Bolotin et al. 1997; Lapidus et al. 1997). It involves mixing a large number of primers that correspond to the ends of contigs. A product will be obtained each time when the distance between two sites on the genome, corresponding to two contig ends, is less than the maximal size that can be synthesised by LR PCR. The probability of producing such pruducts, as a function of the genome size and the number of primers present in LR PCR reaction is shown in Figure 1. For the genome of *L. lactis* size this probability is between 0.5 and 1 if some 20 primers are mixed (that is at least half of PCR reactions containing 20 randomly chosen primers will give an amplification product). It should be noted that the efficiency of gap closure is expected to increase with the progression of the sequencing project, since the number of contigs continuously decreases. This is in sharp contrast with the fully random approach, where the efficiency of gap closure decreases exponentially with sequence acquisition. We now believe that the long-insert library construction can be omitted.

Some statistical data of application of this strategy for the *L. lactis* sequencing are presented in Table 2. A library consisting of 2854 clones with insert sizes in the range from 1 to 2 kb, was constructed by using pBluescript II KS+ (Stratagene) or pSGMU2 (Errington 1986) vectors. 2625 clones were sequenced by forward (M13-21) and 2168 by reverse (M13RP1)

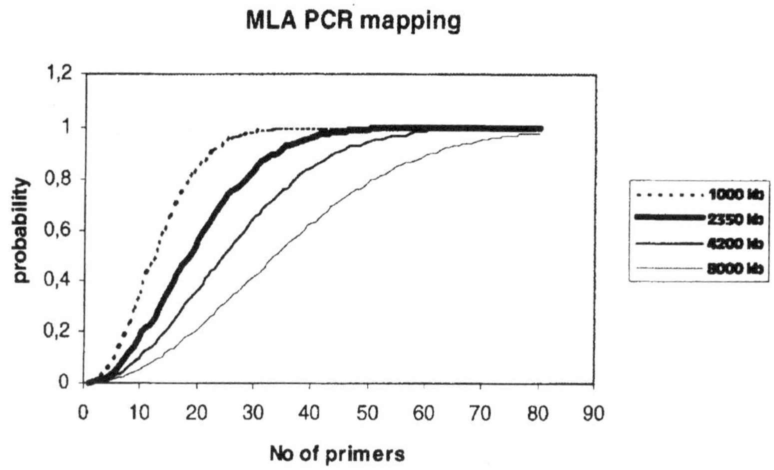

Figure 1. Probability of successful amplification in MLA PCR as a function of the number of oligonucleotides used in the same mixture. The calculation was done by generating the results of 10,000 independent random experiments and averaging these results. The longest distance between primers still allowing the amplification was supposed to be 20 kb. The genome sizes used for calculation are: 1 Mb (dashed line); 2.35 Mb (bold line); 4.2 Mb (thin line); 8 Mb (dotted line).

primers with successful sequencing rate of about 90%. After accumulating about 2100 kb of sequence 2357 oligonucleotides were synthesised to fill the gaps between the forward and reverse sequences. A total of about 3.3 Mb of sequence was thus acquired.

λ-FIXII vector (Stratagene) was used to construct a long insert library. *L. lactis* chromosome was partially digested with *Sau*3A, fractionated by centrifugation in sucrose gradient, treated with Klenow polymerase in the presence of dGTP and dATP, and ligated with λ-FIXII arms cut by *Xho*I and treated with Klenow in the presence of dCTP and dTTP. 262 phages were chosen at random and the ends of inserts sequenced by the T7 primer (Stratagene). Of these, 122 phages which had a unique sequence obtained by the T7 primer, were sequenced by the T3 primer (Stratagene). These 122 clones appeared to constitute a nonredundant set and their inserts were sequenced by primer walking (see Table 2). Some 250 kb of sequence were obtained in this way.

MLA PCR was used to obtain new sequencing substrates. A critical step in the procedure is to identify in the mixture the two primers that give a product useful for sequencing. The previously developed protocol (Sorokin et al. 1996), where two steps were required for the identification, was modified here so that only a single step is required. The principle of the modification is illustrated in Figure 2A. In this theoretical example each reaction contains 8 different primers. A total of 16 primers is used (labelled 1–16 on the left) and there are 16 reactions (labelled 1–16 on the top). To obtain a different primer combination in each reaction one primer is replaced at a time with a new primer (for instance, reaction 8 contains primers 1–8, reaction 9 contains primers 2–9, etc). The primer couples which give a useful product can be identified unambigously, by simply identifying the reaction mixtures in which the product is synthesized, as illustrated in the three lower rows. For instance, if a product were found only in the reaction 8, primers 1 and 8 would be required for it, since only the reaction 8 contains these two primers together. Alternatively, if a product were found in reactions 4 to 10, primers 3 and 4 would be required. A general correlation between positive reactions and primer couples is given in the legend of Figure 2. An example of application of this protocol to the IL1403 genome, using 32 oligonucleotides in each reaction, is shown on Figure 2B (the analysis of 36 reactions out of a total of 96 that were routinely run together is displayed). There are 4 different products in the reaction mixtures, ranging between 1 and 10 kb. The largest product appears in reactions 7 to 34, identifing that the primers 7 and 3 are required for its synthesis (see legend, Figure 2). The primers identified in this way were used to sequence the ends of the PCR products and thus confirm the identification and to extend the sequence. In total 1641 such sequencing reactions on the products with sizes from 1 to 20 kb were performed and about 0.77 Mb of sequence was read (Table 2). This step allowed to finish the entire chromosome assembling, giving a contig of 2.34 Mb. The total redundancy was close to 2 (Table 2).

Table 1. Low-redundancy whole-genome sequencing strategy

Stage	Description
Random small insert and large insert library construction	Shearing DNA randomly to 1–2 kb and 15–20 kb, respectively. Cloning in multicopy plasmids and λ-based FIXII vector.
Random sequencing of a limited number of plasmid inserts	Sequencing of plasmid inserts up to coverage of about one. Using forward (M13-21), reverse (M13RP1) and middle primers. The latter are chosen to sequence the middle of plasmid inserts, when necessary. Assembling is done in parallel with sequencing.
Random sequencing of a number of large inserts by primer walking.	Sequencing of large inserts by using synthetic oligonucleotides. Assembling is done in parallel with sequencing to avoid too high redundancy.
Synthesis of new sequencing substrates by MLA PCR approach.	Application of Multiplex Long Accurate PCR protocol for random generation of new sequencing substrates.
Sequencing of MLA PCR products.	Sequencing of MLA PCR products by primer walking. Assembling is done in parallel with sequencing.
Iteration	Iteration of the last two stages until the whole contigous genome sequence is obtained.
Annotation	On-line annotation is done in parallel with the sequencing progress by using automatic computer tools based on BLASTx and FASTx algorithms.

To test the correctness of the assembling we carried out LR PCR amplification of the entire genome using 266 oligonucleotides, predicted to be separated by distances 10–20 kb. The expected products were obtained (Figure 3), which indicates that the assembling is correct. The sequences of these 266 oligonucleotides are presented at http://spock.jouy.inra.fr.

The average redundancy of the *L. lactis* genome sequence is two. Such low redundancy does not allow to eliminate all errors by comparing multiple sequences corresponding to the same genome region. The error rate of this sequence is therefore determined by individual gel reading quality. We counted the number of ambiguities in the raw sequence data used for the assembling and found 46,954 ambiguities for a total of 4,687,630 bases. From this estimation the accuracy of the *L. lactis* diagnostic sequence is about 1% (one ambiguity/100 bases). However, assembling usually decreases the number of errors in the final consensus sequence and the real error rate may be lower. We therefore carried out a different quality test, aiming to detect the level of frame-shift errors in the open reading frames present in the genome sequence. For this purpose we used the GENMARK programme (Borodovsky & McIninch 1993), which predicts the genome regions statistically fit to encode proteins. Some 3544 protein coding regions were thus detected. If the real density of genes in *L. lactis* were similar to that observed in other bacterial genomes, we would expect only about 2300 such regions. There should therefore be over 1200 frame-shift errors in our sequence (the difference between 3544 regions detected and 2300 expected), corresponding to about 1 error for two genes, or 2 kb. We suggest therefore that the overall error rate of the diagnostic quality sequence after assembling, at a redundance of 2, may be close to 1 in 1000.

Genome analysis

GC skew, replication origin and terminus. *L. lactis* IL1403 chromosome has the size of 2345 ± 5 kb. The average GC-content is 35.4%, its variation over

A

	1	2	3	4	5	6	7	8	9	10	11	12	13	14	15	16
1	+	+	+	+	+	+	+	+								
2		+	+	+	+	+	+	+	+							
3			+	+	+	+	+	+	+	+						
4				+	+	+	+	+	+	+	+					
5					+	+	+	+	+	+	+	+				
6						+	+	+	+	+	+	+	+			
7							+	+	+	+	+	+	+	+		
8								+	+	+	+	+	+	+	+	
9									+	+	+	+	+	+	+	+
10	+									+	+	+	+	+	+	+
11	+	+									+	+	+	+	+	+
12	+	+	+									+	+	+	+	+
13	+	+	+	+									+	+	+	+
14	+	+	+	+	+									+	+	+
15	+	+	+	+	+	+									+	+
16	+	+	+	+	+	+	+									+
Product 1x8	n	n	n	n	n	n	n	y	n	n	n	n	n	n	n	n
Product 3x4	n	n	n	y	y	y	y	y	y	y	n	n	n	n	n	n
Product 12x15	y	y	y	n	n	n	n	n	n	n	n	n	n	n	y	y

B

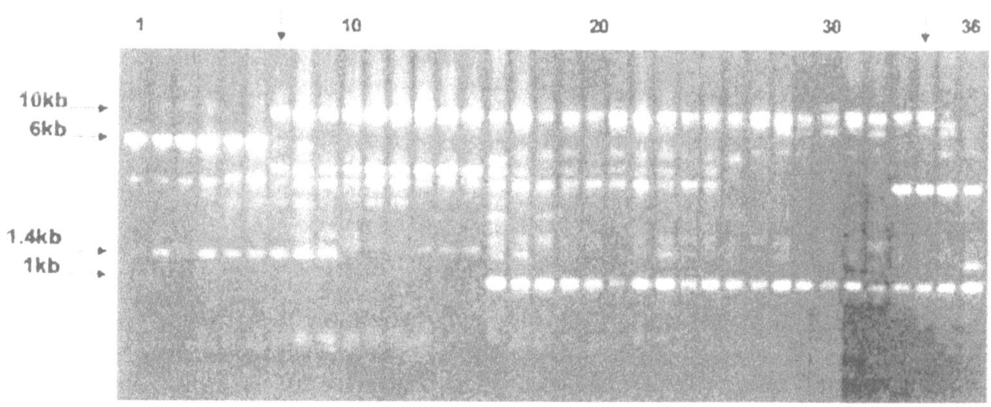

Figure 2. MLA PCR procedure. (A) The table illustrates a theoretical experiment with a total of 16 (M) primers, where each reaction mixture contains 8 (N) primers. The numbers at the top designate reaction mixtures, the numbers on the left designate individual primers, the '+' sign indicates presence of a given primer in a reaction mixture. The three last rows indicate the reaction mixtures in which amplification does (y) or does not (n) take place, if the primer couples 1 and 8, 3 and 4 or 12 and 15, respectively, give a product. To identify the primers which give a product from the analysis of the reactions the following rule is used: If N_1 and N_2 are the first and the last positive reaction, respectively, the primers giving the product will be: $P_1 = N_1$; $P_2 = N_2 - N + 1$, if $N_2 > N_1$ (the cases 1×8 and 3×4), or $P_2 = M + N_2 - N + 1$, if $N_2 < N_1$ (the case 12×15). (B) Part of a real experiment with 96 primers, where each reaction contained 32 primers and *L. lactis* IL1403 chromosome as template. Sizes and positions of identified fragments are shown on the left. Numbers on the top designate the reaction mix. Arrows on the top indicate the reactions that are used to identify the primers which gave the 10 kb fragment. These primers are $P_1 = N_1 = 7$, and $P_2 = N_2 - N + 1 = 34 - 32 + 1 = 3$.

the chromosome is shown on Figure 4. By analogy with the analysis performed on the *B. subtilis* genome (Kunst et al. 1997), the deviation from the mean GC-content may indicate the presence of foreign insertion

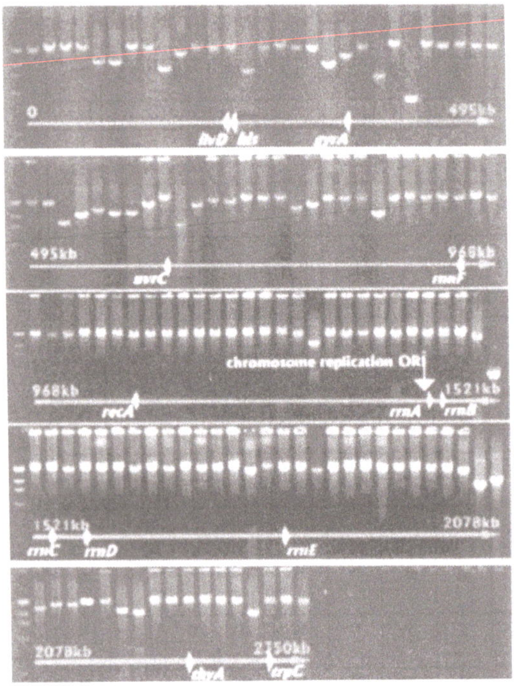

Figure 3. Whole-genome PCR amplification of the *L. lactis* IL1403 chromosome. 266 oligonucleotides were used in 133 LR PCR reactions to amplify the entire genome. Correspondence of the sizes of amplified fragments and the physical map predicted from the sequence (Figure 9) indicates the correctness of sequence assembling. Left slot on each panel shows size marker (DNA of phage λ cut by *Hind*III). The PCR product designations correspond to the ones shown on Figure 9. The first panel from the top, left to the right: BA7, BA8, BB1, BB2, CC1, CC3, BB3, BB4, CC5, CC6, BB5, BB6, BB7, BA1, CB8, BB8, BC1, BC2, CL1, CL2, BC4, BA2, BC5, BC6, BC7, CB3, BC8, BD1, BD2; second panel: BD3, CB4, CC7, CC8, CD1, CD2, CD3, CG4, BD4, CD4, cg7, BD5, BD6, BD7, BD8, BE1, CH4, CE1, BE2, BE3, BE4, BE5, BE6, BE7, BE8, BF1, BF2, BF3, BF4; third panel: BF5, BA3, BF6, BF7, BF8, BG1, BG2, BG3, BG4, BG5, BG6, BG7, BG8, BH1, BH2, BH3, BH4, BA4, BH5, BH6, BH7, BH8, CB5, BI1, BI2, BI3, BI4, DA1, DA2; fourth panel: BI5, BI6, BI7, BI8, BJ1, BJ2, BJ3, BJ4, BJ5, BJ6, BJ7, BJ8, BK1, CB7, BK2, BK3, BK4, BA5, BK5, BK6, BK7, BK8, BL1, BL2, BL3, BL4, BL5, BL6, CE3; fifth panel: CE4, CK5, CK7, BA6, BL8, CE7, CE8, CA1, CA2, CA3, CA4, CA5, CA6, CA7, CA8, CB1, CF3. The sequences of the used oligonucleotides can be found at http://spock.jouy.inra.fr. Chromosome drawing, which is complementary to the one shown on Figure 9, approximately corresponds to the PCR fragments above it.

Table 2. Summary of low-redundancy sequencing of the *L. lactis* IL1403 genome*

Small insert plasmids	2854
Average insert length (bp)	1520
Forward sequence reactions (M13-21 primer)	2625
Average useful sequence length (bp)	462
Reverse sequence reactions (M13RP1 primer)	2168
Average useful sequence length (bp)	422
Sequencing reactions in the middle of insert	2357
Average read length (bp)	494
Total characters from plasmid insert sequencing (bp)	3292004
Number of selected λ-clones	122
Average insert length (kb)	15.6
Number of gel readings from λ-inserts	622
Average read length (bp)	409
Total characters from λ-insert reads (bp)	254398
Number of gel readings from LR PCR products	1641
Average read length (bp)	470
Total characters read from LR PCR products (bp)	771270
Total oligonucleotides used	4620
Total gel readings	10235
Average read length (bp)	458
Total characters read (bp)	4687630
Size of final contig (bp)	2345160
Undefined letters in the final contig (bp)	3563

*The figures were estimated by analysis of the assembling database and are therefore approximate (±5%).

elements, such as transposons or prophages. We did not detect, however, in the *L. lactis* genome, a correlation of GC-content with location of these elements, determined by other approaches (see below). Several regions of highly elevated GC-content were revealed (Figure 4). Six such peaks with GC-content higher than 46%, at positions 540, 1970, 2210, 2250, 2325 and 2340 (Figure 4, bottom) correspond to rRNA operons. Another GC rich region, near position 2150 kb (Fig 4, bottom), corresponds to the cluster of ribosomal protein genes. We conclude that the regions encoding ribosomes have a higher GC content than the other regions in *L. lactis*. In two other AT-rich bacteria, *B. subtilis* and *Rickettsia prowazekii*, the regions encoding ribosomal genes are not particularly GC-rich (Kunst et al. 1997; Andersson et al. 1998).

GC skew, replication origin and terminus. GC or AT skews are useful to predict positions of replication origin and terminus (Lobry 1996a, b). The distributions of the values (C−G)/(C+G) and (A−T)/(A+T) over the chromosomal region in the vicinity of *dnaA* gene is shown in Figure 5, top. A sharp transition between positive and negative values is observed, and indicates the presence of the replication origin (Figure 5A). This region contains four DnaA boxes (Figure 6), which also indicates the presence of the replication origin. We choose the start point for the presentation of the

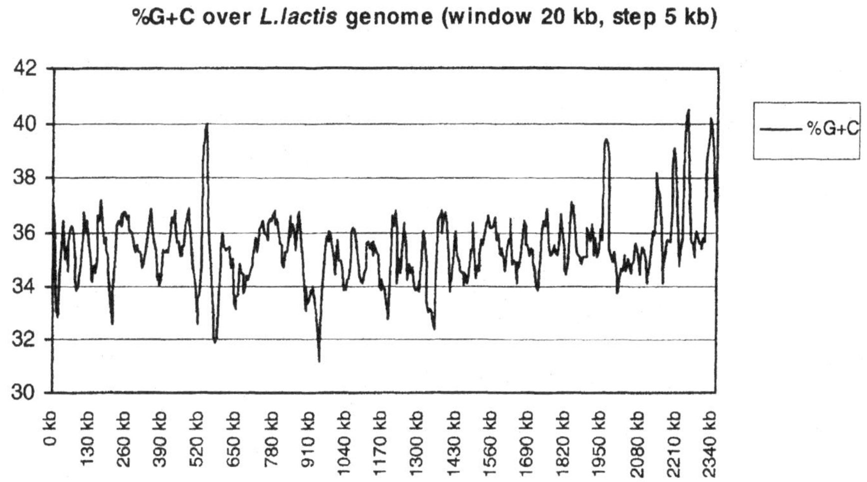

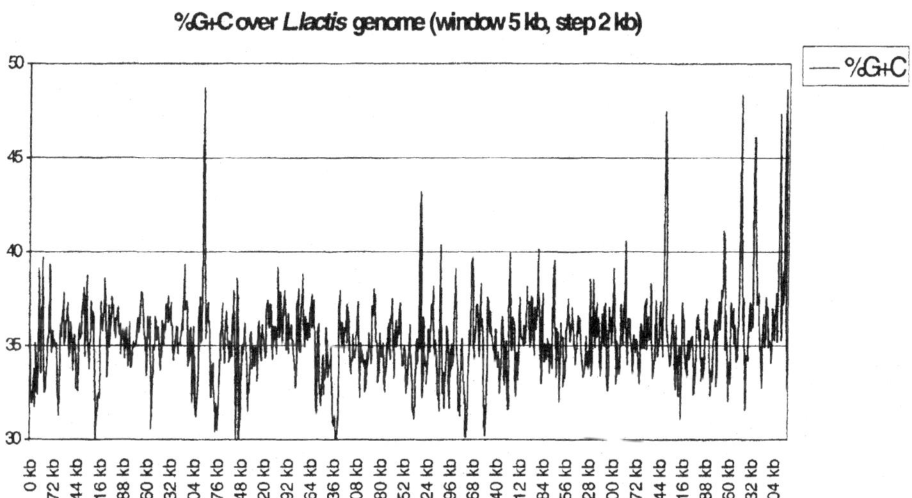

Figure 4. Distribution of the GC-content over the *L. lactis* chromosome. Zero point corresponds to the position of the origin of replication. Top panel: a sliding window of 20 kb and a step of 5 kb were used. Bottom panel: a sliding window of 5 kb and the step of 2 kb were used. The peaks near coordinates 540, 1970, 2210, 2250, 2325 and 2340 correspond to rRNA operons. A peak near 2150 kb corresponds to the cluster of ribosomal protein genes.

circular *L. lactis* genome map at the middle of *Hind*III site near the replication origin and the map is oriented to be colinear with the transcription direction of the *dnaA* and *dnaN* genes (see below, Figure 9).

GC and AT skews may also indicate location of the replication terminus. This is seen for *L. lactis* genome from Figure 5, bottom. The transition between positive and negative values occurs near the position 1260 kb. These data correlate with the replication terminus location based on orientation of transcription of potential genes and distribution of chi-sites over the genome (see below).

IS-elements and prophages in L.lactis IL1403. Three different IS elements in the genome of *L. lactis* IL1403 were known previously, designated IS981, IS982 and IS1076. Their number of copies (ten, one and seven, respectively) and approximate locations were reported

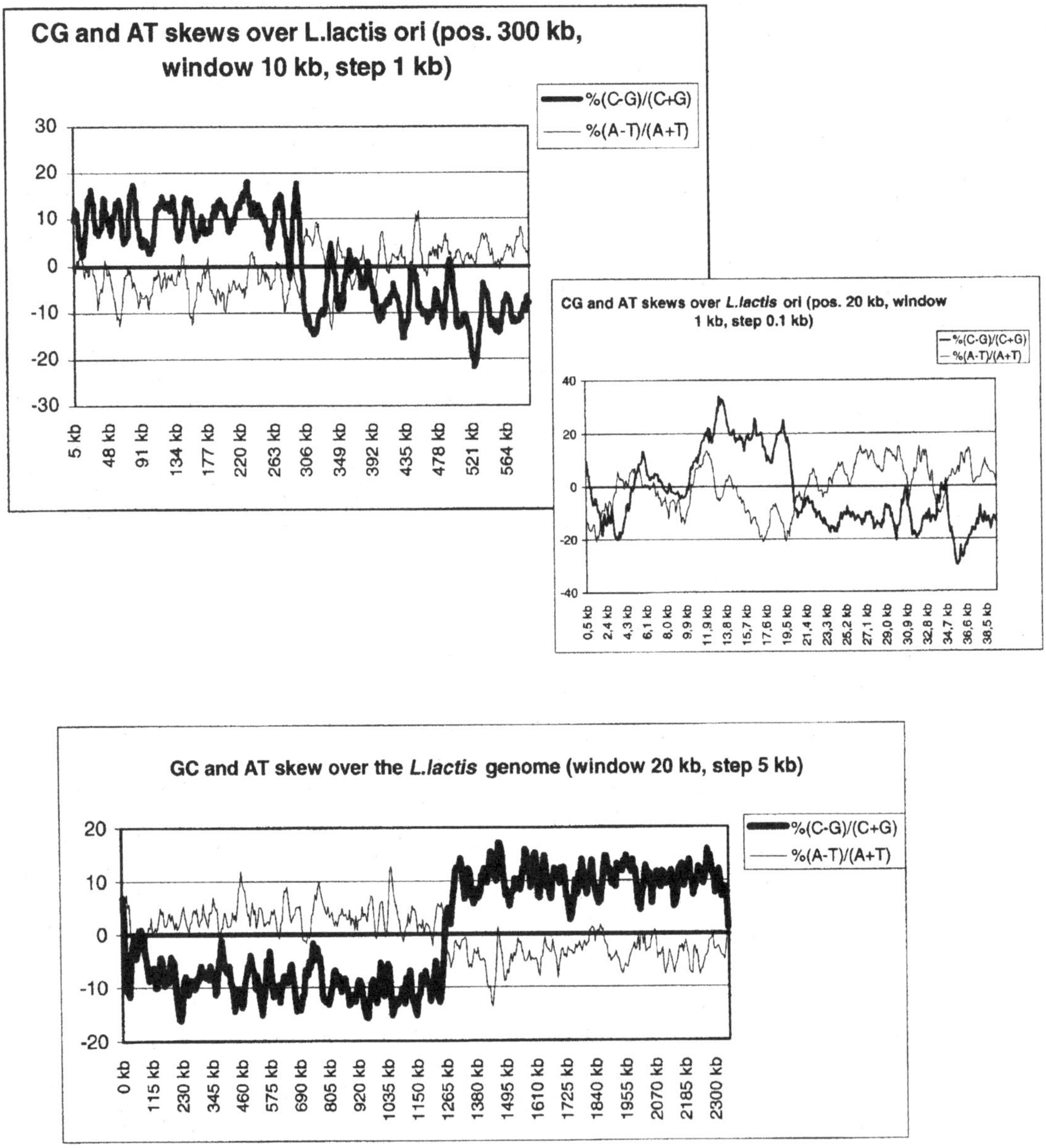

Figure 5. GC and AT skew distribution over the *L. lactis* genome. Top panel: a sliding window of 10 kb and step of 1 kb were used. Inset: a sliding window of 1 kb and step of 0.1 kb, were used. In both cases a region near the *dnaA* gene was analysed. Change of the skew distribution sign indicates the position of replication origin. Bottom panel: a sliding window 20 kb and a step of 5 kb were used. Zero coordinate is placed at the replication origin. The whole genome was analysed. Change of the skew distribution sign indicates the position of replication terminus.

previously (Le Bourgeois et al. 1992, 1995; Polzin et al. 1993; Schouler et al. 1998). The element IS1076 was initially identified in the strain *L. lactis* subsp. *lactis* CNRZ270 on the plasmid pUCL22 (Huang et al. 1991). Nucleotide sequence of this element (Genbank accession X53013) shows complete identity to IS904, known from nisin-producing lactococcus (Engelke et al. 1992). Our sequencing data confirmed that in all chromosome locations where IS1076 was mapped, the nucleotide sequence identical to IS904 is present, and we therefore keep the latter name on the map. We found, however, that another element, which we des-

tattaggcacgccgccagccagtcggtacaagtaccaaccttcagcgctcaacttgcatg 2344620
ataatccgtgcggcggtcggtcagccatgttcatggttggaagtcgcgagttgaacgtac

cgttcgtcctgagccaggatcaaactctcaaataaagtatttaagtcaccttagtgactg 2344680
gcaagcaggactcggtcctagtttgagagtttatttcataaattcagtggaatcactgac
<------| 16S rRNA *rrnA*
gcttgtctttcattattgattgacagattttgatgtatcttactacatcacatgagtca 2344740
cgaacagaaagtaataactaactgtctaaaaactacatagaatgatgtagtgtactcagt

ttcttctttattcagttttcaatggtctagcttatcgcttcaagaccctcgtctcttgcg 2344800
aagaagaaataagtcaaaagttaccagatcgaatagcgaagttctgggagcagagaacgc

acaactattatattctatcaaacctaactaaacttgtcaagaattaactgtgacttttat 2344860
tgttgataatataagatagtttggattgatttgaacagttcttaattgacactgaaaata

accttttttatcttacgttcgctttttccctctcctgtattcgctttatttcatgccac 2344920
tggaaaaaaatagaatgcaagcgaaaaagggagaggacataagcgaaataaagtacggtg

cactagctttggtaaattaaaaaaattagtctttttaacttgctttttttcttttttttaa 2344980
gtgatcgaaaccatttaatttttttaatcagaaaaaattgaacgaaaaaagaaaaaaatt

taagataaggtgcttattttagtcttctttaccttatgagaataaaattctttgattttt 2345040
attctattccacgaataaaatcagaagaaatggaatactcttattttaagaaactaaaaa

acgtagaaatgatgtttttaaacttaactaaattttttattataaatagtagaaaattgaa 2345100
tgcatctttactacaaaaatttgaattgatttaaaaataatatttatcatcttttaactt

ttctcttttattttattgcaaatagtttgtctctatattccgtctttatatagtaga**AAG** 0
aagagaaaataaaataacgtttatcaaacagagatataaggcagaaatatatcatct**TTC**

HindIII

CTTaaaatttaacaatctattttcaaattcttagattctattacattagagataaaatat 60
GAAttttaaattgttagataaaagtttaagaatctaagataatgtaatctctattttata

ccccaaagatagaaaagactgtctaaaagctgtggtaaatttttcttaatttttttctttt 120
ggggtttctatcttttctgacagatttttcgacaccatttaaaaagaattaaaaaagaaaa

tccacagtaaataatctataaatttgactttttttgatttgtggataagtctgtatgcat 180
aggtgtcatttattagatatttaaactgaaaaaaactaa**acacctatt**cagacatacgta
DnaA box
ttgatattactcatttttttctgtggataagaaaacttttacaca**ttatccaca**gttttg 240
aactataatgagtaaaaaaag**acacctatt**cttttgaaaatgtgtaataggtgtcaaaac
DnaA box
ttaataactttttcttatggtggaaaactttcttttgttaatttcaacagactgtggaaa 300
aattattgaaaaagaataccaccttttgaaagaaaacaattaaagttgtctgacaccttt

DnaA box M *dnaA*
aatataatgtgaaatgatataatgag**ttatcaaca**aactaaaaagtaaaggagtaatatg 360
ttatattacactttactatattactcaatagttgtttgatttttcatttcctcattatac

A S L N E N Q K F W A R V T E L A Q Q S *dnaA*
gcatcccttaatgaaaatcaaaaatttgggcccgtgtaaccgaactagctcagcagagt 420
cgtagggaattacttttagtttttaaacccgggcacattggcttgatcgagtcgtctca

I G K Q A Y D F F I E P A Q L M S V E Q *dnaA*
atcggtaaacaagcttatgatttttttattgagcctgctcaattaatgtctgtcgagcaa 480
tagccatttgttcgaatactaaaaaaataactcggacgagttaattacagacagctcgtt

Figure 6. Sequence of the *L. lactis* chromosome near the origin of replication. The *Hind*III site shown in bold was chosen as a zero point to represent the genetic map of *L. lactis* IL1403. Putative DnaA boxes are shown in bold. The begining of the *dnaA* gene and of the *rrnA* genes are shown. Numbering corresponds to that used in the map shown on Figure 9.

ignate IS1077, was also present in each of these seven sites. At two sites the IS1077 element is flanked by two sequences identical to IS904. The orientation of IS904 relative to IS1077 is different in five cases out of seven. A possible explanation for the fact that the two elements always occur together in this genome would be

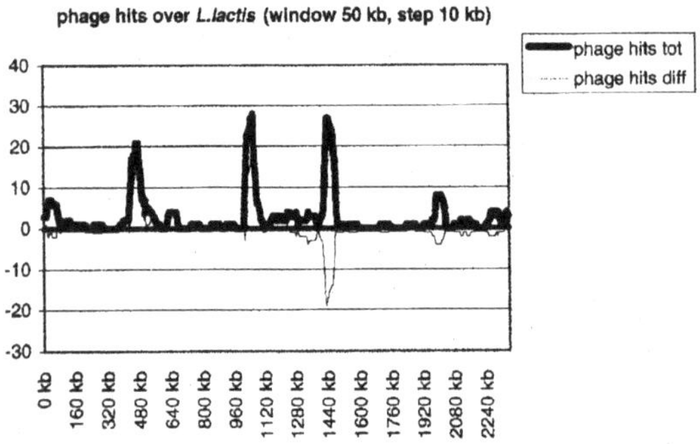

Figure 7. Identification of putative prophages in the *L. lactis* IL1403 genome. The distribution shows the number of phage protein hits detected with a sliding window of 50 kb with a step of 10 kb, over the whole genome. Total number of hits on both strands is shown by bold line. Thin line shows difference of the numbers of hits detected separately on each strand and indicates the orientation of most of the phage genes.

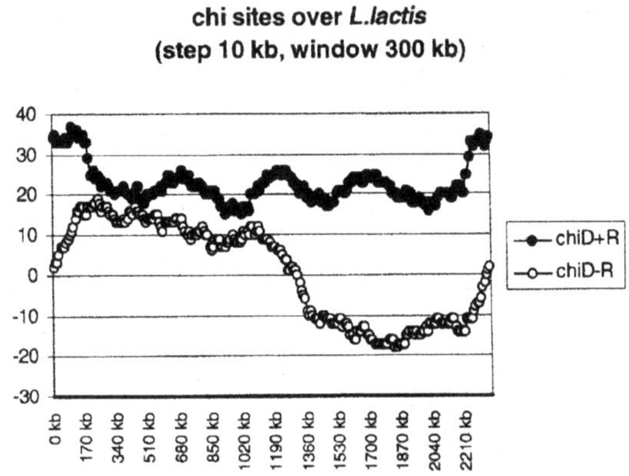

Figure 8. Distribution of chi-sites over the *L. lactis* genome. The distribution represents the number of chi sequences detected with a sliding window of 300 kb with a step of 10 kb, over the whole genome. Filled circles show total numbers of chi-sites in the window, open circles show the difference of the numbers of sites detected separately on each strand, refered to as D and R.

that one of these elements carry a sequence essential for the transposition of the other. Since IS904 can be found without IS1077 in other genomes, the essential sequence must be in IS904. By searching Genbank we identified another copy of IS1077 in the plasmid pK214, (sequence accession number is X92946). IS904 is also present on this plasmid.

Fifteen copies of an element, which was not described earlier for lactococcus species and is designated here IS983, were detected in the IL1403 genome. The closest relative from other species is IS1239 from *Streptococcus pyogenes* (Ashbauch et al. 1998), and, from other lactic acid bacteria, it is IS1070, discovered in the plasmid pNZ63 in *Leuconostoc lactis* NZ6009 (Vaughan & de Vos 1995).

Chopin et al. (1989) detected in the chromosome of IL1403 two different prophages, bIL285 and bIL286, which can be induced for lysogenic development by UV irradiation. Two more prophages, bIL309 and bIL310, can also probably be induced if supplied *in trans* by some functions of the first two phages (Chopin & Perreira, pers. comm.). We tried to identify potential prophages in the chromosome by using homology search against a database containing known phage protein sequences. The database consisted of 1219 protein sequences, including the complete set of 50 putative proteins derived from the sequence of the r1t temperate phage of *L. lactis* (van Sinderen et al. 1996). A distribution of nonredundant homology hits over *L. lactis* genome is shown on Figure 7. This distribution indicates the presence of three regions, around 460, 1060 and 1410 kb, which may contain the prophages identified by Chopin et al. (1989). Two areas, around 40 and 2005 kb, may indicate a fourth and a fifth prophage. More detailed analysis of these areas will help to elucidate the correspondence between the prophages detected by sequencing and by induction of the lysogenic development.

Distribution of chi-sites over the chromosome. Recent analyses suggested that so-called chi-sites, which in *L. lactis* have sequence 5′-GCGCGTG-3′, protect DNA from exonuclease degradation (Biswas et al. 1995). The same sequence was reported to modulate the activities of RexAB exonuclease corresponding to RecBCD of *E. coli* and is thus involved in the regulation of homologous recombination (el Karoui et al. 1998). In *E. coli* and *B. subtilis* such sequences are over-represented and most of them are found to be co-oriented with DNA replication (Burland et al. 1993; Chédin et al. 1998). The latter property was correlated either with the involvement of RecBCD in recombination, and therefore preferring the same DNA structures as the RecA protein (Tracy et al. 1998), or with a statistical bias of base distribution, due to codon usage (Biaudet et al. 1998). The chi sequence is presented in *L. lactis* on average once per 13.2 kb (178/2350 kb) and tends to be co-oriented with replication (Figure 8). Interestingly, the density of chi sequences increases up to one site per 9 kb (34/300 kb) in the region close to the replication origin. Also, chi-site density decreases in the regions near 460, 1060, 1410 and 2000 kb, where the putative prophages were detected (see above).

RNA genes. Six rRNA operons were mapped on the *L. lactis* chromosome (Tulloch et al. 1991; Le Bourgeois et al. 1992) and one was sequenced (Beresford & Condon 1991; Chiaruttini & Milet 1993). We did not resequence all other operons since the diagnostic sequencing would not reveal reliably any difference. However, we confirmed map positions of these operons by PCR amplification of the corresponding genome regions.

tRNA genes were identified with the computer programme tRNAscan-SE (Lowe & Eddy 1997). We detected 62 tRNA genes, corresponding to the 20 aminoacids. Most of the tRNA genes were found to be organised in four large operons, three of them transcribed as parts of *rrnB*, *rrnE* and *rrnF* rRNA operons. The *trnA* operon, encoding seven tRNAs, was described earlier (Nilsson & Johansen 1994). Twenty genes encoding tRNA were found to be isolated from other tRNA genes. RNA component of RNAse P (*rnpB* gene) and 10S RNA (*ssrA*) were detected by homology at 1703 and 1458 kb, respectively.

Protein-coding genes. Prediction of protein-coding regions by methods based on statistical properties of DNA is limited for the sequences of diagnostic quality, since potential frameshifts will hamper the prediction of true stop codons. It is, however, possible to predict genes encoding proteins homologous to proteins from other organisms. We used for the comparison 30757 protein sequences, corresponding to prokaryotic and phage genes, available from SwissProt (release of 17.02.98). Several home-made Perl 5.0 or UNIX shell scripts had been used for automatic extraction of *L. lactis* protein sequences and other parameters from the standard BLASTx or FASTx output files and organising them into clickable tables or genome map drawings. The extracted protein set of *L. lactis* genome was then again used for homology searches, to find all paralogous proteins encoded in the genome. The largest protein from the FASTx comparison, or the one, having BLASTx homology hits to the same Swissprot entry, was considered as really encoded by a particular area of the *L. lactis* genome. The final protein set consists of 1495 proteins. If the true number of genes in the *L. lactis* genome were close to 2300 (a density of one gene per kb), our analysis would have identified about two thirds of them. A version of the automatically generated ORF map of *L. lactis* IL1403 genome is shown on Figure 9. There is a strong bias of gene orientations to be colinear with the direction of replication. The same phenomenon was seen for another AT-rich Gram positive bacterium *B. subtilis* (Kunst et al. 1997), but not for the GC-rich *M. tuberculosis* (Cole et al. 1998). It may be a characteristic of AT-rich gram positive microorganisms.

Another automatic tool, based on the proposed functional classifications of genes in prokaryotic genomes (Riley 1993; Fleischmann et al. 1995) and commentaries from SwissProt, has been used to classify the genes. The programme uses list of functional categories and the relevant keywords with assigned weights corresponding to each category for a given keyword. The advantages of such a classification method are: it is completely automatic, it can be applied to genes with hypothetical functions and needs the interaction with an expert only at the stage of construction of the keyword list. After a keyword list was constructed and the programme had been educated by using several functionally annotated genomes, it provided satisfactory results, which can be used by experts as a draft for more precise functional annotation of the genome and classification of the genes. Any protein database can be automatically classified by using this approach and then used in turn to classify a set of new proteins. Presumably, such method of automatic functional classification of protein data bases can be further developed to give tools, which are completely independent from human expert and use different sources for functional description of pro-

Figure 9. Linear representation of the *L.lactis* IL1403 genome.

The map shows predicted protein encoding genes, IS elements, RNA genes and the ends of regions amplified by PCR (Fig 3). The genes are shown as arrows, with the best BLASTx hit SwissProt name indicated above. Colours correspond to the functional categories assigned by the automatic gene classification method described in the text: amino acid biosynthesis ( ), biosynthesis of cofactors ( ), cell envelope ( ), cellular processes ( ), central intermediary metabolism ( ), energy metabolism ( ), fatty acid and phospholipid metabolism ( ), purines, pyrimidines, nucleosides and nucleotides ( ), regulatory functions ( ), replication ( ), transcription ( ), translation ( ), transport and binding proteins ( ), other categories ( ), hypothetical and unknown ( ). The arrows outlined by bold line correspond to the proteins whose size differ less than 20% of their Swissprot homologue. IS elements and tRNA genes are shown as black arrows with corresponding designation inside. Arrow directions for IS elements are arbitrary, indicating orientations. tRNA genes are shown as small black triangles. The ends of regions amplified by PCR are indicated by open arrows below the scale line. Identifier and size of the region are indicated near the left end

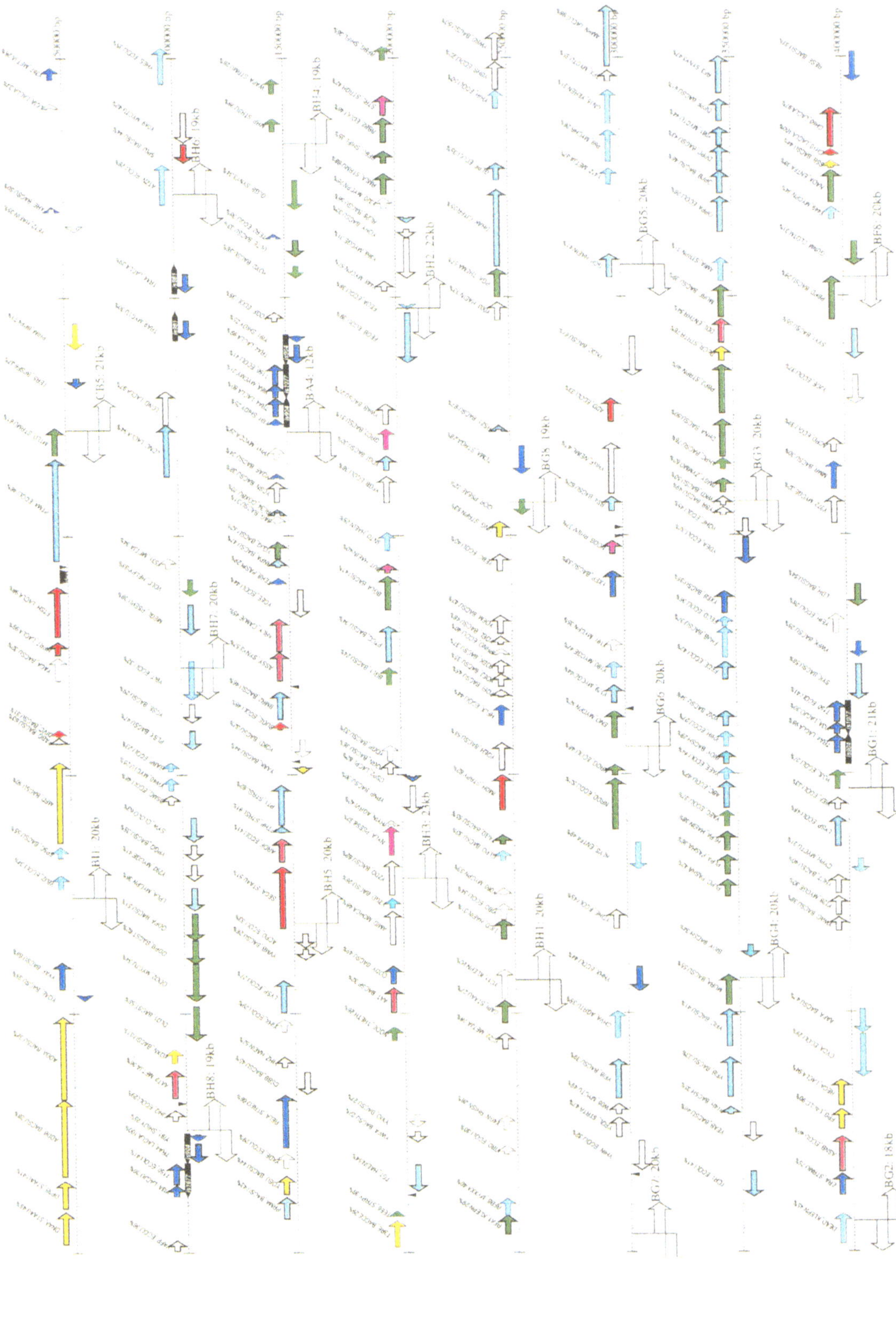

CR5: 21kb
BH1: 20kb
BH6: 19kb
BH7: 20kb
BH8: 19kb
BH4: 19kb
BA4: 12kb
BH5: 20kb
BH2: 22kb
BH3: 23kb
BG8: 19kb
BH1: 20kb
BG5: 20kb
BG6: 20kb
BG7: 20kb
BG3: 20kb
BG4: 20kb
BF8: 20kb
BG1: 21kb
BG2: 18kb

Figure 9.

Figure 9.

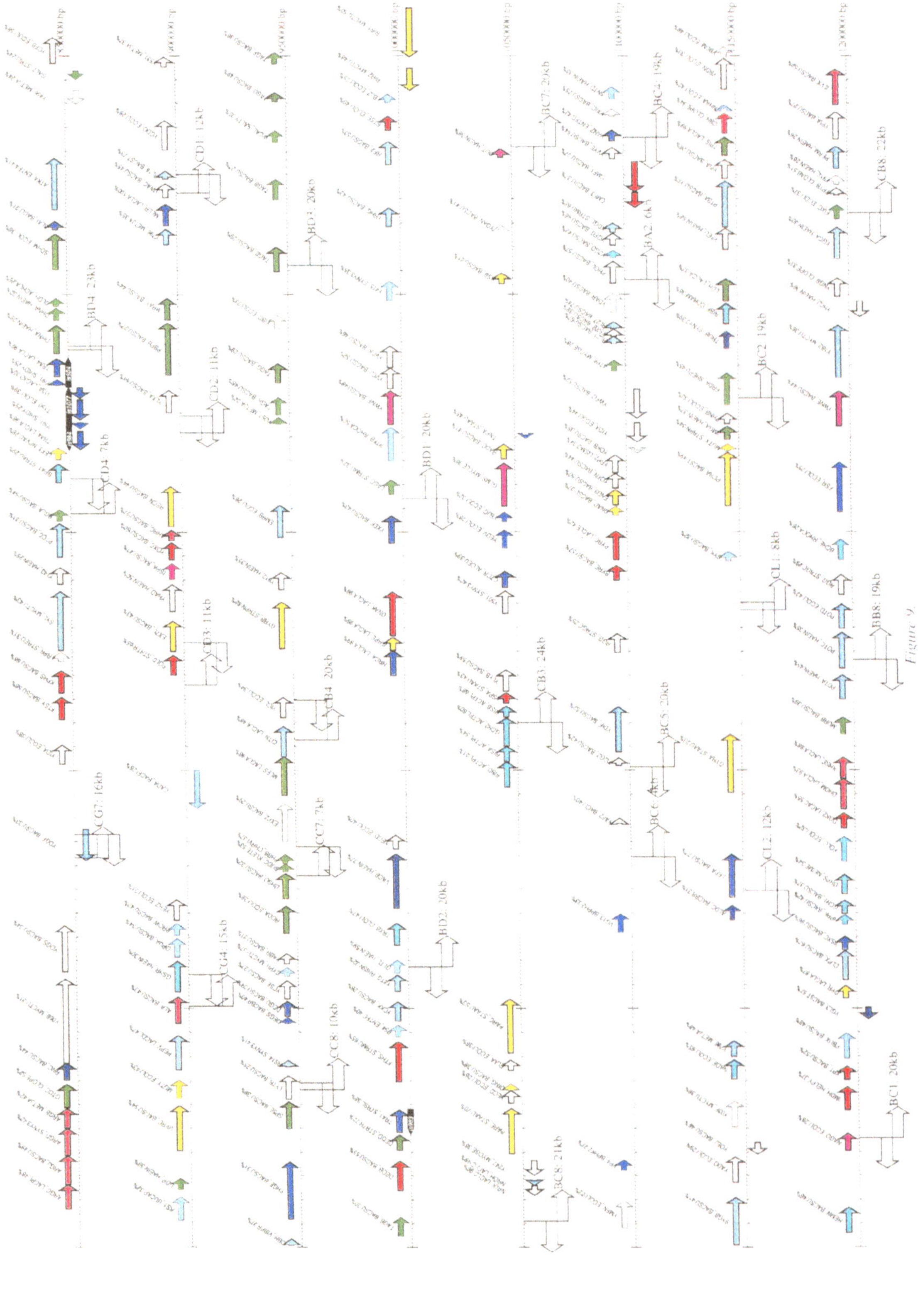

Figure 9

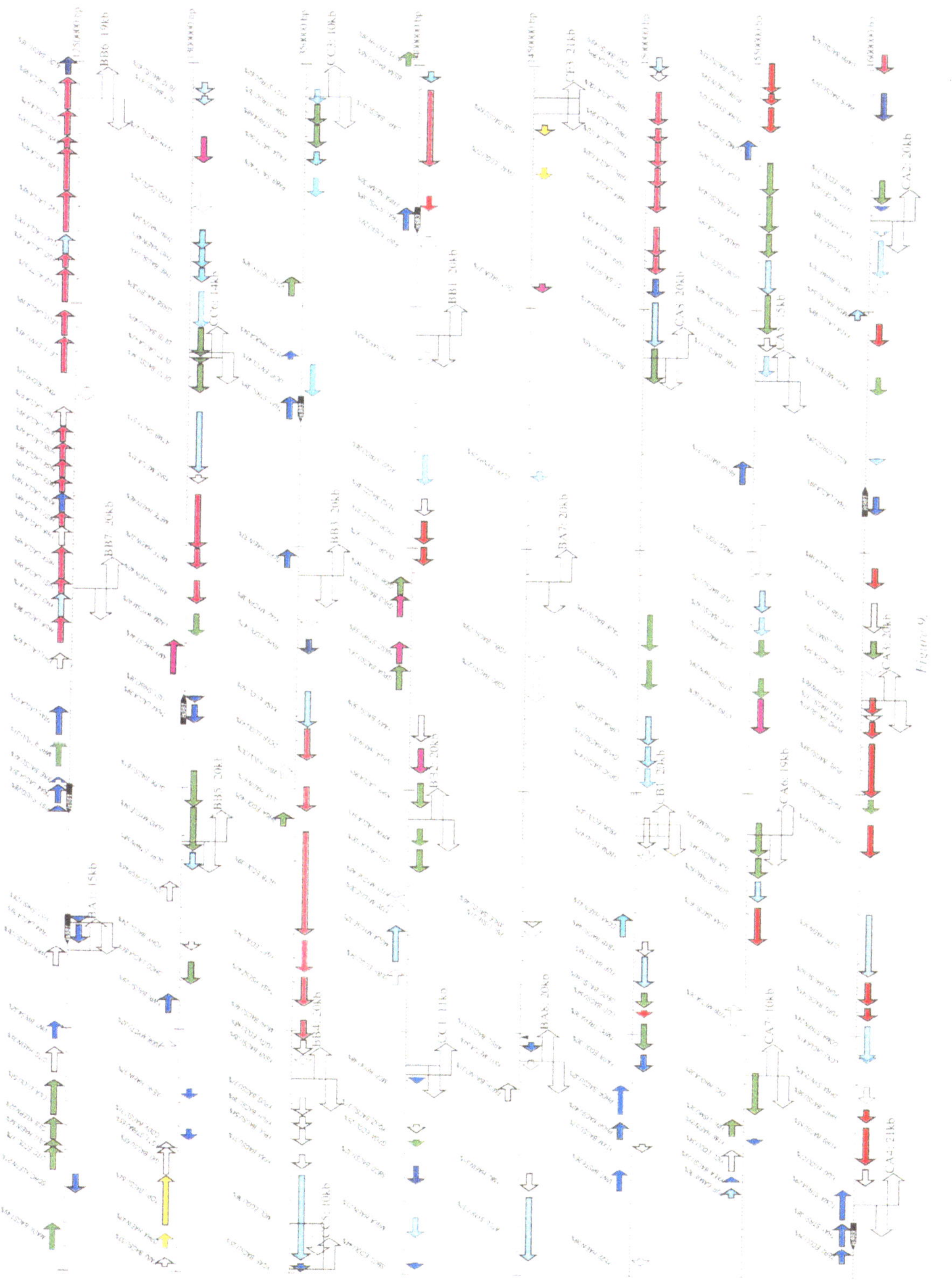
BB6 19kb
BB7 20kb
BA1 15kb
BB5 20kb
CC6 14kb
CC3 10kb
BB3 20kb
BB4 20kb
CC5 10kb
BB1 20kb
BB2 20kb
CC1 11kb
BA7 20kb
BA8 20kb
CA5 20kb
B1 20kb
CA6 19kb
CA7 10kb
CA2 20kb
CA4 21kb

Figure 9.

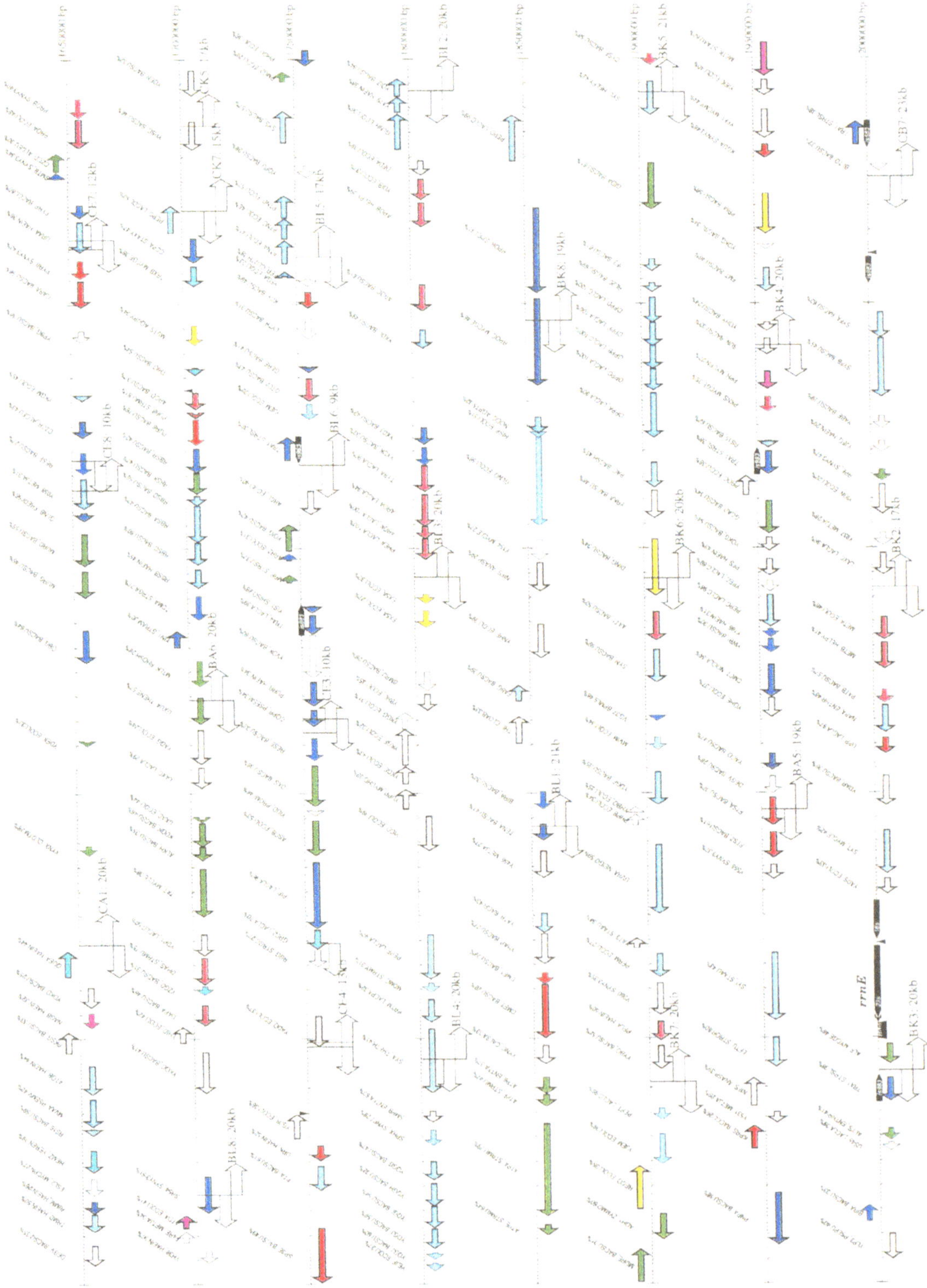

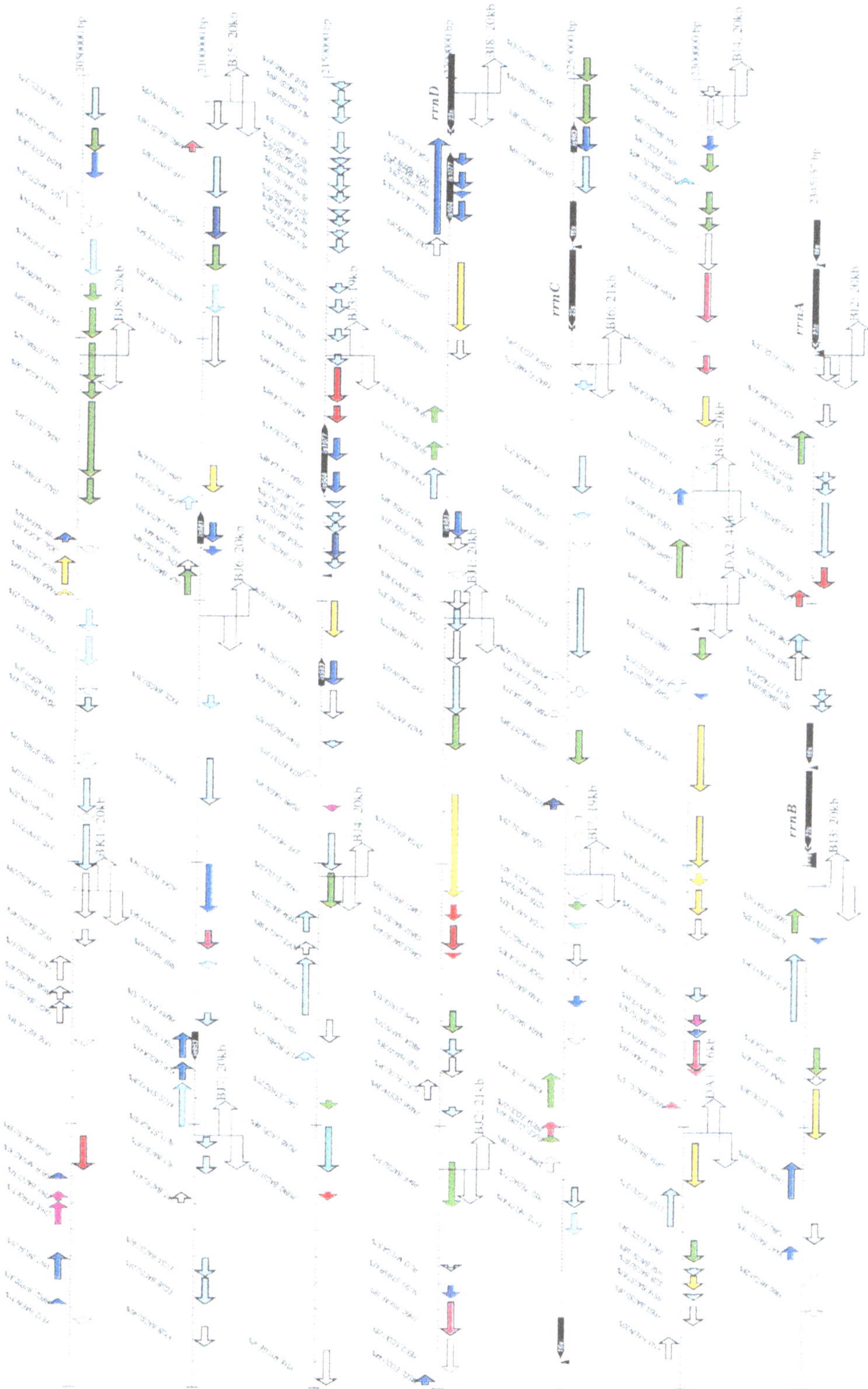

Figure 9.

Table 3. Functional classification of the *L. lactis* genes. Genes are ordered by functional categories and their positions on the chromosome. Position on the chromosome (in kb) and the percentage of identity with the most homologous Swissprot entry is also shown

BIO01. AMINO-ACID BIOSYNTHESIS			**95 genes**
BIO01.01 General			**2 genes**
DHSS_SYNP1	29%	485	SOLUBLE HYDROGENASE, SMALL SUBUNIT (EC 1.12.-.-)
PATB_BACSU	57%	1974	PUTATIVE AMINOTRANSFERASE B (EC 2.6.1.-)
BIO01.02 Aromatic amino-acid family			**19 genes**
AROF_ECOLI	41%	117	PHOSPHO-2-DEHYDRO-3-DEOXYHEPTONATE ALDOLASE, TYR-SENSITIVE (EC 4.1.2.15)
TRPE_THEMA	26%	736	ANTHRANILATE SYNTHASE COMPONENT I (EC 4.1.3.27)
AROG_HAEIN	40%	1278	PHOSPHO-2-DEHYDRO-3-DEOXYHEPTONATE ALDOLASE (EC 4.1.2.15)
TRPA_LACLA	100%	1492	TRYPTOPHAN SYNTHASE ALPHA CHAIN (EC 4.2.1.20)
TRPB_LACLA	100%	1493	TRYPTOPHAN SYNTHASE BETA CHAIN (EC 4.2.1.20)
TRPF_LACLA	99%	1495	N-(5'-PHOSPHORIBOSYL)ANTHRANILATE ISOMERASE (EC 5.3.1.24)
TRPC_LACLA	100%	1495	INDOLE-3-GLYCEROL PHOSPHATE SYNTHASE (EC 4.1.1.48)
TRPD_LACLA	100%	1496	ANTHRANILATE PHOSPHORIBOSYLTRANSFERASE (EC 2.4.2.18)
TRPG_LACLA	99%	1497	ANTHRANILATE SYNTHASE COMPONENT II (EC 4.1.3.27)
TRPE_LACLA	100%	1498	ANTHRANILATE SYNTHASE COMPONENT I (EC 4.1.3.27)
AROD_BACSU	37%	1686	3-DEHYDROQUINATE DEHYDRATASE (EC 4.2.1.10)
PHEA_LACLA	100%	1780	PREPHENATE DEHYDRATASE (EC 4.2.1.51)
AROK_LACLA	100%	1781	SHIKIMATE KINASE (EC 2.7.1.71)
AROA_LACLA	94%	1782	3-PHOSPHOSHIKIMATE 1-CARBOXYVINYLTRANSFERASE (EC 2.5.1.19)
TYRA_LACLA	90%	1783	PREPHENATE DEHYDROGENASE (EC 1.3.1.12)
AROC_BACSU	61%	1790	CHORISMATE SYNTHASE (EC 4.6.1.4)
AROB_HELPY	39%	1794	3-DEHYDROQUINATE SYNTHASE (EC 4.6.1.3)
YDIB_ECOLI	37%	1795	HYPOTHETICAL 31.2 KD PROTEIN IN LPP-AROD INTERGENIC REGION
AROG_BACSU	32%	2261	PHOSPHO-2-DEHYDRO-3-DEOXYHEPTONATE ALDOLASE (EC 4.1.2.15)
BIO01.03 Aspartate family			**19 genes**
AAT2_METJA	36%	57	PUTATIVE ASPARTATE AMINOTRANSFERASE 2 (EC 2.6.1.1)
ASSY_SYNY3	57%	125	ARGININOSUCCINATE SYNTHASE (EC 6.3.4.5)
AAT_BACSP	32%	161	ASPARTATE AMINOTRANSFERASE (EC 2.6.1.1)
ASNB_ECOLI	40%	354	ASPARAGINE SYNTHETASE B (GLUTAMINE-HYDROLYZING) (EC 6.3.5.4)
AK_METJA	30%	746	PROBABLE ASPARTOKINASE (EC 2.7.2.4)
METB_HELPY	65%	789	CYSTATHIONINE GAMMA-SYNTHASE (EC 4.2.99.9)
DHOM_LACLA	93%	1169	HOMOSERINE DEHYDROGENASE (EC 1.1.1.3)
KHSE_LACLA	88%	1170	HOMOSERINE KINASE (EC 2.7.1.39)
METF_HAEIN	32%	1279	5,10-METHYLENETETRAHYDROFOLATE REDUCTASE (EC 1.7.99.5)
METE_HAEIN	46%	1281	5-METHYLTETRAHYDROPTEROYLTRIGLUTAMATE--HOMOCYSTEINE METHYLTRANSFERASE (EC 2.1.1.14)
DAPB_BACSU	50%	1600	DIHYDRODIPICOLINATE REDUCTASE (EC 1.3.1.26)
DAPA_BACSU	46%	1661	DIHYDRODIPICOLINATE SYNTHASE (EC 4.2.1.52)
DHAS_STRMU	75%	1663	ASPARTATE-SEMIALDEHYDE DEHYDROGENASE (EC 1.2.1.11)
AAT1_BACSU	50%	1877	ASPARTATE AMINOTRANSFERASE (EC 2.6.1.1)
METB_HELPY	41%	1976	CYSTATHIONINE GAMMA-SYNTHASE (EC 4.2.99.9)
META_ECOLI	46%	1977	HOMOSERINE O-SUCCINYLTRANSFERASE (EC 2.3.1.46)
PYRH_SYNY3	56%	2067	URIDYLATE KINASE (EC 2.7.4.-)
THRC_PSEAE	38%	2153	THREONINE SYNTHASE (EC 4.2.99.2)
ASNH_MYCTU	37%	2292	PUTATIVE ASPARAGINE SYNTHETASE (GLUTAMINE-HYDROLYZING) (EC 6.3.5.4)
BIO01.04 Branched chain family			**11 genes**
ILVX_BACSU	52%	1199	ACETOLACTATE SYNTHASE (EC 4.1.3.18)
LEU1_SYNY3	45%	1238	2-ISOPROPYLMALATE SYNTHASE (EC 4.1.3.12)
LEU3_LACLA	99%	1239	3-ISOPROPYLMALATE DEHYDROGENASE (EC 1.1.1.85)
LEU2_ACTTI	53%	1241	3-ISOPROPYLMALATE DEHYDRATASE (EC 4.2.1.33)
LEUD_LACLA	100%	1242	3-ISOPROPYLMALATE DEHYDRATASE SMALL SUBUNIT (EC 4.2.1.33)
ILVD_LACLA	98%	1244	DIHYDROXY-ACID DEHYDRATASE (EC 4.2.1.9)
ILVB_LACLA	82%	1246	ACETOLACTATE SYNTHASE LARGE SUBUNIT (EC 4.1.3.18)
ILVN_LACLA	100%	1247	ACETOLACTATE SYNTHASE SMALL SUBUNIT (EC 4.1.3.18)
ILVC_LACLA	97%	1247	KETOL-ACID REDUCTOISOMERASE (EC 1.1.1.86)
THD1_LACLA	99%	1249	THREONINE DEHYDRATASE BIOSYNTHETIC (EC 4.2.1.16)

Table 3.

ILVE_HAEIN	67%	1320	BRANCHED-CHAIN AMINO ACID AMINOTRANSFERASE (EC 2.6.1.42)
BIO01.05 Glutamate family			**21 genes**
ARLY_CAMJE	45%	126	ARGININOSUCCINATE LYASE (EC 4.3.2.1)
ARGE_ECOLI	26%	557	ACETYLORNITHINE DEACETYLASE (EC 3.5.1.16)
ARGC_LACPL	48%	802	N-ACETYL-GAMMA-GLUTAMYL-PHOSPHATE REDUCTASE (EC 1.2.1.38)
ARGJ_BACSU	49%	803	GLUTAMATE N-ACETYLTRANSFERASE (EC 2.3.1.35)
ARGD_SYNY3	42%	804	ACETYLORNITHINE AMINOTRANSFERASE (EC 2.6.1.11)
ARGB_METJA	42%	805	PROBABLE ACETYLGLUTAMATE KINASE (EC 2.7.2.8)
AHRC_BACSU	37%	880	ARGININE HYDROXIMATE RESISTANCE PROTEIN
MURI_BACSU	46%	1310	GLUTAMATE RACEMASE (EC 5.1.1.3)
TABA_PSESZ	40%	1312	TABA PROTEIN
YGFT_ECOLI	34%	1313	HYPOTHETICAL 69.7 KD PROTEIN IN KDUI-LYSS INTERGENIC REGION
GLTB_ECOLI	39%	1316	GLUTAMATE SYNTHASE (NADPH) LARGE CHAIN PRECURSOR (EC 1.4.1.13)
DCEB_ECOLI	43%	1322	GLUTAMATE DECARBOXYLASE BETA (EC 4.1.1.15)
CARY_BACSU	51%	1397	CARBAMOYL-PHOSPHATE SYNTHASE, ARGININE-SPECIFIC, LARGE CHAIN (EC 6.3.5.5)
PROA_ECOLI	46%	1647	GAMMA-GLUTAMYL PHOSPHATE REDUCTASE (GPR) (EC 1.2.1.41)
PROB_SYNY3	40%	1648	GLUTAMATE 5-KINASE (EC 2.7.2.11)
OTC2_BACSU	43%	1736	ORNITHINE CARBAMOYLTRANSFERASE CHAIN F (EC 2.1.3.3)
AHRC_BACSU	36%	2097	ARGININE HYDROXIMATE RESISTANCE PROTEIN
PROA_ECOLI	22%	2210	GAMMA-GLUTAMYL PHOSPHATE REDUCTASE (GPR) (EC 1.2.1.41)
GLNA_STAAU	45%	2262	GLUTAMINE SYNTHETASE (EC 6.3.1.2)
GLNA_HAEIN	41%	2263	GLUTAMINE SYNTHETASE (EC 6.3.1.2)
RACD_STRTR	53%	2289	ASPARTATE RACEMASE (EC 5.1.1.13)
BIO01.06 Histidine family			**9 genes**
PT1P_ECOLI	27%	118	PHOSPHOENOLPYRUVATE-PROTEIN PHOSPHOTRANSFERASE PTSP (EC 2.7.3.9)
HIS8_LACLA	96%	1226	HISTIDINOL-PHOSPHATE AMINOTRANSFERASE (EC 2.6.1.9)
HIS1_LACLA	99%	1228	ATP PHOSPHORIBOSYLTRANSFERASE (EC 2.4.2.17)
HISX_LACLA	98%	1229	HISTIDINOL DEHYDROGENASE (EC 1.1.1.23)
HIS7_LACLA	98%	1231	IMIDAZOLEGLYCEROL-PHOSPHATE DEHYDRATASE (EC 4.2.1.19)
HIS5_LACLA	99%	1232	AMIDOTRANSFERASE HISH (EC 2.4.2.-)
HIS4_LACLA	96%	1233	PHOSPHORIBOSYLFORMIMINO-5-AMINOIMIDAZOLE CARBOXAMIDE RIBOTIDE ISOMERASE (EC 5.3.1.16)
HIS6_LACLA	98%	1234	HISF PROTEIN (CYCLASE)
HIS2_LACLA	99%	1235	PHOSPHORIBOSYL-AMP CYCLOHYDROLASE (EC 3.5.4.19) / PHOSPHORIBOSYL-ATP PYROPHOSPHOHYDROLASE (EC 3.6.1.31)
BIO01.07 Pyruvate family			**2 genes**
DDL_ENTHR	54%	339	D-ALANINE--D-ALANINE LIGASE (EC 6.3.2.4)
ALR_BACSU	43%	860	ALANINE RACEMASE (EC 5.1.1.1)
BIO01.08 Serine family			**12 genes**
SERC_ECOLI	32%	463	PHOSPHOSERINE AMINOTRANSFERASE (EC 2.6.1.52)
GLYA_HELPY	30%	471	SERINE HYDROXYMETHYLTRANSFERASE (EC 2.1.2.1)
CYSK_BACSU	56%	524	CYSTEINE SYNTHASE (EC 4.2.99.8)
GLYA_BACSU	61%	589	SERINE HYDROXYMETHYLTRANSFERASE (EC 2.1.2.1)
SERC_BACCI	47%	591	PHOSPHOSERINE AMINOTRANSFERASE (EC 2.6.1.52)
SERA_HAEIN	32%	592	D-3-PHOSPHOGLYCERATE DEHYDROGENASE (EC 1.1.1.95)
SERB_HAEIN	42%	593	PHOSPHOSERINE PHOSPHATASE (EC 3.1.3.3)
CYSK_BACSU	53%	790	CYSTEINE SYNTHASE (EC 4.2.99.8)
CBH_CLOPE	34%	1147	CHOLOYLGLYCINE HYDROLASE (EC 3.5.1.24)
PFLA_HAEIN	50%	1860	PYRUVATE FORMATE-LYASE 1 ACTIVATING ENZYME (EC 1.97.1.4)
CYSE_BACSU	58%	1900	SERINE ACETYLTRANSFERASE (EC 2.3.1.30)
PKNS_MYCGE	35%	1936	PUTATIVE SERINE/THREONINE-PROTEIN KINASE (EC 2.7.1.-)
BIO02. BIOSYNTHESIS OF COFACTORS, PROSTHETIC GROUPS, AND CARRIERS			**40 genes**
BIO02.01 Biotin			**3 genes**
BIOY_BACSH	34%	313	BIOY PROTEIN PRECURSOR
BIOY_BACSH	28%	321	BIOY PROTEIN PRECURSOR
ACCD_SYNP7	53%	787	ACETYL-COENZYME A CARBOXYLASE CARBOXYL TRANSFERASE SUBUNIT BETA (EC 6.4.1.2)
BIO02.02 Folic acid			**6 genes**
HPPK_BACSU	42%	1164	2-AMINO-4-HYDROXY-6-HYDROXYMETHYLDIHYDROPTERIDINE PYROPHOSPHOKINASE (EC 2.7.6.3)
GCH1_BACSU	37%	1164	GTP CYCLOHYDROLASE I (EC 3.5.4.16)
DHPS_NEIME	34%	1165	DIHYDROPTEROATE SYNTHASE (EC 2.5.1.15)
FOLC_ECOLI	32%	1167	FOLYLPOLYGLUTAMATE SYNTHASE (EC 6.3.2.17)
PABB_SALTY	40%	1345	PARA-AMINOBENZOATE SYNTHASE COMPONENT I (EC 4.1.3.-)
PABA_SALTY	50%	1346	PARA-AMINOBENZOATE SYNTHASE GLUTAMINE AMIDOTRANSFERASE COMPONENT II (EC 4.1.3.-)
BIO02.03 Heme and porphyrin			**5 genes**

Table 3.

HEMK_HAEIN	46%	586	HEMK PROTEIN HOMOLOG
BPA1_STRAU	29%	832	NON-HAEM BROMOPEROXIDASE BPO-A1 (EC 1.11.1.-)
HEMN_BACSU	46%	1151	PROBABLE OXYGEN-INDEPENDENT COPROPORPHYRINOGEN III OXIDASE (EC 1.-.-.-)
BCHE_RHOCA	26%	1179	MAGNESIUM-PROTOPORPHYRIN IX MONOMETHYL ESTER OXIDATIVE CYCLASE 66 KD SUBUNIT
HEMZ_YEREN	39%	1605	FERROCHELATASE (EC 4.99.1.1)
BIO02.04 Lipoate			**0 genes**
BIO02.05 Menaquinone and ubiquinone			**4 genes**
MENB_BACSU	73%	732	NAPHTHOATE SYNTHASE (EC 4.1.3.36)
MEND_BACSU	42%	735	2-SUCCINYL-6-HYDROXY-2,4-CYCLOHEXADIENE-1-CARBOXYLATE SYNTHASE (SHCHC SYNTHASE) / 2-OXOGLUTARATE DECARBOXYLASE (EC 4.1.1.71)
QUEA_HAEIN	44%	1613	S-ADENOSYLMETHIONINE:TRNA RIBOSYLTRANSFERASE-ISOMERASE (EC 5.-.-.-)
GRC2_LACLA	70%	1714	PROBABLE MENAQUINONE BIOSYNTHESIS METHLYTRANSFERASE (EC 2.1.1.-)
BIO02.06 Molybdopterin			**0 genes**
BIO02.07 Pantothenate			**1 genes**
COAA_HAEIN	43%	1464	PANTOTHENATE KINASE (EC 2.7.1.33)
BIO02.08 Pyridoxine			**0 genes**
BIO02.09 Riboflavin			**5 genes**
RIBD_ACTPL	51%	1020	RIBOFLAVIN-SPECIFIC DEAMINASE (EC 3.5.4.-)
RISA_ACTPL	54%	1021	RIBOFLAVIN SYNTHASE ALPHA CHAIN (EC 2.5.1.9)
GCH2_ACTPL	60%	1022	GTP CYCLOHYDROLASE II (EC 3.5.4.25) / 3,4-DIHYDROXY-2-BUTANONE 4- PHOSPHATE SYNTHASE
RISB_ACTPL	68%	1022	6,7-DIMETHYL-8-RIBITYLLUMAZINE SYNTHASE (EC 2.5.1.9)
RIBF_CORAM	38%	1139	RIBOFLAVIN KINASE (EC 2.7.1.26)
BIO02.10 Thioredoxin, glutaredoxin, and glutathione			**5 genes**
GSHR_HAEIN	30%	861	GLUTATHIONE REDUCTASE (EC 1.6.4.2)
TRXB_CLOLI	47%	963	THIOREDOXIN REDUCTASE (EC 1.6.4.5)
BSAA_BACSU	56%	1400	GLUTATHIONE PEROXIDASE HOMOLOG BSAA
THIO_BACSU	54%	1687	THIOREDOXIN (TRX)
TRXB_MYCGE	36%	1691	THIOREDOXIN REDUCTASE (EC 1.6.4.5)
BIO02.11 Thiamine			**10 genes**
YZGD_BACSU	26%	165	HYPOTHETICAL 45.4 KD PROTEIN IN THIAMINASE I 5'REGION
THID_HAEIN	38%	483	PHOSPHOMETHYLPYRIMIDINE KINASE (EC 2.7.4.7)
YZGD_BACSU	42%	1092	HYPOTHETICAL 45.4 KD PROTEIN IN THIAMINASE I 5'REGION
THIE_BACSU	39%	1291	THIAMIN-PHOSPHATE PYROPHOSPHORYLASE (EC 2.5.1.3)
THID_HAEIN	48%	1292	PHOSPHOMETHYLPYRIMIDINE KINASE (EC 2.7.4.7)
THIM_HAEIN	34%	1293	HYDROXYETHYLTHIAZOLE KINASE (EC 2.7.1.50)
DCIP_ENTCL	41%	1337	INDOLE-3-PYRUVATE DECARBOXYLASE (EC 4.1.1.74)
YZGD_BACSU	31%	1662	HYPOTHETICAL 45.4 KD PROTEIN IN THIAMINASE I 5'REGION
POXB_LACPL	46%	2109	PYRUVATE OXIDASE (EC 1.2.3.3)
YZGD_BACSU	40%	2296	HYPOTHETICAL 45.4 KD PROTEIN IN THIAMINASE I 5'REGION
BIO02.12 Pyridine nucleotides			**1 genes**
NADE_ECOLI	65%	1107	NH(3)-DEPENDENT NAD
BIO03. CELL ENVELOPE			**73 genes**
BIO03.01 Membranes, lipoproteins, and porins			**29 genes**
YQJG_BACSU	32%	129	HYPOTHETICAL 30.7 KD LIPOPROTEIN IN GLNQ-ANSR INTERGENIC REGION PRECURSOR
MRP_STRSU	29%	147	MURAMIDASE-RELEASED PROTEIN PRECURSOR (136 KD SURFACE PROTEIN)
WAPA_STRMU	26%	149	WALL-ASSOCIATED PROTEIN PRECURSOR
TEE6_STRPY	39%	152	TRYPSIN-RESISTANT SURFACE T6 PROTEIN PRECURSOR
RFA1_KLEPN	29%	201	O-ANTIGEN EXPORT SYSTEM PERMEASE PROTEIN RFBA
PLPC_PASHA	42%	315	OUTER MEMBRANE LIPOPROTEIN 3 PRECURSOR (PLP3)
PLPA_PASHA	36%	316	OUTER MEMBRANE LIPOPROTEIN 1 PRECURSOR (PLP1)
HLPA_HAEIN	38%	317	28 KD OUTER MEMBRANE PROTEIN PRECURSOR
YAEC_ECOLI	37%	318	HYPOTHETICAL 29.4 KD LIPOPROTEIN IN RCSF-ABC INTERGENIC REGION PRECURSOR
YOJL_ECOLI	27%	370	HYPOTHETICAL 38.5 KD LIPOPROTEIN IN ADA-OMPC INTERGENIC REGION PRECURSOR
CSTA_MYCTU	51%	412	CARBON STARVATION PROTEIN A HOMOLOG
SPR_ECOLI	35%	480	LIPOPROTEIN SPR PRECURSOR
60IM_PSEPU	29%	562	60 KD INNER-MEMBRANE PROTEIN
CRTK_RHOCA	33%	571	CRTK PROTEIN
LGT_ECOLI	25%	604	PROLIPOPROTEIN DIACYLGLYCERYL TRANSFERASE (EC 2.4.99.-)

Table 3.

WAPA_STRMU	26%	682	WALL-ASSOCIATED PROTEIN PRECURSOR
CAIC_ECOLI	26%	731	PROBABLE CROTONOBETAINE/CARNITINE-COA LIGASE (EC 6.3.2.-)
SP5E_BACSU	38%	906	STAGE V SPORULATION PROTEIN E
YD57_METJA	32%	935	PUTATIVE POTASSIUM CHANNEL PROTEIN MJ1357
MRP_STRSU	31%	1221	MURAMIDASE-RELEASED PROTEIN PRECURSOR (136 KD SURFACE PROTEIN)
CD59_HSVSA	32%	1299	SURFACE GLYCOPROTEIN CD59 HOMOLOG PRECURSOR
ADHS_STRPA	68%	1347	ADHESIN B PRECURSOR (SALIVA-BINDING PROTEIN)
P29K_STRGC	76%	1348	29 KD MEMBRANE PROTEIN IN PSAA 5'REGION (ORF1)
TMPC_TREPA	36%	1460	MEMBRANE LIPOPROTEIN TMPC PRECURSOR (MEMBRANE PROTEIN C)
LPLA_BACSU	27%	1526	LIPOPROTEIN LPLA PRECURSOR
YDEK_ECOLI	33%	1622	HYPOTHETICAL 136.5 KD LIPOPROTEIN IN HIPB-UXAB INTERGENIC REGION PRECURSOR (ORFT)
CZCD_ALCEU	30%	1646	CATION EFFLUX SYSTEM PROTEIN CZCD
MRP_STRSU	37%	1729	MURAMIDASE-RELEASED PROTEIN PRECURSOR (136 KD SURFACE PROTEIN)
ADHS_STRSA	35%	2164	ADHESIN B PRECURSOR (SALIVA-BINDING PROTEIN)

BIO03.02 Murein sacculus and peptidoglycan 23 genes

MURA_BACSU	55%	311	PROBABLE UDP-N-ACETYLGLUCOSAMINE 1-CARBOXYVINYLTRANSFERASE (EC 2.5.1.7)
PBP2_STRPN	46%	336	PENICILLIN-BINDING PROTEIN 2B
MURF_BACSU	39%	340	UDP-N-ACETYLMURAMOYLALANYL-D-GLUTAMYL-2,6-DIAMINOPIMELATE--D-ALANYL-D- ALANYL LIGASE (EC 6.3.2.15)
PBPD_BACSU	28%	390	PENICILLIN-BINDING PROTEIN 4 PRECURSOR (PBP 4)
PBPA_STROR	54%	528	PENICILLIN-BINDING PROTEIN 1A (PBP-1A)
MURA_ACICA	49%	532	UDP-N-ACETYLGLUCOSAMINE 1-CARBOXYVINYLTRANSFERASE (EC 2.5.1.7)
RODA_HELPY	32%	661	ROD SHAPE-DETERMINING PROTEIN RODA
PBPB_BACSU	27%	888	PENICILLIN-BINDING PROTEIN 2B (PBP-2B)
MRAY_BACSU	44%	889	PHOSPHO-N-ACETYLMURAMOYL-PENTAPEPTIDE-TRANSFERASE (EC 2.7.8.13)
RODA_ECOLI	28%	914	ROD SHAPE-DETERMINING PROTEIN RODA
MURB_BACSU	39%	1172	UDP-N-ACETYLENOLPYRUVOYLGLUCOSAMINE REDUCTASE (EC 1.1.1.158)
ALYS_ENTHR	39%	1400	AUTOLYSIN PRECURSOR (EC 3.5.1.28)
MURG_BACSU	45%	1629	UDP-N-ACETYLGLUCOSAMINE--N-ACETYLMURAMYL-(PENTAPEPTIDE) PYROPHOSPHORYL-UNDECAPRENOL N-ACETYLGLUCOSAMINE TRANSFERASE (EC 2.4.1.-)
MURD_BACSU	55%	1630	UDP-N-ACETYLMURAMOYLALANINE--D-GLUTAMATE LIGASE (EC 6.3.2.9)
MURE_BACSU	31%	1851	UDP-N-ACETYLMURAMOYLALANYL-D-GLUTAMATE--2,6-DIAMINOPIMELATE LIGASE (EC 6.3.2.13)
GCAD_BACSU	54%	1931	UDP-N-ACETYLGLUCOSAMINE PYROPHOSPHORYLASE (EC 2.7.7.23)
ALYS_ENTHR	41%	1956	AUTOLYSIN PRECURSOR (EC 3.5.1.28)
DAC_STRSQ	25%	2111	D-ALANYL-D-ALANINE CARBOXYPEPTIDASE PRECURSOR (EC 3.4.16.4)
PBPF_BACSU	31%	2158	PENICILLIN-BINDING PROTEIN 1A (PBP-1A)
BACA_ECOLI	34%	2255	BACITRACIN RESISTANCE PROTEIN (PUTATIVE UNDECAPRENOL KINASE) (EC 2.7.1.66)
MRED_BACSU	22%	2294	ROD SHAPE-DETERMINING PROTEIN MRED
MREC_BACSU	30%	2295	ROD SHAPE-DETERMINING PROTEIN MREC
DACA_BACSU	35%	2336	D-ALANYL-D-ALANINE CARBOXYPEPTIDASE PRECURSOR (EC 3.4.16.4)

BIO03.03 Surface polysaccharides, lipopolysaccharides and antigens 21 genes

DLTE_BACSU	33%	142	DLTE PROTEIN
RFBC_SHIFL	35%	196	DTDP-4-DEHYDRORHAMNOSE 3,5-EPIMERASE (EC 5.1.3.13)
RBB2_ECOLI	46%	197	DTDP-GLUCOSE 4,6-DEHYDRATASE (EC 4.2.1.46)
RFBG_SHIFL	36%	200	DTDP-RHAMNOSYL TRANSFERASE RFBG (EC 2.-.-.-)
LICD_HAEIN	31%	213	LICD PROTEIN
TAGD_BACSU	55%	217	GLYCEROL-3-PHOSPHATE CYTIDYLYLTRANSFERASE (EC 2.7.7.39)
TAGH_BACSU	48%	935	TEICHOIC ACID TRANSLOCATION ATP-BINDING PROTEIN TAGH
TAGG_BACSU	29%	937	TEICHOIC ACID TRANSLOCATION PERMEASE PROTEIN TAGG
TAGB_BACSU	29%	941	TEICHOIC ACID BIOSYNTHESIS PROTEIN B PRECURSOR
TAGB_BACSU	27%	944	TEICHOIC ACID BIOSYNTHESIS PROTEIN B PRECURSOR
TAGD_BACSU	49%	948	GLYCEROL-3-PHOSPHATE CYTIDYLYLTRANSFERASE (EC 2.7.7.39)
TAGF_BACSU	38%	950	TEICHOIC ACID BIOSYNTHESIS PROTEIN F
TAGB_BACSU	30%	951	TEICHOIC ACID BIOSYNTHESIS PROTEIN B PRECURSOR
DLTD_BACSU	34%	1287	PROTEIN DLTD PRECURSOR
DLTC_LACCA	43%	1288	D-ALANYL CARRIER PROTEIN (DCP)
DLTB_BACSU	49%	1288	DLTB PROTEIN
TAGE_BACSU	25%	1475	PROBABLE POLY(GLYCEROL-PHOSPHATE) ALPHA-GLUCOSYLTRANSFERASE (EC 2.4.1.52)
TAGE_BACSU	23%	1477	PROBABLE POLY(GLYCEROL-PHOSPHATE) ALPHA-GLUCOSYLTRANSFERASE (EC 2.4.1.52)
RFBU_SALTY	25%	2186	RFBU PROTEIN
RFAK_SALTY	28%	2187	LIPOPOLYSACCHARIDE 1,2-N-ACETYLGLUCOSAMINETRANSFERASE (EC 2.4.1.56)
KDTB_HAEIN	39%	2218	LIPOPOLYSACCHARIDE CORE BIOSYNTHESIS PROTEIN KDTB

BIO03.04 Surface structures 0 genes

BIO04. CELLULAR PROCESSES 29 genes

BIO04.01 Cell division 8 genes

Table 3.

DIVC_BACSU	31%	22	CELL DIVISION PROTEIN DIVIC
FTSH_LACLA	98%	27	CELL DIVISION PROTEIN FTSH HOMOLOG (EC 3.4.24.-)
SP0J_BACSU	44%	96	STAGE 0 SPORULATION PROTEIN J
FTSY_BACSU	58%	823	CELL DIVISION PROTEIN FTSY HOMOLOG
FTSE_ECOLI	40%	997	CELL DIVISION ATP-BINDING PROTEIN FTSE
SP3E_BACSU	49%	1701	STAGE III SPORULATION PROTEIN E
FTSZ_BACSU	61%	1918	CELL DIVISION PROTEIN FTSZ
FTSA_BACSU	37%	1919	CELL DIVISION PROTEIN FTSA
BIO04.02 Cell killing			**3 genes**
HLY3_BACCE	44%	495	HEMOLYSIN III (HLY-III)
YQXC_BACSU	60%	879	HYPOTHETICAL 29.7 KD PROTEIN IN FOLD-AHRC INTERGENIC REGION
FRPA_NEIME	30%	1394	IRON-REGULATED PROTEIN FRPA
BIO04.03 Chaperones			**4 genes**
SUGE_ECOLI	46%	122	SUGE PROTEIN
CH10_LACLA	100%	396	10 KD CHAPERONIN (PROTEIN CPN10)
CH60_LACLA	97%	397	60 KD CHAPERONIN (PROTEIN CPN60)
DNAK_LACLA	99%	977	DNAK PROTEIN (HEAT SHOCK PROTEIN 70)
BIO04.04 Detoxification			**1 genes**
SODM_LACLA	100%	410	SUPEROXIDE DISMUTASE (MN) (EC 1.15.1.1)
BIO04.05 Protein and peptide secretion			**5 genes**
SECA_STAAU	51%	115	PREPROTEIN TRANSLOCASE SECA SUBUNIT
TIG_BACSU	52%	534	TRIGGER FACTOR (TF)
LSPA_STAAU	43%	1023	LIPOPROTEIN SIGNAL PEPTIDASE (EC 3.4.23.36)
SECY_LACLA	98%	2138	PREPROTEIN TRANSLOCASE SECY SUBUNIT
LEPC_BACCL	41%	2330	SIGNAL PEPTIDASE I (EC 3.4.21.89)
BIO04.06 Transformation			**8 genes**
CMA2_BACSU	46%	729	COMA OPERON PROTEIN 2
CMF3_BACSU	37%	1094	COMF OPERON PROTEIN 3
CMF1_BACSU	37%	1095	COMF OPERON PROTEIN 1
CME3_BACSU	28%	1811	COME OPERON PROTEIN 3
CME1_BACSU	39%	1812	COME OPERON PROTEIN 1
CMG3_BACSU	44%	2166	COMG OPERON PROTEIN 3 PRECURSOR
CMG2_BACSU	30%	2167	COMG OPERON PROTEIN 2
CMG1_BACSU	39%	2168	COMG OPERON PROTEIN 1
BIO04.07 Other			**0 genes**
BIO05. CENTRAL INTERMEDIARY METABOLISM			**30 genes**
BIO05.01 General			**7 genes**
GRC3_BACSU	31%	184	PROBABLE HEPTAPRENYL DIPHOSPHATE SYNTHASE COMPONENT II (EC 2.5.1.30)
STRL_STRGR	42%	198	DTDP-4-DEHYDRORHAMNOSE REDUCTASE (EC 1.1.1.133)
ISPA_BACSU	47%	878	GERANYLTRANSTRANSFERASE (EC 2.5.1.10)
YWNE_BACSU	48%	985	HYPOTHETICAL 55.8 KD PROTEIN IN SPOIIQ-MTA INTERGENIC REGION
YWNE_BACSU	44%	1185	HYPOTHETICAL 55.8 KD PROTEIN IN SPOIIQ-MTA INTERGENIC REGION
GRC3_BACSU	35%	1378	PROBABLE HEPTAPRENYL DIPHOSPHATE SYNTHASE COMPONENT II (EC 2.5.1.30)
METK_STAAU	62%	1950	S-ADENOSYLMETHIONINE SYNTHETASE (EC 2.5.1.6)
BIO05.02 Amino sugars			**8 genes**
NYLA_PSES8	32%	167	6-AMINOHEXANOATE-CYCLIC-DIMER HYDROLASE (EC 3.5.2.12)
NODB_RHILV	37%	279	CHITOOLIGOSACCHARIDE DEACETYLASE (EC 3.5.1.-)
GLMS_MYCLE	36%	1032	PROBABLE GLUCOSAMINE--FRUCTOSE-6-PHOSPHATE AMINOTRANSFERASE (ISOMERIZING) (EC 2.6.1.16)
NAGD_ECOLI	28%	1154	NAGD PROTEIN
YVYH_BACSU	47%	1296	PUTATIVE UDP-N-ACETYLGLUCOSAMINE 2-EPIMERASE (EC 5.1.3.14)
NAGA_HAEIN	30%	1371	N-ACETYLGLUCOSAMINE-6-PHOSPHATE DEACETYLASE (EC 3.5.1.25)
HASC_STRPY	58%	1376	UTP--GLUCOSE-1-PHOSPHATE URIDYLYLTRANSFERASE (EC 2.7.7.9)
NAGB_HAEIN	32%	1611	GLUCOSAMINE-6-PHOSPHATE ISOMERASE (EC 5.3.1.10)
BIO05.03 Degradation of polysaccharides			**5 genes**
NEPU_THEVU	30%	701	NEOPULLULANASE (EC 3.2.1.135)
AMY_BACST	44%	1275	ALPHA-AMYLASE PRECURSOR (EC 3.2.1.1)
O16G_BACTR	60%	1523	OLIGO-1,6-GLUCOSIDASE (EC 3.2.1.10)
CHIX_STROI	31%	2007	EXOCHITINASE 1 PRECURSOR (EC 3.2.1.14)
CHI1_BACCI	34%	2007	CHITINASE A1 PRECURSOR (EC 3.2.1.14)
BIO05.04 Phosphorus compounds			**9 genes**
DUT_HAEIN	40%	179	DEOXYURIDINE 5'-TRIPHOSPHATE NUCLEOTIDOHYDROLASE (EC 3.6.1.23)
DUT_HAEIN	38%	456	DEOXYURIDINE 5'-TRIPHOSPHATE NUCLEOTIDOHYDROLASE (EC 3.6.1.23)
SUHB_ECOLI	29%	531	EXTRAGENIC SUPPRESSOR PROTEIN SUHB

Table 3.

APL_LACLA	96%	717	ALKALINE PHOSPHATASE LIKE PROTEIN
DUT_HAEIN	38%	1046	DEOXYURIDINE 5'-TRIPHOSPHATE NUCLEOTIDOHYDROLASE (EC 3.6.1.23)
DUT_HAEIN	37%	1441	DEOXYURIDINE 5'-TRIPHOSPHATE NUCLEOTIDOHYDROLASE (EC 3.6.1.23)
PPH_MYCPN	27%	1937	PUTATIVE PROTEIN PHOSPHATASE (EC 3.1.3.16)
PGPB_HAEIN	36%	2122	PHOSPHATIDYLGLYCEROPHOSPHATASE B (EC 3.1.3.27)
Y328_SYNY3	35%	2264	PUTATIVE LOW MOLECULAR WEIGHT PROTEIN-TYROSINE-PHOSPHATASE (EC 3.1.3.48)

BIO05.05 Polyamine biosynthesis 1 genes

ATDA_ECOLI	41%	1653	SPERMIDINE N1-ACETYLTRANSFERASE (EC 2.3.1.57)

BIO05.06 Other 0 genes

BIO06. ENERGY METABOLISM 148 genes

BIO06.01 Aerobic 32 genes

MTLD_STRMU	61%	34	MANNITOL-1-PHOSPHATE 5-DEHYDROGENASE (EC 1.1.1.17)
ODO2_MYCTU	34%	61	DIHYDROLIPOAMIDE SUCCINYLTRANSFERASE COMPONENT (E2) OF 2-OXOGLUTARATE DEHYDROGENASE COMPLEX (EC 2.3.1.61)
VDLC_HELPY	33%	78	PROBABLE SHORT-CHAIN TYPE DEHYDROGENASE/REDUCTASE VDLC (EC 1.-.-.-)
YOXD_BACSU	36%	141	HYPOTHETICAL OXIDOREDUCTASE IN RTP 5'REGION (EC 1.-.-.-)
NQOE_THETH	26%	159	NADH-UBIQUINONE OXIDOREDUCTASE CHAIN 14 (EC 1.6.5.3)
DHNA_BACSU	50%	334	NADH DEHYDROGENASE (EC 1.6.99.3)
LDH_BACSU	54%	377	L-LACTATE DEHYDROGENASE (EC 1.1.1.27)
NAOX_ENTFA	39%	395	NADH OXIDASE (EC 1.6.-.-)
LDH_LISMO	24%	488	L-LACTATE DEHYDROGENASE (EC 1.1.1.27)
YBBO_ECOLI	29%	617	HYPOTHETICAL OXIDOREDUCTASE IN USHA-TESA INTERGENIC REGION (EC 1.-.-.-)
NAOX_METJA	34%	792	PUTATIVE NADH OXIDASE (EC 1.6.-.-)
YWFD_BACSU	38%	799	HYPOTHETICAL OXIDOREDUCTASE IN ROCC-PTA INTERGENIC REGION (EC 1.-.-.-)
BUDC_KLETE	53%	916	ACETOIN(DIACETYL) REDUCTASE (EC 1.1.1.5)
PHBB_CHRVI	37%	916	ACETOACETYL-COA REDUCTASE (EC 1.1.1.36)
FRDA_SHEPU	45%	1136	FUMARATE REDUCTASE FLAVOPROTEIN SUBUNIT PRECURSOR (EC 1.3.99.1)
LDH_LACCA	37%	1140	L-LACTATE DEHYDROGENASE (EC 1.1.1.27)
GLPD_MYCLE	34%	1268	GLYCEROL-3-PHOSPHATE DEHYDROGENASE (EC 1.1.99.5)
LA2M_MYCSM	40%	1277	LACTATE 2-MONOOXYGENASE (EC 1.13.12.4)
VDLC_HELPY	36%	1341	PROBABLE SHORT-CHAIN TYPE DEHYDROGENASE/REDUCTASE VDLC (EC 1.-.-.-)
GPDA_ECOLI	25%	1355	GLYCEROL-3-PHOSPHATE DEHYDROGENASE (NAD+) (EC 1.1.1.8)
LDH_LACLA	100%	1367	L-LACTATE DEHYDROGENASE (EC 1.1.1.27)
GPDA_BACSU	34%	1375	GLYCEROL-3-PHOSPHATE DEHYDROGENASE (NAD
YINL_LISMO	36%	1506	HYPOTHETICAL OXIDOREDUCTASE IN INLA 5'REGION (EC 1.-.-.-)
YQJQ_BACSU	38%	1569	HYPOTHETICAL OXIDOREDUCTASE IN GLNQ-ANSR INTERGENIC REGION (EC 1.-.-.-)
YINL_LISMO	72%	1576	HYPOTHETICAL OXIDOREDUCTASE IN INLA 5'REGION (EC 1.-.-.-)
MTLK_RHOSH	28%	1675	MANNITOL 2-DEHYDROGENASE (EC 1.1.1.67)
YQIG_BACSU	42%	1730	PROBABLE NADH-DEPENDENT FLAVIN OXIDOREDUCTASE YQIG (EC 1.-.-.-)
YPRA_CORGL	29%	2048	HYPOTHETICAL 33.0 KD PROTEIN IN PROB-PROA INTERGENIC REGION
YKGC_ECOLI	39%	2119	PROBABLE PYRIDINE NUCLEOTIDE-DISULFIDE OXIDOREDUCTASE IN EAEH-BETA INTERGENIC REGION
NAOX_ENTFA	35%	2175	NADH OXIDASE (EC 1.6.-.-)
YBBQ_ECOLI	33%	2278	HYPOTHETICAL 30.8 KD PROTEIN IN GIP-FDRA INTERGENIC REGION
OXIR_STRAT	40%	2318	PROBABLE OXIDOREDUCTASE (EC 1.-.-.-)

BIO06.02 Amino acids and amines 5 genes

DHL2_LACCO	57%	488	L-2-HYDROXYISOCAPROATE DEHYDROGENASE (EC 1.1.1.-)
ASPG_BACLI	34%	741	L-ASPARAGINASE (EC 3.5.1.1)
OTCC_CLOPE	52%	806	ORNITHINE CARBAMOYLTRANSFERASE, CATABOLIC (EC 2.1.3.3)
DHCO_LACLA	86%	1262	N5-(CARBOXYETHYL)ORNITHINE SYNTHASE (EC 1.5.1.24)
OTCC_CLOPE	54%	2093	ORNITHINE CARBAMOYLTRANSFERASE, CATABOLIC (EC 2.1.3.3)

BIO06.03 Anaerobic 2 genes

NRDD_ECOLI	52%	269	ANAEROBIC RIBONUCLEOSIDE-TRIPHOSPHATE REDUCTASE (EC 1.17.4.2)
NRDG_ECOLI	46%	270	ANAEROBIC RIBONUCLEOSIDE-TRIPHOSPHATE REDUCTASE ACTIVATING PROTEIN (EC 1.97.1.-)

BIO06.04 ATP-proton motive force interconversion 6 genes

ATP6_MYCGE	28%	1087	ATP SYNTHASE A CHAIN (EC 3.6.1.34)
ATPE_STRMU	64%	1802	ATP SYNTHASE EPSILON CHAIN (EC 3.6.1.34)
ATPA_STRMU	79%	1804	ATP SYNTHASE ALPHA CHAIN (EC 3.6.1.34)
ATPA_SYNP1	65%	1806	ATP SYNTHASE ALPHA CHAIN (EC 3.6.1.34)
ATPF_STRMU	47%	1807	ATP SYNTHASE B CHAIN (EC 3.6.1.34)
ATP6_ENTFA	37%	1808	ATP SYNTHASE A CHAIN (EC 3.6.1.34)

BIO06.05 Electron transport 6 genes

Table 3.

NIFJ_ANASP	44%	428	PYRUVATE-FLAVODOXIN OXIDOREDUCTASE (EC 1.-.-.-)
CYDA_BACSU	49%	706	CYTOCHROME D UBIQUINOL OXIDASE SUBUNIT I (EC 1.10.3.-)
CYDB_BACSU	43%	707	CYTOCHROME D UBIQUINOL OXIDASE SUBUNIT II (EC 1.10.3.-)
DHNA_HAEIN	28%	838	NADH DEHYDROGENASE (EC 1.6.99.3)
DHNA_HAEIN	32%	839	NADH DEHYDROGENASE (EC 1.6.99.3)
FIXC_ECOLI	37%	1193	FIXC PROTEIN
BIO06.06 Entner-Doudoroff			**1 genes**
KDGK_BACSU	46%	1668	2-DEHYDRO-3-DEOXYGLUCONOKINASE (EC 2.7.1.45)
BIO06.07 Fermentation			**8 genes**
QOR_PSEAE	32%	231	QUINONE OXIDOREDUCTASE (EC 1.6.5.5)
QOR_PSEAE	32%	721	QUINONE OXIDOREDUCTASE (EC 1.6.5.5)
DHSO_BACSU	32%	915	SORBITOL DEHYDROGENASE (EC 1.1.1.14)
FADH_METMR	37%	1587	GLUTATHIONE-DEPENDENT FORMALDEHYDE DEHYDROGENASE (EC 1.2.1.1)
YBDH_ECOLI	31%	1595	HYPOTHETICAL OXIDOREDUCTASE IN CSTA-AHPC INTERGENIC REGION
ADH1_ZYMMO	60%	1852	ALCOHOL DEHYDROGENASE I (EC 1.1.1.1)
SUCD_CLOKL	40%	2210	SUCCINATE-SEMIALDEHYDE DEHYDROGENASE (NAD
ADHE_ECOLI	44%	2211	ALCOHOL DEHYDROGENASE (EC 1.1.1.1)
BIO06.08 Gluconeogenesis			**3 genes**
DCOA_KLEPN	40%	664	OXALOACETATE DECARBOXYLASE ALPHA CHAIN (EC 4.1.1.3)
SDHM_ECOLI	36%	842	L-SERINE DEHYDRATASE 2 (EC 4.2.1.13)
G6PB_BACST	39%	2224	GLUCOSE-6-PHOSPHATE ISOMERASE B (GPI B) (EC 5.3.1.9)
BIO06.09 Glycolysis			**14 genes**
PGK_THEMA	57%	241	PHOSPHOGLYCERATE KINASE (EC 2.7.2.3)
ENO_MYCPN	52%	272	ENOLASE (EC 4.2.1.11)
PMGY_ZYMMO	62%	332	PHOSPHOGLYCERATE MUTASE (EC 5.4.2.1)
PGMB_LACLA	100%	440	BETA-PHOSPHOGLUCOMUTASE (EC 5.4.2.6)
G3P_LACLA	99%	552	GLYCERALDEHYDE 3-PHOSPHATE DEHYDROGENASE (EC 1.2.1.12)
ENO_BACSU	70%	631	ENOLASE (EC 4.2.1.11)
ENO_HAEIN	55%	649	ENOLASE (EC 4.2.1.11)
PGMB_LACLA	32%	748	BETA-PHOSPHOGLUCOMUTASE (EC 5.4.2.6)
TPIS_LACLA	99%	1146	TRIOSEPHOSPHATE ISOMERASE (EC 5.3.1.1)
KPYK_LACLA	54%	1368	PYRUVATE KINASE (EC 2.7.1.40)
K6PF_LACLA	99%	1370	6-PHOSPHOFRUCTOKINASE (EC 2.7.1.11)
PMG2_ECOLI	28%	1749	PROBABLE PHOSPHOGLYCERATE MUTASE 2 (EC 5.4.2.1)
ALF_MYCGE	46%	1959	FRUCTOSE-BISPHOSPHATE ALDOLASE (EC 4.1.2.13)
G3P_LACLA	81%	2313	GLYCERALDEHYDE 3-PHOSPHATE DEHYDROGENASE (EC 1.2.1.12)
BIO06.10 Pentose phosphate pathway			**5 genes**
DXS_RHOCA	38%	1508	PROBABLE 1-DEOXYXYLULOSE-5-PHOSPHATE SYNTHASE (DXP SYNTHASE)
TKT_MYCLE	38%	1666	TRANSKETOLASE (EC 2.2.1.1)
DXS_BACSU	38%	1720	PROBABLE 1-DEOXYXYLULOSE-5-PHOSPHATE SYNTHASE (DXP SYNTHASE)
G6PD_BACSU	48%	2282	GLUCOSE-6-PHOSPHATE 1-DEHYDROGENASE (EC 1.1.1.49)
RPIA_ECOLI	39%	2297	RIBOSE 5-PHOSPHATE ISOMERASE A (EC 5.3.1.6)
BIO06.11 Pyruvate dehydrogenase			**4 genes**
DLD1_BACST	55%	60	LIPOAMIDE DEHYDROGENASE COMPONENT (E3) OF PYRUVATE DEHYDROGENASE COMPLEX (EC 1.8.1.4)
ODPB_BACST	62%	63	PYRUVATE DEHYDROGENASE E1 COMPONENT, BETA SUBUNIT (EC 1.2.4.1)
ODPA_BACSU	51%	64	PYRUVATE DEHYDROGENASE E1 COMPONENT, ALPHA SUBUNIT (EC 1.2.4.1)
DLDH_ACHLA	26%	840	LIPOAMIDE DEHYDROGENASE COMPONENT (E3) OF PYRUVATE DEHYDROGENASE COMPLEX (EC 1.8.1.4)
BIO06.12 Sugars			**53 genes**
GLGB_SYNY3	34%	144	1,4-ALPHA-GLUCAN BRANCHING ENZYME (EC 2.4.1.18)
BGL1_BACSU	24%	174	BETA-GLUCOSIDASE (EC 3.2.1.21)
BGLA_BACSU	41%	178	6-PHOSPHO-BETA-GLUCOSIDASE (EC 3.2.1.86)
RMLA_STRMU	88%	195	GLUCOSE-1-PHOSPHATE THYMIDYLYLTRANSFERASE (EC 2.7.7.24)
CAPI_STAAU	27%	210	CAPI PROTEIN
AHPC_BACSU	76%	333	ALKYL HYDROPEROXIDE REDUCTASE C22 PROTEIN (EC 1.6.4.-)
GUNM_CLOTM	31%	392	ENDOGLUCANASE M (EC 3.2.1.4)
CASB_KLEOX	51%	420	PHOSPHO-CELLOBIASE (EC 3.2.1.-)
XYNB_BUTFI	29%	440	ENDO-1,4-BETA-XYLANASE B (EC 3.2.1.8)
6PGD_SALTY	68%	606	6-PHOSPHOGLUCONATE DEHYDROGENASE, DECARBOXYLATING (EC 1.1.1.44)
GAL7_LACHE	36%	614	GALACTOSE-1-PHOSPHATE URIDYLYLTRANSFERASE (EC 2.7.7.10)
PFL_STRMU	83%	656	FORMATE ACETYLTRANSFERASE (EC 2.3.1.54)
MALQ_ECOLI	29%	692	4-ALPHA-GLUCANOTRANSFERASE (EC 2.4.1.25)
GLGC_BACST	57%	694	GLUCOSE-1-PHOSPHATE ADENYLYLTRANSFERASE (EC 2.7.7.27)
GLGD_BACST	31%	695	GLYCOGEN BIOSYNTHESIS PROTEIN GLGD
GLGA_BACSU	46%	697	GLYCOGEN SYNTHASE (EC 2.4.1.21)
PHSG_BACSU	44%	699	GLYCOGEN PHOSPHORYLASE (EC 2.4.1.1)
BGLA_BACSU	54%	831	6-PHOSPHO-BETA-GLUCOSIDASE (EC 3.2.1.86)
GAL7_STRLI	24%	849	GALACTOSE-1-PHOSPHATE URIDYLYLTRANSFERASE (EC 2.7.7.10)
UDP_HAEIN	39%	853	URIDINE PHOSPHORYLASE (EC 2.4.2.3)

Table 3.

WCAL_SALTY	35%	947	PUTATIVE COLANIC ACID BIOSYNTHESIS GLYCOSYL TRANSFERASE WCAL
DEOD_STRTR	72%	954	PURINE NUCLEOSIDE PHOSPHORYLASE (EC 2.4.2.1)
LACC_STRMU	33%	982	TAGATOSE-6-PHOSPHATE KINASE (EC 2.7.1.-)
3MGA_HAEIN	37%	1134	DNA-3-METHYLADENINE GLYCOSIDASE (EC 3.2.2.20)
GLPK_BACSU	61%	1270	GLYCEROL KINASE (EC 2.7.1.30)
THGA_ECOLI	40%	1319	GALACTOSIDE ACETYLTRANSFERASE (EC 2.3.1.18)
GIDB_BACSU	60%	1378	GLUCOSE INHIBITED DIVISION PROTEIN B
DEOC_BACSU	60%	1461	DEOXYRIBOSE-PHOSPHATE ALDOLASE (EC 4.1.2.4)
BGL1_BACSU	56%	1488	BETA-GLUCOSIDASE (EC 3.2.1.21)
GLK_BACSU	34%	1517	GLUCOKINASE (EC 2.7.1.2)
BGLA_THEMA	33%	1518	BETA-GLUCOSIDASE A (EC 3.2.1.21)
STRH_STRPN	29%	1524	BETA-N-ACETYLHEXOSAMINIDASE PRECURSOR (EC 3.2.1.52)
XYNB_BACPU	48%	1540	BETA-XYLOSIDASE (EC 3.2.1.37)
GALM_ACICA	29%	1543	ALDOSE 1-EPIMERASE PRECURSOR (EC 5.1.3.3)
XYLB_BACSU	49%	1544	XYLULOSE KINASE (EC 2.7.1.17)
XYLA_THETU	50%	1545	XYLOSE ISOMERASE (EC 5.3.1.5)
XYNX_CLOTM	28%	1618	EXOGLUCANASE XYNX PRECURSOR (EC 3.2.1.91)
ALKH_BACSU	35%	1668	4-HYDROXY-2-OXOGLUTARATE ALDOLASE (EC 4.1.3.16)
UXAC_ECOLI	44%	1669	URONATE ISOMERASE (EC 5.3.1.12)
UXUA_THENE	57%	1673	MANNONATE DEHYDRATASE (EC 4.2.1.8)
RBSK_BACSU	47%	1683	RIBOKINASE (EC 2.7.1.15)
ASCB_ECOLI	53%	1718	6-PHOSPHO-BETA-GLUCOSIDASE ASCB (EC 3.2.1.86)
GIDA_BACSU	61%	1895	GLUCOSE INHIBITED DIVISION PROTEIN A
RPE_SYNY3	47%	1983	RIBULOSE-PHOSPHATE 3-EPIMERASE (EC 5.1.3.1)
GALE_STRMU	60%	2034	UDP-GLUCOSE 4-EPIMERASE (EC 5.1.3.2)
BGA2_ECOLI	34%	2036	EVOLVED BETA-GALACTOSIDASE ALPHA-SUBUNIT (EC 3.2.1.23)
THGA_LACLA	100%	2038	GALACTOSIDE O-ACETYLTRANSFERASE (EC 2.3.1.18)
GAL7_STRMU	52%	2039	GALACTOSE-1-PHOSPHATE URIDYLYLTRANSFERASE (EC 2.7.7.10)
GAL1_STRMU	58%	2041	GALACTOKINASE (EC 2.7.1.6)
GALM_HAEIN	36%	2042	ALDOSE 1-EPIMERASE (EC 5.1.3.3)
GLK_BACSU	47%	2081	GLUCOKINASE (EC 2.7.1.2)
GNTK_BACSU	49%	2249	GLUCONOKINASE (EC 2.7.1.12)
YQEC_BACSU	53%	2250	HYPOTHETICAL 32.8 KD PROTEIN IN NUCB-AROD INTERGENIC REGION
BIO06.13 TCA cycle			**9 genes**
CISY_THIFE	29%	665	CITRATE SYNTHASE (EC 4.1.3.7)
ACON_BRAJA	51%	667	ACONITATE HYDRATASE (EC 4.2.1.3)
IDH_STRMU	60%	669	ISOCITRATE DEHYDROGENASE (NADP) (EC 1.1.1.42)
MLES_LACLA	98%	920	MALOLACTIC ENZYME (EC 1.-.-.-)
MAOX_BACST	49%	1201	MALATE OXIDOREDUCTASE (NAD) (EC 1.1.1.38)
CITC_ECOLI	37%	1205	[CITRATE (PRO-3S)-LYASE] LIGASE (EC 6.2.1.22)
CILG_HAEIN	52%	1205	CITRATE LYASE ACYL CARRIER PROTEIN (CITRATE LYASE GAMMA CHAIN)
CILB_KLEPN	52%	1206	CITRATE LYASE BETA CHAIN (EC 4.1.3.6)
CILA_ECOLI	56%	1207	CITRATE LYASE ALPHA CHAIN (EC 4.1.3.6)
BIO07. FATTY ACID AND PHOSPHOLIPID METABOLISM			**20 genes**
BIO07.01 General			**20 genes**
LPLA_MYCPN	36%	65	PROBABLE LIPOATE-PROTEIN LIGASE A (EC 6.-.-.-)
PLSX_BACSU	47%	71	FATTY ACID/PHOSPHOLIPID SYNTHESIS PROTEIN PLSX HOMOLOG
GLPQ_BACSU	33%	217	GLYCEROPHOSPHORYL DIESTER PHOSPHODIESTERASE (EC 3.1.4.46)
FABZ_ECOLI	42%	558	(3R)-HYDROXYMYRISTOYL-[ACYL CARRIER PROTEIN] DEHYDRATASE (EC 4.2.1.-)
FABI_HAEIN	42%	559	ENOYL-[ACYL-CARRIER-PROTEIN] REDUCTASE (NADH) (EC 1.3.1.9)
DLTA_LACCA	24%	653	D-ALANINE-ACTIVATING ENZYME (EC 6.3.2.-)
ACCC_METJA	48%	662	BIOTIN CARBOXYLASE (EC 6.3.4.14)
FABD_BACSU	45%	781	MALONYL COA-ACYL CARRIER PROTEIN TRANSACYLASE (EC 2.3.1.39)
FABG_BACSU	46%	782	3-OXOACYL-[ACYL-CARRIER PROTEIN] REDUCTASE (EC 1.1.1.100)
FABF_VIBHA	42%	783	3-OXOACYL-[ACYL-CARRIER-PROTEIN] SYNTHASE II (EC 2.3.1.41)
BCCP_ANASP	51%	784	BIOTIN CARBOXYL CARRIER PROTEIN OF ACETYL-COA CARBOXYLASE (EC 6.4.1.2)
FABZ_HAEIN	50%	784	(3R)-HYDROXYMYRISTOYL-[ACYL CARRIER PROTEIN] DEHYDRATASE (EC 4.2.1.-)
ACCC_BACSU	59%	786	BIOTIN CARBOXYLASE (EC 6.3.4.14)
ACCA_HAEIN	49%	788	ACETYL-COENZYME A CARBOXYLASE CARBOXYL TRANSFERASE SUBUNIT ALPHA (EC 6.4.1.2)
KDGL_STRMU	68%	1093	DIACYLGLYCEROL KINASE (EC 2.7.1.107)
ATOB_HAEIN	44%	1608	ACETYL-COA ACETYLTRANSFERASE (EC 2.3.1.9)
FABG_BACSU	39%	1824	3-OXOACYL-[ACYL-CARRIER PROTEIN] REDUCTASE (EC 1.1.1.100)
PAC_BACSH	45%	1883	PENICILLIN ACYLASE (EC 3.5.1.11)
PGSA_BACSU	43%	2026	CDP-DIACYLGLYCEROL--GLYCEROL-3-PHOSPHATE 3-PHOSPHATIDYLTRANSFERASE (EC 2.7.8.5)
CDSA_PSEAE	32%	2179	PHOSPHATIDATE CYTIDYLYLTRANSFERASE (EC 2.7.7.41)
BIO08. PURINES, PYRIMIDINES, NUCLEOSIDES AND NUCLEOTIDES			**47 genes**
BIO08.01 2'-deoxyribonucleotide metabolism			**0 genes**

Table 3.

BIO08.02 Nucleotide and nucleoside interconversions **4 genes**

DGK2_LACAC	62%	491	DEOXYGUANOSINE KINASE (EC 2.7.1.113) / DEOXYADENOSINE KINASE (EC 2.7.1.76) SUBUNIT 2
KITH_STRGC	79%	583	THYMIDINE KINASE (EC 2.7.1.21)
DGK2_LACAC	55%	1168	DEOXYGUANOSINE KINASE (EC 2.7.1.113) / DEOXYADENOSINE KINASE (EC 2.7.1.76) SUBUNIT 2
KCY_BACSU	49%	1740	PROBABLE CYTIDYLATE KINASE (EC 2.7.4.14)

BIO08.03 Purine ribonucleotide biosynthesis **22 genes**

IMDH_STRPY	82%	219	INOSINE-5'-MONOPHOSPHATE DEHYDROGENASE (EC 1.1.1.205)
ADD_ECOLI	33%	285	ADENOSINE DEAMINASE (EC 3.5.4.4)
FOLD_STRTR	65%	874	METHYLENETETRAHYDROFOLATE DEHYDROGENASE (EC 1.5.1.5) / METHENYLTETRAHYDROFOLATE CYCLOHYDROLASE (EC 3.5.4.9)
FTHS_STRMU	65%	958	FORMATE--TETRAHYDROFOLATE LIGASE (EC 6.3.4.3)
IMDH_HELPY	37%	1156	INOSINE-5'-MONOPHOSPHATE DEHYDROGENASE (EC 1.1.1.205)
GUAA_BACSU	67%	1514	GMP SYNTHASE (GLUTAMINE-HYDROLYZING) (EC 6.3.5.2)
PURK_SYNY3	33%	1548	PHOSPHORIBOSYLAMINOIMIDAZOLE CARBOXYLASE ATPASE SUBUNIT (EC 4.1.1.21)
PUR6_HAEIN	59%	1549	PHOSPHORIBOSYLAMINOIMIDAZOLE CARBOXYLASE CATALYTIC SUBUNIT (EC 4.1.1.21)
PUR2_BACSU	55%	1550	PHOSPHORIBOSYLAMINE--GLYCINE LIGASE (EC 6.3.4.13)
PUR9_BACSU	55%	1555	PHOSPHORIBOSYLAMINOIMIDAZOLECARBOXAMIDE FORMYLTRANSFERASE (EC 2.1.2.3)
PUR3_BACSU	45%	1561	PHOSPHORIBOSYLGLYCINAMIDE FORMYLTRANSFERASE (EC 2.1.2.2)
PUR5_BACSU	60%	1562	PHOSPHORIBOSYLFORMYLGLYCINAMIDINE CYCLO-LIGASE (EC 6.3.3.1)
PUR1_BACSU	61%	1568	AMIDOPHOSPHORIBOSYLTRANSFERASE PRECURSOR (EC 2.4.2.14)
PURL_BACSU	58%	1571	PHOSPHORIBOSYLFORMYLGLYCINAMIDINE SYNTHASE II (EC 6.3.5.3)
PURQ_BACSU	61%	1573	PHOSPHORIBOSYLFORMYLGLYCINAMIDINE SYNTHASE I (EC 6.3.5.3)
PUR7_STRPN	66%	1574	PHOSPHORIBOSYLAMINOIMIDAZOLE-SUCCINOCARBOXAMIDE SYNTHASE (EC 6.3.2.6)
PUR8_BACSU	76%	1685	ADENYLOSUCCINATE LYASE (EC 4.3.2.2)
PUR8_STRMU	81%	1685	ADENYLOSUCCINATE LYASE (EC 4.3.2.2)
KGUA_SYNY3	48%	1946	GUANYLATE KINASE (EC 2.7.4.8)
PURA_BACSU	59%	2009	ADENYLOSUCCINATE SYNTHETASE (EC 6.3.4.4)
PURQ_BACSU	31%	2107	PHOSPHORIBOSYLFORMYLGLYCINAMIDINE SYNTHASE I (EC 6.3.5.3)
KAD_LACLA	98%	2137	ADENYLATE KINASE (EC 2.7.4.3)

BIO08.04 Pyrimidine ribonucleotide biosynthesis **10 genes**

PYRG_BACSU	64%	486	CTP SYNTHASE (EC 6.3.4.2)
PYRE_BACSU	57%	1078	OROTATE PHOSPHORIBOSYLTRANSFERASE (EC 2.4.2.10)
PYRC_LACLE	42%	1079	DIHYDROOROTASE (EC 3.5.2.3)
DCOP_LACLC	97%	1380	OROTIDINE 5'-PHOSPHATE DECARBOXYLASE (EC 4.1.1.23)
PYDB_LACLC	93%	1381	DIHYDROOROTATE DEHYDROGENASE B (EC 1.3.3.1)
CDD_BACSU	54%	1461	CYTIDINE DEAMINASE (EC 3.5.4.5)
TYSY_LACLA	98%	1579	THYMIDYLATE SYNTHASE (EC 2.1.1.45)
PYDA_LACLC	98%	1589	DIHYDROOROTATE DEHYDROGENASE A (EC 1.3.3.1)
CARA_BACSU	56%	1640	CARBAMOYL-PHOSPHATE SYNTHASE, PYRIMIDINE-SPECIFIC, SMALL CHAIN (EC 6.3.5.5)
PYRB_SYNY3	42%	1641	ASPARTATE CARBAMOYLTRANSFERASE (EC 2.1.3.2)

BIO08.05 Salvage of nucleosides and nucleotides **11 genes**

HPRT_LACLA	99%	25	HYPOXANTHINE-GUANINE PHOSPHORIBOSYLTRANSFERASE (EC 2.4.2.8)
KTHY_BACSU	48%	401	THYMIDYLATE KINASE (EC 2.7.4.9)
APT_ECOLI	54%	621	ADENINE PHOSPHORIBOSYLTRANSFERASE (EC 2.4.2.7)
KPRS_BACSU	66%	824	RIBOSE-PHOSPHATE PYROPHOSPHOKINASE (EC 2.7.6.1)
DEOB_BACSU	55%	953	PHOSPHOPENTOMUTASE (EC 5.4.2.7)
XPT_BACSU	52%	1157	XANTHINE PHOSPHORIBOSYLTRANSFERASE (EC 2.4.2.-)
HPRT_BACSU	56%	1557	HYPOXANTHINE-GUANINE PHOSPHORIBOSYLTRANSFERASE (EC 2.4.2.8)
URK_HAEIN	52%	1705	URIDINE KINASE (EC 2.7.1.48)
KPRS_BACCL	26%	1906	RIBOSE-PHOSPHATE PYROPHOSPHOKINASE (EC 2.7.6.1)
UPP_LACLA	97%	1972	URACIL PHOSPHORIBOSYLTRANSFERASE (EC 2.4.2.9)
PURR_BACSU	50%	2331	PUR OPERON REPRESSOR

BIO08.06 Sugar-nucleotide biosynthesis and interconversions **0 genes**

BIO09. REGULATORY FUNCTIONS **94 genes**

BIO09.01 General **41 genes**

YQAE_BACSU	35%	11	HYPOTHETICAL TRANSCRIPTIONAL REGULATOR IN SPOIIIC-CWLA INTERGENIC REGION (ORF7)
XRE_BACSU	50%	44	PUTATIVE PBSX REPRESSOR
RELA_STREQ	66%	106	PUTATIVE GTP PYROPHOSPHOKINASE (EC 2.7.6.5)
TER8_PASPI	29%	128	TETRACYCLINE REPRESSOR PROTEIN CLASS H
YQAE_BACSU	34%	133	HYPOTHETICAL TRANSCRIPTIONAL REGULATOR IN SPOIIIC-CWLA INTERGENIC REGION (ORF7)
TER2_ECOLI	38%	143	TETRACYCLINE REPRESSOR PROTEIN CLASS B (TRANSPOSON TN10)

Table 3.

CODY_BACSU	49%	162	CODY PROTEIN (VEGETATIVE PROTEIN 286B)
MECB_BACSU	51%	549	NEGATIVE REGULATOR OF GENETIC COMPETENCE MECB
CTSR_BACSU	45%	627	TRANSCRIPTIONAL REGULATOR CTSR
MECB_BACSU	50%	629	NEGATIVE REGULATOR OF GENETIC COMPETENCE MECB
YP20_BACLI	34%	737	HYPOTHETICAL P20 PROTEIN
SLYA_SALTY	26%	747	TRANSCRIPTIONAL REGULATOR SLYA (SALMOLYSIN)
YP20_BACLI	33%	795	HYPOTHETICAL P20 PROTEIN
BLAI_BACLI	31%	843	PENICILLINASE REPRESSOR (REGULATORY PROTEIN BLAI)
DEGU_BACSU	39%	910	TRANSCRIPTIONAL REGULATORY PROTEIN DEGU (PROTEASE PRODUCTION ENHANCER PROTEIN)
HRCA_LACLA	99%	975	HEAT-INDUCIBLE TRANSCRIPTION REPRESSOR HRCA
SORC_KLEPN	21%	1204	SORBITOL (SOR) OPERON REGULATOR
XRE_BACSU	42%	1220	PUTATIVE PBSX REPRESSOR
YQAE_BACSU	35%	1300	HYPOTHETICAL TRANSCRIPTIONAL REGULATOR IN SPOIIIC-CWLA INTERGENIC REGION (ORF7)
Y186_HAEIN	39%	1506	HYPOTHETICAL TRANSCRIPTIONAL REGULATOR HI0186
TER2_ECOLI	29%	1551	TETRACYCLINE REPRESSOR PROTEIN CLASS B (TRANSPOSON TN10)
TFDT_ALCEU	40%	1594	TRANSCRIPTIONAL REGULATORY PROTEIN TFDT
GLNB_FREDI	56%	1631	NITROGEN REGULATORY PROTEIN P-II (PII SIGNAL TRANSDUCING PROTEIN)
PYRR_BACCL	60%	1644	PYRIMIDINE OPERON REGULATORY PROTEIN PYRR
SMTB_SYNY3	34%	1645	TRANSCRIPTIONAL REPRESSOR SMTB HOMOLOG
ARR2_ECOLI	31%	1730	ARSENICAL RESISTANCE OPERON REPRESSOR
GERE_BACSU	41%	1737	GERMINATION PROTEIN GERE
PHOU_ECOLI	36%	1750	PHOSPHATE TRANSPORT SYSTEM REGULATORY PROTEIN
TENA_BACSU	41%	1818	TRANSCRIPTIONAL ACTIVATOR TENA
BIRA_BACSU	34%	1820	BIRA BIFUNCTIONAL PROTEIN (BIOTIN OPERON REPRESSOR)
Y186_HAEIN	31%	1927	HYPOTHETICAL TRANSCRIPTIONAL REGULATOR HI0186
BIRA_BACSU	33%	1953	BIRA BIFUNCTIONAL PROTEIN (BIOTIN OPERON REPRESSOR)
RPC2_BPP22	30%	2003	REPRESSOR PROTEIN C2
Y186_HAEIN	34%	2032	HYPOTHETICAL TRANSCRIPTIONAL REGULATOR HI0186
NADR_ECOLI	30%	2047	TRANSCRIPTIONAL REGULATOR NADR
FLP_LACCA	41%	2062	PROBABLE TRANSCRIPTIONAL REGULATOR FLP
ACKA_BACSU	54%	2069	ACETATE KINASE (EC 2.7.2.1)
GLNR_BACSU	50%	2264	REGULATORY PROTEIN GLNR
YQAE_BACSU	41%	2276	HYPOTHETICAL TRANSCRIPTIONAL REGULATOR IN SPOIIIC-CWLA INTERGENIC REGION (ORF7)
FNR_BACSU	28%	2298	ANAEROBIC REGULATORY PROTEIN
ACRR_ECOLI	32%	2317	POTENTIAL ACRAB OPERON REPRESSOR

BIO09.02 Two-component systems 13 genes

RESE_BACSU	33%	400	SENSOR PROTEIN RESE (EC 2.7.3.-)
YYCF_BACSU	69%	401	HYPOTHETICAL 27.2 KD SENSORY TRANSDUCTION PROTEIN IN ROCR-PURA INTERGENIC REGION
DEGS_BACBR	40%	909	SENSOR PROTEIN DEGS (EC 2.7.3.-)
CZCR_ALCEU	39%	1028	TRANSCRIPTIONAL ACTIVATOR PROTEIN CZCR
YEDV_ECOLI	28%	1030	PROBABLE SENSOR PROTEIN YEDV (EC 2.7.3.-)
PHOP_BACSU	40%	1456	ALKALINE PHOSPHATASE SYNTHESIS TRANSCRIPTIONAL REGULATORY PROTEIN PHOP
PHOR_BACSU	32%	1457	ALKALINE PHOSPHATASE SYNTHESIS SENSOR PROTEIN PHOR (EC 2.7.3.-)
RESE_BACSU	27%	1633	SENSOR PROTEIN RESE (EC 2.7.3.-)
CZCR_ALCEU	43%	1635	TRANSCRIPTIONAL ACTIVATOR PROTEIN CZCR
RESE_BACSU	33%	1722	SENSOR PROTEIN RESE (EC 2.7.3.-)
COPR_PSESM	34%	1723	TRANSCRIPTIONAL ACTIVATOR PROTEIN COPR
YXDK_BACSU	33%	1784	HYPOTHETICAL SENSOR-LIKE HISTIDINE KINASE IN IDH 3'REGION (EC 2.7.3.-)
YXDJ_BACSU	42%	1785	HYPOTHETICAL 26.6 KD SENSORY TRANSDUCTION PROTEIN IN IDH 3'REGION

BIO09.03 LacI-family regulators 6 genes

CYTR_ECOLI	28%	728	TRANSCRIPTIONAL REPRESSOR CYTR
RBSR_BACSU	31%	1533	RIBOSE OPERON REPRESSOR
RBSR_BACSU	42%	1684	RIBOSE OPERON REPRESSOR
CCPA_STAXY	47%	1692	PROBABLE CATABOLITE CONTROL PROTEIN A
PURR_HAEIN	34%	1724	PURINE NUCLEOTIDE SYNTHESIS REPRESSOR
YYAG_BACSU	36%	2196	HYPOTHETICAL TRANSCRIPTIONAL REGULATOR IN TETB-EXOA INTERGENIC REGION

BIO09.04 LysR-family regulators 4 genes

YWFK_BACSU	25%	375	HYPOTHETICAL TRANSCRIPTIONAL REGULATOR IN PTA 3'REGION
MLER_LACLA	100%	893	MALOLACTIC FERMENTATION SYSTEM TRANSCRIPTIONAL ACTIVATOR
YWBI_BACSU	29%	1261	HYPOTHETICAL TRANSCRIPTIONAL REGULATOR IN EPR-GALK INTERGENIC REGION
YWBI_BACSU	30%	1926	HYPOTHETICAL TRANSCRIPTIONAL REGULATOR IN EPR-GALK INTERGENIC REGION

BIO09.05 AraC-family regulators 4 genes

Table 3.

SLAP_LACBR	30%	480	S-LAYER PROTEIN PRECURSOR (SURFACE LAYER PROTEIN)
ADAA_BACSU	44%	517	METHYLPHOSPHOTRIESTER-DNA ALKYLTRANSFERASE
LUMQ_PHOLE	32%	1547	PROBABLE TRANSCRIPTIONAL REGULATOR LUMQ
TOXA_CLODI	30%	2284	TOXIN A
BIO09.06 GntR-family regulators			**6 genes**
YHHX_ECOLI	44%	261	HYPOTHETICAL 38.8 KD PROTEIN IN GNTR-GGT INTERGENIC REGION (F345)
TRER_BACSU	44%	434	TREHALOSE OPERON TRANSCRIPTIONAL REPRESSOR
FARR_ECOLI	30%	1459	FATTY ACYL RESPONSIVE REGULATOR (P30 PROTEIN)
YIN1_STRAM	30%	1676	HYPOTHETICAL TRANSCRIPTIONAL REGULATOR IN INSTABLE DNA LOCUS (ORF 1)
YXAM_BACSU	24%	2215	HYPOTHETICAL 44.7 KD PROTEIN IN GNTR-HTPG INTERGENIC REGION
YXAT_BACSU	18%	2305	HYPOTHETICAL 44.3 KD PROTEIN IN GNTR-HTPG INTERGENIC REGION
BIO09.07 DeoR-family regulators			**8 genes**
YXEP_BACSU	33%	278	HYPOTHETICAL 41.6 KD PROTEIN IN IDH-DEOR INTERGENIC REGION
YXEB_BACSU	35%	327	PROBABLE ABC TRANSPORTER BINDING PROTEIN IN IDH-DEOR INTERGENIC REGION PRECURSOR
YXEA_BACSU	37%	575	HYPOTHETICAL 13.0 KD PROTEIN IN IDH-DEOR INTERGENIC REGION PRECURSOR
GATR_ECOLI	31%	794	GALACTITOL UTILIZATION OPERON REPRESSOR
YXEP_BACSU	43%	980	HYPOTHETICAL 41.6 KD PROTEIN IN IDH-DEOR INTERGENIC REGION
YBJG_ECOLI	32%	1031	HYPOTHETICAL 22.4 KD PROTEIN IN DEOR-CMR INTERGENIC REGION
GLPR_HAEIN	33%	1330	GLYCEROL-3-PHOSPHATE REGULON REPRESSOR
YXEA_BACSU	39%	1504	HYPOTHETICAL 13.0 KD PROTEIN IN IDH-DEOR INTERGENIC REGION PRECURSOR
BIO09.08 MarR-family regulators			**2 genes**
YDEA_ECOLI	32%	329	HYPOTHETICAL 42.5 KD PROTEIN IN UXAB-MARR INTERGENIC REGION
PETP_RHOCA	34%	1338	PETP PROTEIN
BIO09.09 BglG-family regulators			**1 genes**
LICT_BACSU	41%	1491	TRANSCRIPTION ANTITERMINATOR LICT
BIO09.10 GTP-binding proteins			**9 genes**
YYAF_BACSU	68%	12	HYPOTHETICAL 40.1 KD GTP-BINDING PROTEIN IN RPSF-SPO0J INTERGENIC REGION
HFLX_ECOLI	44%	222	GTP-BINDING PROTEIN HFLX
ERA_STRMU	75%	353	GTP-BINDING PROTEIN ERA HOMOLOG
YPHC_BACSU	62%	760	HYPOTHETICAL 48.8 KD GTP-BINDING PROTEIN IN CMK-GPSA INTERGENIC REGION
LEPA_BACSU	77%	1116	GTP-BINDING PROTEIN LEPA
YSXC_BACSU	66%	1163	HYPOTHETICAL GTP-BINDING PROTEIN IN LONA-HEMA INTERGENIC REGION (ORFX)
OBG_BACSU	64%	1626	SPO0B-ASSOCIATED GTP-BINDING PROTEIN
SR54_SYNY3	51%	1654	SIGNAL RECOGNITION PARTICLE PROTEIN (FIFTY-FOUR HOMOLOG)
THDF_BACSU	60%	2308	POSSIBLE THIOPHENE AND FURAN OXIDATION PROTEIN THDF
BIO10. REPLICATION			**77 genes**
BIO10.01 Degradation of DNA			**11 genes**
ADDB_BACSU	28%	5	ATP-DEPENDENT NUCLEASE SUBUNIT B
ADDA_BACSU	32%	7	ATP-DEPENDENT NUCLEASE SUBUNIT A
ADDA_BACSU	33%	9	ATP-DEPENDENT NUCLEASE SUBUNIT A
UVRB_HAEIN	40%	555	EXCINUCLEASE ABC SUBUNIT B
RECJ_ERWCH	36%	619	SINGLE-STRANDED-DNA-SPECIFIC EXONUCLEASE RECJ (EC 3.1.-.-)
EXOA_BACSU	40%	796	EXODEOXYRIBONUCLEASE (EC 3.1.11.2)
UVRC_BACSU	54%	855	EXCINUCLEASE ABC SUBUNIT C
EX7L_BACSU	43%	876	PUTATIVE EXODEOXYRIBONUCLEASE LARGE SUBUNIT (EC 3.1.11.6)
END3_BACSU	50%	1081	PROBABLE ENDONUCLEASE III (EC 4.2.99.18)
EX5A_ECOLI	27%	1777	EXODEOXYRIBONUCLEASE V 67 KD POLYPEPTIDE (EC 3.1.11.5)
EX5A_ECOLI	32%	1778	EXODEOXYRIBONUCLEASE V 67 KD POLYPEPTIDE (EC 3.1.11.5)
BIO10.02 DNA replication, restriction, modification, recombination, and repair			**66 genes**
DNAA_STAAU	45%	1	CHROMOSOMAL REPLICATION INITIATOR PROTEIN DNAA
DP3B_STAAU	41%	2	DNA POLYMERASE III, BETA CHAIN (EC 2.7.7.7)
MFD_BACSU	50%	19	TRANSCRIPTION-REPAIR COUPLING FACTOR (TRCF)
PRIM_BPP4	21%	38	PUTATIVE P4-SPECIFIC DNA PRIMASE (EC 2.7.7.-)
YQXN_BACSU	31%	58	HYPOTHETICAL 29.3 KD PROTEIN IN BEX-DNAG/DNAE INTERGENIC REGION (ORF3)
YQEU_BACSU	44%	103	HYPOTHETICAL 28.8 KD PROTEIN IN DNAJ-RPSU INTEREGENIC REGION
YAAK_BACSU	44%	120	HYPOTHETICAL 11.8 KD PROTEIN IN DNAZ-RECR INTERGENIC REGION
T3RE_BACCE	29%	151	TYPE III RESTRICTION-MODIFICATION SYSTEM ENZYME RES (EC 3.1.21.5)
UNG_STRPN	53%	230	URACIL-DNA GLYCOSYLASE (EC 3.2.2.-)

Table 3.

RECR_STRTR	76%	338	RECOMBINATION PROTEIN RECR
FPG_LACLC	90%	356	FORMAMIDOPYRIMIDINE-DNA GLYCOSYLASE (EC 3.2.2.23)
RECA_LACLA	94%	357	RECA PROTEIN
SSB_BACSU	44%	395	SINGLE-STRAND BINDING PROTEIN (SSB)
HOLB_BACSU	35%	402	DNA POLYMERASE III, DELTA' SUBUNIT (EC 2.7.7.7)
DNLJ_RHOMR	43%	423	DNA LIGASE (EC 6.5.1.2)
DNAC_ECOLI	27%	454	DNA REPLICATION PROTEIN DNAC
DP3A_HELPY	26%	494	DNA POLYMERASE III, ALPHA CHAIN (EC 2.7.7.7)
PRIM_LACLA	100%	542	DNA PRIMASE (EC 2.7.7.-)
DFP_ECOLI	30%	565	DNA/PANTOTHENATE METABOLISM FLAVOPROTEIN
T1R1_ECOLI	32%	640	TYPE I RESTRICTION ENZYME ECOR124II R PROTEIN (EC 3.1.21.3)
YC18_METJA	27%	644	HYPOTHETICAL PROTEIN MJ1218
YABF_BACSU	47%	687	HYPOTHETICAL 20.7 KD PROTEIN IN METS-KSGA INTERGENIC REGION
YAAJ_BACSU	51%	725	HYPOTHETICAL 17.8 KD PROTEIN IN SERS-DNAH INTERGENIC REGION
DNAC_BACSU	56%	752	REPLICATIVE DNA HELICASE (EC 3.6.1.-)
DNAB_BACSU	27%	756	REPLICATION INITIATION AND MEMBRANE ATTACHMENT PROTEIN
DNAI_BACSU	38%	757	PRIMOSOMAL PROTEIN DNAI
LEF3_NPVAC	26%	833	LATE EXPRESSION FACTOR 3
MUTY_ECOLI	43%	857	A/G-SPECIFIC ADENINE GLYCOSYLASE (EC 3.2.2.-)
RECN_BACSU	44%	881	DNA REPAIR PROTEIN RECN (RECOMBINATION PROTEIN N)
GYRB_STRPN	82%	926	DNA GYRASE SUBUNIT B (EC 5.99.1.3)
GRPE_LACLA	99%	975	GRPE PROTEIN
RIR2_MYCTU	48%	999	RIBONUCLEOSIDE-DIPHOSPHATE REDUCTASE BETA CHAIN (EC 1.17.4.1)
RIR1_MYCTU	50%	1001	RIBONUCLEOSIDE-DIPHOSPHATE REDUCTASE ALPHA CHAIN (EC 1.17.4.1)
PARE_STAAU	70%	1005	TOPOISOMERASE IV SUBUNIT B (EC 5.99.1.-)
DING_BACSU	38%	1007	PROBABLE ATP-DEPENDENT HELICASE DING HOMOLOG
PARC_STAAU	53%	1009	TOPOISOMERASE IV SUBUNIT A (EC 5.99.1.-)
RADC_BACSU	41%	1033	DNA REPAIR PROTEIN RADC HOMOLOG (ORFB)
SSB_BACSU	61%	1041	SINGLE-STRAND BINDING PROTEIN (SSB)
DNAD_BACSU	37%	1081	DNA REPLICATION PROTEIN DNAD
GYRA_STAAU	23%	1120	DNA GYRASE SUBUNIT A (EC 5.99.1.3)
PCRA_BACST	49%	1132	ATP-DEPENDENT HELICASE PCRA (EC 3.6.1.-)
MUTX_STRPN	34%	1134	MUTATOR MUTT PROTEIN (7,8-DIHYDRO-8-OXOGUANINE-TRIPHOSPHATASE)
DYR_LACLA	87%	1161	DIHYDROFOLATE REDUCTASE (EC 1.5.1.3)
DPRA_HAEIN	43%	1251	DNA PROCESSING CHAIN A
TOP1_BACSU	64%	1253	DNA TOPOISOMERASE I (EC 5.99.1.2)
DNAC_ECOLI	27%	1446	DNA REPLICATION PROTEIN DNAC
SSB_BACSU	68%	1447	SINGLE-STRAND BINDING PROTEIN (SSB)
MUTS_AQUPY	34%	1689	DNA MISMATCH REPAIR PROTEIN
RECQ_ECOLI	35%	1854	ATP-DEPENDENT DNA HELICASE RECQ (EC 3.6.1.-)
DING_BACSU	34%	1878	PROBABLE ATP-DEPENDENT HELICASE DING HOMOLOG
DING_BACSU	28%	1880	PROBABLE ATP-DEPENDENT HELICASE DING HOMOLOG
PRIA_BACSU	58%	1944	PRIMOSOMAL PROTEIN N' (REPLICATION FACTOR Y)
YAAA_BACSU	49%	2030	HYPOTHETICAL 7.9 KD PROTEIN IN DNAN-RECF INTERGENIC REGION
RECF_LACLC	99%	2031	RECF PROTEIN
DINP_ECOLI	42%	2085	DNA-DAMAGE-INDUCIBLE PROTEIN P
RADA_BACSU	60%	2129	DNA REPAIR PROTEIN RADA HOMOLOG (DNA REPAIR PROTEIN SMS HOMOLOG)
DP3A_BACSU	42%	2171	DNA POLYMERASE III, ALPHA CHAIN (EC 2.7.7.7)
DPO1_STRPN	68%	2192	DNA POLYMERASE I (EC 2.7.7.7)
SSB_BACSU	60%	2254	SINGLE-STRAND BINDING PROTEIN (SSB)
DP3X_BACSU	43%	2259	DNA POLYMERASE III SUBUNITS GAMMA AND TAU (EC 2.7.7.7)
RUVB_PSEAE	59%	2269	HOLLIDAY JUNCTION DNA HELICASE RUVB
RUVA_HAEIN	40%	2269	HOLLIDAY JUNCTION DNA HELICASE RUVA
HEXB_STRPN	57%	2271	DNA MISMATCH REPAIR PROTEIN HEXB
HEXA_STRPN	79%	2274	DNA MISMATCH REPAIR PROTEIN HEXA
DNAJ_LACLA	99%	2287	DNAJ PROTEIN
RECG_ECOLI	36%	2311	ATP-DEPENDENT DNA HELICASE RECG (EC 3.6.1.-)
BIO11. TRANSCRIPTION			**27 genes**
BIO11.01 Degradation of RNA			**18 genes**
TRUA_ECOLI	34%	482	PSEUDOURIDYLATE SYNTHASE I (EC 4.2.1.70)
YACO_BACSU	26%	595	HYPOTHETICAL RRNA METHYLASE IN CYSS 3'REGION
MIAA_MYCLE	32%	613	PROBABLE TRNA DELTA(2)-ISOPENTENYLPYROPHOSPHATE TRANSFERASE (EC 2.5.1.8)
RNC_BACSU	44%	807	RIBONUCLEASE III (EC 3.1.26.3)
VACB_HAEIN	37%	965	VACB PROTEIN HOMOLOG
ALDC_BACBR	31%	1114	ALPHA-ACETOLACTATE DECARBOXYLASE PRECURSOR (EC 4.1.1.5)
TRUB_SYNY3	39%	1138	TRNA PSEUDOURIDINE 55 SYNTHASE (PSI55 SYNTHASE)
YGL3_BACST	57%	1160	HYPOTHETICAL 18.2 KD PROTEIN IN GLDA 3'REGION (ORF3)
ALDC_BACSU	35%	1250	ALPHA-ACETOLACTATE DECARBOXYLASE (EC 4.1.1.5)
RNH2_ECOLI	47%	1326	RIBONUCLEASE HII (EC 3.1.26.4)
SBCD_BACSU	32%	1354	EXONUCLEASE SBCD HOMOLOG (FRAGMENT)
PAPS_BACSU	39%	1598	POLY(A) POLYMERASE (EC 2.7.7.19)
RIMM_HAEIN	26%	1603	16S RRNA PROCESSING PROTEIN RIMM
PNPA_BACSU	58%	1903	POLYRIBONUCLEOTIDE NUCLEOTIDYLTRANSFERASE (EC 2.7.7.8)

Table 3.

YACO_BACSU	47%	1921	HYPOTHETICAL RRNA METHYLASE IN CYSS 3'REGION
SPLR_NPVAC	30%	2008	SPHEROIDIN-LIKE PROTEIN PRECURSOR (SPINDOLIN-LIKE PROTEIN)
SAGP_STRPY	47%	2094	STREPTOCOCCAL ACID GLYCOPROTEIN
NUSG_STRGR	43%	2154	TRANSCRIPTION ANTITERMINATION PROTEIN NUSG

BIO11.02 RNA synthesis, modification, and DNA transcription **8 genes**

RPSD_NEIGO	24%	466	RNA POLYMERASE SIGMA FACTOR RPOD (SIGMA-70)
RPSD_LACLA	100%	544	RNA POLYMERASE SIGMA FACTOR RPOD
RPOE_BACSU	33%	621	DNA-DIRECTED RNA POLYMERASE DELTA SUBUNIT (EC 2.7.7.6)
GREA_RICPR	47%	624	TRANSCRIPTION ELONGATION FACTOR GREA (TRANSCRIPT CLEAVAGE FACTOR GREA)
RPOC_MYCGE	46%	1838	DNA-DIRECTED RNA POLYMERASE BETA' CHAIN (EC 2.7.7.6)
RPOB_SPICI	31%	1842	DNA-DIRECTED RNA POLYMERASE BETA CHAIN (EC 2.7.7.6)
RPOA_BACSU	57%	2132	DNA-DIRECTED RNA POLYMERASE ALPHA CHAIN (EC 2.7.7.6)
SIGV_BACSU	23%	2222	RNA POLYMERASE SIGMA FACTOR SIGV

BIO11.03 RNA processing **1 genes**

YQFR_BACSU	48%	414	PROBABLE RNA HELICASE IN CCCA-SODA INTERGENIC REGION

BIO12. TRANSLATION **138 genes**

BIO12.01 Amino acyl tRNA synthetases **26 genes**

SYW_CLOLO	63%	68	TRYPTOPHANYL-TRNA SYNTHETASE (EC 6.1.1.2)
TGT_HAEIN	54%	153	QUEUINE TRNA-RIBOSYLTRANSFERASE (EC 2.4.2.29)
SYK_BACSU	58%	374	LYSYL-TRNA SYNTHETASE (EC 6.1.1.6)
SYY1_BACSU	55%	388	TYROSYL-TRNA SYNTHETASE 1 (EC 6.1.1.1)
Y449_MYCPN	34%	393	HYPOTHETICAL PROTEIN MG449 HOMOLOG
SYM_BACSU	62%	798	METHIONYL-TRNA SYNTHETASE (EC 6.1.1.10)
SYL_MYCTU	43%	826	LEUCYL-TRNA SYNTHETASE (EC 6.1.1.4)
SYGB_BACSU	41%	1101	GLYCYL-TRNA SYNTHETASE BETA CHAIN (EC 6.1.1.14)
SYK_BACSU	32%	1129	LYSYL-TRNA SYNTHETASE (EC 6.1.1.6)
YHI3_LACLA	97%	1228	HYPOTHETICAL 38.0 KD PROTEIN IN HISC-HISG INTERGENIC REGION (ORF3)
SYS_BACSU	57%	1747	SERYL-TRNA SYNTHETASE (EC 6.1.1.11)
SYA_THETH	41%	1759	ALANYL-TRNA SYNTHETASE (EC 6.1.1.7)
SYN_BACSU	59%	1875	ASPARAGINYL-TRNA SYNTHETASE (EC 6.1.1.22)
SYC_HELPY	43%	1898	CYSTEINYL-TRNA SYNTHETASE (EC 6.1.1.16)
SYI_STAAU	28%	1911	ISOLEUCYL-TRNA SYNTHETASE (EC 6.1.1.5)
SYI_CAMJE	43%	1912	ISOLEUCYL-TRNA SYNTHETASE (EC 6.1.1.5)
SYT_MYCLE	42%	1968	THREONYL-TRNA SYNTHETASE (EC 6.1.1.3)
SYFB_BACSU	45%	1987	PHENYLALANYL-TRNA SYNTHETASE BETA CHAIN (EC 6.1.1.20)
SYFA_BACSU	62%	1989	PHENYLALANYL-TRNA SYNTHETASE ALPHA CHAIN (EC 6.1.1.20)
SYD_SYNY3	51%	2021	ASPARTYL-TRNA SYNTHETASE (EC 6.1.1.12)
SYH_STREQ	63%	2023	HISTIDYL-TRNA SYNTHETASE (EC 6.1.1.21)
SYR_SYNY3	36%	2096	ARGINYL-TRNA SYNTHETASE (EC 6.1.1.19)
SYE_HELPY	45%	2120	GLUTAMYL-TRNA SYNTHETASE (EC 6.1.1.17)
SYP_HAEIN	45%	2177	PROLYL-TRNA SYNTHETASE (EC 6.1.1.15)
SYV_THETH	43%	2229	VALYL-TRNA SYNTHETASE (EC 6.1.1.9)
SYFB_MYCGE	29%	2233	PHENYLALANYL-TRNA SYNTHETASE BETA CHAIN (EC 6.1.1.20)

BIO12.02 Degradation of proteins, peptides, and glycopeptides **22 genes**

GCP_HAEIN	41%	291	PROBABLE O-SIALOGLYCOPROTEIN ENDOPEPTIDASE (EC 3.4.24.57)
AMPN_LACLC	98%	301	AMINOPEPTIDASE N (EC 3.4.11.2)
AMP1_SYNY3	45%	599	PUTATIVE METHIONINE AMINOPEPTIDASE A (EC 3.4.11.18)
CLPP_STRSL	87%	670	ATP-DEPENDENT CLP PROTEASE PROTEOLYTIC SUBUNIT (EC 3.4.21.92)
YQHT_BACSU	41%	688	PUTATIVE PEPTIDASE IN GCVT-SPOIIIAA INTERGENIC REGION (EC 3.4.-.-)
PEPV_LACDL	47%	858	XAA-HIS DIPEPTIDASE (EC 3.4.13.3)
P54_ENTFC	40%	959	P54 PROTEIN PRECURSOR
CLPX_BACSU	67%	1162	ATP-DEPENDENT CLP PROTEASE ATP-BINDING SUBUNIT CLPX
CLPP_SYNP7	29%	1433	ATP-DEPENDENT CLP PROTEASE PROTEOLYTIC SUBUNIT (EC 3.4.21.92)
CLPB_HAEIN	56%	1564	CLPB PROTEIN
PEPQ_LACDL	50%	1693	XAA-PRO DIPEPTIDASE (EC 3.4.13.9)
PRTM_LACPA	32%	1761	PROTEASE MATURATION PROTEIN PRECURSOR
PEPF_LACLA	97%	1763	OLIGOENDOPEPTIDASE F (EC 3.4.24.-)
PEPO_LACLC	99%	1847	NEUTRAL ENDOPEPTIDASE (EC 3.4.24.-)
PEPC_LACLC	98%	1927	AMINOPEPTIDASE C (EC 3.4.22.-)
PTR_ECOLI	22%	2028	PROTEASE III PRECURSOR (EC 3.4.24.55)
YMXG_BACSU	23%	2029	HYPOTHETICAL PROCESSING PROTEASE (EC 3.4.-.-)
YEGQ_ECOLI	37%	2049	PUTATIVE PROTEASE IN BAER-OGRK INTERGENIC REGION (EC 3.4.-.-)
LEP3_BACSU	34%	2084	TYPE 4 PREPILIN-LIKE PROTEIN SPECIFIC LEADER PEPTIDASE (EC 3.4.99.-)
PEPX_LACLC	79%	2115	XAA-PRO DIPEPTIDYL-PEPTIDASE (EC 3.4.14.11)
YYXA_BACSU	47%	2185	HYPOTHETICAL PROTEASE IN ROCR-PURA INTERGENIC REGION (EC 3.4.21.-)
HTOA_HAEIN	33%	2218	PROBABLE PERIPLASMIC SERINE PROTEASE DO/HHOA-LIKE PRECURSOR

Table 3.

BIO12.03 Nucleoproteins			**2 genes**
DBH_BACST	67%	500	DNA-BINDING PROTEIN II (HB)
DBH_VIBPR	37%	900	DNA-BINDING PROTEIN HU
BIO12.04 Protein modification			**2 genes**
CYPH_MYCTU	31%	366	PROBABLE PEPTIDYL-PROLYL CIS-TRANS ISOMERASE (EC 5.2.1.8)
CYPH_MYCTU	37%	912	PROBABLE PEPTIDYL-PROLYL CIS-TRANS ISOMERASE (EC 5.2.1.8)
BIO12.05 Ribosomal proteins: synthesis and modification			**55 genes**
PRMA_BACSU	43%	102	PROBABLE METHYLTRANSFERASE (EC 2.1.1.-)
RL34_COXBU	70%	131	50S RIBOSOMAL PROTEIN L34
RL28_BACSU	56%	193	50S RIBOSOMAL PROTEIN L28
RS21_BACSU	87%	235	30S RIBOSOMAL PROTEIN S21 (BS-B)
RS4_BACSU	62%	281	30S RIBOSOMAL PROTEIN S4 (BS4)
R14H_BACSH	35%	306	30S RIBOSOMAL PROTEIN S14 HOMOLOG
DEAD_KLEPN	45%	351	ATP-DEPENDENT RNA HELICASE DEAD
YVYD_BACSU	51%	630	HYPOTHETICAL 22.0 KD PROTEIN IN FLIT-SECA INTERGENIC REGION
RL33_LACLA	100%	659	50S RIBOSOMAL PROTEIN L33
RL9_BACST	52%	751	50S RIBOSOMAL PROTEIN L9 (BL17)
YLXQ_ENTFC	48%	773	PROBABLE RIBOSOMAL PROTEIN IN INFB 5'REGION
RS1_BUCAP	32%	852	30S RIBOSOMAL PROTEIN S1
RL19_BACST	73%	895	50S RIBOSOMAL PROTEIN L19
RS14_SYNY3	41%	908	30S RIBOSOMAL PROTEIN S14
RL21_BACSU	67%	1088	50S RIBOSOMAL PROTEIN L21 (BL20)
RL27_BACSU	74%	1089	50S RIBOSOMAL PROTEIN L27 (BL30)
RL7_BACSU	61%	1298	50S RIBOSOMAL PROTEIN L7/L12 (BL9)
RL7_STRPN	76%	1299	50S RIBOSOMAL PROTEIN L7/L12 (FRAGMENT)
RL10_BACSU	63%	1299	50S RIBOSOMAL PROTEIN L10 (BL5)
RS16_BACSU	59%	1606	30S RIBOSOMAL PROTEIN S16 (BS17)
YKGM_ECOLI	45%	1636	HYPOTHETICAL 9.9 KD PROTEIN IN INTF-EAEH INTERGENIC REGION
RL20_BACSU	69%	1891	50S RIBOSOMAL PROTEIN L20
RS15_BACSU	62%	1934	30S RIBOSOMAL PROTEIN S15 (BS18)
RL1_BACSU	67%	2058	50S RIBOSOMAL PROTEIN L1 (BL1)
RL11_STACA	80%	2059	50S RIBOSOMAL PROTEIN L11
R14H_BACSH	30%	2125	30S RIBOSOMAL PROTEIN S14 HOMOLOG
RL17_BACSU	72%	2131	50S RIBOSOMAL PROTEIN L17 (BL15)
RS11_BACSU	82%	2133	30S RIBOSOMAL PROTEIN S11 (BS11)
RS13_BACSU	70%	2133	30S RIBOSOMAL PROTEIN S13 (BS14)
RL15_STAAU	67%	2139	50S RIBOSOMAL PROTEIN L15
RS5_BACSU	77%	2140	30S RIBOSOMAL PROTEIN S5 (BS5)
RL6_BACST	63%	2141	50S RIBOSOMAL PROTEIN L6 (BL10)
RS8_BACSU	72%	2142	30S RIBOSOMAL PROTEIN S8 (BS8)
RL5_BACST	80%	2144	50S RIBOSOMAL PROTEIN L5
RL24_BACST	62%	2144	50S RIBOSOMAL PROTEIN L24
RL14_BACSU	80%	2144	50S RIBOSOMAL PROTEIN L14
RS17_BACSU	77%	2145	30S RIBOSOMAL PROTEIN S17 (BS16)
RL16_BACSU	76%	2145	50S RIBOSOMAL PROTEIN L16
RS3_BACSU	73%	2146	30S RIBOSOMAL PROTEIN S3 (BS3)
RL22_BACSU	62%	2146	50S RIBOSOMAL PROTEIN L22
RS19_BACST	77%	2147	30S RIBOSOMAL PROTEIN S19 (BS19)
RL2_BACSU	76%	2147	50S RIBOSOMAL PROTEIN L2 (BL2)
RL4_BACSU	60%	2149	50S RIBOSOMAL PROTEIN L4
RL3_BACSU	69%	2149	50S RIBOSOMAL PROTEIN L3 (BL3)
RS10_STRMU	81%	2150	30S RIBOSOMAL PROTEIN S10
RL33_MYCCA	57%	2155	50S RIBOSOMAL PROTEIN L33
RS2_PEDAC	71%	2207	30S RIBOSOMAL PROTEIN S2
RS2_BACSU	100%	2208	30S RIBOSOMAL PROTEIN S2 (BS1)
RS18_BACST	63%	2254	30S RIBOSOMAL PROTEIN S18 (BS21)
RS6_BACSU	58%	2254	30S RIBOSOMAL PROTEIN S6 (BS9)
RS9_BACSU	65%	2326	30S RIBOSOMAL PROTEIN S9 (BS10)
RL13_STACA	63%	2327	50S RIBOSOMAL PROTEIN L13
RS7_ECOLI	64%	2334	30S RIBOSOMAL PROTEIN S7
RS7_BACSU	71%	2334	30S RIBOSOMAL PROTEIN S7 (BS7)
RS12_SYNY3	67%	2335	30S RIBOSOMAL PROTEIN S12
BIO12.06 tRNA modification			**8 genes**
SP5C_BACSU	55%	17	PROBABLE PEPTIDYL-TRNA HYDROLASE (EC 3.1.1.29)
RNPA_BACSU	47%	129	RIBONUCLEASE P PROTEIN COMPONENT (EC 3.1.26.5)
KSGA_BACSU	52%	687	DIMETHYLADENOSINE TRANSFERASE (EC 2.1.1.-)
YGCA_HAEIN	30%	1192	HYPOTHETICAL RNA METHYLTRANSFERASE HI0333 (EC 2.1.1.-)
TRMD_HAEIN	50%	1603	TRNA (GUANINE-N1)-METHYLTRANSFERASE (EC 2.1.1.31)
MDMC_STRMY	31%	1762	O-METHYLTRANSFERASE (EC 2.1.1.-)
FMT_BACSU	46%	1941	METHIONYL-TRNA FORMYLTRANSFERASE (EC 2.1.2.9)
YGCA_HAEIN	27%	2235	HYPOTHETICAL RNA METHYLTRANSFERASE HI0333 (EC 2.1.1.-)
BIO12.07 Translation factors			**23 genes**
YDGI_BACSU	30%	183	PUTATIVE NAD(P)H NITROREDUCTASE (EC 1.-.-.-)

Table 3.

RF3_SYNY3	47%	349	PEPTIDE CHAIN RELEASE FACTOR 3 (RF-3)
RF2_HELPY	25%	451	PEPTIDE CHAIN RELEASE FACTOR 2 (RF-2)
IF2_ENTFC	36%	543	TRANSLATION INITIATION FACTOR IF-2
RF1_BACSU	54%	584	PEPTIDE CHAIN RELEASE FACTOR 1 (RF-1)
FLAV_DESVH	38%	595	FLAVODOXIN
EFP_BACSU	48%	689	ELONGATION FACTOR P (EF-P)
IF2_ENTFC	69%	775	TRANSLATION INITIATION FACTOR IF-2
RBFA_BACSU	37%	777	RIBOSOME-BINDING FACTOR A (P15B PROTEIN)
RF2_BACSU	53%	996	PEPTIDE CHAIN RELEASE FACTOR 2 (RF-2)
YDGI_BACSU	49%	1500	PUTATIVE NAD(P)H NITROREDUCTASE (EC 1.-.-.-)
MVIM_ECOLI	34%	1872	VIRULENCE FACTOR MVIM HOMOLOG
IF3_BACSU	61%	1892	TRANSLATION INITIATION FACTOR IF-3
EFTU_STROR	80%	1909	ELONGATION FACTOR TU (EF-TU)
RRF_HAEIN	43%	2066	RIBOSOME RECYCLING FACTOR (RIBOSOME RELEASING FACTOR)
YIHK_ECOLI	55%	2073	65.4 KD GTP-BINDING PROTEIN IN GLNA-RBN INTERGENIC REGION
IF1_LACLA	100%	2134	TRANSLATION INITIATION FACTOR IF-1
FMNB_DESVM	39%	2161	FMN-BINDING PROTEIN
EFTS_HELPY	43%	2206	ELONGATION FACTOR TS (EF-TS)
EFTS_STRPN	71%	2207	ELONGATION FACTOR TS (EF-TS)
EFTS_BACSU	77%	2207	ELONGATION FACTOR TS (EF-TS)
EFG_BACSU	72%	2333	ELONGATION FACTOR G (EF-G)
EFG_THEMA	55%	2333	ELONGATION FACTOR G (EF-G)
BIO13. TRANSPORT AND BINDING PROTEINS			**185 genes**
BIO13.01 General			**35 genes**
LCNC_LACLA	94%	84	LACTOCOCCIN A TRANSPORT ATP-BINDING PROTEIN LCNC
RFBB_MYXXA	40%	202	O-ANTIGEN EXPORT SYSTEM ATP-BINDING PROTEIN RFBB
CHVA_AGRTU	35%	260	BETA-(1-->2)GLUCAN EXPORT ATP-BINDING PROTEIN CHVA
P69_MYCHR	26%	296	TRANSPORT SYSTEM PERMEASE PROTEIN P69
ABC_ECOLI	42%	319	ATP-BINDING PROTEIN ABC
LCN3_LACLA	34%	517	LACTICIN 481/LACTOCOCCIN TRANSPORT ATP-BINDING PROTEIN LCNDR3
UUP1_HAEIN	29%	560	ABC TRANSPORTER ATP-BINDING PROTEIN UUP-1
LMRA_STRLN	34%	597	LINCOMYCIN RESISTANCE PROTEIN
KUP_ECOLI	36%	608	KUP SYSTEM POTASSIUM UPTAKE PROTEIN
KUP_ECOLI	29%	610	KUP SYSTEM POTASSIUM UPTAKE PROTEIN
CYDC_BACSU	48%	709	TRANSPORT ATP-BINDING PROTEIN CYDC
CYDD_BACSU	45%	711	TRANSPORT ATP-BINDING PROTEIN CYDD
NRDH_LACLC	99%	1003	GLUTAREDOXIN-LIKE PROTEIN NRDH
PERM_HAEIN	26%	1196	PUTATIVE PERMEASE PERM HOMOLOG
ATMB_SALTY	51%	1284	MG(2+) TRANSPORT ATPASE, P-TYPE 2 (EC 3.6.1.-)
MGLA_HAEIN	23%	1352	GALACTOSIDE TRANSPORT ATP-BINDING PROTEIN MGLA
MGLA_MYCGE	32%	1364	GALACTOSIDE TRANSPORT ATP-BINDING PROTEIN MGLA HOMOLOG
AC22_STRCO	26%	1383	PROBABLE ACTINORHODIN TRANSPORTER
LPLC_BACSU	42%	1527	LPLC PROTEIN
LPLB_BACSU	27%	1528	LPLB PROTEIN
LCN3_LACLA	41%	1559	LACTICIN 481/LACTOCOCCIN TRANSPORT ATP-BINDING PROTEIN LCNDR3
COMA_STRPN	24%	1560	TRANSPORT ATP-BINDING PROTEIN COMA
AC22_STRCO	28%	1584	PROBABLE ACTINORHODIN TRANSPORTER
UUP_ECOLI	31%	1592	ABC TRANSPORTER ATP-BINDING PROTEIN UUP
YJJK_HAEIN	38%	1593	ABC TRANSPORTER ATP-BINDING PROTEIN HI1252
MVAA_PSEMV	42%	1607	3-HYDROXY-3-METHYLGLUTARYL-COENZYME A REDUCTASE (EC 1.1.1.88)
YQGI_BACSU	54%	1753	PROBABLE ABC TRANSPORTER PERMEASE PROTEIN IN SODA-COMGA INTERGENIC REGION (ORF73)
YQGH_BACSU	32%	1754	PROBABLE ABC TRANSPORTER PERMEASE PROTEIN IN SODA-COMGA INTERGENIC REGION (ORF72)
YQGG_BACSU	47%	1755	PROBABLE ABC TRANSPORTER BINDING PROTEIN IN SODA-COMGA INTERGENIC REGION PRECURSOR (ORF108)
NOCQ_AGRT5	32%	1835	NOPALINE TRANSPORT SYSTEM PERMEASE PROTEIN NOCQ
PERM_ECOLI	27%	1863	PUTATIVE PERMEASE PERM
UVRA_NEIGO	59%	1866	EXCINUCLEASE ABC SUBUNIT A
ECSB_BACSU	28%	2054	PROTEIN ECSB
ECSA_BACSU	60%	2055	ABC-TYPE TRANSPORTER ATP-BINDING PROTEIN ECSA
FLIP_BORBU	27%	2113	FLAGELLAR BIOSYNTHETIC PROTEIN FLIP
BIO13.02 Amino acids, peptides and amines			**52 genes**
YW40_MYCTU	25%	70	HYPOTHETICAL TRANSPORT PROTEIN CY39.40 (FRAGMENT)
GABP_ECOLI	23%	70	GABA PERMEASE (4-AMINO BUTYRATE TRANSPORT CARRIER)
LYSP_ECOLI	27%	111	LYSINE-SPECIFIC PERMEASE
PTHP_STRSL	81%	118	PHOSPHOCARRIER PROTEIN HPR (HISTIDINE-CONTAINING PROTEIN)
PT1_STRSL	82%	119	PHOSPHOENOLPYRUVATE-PROTEIN PHOSPHOTRANSFERASE (EC 2.7.3.9)
ALYS_ENTFA	49%	267	AUTOLYSIN (EC 3.5.1.28)
YQIZ_BACSU	34%	322	PROBABLE AMINO-ACID ABC TRANSPORTER ATP-BINDING PROTEIN IN BMRU-ANSR INTERGENIC REGION
AMIA_STRPN	31%	341	OLIGOPEPTIDE-BINDING PROTEIN AMIA PRECURSOR
OPPA_ECOLI	28%	343	PERIPLASMIC OLIGOPEPTIDE-BINDING PROTEIN PRECURSOR
OPPB_BACSU	46%	345	OLIGOPEPTIDE TRANSPORT SYSTEM PERMEASE PROTEIN OPPB

Table 3.

DPPC_BACSU	43%	346	DIPEPTIDE TRANSPORT SYSTEM PERMEASE PROTEIN DPPC
OPPF_BACSU	61%	348	OLIGOPEPTIDE TRANSPORT ATP-BINDING PROTEIN OPPF
CYCA_ECOLI	28%	358	D-SERINE/D-ALANINE/GLYCINE TRANSPORTER
AAPA_BACSU	47%	360	AMINO ACID PERMEASE AAPA
LYSP_ECOLI	48%	368	LYSINE-SPECIFIC PERMEASE
OPPA_BACSU	30%	523	OLIGOPEPTIDE-BINDING PROTEIN OPPA PRECURSOR
ALYS_ENTFA	35%	525	AUTOLYSIN (EC 3.5.1.28)
DEF_HELPY	39%	553	POLYPEPTIDE DEFORMYLASE (EC 3.5.1.31)
YQIX_BACSU	36%	556	PROBABLE AMINO-ACID ABC TRANSPORTER BINDING PROTEIN IN BMRU-ANSR INTERGENIC REGION PRECURSOR
BRNQ_LACDL	47%	683	BRANCHED-CHAIN AMINO ACID TRANSPORT SYSTEM CARRIER PROTEIN (BRANCHED- CHAIN AMINO ACID UPTAKE CARRIER)
DTPT_LACLA	97%	703	DI-/TRIPEPTIDE TRANSPORTER
YDGF_BACSU	33%	817	HYPOTHETICAL TRANSPORT PROTEIN IN EXPZ-DINB INTERGENIC REGION
OPUA_BACSU	34%	863	GLYCINE BETAINE TRANSPORT ATP-BINDING PROTEIN OPUAA
PROW_BACSU	42%	863	GLYCINE BETAINE/L-PROLINE TRANSPORT SYSTEM PERMEASE PROTEIN PROW
YCKK_BACSU	28%	960	PROBABLE AMINO-ACID ABC TRANSPORTER BINDING PROTEIN PRECURSOR
Y4TG_RHISN	32%	961	PROBABLE AMINO-ACID ABC TRANSPORTER PERMEASE PROTEIN Y4TG
GLTL_HAEIN	59%	962	GLUTAMATE/ASPARTATE TRANSPORT ATP-BINDING PROTEIN GLTL
POTA_HAEIN	45%	1173	SPERMIDINE/PUTRESCINE TRANSPORT ATP-BINDING PROTEIN POTA
POTC_HAEIN	35%	1175	SPERMIDINE/PUTRESCINE TRANSPORT SYSTEM PERMEASE PROTEIN POTC
POTD_ECOLI	43%	1176	SPERMIDINE/PUTRESCINE-BINDING PERIPLASMIC PROTEIN PRECURSOR (SPBP)
GRSB_BACBR	30%	1290	GRAMICIDIN S SYNTHETASE II (GRAMICIDIN S BIOSYNTHESIS GRSB PROTEIN) (EC 6.-.-.-)
XASA_ECOLI	50%	1323	AMINO ACID ANTIPORTER (EXTREME ACID SENSITIVITY PROTEIN)
OPUC_BACSU	47%	1471	GLYCINE BETAINE-BINDING PROTEIN PRECURSOR
OPUB_BACSU	51%	1471	GLYCINE BETAINE TRANSPORT SYSTEM PERMEASE PROTEIN OPUAB
OPUA_BACSU	58%	1473	GLYCINE BETAINE TRANSPORT ATP-BINDING PROTEIN OPUAA
PMSR_BACSU	54%	1590	PUTATIVE PEPTIDE METHIONINE SULFOXIDE REDUCTASE (PEPTIDE MET
GLNH_ECOLI	36%	1797	GLUTAMINE-BINDING PERIPLASMIC PROTEIN PRECURSOR (GLNBP)
YA79_HAEIN	38%	1798	HYPOTHETICAL AMINO-ACID ABC TRANSPORTER PERMEASE PROTEIN HI0179
YQIZ_BACSU	64%	1799	PROBABLE AMINO-ACID ABC TRANSPORTER ATP-BINDING PROTEIN IN BMRU-ANSR INTERGENIC REGION
GLNQ_ECOLI	56%	1833	GLUTAMINE TRANSPORT ATP-BINDING PROTEIN GLNQ
HISP_ECOLI	51%	1835	HISTIDINE TRANSPORT ATP-BINDING PROTEIN HISP
PEPT_LACLC	95%	1857	PEPTIDASE T (EC 3.4.11.-)
YDGF_BACSU	35%	1870	HYPOTHETICAL TRANSPORT PROTEIN IN EXPZ-DINB INTERGENIC REGION
OPPA_LACLA	90%	1885	OLIGOPEPTIDE-BINDING PROTEIN OPPA PRECURSOR
OPPC_LACLA	100%	1887	OLIGOPEPTIDE TRANSPORT SYSTEM PERMEASE PROTEIN OPPC
OPPB_LACLC	100%	1888	OLIGOPEPTIDE TRANSPORT SYSTEM PERMEASE PROTEIN OPPB
OPPF_LACLA	100%	1889	OLIGOPEPTIDE TRANSPORT ATP-BINDING PROTEIN OPPF
OPPD_LACLC	100%	1890	OLIGOPEPTIDE TRANSPORT ATP-BINDING PROTEIN OPPD
PMSR_BACSU	53%	2064	PUTATIVE PEPTIDE METHIONINE SULFOXIDE REDUCTASE (PEPTIDE MET
ARCD_PSEAE	35%	2092	ARGININE/ORNITHINE ANTIPORTER
YXEM_BACSU	33%	2118	PROBABLE AMINO-ACID ABC TRANSPORTER BINDING PROTEIN IN IDH-DEOR INTERGENIC REGION PRECURSOR
LYSP_ECOLI	37%	2257	LYSINE-SPECIFIC PERMEASE

BIO13.03 Anions **13 genes**

MEGL_PSEPU	38%	77	METHIONINE GAMMA-LYASE (EC 4.4.1.11)
ATZN_ECOLI	28%	95	ZINC-TRANSPORTING ATPASE (EC 3.6.1.-)
5NTD_HAEIN	25%	180	PROBABLE 5'-NUCLEOTIDASE PRECURSOR (EC 3.1.3.5)
DHAK_CITFR	35%	243	DIHYDROXYACETONE KINASE (EC 2.7.1.29)
CN16_YEREN	31%	298	2',3'-CYCLIC-NUCLEOTIDE 2'-PHOSPHODIESTERASE PRECURSOR (EC 3.1.4.16)
GLPT_BACSU	34%	546	GLYCEROL-3-PHOSPHATE TRANSPORTER (G-3-P TRANSPORTER)
MANA_STRMU	65%	777	MANNOSE-6-PHOSPHATE ISOMERASE (EC 5.3.1.8)
ATKA_ENTFA	41%	845	POTASSIUM/COPPER-TRANSPORTING ATPASE A (EC 3.6.1.36)
CADA_BACFI	28%	869	PROBABLE CADMIUM-TRANSPORTING ATPASE (EC 3.6.1.-)
5NTD_HAEIN	33%	1098	PROBABLE 5'-NUCLEOTIDASE PRECURSOR (EC 3.1.3.5)
SCRK_STRMU	62%	1516	FRUCTOKINASE (EC 2.7.1.4)
PTA_BACSU	61%	1704	PROBABLE PHOSPHATE ACETYLTRANSFERASE (EC 2.3.1.8)
SPHX_SYNP7	28%	1756	SPHX PROTEIN PRECURSOR

BIO13.04 Carbohydrates, organic alcohols and acids **13 genes**

GLPF_ECOLI	35%	245	GLYCEROL UPTAKE FACILITATOR PROTEIN
FECE_ECOLI	42%	324	IRON(III) DICITRATE TRANSPORT ATP-BINDING PROTEIN FECE
FECD_ECOLI	35%	326	IRON(III) DICITRATE TRANSPORT SYSTEM PERMEASE PROTEIN FECD
CITN_LACLA	48%	921	CITRATE-SODIUM SYMPORT (CITRATE TRANSPORTER)
GLPF_STRPN	58%	1267	GLYCEROL UPTAKE FACILITATOR PROTEIN
RBSB_HAEIN	45%	1679	D-RIBOSE-BINDING PERIPLASMIC PROTEIN PRECURSOR
RBSC_BACSU	60%	1680	RIBOSE TRANSPORT SYSTEM PERMEASE PROTEIN RBSC
RBSA_BACSU	42%	1681	RIBOSE TRANSPORT ATP-BINDING PROTEIN RBSA
RBSD_BACSU	56%	1682	HIGH AFFINITY RIBOSE TRANSPORT PROTEIN RBSD
LACY_STRTR	43%	2043	LACTOSE PERMEASE (LACTOSE-PROTON SYMPORT)

Table 3.

RBSA_HAEIN	24%	2163	RIBOSE TRANSPORT ATP-BINDING PROTEIN RBSA
EBA2_FLAME	27%	2238	ENDO-BETA-N-ACETYLGLUCOSAMINIDASE F2 PRECURSOR (EC 3.2.1.96)
GNTP_BACLI	59%	2246	GLUCONATE PERMEASE
BIO13.05 Cations			**18 genes**
FEOB_ECOLI	39%	188	FERROUS IRON TRANSPORT PROTEIN B
FEOA_ECOLI	38%	190	FERROUS IRON TRANSPORT PROTEIN A
TCR1_ECOLI	41%	303	TETRACYCLINE RESISTANCE PROTEIN, CLASS A (TRANSPOSON TN1721)
FHUB_BACSU	37%	325	FERRICHROME TRANSPORT PERMEASE PROTEIN FHUB
Y1B2_MYCTU	36%	567	PUTATIVE CATION-TRANSPORTING ATPASE CY21C12.02 (EC 3.6.1.-)
ATA1_SYNY3	40%	675	CATION-TRANSPORTING ATPASE PMA1 (EC 3.6.1.-)
NHAA_ECOLI	42%	1148	NA(+)/H
PBUX_BACSU	48%	1158	XANTHINE PERMEASE
Y1B2_MYCTU	35%	1188	PUTATIVE CATION-TRANSPORTING ATPASE CY21C12.02 (EC 3.6.1.-)
ATCL_SYNP7	37%	1402	CATION-TRANSPORTING ATPASE PACL (EC 3.6.1.-)
FUR_CAMJE	30%	1503	FERRIC UPTAKE REGULATION PROTEIN
UIDB_ECOLI	33%	1541	GLUCURONIDE CARRIER PROTEIN (GLUCURONIDE PERMEASE)
Y058_METJA	31%	1632	PUTATIVE AMMONIUM TRANSPORTER MJ0058
URAA_HAEIN	36%	1643	PROBABLE URACIL PERMEASE (URACIL TRANSPORTER)
FER_CLOTH	38%	1741	FERREDOXIN
NAPA_ENTHR	48%	1973	NA(+)/H
ATCS_SYNY3	36%	2061	CATION-TRANSPORTING ATPASE PACS (EC 3.6.1.-)
ATA1_SYNY3	36%	2315	CATION-TRANSPORTING ATPASE PMA1 (EC 3.6.1.-)
BIO13.06 Nucleosides, purines and pyrimidines			**1 genes**
PDP_BACST	56%	1463	PYRIMIDINE-NUCLEOSIDE PHOSPHORYLASE (EC 2.4.2.2)
BIO13.07 PTS system			**13 genes**
PTMA_ECOLI	46%	31	PTS SYSTEM, MANNITOL-SPECIFIC IIABC COMPONENT (EIIABC-MTL)
PTCC_BACSU	34%	175	PTS SYSTEM, CELLOBIOSE-SPECIFIC IIC COMPONENT (EIIC-CEL)
PTCB_BACSU	52%	416	PTS SYSTEM, CELLOBIOSE-SPECIFIC IIB COMPONENT (EIIB-CEL)
PTCA_BACSU	48%	417	PTS SYSTEM, CELLOBIOSE-SPECIFIC IIA COMPONENT (EIIA-CEL)
PTSA_STRMU	48%	435	PTS SYSTEM, SUCROSE-SPECIFIC IIABC COMPONENT (EIIABC-SCR)
PTSA_STRMU	31%	436	PTS SYSTEM, SUCROSE-SPECIFIC IIABC COMPONENT (EIIABC-SCR)
PTCC_BACSU	31%	830	PTS SYSTEM, CELLOBIOSE-SPECIFIC IIC COMPONENT (EIIC-CEL)
PTFB_RHOCA	52%	984	PTS SYSTEM, FRUCTOSE-SPECIFIC IIBC COMPONENT (EIIBC-FRU)
PTBA_BACSU	37%	1144	PTS SYSTEM, BETA-GLUCOSIDES-SPECIFIC IIABC COMPONENT (EIIABC-BGL)
PTBA_ERWCH	41%	1489	PTS SYSTEM, BETA-GLUCOSIDES-SPECIFIC IIABC COMPONENT (EIIABC-BGL)
PTNA_ECOLI	41%	1742	PTS SYSTEM, MANNOSE-SPECIFIC IIAB COMPONENT (EIIAB-MAN)
PTNC_ECOLI	46%	1743	PTS SYSTEM, MANNOSE-SPECIFIC IIC COMPONENT (EIIC-MAN)
PTND_ECOLI	57%	1744	PTS SYSTEM, MANNOSE-SPECIFIC IID COMPONENT (EIID-MAN)
BIO13.08 Multidrug resistance			**3 genes**
BMR2_BACSU	46%	123	MULTIDRUG RESISTANCE PROTEIN 2 (MULTIDRUG-EFFLUX TRANSPORTER 2)
EMRB_ECOLI	26%	930	MULTIDRUG RESISTANCE PROTEIN B
MDL_ECOLI	36%	1302	MULTIDRUG RESISTANCE-LIKE ATP-BINDING PROTEIN MDL
BIO13.09 Other			**37 genes**
YBJZ_ECOLI	34%	15	HYPOTHETICAL ABC TRANSPORTER ATP-BINDING PROTEIN YBJZ
YCHM_ECOLI	37%	53	HYPOTHETICAL 58.4 KD PROTEIN IN PTH-PRSA INTERGENIC REGION
YBIT_ECOLI	32%	74	HYPOTHETICAL ABC TRANSPORTER ATP-BINDING PROTEIN YBIT
YHES_ECOLI	35%	100	HYPOTHETICAL ABC TRANSPORTER ATP-BINDING PROTEIN YHES
YHIH_ECOLI	50%	248	HYPOTHETICAL ABC TRANSPORTER ATP-BINDING PROTEIN YHIH
YFIB_BACSU	39%	257	HYPOTHETICAL ABC TRANSPORTER ATP-BINDING PROTEIN 1 IN GLVBC 3'REGION
Y179_MYCGE	44%	273	HYPOTHETICAL ABC TRANSPORTER ATP-BINDING PROTEIN MG179
Y180_MYCGE	47%	274	HYPOTHETICAL ABC TRANSPORTER ATP-BINDING PROTEIN MG180
Y412_METJA	37%	295	HYPOTHETICAL ABC TRANSPORTER ATP-BINDING PROTEIN MJ0412
YFIB_BACSU	33%	307	HYPOTHETICAL ABC TRANSPORTER ATP-BINDING PROTEIN 1 IN GLVBC 3'REGION
YFIC_BACSU	41%	309	HYPOTHETICAL ABC TRANSPORTER ATP-BINDING PROTEIN 2 IN GLVBC 3'REGION
YAEE_ECOLI	44%	320	HYPOTHETICAL ABC TRANSPORTER PERMEASE PROTEIN YAEE
YHIH_ECOLI	27%	321	HYPOTHETICAL ABC TRANSPORTER ATP-BINDING PROTEIN YHIH
Y081_MYCTU	44%	347	HYPOTHETICAL ABC TRANSPORTER ATP-BINDING PROTEIN CY50.01
YCJV_ECOLI	60%	426	HYPOTHETICAL ABC TRANSPORTER ATP-BINDING PROTEIN YCJV
YCFV_ECOLI	41%	650	HYPOTHETICAL ABC TRANSPORTER ATP-BINDING PROTEIN YCFV
YWJA_BACSU	35%	715	HYPOTHETICAL ABC TRANSPORTER ATP-BINDING PROTEIN IN ACDA 5'REGION
YBJZ_ECOLI	40%	718	HYPOTHETICAL ABC TRANSPORTER ATP-BINDING PROTEIN YBJZ
Y796_METJA	35%	892	HYPOTHETICAL ABC TRANSPORTER ATP-BINDING PROTEIN MJ0796
Y415_SYNY3	35%	990	HYPOTHETICAL ABC TRANSPORTER ATP-BINDING PROTEIN SLL0415
YRHG_BACSU	43%	993	HYPOTHETICAL 28.5 KD PROTEIN IN SIGV-GREA INTERGENIC REGION
YBJZ_ECOLI	25%	998	HYPOTHETICAL ABC TRANSPORTER ATP-BINDING PROTEIN YBJZ
YDIF_BACSU	52%	1072	HYPOTHETICAL ABC TRANSPORTER ATP-BINDING PROTEIN YDIF

Table 3.

Y796_METJA	48%	1108	HYPOTHETICAL ABC TRANSPORTER ATP-BINDING PROTEIN MJ0796
YLUD_LACLA	100%	1242	HYPOTHETICAL ABC TRANSPORTER ATP-BINDING PROTEIN IN LEUD 3'REGION
YSC1_STRGC	62%	1349	HYPOTHETICAL ABC TRANSPORTER ATP-BINDING PROTEIN IN SCAA 5'REGION (ORF1)
YXBC_BACSU	51%	1538	HYPOTHETICAL METABOLITE TRANSPORT PROTEIN IN HTPG-IOLR INTERGENIC REGION
YJEM_ECOLI	22%	1736	HYPOTHETICAL 56.3 KD PROTEIN IN GENX-PSD INTERGENIC REGION (O514)
YEJF_ECOLI	27%	1751	HYPOTHETICAL ABC TRANSPORTER ATP-BINDING PROTEIN YEJF
YQGJ_BACSU	60%	1751	HYPOTHETICAL ABC TRANSPORTER ATP-BINDING PROTEIN IN SODA-COMGA INTERGENIC REGION (ORF74)
YQGJ_BACSU	59%	1752	HYPOTHETICAL ABC TRANSPORTER ATP-BINDING PROTEIN IN SODA-COMGA INTERGENIC REGION (ORF74)
YXDL_BACSU	49%	1789	HYPOTHETICAL ABC TRANSPORTER ATP-BINDING PROTEIN IN IDH 3'REGION
YATR_BACFI	43%	1815	HYPOTHETICAL ABC TRANSPORTER ATP-BINDING PROTEIN
YJEM_ECOLI	39%	1855	HYPOTHETICAL 56.3 KD PROTEIN IN GENX-PSD INTERGENIC REGION (O514)
YXDL_BACSU	35%	2076	HYPOTHETICAL ABC TRANSPORTER ATP-BINDING PROTEIN IN IDH 3'REGION
YYBJ_BACSU	28%	2265	HYPOTHETICAL ABC TRANSPORTER ATP-BINDING PROTEIN IN COTF-TETB INTERGENIC REGION
Y796_METJA	42%	2329	HYPOTHETICAL ABC TRANSPORTER ATP-BINDING PROTEIN MJ0796

BIO14. OTHER CATEGORIES 94 genes

BIO14.01 Adaptations and atypical conditions 13 genes

CSP2_LACPL	82%	170	COLD SHOCK PROTEIN 2
TCMA_STRGA	28%	233	TETRACENOMYCIN C RESISTANCE AND EXPORT PROTEIN
MMR_BACSU	30%	382	METHYLENOMYCIN A RESISTANCE PROTEIN (MMR PEPTIDE)
CSPC_BACSU	75%	515	COLD SHOCK PROTEIN CSPC
TCMA_STRGA	25%	596	TETRACENOMYCIN C RESISTANCE AND EXPORT PROTEIN
VANZ_ENTFC	42%	1097	VANZ PROTEIN
FSR_ECOLI	25%	1182	FOSMIDOMYCIN RESISTANCE PROTEIN
ARSC_BACSU	51%	1409	PUTATIVE ARSENATE REDUCTASE
TCMA_STRGA	29%	1553	TETRACENOMYCIN C RESISTANCE AND EXPORT PROTEIN
TCMA_STRGA	27%	1678	TETRACENOMYCIN C RESISTANCE AND EXPORT PROTEIN
CMCT_NOCLA	34%	1925	CEPHAMYCIN EXPORT PROTEIN CMCT
FRI_LISIN	44%	2082	NON-HEME IRON-CONTAINING FERRITIN
MSCL_ECOLI	44%	2150	LARGE CONDUCTANCE MECHANOSENSITIVE CHANNEL

BIO14.02 Drug and analog sensitivity 1 genes

KKTH_LACLA	100%	1232	PROBABLE AMINOGLYCOSIDE 3'-PHOSPHOTRANSFERASE (EC 2.7.1.95)

BIO14.03 Phage related functions and prophages 21 genes

TERS_BPSF6	37%	36	TERMINASE SMALL SUBUNIT (G1P)
Y367_METJA	26%	49	PROBABLE INTEGRASE/RECOMBINASE PROTEIN MJ0367
VINT_BPL54	25%	445	INTEGRASE
VG64_BPML5	45%	460	GENE 64 PROTEIN (GP64)
VP3_BPHK7	21%	464	PORTAL PROTEIN (GP3)
VP4_BPHK7	31%	465	PUTATIVE PROHEAD PROTEASE (GP4)
VINT_BPL54	31%	501	INTEGRASE
XERC_SALTY	32%	633	INTEGRASE/RECOMBINASE XERC
YHGE_BACSU	31%	902	HYPOTHETICAL 84.1 KD PROTEIN IN HEMY-GLTT INTERGENIC REGION (ORFB)
TNPA_STAAU	45%	1034	TRANSPOSASE A (TRANSPOSON TN554)
VP4_BPHK7	27%	1053	PUTATIVE PROHEAD PROTEASE (GP4)
VG13_BPPH2	33%	1064	MORPHOGENESIS PROTEIN 1 (LATE PROTEIN GP13)
VINT_BPL54	32%	1210	INTEGRASE
CODV_BACSU	31%	1256	PROBABLE INTEGRASE/RECOMBINASE CODV
XERC_HAEIN	37%	1257	INTEGRASE/RECOMBINASE XERC
SBCC_ECOLI	44%	1350	EXONUCLEASE SBCC
MCP_BPF41	48%	1358	MAJOR CAPSID PROTEIN (MCP)
VINT_BPSF6	26%	1454	INTEGRASE
PIP_LACLA	96%	1716	PHAGE INFECTION PROTEIN
VG37_BPML5	48%	1873	GENE 37 PROTEIN (GP37)
VINT_BPL54	25%	2005	INTEGRASE

BIO14.04 Transposon related functions 59 genes

YI3A_LACJO	30%	53	INSERTION ELEMENT IS1223 HYPOTHETICAL 20.7 KD PROTEIN (ORFA)
T150_ECOLI	41%	53	PUTATIVE TRANSPOSASE FOR INSERTION SEQUENCE IS150
TRA4_LACLA	100%	54	TRANSPOSASE FOR INSERTION SEQUENCE ELEMENT IS904
YI91_SHIDY	25%	55	INSERTION ELEMENT IS911 HYPOTHETICAL 12.7 KD PROTEIN
TRA9_MYCTU	30%	89	PUTATIVE TRANSPOSASE FOR INSERTION SEQUENCE ELEMENT IS986 (ORFB)
TRA4_LACLA	39%	91	TRANSPOSASE FOR INSERTION SEQUENCE ELEMENT IS904

Table 3.

YI91_SHIDY	25%	135	INSERTION ELEMENT IS911 HYPOTHETICAL 12.7 KD PROTEIN
TRA4_LACLA	80%	135	TRANSPOSASE FOR INSERTION SEQUENCE ELEMENT IS904
YI6A_MYCMY	31%	136	INSERTION ELEMENT IS1296 HYPOTHETICAL 21.4 KD PROTEIN (ORFA)
T150_ECOLI	41%	137	PUTATIVE TRANSPOSASE FOR INSERTION SEQUENCE IS150
TRA4_LACLA	99%	138	TRANSPOSASE FOR INSERTION SEQUENCE ELEMENT IS904
YI91_SHIDY	25%	138	INSERTION ELEMENT IS911 HYPOTHETICAL 12.7 KD PROTEIN
TRA4_LACLA	98%	371	TRANSPOSASE FOR INSERTION SEQUENCE ELEMENT IS904
YI3A_LACJO	30%	372	INSERTION ELEMENT IS1223 HYPOTHETICAL 20.7 KD PROTEIN (ORFA)
T150_ECOLI	41%	373	PUTATIVE TRANSPOSASE FOR INSERTION SEQUENCE IS150
TRA4_LACLA	100%	624	TRANSPOSASE FOR INSERTION SEQUENCE ELEMENT IS904
YI91_SHIDY	26%	625	INSERTION ELEMENT IS911 HYPOTHETICAL 12.7 KD PROTEIN
T150_ECOLI	41%	626	PUTATIVE TRANSPOSASE FOR INSERTION SEQUENCE IS150
YI3A_LACJO	30%	626	INSERTION ELEMENT IS1223 HYPOTHETICAL 20.7 KD PROTEIN (ORFA)
TRA4_LACLA	39%	648	TRANSPOSASE FOR INSERTION SEQUENCE ELEMENT IS904
TRA1_STRSL	38%	680	PROBABLE TRANSPOSASE FOR INSERTION-LIKE SEQUENCE ELEMENT IS1161
TRA1_STRSL	38%	705	PROBABLE TRANSPOSASE FOR INSERTION-LIKE SEQUENCE ELEMENT IS1161
TRA4_LACLA	39%	726	TRANSPOSASE FOR INSERTION SEQUENCE ELEMENT IS904
YIS1_SHISO	28%	727	INSERTION ELEMENT IS600 (ISO-S3) HYPOTHETICAL 11 KD PROTEIN
TRA4_LACLA	98%	834	TRANSPOSASE FOR INSERTION SEQUENCE ELEMENT IS904
YI91_SHIDY	25%	834	INSERTION ELEMENT IS911 HYPOTHETICAL 12.7 KD PROTEIN
T150_ECOLI	39%	835	PUTATIVE TRANSPOSASE FOR INSERTION SEQUENCE IS150
YI3A_LACJO	30%	836	INSERTION ELEMENT IS1223 HYPOTHETICAL 20.7 KD PROTEIN (ORFA)
YI91_SHIDY	25%	836	INSERTION ELEMENT IS911 HYPOTHETICAL 12.7 KD PROTEIN
TRA4_LACLA	99%	837	TRANSPOSASE FOR INSERTION SEQUENCE ELEMENT IS904
TRA1_STRSL	38%	955	PROBABLE TRANSPOSASE FOR INSERTION-LIKE SEQUENCE ELEMENT IS1161
TRA4_LACLA	39%	1214	TRANSPOSASE FOR INSERTION SEQUENCE ELEMENT IS904
YIS1_SHISO	28%	1215	INSERTION ELEMENT IS600 (ISO-S3) HYPOTHETICAL 11 KD PROTEIN
YIS1_SHISO	28%	1219	INSERTION ELEMENT IS600 (ISO-S3) HYPOTHETICAL 11 KD PROTEIN
TRA4_LACLA	39%	1220	TRANSPOSASE FOR INSERTION SEQUENCE ELEMENT IS904
TRA5_LACLA	97%	1223	TRANSPOSASE FOR INSERTION SEQUENCE ELEMENT IS905
TRA4_LACLA	39%	1273	TRANSPOSASE FOR INSERTION SEQUENCE ELEMENT IS904
YIS1_SHISO	28%	1274	INSERTION ELEMENT IS600 (ISO-S3) HYPOTHETICAL 11 KD PROTEIN
TRA1_STRSL	38%	1336	PROBABLE TRANSPOSASE FOR INSERTION-LIKE SEQUENCE ELEMENT IS1161
TRA1_STRSL	38%	1394	PROBABLE TRANSPOSASE FOR INSERTION-LIKE SEQUENCE ELEMENT IS1161
TRA1_STRSL	38%	1552	PROBABLE TRANSPOSASE FOR INSERTION-LIKE SEQUENCE ELEMENT IS1161
TRA4_LACLA	39%	1582	TRANSPOSASE FOR INSERTION SEQUENCE ELEMENT IS904
TRA4_LACLA	39%	1727	TRANSPOSASE FOR INSERTION SEQUENCE ELEMENT IS904
YIS1_SHISO	28%	1728	INSERTION ELEMENT IS600 (ISO-S3) HYPOTHETICAL 11 KD PROTEIN
TRA1_STRSL	38%	1734	PROBABLE TRANSPOSASE FOR INSERTION-LIKE SEQUENCE ELEMENT IS1161
TRA1_STRSL	38%	1933	PROBABLE TRANSPOSASE FOR INSERTION-LIKE SEQUENCE ELEMENT IS1161
TRA1_STRSL	38%	1958	PROBABLE TRANSPOSASE FOR INSERTION-LIKE SEQUENCE ELEMENT IS1161
TRA1_STRSL	38%	1997	PROBABLE TRANSPOSASE FOR INSERTION-LIKE SEQUENCE ELEMENT IS1161
TRA1_STRSL	42%	2063	PROBABLE TRANSPOSASE FOR INSERTION-LIKE SEQUENCE ELEMENT IS1161
TRA4_LACLA	39%	2083	TRANSPOSASE FOR INSERTION SEQUENCE ELEMENT IS904
TRA1_STRSL	38%	2127	PROBABLE TRANSPOSASE FOR INSERTION-LIKE SEQUENCE ELEMENT IS1161
TRA4_LACLA	98%	2135	TRANSPOSASE FOR INSERTION SEQUENCE ELEMENT IS904
T150_ECOLI	41%	2136	PUTATIVE TRANSPOSASE FOR INSERTION SEQUENCE IS150
TRA1_STRSL	38%	2183	PROBABLE TRANSPOSASE FOR INSERTION-LIKE SEQUENCE ELEMENT IS1161
TRA4_LACLA	100%	2195	TRANSPOSASE FOR INSERTION SEQUENCE ELEMENT IS904
YI91_SHIDY	25%	2196	INSERTION ELEMENT IS911 HYPOTHETICAL 12.7 KD PROTEIN
YH21_HAEIN	47%	2196	HYPOTHETICAL TRANSPOSASE-LIKE PROTEIN HI1721
YI3A_LACJO	31%	2197	INSERTION ELEMENT IS1223 HYPOTHETICAL 20.7 KD PROTEIN (ORFA)
TRA1_STRSL	38%	2248	PROBABLE TRANSPOSASE FOR INSERTION-LIKE SEQUENCE ELEMENT IS1161
BIO14.05 Other			**0 genes**
BIO15. HYPOTHETICAL			**315 genes**
BIO15.01 General			**315 genes**
YABO_BACSU	63%	21	HYPOTHETICAL 9.7 KD PROTEIN IN MFD-DIVIC INTERGENIC REGION
YACA_BACSU	32%	24	HYPOTHETICAL 55.1 KD PROTEIN IN SPOIIE-HPT INTERGENIC REGION
YE12_HAEIN	35%	43	HYPOTHETICAL PROTEIN HI1412
YAFP_ECOLI	36%	50	HYPOTHETICAL 17.6 KD PROTEIN IN DINP-PRFH INTERGENIC REGION
YJHQ_ECOLI	29%	56	HYPOTHETICAL 20.0 KD PROTEIN IN FECI-FIMB INTERGENIC REGION
(F181)			
Y293_MYCGE	41%	66	HYPOTHETICAL PROTEIN MG293
YPGQ_BACSU	35%	67	HYPOTHETICAL 23.1 KD PROTEIN IN BSAA-ILVD INTERGENIC REGION
YCSE_BACSU	29%	73	HYPOTHETICAL 28.1 KD PROTEIN IN SIPU 3'REGION
Y531_METJA	34%	79	HYPOTHETICAL PROTEIN MJ0531
YOA9_MYCTU	42%	97	HYPOTHETICAL 47.5 KD PROTEIN CY9C4.09
YADE_ECOLI	29%	104	HYPOTHETICAL 46.3 KD PROTEIN IN HPT-PAND INTERGENIC REGION
PRECURSOR			
YIHZ_HAEIN	52%	108	HYPOTHETICAL PROTEIN HI0670
YIHX_ECOLI	29%	109	HYPOTHETICAL 23.5 KD PROTEIN IN GLNA-RBN INTERGENIC REGION
(O206)			
YVMB_BACSU	26%	112	HYPOTHETICAL 19.7 KD PROTEIN IN CYPX-UVRA INTERGENIC REGION
YKXH_BACSU	58%	118	HYPOTHETICAL 7.4 KD PROTEIN IN PTSX OPERON (PROTEIN K)

Table 3.

YQKD_BACSU	32%	121	HYPOTHETICAL 34.6 KD PROTEIN IN GLNQ-ANSR INTERGENIC REGION
YCEE_ECOLI (ORF1)	44%	127	HYPOTHETICAL 43.9 KD PROTEIN IN MSYB-HTRB INTERGENIC REGION
YCSN_BACSU	38%	132	HYPOTHETICAL 34.1 KD PROTEIN IN PBPC 3'REGION
YPH1_MYCCA	29%	133	HYPOTHETICAL 37.9 KD PROTEIN IN PTSH 3'REGION (ORF1)
YQGF_ECOLI (O180)	35%	139	HYPOTHETICAL 15.2 KD PROTEIN IN GSHB-ANSB INTERGENIC REGION
YWFA_BACSU	22%	155	HYPOTHETICAL 44.5 KD PROTEIN IN ROCC-PTA INTERGENIC REGION
YXIO_BACSU	27%	156	HYPOTHETICAL 47.3 KD PROTEIN IN WAPA-LICT INTERGENIC REGION
YPNP_BACSU	30%	169	HYPOTHETICAL 48.5 KD PROTEIN IN ILVA 3'REGION
YQGN_BACSU	36%	170	HYPOTHETICAL 21.4 KD PROTEIN IN SODA-COMGA INTERGENIC REGION
YQGP_BACSU	33%	171	HYPOTHETICAL 56.4 KD PROTEIN IN SODA-COMGA INTERGENIC REGION
YFGB_ECOLI	36%	182	HYPOTHETICAL 43.1 KD PROTEIN IN NDK-GCPE INTERGENIC REGION
YWAB_BACSU	27%	185	HYPOTHETICAL 33.8 KD PROTEIN IN DAE-TYRZ INTERGENIC REGION
Y448_MYCPN	57%	190	HYPOTHETICAL PROTEIN MG448 HOMOLOG
Y369_MYCGE	31%	192	HYPOTHETICAL PROTEIN MG369
YQHY_BACSU	32%	193	HYPOTHETICAL 14.7 KD PROTEIN IN ACCC-FOLD INTERGENIC REGION
Y320_MYCPN	26%	194	HYPOTHETICAL PROTEIN MG320 HOMOLOG
YIBD_ECOLI	30%	204	HYPOTHETICAL 40.5 KD PROTEIN IN SECB-TDH INTERGENIC REGION
Y4GN_RHISN	28%	205	HYPOTHETICAL 45.0 KD PROTEIN Y4GN
YC22_METJA	28%	209	HYPOTHETICAL PROTEIN MJ1222
YC11_KLEPN	20%	211	HYPOTHETICAL 55.0 KD PROTEIN IN CPS REGION (ORF11)
YIBD_ECOLI	34%	214	HYPOTHETICAL 40.5 KD PROTEIN IN SECB-TDH INTERGENIC REGION
Y060_MYCPN	33%	215	HYPOTHETICAL PROTEIN MG060 HOMOLOG
YQEH_BACSU	55%	221	HYPOTHETICAL 41.0 KD PROTEIN IN NUCB-AROD INTERGENIC REGION
YQEI_BACSU	45%	223	HYPOTHETICAL 10.8 KD PROTEIN IN AROD-COMER INTERGENIC REGION
YQEJ_BACSU	35%	224	HYPOTHETICAL 22.2 KD PROTEIN IN AROD-COMER INTERGENIC REGION
YQEK_BACSU	37%	225	HYPOTHETICAL 21.3 KD PROTEIN IN AROD-COMER INTERGENIC REGION
YQEL_BACSU	46%	226	HYPOTHETICAL 13.3 KD PROTEIN IN AROD-COMER INTERGENIC REGION
YQEM_BACSU	42%	226	HYPOTHETICAL 28.3 KD PROTEIN IN AROD-COMER INTERGENIC REGION
YEBC_ECOLI	40%	229	HYPOTHETICAL 26.4 KD PROTEIN IN RUVC-ASPS INTERGENIC REGION
Y003_HAEIN	33%	239	HYPOTHETICAL PROTEIN HI0003
YBHS_ECOLI	22%	249	HYPOTHETICAL 42.1 KD PROTEIN IN MOAE-RHLE INTERGENIC REGION
YWBD_BACSU	50%	250	HYPOTHETICAL 44.4 KD PROTEIN IN EPR-GALK INTERGENIC REGION
YHIN_ECOLI	32%	254	HYPOTHETICAL 43.8 KD PROTEIN IN RHSB-PIT INTERGENIC REGION
YFOL_STRTR (ORF1091)	47%	255	HYPOTHETICAL 29.9 KD PROTEIN IN FOLD-PBP2B INTERGENIC REGION
YW05_MYCTU	49%	256	HYPOTHETICAL 18.2 KD PROTEIN CY39.05C
YQHE_ECOLI	45%	264	HYPOTHETICAL OXIDOREDUCTASE IN METC-SUFI INTERGENIC REGION
Y181_MYCPN	35%	275	HYPOTHETICAL PROTEIN MG181 HOMOLOG (GT9_ORF434)
YHGF_NEIME	47%	283	HYPOTHETICAL 83.1 KD PROTEIN IN REGION E
YKQC_BACSU	61%	287	HYPOTHETICAL 61.5 KD PROTEIN IN ADEC-PDHA INTERGENIC REGION
YEAB_BACSU (ORFX)	46%	305	HYPOTHETICAL 31.8 KD PROTEIN IN GABP-GUAA INTERGENIC REGION
YQHE_ECOLI	45%	330	HYPOTHETICAL OXIDOREDUCTASE IN METC-SUFI INTERGENIC REGION
YYBR_BACSU	49%	331	HYPOTHETICAL 14.7 KD PROTEIN IN COTF-TETB INTERGENIC REGION
YWKD_BACSU	50%	331	HYPOTHETICAL 14.8 KD PROTEIN IN TDK-PRFA INTERGENIC REGION
YWAC_BACSU	38%	364	HYPOTHETICAL 24.6 KD PROTEIN IN DAE-TYRZ INTERGENIC REGION
Y128_MYCGE	30%	364	HYPOTHETICAL PROTEIN MG128
YHCT_BACSU	40%	365	HYPOTHETICAL 33.7 KD PROTEIN IN CSPB-GLPP INTERGENIC REGION
YIEF_ECOLI	33%	369	HYPOTHETICAL 20.4 KD PROTEIN IN TNAB-BGLB INTERGENIC REGION
YEIH_ECOLI	28%	376	HYPOTHETICAL 36.9 KD PROTEIN IN LYSP-NFO INTERGENIC REGION
Y372_MYCGE	32%	381	HYPOTHETICAL PROTEIN MG372
YJCE_ECOLI (O549)	37%	386	HYPOTHETICAL 60.5 KD PROTEIN IN SOXR-ACS INTERGENIC REGION
YAAT_BACSU	44%	403	HYPOTHETICAL 31.2 KD PROTEIN IN XPAC-ABRB INTERGENIC REGION
Y552_HELPY	35%	404	HYPOTHETICAL PROTEIN HP0552
YF56_SYNY3	37%	409	HYPOTHETICAL 37.5 KD PROTEIN SLL1556
YWBA_BACSU	40%	419	HYPOTHETICAL 47.6 KD PROTEIN IN EPR-GALK INTERGENIC REGION
Y105_MYCPN	35%	431	HYPOTHETICAL PROTEIN MG105 HOMOLOG
YW11_MYCTU	38%	438	HYPOTHETICAL 145.8 KD PROTEIN CY39.11C
Y097_MYCTU	39%	442	HYPOTHETICAL 18.2 KD PROTEIN CY78.07
LYTR_BACSU	31%	443	MEMBRANE-BOUND PROTEIN LYTR
YQAK_BACSU	46%	451	HYPOTHETICAL 32.2 KD PROTEIN IN SPOIIIC-CWLA INTERGENIC REGION
YMCP_BPF41	27%	471	HYPOTHETICAL PROTEIN IN MCP 3' REGION (FRAGMENT)
YQBO_BACSU	35%	473	HYPOTHETICAL 171.0 KD PROTEIN IN SPOIIIC-CWLA INTERGENIC REGION
YWLG_BACSU	48%	484	HYPOTHETICAL 19.4 KD PROTEIN IN SPOIIR-GLYC INTERGENIC REGION
DEGV_BACSU	34%	496	DEGV PROTEIN
YPMR_BACSU	29%	497	HYPOTHETICAL 28.9 KD PROTEIN IN ILVA 3'REGION
YBBD_BACSU	25%	499	HYPOTHETICAL 70.6 KD PROTEIN IN FEUA 5'REGION PRECURSOR (ORF1)
YQGX_BACSU	44%	526	HYPOTHETICAL 23.2 KD PROTEIN IN SODA-COMGA INTERGENIC REGION
YPOA_STROR	73%	529	HYPOTHETICAL 23.1 KD PROTEIN IN PONA 5'REGION
YPSA_BACSU	40%	530	HYPOTHETICAL 21.1 KD PROTEIN IN COTD-KDUD INTERGENIC REGION
YQGZ_BACSU	42%	530	HYPOTHETICAL 14.8 KD PROTEIN IN SODA-COMGA INTERGENIC REGION
YNGB_CLOPE	25%	545	HYPOTHETICAL 31.2 KD PROTEIN IN NAGH 5'REGION (ORFB)
YJEB_ECOLI	28%	551	HYPOTHETICAL 15.6 KD PROTEIN IN PURA-VACB INTERGENIC REGION
YGDP_ECOLI	33%	553	HYPOTHETICAL 20.8 KD PROTEIN IN PTPH-MUTH INTERGENIC REGION

Table 3.

YJAB_ECOLI (F147)	41%	569	HYPOTHETICAL 16.4 KD PROTEIN IN RRFE-META INTERGENIC REGION
Y369_MYCGE	28%	572	HYPOTHETICAL PROTEIN MG369
DEGV_BACSU	30%	573	DEGV PROTEIN
YQJW_BACSU	35%	578	HYPOTHETICAL 45.9 KD PROTEIN IN GLNQ-ANSR INTERGENIC REGION
YF63_METJA	31%	581	HYPOTHETICAL PROTEIN MJ1563
YWCG_BACSU	41%	585	HYPOTHETICAL 28.3 KD PROTEIN IN QOXD-VPR INTERGENIC REGION
YAT3_STAAU	28%	587	HYPOTHETICAL 16.6 KD PROTEIN IN ATL 5'REGION (ORF3)
YWLC_BACSU	33%	588	HYPOTHETICAL 37.0 KD PROTEIN IN SPOIIR-GLYC INTERGENIC REGION
Y553_METJA	40%	594	HYPOTHETICAL PROTEIN MJ0553
YWCD_BACSU	33%	601	HYPOTHETICAL 14.3 KD PROTEIN IN EPR-GALK INTERGENIC REGION
Y085_MYCGE	33%	603	HYPOTHETICAL PROTEIN MG085
YQFU_BACSU	29%	612	HYPOTHETICAL 32.5 KD PROTEIN IN CCCA-SODA INTERGENIC REGION
Y119_MYCTU	25%	615	HYPOTHETICAL 41.2 KD PROTEIN CY277.09
YQJK_BACSU	51%	617	HYPOTHETICAL 34.0 KD PROTEIN IN GLNQ-ANSR INTERGENIC REGION
YCEG_ECOLI	30%	623	HYPOTHETICAL 38.2 KD PROTEIN IN PABC-HOLB INTERGENIC REGION
YZ42_METJA	32%	642	HYPOTHETICAL PROTEIN MJECL42
Y0A5_MYCTU	22%	651	HYPOTHETICAL 40.2 KD PROTEIN CY9C4.05C
Y553_SYNY3	47%	658	HYPOTHETICAL 22.5 KD PROTEIN SLR0553
YCEE_ECOLI (ORF1)	42%	659	HYPOTHETICAL 43.9 KD PROTEIN IN MSYB-HTRB INTERGENIC REGION
YRG2_LACLA	95%	660	HYPOTHETICAL PROTEIN IN RPMG 3'REGION (ORF2)
YQGZ_BACSU	38%	672	HYPOTHETICAL 14.8 KD PROTEIN IN SODA-COMGA INTERGENIC REGION
YCDQ_ECOLI	35%	678	HYPOTHETICAL 50.8 KD PROTEIN IN PHOH-CSGG INTERGENIC REGION
YADE_ECOLI PRECURSOR	28%	680	HYPOTHETICAL 46.3 KD PROTEIN IN HPT-PAND INTERGENIC REGION
YJDJ_ECOLI (O90A)	35%	684	HYPOTHETICAL 10.5 KD PROTEIN IN DCUB-LYSU INTERGENIC REGION
YJDI_ECOLI (O76)	56%	684	HYPOTHETICAL 8.6 KD PROTEIN IN DCUB-LYSU INTERGENIC REGION
YABD_BACSU	43%	685	HYPOTHETICAL 29.2 KD PROTEIN IN METS-KSGA INTERGENIC REGION
YQHY_BACSU	39%	690	HYPOTHETICAL 14.7 KD PROTEIN IN ACCC-FOLD INTERGENIC REGION
YXJH_BACSU	43%	722	HYPOTHETICAL 38.3 KD PROTEIN IN PEPT-KATB INTERGENIC REGION
YICI_ECOLI	33%	727	HYPOTHETICAL 88.1 KD PROTEIN IN GLTS-SELC INTERGENIC REGION
YFBB_HAEIN	26%	734	HYPOTHETICAL PROTEIN HI0282
Y188_METJA	24%	739	HYPOTHETICAL PROTEIN MJ0188
Y371_MYCGE	28%	740	HYPOTHETICAL PROTEIN MG371
YCSE_BACSU	40%	742	HYPOTHETICAL 28.1 KD PROTEIN IN SIPU 3'REGION
YYBT_BACSU	38%	750	HYPOTHETICAL 74.3 KD PROTEIN IN RPLI-COTF INTERGENIC REGION
Y347_MYCPN	26%	754	HYPOTHETICAL PROTEIN MG347 HOMOLOG
YCND_BACSU	36%	758	HYPOTHETICAL 27.9 KD PROTEIN IN PHRC-GDH INTERGENIC REGION
YHCS_BACSU	31%	768	HYPOTHETICAL 22.0 KD PROTEIN IN CSPB-GLPP INTERGENIC REGION
YWBG_BACSU	36%	770	HYPOTHETICAL 25.8 KD PROTEIN IN EPR-GALK INTERGENIC REGION
YWBH_BACSU	27%	770	HYPOTHETICAL 14.3 KD PROTEIN IN EPR-GALK INTERGENIC REGION
YLXS_BACSU	38%	771	HYPOTHETICAL 17.6 KD PROTEIN IN NUSA 5'REGION (P15A)
YLXR_BACSU (ORF3)	51%	773	HYPOTHETICAL 10.4 KD PROTEIN IN NUSA-INFB INTERGENIC REGION
YHI3_RHOCA	44%	779	HYPOTHETICAL 33.7 KD PROTEIN IN HIMA 5'REGION (ORF3)
Y06B_MYCTU	31%	809	HYPOTHETICAL 139.6 KD PROTEIN CY338.11C PRECURSOR
YQGS_BACSU	34%	813	HYPOTHETICAL 73.2 KD PROTEIN IN SODA-COMGA INTERGENIC REGION
YIDA_ECOLI	38%	821	HYPOTHETICAL 29.7 KD PROTEIN IN IBPA-GYRB INTERGENIC REGION
Y143_HAEIN	25%	828	HYPOTHETICAL PROTEIN HI0143
Y456_METJA	24%	848	HYPOTHETICAL PROTEIN MJ0456
YCFB_ECOLI (ORF-15)	58%	850	HYPOTHETICAL 42.6 KD PROTEIN IN PURB-ICDA INTERGENIC REGION
YEHZ_ECOLI PRECURSOR	31%	864	HYPOTHETICAL 32.6 KD PROTEIN IN MOLR-BGLX INTERGENIC REGION
YHAD_HAEIN	50%	877	HYPOTHETICAL PROTEIN HI0091
YLXA_BACSU	64%	885	HYPOTHETICAL 35.3 KD PROTEIN IN FTSL 5'REGION (ORFB)
YWAC_BACSU	45%	894	HYPOTHETICAL 24.6 KD PROTEIN IN DAE-TYRZ INTERGENIC REGION
YCDQ_ECOLI	28%	897	HYPOTHETICAL 50.8 KD PROTEIN IN PHOH-CSGG INTERGENIC REGION
Y531_METJA	37%	900	HYPOTHETICAL PROTEIN MJ0531
LYTR_BACSU	25%	907	MEMBRANE-BOUND PROTEIN LYTR
YCSE_BACSU	27%	911	HYPOTHETICAL 28.1 KD PROTEIN IN SIPU 3'REGION
YABR_BACSU	31%	912	HYPOTHETICAL 14.2 KD PROTEIN IN DIVIC-SPOIIE INTERGENIC REGION
YICL_ECOLI	34%	923	HYPOTHETICAL 33.1 KD PROTEIN IN SELC-NLPA INTERGENIC REGION
Y912_HAEIN	35%	928	HYPOTHETICAL PROTEIN HI0912
YIBD_ECOLI	33%	939	HYPOTHETICAL 40.5 KD PROTEIN IN SECB-TDH INTERGENIC REGION
YECD_ECOLI	40%	967	HYPOTHETICAL 21.7 KD PROTEIN IN ASPS-BISZ INTERGENIC REGION
YVCJ_BACSU	52%	986	HYPOTHETICAL 33.9 KD PROTEIN IN CRH-TRXB INTERGENIC REGION
YVCK_BACSU	46%	987	HYPOTHETICAL 34.7 KD PROTEIN IN CRH-TRXB INTERGENIC REGION
Y247_MYCGE	36%	1003	HYPOTHETICAL PROTEIN MG247
YECD_ECOLI	26%	1006	HYPOTHETICAL 21.7 KD PROTEIN IN ASPS-BISZ INTERGENIC REGION
YLYB_BACSU (ORF-X)	59%	1024	HYPOTHETICAL 33.7 KD PROTEIN IN LSP-PYRR INTERGENIC REGION
Y501_SYNY3	42%	1027	HYPOTHETICAL 36.7 KD PROTEIN SLL0501
YQAN_BACSU	41%	1043	HYPOTHETICAL 16.1 KD PROTEIN IN SPOIIIC-CWLA INTERGENIC REGION

Table 3.

YMFN_ECOLI	23%	1051	HYPOTHETICAL 50.9 KD PROTEIN IN INTE-PIN INTERGENIC REGION
YAF2_BACLI	40%	1068	HYPOTHETICAL 9.7 KD PROTEIN IN CWLL 5'REGION
YQGZ_BACSU	42%	1070	HYPOTHETICAL 14.8 KD PROTEIN IN SODA-COMGA INTERGENIC REGION
YQFN_BACSU	44%	1082	HYPOTHETICAL 23.7 KD PROTEIN IN CCCA-SODA INTERGENIC REGION
YQKB_BACSU	35%	1083	HYPOTHETICAL 11.8 KD PROTEIN IN GLNQ-ANSR INTERGENIC REGION
YIDA_ECOLI	44%	1084	HYPOTHETICAL 29.7 KD PROTEIN IN IBPA-GYRB INTERGENIC REGION
YWFO_BACSU	52%	1085	HYPOTHETICAL 51.0 KD PROTEIN IN PTA 3'REGION
YSXB_BACSU (ORF X)	38%	1088	HYPOTHETICAL 12.3 KD PROTEIN IN RPLU-RPMA INTERGENIC REGION
YDAR_BACSU	40%	1090	HYPOTHETICAL 45.7 KD PROTEIN IN MUTT-GSIB INTERGENIC REGION
YQFG_BACSU	46%	1092	HYPOTHETICAL 17.8 KD PROTEIN IN PHOH-DGKA INTERGENIC REGION
YVYE_BACSU	54%	1096	HYPOTHETICAL PROTEIN IN DEGS-TAGO INTERGENIC REGION (FRAGMENT)
YXIQ_BACSU	25%	1098	HYPOTHETICAL 45.5 KD PROTEIN IN BGLS-KATB INTERGENIC REGION
YADQ_ECOLI	29%	1103	HYPOTHETICAL 46.0 KD PROTEIN IN HEML-PFS INTERGENIC REGION
YQEJ_BACSU	46%	1104	HYPOTHETICAL 22.2 KD PROTEIN IN AROD-COMER INTERGENIC REGION
Y03F_MYCTU	36%	1106	HYPOTHETICAL 54.3 KD PROTEIN CY130.15C
YFEU_HAEIN	52%	1142	HYPOTHETICAL PROTEIN HI0754
YFIA_BACSU	26%	1145	HYPOTHETICAL 29.3 KD PROTEIN IN GLVG-GLVBC INTERGENIC REGION
YBDN_ECOLI	33%	1149	HYPOTHETICAL 47.8 KD PROTEIN IN CSTA-DSBG INTERGENIC REGION
YBDM_ECOLI	46%	1150	HYPOTHETICAL 23.9 KD PROTEIN IN CSTA-DSBG INTERGENIC REGION
YHCJ_HAEIN	36%	1189	HYPOTHETICAL PROTEIN HI0145
YNGB_CLOPE	33%	1190	HYPOTHETICAL 31.2 KD PROTEIN IN NAGH 5'REGION (ORFB)
YPTB_CLOAB	51%	1194	HYPOTHETICAL PROTEIN IN PTB 5'REGION (ORF1)
Y912_HAEIN	28%	1195	HYPOTHETICAL PROTEIN HI0912
YPIA_BACSU	27%	1197	HYPOTHETICAL 48.3 KD PROTEIN IN QCRA 5'REGION
YWKB_BACSU	26%	1213	HYPOTHETICAL 33.6 KD PROTEIN IN TDK-PRFA INTERGENIC REGION
YHI1_LACLA	92%	1225	HYPOTHETICAL PROTEIN IN HISC 5'REGION (ORF1)
YHI6_LACLA (ORF6)	98%	1230	HYPOTHETICAL 30.7 KD PROTEIN IN HISD-HISB INTERGENIC REGION
YHIC_LACLA	97%	1235	HYPOTHETICAL 31.3 KD PROTEIN IN HISIE 3'REGION (ORF13)
YABJ_BACSU (ORF2)	53%	1250	HYPOTHETICAL 13.7 KD PROTEIN IN PURR-SPOVG INTERGENIC REGION
YGT1_STRMU	65%	1255	HYPOTHETICAL PROTEIN 1 IN GTFB REGION (FRAGMENT)
Y0D8_MYCTU	32%	1259	HYPOTHETICAL 69.9 KD PROTEIN CY1A11.08
YQHY_BACSU	24%	1263	HYPOTHETICAL 14.7 KD PROTEIN IN ACCC-FOLD INTERGENIC REGION
Y568_METJA	35%	1283	HYPOTHETICAL PROTEIN MJ0568
YCDQ_ECOLI	25%	1294	HYPOTHETICAL 50.8 KD PROTEIN IN PHOH-CSGG INTERGENIC REGION
YPAA_BACSU	31%	1305	HYPOTHETICAL 20.5 KD PROTEIN IN SERA-FER INTERGENIC REGION
YPUL_BACSU (ORFX13)	54%	1306	HYPOTHETICAL 26.0 KD PROTEIN IN SPMB-AROC INTERGENIC REGION
YPUH_BACSU (ORFX8)	39%	1306	HYPOTHETICAL 22.0 KD PROTEIN IN RIBT-DACB INTERGENIC REGION
YPUG_BACSU (ORFX7)	37%	1307	HYPOTHETICAL 29.6 KD PROTEIN IN RIBT-DACB INTERGENIC REGION
YSNB_BACSU	39%	1309	HYPOTHETICAL 19.2 KD PROTEIN IN RPH-ILVB INTERGENIC REGION
YGGV_ECOLI (O197)	48%	1309	HYPOTHETICAL 21.0 KD PROTEIN IN G3HD ANSB INTERGENIC REGION
YBIO_ECOLI	32%	1321	HYPOTHETICAL 86.8 KD PROTEIN IN DING-GLNQ INTERGENIC REGION
Y442_MYCPN	36%	1327	HYPOTHETICAL PROTEIN MG442 HOMOLOG
YQGZ_BACSU	37%	1356	HYPOTHETICAL 14.8 KD PROTEIN IN SODA-COMGA INTERGENIC REGION
YEBH_ECOLI (ORF30)	34%	1362	HYPOTHETICAL 30.4 KD PROTEIN IN MANZ-CSPC INTERGENIC REGION
Y120_MYCGE	30%	1365	HYPOTHETICAL PROTEIN MG120
Y121_MYCGE	48%	1366	HYPOTHETICAL PROTEIN MG121
YAAN_BACSU	50%	1373	HYPOTHETICAL 43.8 KD PROTEIN IN XPAC-ABRB INTERGENIC REGION
YLXD_BACCL (ORF2)	46%	1382	HYPOTHETICAL 27.6 KD PROTEIN IN PYRAB-PYRD INTERGENIC REGION
YMFC_HAEIN	62%	1388	HYPOTHETICAL PROTEIN HI0694
Y882_HAEIN	35%	1404	HYPOTHETICAL PROTEIN HI0882
Y531_METJA	34%	1409	HYPOTHETICAL PROTEIN MJ0531
YQBO_BACSU	27%	1424	HYPOTHETICAL 171.0 KD PROTEIN IN SPOIIIC-CWLA INTERGENIC REGION
YQBO_BACSU	29%	1426	HYPOTHETICAL 171.0 KD PROTEIN IN SPOIIIC-CWLA INTERGENIC REGION
YE22_HAEIN	39%	1451	HYPOTHETICAL PROTEIN HI1422
YLXM_BACSU	53%	1455	HYPOTHETICAL 13.2 KD PROTEIN IN FFH 5'REGION
YBXB_BACSU (P23)	44%	1464	HYPOTHETICAL 22.5 KD PROTEIN IN RPLL-RPOB INTERGENIC REGION
YBDM_ECOLI	47%	1467	HYPOTHETICAL 23.9 KD PROTEIN IN CSTA-DSBG INTERGENIC REGION
YBDN_ECOLI	32%	1468	HYPOTHETICAL 47.8 KD PROTEIN IN CSTA-DSBG INTERGENIC REGION
YTRP_LACLA	99%	1500	HYPOTHETICAL 13.3 KD PROTEIN IN TRPE 5'REGION
YICL_ECOLI	34%	1505	HYPOTHETICAL 33.1 KD PROTEIN IN SELC-NLPA INTERGENIC REGION
Y709_METJA	26%	1511	HYPOTHETICAL PROTEIN MJ0709
YBGG_ECOLI	29%	1530	HYPOTHETICAL 100.0 KD PROTEIN IN HRSA-CYDA INTERGENIC REGION
YYAI_BACSU (ORFG)	33%	1539	HYPOTHETICAL 20.2 KD PROTEIN IN TETB-EXOA INTERGENIC REGION
YJJG_ECOLI	27%	1554	HYPOTHETICAL 25.3 KD PROTEIN IN RIMI-PRFC INTERGENIC REGION
YEXA_BACSU	37%	1573	HYPOTHETICAL 9.7 KD PROTEIN IN PUR OPERON
YEGB_ECOLI	27%	1577	HYPOTHETICAL 50.9 KD PROTEIN IN ALKA-BAES INTERGENIC REGION

Table 3.

YMO1_RHIME	30%	1591	HYPOTHETICAL 36.4 KD PROTEIN IN MOCC-MOCA INTERGENIC REGION
(ORF334)			
DEGV_BACSU	35%	1601	DEGV PROTEIN
YQKD_BACSU	25%	1612	HYPOTHETICAL 34.6 KD PROTEIN IN GLNQ-ANSR INTERGENIC REGION
YPBG_BACSU	30%	1639	HYPOTHETICAL 28.6 KD PROTEIN IN RECQ-CMK INTERGENIC REGION
PRECURSOR			
YH38_HAEIN	30%	1651	HYPOTHETICAL PROTEIN HI1738
YKQC_BACSU	41%	1659	HYPOTHETICAL 61.5 KD PROTEIN IN ADEC-PDHA INTERGENIC REGION
YHID_ECOLI	42%	1660	HYPOTHETICAL 23.2 KD PROTEIN IN SLP-HDEB INTERGENIC REGION
(ORF-C)			
YDP1_LACLC	50%	1664	HYPOTHETICAL 30.9 KD PROTEIN IN PEPX 5'REGION (ORF1)
YAGG_ECOLI	24%	1672	HYPOTHETICAL 50.6 KD PROTEIN IN PERR-ARGF INTERGENIC REGION
YPSC_BACSU	46%	1697	HYPOTHETICAL 43.5 KD PROTEIN IN COTD-KDUD INTERGENIC REGION
PRECURSOR			
YQEN_BACSU	32%	1699	HYPOTHETICAL 40.5 KD PROTEIN IN COMEC-RPST INTERGENIC REGION
YGJR_ECOLI	36%	1706	HYPOTHETICAL 37.0 KD PROTEIN IN EBGC-UXAA INTERGENIC REGION
YADQ_ECOLI	37%	1710	HYPOTHETICAL 46.0 KD PROTEIN IN HEML-PFS INTERGENIC REGION
YIDD_PROMI	46%	1719	HYPOTHETICAL 9.6 KD PROTEIN IN RNPA 3'REGION
YVDK_BACSU	60%	1726	HYPOTHETICAL 88.3 KD PROTEIN IN CLPP-CRH INTERGENIC REGION
LYTR_BACSU	31%	1739	MEMBRANE-BOUND PROTEIN LYTR
YFGA_ECOLI	23%	1741	HYPOTHETICAL 36.2 KD PROTEIN IN NDK-GCPE INTERGENIC REGION
YQFU_BACSU	28%	1745	HYPOTHETICAL 32.5 KD PROTEIN IN CCCA-SODA INTERGENIC REGION
YICO_ECOLI	39%	1768	HYPOTHETICAL 49.9 KD PROTEIN IN NLPA-UHPT INTERGENIC REGION
YUPP_MYCHO	32%	1770	HYPOTHETICAL 31.6 KD PROTEIN IN UPP 3'REGION
YKGE_ECOLI	43%	1771	HYPOTHETICAL 26.0 KD PROTEIN IN EAEH-BETA INTERGENIC REGION
YKGF_ECOLI	32%	1772	HYPOTHETICAL 53.1 KD PROTEIN IN EAEH-BETA INTERGENIC REGION
YKGG_ECOLI	31%	1773	HYPOTHETICAL 31.1 KD PROTEIN IN EAEH-BETA INTERGENIC REGION
YBHL_ECOLI	35%	1774	HYPOTHETICAL 25.9 KD PROTEIN IN MOAE-RHLE INTERGENIC REGION
YHAP_BACSU	27%	1814	HYPOTHETICAL 45.4 KD PROTEIN IN SSPB-PRSA INTERGENIC REGION
Y456_METJA	27%	1817	HYPOTHETICAL PROTEIN MJ0456
YNHE_ECOLI	39%	1826	HYPOTHETICAL 56.3 KD PROTEIN IN LPP-AROD INTERGENIC REGION
Y154_MYCLE	27%	1830	HYPOTHETICAL 42.2 KD PROTEIN U1496A/B1496_C1_154
YYBQ_BACSU	52%	1859	HYPOTHETICAL 34.0 KD PROTEIN IN COTF-TETB INTERGENIC REGION
Y260_SYNY3	35%	1862	HYPOTHETICAL 49.2 KD PROTEIN SLL0260
YAT3_STAAU	34%	1864	HYPOTHETICAL 16.6 KD PROTEIN IN ATL 5'REGION (ORF3)
YHBQ_ECOLI	35%	1869	HYPOTHETICAL 11.3 KD PROTEIN IN SOHA-MTR INTERGENIC REGION
(O100)			
YRKA_BACSU	48%	1882	HYPOTHETICAL 49.0 KD PROTEIN IN BLTD-TRKA INTERGENIC REGION
Y531_METJA	26%	1907	HYPOTHETICAL PROTEIN MJ0531
Y556_SYNY3	37%	1917	HYPOTHETICAL 24.0 KD PROTEIN SLR0556
DEGV_BACSU	28%	1920	DEGV PROTEIN
YQHE_ECOLI	37%	1924	HYPOTHETICAL OXIDOREDUCTASE IN METC-SUFI INTERGENIC REGION
YPEC_LACLC	88%	1928	HYPOTHETICAL PROTEIN IN PEPC 5'REGION (ORF 2)
YQKG_BACSU	54%	1930	HYPOTHETICAL 21.0 KD PROTEIN IN GLNQ-ANSR INTERGENIC REGION
YQHY_BACSU	24%	1939	HYPOTHETICAL 14.7 KD PROTEIN IN ACCC-FOLD INTERGENIC REGION
YQKG_BACSU	34%	1942	HYPOTHETICAL 21.0 KD PROTEIN IN GLNQ-ANSR INTERGENIC REGION
YIEK_ECOLI	40%	1949	HYPOTHETICAL 23.3 KD PROTEIN IN TNAB-BGLB INTERGENIC REGION
YLP3_PSEPU	40%	1951	HYPOTHETICAL 44.7 KD PROTEIN IN LPD-3 5'REGION (ORF3)
YADS_ECOLI	33%	1966	HYPOTHETICAL 22.1 KD PROTEIN IN HEML-PFS INTERGENIC REGION
YDAR_BACSU	30%	1970	HYPOTHETICAL 45.7 KD PROTEIN IN MUTT-GSIB INTERGENIC REGION
Y837_METJA	28%	1980	HYPOTHETICAL PROTEIN MJ0837
YIGN_ECOLI	22%	1982	HYPOTHETICAL 54.7 KD PROTEIN IN UDP-UBIE INTERGENIC REGION
PRECURSOR			
YJEQ_HAEIN	29%	1984	HYPOTHETICAL PROTEIN HI1714
YABE_BACSU	39%	1985	HYPOTHETICAL 47.7 KD PROTEIN IN METS-KSGA INTERGENIC REGION
YE12_HAEIN	31%	2002	HYPOTHETICAL PROTEIN HI1412
YF56_METJA	38%	2013	HYPOTHETICAL PROTEIN MJ1556
YACC_BACSU	45%	2014	HYPOTHETICAL 31.8 KD PROTEIN IN FTSH-CYSK INTERGENIC REGION
YWJB_BACSU	40%	2015	HYPOTHETICAL 19.6 KD PROTEIN IN ACDA 5'REGION
YACF_BACSU	57%	2016	HYPOTHETICAL 37.1 KD PROTEIN IN FOLK-LYSS INTERGENIC REGION
YPJC_BACSU	45%	2017	HYPOTHETICAL 23.6 KD PROTEIN IN QCRC-DAPB INTERGENIC REGION
YQFU_BACSU	29%	2019	HYPOTHETICAL 32.5 KD PROTEIN IN CCCA-SODA INTERGENIC REGION
YHIT_MYCPN	37%	2022	HYPOTHETICAL 16.1 KD PROTEIN HIT-LIKE PROTEIN
Y143_HAEIN	24%	2044	HYPOTHETICAL PROTEIN HI0143
YU12_MYCTU	30%	2045	HYPOTHETICAL 34.9 KD PROTEIN CY339.12C
YXJB_BACSU	40%	2052	HYPOTHETICAL 31.5 KD PROTEIN IN KATB 3'REGION
YQHL_BACSU	38%	2081	HYPOTHETICAL 14.6 KD PROTEIN IN GCVT-SPOIIIAA INTERGENIC REGION
YJFG_HAEIN	29%	2099	HYPOTHETICAL PROTEIN HI0121
Y018_MYCGE	36%	2101	HYPOTHETICAL HELICASE MG018
YDP1_LACLC	95%	2114	HYPOTHETICAL 30.9 KD PROTEIN IN PEPX 5'REGION (ORF1)
YDP3_LACLA	88%	2117	HYPOTHETICAL 18.7 KD PROTEIN IN PEPX 3'REGION (ORF3)
YACL_BACSU	42%	2126	HYPOTHETICAL 40.9 KD PROTEIN IN MECB-GLTX INTERGENIC REGION
YEEO_ECOLI	28%	2151	HYPOTHETICAL 60.1 KD PROTEIN IN AMN-CBL INTERGENIC REGION
YCEC_ECOLI	38%	2161	HYPOTHETICAL 36.0 KD PROTEIN IN RNE-RPMF INTERGENIC REGION
(ORFX)			
YEBI_HAEIN	28%	2162	HYPOTHETICAL PROTEIN HI0407
YAEL_HAEIN	27%	2178	HYPOTHETICAL PROTEIN HI0918

Table 3.

Y506_SYNY3	46%	2180	HYPOTHETICAL 28.8 KD PROTEIN SLL0506
Y0BD_MYCTU	31%	2181	HYPOTHETICAL 13.0 KD PROTEIN CY227.13 PRECURSOR
YBEA_ECOLI	31%	2182	HYPOTHETICAL 17.3 KD PROTEIN IN MRDA-PHPB INTERGENIC REGION
YABB_BACSU	47%	2190	HYPOTHETICAL 28.3 KD PROTEIN IN XPAC-ABRB INTERGENIC REGION
Y003_HAEIN	28%	2194	HYPOTHETICAL PROTEIN HI0003
YBHE_ECOLI	28%	2209	HYPOTHETICAL 36.3 KD PROTEIN IN MODC-BIOA INTERGENIC REGION
YQJV_BACSU	25%	2221	HYPOTHETICAL 44.7 KD PROTEIN IN GLNQ-ANSR INTERGENIC REGION
Y661_METJA	31%	2226	HYPOTHETICAL PROTEIN MJ0661
YTFG_ECOLI (F286)	45%	2227	HYPOTHETICAL 29.7 KD PROTEIN IN RPLI-CPDB INTERGENIC REGION
YYBR_BACSU	42%	2227	HYPOTHETICAL 14.7 KD PROTEIN IN COTF-TETB INTERGENIC REGION
YJBR_ECOLI	40%	2232	HYPOTHETICAL 13.5 KD PROTEIN IN APHA-UVRA INTERGENIC REGION
Y143_HAEIN	30%	2252	HYPOTHETICAL PROTEIN HI0143
YPBG_BACSU PRECURSOR	26%	2253	HYPOTHETICAL 28.6 KD PROTEIN IN RECQ-CMK INTERGENIC REGION
YCEE_ECOLI (ORF1)	25%	2277	HYPOTHETICAL 43.9 KD PROTEIN IN MSYB-HTRB INTERGENIC REGION
Y441_METJA	38%	2280	HYPOTHETICAL PROTEIN MJ0441
YQEG_BACSU	42%	2283	HYPOTHETICAL 20.1 KD PROTEIN IN NUCB-AROD INTERGENIC REGION
YWFA_BACSU	28%	2299	HYPOTHETICAL 44.5 KD PROTEIN IN ROCC-PTA INTERGENIC REGION
Y531_METJA	32%	2300	HYPOTHETICAL PROTEIN MJ0531
Y800_METJA	28%	2304	HYPOTHETICAL PROTEIN MJ0800
YJBC_ECOLI	30%	2306	HYPOTHETICAL 32.5 KD PROTEIN IN PEPE-LYSC INTERGENIC REGION
Y0A5_MYCTU	24%	2328	HYPOTHETICAL 40.2 KD PROTEIN CY9C4.05C
YHA2_EIKCO	29%	2333	HYPOTHETICAL 66.3 KD PROTEIN IN HAG2 5'REGION
YCXE_BACME	41%	2337	HYPOTHETICAL 30.5 KD PROTEIN IN GDHI 5'REGION (ORF 2)
YHCC_ECOLI	37%	2339	HYPOTHETICAL 34.6 KD PROTEIN IN ARCB-GLTB INTERGENIC REGION

BIO16. UNKNOWN **83 genes**

BIO16.01 General **5 genes**

NIFS_LACDE	53%	520	NIFS PROTEIN HOMOLOG (FRAGMENT)
NIFS_ANASP	29%	1829	NIFS PROTEIN
NIFS_ANASP	39%	1908	NIFS PROTEIN
US45_LACLA	39%	1956	SECRETED 45 KD PROTEIN PRECURSOR
US45_LACLA	99%	2293	SECRETED 45 KD PROTEIN PRECURSOR

BIO16.02 Unclassified **78 genes**

LCIA_LACLA	33%	48	LACTOCOCCIN A IMMUNITY PROTEIN
OSMC_ECOLI	48%	69	OSMOTICALLY INDUCIBLE PROTEIN C
LCND_LACLA	97%	85	LACTOCOCCIN A SECRETION PROTEIN LCND
CSBB_BACSU	45%	107	CSBB PROTEIN
ACPD_ECOLI	33%	113	ACYL CARRIER PROTEIN PHOSPHODIESTERASE (ACP PHOSPHODIESTERASE)
JAG_BACSU	31%	131	JAG PROTEIN (SPOIIIJ ASSOCIATED PROTEIN)
AMID_MORCA	48%	164	PUTATIVE AMIDASE (EC 3.5.1.4)
YZDD_BACSU	62%	166	PET112-LIKE PROTEIN
HYIN_AGRVI	42%	168	INDOLEACETAMIDE HYDROLASE (EC 3.5.1.-)
PHNO_ECOLI	40%	225	PHNO PROTEIN
TPX_MYCTU	35%	299	PROBABLE THIOL PEROXIDASE (EC 1.11.1.-)
ACPD_ECOLI	39%	384	ACYL CARRIER PROTEIN PHOSPHODIESTERASE (ACP PHOSPHODIESTERASE)
KIME_METJA	31%	406	MEVALONATE KINASE (EC 2.7.1.36)
KIME_METJA	29%	408	MEVALONATE KINASE (EC 2.7.1.36)
BMRU_BACSU	27%	424	BMRU PROTEIN
MRSA_HAEIN	46%	433	MRSA PROTEIN HOMOLOG
COLL_HSVS7	37%	478	COLLAGEN-LIKE PROTEIN
FAS6_RHOFA	30%	491	HYPOTHETICAL PROTEIN IN AZU 5'REGION (ORF1)
EPUA_STRPN	40%	533	EPUA PROTEIN
XYLH_PSEPU	32%	566	4-OXALOCROTONATE TAUTOMERASE (EC 5.3.2.-)
STA_ECOLI	31%	580	STREPTOTHRICIN ACETYLTRANSFERASE (EC 2.3.-.-)
HPRK_ENTFA	43%	602	HPR-KINASE (EC 2.7.1.-)
NUSB_BACSU	46%	691	N UTILIZATION SUBSTANCE PROTEIN B HOMOLOG (NUSB PROTEIN)
CLCB_PSEPU	26%	730	CHLOROMUCONATE CYCLOISOMERASE (EC 5.5.1.7)
VEG_BACSU	30%	753	VEG PROTEIN
YORV_CLOAB	52%	755	HYPOTHETICAL PROTEIN HI0943
NUSA_BACSU	48%	772	N UTILIZATION SUBSTANCE PROTEIN A HOMOLOG (NUSA PROTEIN)
LRPR_STREQ	31%	825	LEUCINE RICH PROTEIN
EXPZ_BACSU	29%	918	NUCLEOTIDE BINDING PROTEIN EXPZ
NRDI_LACLC	96%	1002	NRDI PROTEIN
ELAA_ECOLI	38%	1008	ELAA PROTEIN
RGG_STRGC	25%	1075	RGG PROTEIN
YRP2_LISMO	34%	1083	HYPOTHETICAL PROTEIN MJ0927
PHOL_BACSU	57%	1091	PHOH-LIKE PROTEIN
PHNB_ECOLI	33%	1135	PHNB PROTEIN
RGG_STRGC	29%	1178	RGG PROTEIN
CITG_HAEIN	35%	1209	CITG PROTEIN
Y002_RICPR (ORF39.9)	33%	1236	HYPOTHETICAL 40.0 KD PROTEIN IN HTRB-DINI INTERGENIC REGION

Table 3.

GID_BACSU	69%	1255	GID PROTEIN (FRAGMENT)
2DKG_CORSP	46%	1266	2,5-DIKETO-D-GLUCONIC ACID REDUCTASE (EC 1.1.1.-)
FIBP_ADEB3	29%	1393	FIBER PROTEIN
FMCH_BACNO	31%	1408	POSSIBLE FIMBRIAL ASSEMBLY PROTEIN FIMC (SEROGROUP H1)
P60_LISSE	31%	1414	PROTEIN P60 PRECURSOR (INVASION-ASSOCIATED PROTEIN)
XHLB_BACSU	40%	1414	HOLIN
DFRA_SYNY3	30%	1558	PUTATIVE DIHYDROFLAVONOL-4-REDUCTASE (EC 1.1.1.219)
DC4C_ACICA	34%	1575	4-CARBOXYMUCONOLACTONE DECARBOXYLASE (EC 4.1.1.44)
ESL2_MYCPN	27%	1604	PUTATIVE ESTERASE/LIPASE 2 (EC 3.1.-.-)
PKSG_BACSU	33%	1610	PUTATIVE POLYKETIDE BIOSYNTHESIS PROTEIN PKSG
F4RE_METJA	34%	1652	PUTATIVE F420-DEPENDENT NADP REDUCTASE (EC 1.-.-.-)
LAXC_LACLA	29%	1671	LACX PROTEIN, CHROMOSOMAL
RGG_STRGC	27%	1713	RGG PROTEIN
ARCL_ECOLI	40%	1732	CARBAMATE KINASE-LIKE PROTEIN 1
SMPB_ENTFA	63%	1757	SMALL PROTEIN B HOMOLOG
BMRU_BACSU	28%	1775	BMRU PROTEIN
DEDA_ECOLI	39%	1796	DEDA PROTEIN (DSG-1 PROTEIN)
XYNC_CALSA	33%	1809	ACETYL ESTERASE (EC 3.1.-.-)
THL_CLOAB	33%	1823	ACETYL-COA ACETYLTRANSFERASE (EC 2.3.1.9)
PHPB_ECOLI	34%	1869	PHPB PROTEIN
PFS_HAEIN	43%	1929	PFS PROTEIN HOMOLOG
PROC_ECOLI	38%	1933	PYRROLINE-5-CARBOXYLATE REDUCTASE (EC 1.5.1.2)
SUN_BACSU	33%	1938	SUN PROTEIN (FRAGMENT)
YFF1_MYCMY	44%	1947	HYPOTHETICAL PROTEIN MG130
LAXC_LACLA	36%	1979	LACX PROTEIN, CHROMOSOMAL
BLTD_BACSU	33%	1996	TRANSACETYLASE BLTD (EC 2.3.1.-)
RGG_STRGC	19%	2024	RGG PROTEIN
FIB1_ADE41	30%	2027	FIBER PROTEIN 1
DC4C_ACICA	35%	2032	4-CARBOXYMUCONOLACTONE DECARBOXYLASE (EC 4.1.1.44)
HIT_BACSU	41%	2057	HIT PROTEIN
ARCL_ECOLI	45%	2090	CARBAMATE KINASE-LIKE PROTEIN 1
STA_ECOLI	31%	2123	STREPTOTHRICIN ACETYLTRANSFERASE (EC 2.3.-.-)
NARA_BACSU	27%	2213	NARA PROTEIN
PQQE_ACICA	30%	2216	COENZYME PQQ SYNTHESIS PROTEIN E (COENZYME PQQ SYNTHESIS PROTEIN III)
RGG_STRGC	31%	2217	RGG PROTEIN
YHHF_ECOLI	30%	2219	21.7 KD PROTEIN IN FTSY-NIKA INTERGENIC REGION
GSPA_ECOLI	28%	2239	PUTATIVE GENERAL SECRETION PATHWAY PROTEIN B
RGG_STRGC	19%	2268	RGG PROTEIN
TOXB_CLODI	27%	2285	TOXIN B
PHNA_ECOLI	47%	2312	PHNA PROTEIN
	1495 genes		

teins. Parameters of homology with several proteins, rather than with the best one, can be used to increase the performance of the approach for classification of hypothetical proteins. The functional classification of predicted genes is colour-encoded on Figure 9 and summarised in Figure 10 and in Table 3. A more detailed and systematically updated version of Table 3 can be found at http://spock.jouy.inra.fr.

Metabolic potential of *L. lactis* IL1403.

The automatic procedure for protein finding, function assignment and classification, described above, allows to estimate the potential of *L. lactis* as a free-living microorganism. Several of the detected systems were characterised experimentally earlier, by biochemical, genetic or gene cloning means.

Amino acid, vitamin and nucleotide biosynthesis

Lactic acid bacteria are nutritionally fastidious. The presence of a gene for synthesis of a metabolite does not mean that a strain of *L. lactis* where the gene was found will not require this metabolite in the growth medium. The studies performed long ago (Anderson & Elliker 1953) and recently (Jensen & Hammer 1993; Cocaign-Bousquet et al. 1995) showed that synthetic medium for *L. lactis* should contain at least six amino acids (isoleucine, valine, leucine, histidine, methionine and glutamic acid) and seven vitamins (biotin, pyridoxal, folic acid, riboflavin, niacinamide, thiamine and pantothenic acid). Our analysis showed that *L. lactis* has a genetic potential to synthesise 20 standard amino acids and at least four cofactors (folic acid, menaquinone, riboflavin and thioredoxin). The genes for biosynthesis of aromatic amino acids, threonine, branched chain amino acids and histidine were cloned and characterised earlier (Bardowski et al. 1992; Delorme et al. 1992; Godon et al. 1992; Griffin & Gasson 1995; Madsen et al. 1996). The regulation of transcription of the tryptophane operon was studied in detail (Raya et al. 1998; van de Guchte et al. 1998). The problem of fastidious nutritional requirements of *L. lactis* strains has recently been addressed by application of the single-omission technique (Cocaign-Bousquet et al. 1995; Novak et al. 1997) and genetic approaches (Lapujade et al. 1998). It was shown also that auxotrophy of IL1403, used as a dairy strain, for histidine and branched-chain amino acids is due to the recently acquired mutations (Delorme et al. 1993; Godon et al. 1993).

Biosynthesis and salvage of nucleotides and nucleosides in *L. lactis* were explored earlier and several pyrimidine and purine biosynthetic genes were cloned and sequenced (Martinussen et al. 1994; Andersen et al. 1996; Nilsson & Kilstrup 1998). It was demonstrated that milk does not contain sufficient purine compounds to support growth and therefore *de novo* biosynthesis is necessary (Dickely et al. 1995). Nucleoside compounds are not needed in the minimal synthetic medium to maintain exponential growth of *L. lactis* (Jensen & Hammer 1993). We detected 47 genes implicated in this metabolism. Most of the genes for biosynthesis of purines are clustered near the *purDEK* operon, which was recently characterised (Nilsson & Kilstrup 1998). A copy of IS983 is detected between the *purDEK* operon and other genes of purine biosynthesis.

Energy metabolism and transporters

The main feature of the energy metabolism in *L. lactis* is that fermented carbon sources are used mostly to maintain the energetic potential of the cells (Poolman 1993; Cocaign-Bousque et al. 1996). About 5% of the fermented sugar is used as the source of substrates for synthetic reactions (Novak et al. 1998). *L. lactis* does not have a functional citric acid cycle and gluconeogenesis enzymes and is unable to carry out many anaplerotic reactions. The exponential growth of this bacterium appears to be 'programmed' to happen only in sugar rich complex media, like milk. As mentioned above, many metabolites have to be added into the synthetic medium (Jensen & Hammer 1993), although the corresponding biosynthetic genes seem to be present.

It is interesting to compare the genetic potential of *L. lactis* to grow on different carbon sources with the available experimental data. The genetic potential can be estimated from the presence of the biodegradation genes and the relevant transporters. IL1403 has genes which can be used to grow on different carbon sources: glucose (glycolysis genes), fructose (positions 1519 and 2230 kb), N-acetyl glucosamine (1032 kb), xylose (1550 kb), ribose (1685 kb), mannose (777 kb), gluconate (606 and 2253 kb), maltose (692, 700 and 1526 kb), lactose (2040 kb), galactose (2045 kb), mannitol (33 kb), different β-glucosides (186, 419, 830, 1490 and 1520 kb). Glucuronate or galacturonate catabolic operon (1670 kb) can be used for utilisation of pectin degradation products as a supplementary source of energy and carbon. Components of sugar-specific enzyme II of phosphoenol-pyruvate dependent transport systems were found for mannitol (30 kb), sucrose or trehalose (435 kb), fructose (984 and 1747kb), mannose (1748 kb) and β-glucosides (175, 416, 830, 1144 and 1489 kb). These findings are summarised in Table 4, along with references to experimental data on the growth of *L. lactis* on different carbon sources.

The ability to grow more efficiently in particular conditions can be acquired by *L. lactis* by means of plasmid transfer. This is well illustrated by experimental studies of the system needed for efficient growth of *L. lactis* in milk. To overgrow other bacteria in this niche *L. lactis* needs efficient systems for utilisation of lactose and peptides. Such genes, which include PEP-dependent transporter for lactose and proteinase for peptides, were found in the plasmid pLP712 isolated from *Streptococcus lactis* NCDO 712 (Gasson 1983; Maeda & Gasson 1986). The system encoded by pLP712 contains a PEP-dependent transporter for lactose. This allows efficient entry of lactose into glycolytic pathways through glucose, which is phosphorylated by glucokinase, and galactose-6P, which is metabolised through the tagatose-6P pathway, also encoded by pLP712. Sequence analysis of the IL1403 chromosome revealed that the genes encoding the PTS-dependent pathway of lactose utilisation are absent in this strain. The chromosome contains, however, another system for lactose utilisation, dependent on the transport by the product of the *lacS* gene, encoding an H^+ symporter or galactose-lactose antiporter. This system was characterised earlier for *Streptococcus thermophilus*. Special selection is needed for *S. thermophilus* strains to be able to ferment galactose (Thomas & Craw, 1984; Hutkins et al. 1985). Strain IL1403 was isolated from

Table 4. Ability of of *L. lactis* for growth on different carbon sources

Carbon source	Catabolic genes (position on genome map)	Transport genes (position on genome map)	Experimental growth (reference)
glucose	glycolysis genes		+ (Bardowski et al. 1994) + (de Vos 1996)#
fructose	fructokinase (1519 kb), glucoso-6P isomerase (2230 kb)	PTS (984 kb and 1747 kb)	+ (Bardowski et al. 1994) + (de Vos 1996)#
N-acetyl glucosamine	glucosamine-fructoso-6P aminotransferase (1032 kb)		+ (Bardowski et al. 1994) + (de Vos 1996)#
sucrose		PTS (435 kb)*	- (de Vos 1996)#
mannose	mannose-6P isomerase (777 kb)	PTS (1748 kb)	+ (Bardowski et al. 1994) + (de Vos 1996)#
maltose	maltodextrin glucosidase (700 kb and 1526 kb), 4-α-glucanotransferase (692 kb)		+ (Bardowski et al. 1994) + (de Vos 1996)#
trehalose		PTS (435 kb)*	+ (Bardowski et al. 1994) + (de Vos 1996)#
lactose	β-galactosidase (2040 kb)	lactose permease (2048 kb)	+ (de Vos 1996)#
galactose	gal operon (2045 kb)		+ (Bardowski et al. 1994) + (de Vos 1996)#
ribose	rbs operon (1685 kb)	ribose transport protein (1686 kb)	+ (Bardowski et al. 1994) - (de Vos 1996)#
xylose	xyl operon (1550 kb)		
cellobiose	glucosidase (186 kb and 1521 kb), 6-P β-glucosidase (419 kb, 830 kb and 1490 kb)	PTS (175 kb, 416 kb and 830 kb)	+ (Bardowski et al. 1994) + (de Vos 1996)#
salicin	6-P β-glucosidase (1722 kb)		+(Bardowski et al. 1994) + (de Vos 1996)#
arbutin	See cellobiose	See cellobiose	+ (Bardowski et al. 1994)
esculin	See cellobiose	See cellobiose	+ (Bardowski et al. 1994)
gluconate	6-P-gluconate gehydrogenase (606 kb), gluconate kinase (2253 kb)	gluconate permease (2246 kb)	
mannitol	mannitol-1P 5-dehydrogenase (33 kb)	PTS (29 kb and 32 kb)	

#For the strain mg1363 with the plasmid pMG820.
*Presumably a trehalose specific system.

cheese starter culture and cured from plasmids for the purpose to be used as a general genetic engineering host (Chopin et al. 1984; Simon et al. 1985). Probably one of the eliminated plasmids contained genes for efficient utilisation of lactose, similar to the ones encoded by pLP712.

Chromosomal genes for galactose and lactose utilisation were cloned and characterised recently from different *L. lactis* strains (Grossiord et al. 1998; Vaughan et al. 1998). It appeared that the strains can be divided in two groups by their possession or lack of the inducible β-galactosidase gene. Since we found the gene in IL1403, this strain belongs to the former group. It is different in this respect from another model strain cured of plasmids, MG1363 (Gasson 1983), which does not have β-galactosidase.

Automatic analysis detected 14 genes involved in glycolysis, completing the description of this system and revealed a second glyceraldehyde-phosphate dehydrogenase gene. It also confirmed the absence of a complete citric acid cycle, revealing only the genes working around pyruvate, like citrate lyase (5 genes), malate oxidoreductase and pyruvate dehydrogenase (4 genes). Close analysis of the genes automatically classified as relevant to glyconeogenesis confirmed that this process is not present in *L. lactis*. No genes for fructose 1,6 bisphosphatase, phosphoenolpyruvate carboxykinase or phosphoenolpyruvate synthetase were found.

It is interesting to mention that all genes of the pentose-phosphate pathway were detected, but not a putative trans-aldolase. It is not yet clear which gene is encoding this function in *B. subtilis*. It seems that the structure of the corresponding enzyme in AT-rich gram positive bacteria is different from that in gram negative and GC-rich gram positive bacteria. The corresponding gene in AT-rich gram positive bacteria still has to be identified.

Cell envelope

The cell surface structure has a large impact on the properties of lactic acid bacteria relevant for their application in milk industry, such as resistance to stress, cell lysis, sencsitivity to phages antimicrobial agents and influence on the product texture. However, very few studies, mostly related to the potential application of lactic acid bacteria as a live vehicle for vaccine delivery, were reported (Pouwels et al. 1998). Our analysis revealed 73 genes implicated in cell envelope function, including 29 membrane proteins, 23 murein succulus and peptidoglycan biosynthesis genes and 21 surface polysaccharide biosynthesis genes. Operons for the biosynthesis of extracellular polysacharides were identified recently in the chromosome of *Streptococcus thermophilus* (Stingele et al. 1996) and on a plasmid in *L. lactis* (van Kranenburg et al. 1997). We did not find similar genes in the chromosome of IL1403. This may be another property of industrial interest which is encoded and propagated predominantly by plasmids in *L. lactis*.

Cellular processes

L. lactis is not known as a good secretor of proteins compared to, for example, Bacillus species. Optimised gene constructions allow *L. lactis* to produce up to 20 mg of foreign protein per litre (Wells et al. 1993; Le Loir et al. 1998), compared to 800 mg produced by *B. subtilis* (Hemila et al. 1992). The reasons may be the lack of some components of the secretion machinery or the inefficiency of energy metabolism. We did not detect components such as *secDF* genes, which are known to improve the secretion efficiency in gram negative and gram positive bacteria (Pogliano & Beckwith 1994; Duong & Wickner 1997; Bolhuis et al. 1998). Also, only one signal peptidase gene was found in IL1403, compared to five detected in the *B. subtilis* genome (Kunst et al. 1997). Probably, the useful genes

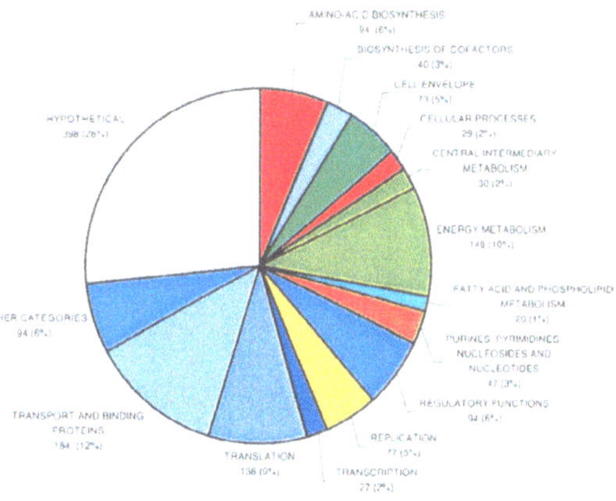

Figure 10. Distribution of the detected genes by functional categories. Colours designating different categories are defined in the legend of Figure 9. The number and the percentage of homology hits to SwissProt are indicated for each category. Classification method is presented in the text.

for protein secretion can be acquired by lactococci from closely related species through plasmid transfer, if selective conditions for this are created by industrial applications.

Natural competence for DNA-mediated transformation was not described for *L. lactis* species. However, we found that the *L. lactis* genome contains four operons, *comE*, *comF*, *comC* and *comG*, which encode nine genes that share strong similarity with the late competence genes from *B. subtilis* and *S. pneumoniae*. Since *L. lactis* seems to possess a complete set of these genes and since no function other than the involvement in DNA-mediated transformation is known for them, it is possible that *L. lactis* can acquire natural competence. Sequence studies of late competence genes showed that they are quite similar in different AT-rich gram positive bacteria, like *B. subtilis* or Streptococci (Dubnau 1997; Campbell et al. 1998). It is apparent also that a big difference exists in the molecular mechanisms which regulate the development of this process in Streptococci and Bacilli. In *B. subtilis* the regulator ComK assembles the signals from earlier stages of competence development (van Sinderen et al. 1995; Hahn et al. 1996). A functional counterpart of such a regulator was not yet discovered for Streptococci. A way to discover the conditions for acquiring the competence may be the study of regulation of the late genes. We analysed the regions preceding the first ORF in the four late competence operons. The corresponding alignment is shown in Figure 11. It is apparent that a common feature exists for these re-

```
A

aaattttcggcttttttGTTACATTaactgAttaTTTgCGTATAgtct AGTATG. . . comGA

tgattttaacaaattttGTTACATTttaacAgcaTTTTCGTATAgaat AGTATG. . . comFA

ttataattcaactcttgGaaACtTTccaagAgttTTTTCGTATAtaaa AGTATG. . . comEA

tttcgctatatttaaaGTTACAaTtcttcAtttaTTTCGTATAgtattAGTATG. . . comC

consensus          GTTACATTnnnnnAnnnTTTTCGTATAnnnn AGT

B

tgtcagtaatttttttgGTTACAagtcagtctttTTTaCGTATAgaaa AtTataagccaattttaaaattg

                                   M  I  T  N  F  D  L  Y  R  W  K  K  A
ctgacaaagctgtcagtaaaaaatGGAGGaattATGataacaaattttgatttatatcgttggaaaaaagct

 G  M  T  N  L  G  V  N  K  L  L  K  F  F  H  R  Y  D  K  K  stop
ggcatgaccaatttaggagtaaataaacttttaaatttttttcatagatatgataaaaaaTAAgtttaagac
```

Figure 11. Regulatory regions of the *L. lactis* competence genes. (A) Sequences of the regions upstream of the first gene in the four putative competence operons. Sequences corresponding to the recognition site of a hypothetical regulator are shown in bold capital letters. Start codons of the competence genes are italicised. Consensus sequence of the recognition site is shown at the bottom. (B) Sequence of the upstream region of the *dprA* gene. Putative recognition site of the hypothetical competence regulator is shown in bold capital letters. Conceptual translation of a small ORF upstream of the *dprA* gene is shown.

gions, which suggests that they may be recognised by a common transcription regulator. An interesting feature, which seems to be unique for these genes, is the presence of a strictly conserved sequence AGTATG, in which ATG always corresponds to the start codon of the first gene in each operon. This is different from the usual ribosome binding site in AT-rich gram positive bacteria, which has a sequence (SD) homologous to the 3′-end of 16S rRNA (Vollenoweth & Rabinowitz 1992). Several recent reports show that the efficiency of translation may be not much lower for mRNAs that start directly from ATG codon than for mRNAs with long leaders and strong SD-sequences (Resch et al. 1996; Moll et al. 1998; Van Etten & Janssen 1998), and that the ribosomal protein S1 is not needed for the translation of unleadered mRNA (Tedin et al. 1997). It appears that the absence of the leader region is a common feature of mRNAs of late competence genes in *L. lactis*.

Studies of the regulation of competence in Streptococci revealed the presence of a gene, encoding a small peptide, that stimulates the induction of expression of the whole competence regulon (Havarstein et al. 1995, 1997). In Streptococci the gene encoding competence-stimulating peptide (CSP) precursor was always found to be the first gene of an operon encoding a two-component regulatory system, which is involved in the induction of competence (Havarstein et al. 1996). Homology search over the *L. lactis* genome sequence did not reveal any two-component system which resembles the one regulating competence in Streptococci. However, we found an operon encoding proteins homologous to DprA or Smf protein, supposed to be required for chromosomal transformation in *H. influenzae*, and DNA topoisomerase I from *B. subtilis*. Interestingly, upstream of the *dprA* gene there is a sequence resembling that found in the putative *com*-promoters, described above (Figure 11B). A small ORF between this putative promoter and initiation codon of the *dprA* gene, may be a candidate for the gene of CSP precursor, since the C-terminal part of the encoded peptide resembles that of CSP from Streptococci. This potential CSP precursor does not have, however, a Gly-Gly pattern that is always found in the Streptococcal CSP precursors (Havarstein et al. 1997).

Regulatory functions

Regulation of gene expression in *L. lactis* was not yet thoroughly explored. A few examples of experimental study of chromosomal gene regulation are known (Bardowski et al. 1994; Kilstrup & Martinussen 1998; Luesink et al. 1998; Raya et al. 1998; van de Guchte et al. 1998; Vaughan et al. 1998; Wouters et al. 1998). Automatic analysis revealed 83 genes potentially involved in regulation, which is about 6% of the total number of the identified ORFs. This figure is about two times higher than that for *M. genitalium* (Fraser et al. 1995), coincides with that found in *H. influenzae* (Fleischmann et al. 1995) and is about two-fold lower than that found in *B. subtilis* using the same classification criteria (Lapidus et al. 1997). These ratios support the hypothesis that the increase of genome size in bacteria results in the proportional increase of the number of different regulatory systems.

Seven two-component systems were detected. They are all organised in operons, consisting of a cognate sensor kinase and a responce regulator. The genes of five two-component systems were cloned recently from the strain MG1363 by complementation of *E. coli* mutants (O'Connell-Motherway et al. 1997). All of them were identified in our sequence.

An interesting class of regulatory genes are the RAS-like GTP-binding proteins. Several such regulators were found in other bacteria, including the most intensively studied Era protein of *E. coli*, Obg of *B. subtilis* and several others, usually revealed by genome sequencing. Exact biological functions of most of these proteins are unknown, although they are sometimes essential and have pleiotropic functions. Interestingly, although all these proteins are homologous and can be considered as a family of paralogs. Each of the nine *L. lactis* proteins has its clear counterpart in *B. subtilis* or other bacteria. This suggests that each of these proteins may have a unique biological function in the cell and that revealing it in one microorganism will be directly applicable to the others. It would be interesting to test whether the counterparts of the *B. subtilis* genes from *L. lactis* can complement the corresponding mutations in *B. subtilis* and *vice versa*.

A gene encoding the extracytoplasmic function (ECF) σ-factor was detected in *L. lactis*. This finding can be interesting for studies of regulation of stress response in *L. lactis*. It is not clear if any of the 11 ECF σ-factors from *B. subtilis* has the same biological function as the one detected in *L. lactis*. The most similar proteins are SigV, SigX and YlaC, but the biological function of none of them has yet been established in *B. subtilis*.

Replication, transcription and translation

Seventy-seven, 30 and 134 genes were assigned to replication, transcription and translation functional categories, respectively. It is apparent that the replication system of *L. lactis* is very similar to that of *B. subtilis*. Counterparts of *dnaB* and *dnaD* genes, essential for DNA replication in *B. subtilis* and not present in gram negative bacteria were detected. Two DNA-polymerase III α-chain genes, one corresponding to *polC* and another to *dnaE* of *B. subtilis*, were also detected in *L. lactis*. *E. coli* has only the latter gene.

The transcription and translation machinery does not seem to have any remarkable difference compared to *B. subtilis*. Presumably *B. subtilis* with its well-developed genetic tools can be a convenient host organism to study gene regulation of *L. lactis* systems.

Conclusions and perspectives

We present here the data on a diagnostic sequencing study of the *L. lactis* IL1403 genome. The sequencing data with a low average redundancy (about 2) allowed to reveal 1495 proteins, encoded by the *L. lactis* genome, and to classify them by functions predicted on the basis of homology to known proteins. Some statistical properties of the *L. lactis* genome sequence were also calculated.

The data presented here allow to evaluate *L. lactis* as a model organism for studies of bacterial metabolism, physiology and regulation. Compared to the well established models, like *E. coli* or *B. subtilis*, *L. lactis* represents a simplified system and is, therefore, potentially more amenable for detailed understanding. The absence of numerous genes, relative to *E. coli* or *B. subtilis*, should facilitate, in the case of *L. lactis*, detecting of phenotypes associated with genes of unknown function. With regard to even more simple microorganisms, such as Mycoplasma, the advantage of *L. lactis* is the ease of manipulation and the absence of any pathogenicity for humans. These properties, with the complete genome information available, place *L. lactis* among the best model bacteria, to be used for fundamental biological research.

Acknowledgements

We are grateful to the *L. lactis* community of this lab, and especially to the team leaders Marie-Christine and Alain Chopin, Emmanuelle Maguin, Pierre Renault and Alexandra Gruss for the interest to this work and stimulating discussions. Pierre Renault, Patrick Duwat, Meriem El Karoui, Isabelle Poquet, Catherine Schouler are gratefully acknowledged for the use of their unpublished sequence data in the early stages of this project. We thank Philippe Bessieres, Franck Samson, Véronique Biaudet-Brunaud and Nicole Vermeire (INRA, Jouy en Josas), and Gregory Kucherov and Vladimir Grebinskii (INRIA, Nancy) for the discussions and advice on programming and computing.

References

Altschul SF, Gish W, Miller W, Myers EW, & Lipman DJ (1990) Basic local alignment search tool. J. Mol. Biol. 215(3): 403–410

Andersen PS, Martinussen J & Hammer K (1996) Sequence analysis and identification of the *pyrKDbF* operon from *Lactococcus lactis* including a novel gene, *pyrK*, involved in pyrimidine biosynthesis. J Bacteriol 178(16): 5005–5012

Anderson AW & Elliker PR (1953) The nutritional requirements of lactic streptococci isolated from starter cultures. I. Growth in a synthetic medium. J Dairy Science 36: 161–167

Andersson SGE, Zomorodipour A, Andersson JO, Sicheritz-Pontén T, Alsmark UC, Podowski RM, Näslund AK, Eriksson A-S, Winkler HH & Kurland CG (1998) The genome sequence of *Rickettsia prowazekii* and the origin of mitochondria. Nature 396: 133–140

Bardowski J, Ehrlich SD & Chopin A (1992) Tryptophan biosynthesis genes in *Lactococcus lactis* subsp. *lactis*. J. Bacteriol. 174(20): 6563–7650

Bardowski J, Ehrlich SD & Chopin A (1994) BglR protein, which belongs to the BglG family of transcriptional antiterminators, is involved in beta-glucoside utilization in *Lactococcus lactis*. J. Bacteriol. 176(18): 5681–5685

Beresford T & Condon S (1991) Cloning and partial characterization of genes for ribosomal ribonucleic acid in *Lactococcus lactis* subsp. *lactis*. FEMS Microbiol. Lett. 62(2–3): 319–323

Biaudet V, el Karoui M & Gruss A (1998) Codon usage can explain GT-rich islands surrounding Chi sites on the *Escherichia coli* genome. Mol. Microbiol. 29(2): 666–669

Biswas I, Maguin E, Ehrlich SD & Gruss A (1995) A 7-base-pair sequence protects DNA from exonucleolytic degradation in *Lactococcus lactis*. Proc. Natl. Acad. Sci. USA 92(6): 2244–2248

Bolhuis A, Broekhuizen CP, Sorokin A, van Roosmalen ML, Venema G, Bron S, Quax WJ & van Dijl JM (1998) SecDF of *Bacillus subtilis*, a molecular Siamese twin required for the efficient secretion of proteins. J. Biol. Chem. 273(33): 21217–21224

Bolotin A, Sorokin A & Ehrlich SD (1996) Mapping of the 150 kb *spoIIIC-pheA* region of the *Bacillus subtilis* chromosome using Long Accurate PCR and three yeast artificial chromosomes. *Microbiology* 142: 3017–3020

Borodovsky M & McIninch J (1993) GENMARK: a parallel gene recognition for both DNA strands. Comput. Chem. 17: 123–133

Burland V, Plunkett G, Daniels DL & Blattner F R (1993) DNA sequence and analysis of 136 kilobases of the *Escherichia coli* genome: organizational symmetry around the origin of replication. Genomics 16(3): 551–561

Campbell EA, Choi SY & Masure HR (1998) A competence regulon in *Streptococcus pneumoniae* revealed by genomic analysis. Mol. Microbiol. 27(5): 929–939

Chedin F, Noirot P, Biaudet V & Ehrlich SD (1998) A five-nucleotide sequence protects DNA from exonucleolytic degradation by AddAB, the RecBCD analogue of *Bacillus subtilis*. Mol. Microbiol. 29(6): 1369–1377

Cheng S, Fockler C, Barnes WM & Higuchi R (1994) Effective amplification of long targets from cloned inserts and human genomic DNA. Proc. Natl. Acad. Sci. U.S.A. 91(12): 5695–5699

Chiaruttini C & Milet M (1993) Gene organization, primary structure and RNA processing analysis of a ribosomal RNA operon in *Lactococcus lactis*. J. Mol. Biol. 230(1): 57–76

Chopin A, Chopin MC, Moillo-Batt A & Langella P (1984) Two plasmid-determined restriction and modification systems in *Streptococcus lactis*. Plasmid 11(3): 260–263

Chopin MC, Chopin A, Rouault A & Galleron N (1989) Insertion and amplification of foreign genes in the *Lactococcus lactis* subsp. *lactis* chromosome. Appl. Environ. Microbiol. 55(7): 1769–1774

Cocaign-Bousquet M, Garrigues C, Novak L, Lindley ND & Loubiere P (1995) Rational development of a simple synthetic medium for the sustained growth of *Lactococcus lactis*. J. Appl. Bacteriol. 79: 108–116

Cocaign-Bousquet M, Garrigues C, Loubiere P & Lindley ND (1996) Physiology of pyruvate metabolism in *Lactococcus lactis*. Antonie Van Leeuwenhoek 70(2–4): 253–267

Cole ST, Brosch R, Parkhill J, Garnier T, Churcher C, Harris D, Gordon SV, Eiglmeier K, Gas S & Barry C.E. et al (1998) Deciphering the biology of Mycobacterium tuberculosis from the complete genome sequence. *Nature* 393(6685): 537–544

Delorme C, Ehrlich SD & Renault P (1992) Histidine biosynthesis genes in *Lactococcus lactis* subsp. *lactis*. J. Bacteriol. 174(20): 6571–6579

Delorme C, Godon J-J, Ehrlich SD & Renault P (1993) Gene inactivation in *Lactococcus lactis*: Histidine biosynthesis. J. Bacteriol. 175(14): 4391–4399

De Vos WM & Vaughan EE (1994) Genetics of lactose utilisation in lactic acid bacteria. *FEMS Microbiol. Rev.* 15: 217–237

De Vos WM (1996) Metabolic engineering of sugar catabolism in lactic acid bacteria. *Antonie Van Leeuwenhoek* 70(2–4): 223–242

Dickely F, Nilsson D, Hansen EB & Johansen E (1995) Isolation of *Lactococcus lactis* nonsense suppressors and construction of a food-grade cloning vector. Mol. Microbiol. 15(5): 839–847

Dubnau D (1997) Binding and transport of transforming DNA by *Bacillus subtilis*: the role of type-IV pilin-like proteins – a review. Gene 192(1): 191–198

Duong F & Wickner W (1997) The SecDFyajC domain of preprotein translocase controls preprotein movement by regulating SecA membrane cycling. EMBO J. 16(16): 4871–4879

El Karoui M, Ehrlich D & Gruss A (1998) Identification of the lactococcal exonuclease/recombinase and its modulation by the putative Chi sequence. Proc. Natl. Acad. Sci. USA 95(2): 626–631

Engelke G, Gutowski-Eckel Z, Hammelmann M & Entian KD (1992) Biosynthesis of the lantibiotic nisin: genomic organization and membrane localization of the NisB protein. Appl. Environ. Microbiol. 58(11): 3730–3743

Errington J (1986) A general method for fusion of the *Escherichia coli lacZ* gene to chromosomal genes in *Bacillus subtilis*. J. Gen. Microbiol. 132: 2953–2966

Fleischmann RD, Adams MD, White O, Clayton RA, Kirkness EF, Kerlavage AR, Bult CJ, Tomb JF, Dougherty BA, Merrick JM et al. (1995) Whole-genome random sequencing and assembly of *Haemophilus influenzae* Rd. Science 269(5223): 496–512

Gasson MJ (1983) Plasmid complements of *Streptococcus lactis* NCDO 712 and other lactic streptococci after protoplast-induced curing. J. Bacteriol. 154(1): 1–9

Godon JJ, Chopin MC & Ehrlich SD (1992) Branched-chain amino acid biosynthesis genes in *Lactococcus lactis* subsp. *lactis*. J. Bacteriol. 174(20): 6580–6589

Godon JJ, Delorme C, Bardowski J, Chopin MC, Ehrlich SD & Renault P (1993) Gene inactivation in *Lactococcus lactis*:

Branched-chain amino acid biosynthesis. J. Bacteriol. 175(14): 4383–4390

Griffin HG & Gasson MJ (1995) Genetic aspects of aromatic amino acid biosynthesis in *Lactococcus lactis*. Mol. Gen. Genet. 246(1): 119–127

Grossiord B, Vaughan EE, Luesink E & de Vos WM (1998) Genetics of galactose utilisation via the Leloir pathway in lactic acid bacteria. Le Lait 78: 77–84

Hahn J, Luttinger A & Dubnau D (1996) Regulatory inputs for the synthesis of ComK, the competence transcription factor of *Bacillus subtilis*. Mol. Microbiol. 21(4): 763–775

Havarstein LS, Coomaraswamy G & Morrison DA (1995) An unmodified heptadecapeptide pheromone induces competence for genetic transformation in *Streptococcus pneumoniae*. Proc. Natl. Acad. Sci. USA 92(24): 11140–11144

Havarstein LS, Gaustad P, Nes IF & Morrison DA (1996) Identification of the streptococcal competence-pheromone receptor. Mol. Microbiol. 21(4): 863–869

Havarstein LS, Hakenbeck R & Ganstad P (1997) Natural competence in the genus *Streptococcus*: Evidence that Streptococci can change phenotype by interspecies recombinational exchanges. J. Bacteriol. 179: 6589–6594

Hemila H, Pakkanen R, Heikinheimo R, Palva ET & Palva I (1992) Expression of the *Erwinia carotovora* polygalacturonase-encoding gene in *Bacillus subtilis*: role of signal peptide fusions on production of a heterologous protein. Gene 116(1): 27–33

Huang DC, Novel M & Novel G (1991) A transposon-like element on the lactose plasmid of *Lactococcus lactis* subsp. *lactis* Z270. FEMS Microbiol. Lett. 61(1): 101–106

Hutkins RW, Morris HA & McKay LL (1985) Galactokinase activity in *Streptococcus thermophilus*. Appl. Env. Microbiol. 50: 777–780

Jensen PR & Hammer K (1993) Minimal requirements for exponential growth of *Lactococcus lactis*. Appl. Env. Microbiol. 59: 4363–4366

Kilstrup M & Martinussen J (1998) A transcriptional activator, homologous to the *Bacillus subtilis* PurR repressor, is required for expression of purine biosynthetic genes in *Lactococcus lactis*. J. Bacteriol. 180(15): 3907–3916

Kunst F, Ogasawara N, Moszer I, Albertini AM, Alloni G et al. (1997) The complete genome sequence of the Gram-positive bacterium *Bacillus subtilis*. Nature 390: 249–256

Lander ES & Waterman MS (1988) Genomic mapping by fingerprinting random clones: a mathematical analysis. Genomics 2(3): 231–239

Lapidus A, Galleron N, Sorokin A & Ehrlich SD (1997) Sequencing and functional annotation of the *Bacillus subtilis* genes in the 200 kb *rrnB-dnaB* region. Microbiology 143: 3431–3441

Lapujade P, Cocaign-Bousquet M & Loubiere P (1998) Glutamate biosynthesis in *Lactococcus lactis* subsp lactis NCDO 2118. Appl. Environ. Microbiol 64: 2485–2489

Le Bourgeois P, Lautier M, Mata M & Ritzenthaler P (1992) Physical and genetic map of the chromosome of *Lactococcus lactis subsp. lactis* IL1403. J. Bacteriol. 174(21): 6752–6762

Le Bourgeois P, Lautier M, van den Berghe L, Gasson MJ & Ritzenthaler P (1995) Physical and genetic map of the *Lactococcus lactis* subsp. *cremoris* MG1363 chromosome: comparison with that of *Lactococcus lactis* subsp. *lactis* IL 1403 reveals a large genome inversion. J. Bacteriol. 177(10): 2840–2850

Le Loir Y, Gruss A, Ehrlich SD & Langella P (1998) A nine-residue synthetic propeptide enhances secretion efficiency of heterologous proteins in *Lactococcus lactis*. J. Bacteriol. 180(7): 1895–1903

Lobry JR (1996a) Origin of replication of *Mycoplasma genitalium*. Nature 272: 745–746

Lobry JR (1996b) Asymmetric substitution patterns in the two DNA strands of bacteria. Mol. Biol. Evol. 13: 660–665

Lowe TM & Eddy SR (1997) tRNAscan-SE: a program for improved detection of transfer RNA genes in genomic sequence. Nucleic Acids Res. 25(5): 955–964

Luesink EJ, van Herpen REMA, Grossiord BP, Kuipers OP & de Vos WM (1998) Transcriptional activation of the glycolytic *las* operon and catabolite repression of the *gal* operon in *Lactococcus lactis* are mediated by the catabolite control protein CcpA. Mol. Microbiol. 30(4): 789–798

Madsen SM, Albrechtsen B, Hansen EB & Israelsen H (1996) Cloning and transcriptional analysis of two threonine biosynthetic genes from *Lactococcus lactis* MG1614. J. Bacteriol. 178(13): 3689–3694

Maeda S & Gasson MJ (1986) Cloning, expression and location of the *Streptococcus lactis* gene for phospho-beta-D-galactosidase. J. Gen. Microbiol. 132 (2): 331–340

Martinussen J, Andersen PS & Hammer K (1994) Nucleotide metabolism in *Lactococcus lactis*: salvage pathways of exogenous pyrimidines. J. Bacteriol. 176(5): 1514–1516

Moll I, Resch A & Blasi U (1998) Discrimination of 5′-terminal start codons by translation initiation factor 3 is mediated by ribosomal protein S1. FEBS Lett. 436(2): 213–217

Nilsson D & Johansen E (1994) A conserved sequence in tRNA and rRNA promoters of *Lactococcus lactis*. *Biochim Biophys Acta* 1219(1): 141–144

Nilsson D & Kilstrup M (1998) Cloning and expression of the *Lactococcus lactis purDEK* genes, required for growth in milk. Appl. Environ. Microbiol. 64: 4321–4327

Novak L, Cocaign-Bousquet M, Lindley ND & Loubiere P (1997) Metabolism and energetics of *Lactococcus lactis* during growth in complex or synthetic media. Appl. Environ. Microbiol. 63: 2665–2670

Novak L, Cocaign-Bousquet M, Lindley ND & Loubiere P (1998) Cométabolisme sucre-acides aminés chez *Lactococcus lactis* (in French). Le Lait 78: 17–22

O'Connell-Motherway M, Fitzgerald GF & van Sinderen D (1997) Cloning and sequence analysis of putative histidine protein kinases isolated from *Lactococcus lactis* MG1363. Appl. Environ. Microbiol. 63(6): 2454–2459

Pearson WR, Wood T, Zhang Z & Miller W (1997) Comparison of DNA sequences with protein sequences. Genomics 46(1): 24–36

Pogliano JA & Beckwith J (1994) SecD and SecF facilitate protein export in *Escherichia coli*. EMBO J. 13(3): 554–561

Polzin KM, Romero D, Shimizu-Kadota M, Klaenhammer TR & McKay LL (1993) Copy number and location of insertion sequences ISS1 and IS981 in lactococci and several other lactic acid bacteria. J. Dairy Sci. 76(5): 1243–1252

Poolman B (1993) Energy transduction in lactic acid bacteria. FEMS Microbiol. Rev. 12(1-3): 125–147

Pouwels PH, Leer RJ, Shaw M, Heijne den Bak-Glashouwer MJ, Tielen FD, Smit E, Martinez B, Jore J & Conway PL (1998) Lactic acid bacteria as antigen delivery vehicles for oral immunization purposes. Int. J. Food. Microbiol. 41(2): 155–167

Raya R, Bardowski J, Andersen PS, Ehrlich SD & Chopin A (1998) Multiple transcriptional control of the *Lactococcus lactis trp* operon. J. Bacteriol. 180(12): 3174–3180

Resch A, Tedin K, Grundling A, Mundlein A & Blasi U (1996) Downstream box-anti-downstream box interactions are dispensable for translation initiation of leaderless mRNAs. EMBO J. 15(17): 4740–4748

Riley M (1993) Functions of the gene products of *Escherichia coli*. Microbiol. Rev. 57(4): 862–952

Schouler C, Gautier M, Ehrlich SD & Chopin MC (1998) Combinational variation of restriction modification specificities in *Lactococcus lactis*. Mol. Microbiol. 28(1): 169–178

Simon D, Rouault A & Chopin MC (1985) Protoplast transformation of group N streptococci with cryptic plasmids. FEMS Microbiol. Lett. 26: 239–241

Sorokin A, Lapidus A, Capuano V, Galleron N, Pujic P & Ehrlich SD (1996) A new approach using multiplex long accurate PCR and yeast artificial chromosomes for bacterial chromosome mapping and sequencing. Genome Res. 6(5): 448–453

Stingele F, Neeser JR & Mollet B (1996) Identification and characterization of the *eps* (Exopolysaccharide) gene cluster from *Streptococcus thermophilus* Sfi6. J. Bacteriol. 178(6): 1680–1690

Tedin K, Rosch A & Blösi U (1997) Requirements for ribosomal protein S1 for translation initiation of mRNAs with and without a 5' leader sequence. Mol. Microbiol. 25(1): 189–199

Thomas TD & Craw VF (1984) Selection of galactose-fermenting *Streptococcus thermophilus* in lactose-limited chemostat cultures. Appl. Env. Microbiol. 48: 186–191

Tracy RB, Chedin F & Kowalczykowski SC (1997) The recombination hot spot chi is embedded within islands of preferred DNA pairing sequences in the *E. coli* genome. Cell 90(2): 205–206

Tulloch DL, Finch LR, Hillier AJ & Davidson BE (1991) Physical map of the chromosome of *Lactococcus lactis* subsp. *lactis* DL11 and localization of six putative rRNA operons. J. Bacteriol. 173(9): 2768–7275

Van de Guchte M, Ehrlich DS & Chopin A (1998) tRNATrp as a key element of antitermination in the *Lactococcus lactis trp* operon. Mol. Microbiol. 29(1): 61–74

Van Etten WJ & Janssen GR (1998) An AUG initiation codon, not codon-anticodon complementarity, is required for the translation of unleadered mRNA in *Escherichia coli*. Mol. Microbiol. 27(5): 987–1001

Van Kranenburg R, Marugg JD, van Swam II, Willem NJ & de Vos WM (1997) Molecular characterization of the plasmid-encoded eps gene cluster essential for exopolysaccharide biosynthesis in *Lactococcus lactis*. Mol. Microbiol. 24(2): 387–397

Van Sinderen D, Karsens H, Kok J, Terpstra P, Ruiters MH, Venema G & Nauta A (1996) Sequence analysis and molecular characterization of the temperate lactococcal bacteriophage r1t. Mol. Microbiol. 19(6): 1343–1355

Van Sinderen D, Luttinger A, Kong L, Dubnau D, Venema G & Hamoen L (1995) *comK* encodes the competence transcription factor, the key regulatory protein for competence development in *Bacillus subtilis*. Mol. Microbiol. 15(3): 455–462

Vaughan EE & de Vos WM (1995) Identification and characterization of the insertion element IS1070 from *Leuconostoc lactis* NZ6009. Gene 155 (1): 95–100

Vaughan EE, Pridmore RD & Mollet B (1998) Transcriptional regulation and evolution of lactose genes in the galactose-lactose operon of *Lactococcus lactis* NCDO 2054. J. Bacteriol. 180(18): 4893–4902

Vellanoweth RL & Rabinowitz JC (1992) The influence of ribosome-binding-site elements on translational efficiency in *Bacillus subtilis* and *Escherichia coli in vivo*. Mol. Microbiol. 6(9): 1105–1114

Venter JC, Smith HO & Hood L (1996) A new strategy for genome sequencing. Nature 381: 364–366

Wells JM, Wilson PW, Norton PM & Le Page RW (1993) A model system for the investigation of heterologous protein secretion pathways in *Lactococcus lactis*. Appl. Environ. Microbiol. 59(11): 3954–3959

Wouters JA, Sanders JW, Kok J, de Vos WM, Kuipers OP & Abee T (1998) Clustered organization and transcriptional analysis of a family of five *csp* genes of *Lactococcus lactis* MG1363. Microbiology 144 (10): 2885–2893

Antonie van Leeuwenhoek **76:** 77–88, 1999.

Group II introns and expression of conjugative transfer functions in lactic acid bacteria

Gary M. Dunny[1,*] & Larry L. McKay[2]
[1]*Department of Microbiology, University of Minnesota Medical School, Minneapolis, MN, USA and*
[2]*Department of Food Science and Nutrition, University of Minnesota, St. Paul, MN, USA. (*Author for correspondence; E-mail: gary-d@biosci.cbs.umn.edu)*

Key words: conjugation, relaxase, RNA splicing, retroelement

Abstract

The homologous lactococcal conjugative elements pRS01 and the sex factor of *Lactococcus lactis* strain 712 both contain a Group II intron within a gene believed to encode a conjugative relaxase enzyme. This enzyme is responsible for nicking of DNA at the origin of transfer (*oriT*) sequence of the sex factor DNA to initiate the strand transfer process. Group II introns have been studied in eukaryotes, and several of these elements in yeast mitochondrial genes have received considerable attention. These introns are relatively large in size and generally encode a protein within the intron sequence. In addition to splicing activity, Group II introns are mobile genetic elements. The intron-encoded proteins (IEPs) contain endonuclease and reverse transcriptase domains believed to play an enzymatic role in genetic mobility reactions, while a putative maturase domain is thought to promote splicing by stabilizing the folding of the intron RNA into an active ribozyme structure which carries out the splicing reaction. The lactococcal introns represent the first examples of Group II introns shown to be functional *in vivo* in prokaryotes. Because of the advantages of a bacterial system for genetic and molecular studies, the Ll.ltrB intron from pRS01 has attracted the attention of several laboratories interested in Group II intron biology. Recently, it has been shown that the system can be adapted to function in *Escherichia coli* (although at somewhat reduced efficiency). In addition, it has been recently proven that the best studied form of mobility, the homing of the intron into an intronless allele of the cognate exon gene, occurs via an RNA intermediate and does not require DNA homology or generalized host recombination functions. Current efforts are analysis of the role Ll.ltrB splicing in regulating expression of pRS01 conjugation functions. The lactococcal Group II introns represent the first demonstrated genetically mobile prokaryotic retroelements, and they also have considerable potential as genetic engineering tools for Lactic Acid Bacteria (LAB) and other organisms.

Introduction

One of the most rewarding benefits of carrying out basic research is that investigators can sometimes obtain completely unexpected results which dramatically change the ways in which they think about their research topic, and which throw open completely new directions for future investigation. This type of unpredictable discovery was recently made during analysis of conjugative transfer in LAB. In the course of analyzing a lactococcal genetic transfer system in order to develop improved mobilizable cloning vectors, an unusual bacterial intron was discovered within an essential transfer gene. Not only did this finding reveal a novel genetic element in prokaryotes, but it also changed our perceptions of evolutionary mechanisms in these bacteria. In addition, it has allowed us to envision new approaches to functional genomic analysis and other types of genetic manipulation that could not have been foreseen a few years ago.

Results obtained from two diverse research areas form the basis for the work summarized in this review. The first of these is the analysis of Group II introns (until recently, carried out almost exclusively in eukaryotes), and the second area is the study of conjugative plasmid transfer in lactic acid bacteria. Our work in the latter area led to the discovery of the Ll.ltrB Group II intron within an essential gene (*ltrB*) required for conjugative transfer of the lactococcal sex factor pRS01. Some of the relevant published work in each of these two areas is summarized below, followed by an overview of recent experiments involving the use of conjugation-based genetic assays to study Group II intron splicing in lactococci. The review will conclude with some comments on the significance of Group II introns in LAB from basic and applied points of view.

Lactococcal conjugation and the pRS01 system

The ability to transfer genetic material via conjugation is widespread among lactococci. Detailed analysis of conjugation initially focused on the conjugative plasmid pRS01 from *L. lactis* ML3 and the sex factor of *L. lactis* 712 (Anderson & McKay 1984; Gasson et al. 1995). More recently, a number of additional conjugative elements, such as a sucrose-nisin conjugative transposon, have been discovered in these organisms (Rauch & DeVos 1992).

Conjugal transfer of lactose utilization (Lac$^+$) from ML3 was first described by McKay et al. (1980). Walsh & McKay (1981) then showed that Lac$^+$ transconjugants resulting from matings between *L. lactis* ML3 and a plasmid-free recipient possessed a single plasmid of 104-kb which was nearly twice the size of the 55-kb Lac plasmid in ML3. Subsequently, Anderson & McKay (1984) demonstrated that this transfer involved the co-integration of a conjugative plasmid, pRS01, with the non-conjugative Lac plasmid, pSK08, resulting in the 104-kb hybrid plasmid in the recipient cell. The co-integration event interrupted pRS01 randomly and was later shown to be directed by one of two ISS1 sequences present within pSK08 (Polzin & Shimizu-Kadota 1987). By analyzing the junction sites of pSK08::pRS01 co-integrate plasmids in various transfer-deficient (Tra$^-$) transconjugants, a crude map of the pRS01 regions responsible for the Tra phenotype was generated (Anderson & McKay 1984). The region of pRS01 responsible for Tra was localized to a 15-kb segment. The large size, variable copy number, and relative instability of the pSK08::pRS01 cointegrate plasmids seriously hampered further efforts to analyze the Tra regions of pRS01.

Romero & Klaenhammer (1991) then demonstrated the ability of the 12 kb plasmid pTRK28 to integrate randomly within pRS01. Like that of pSK08, cointegration of pTRK28 with pRS01 was shown to be a Rec-independent recombination event with a strict requirement for the presence of an IS element. The pTRK28::pRS01 cointegrates were stable and were maintained at a higher copy number. We then used pTRK28 for insertional mutagenesis to further localize regions of pRS01 involved in conjugal transfer (Mills et al. 1994). Mapping of the site of pTRK28 integration within 34 pTRK28::pRS01 cointegrates exhibiting altered Tra phenotypes identified four regions of pRS01 involved in cointegrate plasmid transfer (Tra1, Tra2, Tra3, and Tra4). Two of these regions (Tra3 and Tra4) correspond to those previously identified (Anderson & McKay 1984). Tra1 and Tra2, two previously unidentified regions, were located about 9-kb from Tra3. A map of a transfer-proficient pRS01::pTRK28 cointegrate plasmid, pM2036, indicating the positions of the transfer regions is shown below in Figure 1. Subsequent analysis of the pRS01 transfer genes focused on the identification of the origin of transfer (*oriT*) region (Clewell 1993), in order to develop vectors that could be used to introduce genetic material into many gram-positive organisms of industrial and medical importance (Fitzgerald & Gasson 1988; Gasson et al. 1995). This resulted in identification of a 446 bp region containing the functional *oriT* (Figure 2) (Mills et al. 1998). Results of the Tra 1-2 analysis indicate that this region of the plasmid probably encodes both the target sites and the gene products for the DNA processing functions of conjugative transfer, whereas the Tra3-4 region seems to encode mating pair formation functions.

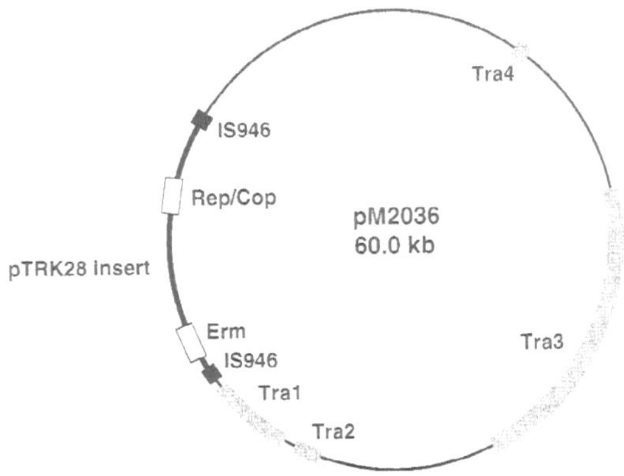

Figure 1. Map of pM2036 indicating the positions of the transfer regions of pRS01 (Tra 1-4), and the site of pTRK28 insertion. Based on the results of Mills et al. (1994).

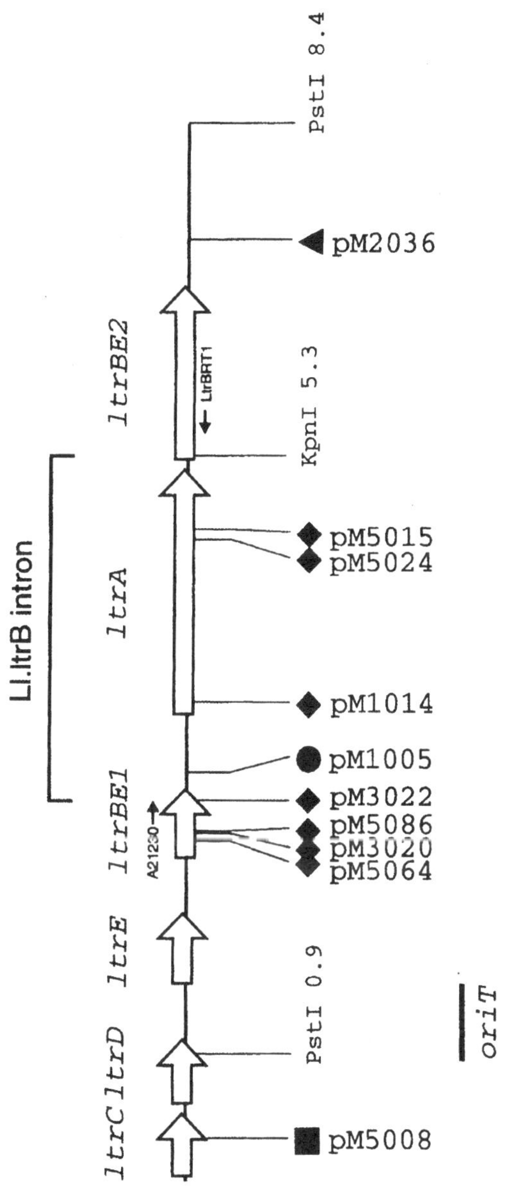

Figure 2. A. Map of the Tra 1-2 region of pRS01, showing the location of Ll.ltrB, the positions of pTRK28 insertions affecting conjugative transfer (Tra$^+$, squares and triangles; Tra-low, circle; Tra$^-$, diamonds), and the *oriT* region. Based on data from Mills et al. (1994, 1996, 1998).

Identification and analysis of the Ll.ltrB intron in *L. lactis*

Sequence analysis of the region encompassing Tra1 and Tra2 revealed six open reading frames and indicated the presence of a bacterial Group II intron, termed Ll.ltrB, within the gene (*ltrB*) encoding a putative conjugative relaxase (Mills et al. 1996). The likely function of this enzyme is to nick the DNA at the origin of transfer (*oriT*) sequence to initiate single strand transfer from donor to recipient (Clewell 1993). A map of the Tra 1-2 regions of pRS01, including the intron, is shown in Figure 2.

Subsequent genetic and molecular analysis demonstrated highly efficient splicing of the intron from the mRNA *in vivo*, and sequencing of the Tra1 region cDNA amplified from cellular RNA showed that splicing joined the two exon sequences (*ltrBE1* and *ltrBE2*) in frame. In addition pTRK28 insertions within the *ltrA* (Intron-encoded protein, IEP) gene completely abolished detectable splicing and were transfer deficient, whereas another insertion near the 5′ end of Ll.ltrB (pM1005, see Figure 2) retained some residual transfer ability. Transfer of plasmids containing insertions in *ltrA* was restored when an intact copy of the Tra 1 region was supplied *in trans*; this result implied that intron splicing (to generate a functional *ltrB*-encoded relaxase) was required for conjugation.

Prior to the identification of Ll.ltrB, no functional analyses of Group II introns in bacteria had been carried out. By searching for homology to intron RNA and IEP sequences, putative Group II introns were identified in cyanobacteria (Ferat & Michel 1993) and in proteobacteria (Ferat et al. 1994; Knoop & Brennicke 1994), but no demonstration of functionality of these elements *in vivo* has been forthcoming. At nearly the same time as the existence of Ll.ltrB in pRS01 was reported, Shearman et al. (1996) published results of a study of the sex factor of *L. lactis* strain 712 indicating the presence of an intron/exon structure in the corresponding location, relative to pRS01. This finding is not surprising in view of the extensive homology between these two elements. Our search for similar elements in other lactococcal strains indicated the presence of the intron only in strains carrying pRS01-like sex factors (Mills et al. 1996). Subsequently, an additional Group II intron was reported in *Clostridium difficile* by Mullany et al. (1996). In this case, the intron, which is highly related to Ll.ltrB, is located in a different gene, one involved in mating pair formation within the conjugative transposon Tn*5397*. Recently, a report of a

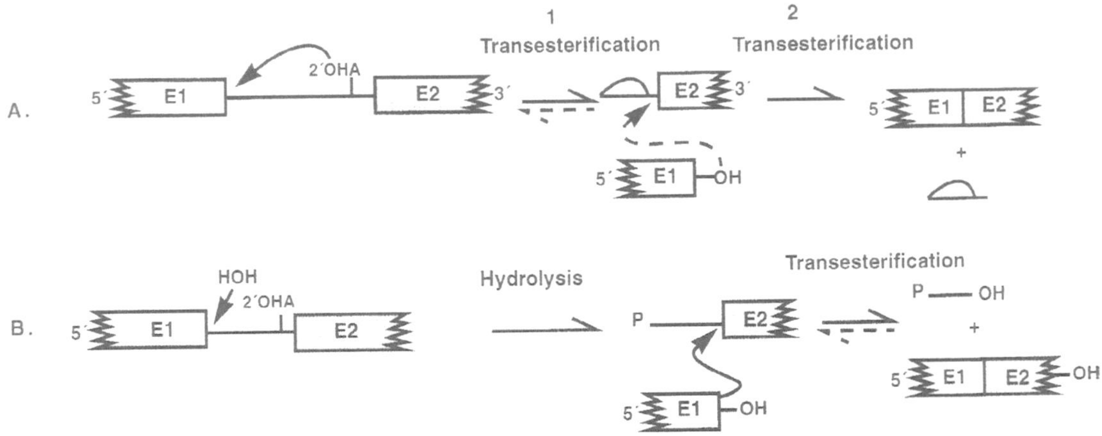

Figure 3. Simplified diagrams of Group II intron splicing pathways. A. Splicing via the branching pathway, releasing an intron lariat, and believed to be the primary pathway. B. The alternative pathway, where splicing is initiated by hydrolysis, and a linear intron RNA is released. Adapted from Podar et al. (1998a).

Group II intron in *Pseudomonas* appeared (Yeo et al. 1997).

Ribozyme activity and RNA structural features of Group II introns

The Group II introns are a family of large self-splicing RNAs, which often function as mobile genetic elements (Lambowitz & Belfort 1993; Michel & Ferat 1995). Over 100 different Group II introns have been identified in organisms ranging from plants to bacteria. They are very commonly found in the organellar genes of lower eukaryotes and have been studied extensively in these organisms. Group II introns generally can be shown to self-splice *in vitro*, although efficient *in vivo* splicing seems to involve participation of proteins (Lambowitz & Belfort 1993; Michel & Ferat 1995; Qin & Pyle 1998). The primary splicing pathway (shown in Figure 3A) involves a nucleophilic attack on the 5′ splice junction site by an adenosine residue within the intron (the 'bulging A' residue; see below). This initial step results in formation of a 2′-5′ phosphodiester branch in the intron. A mechanistically distinct second step (Podar et al. 1998b) joins the two exons via a 3′-5′ phosphodiester bond and releases the excised intron as a 'lariat'. An alternative pathway is possible where the first step occurs via hydrolysis rather than transesterification and excises a linear intron RNA (Figure 3B). Interestingly, it has been shown recently that yeast intron mutants completely deficient in the branching pathway can splice *in vivo* via hydrolysis at a level sufficient for normal phenotypic expression of the exon gene (Podar et al. 1998a). Because the linear intron product released in the hydrolysis pathway is extremely labile relative to intron lariats, it is difficult to assess the extent to which the hydrolysis pathway occurs in fully functional Group II introns. Group II intron splicing and mobility mechanisms are very different from those employed by Group I introns, which include most of the bacterial introns studied previously, as well as numerous eukaryotic introns (Belfort et al. 1995). The considerable similarity between splicing mechanisms of Group II introns and spliceosomal introns of eukaryotic nuclear genes has led to suggestions that the latter class of introns arose form the former (Lambowitz & Belfort 1993; Michel & Ferat 1995).

While there is considerable nucleotide sequence variation over the length of various Group II introns, they all seem to be capable of folding into a similar set of subdomains that show a remarkable similarity. Figure 4 shows the sequence of the lactococcal Ll.ltrB intron folded into this structure. In addition to the phylogenetic conservation of the illustrated domain structure, numerous genetic studies showing predicted effects of mutations altering these domains, and biochemical studies showing that some of the domains can function as independent RNAs provide strong support for the general validity of the model shown in Figure 4. Some of the main features of intron structure will be summarized here, with the reader referred to a recent review by Qin & Pyle (1998), and previous reviews (Lambowitz & Belfort 1993; Belfort

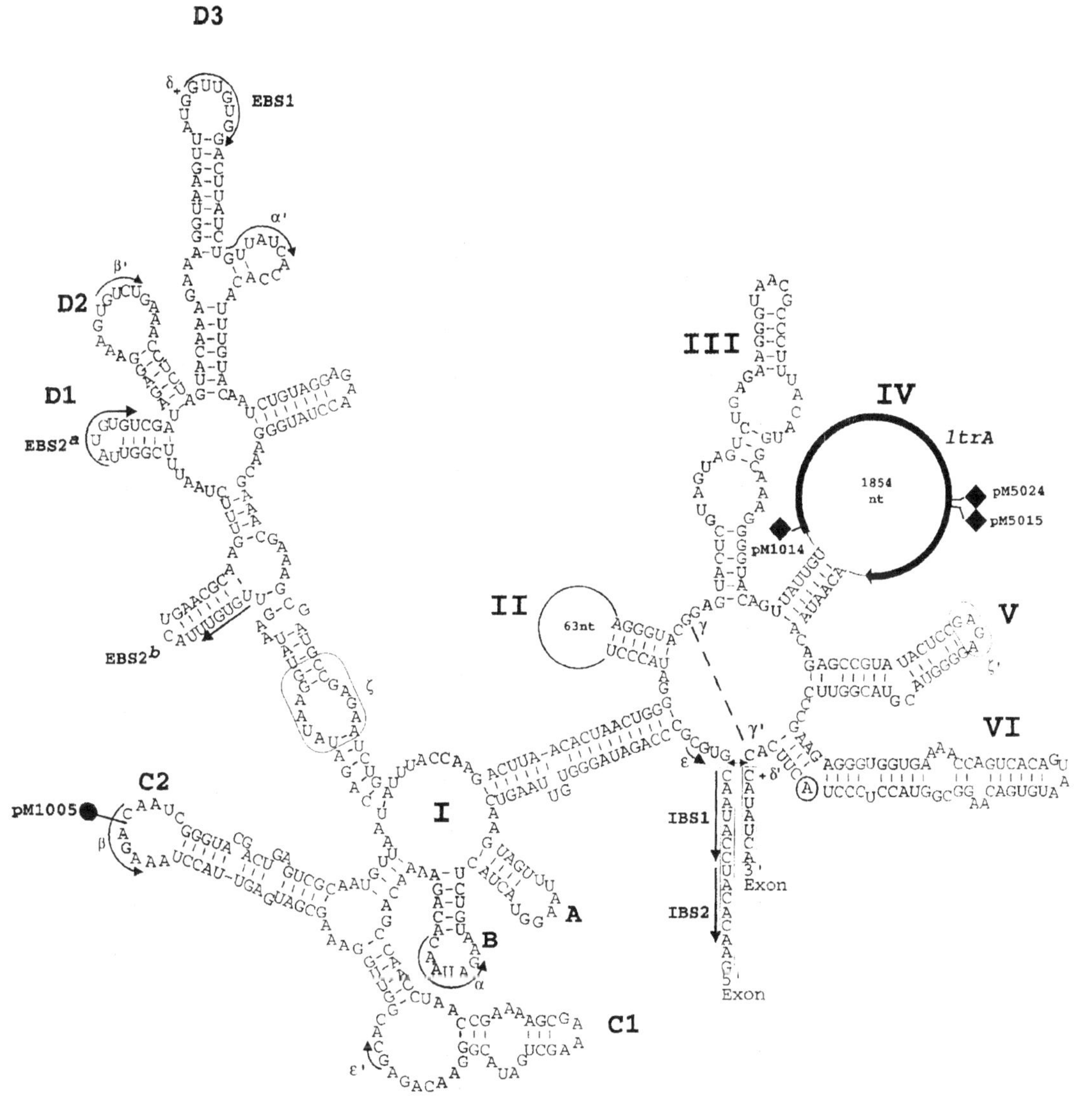

Figure 4. Predicted folded structure of Ll.ltrB RNA indicating structural domains based on comparison with other Group II introns. Major domains are indicated by Roman numerals, and complementary sequences believed to be paired are indicated by Greek letters. The 'bulging A' residue involved in splicing is circled and the position of the *ltrA* gene, encoding the IEP is indicated (heavy line), as are the locations of pTRK28 insertions affecting conjugation. Adapted from Mills et al. (1996).

et al. 1995; Michel & Ferat 1995) for more detailed information. Group II introns generally contain 6 domains, with domains 1, 5, and 6 playing crucial roles in the splicing reaction, domains 2 and 3 probably involved in stabilization of proper folded structures, and domain 4 being dispensable in terms of folding or catalytic activity. The latter domain typically contains the intron-encoded protein (IEP) gene, whose functions are discussed in the next section. Domain 4 is probably looped out from the functional catalytic RNA molecule.

Domain 1 is the largest and most complex. The EBS1 and EBS2 sequences interact with specific complementary exon sequences (IBS1 and IBS2). This binding is important for proper interaction with the 5′ exon and also in intron mobility reactions (see following paragraphs). Several short sequences (designated by Greek letters in Figure 4) interact specifically

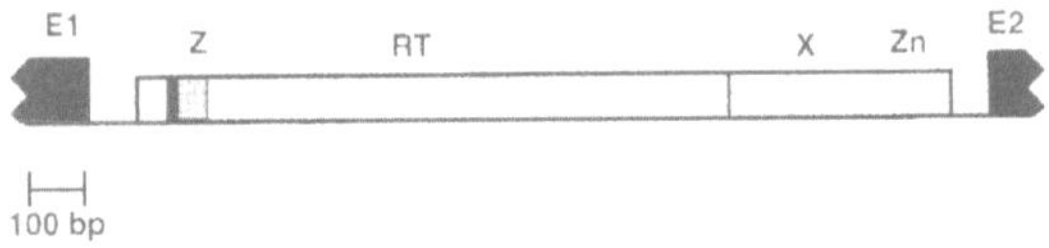

Figure 5. Conserved domains in Group II IEPs. See Michel & Ferat (1995) and Lambowitz & Belfort (1993) for additional information.

within the intron RNA to stabilize important structures within this domain, as well as effecting important interactions between domains 1 and 5. Domain 5 is a small hairpin with a small bulge in the stem and a short loop containing sequences essential for interaction with domain 1. There is considerable evidence that domain 5 is an essential catalytic component of the active ribozyme, including the finding that this domain can initiate a splicing reaction when added *in trans* to molecules containing a 5′ exon joined to a deleted intron comprised of domains 1-3 (Qin & Pyle 1998). Domain 6, containing the 'bulging A' residue, is near the 3′ intron/exon junction and presents the branch point to the 5′ exon in the first step in the reaction.

Intron-encoded proteins and their functions

Group II introns known to be functional *in vivo* often encode a protein (IEP) within domain 4. IEPs have multiple domains (Figure 5), identified by both sequence homology and biochemical analysis, including reverse transcriptase, endonuclease, and Zn finger domains (Lambowitz & Belfort 1993; Michel & Ferat 1995). The IEPs are known to play an essential catalytic role in mobility (Belfort & Perlman 1995; Zimmerly et al. 1995), and are also required for efficient splicing *in vivo*. The IEP 'maturase' function is felt to involve a type of RNA chaperone activity where the intron RNA is stabilized in a conformation such that the ribozyme efficiently catalyzes the splicing reaction (Moran et al. 1994; Qin & Pyle, 1998). A domain termed 'X' is highly conserved in many Group II IEPs, and may be important in maturase activity. In eukaryotic Group II introns, the IEP coding sequence is often in-frame with the 5′ exon. It is believed that the protein is translated from the unspliced 'pre-mRNA' and proteolytically processed to the mature protein (Anziano et al. 1982; Carignani et al. 1983). In this scenario, the protein would essentially autoregulate its synthesis by facilitating the processing of its own functional mRNA as the mature protein accumulated in the cell.

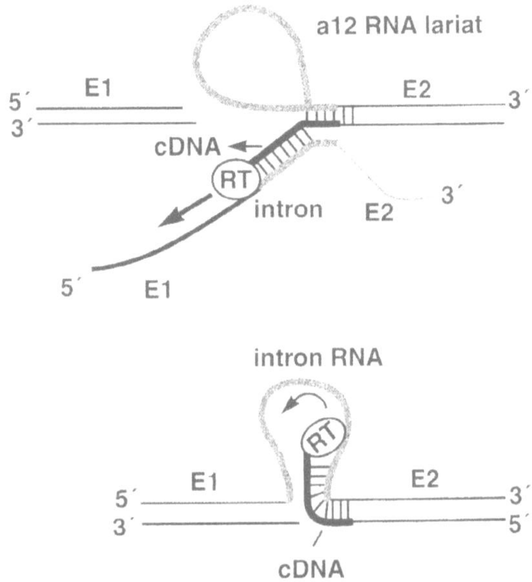

Figure 6. Proposed mechanisms for Intron homing via an RNA intermediate. A. In partial reverse splicing an unspliced pre-mRNA serves as template for reverse transcriptase, and DNA recombination with co-conversion of 5′ flanking exon sequences occurs. B. In complete reverse splicing, the intron lariat ligated to the splice junction is the only template for reverse transcription. DNA recombination is not required and no co-conversion occurs. The thin line represents RNA, the very thick line represents newly synthesized cDNA, and the intermediate line represents the target DNA. Based on models of Zimmerly et al. (1995).

Group II introns as mobile genetic elements

A remarkable property of Group II introns is their genetic mobility. The most efficient form of mobility is termed homing. In this situation, two or more alleles of the same exon exist in a cell. If one of these alleles lacks the intron, a copy of the intron sequence is rapidly acquired. It is also possible, at much lower frequency, for the intron to transpose to ectopic sites. Any mechanistic differences that may exist between these two forms of mobility have not been elucidated. Group II intron homing has been studied extensively in yeast mitochondria, as reviewed by Lambowitz & Belfort (1993) and Belfort & Perlman (1995). An elegant series of biochemical studies of the yeast mitochondrial Group II intron aI2 have been carried out. The results suggest that mobility involves a novel reaction where two nuclease activities, one conferred by the intron RNA and a second by the IEP, generate a staggered cleavage in the target DNA (Zimmerly et al. 1995; Yang et al. 1996). The available biochemical evidence suggest that both the RNA and protein components of the endonuclease activities are located in ribonucleoprotein (RNP) particles formed by spontaneous association of the intron RNA and IEP. DNA target site cleavage is presumably followed by a reverse splicing reaction in which the intron RNA is copied into a DNA, which is, turn, integrated into the DNA target site. Possible pathways for homing involving RNA intermediates ('retrohoming') have been proposed, including those shown in Figure 6. In partial reverse splicing, a pre-mRNA serves as template for a portion of the reverse transcription reaction (in Figure 6A, the initial RT template is an intron lariat joined to the opposite strand of the target DNA). Following reverse transcription, the 3′ end of the newly synthesized DNA is integrated into the target by DNA recombination. This may result in co-conversion of the 5′ exon flanking sequence in the target to that of the template molecule. In complete reverse splicing (Figure 6B), both ends of the invading intron are ligated to the target DNA sense strand. Reverse transcription (primed by the 3′ end of the cleaved antisense strand) results in precise insertion of the intron sequence with no co-conversion or dependence on homologous DNA recombination predicted. In the case of bacterial Group I introns, mobility clearly occurs at the DNA level (Mueller et al. 1996), whereas genetic evidence for co-conversion 5′ exon sequences in some yeast Group II intron homing events is suggestive of a partial reverse splicing homing mechanism (Figure 6A). Recent studies provided evidence for multiple pathways of homing in yeast mitochondrial Group II introns (Yang et al. 1996; Eskes et al. 1997). The available information suggests that the two short intron binding sequences (IBSs) in the 5′ exon with complementarity to exon binding sequences (EBSs) in the intron are important for target activity in homing (Zimmerly et al. 1995; Yang et al. 1996; Eskes et al. 1997).

Following the discovery of Ll.ltrB (Mills et al. 1996), we initiated genetic analyses of the mobility of Ll.ltrB in lactococci. The results obtained in these studies showed that homing occurs efficiently in lactococci (Mills et al. 1997). When homing was assayed by agarose gel electrophoresis of plasmid preparations from strains transformed with both an intron-donor plasmid and a target plasmid, abundant products were detected without genetic selection. The homing event showed target specificity, and a 250 bp exon segment flanking the splice site was equally efficient as a target as compared with a full-length intronless *ltrB* allele. Interestingly, homing was not only *rec*-independent, but was actually somewhat more efficient in a recombination-deficient host strain.

Ll.ltrB retrohoming via an homologous recombination-independent mechanism

An inherent limitation to both genetic and molecular analyses of Group II introns characterized in eukaryotes is their organellar location. The techniques for introduction of mutant constructs into a functional mitochondrial genome for *in vivo* studies are cumbersome and can effectively limit the number of mutations analyzed. In addition, it is not feasible to do random mutagenesis. Purification of the intron-containing RNP (ribonucleoprotein) particles, for *in vitro* assays of biochemical reactions involved in splicing and mobility, has been complicated by the requirement for their isolation from mitochondria and by the lack of appropriate expression systems (Zimmerly et al. 1995; Yang et al. 1996). Therefore, several investigators who have worked extensively on yeast Group II introns have become interested in Ll.ltrB. Our recent investigations of Ll.ltrB have been carried out as a multi-laboratory collaboration involving the laboratories of Drs Marlene Belfort (Wadsworth Center, NYS Dept. of Health), Alan Lambowitz (Univ. of Texas at Austin), and Craig Peebles (Univ. of Pittsburgh).

Dr Peebles work has focused on molecular and genetic analysis of domain 5 and on the 'joiner' sequences between domains 5 and 6 (Peebles et al. 1995; Podar et al. 1998c). Dr Lambowitz's group has focused most of their effort on expression and purification of the intron RNA, the LtrA protein, and RNP particles containing both components for use in biochemical assays. Dr Belfort's lab has developed plasmid vectors and probes for molecular analysis of *in vivo* splicing and homing of Ll.ltrB in *E. coli* and has worked with us to develop similar methods for *L.lactis*. Thus far, it has been possible to express Ll.ltrB at high level in a T7-based expression system, to isolate RNP particles from the induced cell extracts, and to use these to demonstrate the *in vitro* biochemical activities previously associated with fungal Group II intron-derived RNP particles. In addition, functional RNPs can be reconstituted from the individually purified intron RNA and LtrA protein components. The first collaborative paper (Matsuura et al. 1997) reported functionality of Ll.ltrB *in vitro* and in *E. coli*. In addition, it was demonstrated that changing the EBS of the intron could alter the endonuclease specificity of Ll.ltrB-derived RNPs. This allowed for targeted insertion of the intron into novel sites, and has important implications for the use of these elements as genetic engineering tools.

A second paper (Cousineau et al. 1998) by the same 3 groups reported analysis of the *in vivo* homing process in *L. lactis* and in *E. coli* by molecular techniques and by genetic screens (using genetically marked introns, as well as 'Twintron' constructs containing small Group I intron derivatives inserted in non-essential regions of Ll.ltrB). The data provided compelling genetic evidence for intron homing in bacteria via a complete reverse splicing mechanism involving an RNA intermediate. These studies also confirmed previous results that homing is completely independent of the host-encoded homologous recombination machinery. Interestingly, Ll.ltrB homing in either *L. lactis* or *E. coli*, did not result in co-conversion of either 3′ or 5′ sequences flanking the homing target. This indicates that the major homing mechanism of Ll.ltrB in bacteria is a single pathway.

Development of the Ll.ltrB system for conjugation-based genetic analysis of splicing

We recently shifted the major focus of our own work to the analysis of Ll.ltrB splicing in *L. lactis* and its role in conjugative transfer of pRS01. In order to carry out the detailed genetic analysis of Ll.ltrB, we developed a two-plasmid system for genetic analysis of splicing-dependent conjugation. In this system, Rec^- lactococcal donor cells carry the pRS01-derived plasmid pM1014. This plasmid contains a pTRK28 insertion in the 5′ end of *ltrA* (Figure 2) and neither splicing nor conjugation (donor) ability are detectable in strains carrying only this plasmid. Splicing-dependent production of *ltrB* gene product is provided *in trans* by pCOM9, which contains the wild-type *ltrBE1*-intron-*ltrBE2* region and about 300 bp of non-coding upstream DNA containing the functional *ltrB* promoter (Figure 7A). The vector portion of pCOM9 is pDL281, a pDL278-based shuttle vector (LeBlanc et al. 1992) which replicates in both *E. coli* and *L. lactis*. This construct allows for fairly simple insertion of either specific or random mutations into the intron in a genetic context that allows for phenotypic analysis of splicing function in lactococci.

To assess the utility of the pCOM system, we introduced a set of defined mutations obtained from our collaborators in regions of the intron known to be functionally important. We compared the frequency of pM1014 transfer in the presence of pCOM9 to that observed with a vector derivative lacking intron or exon, to that observed with pCOM-ex (Figure 7B; a derivative of pCOM9 containing an intronless allele of *ltrB*), and to various mutant derivatives of the intron

Table 1. Analysis of splicing of Ll.ltrB and mutant derivatives using a conjugation assay in *L. lactis*[1]

A. Frequency of transfer of pRS01-derivative plasmid.

Plasmids in donor[2]	Transfer frequency of frequency cointegrate plasmid[3]
pM2036	1.6×10^{-1}
pM1014, pDL281	$<2.0\times10^{-9}$
pM1014, pCOM9	4.9×10^{-3}
pM1014, pCOM-ex	1.3×10^{-2}

B. Effects of various intron mutations in pCOM9 on transfer frequency.

Intron mutation	Approximate transfer relative to wild-type
G471→A	10^{-2}
G2399→A	5×10^{-2}
ΔOrf	10^{-4}
PltrA-35	10^{-3}
PltrA-10	10^{-3}

[1]Matings were carried out on GM17 agar plates for 12–16 h, similar to techniques described by Mills et al. (1994). All donor strains were derived by insertion of plasmids into the Rec^- MMS370 strain. The recipient was LM2301.
[2]The plasmids pM2036 and pM1014 are both cointegrates between pTRK28 and pRS01 (Mills et al. 1994). As discussed in the text, in pM2036 the cointegration took place outside the Tra sequences, and this plasmid serves as a positive control for conjugation. In pM1014 the *ltrA* gene was interrupted, and no splicing of the intron from the *ltrB* gene occurs, resulting in a Tra-phenotype. The shuttle vector pDL281 was used to construct all complementing plasmids in the pCOM series (Figure 7). The wild-type intron/exon region is cloned in pDL281 to generate pCOM9. The isogenic intronless allele of ltrB is contained in pCOM-ex. The G471→A, and G2399→A derivatives generated by C. Peebles contain single (G→A) substitutions at the intron nucleotides indicated by the number. The pCOMΔOrf mutant is deleted for nearly all the *ltrA* coding sequence. The pCOM-35, and pCOM-10 plasmids contain mutations in the -35 or -10 regions of the *ltrA* promoter recently identified within the intron, as discussed in the text.
[3]The frequencies represent the numbers of transconjugants/donor for the plasmid selected.

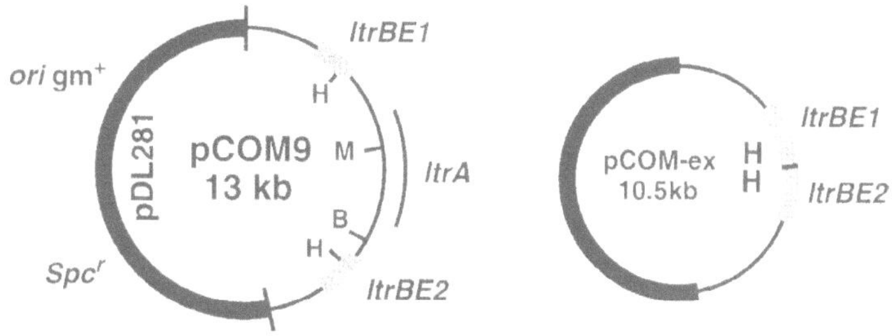

Figure 7. Plasmids used for Ll.ltrB analysis. A. pCOM9, in which the wild-type exon/intron region is expressed from native promoter sequences. B. pCOM-ex, in which the intron sequences are precisely deleted from the DNA.

inserted into the context of pCOM9. In addition, we used a pRS01 derivative, pM2036, where the cointegration of pTRK28 was outside the Tra regions, as a positive control for mating competence of the recipient strain. Some results of these matings are summarized in Table 1. The difference between the frequency of transfer observed with pCOM9 and the lower limit of detection in the system is nearly 6 orders of magnitude. The mutant intron derivatives tested include two point mutants constructed by Dr Peebles. Mutant G471A contains a single base substitution at a highly conserved G residue (adjacent to the 'γG' residue) in the 2-3 joiner region (Podar et al. 1998c). Mutant G2399A contains a single substitution in Domain 5. Each of these mutations in the corresponding residues in the yeast intron aI5 gamma are defective in splicing. In the conjugation assay, these mutants displayed complementing activity ranging from about 50–100 fold below the wild-type frequency. For an intron derivative, ΔOrf (Cousineau et al. 1998) which has been deleted for nearly all of LtrA, a more drastic reduction (10000-fold) in frequency of transfer was observed. Interestingly, the latter mutant still provided sufficient complementation to mediate a level of transfer over 100 fold above background. This result suggests the possibility of maturase-independent Group II intron splicing *in vivo*.

In order to initiate analysis of regulatory mechanisms in the pRS01 system, we recently began to identify transcription initiation sites in the *ltrB* region. The most interesting result obtained thus far in these studies, is the identification of a transcription start site within the intron, just upstream from *ltrA*, in the loop region of domain 2 (Figure 4). Northern blot analysis suggests that in lactococcal cells carrying the wild-type *ltrB* region, there are two *ltrA* transcripts. Their sizes correspond to those expected if both transcripts terminated at the 3′ end of *ltrBE2*, with one initiating at a promoter upstream from *ltrBE1*, and the second initiating at the putative *ltrA* promoter within the intron. The start site of this internal transcript is preceded by reasonable '−10' and '−35' sequences. We have introduced mutations into both of these putative promoter domains, and tested them in the conjugation assay. Each mutation resulted in a the conjugation frequency, about 10-fold below that observed with the G471 and G2399 point mutations, and about 10-fold above that of the ΔOrf mutation. The simplest interpretation of the data is that this *ltrA* promoter is important for intron function. However, the mutations could also affect the structural integrity of the intron. This may be especially true for the '−35' mutation, which could significantly alter the stability of the base pairing in the stem between domains 1 and 2. To examine this possibility, we introduced compensatory mutations in the stem sequence complementary to the mutated −35 region. While the compensatory mutations appear to increase plasmid transfer about 10-fold relative to the original mutant, the level of transfer is still about 100-fold lower than wild type. We are interested in testing the idea that RNA folding inhibits efficient translation of the *ltrA* Orf in the context of the full-length pre-mRNA or the intron lariat in *L. lactis*. Interestingly, examination of the sequence between ltrA and ltrBE1 suggests that many mutations would need to be introduced to put these two genes in frame, also suggesting that the mechanism of LtrA synthesis is different from eukaryotic group II IEPs. Future work on the system will focus on selection of secondary intron mutations that result in phenotypic reversion of the primary mutations listed in Table 1. This type of suppression/reversion analysis should be useful in elucidation of structural domains of the RNA important in ribozyme function, as well as in interaction with the LtrA protein. In addition, we hope to gain further insights into the mechanisms by which LtrA synthesis is regulated, and how this impacts on intron activity *in vivo*.

Significance of Group II Introns in evolution of LAB

Ll.ltrB was the first bacterial Group II intron shown to be functional *in vivo*. The fact that these elements seem to be relatively rare in bacteria suggests that they may not be major evolutionary players in prokaryotes. Interestingly, the amino acid sequence of the LtrA, the IEP of Ll.ltrB is more highly related to IEPs of fungal introns than to the putative IEPs identified by sequence searching in other bacteria (including the putative endosymbiont organisms from which organelles are believed to have evolved). This suggests the possibility of horizontal genetic transfer between LAB and yeasts subsequent to the evolution of organelles. However, at the DNA sequence level, both the%G+C, as well as the codon usage within the Ll.ltrb sequence, are clearly lactococcal, indicating that this element has resided within the 'low GC' branch of the gram positive eubacteria for a considerable length of time.

It is quite possible that functional bacterial Group II introns are more common in bacteria than has been recognized previously. Since they are self-splicing, their insertion into a gene does not generally inactivate its function. Interestingly, searches of the genomic sequences of both *Streptococcus pneumoniae*, and *Enterococcus faecalis* deposited in The Institute for Genomic Research (TIGR) databases indicated the presence of putative Group II introns. The functionality of these elements has not been tested. Previous searches for homologous Group II intron sequences (generally based on the IEP amino acid sequence) may have missed elements that show sequence divergence. Possibly, more of these introns will be found as more bacterial genomes are sequenced completely. Given the large number of individual bacterial genes already sequenced, it seems safe to say that Group II introns are not commonly present in bacterial genes. However, they may turn out to be present (in low copy number) in most genomes. Since very little is known about intron mobility reactions other than homing, it is possible that, in the absence of a consensus homing target site, they may show preferences for certain targets, e.g. intergenic regions, that have made them more difficult to detect.

Potential applications of Group II introns in genetic manipulation of LAB

Our present understanding of Group II introns and their mobility mechanisms suggests that these elements may have great potential as genetic engineering tools for all organisms, including bacteria. Perhaps the most obvious application would be the use of these elements as specialized transposons, which have the novel property of not abolishing the function of the gene into which the element is inserted. Studies of intron homing suggest that the specificity of insertion is determined by very short sequences (EBSs) within the element which interact with complementary sequences (IBSs) in the target exon. Target specificity can be changed by very minor changes in the EBSs (Matsuura et al. 1997). It would therefore be feasible to either engineer a specific EBS to direct an intron to the genomic target of choice, or to randomize the EBS for random insertion. Both approaches could be very useful. The demonstration that domain 4 is dispensable for intron function, and can accommodate significant amounts of extra genetic material, provides the foundation to use these elements to introduce genes of interest into new hosts. In addition, various manipulations of either the IEP or RNA sequences of these introns to generate conditionally functional elements, can be readily envisioned. This could be very useful in genetic and physiological analysis of organisms not currently amenable to such studies. The pRS01 conjugation system has considerable potential for introduction of genetic material across a broad phylogenetic range of bacterial hosts, and Ll.ltrB has already been shown to function in both gram positive and gram negative organisms (Matsuura et al. 1997; Cousineau et al. 1998). In addition, it may well be feasible to prepare RNP particles containing Ll.ltrB lariats complexed with LtrA protein, and introduce these directly into either prokaryotic or eukaryotic cells by electroporation or other physical methods. Such particles may be active in genomic insertion in target cells. Of course, a great deal of effort will be required to bring any of these ideas to fruition, but they clearly have great potential.

Concluding thoughts

A central question relating to Ll.ltrb is how it impacts on the biology of pRS01 conjugation in lactococci. Splicing is clearly required for conjugation. Is it a rate limiting step? Alternatively do the physiological factors that regulate expression of conjugation functions also affect intron splicing or mobility? What happens to expression of intron function when pRS01 transfers into a new host by conjugation? These questions will keep investigators interested in the basic biology and applications of lactococcal conjugation very busy for a considerable length of time.

Acknowledgements

We thank David Mills, Dawn Manias, and Liang Zhou for carrying out most of the work done in our laboratories on Ll.ltrB. We also thank Marlene Belfort, Alan Lambowitz, and Craig Peebles for excellent discussions and exchange of strains and plasmids. Initial work on Ll.ltrB was supported by the Minnesota-South Dakota Dairy Research Center.

References

Anderson DG & McKay LL (1984) Genetic and physical characterization of recombinant plasmids associated with cell aggregation and high-frequency conjugal transfer in *Streptococcus lactis* ML3. J. Bacteriol 158: 954–962

Anziano PQ, Hanson DK, Mahler HR & Perlman PS (1982) Functional domains in introns: trans-acting and cis-acting regions of intron 4 of the *cob* gene. Cell 30: 925–932

Belfort M & Perlman PS (1995) Mechanisms of Intron Mobility. J. Biol. Chem. 270: 30237–30240

Belfort M, Reaban ME, Coetzee T & Dalgaard JZ (1995) Prokaryotic Introns and Inteins: a Panoply of Form and Function. J. Bacteriol. 177: 3897–3903

Carignani G, Groudinsky O, Frezza D, Schiavon E, Bergantino E & Slonimski PP (1983) An mRNA maturase is encoded by the first intron of the mitochondrial gene for subunit I of cytochrome oxidase in *S. cerevisiae*. Cell 35: 733–745

Clewell DB (1993) Bacterial Conjugation, New York: Plenum

Cousineau B, Smith D, Lawrence-Cavanagh S, Mueller JE, Yang J, Mills D, Manias D, Dunny G, Lambowitz AM & Belfort M (1998) Retrohoming of a bacterial group II intron: mobility via complete reverse splicing, independent of homologous DNA recombination. Cell 94: 451–462

Eskes RJ, Yang J, Lambowitz AM & Perlman P (1997) Mobility of yeast mitochondrial group II introns: engineering new site-specificity and retrohoming via full reverse splicing. Cell 88: 865–874

Ferat J & Michel F (1993) Group II self-splicing introns in bacteria. Nature 364: 358–361

Ferat J, Gouar ML & Michel F (1994) Multiple group II self-splicing introns in mobile DNA from *Escherichia coli*. Life Sciences 317: 141–148

Fitzgerald GF & Gasson MJ (1988) *In vivo* gene transfer systems. Biochimie 70: 489–502

Gasson MJ, Godon J-J, Pillidge CJ, Eaton TJ, Jury K & Shearman CA (1995) Characterization and exploitation of conjugation in *Lactococcus lactis*. Int. Dairy J. 5: 757–762

Knoop V & Brennicke A (1994) Evidence for a group II intron in *Escherichia coli* inserted into a highly conserved reading frame associated with mobile DNA sequences. Nucleic Acids Research 22: 1167–1171

Lambowitz AM & Belfort M (1993) Introns as mobile genetic elements. Ann. Rev. Biochem. 62: 587–622

LeBlanc DJ, Lee LN & Abu-Al-Jaibat A (1992) Molecular, genetic, and functional analysis of the basic replicon of pVA380-1, a plasmid of oral streptococcal origin. Plasmid 28: 130–145

Matsuura M, Saldanha R, Ma H, Wank H, Yang J, Mohr G, Cavanagh S, Dunny GM, Belfort M & Lambowitz AM (1997) A bacterial group II intron encoding reverse transcriptase, maturase, and DNA endonuclease activities: biochemical demonstration of maturase activity and insertion of new genetic information within the intron. Genes Dev. 11: 3333–3345

McKay LL, Baldwin KA & Walsh PM (1980) Conjugal transfer of genetic information in Group N streptococci. Appl. Environ. Microbiol. 40: 1090–1096

Michel F & Ferat J-L (1995) Structure and activities of Group II introns. Ann. Rev. Biochem. 64: 35–61

Mills DA, Choi CK, Dunny GM & McKay LL (1994) Genetic analysis of regions of the *Lactococcus lactis* subsp. *lactis* plasmid pRS01 involved in conjugative transfer. Appl. Environ. Microbiol. 60: 4413–4420

Mills DA, McKay LL & Dunny GM (1996) Splicing of a group II intron involved in the conjugative transfer of pRS01 in lactococci. J. Bacteriol. 178: 3531–3538

Mills DA, Manias DA, McKay LL & Dunny GM (1997) Homing of a Group II Intron from *Lactococcus lactis* subsp. *lactis* ML3. J. Bacteriol 179: 6107–6111

Mills DA, Phister TG, Dunny GM & McKay LL (1998) An origin of transfer (*oriT*) of the conjugative element pRS01 from *Lactococcus lactis* subsp. *lactis* ML3. Appl. Environ. Microbiol. 64: 1541–1544

Moran JV, Mecklenburg KL, Sass P, Belcher SM, Mahnke D, Lewin A & Perlman P (1994) Splicing defective mutants of the *COXI* gene of yeast mitochondrial DNA: initial definition of the maturase domain of the group II intron AI2. Nucl. Acids Res. 22: 2057–2064

Mueller JE, Clyman J, Huang Y-J, Parker MM & Belfort M (1996) Intron mobility in phage T4 occurs in the context of recombination-dependent DNA replication by way of multiple pathways. Genes Dev. 10: 351–364

Mullany P, Pallen M, Wilks M, Stephen JR & Tabaqchali S (1996) A Group II intron in a conjugative transposon from the Gram-positive bacterium, *Clostridium difficile*. Gene 174: 145–150

Peebles CL, Zhang M, Perlman PS & Franzen JS (1995) Catalytically critical nucleotides in domain 5 of a group II intron. Proc. Natl. Acad. Sci. USA 92: 4422–4426

Podar M, Chu VT, Pyle AM & Perlman PS (1998a) Group II intron splicing *in vivo* by first-step hydrolysis. Nature 391: 915–918

Podar M, Perlman PS & Padgett RA (1998b) The two steps of group II intron self-splicing are mechanistically distinguishable. RNA 4: 890–900

Podar M, Zhuo J, Zhang M, Franzen JS, Perlman PS & Peebles CL (1998c) Domain 5 binds near a highly conserved dinucleotide residue in the joiner linking domains 2 and 3 of a group II intron. RNA 4: 151–166

Polzin KM & Shimizu-Kadota M (1987) Identification of a new insertion element, similar to gram-negative IS26, on the lactose plasmid of *Streptococcus lactis* ML3. J. Bacteriol 169: 5481–5488

Qin PZ & Pyle AM (1998) The architectural organization and mechanistic function of group II intron structural elements. Curr. Opinion Struct. Biol. 8: 301–308

Rauch PJG & De Vos WM (1992) Characterization of the novel nisin-sucrose conjugative transposon Tn*5276* and its insertion in *Lactococcus lactis*. J. Bacteriol. 174: 1280–1287

Romero DA & Klaenhammer TR (1991) Construction of an IS*946*-based composite transposon in *Lactococcus lactis* subsp. *lactis*. J. Bacteriol. 173: 7599–7606

Shearman C, Godon J-J & Gasson M (1996) Splicing of a group II intron in a functional transfer gene of *Lactococcus lactis*. Mol. Microbiol. 21: 45–53

Walsh PM & McKay LL (1981) Recombinant plasmid associated with cell aggregation and high-frequency conjugation of *Streptococcus lactis* ML3. J. Bacteriol 146: 937–944

Yang J, Zimmerly S, Perlman PS & Lambowitz AM (1996) Efficient integration of an intron RNA into double-stranded DNA by reverse splicing. Nature 381: 332–335

Yeo CC, Tham JM, Yap MW-C & Poh CL (1997) Group II intron from *Pseudomonas alcaligenes* NCIB 9867(p25X): entrapment in plasmid RP4 and sequence analysis. Microbiology 143: 2833–2840

Zimmerly S, Guo H, Eskes R, Yang J, Perlman PS & Lambowitz AM (1995) A Group II Intron RNA Is a Catalytic Component of a DNA Endonuclease Involved in Intron Mobility. Cell 83: 529–538

Antonie van Leeuwenhoek **76:** 89–113, 1999.

Bacteriophage defence systems in lactic acid bacteria

Amanda Forde[1,3] & Gerald F. Fitzgerald[1,2,3,*]
[1]*Departments of Microbiology and* [2]*Food Science and Technology and* [3]*National Food Biotechnology Centre, University College, Cork, Ireland (*Author for correspondence: E-mail: g.fitzgerald@ucc.ie)*

Abstract

The study of the interactions between lactic acid bacteria and their bacteriophages has been a vibrant and rewarding research activity for a considerable number of years. In the more recent past, the application of molecular genetics for the analysis of phage-host relationships has contributed enormously to the unravelling of specific events which dictate insensitivity to bacteriophage infection and has revealed that while they are complex and intricate in nature, they are also extremely effective. In addition, the strategy has laid solid foundations for the construction of phage resistant strains for use in commercial applications and has provided a sound basis for continued investigations into existing, naturally-derived and novel, genetically-engineered defence systems. Of course, it has also become clear that phage particles are highly dynamic in their response to those defence systems which they do encounter and that they can readily adapt to them as a consequence of their genetic flexibility and plasticity. This paper reviews the exciting developments that have been described in the literature regarding the study of phage-host interactions in lactic acid bacteria and the innovative approaches that can be taken to exploit this basic information for curtailing phage infection.

Introduction

The significance of bacteriophage infection of Lactic Acid Bacteria (LAB), most notably of members of the genus *Lactococcus*, is well recognised by both academic and industrial organisations worldwide. Their negative impact on several dairy fermentation processes was established early this century (Whitehead & Cox 1935) and to the present day phages represent one of the most persistent and stubborn problems in commercial practice. This has stimulated researchers and manufacturers alike to continually devise innovative and effective procedures to alleviate the devastating consequences imposed by lytic bacteriophage attack. The dairy industry has traditionally employed several physical and biological methods to deal with phage related problems, ranging from the implementation of aseptic processing conditions to the utilisation of strain rotation regimes (Moineau et al. 1994a, Limsowtin et al. 1996); yet the effectiveness of such precautions is limited by their protracted use and by the dynamic evolutionary properties of the phages themselves (Hill et al. 1991; Moineau et al. 1994b).

The generation of phage-host range data, classification schemes and the complete genomic sequences of bacteriophages of a number of LAB (Table 1) has not only provided an insight into the origins of these phages and the relationships between them, but has also enabled scientists to develop information-based strategies to curb their successful proliferation. Concurrently, the discovery that natural phage resistance mechanisms reside within many LAB has provided additional new approaches to countering phages and these mechanisms have been exploited in strain improvement programmes. The most extensively studied systems are those of the lactococci where the majority are plasmid encoded and may exist either singly or in combinations on a particular plasmid or number of plasmids within the same strain. This topic has been elaborately reviewed by a number of authors, most recently by Dinsmore & Klaenhammer (1995), Garvey et al. (1995a), Daly et al. (1996) and Allison & Klaenhammer (1998), in which they categorise phage resistance systems based on their point of action in the lytic cycle. Four distinct groups are currently recognised, comprising those mechnisms which are directed towards inhibition of phage adsorption and DNA penetration, and those which hinder any further intracellular phage development (restriction/modification and abortive infection). It has been advantageous that several phage resistance plasmids are transferrable by conjugation, thereby allowing the application of this food grade technology for introducing phage resistance to industrially significant and commercially valuable strains of *Lactococcus* (Sanders et al. 1986; Sing & Klaenhammer 1986; Jarvis 1988; Coffey et al. 1989; Kelly et al. 1990; Klaenhammer 1991; Harrington & Hill 1991; Coakley et al. 1997; O'Sullivan et al. 1998).

Research on phage-host interactions and phage resistance systems, in particular, is very dynamic and is constantly providing increasingly instructive insights into the molecular basis for natural phage insensitivity. In parallel, it is expanding our knowledge of the relationship between both components of the interaction at a very fundamental level. This review will summarise the accumulated information available on the different categories of natural phage resistance in LAB and will allude briefly to 'engineered' systems based on knowledge of phage genomes (this latter topic will be covered more explicitly elsewhere in this volume). More recent developments elucidating the mode of action of some of these systems at the molecular level will also be presented.

Bacteriophages of lactic acid bacteria

The understanding of novel phage resistance systems in LAB is clearly linked to a thorough knowledge of the phages themselves and the manner in which they interact with the defence mechanisms which they may encounter. The earlier efforts to develop research programmes on the genetics of LAB phages can be traced to the generation of classification schemes based on genome relatedness. While early classification schemes grouped these phages according to their host range and morphology (Bradley 1967; Ackerman & DuBow 1987; Jarvis 1989), the later application of hybridisation analysis proved more influential in taxonomic classification, especially in the case of lactococcal phages. Jarvis et al. (1991) defined 12 lactococcal phage species based on DNA relatedness as well as morphology and noted that the phages which were most frequently encountered in dairy fermentations were most commonly of the small isometric-headed 936 species (particularly in New Zealand, the US and Ireland); less frequently of the prolate-headed c2 species and the P335 species which includes virulent and temperate phages (Jarvis & Klaenhammer 1986; Prevots et al. 1990; Moineau et al. 1992; Casey et al. 1993; Jarvis 1995). Reviews by Klaenhammer & Fitzgerald (1994), Neve (1996) and Josephsen & Neve (1998) have also highlighted the morphological and genetic relationships that exist between other phages including those which infect *Streptococcus thermophilus* and *Lactobacillus* species and to a lesser extent *Leuconostoc* species. The generally poor correlation between groupings based on host range and other classification criteria may be attributable to the ease with which phages mutate, often through acquisition of DNA fragments from unrelated sources (Hill et al. 1991). The G+C content of LAB phages usually reflects that of their hosts, ranging from 37% for phages of *Lactococcus* to approximately 48% for phages of *Lactobacillus casei*. Genome sizes can vary from between 18–55 kb (although sizes up to 134 kb have been measured for some lactococcal phages) with the smaller genomes being so compact that the genes may overlap (Josephsen & Neve 1998).

Within the last five years, the study of the molecular genetics of LAB phages has advanced to an impressive level leading to a greater understanding of their genetic organisation and evolutionary origins. The emergence of nucleotide sequence data for complete phage genomes and the characterisation of several individual phage genes (Table 1) has permitted comparative structural analysis as well as assessment of gene regulation (Garvey et al. 1995a; Djordjevic & Klaenhammer 1997). In addition, it has presented the opportunity to develop a range of new applications; for example, LAB phages are being exploited to accelerate lysis and hence ripening of cheese (Shearman et al. 1992); to construct integration vectors (Raya et al. 1992; Fremaux et al. 1993; Dupont et al. 1995; Christiansen et al. 1996; Lillehaug et al. 1997; Auvray et al. 1997) and to generate inducible expression systems (Nauta et al. 1996; O'Sullivan et al. 1996; de Ruyter et al. 1997). Although the nucleotide sequence may differ between LAB phages, their overall compact genome organisation appears to be similar, consisting of functional units involved in replication, morphogenesis, integration and lysis, and they have very little non-coding DNA. The transcription maps of some of these phages have also been determined (Beresford et al. 1993; Chandry et al. 1994; Madsen & Hammer 1998) revealing that genes in the same expression phase (ie. early, middle or late) generally share similar functions and are transcribed in the same direction, often from a common promoter (Allison & Klaenhammer 1998). Homology searches of putative open reading frames (ORFs) have allowed

Table 1. LAB phages for which the genome has been completely sequenced or for which individual genes have been identified

Phage	Genomic region	Host	Reference
bIL67 (lytic)	entire genome	*L. lactis*	Schouler et al. (1994)
c2 (lytic)	entire genome	*L. lactis*	Lubbers et al. (1995)
sk1 (lytic)	entire genome	*L. lactis*	Chandry et al. (1997)
øbIL70 (lytic)	entire genome	*L. lactis*	Unpublished; M.-C. Chopin, personal communication
r1t (temperate)	entire genome	*L. lactis*	van Sinderen et al. (1996)
Tuc2009 (temperate)	entire genome	*L. lactis*	Unpublished; D. van Sinderen, M. Creaven, C. Daly, M. van de Guchte, E. Arendt & G.F. Fitzgerald
	lysin, holin, structural genes; *int*, *attP*, repressor		Arendt et al. (1994); van de Guchte et al. (1994a, b) Sheehan et al. (1996)
	phage encoded resistance (PER)		Mc Grath et al. (1999)
TP-901-1 (temperate)	entire genome	*L. lactis*	Unpublished; F.K. Vogensen, personal communication
	mhp, *mtp* (structural proteins);		Johnsen et al. (1995, 1996)
	attP, *orf1* (resolvase)		Christiansen et al. (1996) Brondsted & Hammer (1999)
	repressor		Madsen & Hammer (1998)
øLL-H (lytic)	entire genome	*Lb. delbrueckii* ssp. *lactis*	Mikkonen et al. (1996)
øgle (temperate)	entire genome	*Lactobacillus*	Kodaira et al. (1997)
ø01205 (temperate)	entire genome	*S. thermophilus*	Stanley et al. (1997)
ø7201 (temperate)	entire genome	*S. thermophilus*	Unpublished; E. Stanley, L. Walsh, D. van Sinderen & G. F. Fitzgerald
øDT1 (temperate)	entire genome	*S. thermophilus*	Tremblay & Moineau (1999)
BK5-T (temperate)	half genome (32 ORFs) sequenced	*L. lactis*	Boyce et al. (1995)
øvML3 (lytic)	lysin, holin	*L. lactis*	Shearman et al. (1989)
F4-1 (lytic)	*mcp*, *p35*, *p43* (structural proteins)	*L. lactis*	Chung et al. (1991); Kim & Batt (1991a)
ø7-9 (lytic)	*orf1356* (possible regulator)	*L. lactis*	Kim & Batt (1991b)
ø50 (lytic)	*per50* (origin of replication)	*L. lactis*	Hill et al. (1990a)
	*Lla*I (methylase)	*L. lactis*	Hill et al. (1991)
øLC3 (temperate)	*cos* (packaging)	*L. lactis*	Lillehaug et al. (1991)
	int; *attP/attB* (integrase and attachment site)		Lillehaug & Birkeland (1993)
	lysA, *lysB* (lysin and holin)		Birkeland (1994)
øUS3 (lytic)	*lytA* (lysin); *orf66* (holin)	*L. lactis*	Platteeuw & de Vos (1992)
ø197 (lytic)	*poa17* (structural protein)	*L. lactis*	Schouler et al. (1992)
bIL66 (lytic)	*orf1-4* ('M' operon-sensititivity to AbiD1)	*L. lactis*	Bidnenko et al. (1995)
bIL41 (lytic)	*orf1*; *orf2* (terminase subunits) *orf6-9, 13* (structural proteins)	*L. lactis*	Parreira et al. (1996a)
ø31 (lytic)	*per31* (origin of replication)	*L. lactis*	O'Sullivan et al. (1993)
	ø31P (phage-specific promoter)		O'Sullivan et al. (1996)
	orf2 (transcriptional activator)		Walker et al. (1996)
	ORF245 (sensitivity to AbiA)		Dinsmore & Klaenhammer (1997)

Table 1. continued

Phage	Genomic region	Host	Reference
øadh (temperate)	*int*; *attP*	*Lb. gasseri*	Raya et al. (1992)
			Fremaux et al. (1993)
	repressor		Engel et al. (1998)
mv4 (temperate)	*int*/*attP*	*Lb. delbrueckii* ssp. *lactis*	Dupont et al. (1995)
			Auvray et al. (1997)
øSfi21 (temperate)	lysogeny module	*S. thermophilus*	Bruttin et al. (1997)
			Neve et al. (1998)
	origin of replication (Per based phage resistance)		Foley et al. (1998)

the deduction of proposed functions which may be conserved among several phages or unique to an individual species. Of exceptional importance was the identification of genetic elements involved in the generation of bacteriophage resistance, such as phage origins of replication Per50 (Hill et al. 1990a) and Per31 (O'Sullivan et al. 1993) and the introduction of a phage inducible promoter upstream of the *Lla*I restriction cassette to create an 'engineered' form of abortive infection (Djordjevic et al. 1997).

These features have proved rewarding in terms of devising innovative phage defence systems. However, the one feature that cannot go unnoticed is the remarkable evolutionary capacity of phages which is based on the exchange of genetic modules (Botstein, 1980) allowing them to overcome natural systems of phage resistance. This genetic flexibility, in combination with the utilisation of lysogenic starter cultures, represents the most likely source of 'new' emerging phages which subsequently prove problematic for the cheese-maker. It stresses the necessity for retaliation through the development of novel insensitivity mechanisms and strains which must ultimately be capable of surviving the pressures of the industrial arena.

Mechanisms of host-mediated defence against phage proliferation in lactic acid bacteria

The dynamic research activity relating to phage-host interactions in LAB has revealed a multiplicity of natural phage insensitivity mechanisms which reside within these hosts. Those which have been studied in greatest detail originate in lactococci, but phage resistance systems have also been identified in other LAB species. In recent years scientists have exploited the molecular biological tools available to them to genetically dissect these organisms for the purpose of analysing their phage resistance systems at fundamental and molecular levels, and also for the construction of strains with stronger defences, either by employing natural resistance systems or by engineering novel types. The natural defence mechanisms are conveniently segregated into four broad categories based on their generalised mechanism of action. In addition, they share the common property of being predominantly plasmid-encoded, with a few exceptions where the insensitivity determinants have been localised on the bacterial chromosome.

Interference with bacteriophage adsorption

Attachment of lactococcal phage particles to the exterior of the bacterial cell usually occurs in a tail-first orientation, either at randomly distributed specific 'hot spots' or more uniformly to receptors which are evenly distributed over the cell surface (Budde-Niekiel & Teuber 1987). It is not possible to state definitively the specific requirement(s) for adsorption since the process is likely to be phage-, strain- and receptor-dependent. However, inhibition of phage adsorption to lactococcal host cells has been observed on a number of occasions, and is generally due to either the lack of, or physical masking of, the receptors. According to the literature available to date, two broad mechanisms have been implicated which lead to the phenomenon; the development of *b*acteriophage *i*nsensitive *m*utants (BIMs) following spontaneous chromosomal mutation, and plasmid systems which mediate inhibition of adsorption. BIMs have been pivotal in generating information regarding the nature of phage receptors, with most investigations indicating that carbohydrate and/or protein moieties might play a key role in adsorption.

In the first studies of adsorption in lactococci, Oram & Reiter (1968) demonstrated the presence of a lipoprotein in the plasma membrane receptor for phage ml3. In most cases, however, it appears that phages adhere to a carbohydrate component in the cell wall. Valeyasevi and colleagues (1990) correlated the inability of five phages to bind to cell walls with the loss of galactose-associated receptor material, and rhamnose was also found to be essential for the adsorption of two of the five phages. Indeed, galactose and/or rhamnose were also implicated by other groups (Keogh & Pettinghill 1983; Monteville et al. 1994; Valeyasevi et al. 1994; Gopal & Reilly 1995) as the most common carbohydrates necessary for adsorption to occur, while Schafer and co-workers (1991) suggested the involvement of a complex array of polysaccharide components in the receptors of *L. lactis* ssp. *cremoris* Wg2-1 and *L. lactis* ssp. *lactis* biovar. *diacetylactis* F7/2. Vlegels et al. (1988) isolated bacteriophage resistant mutants from strains of *L. lactis* ssp. *cremoris* which exhibited a lower adsorbing efficiency than did their wild-type strains due to reduced receptor accessibility. A shielding of the receptors on the cell wall of the mutants by their non-peptidoglycan polymers was likely to be involved in mediating the resistance.

In recent years there has been a renewed interest in defining the mechanisms of phage adsorption in starter lactococci. A study performed by Crow & Gopal (1995) compared the cell surface properties of 15 pairs of lactococcal strains which were comprised of wild-type and phage resistant derivatives, some of which exhibited an adsorption inhibition phenotype. The strains differed principally in hydrophobicity patterns, protein profiles and the concentrations of extracellular lipoteichoic acid, protein, hexose and rhamnose in the non-covalently associated loose material extracted from the cell surface. However, no unifying theme was observed to explain the nature of the cell surface alterations in the phage resistant derivatives. In an earlier study, Gopal & Crow (1993) had characterised in detail this *l*oosely *a*ssociated *m*aterial (LAM) from one of these strain pairs, *L. lactis* ssp. *cremoris* E8 and its phage insensitive variant 398, and demonstrated the inability of three phages, ø852, ø618 and ø833 to bind to the insensitive derivative. It was assumed that the LAM from the 398 variant acted in a fashion similar to the masking polysaccharide produced by the adsorption blocking plasmid pCI528 reported previously by Lucey et al. (1992), and both polymers were indeed found to consist of galactose and rhamnose. In addition, the LAM from 398, compared to that of the parent, contained almost five-fold more rhamnose, twice the amount of galactose and lacked a 21 kDa protein but it was unclear whether this protein played a role in phage adsorption. It is interesting that Gopal & Reilly (1995) later observed two extra plasmid bands in the profile of 398 which were assumed to be due to the rearrangement of a replicon. However, these plasmid bands have not been shown to provide phage resistance and could simply have been lost from the parent strain E8, since the isolation of the resistant derivative occurred approximately two decades prior to these investigations (Limsowtin & Terzaghi 1976).

In at least one instance the molecular events underlying the formation of a BIM have been analysed (Harrington & Hill 1992). The BIM, DPC721, was isolated after challenging the parent strain *L. lactis* ssp. *lactis* biovar. *diacetylactis* DPC220 with its homologous phage, øD1. The activation of a silent adsorption blocking mechanism was established following a precise co-integration event between two endogenous plasmids from DPC220 (pAH82 and pAH33) which had occurred in the formation of DPC721. In fact, a more complicated 'switch' in phage resistance phenotypes had accompanied the reorganisation. While adsorption of øD1 to DPC721 appeared not to be inhibited, the recombinant molecule (pAH90) did encode adsorption inhibition against øc2 when it was introduced to a plasmid-free background (*L. lactis* ssp. *cremoris* MG1614) for analysis. In addition, a restriction-modification phenotype (R/M_{90}) was also detectable against ø712, which incidentally was different in specificity and efficiency to the R/M located on the original plasmid pAH82 (R/M_{82}) (Hill 1993).

The role of plasmid DNA in mediating adsorption inhibition has also been established in a number of strains, and the molecule generally either directs the synthesis of cell surface antigens (Tortorello et al. 1990; Akcelik & Tunail 1992) or mediates the production of an extracellular carbohydrate which shields the receptors against phage attachment (Sijtsma et al. 1988, 1990; Lucey et al. 1992). Since the first description by Sanders & Klaenhammer (1983) of a 31 MDa plasmid, pME0030, which mediated adsorption inhibition against ø18 in the prototype phage resistant strain *L. lactis* ssp. *lactis* ME2, at least seven other plasmids encoding this type of mechanism have been identified (Table 2). Tortorello et al. (1990) attributed the phenotype to the presence of an 80 kb plasmid, pKC50, from *L. lactis* ssp. *lactis* 57150 and indicated

Table 2. Lactococcal plasmids involved in inhibition of phage adsorption

Plasmid	Size (kb)	Host	Mechanism	Source
pME0030	46	*L. lactis* ME2	Undetermined	Sanders & Klaenhammer (1983)
pSK112	52	*L. cremoris* SK11	EPS	de Vos et al. (1984)
p1149-3	12.8	*L. cremoris* 11/49	Undetermined	Geis et al. (1987)
pCI528	46	*L. cremoris* UC503	EPS	Costello (1988)
pKC50	80	*L. lactis* 57150	Surface antigen	Tortorello et al. (1990)
p2520L	37.5	*L. lactis* P25	Surface antigen	Akcelik & Tunail (1992)
pAH90	90	*L. lactis* biovar. *diacetylactis*	BIM	Harrington & Hill (1992)
pCI658	58	*L. cremoris* HO2	EPS	Forde et al. (1999)

Modified from Garvey et al. (1995a)

the possible role of a 100 kDa cell surface antigen in the process. Akcelik & Tunail (1992) demonstrated that a 37.5 kb conjugative plasmid, p2520L, from *L. lactis* ssp. *lactis* P25 was responsible for adsorption inhibition. They found that it directed the synthesis of a 30 kDa cell surface antigen but did not definitively link this protein with the resistance phenotype. The plasmid pSK112 from *L. lactis* ssp. *cremoris* SK110 was studied extensively by Sijtsma and co-workers (1988, 1990, 1991) who associated it with the production of a galactosyl-containing lipoteichoic acid which protected the cell against infection by øsk11G. In similar experiments performed by Lucey et al. (1992) the 46 kb plasmid, pCI528, was shown to be responsible for inhibiting phage attachment through the synthesis of a non-covalently bound hydrophilic polymer rich in galactose and rhamnose. The most recently identified adsorption blocking plasmid, pCI658 (Forde et al. 1999), mirrors the previous examples in that it directs the production of an alkali-soluble hydrophilic polymer, minimising adsorption of phages ø712 and øc2 to their receptors. In this case however, galactose and glucuronic acid were identified as the putative components of the exopolysaccharide (EPS) which was depicted as strands of fibrillar material extending outwards from the cell surface (Forde & Fitzgerald 1999). This feature was also recorded for the previously mentioned phage resistant strain 398 (Gopal & Reilly 1995) and according to an earlier study by Brooker (1976), these filamentous extensions might actually permit adhesion of cheese starter cultures to the curd, reducing the possibility of their loss in the whey during syneresis. The production of a soft 'fluffy' cell pellet which was easily resuspended following centrifugation was recognised for cells harbouring pCI528 and pCI658 and for strain 398. In other LAB, the only instance where an extracellular substance was shown to affect adsorption was in *Lb. helveticus* CNRZ 892. In this case, the purified surface layer (S-layer) protein surrounding the cell was observed to drastically reduce adsorption of the lytic phage CNRZ 832-B1 to the host (Callegari et al. 1996, 1998).

Adsorption inhibition remains poorly characterised with regard to the genetic basis of the phenomenon. The recent reports illuminating exopolysaccharide expression in LAB (Diersken et al. 1995; Stingele et al. 1996, van Kranenburg et al. 1997) suggest that it is a complex process involving large contiguous DNA fragments with the overall organisation, transcriptional direction and deduced functions of the genes being highly conserved between different *eps* clusters. It is likely that the clustering of the *eps* biosynthesis genes is the consequence of a selective process; according to Wilson & Murray (1991), interdependent functions are more efficiently co-regulated and inherited as a unit if their genes are closely linked rather than arranged separately.

In general, the adsorption inhibition shown by BIMs is highly specific and reversion to phage sensitivity can occur at high frequency, while slow acid production has also been recognised as an associated drawback (Lawrence 1978; King et al. 1983). It remains unclear whether plasmid-mediated adsorption blocking is a true defence mechanism or a secondary effect of some other cellular function. In addition, the inherent instability of these molecules can lead to the development of mixed cultures consisting of phage sensitive and phage resistant strains. Therefore, although the phage resistant cell is protected against attachment by the phage, the particle is not removed or destroyed and remains unaffected in the environment where it can subsequently attack a sensitive cell. Perpetuation of bacteriophages in resistant cultures can promote pseudolysogeny or the 'phage carrier state' (whereby bacteria and lytic phages can co-exist in a culture) which has been described by Vlegels et al. (1988) and de Vos (1989) among others. It is this state which may limit the significance of adsorption inhibition as a potent defence mechanism.

Prevention of phage DNA injection

Relatively little information is available regarding phage injection blocking, although recent investigations have provided some compelling evidence for its existence in LAB. Following successful adsorption to the receptors in the cell wall, an energy -requiring, calcium-dependent, irreversible interaction between the phage or its DNA and the plasma membrane occurs, allowing DNA translocation into the cytoplasm (Monteville et al. 1994). When Marshall & Berridge (1976) isolated mutants of *L. lactis* spp. *lactis* NCDO 510 which were recalcitrant to breakdown by the phage tail lysin of ø510, they presented the first example of lactococcal protection against phage DNA injection. In a study by Watanabe et al. (1984), the phenomenon was exemplified in a derivative of *Lactobacillus casei* YIT9002, in which cellular adsorption by phage PL-1 was normal, but the phage DNA was not internalised. Electron micrographs demonstrated that the DNA remained in the phage capsid while an increasing number of empty phage heads was detected as infection of the sensitive host proceeded undeterred.

In *L. lactis* ssp. *lactis* C2, a membrane protein known as Pip (*p*hage *i*nfection *p*rotein) and an unnamed 32 kDa protein have been implicated in øc2 adsorption (Valeyasevi et al. 1991; Geller et al. 1993; Monteville et al. 1994) but whether or not they play a direct role in triggering the ejection of phage DNA has yet to be determined. The Pip protein is predicted to contain transmembrane regions and a putative secretion signal in the amino terminus (Geller et al. 1993). Monteville et al. (1994) demonstrated that ejection of phage DNA was temperature dependent, occurring rapidly at 30 °C but not at 4 °C, which was most likely a reflection of membrane fluidity functions. Spontaneous mutations in *pip* rendered the host resistant to phage and a recent significant step in the genetic engineering of phage resistance in *L. lactis* LM2301 was performed by Garbutt et al. (1997) where the wild-type chromosomal *pip* was replaced with a mutant allele that was engineered *in vitro*. The exchange left no antibiotic resistance genes or non-lactococcal DNA in the newly modified phage resistant strain, JK101, which was significant in terms of its potential use in the manufacture of foods intended for human consumption.

The first plasmid-encoded injection blocking system recognised in *Lactococcus* was that described by Garvey et al. (1996) who demonstrated the presence of an early acting resistance mechanism against øc2 on the plasmid pNP40, which also encodes two independent abortive infection mechanisms *abiE* and *abiF* (Garvey et al. 1995b). The investigators noticed no impairment in cellular adsorption, yet only 10% of cells were killed following infection, indicating the operation of a mechanism prior to the corruption of host functions. In addition, internalisation of the phage genome was delayed until at least 30 min post-infection as opposed to 5 min in the sensitive host. Moreover, circumvention of the early stages of infection by electroporation of øc2 DNA into resistant hosts resulted in an increase in the number of infected cells releasing progeny phages. The authors concluded that the resistance was due to an alteration in plasma membrane components required for øc2 infection, preventing translocation of its DNA into the cell. The inferior resistance observed at 37 °C accommodated the concept of a system acting at the level of the cell membrane since the temperature effect was presumably related to altered membrane fluidity. However, while there is ample physical evidence to support a phage resistance mechanism which involves phage DNA injection blocking, genetic evidence remains elusive and the loci responsible for this phenotype await identification and characterisation.

Restriction and modification

Following successful adsorption and penetration of the phage particle, the ability to complete an infection can be hindered by the presence of restriction-modification (R/M) systems involving the operation of at least two distinct enzymatic activities. The majority of incoming phage DNA is digested by a host-encoded, site-specific restriction endonuclease, but the cell itself remains protected by modifying its own DNA through the action of a methylase. The severity of restriction is dependent on the system and the phage; in general the efficiency of plaquing (EOP) of the phage decreases logarithmically as the number of sites in the viral DNA molecule increases (Wilson & Murray 1991). While the cellular DNA is prevented from autodegradation, the small number of phage genomes that escape restriction will also become modified. These are no longer recognised as foreign and therefore are not degraded and continue to manipulate cellular machinery for their own successful replication. This reiterates the view of Hill (1993) that R/M is the most powerful (since the phage genome is degraded) yet the most fragile (due to the inherent 'leakiness') of defence systems. Indeed phages have responded to the threat of restriction enzymes through the evolution of many other strategies, such as the elimination of certain restriction sites from the genome, the modification of bases, the production of proteins which inhibit host endonucleases and even the acquisition of methylase genes (Klaenhammer & Fitzgerald 1994).

Many R/M systems have been identified in lactococci and have a common plasmid location with a few notable exceptions where the determinants are chromosomally-derived (see Table 3). Early characterisation of R/M systems concentrated on their effects on the EOP, the variation of restriction activity at different temperatures and the ability to isolate/purify a particular endonuclease. Regardless of their abundance, molecular explanations underlying the regulation of these systems are limited. In those instances where the information is available, it appears that expression of the genes is tightly controlled, and close linkage of restriction and modification genes is a feature common to all R/M systems. Four groups of R/M systems (Table 4) can be distinguished based on the complexity of their enzyme structures and the DNA sequences that are recognised and subsequently cleaved (Wilson & Murray 1991).

Type II R/M systems

The vast majority of the known R/M systems in LAB are of type II which is explained by the relative ease with which their activities can be detected biochemically. These systems usually have simple co-factor requirements and molecular organisations, cleaving at or near the site of recognition which renders them easier to study than the other groups. To date, the recognition sites of eight lactococcal type II systems have been established, while nucleotide sequence data are available for seven of these, in addition to one type IIS (where S refers to *s*hifted cleavage, up to 20 bp on one side of the target). In general, type II systems are composed of two structural genes, an endonuclease (ENase) and a methyltransferase (MTase) with the exception of the *Lla*DCHI (Moineau et al. 1995a) and *Scr*FI (Davis et al. 1993, Twomey et al. 1993, 1997) systems, each of which encode two MTases, which, in the latter case, are known to be independently functional.

*Scr*FI, the activity of which was originally detected in *Lactococcus lactis* ssp. *cremoris* UC503 and other closely related strains, is somewhat unusual for lactococcal type II systems in that it is chromosomally encoded. The genetic organisation of the system, illustrated in Figure 1, reveals the presence of the ENase gene (*scrFIR*) flanked by two MTase genes (*scrFIAM* and *scrFIBM*). While the exact role of the dual MTases is as yet undetermined it has been suggested that the second *Scr*FI MTase may have evolved to counteract the degeneracy of the restriction site (5′-CC/NGG-3′) (Twomey et al. 1997). The deduced amino acid sequence of the MTases revealed ten consensus motifs characteristic of 5-methylcytosine (m^5C) MTases, but surprisingly, given that they both methylate the second cytosine in the target sequence, they differed from each other in their target recognition domains (TRDs) (Twomey et al. 1993). There is evidence to suggest that M.*Scr*FIA is the 'primary' MTase of the system since it protects phages from *in vivo* *Scr*FI digestion and the presence of *scrFIAM* was found to be essential for stable maintenance of the ENase gene. In contrast, *scrFIBM* appeared to be dispensable with respect to maintenace of the ENase gene. In addition, the purified M.*Scr*FIA protein was demonstrated by gel retardation assays to bind to the promoter region directly upstream of the *scrFIAM* gene (Figure 1), suggesting that its expression is autoregulated (D. Butler, personal communication). The ENase gene, *scrFIR*, displayed a unique amino acid sequence even though it is an isoschizomer of *Sso*II.R. *Scr*FIR also represents the first lactococcal ENase to be purified and is the only one which is commercially available. Although it was isolated from a mesophilic strain it is noteworthy that its optimum temperature of activity is 40 °C.

The *Lla*I R/M system encoded by pTR2030 is even more complex than the *Scr*FI system since at least six open reading frames (ORFs) are required for expression of the operon. Anomalous characteristics are associated with both the restriction and methylation components and the putative energy requirements seem more reminiscent of type I systems. M.*Lla*I, capable of performing two distinct MTase activities, appears to be a natural fusion of two N^6-methyladenine (m^6A) MTases and it shares homology with the type

Table 3. Restriction/Modification systems in *Lactococcus*

R/M system	Type	Recognition sequence[a,b]	Source organism and genomic location	Characteristics	Reference
***Lla*1403I**	I	Undetermined	pIL2614 and chromosome of *L. lactis* ssp. *lactis* IL1403	3 ORFs encoding ENase, MTase and specificity subunits, adjacent to Rep protein	Schouler et al. (1998a, b)
Lla*DCHI (previously *Lla*II)	II	5′-GATC-3′	pSRQ700 from *L. lactis* ssp. *cremoris* DCH-4	3 ORFs; 2 m^{6}A MTases *Lla*IIAM and *Lla*IIBM ENase *Lla*IIC – isoschizomer of *Dpn*II and *Mbo*I	Moineau et al. (1995a)
Lla*AI	II	5′-/GATC-3′	pFW094 from *L. lactis* ssp. *cremoris* W9	3 ORFs; 2 m^{6}A MTases, 1 ENase, Isoschizomer of *Dpn*II, *Mbo*I, *Lla*DCHI	Nyengaard et al. (1993, 1995)
***Scr*FI**	II	5′-CC/NGG-3′	Chromosome of *L. lactis* ssp. *cremoris* UC503	4 ORFs; *orfX* of unknown function; ENase gene *scr*FIR; 2 independently functional m^{5}C MTases, *Scr*FIBM, *Scr*FIAM (latter is transcribed separately)	Fitzgerald et al. (1982) Davis et al. (1993) Twomey et al. (1993, 1997)
***Lla*497I** (formerly *Lla*I)	II	5′-CC/WGG-3′	*L. lactis* ssp. *lactis* NCD0497 Location unspecified	Isoschizomer of *Scr*FI Genetics undetermined	Mayo et al. (1991)
***Lla*BI**	II	5′-C/TRYAG-3′	pJW553 from *L. lactis* ssp. *cremoris* W56	Divergently transcribed ENase and MTase genes	Nyengaard et al. (1993; 1995, 1996)
***Lla*CI**	II	5′-A/AGCTT-3′	pAW153 from *L. lactis* ssp. *cremoris* W15	Convergently transcribed ENase and MTase genes; similarity with *Hin*dIII system	Madsen & Josephsen (1998a)
***Lla*DII**	II	5′-GC/NGC-3′	pAW393 from *L. lactis* ssp. *cremoris* W39	Tandemly arranged ENase and MTase genes	Madsen & Josephsen (1998b)
***Lla*KR2I**	II	5′-/GATC-3′	pKR223 from *L. lactis* ssp. *lactis* var. *diacetylactis* KR2	Divergently transcribed ENase and MTase (m^{5}C) genes separated by IS*982* Isoschizomer of *Sau*3AI, but no extensive similarity to *Dpn*II, *Mbo*I, *Lla*DCHI	Twomey et al. (1998)
***Lla*I**	IIS	Undetermined	pTR2030 from *L. lactis* ssp. *lactis* ME2	Operon of 6 ORFs; LlaIC – regulatory protein; *Lla*IM- MTase; R. *Lla*I 1, 2, 3 for restriction activity (possible multisubunit complex); *orf4* – unknown function	Hill et al. (1989) O'Sullivan et al. (1995) O'Sullivan & Klaenhammer (1998)
***Lla*FI**	III	Undetermined	pND801 from *L. lactis* ssp. *lactis* LL42-1	2 ORFs encoding Res and Mod subunits Require Mg^{2+} and ATP; stimulated by AdoMet	Su et al. (1999)

(Modified from Allison & Klaenhammer, 1998).
[a]The cleavage point where known is indicated by /.
[b]N = any nucleotide; W = A or T; R = A or G; and Y = C or T.
ENase = Endonuclease.
MTase = Methylase.
**Lla*DCHI and *Lla*I are highly genetically related.

Table 4. Distinguishing features between Type I, II, IIS, III and IV restriction and modification systems

Type	Structural and genetic organisation	Co-factor requirements and cleavage characteristics
I	Multifunctional, multisubunit complex consisting of restriction, modification and specificity subunits Three separate structural genes (*hsdR*, *M* and *S*)	Mg^{2+}, ATP, AdoMet Cut randomly far from bipartite asymmetric recognition site
II	Separate enzymes for restriction and modification Two separate structural genes	Mg^{2+}, AdoMet Cleavage occurs within symmetric recognition site
IIS[a]	As for type II	Mg^{2+}, AdoMet Fixed distance, up to 20 bp from asymmetric recognition site
III	Multifunctional, multisubunit complex composed of restriction and modification subunits Two separate structural genes	Mg^{2+}, ATP, stimulated by AdoMet Cleavage between 24–27 bp to 3′ side of an asymmetric recognition site
IV	Restriction subunit also has methylase activity and individual methylase Two separate structural genes	Mg^{2+}, Ca^{2+}, AdoMet Fixed distance 14–16 bp from the 3′ side of an asymmetric target site

[a] S refers to shifted cleavage

IIS modification components of the *Fok*I and the *Sts*I R/M systems (Hill et al. 1991; O'Sullivan et al. 1995). Therefore, while in essence it is classified as type IIS, it may actually represent an evolutionary intermediate between type IIS and type I systems. The dual action of M.*Lla*I is deemed necessary for methylation of both strands of an asymmetric target recognition sequence. Restriction of phage DNA requires the operation of three ORFs immediately following *llaIM* and the function of the fourth ORF (ORF4) downstream of *llaIM* remains unclear; however O'Sullivan et al. (1995) proposed that it may promote lateral transfer of the R/M system, based on homology with a motif found in DNA relaxases involved in conjugal transfer. Temperature elevation to 40 °C was found to interfere with the action of the regulatory component (LlaIC) of the *Lla*I system which correlated with a dramatic reduction in restriction activity (O'Sullivan et al. 1995; O'Sullivan & Klaenhammer 1998).

Seven other lactococcal type II systems have been characterised in detail, including the highly related and essentially genetically identical *Lla*DCHI (formerly termed *Lla*II, Moineau et al. 1995a) and *Lla*AI (Nyengaard et al. 1995) systems, isolated from hosts which were genetically distinct and of different geographical origins. Both share a common palindromic target sequence (5′-/GATC-3′) and exhibit extensive similarity to the *Dpn*II system. The molecular organisation consists of three genes transcribed in the same direction, the first two of which encode MTases. The *Lla*DCHI system was successfully expressed in a heterologous food-grade organism, *Streptococcus thermophilus*, conferring resistance to phages isolated from yoghurt and mozzarella wheys (Moineau et al. 1995b).

Recently, Twomey et al. (1998) demonstrated that the recognition site of a novel type II R/M system, designated *Lla*KR2I, is also highly likely to be 5′-/GATC-3′, but interestingly the system did not share any significant sequence similarity with the *Lla*DCHI, *Lla*AI and *Dpn*II systems. While these latter systems methylate DNA using a cognate N^6-methyladenine (m^6A) MTase, M.*Lla*KR2I displayed homology with the 5-methylcytosine (m^5C) MTase family of proteins, exhibiting greatest identity with M.*Sau*3AI. The deduced sequence of the *Lla*KR2I ENase also shared homology with the *Sau*3AI ENase and with *Mut*H, a mismatch repair protein. The organisation of the

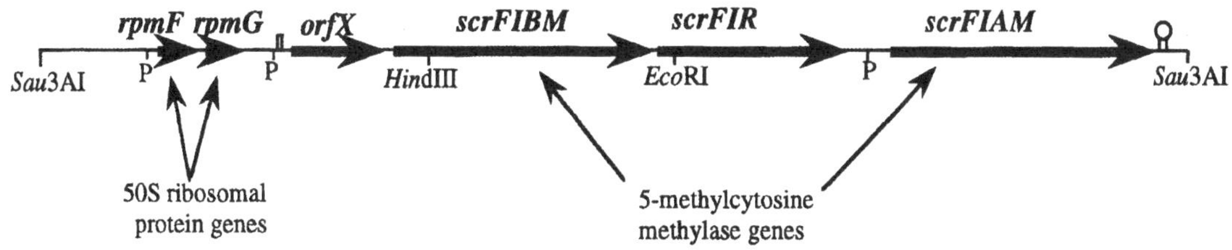

Figure 1. Schematic representation of the molecular organisation of the *Scr*FI restriction-modification locus indicating the presence of the restriction endonuclease gene (*scrFIR*) flanked by two 5′-methylcytosine methylase genes. An open reading frame of unknown function (*orfX*) and two ribosomal protein genes (*rpmF* and *rpmG*) are situated upstream of *scrFIBM*. Promoter sequences are indicated by P. [modified from Garvey et al. (1995a)].

*Lla*KR2I R/M system is unique for type II systems, having an IS element (IS*982*) inserted between the divergently transcribed ENase and MTase genes. The reason for the presence of IS*982* in the operon is unclear as the system was found to operate independently of it and the authors speculated that restriction may originally have been more efficient in its absence. It is generally accepted that the association of IS elements with phage resistance mechanisms may facilitate their dissemination throughout other strains, but their association may also allow some phages to more readily acquire the modification components of R/M systems, due to the increased mobility conferred by the elements. Twomey and co-workers concluded that the removal of IS*982* from the *Lla*KR2I R/M system would provide it with enhanced stability and integrity.

Another pair of isoschizomers exists among lactococcal type II enzymes: the previously described *Scr*FI and the renamed *Lla*497I (formerly *Lla*I; Mayo et al. 1991) with the recognition sequences 5′-CC/NGG-3′ and 5′-CC/WGG-3′, respectively. Although the genetics of the latter system are unknown it would be of interest to determine whether its molecular organisation is also similar to that of *Scr*FI. Nyengaard et al. (1996) reported the cloning and sequencing of the genes coding for the type II system *Lla*BI, which recognises the site 5′-C/TRYAG-3′. This system consists of divergently transcribed ENase and MTase genes, with the MTase showing characteristics of subclass γ of the m^6A group. Josephsen's group also recently analysed the *Lla*CI and *Lla*DII systems which recognise the sequence 5′-A/AGCTT-3′ and 5′-GC/NGC-3′, respectively (Madsen & Josephsen 1998a, b). In both cases restriction was more effective on the P335 phage species than on the 936 species, presumably due to the higher number of type II endonuclease sites in the former (Moineau et al. 1993a). *Lla*CI shows a high degree of similarity with the *Hin*dIII system and is also composed of convergently transcribed ENase and MTase genes. The level of restriction was noted to vary according to gene dosage, being greater when expressed on a high copy number vector. The *Lla*DII ENase and MTase are arranged tandemly, with the MTase displaying typical amino acid sequence motifs indicative of a m^5C MTase.

Nyengaard et al. (1996) suggested that since the overall G+C content of the some R/M systems (e.g. *Lla*BI, *Lla*DII) is lower than the average lactococcal value of 37%, they may have originated from a genus other than *Lactococcus* or this could be an inherent feature of particular R/M systems (low G+C values are also common for the genes encoding abortive infection, which will be discussed later). It is somewhat surprising, considering the relatively low percentage G+C of lactococcal DNA, that the recognition sites of the *Scr*FI, *Lla*497I and *Lla*DII systems is quite high. This implies that these sequences would be encountered less frequently by restriction endonucleases which may impose a practical limitation on the use of such systems as phage defence mechanisms.

R/M systems have also been detected in other members of the lactic acid bacteria. The presence of plasmid-linked R/M systems was demonstrated in three *Lactobacillus helveticus* strains (Reyes-Gavilan et al. 1990) and the activities were not impaired at higher temperatures (60 °C). Takano *et al.* (1990) purified a type II restriction endonuclease designated *Lfe*I from *Lb. fermentum*, which recognised the sequence 5′-GC(A/T)GC-3′ and Auad et al. (1998) implicated the operation of a chromosomally encoded R/M system in *Lb. delbrueckii* ssp. *lactis* CNRZ 326. Biological evidence for chromosomally-linked R/M activity in *S. thermophilus* has been presented in several reports (Benbadis et al. 1990, 1991; Solaiman & Somkuti 1990, 1991; Guimont et al. 1993; Poch et al. 1997) with optimum activity temperatures ranging between 37 and 55 °C. The availability of heat tolerant R/M systems in thermophilic dairy strains could be potentially useful for genetically engineering heat stable R/M systems in mesophilic lactococci, such that their activities would not be affected by the scald temperatures of the various cheese manufacturing processes in which they would be used (Klaenhammer & Fitzgerald 1994).

Type I R/M systems

Type I R/M systems are the most intricate known both biochemically and genetically, which makes them less amenable to analysis than the type II systems. They consist of large multimeric proteins with three subunits responsible for restriction, modification and recognition of the sequence specific for the system (Bickle & Kruger 1993). They require complex co-factors (ATP, Mg^{2+} and AdoMet) for activity and cleave at variable distances (greater than 1000 bp) from the recognition site. Therefore, they do not produce discrete fragments following gel electrophoresis rendering them relatively redundant for routine molecular biology applications and as a consequence, type I systems are frequently detected through their *in vivo* activity but not *in vitro*. They were initially identified and exclusively limited to members of the *Enterobactericiae* such as *Salmonella* and *Escherichia coli* but have been recently identified in the Gram-positive *Mycoplasma pulmonis* (Dybvig & Yu 1994) *Bacillus subtilis* (Xu et al. 1995), *Lactococcus lactis* (Y.-M. Deng & N. W. Dunn, 1997, unpublished, Genbank accession no. AF034786; Schouler et al. 1998a, b) and *S. thermophilus* (O'Sullivan et al. 1999) organisms. These discoveries have stimulated concerted screening programmes to search more extensively for the existence of plasmid and/or chromosomally-located type I R/M systems in LAB. It is proposed that more than 50% of lactococcal R/M systems may be type I (Schouler et al. 1998b), but these may have remained unrecognised because of their complexity. O'Sullivan et al. (1999) located one of the determinants of a type I system on a 6.5 kb plasmid from *S. thermophilus* and while they were unable to attribute any EOP effect to the system, lysis in broth experiments did show a plasmid-free mutant to be more phage-sensitive than the parent.

Type I systems are distinguishable by the manner in which they cleave DNA and the 'hyphenated' nature of the target sequence (Table 4) (for a review see Bickle 1993). Other distuingishing features exist which permit four subclasses or families to be assigned, known as IA, IB, IC (plasmid-borne) and ID. The principal empirical data used to assign members to a family include the biochemical properties of the enzyme, sequence similarities, subunit complementation and transcriptional organisation of the constituent components, namely the *hsdR*, *hsdM* and *hsdS* genes (where *hsd* derives from *h*ost *s*pecificity for *D*NA). The latter two are required to form the methyltransferase while all three are required for activity of the endonuclease. In the system isolated from *Lactococcus*, the gene order appears to be *hsdR*, *hsdM* and *hsdS*, arranged in an operon that frequently includes other genes such as those which specify a replication protein (Schouler et al. 1998a, b), an association which may improve genetic stability of the R/M system. The DNA and therefore the amino acid sequences of the *hsdR* and *hsdM* genes are quite highly conserved within, but not between, a given family. In particular, the predicted R polypeptide sequences share conserved regions with a superfamily of putative helicases, the so-called 'DEAD-box' proteins. These conserved sequences may be associated with the ATP-dependent translocation of DNA that precedes restriction (Murray et al. 1993). The *hsdM* genes display *c*onsensus *m*otifs designated CMI and CMII (NPPY), having a role in co-factor binding and catalysis respectively, and two additional motifs, CMIs and CMIII which are typical of the N^6-adenine class of methylases (Malone et al. 1995; Timinkas et al. 1995; Schouler et al. 1998a).

One of the hallmarks of type I systems is their malleability, manifested by their capacity to adopt new specificities and their potential for evolutionary diversification. These features are dictated by the HsdS subunit whereby merely substituting one for another generates an enzyme that recognises a different sequence of nucleotides. The S subunit contains three consensus core regions that are conserved within a given family (Kannan et al. 1989; Bickle & Kruger 1993) with the notable exception of the lactococcal type IC S subunits which lack the characteristic TAEL direct repeats (Schouler et al. 1998a, b). Two highly variable regions form independent target recognition domains of 150–180 amino acids, each responsible for recognition of one half of the split target sequence and differences in these variable domains imply differences in specificity. It is believed that the HsdS subunit adopts a circular arrangement when complexed with two copies of the the HsdM subunit (Kneale 1994). Schouler et al. (1998b) recently described the acquisiton of new R/M specificities by transferring plasmids bearing only the genetic determinants for the HsdS subunit to a lactococcal strain with a resident, but different, *hsd* locus. In this manner, 'combinational variation' of type I R/M systems may represent a general strategy in which lactococci can acquire R/M systems with novel specificities and may thus be more suitable to deal with the evolution of phage mutants than the type II systems.

Type III R/M systems

While type III R/M enzymes resemble those of the type I category in that they are multimeric, they are composed of just two contiguous structural genes known as *res* and *mod*. The product of the latter, in addition to its role as a methyltransferase, confers the sequence specificity for both the restriction and modification subunits (Bickle 1993). Type III systems require the presence of two inversely oriented asymmetric recognition sites and cleavage occurs approximately 25 nucleotides to one side of the recognition sequence. The reaction has an absolute requirement for Mg^{2+} and ATP and is stimulated by AdoMet. In the presence of both ATP and AdoMet, restriction and modification are competing activities within the same reaction with the result that digestions do not reach completion. One of the distinctive features of type III systems is that methylation takes place only on one strand of the DNA, generating daughter molecules where only half of the the sites are modified following replication. However, since cleavage occurs only if neither strand is modified, protection of the nascent DNA is preserved. Su et al. (1999) described the first Gram positive type III R/M system, designated *Lla*FI, encoded by a 12 kb plasmid (pND801) in *L. lactis* ssp. *lactis* LL42-1. Sequence analysis revealed an operon consisting of two adjacent ORFs exhibiting high homology with previously described type III restriction and modification subunits. *Lla*FI also displayed the co-factor requirements characteristic of these systems being dependent on Mg^{2+} and ATP and stimulated by, but not essentially relying upon, the presence of AdoMet.

Functions of R/M systems

Bickle (1993) believes that the prevalence of DNA restriction systems is a sign of genetic selection at the population level, since a population whose individual members can prevent phage propagation is fitter than one that cannot. The diversity of R/M systems and their common plasmid association demonstrates their effectiveness in natural ecosystems and combinations of systems can be generated which impart higher phage resistance levels. Indeed, in some instances more than one R/M system has been identified within a single strain (Chopin et al. 1984; Josephsen & Vogensen 1989) and they have also been found in combination with other phage resistance systems such as abortive infection (Froseth et al. 1988; Durmaz et al. 1992). In this latter instance, the combination contributes to the overall survival value conferred on the cell; the R/M system acts as the first line of defence, and the abortive infection mechanism will afford protection against the proliferation of modified phages. Since R/M systems are leaky and because phages themselves can acquire a panopoly of mechanisms to escape restriction, only carefully considered combinations of phage resistance systems will counter the genetic flexibility of the phage, increase the degree of insensitivity, and broaden the range of phage types against which the host is resistant (Josephsen & Klaenhammer 1990; Sing & Klaenhammer 1993; Durmaz & Klaenhammer 1995; O'Sullivan et al. 1998). The adaptation of phage defence rotation strategies (PDRS) is particularly effective in the protection of new specialised strains designed for extended use and for meeting specific product requirements in industrial practice.

While R/M systems represent essential elements in strain modification strategies, they are also biologically important in many other situations where they may influence the frequency of conjugation and protoplast transformation (Langella & Chopin 1989; Steenson & Klaenhammer 1986). Their involvement in recombination and plasmid maintenance has been reported for other genera such as *E. coli* (Kulakauskas et al. 1995; Naito et al. 1998). These functions may apply for many lactococcal R/M systems, where their presence could facilitate plasmid stability. It also appears that the evolution and persistence of R/M systems follow as a result of the cell's necessity to protect itself against infection by foreign DNA.

Abortive infection

When a phage has successfully bypassed the early acting resistance mechanisms described above, the last crucial steps involve its intracellular development and the release of progeny phage. Abortive infection (Abi) is a collective term which encompasses a broad range of defences which interfere with these processes, whereby genome replication, transcription, translation, packaging or assembly of phages may be impeded. Abi-mediated resistance typically culmin-

Table 5. Lactococcal *abi* genes for which DNA sequence data is available

Abi	Location	Size (aa)	% G+C	Phage affected	Mechanism	Reference
AbiA	pTR2030; pCI829	628	27	936 sp. (sk1), P335 sp. (31), c2 sp. (c2)	Early	Hill et al. (1990b); Coffey et al. (1991) Dinsmore & Klaenhammer (1994, 1997) O'Sullivan et al. (1995); Dinsmore et al. (1998)
AbiB	unspecified; pCI642	250	27	936 sp. (bIL66, bIL70)	Late	Cluzel et al. (1991); Parreira et al. (1996b); Forde et al. (1999)
AbiC	pTN20	344	27	936 sp (p2, jj50, sk1), P335 sp. (u136, m12r)	Late	Durmaz et al. (1992)
AbiD*	pBF61	366	29	936 sp. (sk1), c2 sp (c2)	Late	McLandsborough et al. (1995)
AbiD1*	pIL105	351	26	936 sp. (bIL66, bIL70), c2 sp. (c6A, bIL67)	Late	Anba et al. (1995); Bidnenko et al. (1995)
AbiE	pNP40	286, 297	29, 29	936 sp. (712)	Late	Garvey et al. (1995b)
AbiF*	pNP40, pAJ2074	341	26	936 sp. (712), c2 sp. (c2)	Early	Garvey et al. (1995b); Jarvis (1988, 1992)
AbiG	pCI750	249, 397	29, 27	936 sp. (712, sk1), c2 sp. (c2)	Late	O'Connor et al. (1996, 1999)
AbiH	chromosome	346	26	936 sp (59), c2 sp. (53)	–	Prevots et al. (1996)
AbiI	pND852	331	29	936 sp. (712), c2 sp. (c2)	Late	Su et al. 1997
AbiJ	pND859	282	30	936 sp. (712)	–	Deng et al. (1997)
AbiK	pSRQ800	599	24	936 sp. c2 sp. P335 sp.	Early	Emond et al. (1997)
AbiL	pND861	458, 297	28, 29	936 sp. (712), c2 sp. (c2)	Late	Deng et al. (1999)
AbiN	chromosome	178	31	936 sp. (59), c2 sp. (53)	–	Prevots et al. (1998)
AbiO	pPF144	540	26	936 sp. (59), c2 sp. (53)	–	Prevots & Ritzenthaler (1998)
AbiP	pIL2614	244	27.5		–	Portugal et al. – Unpublished; Schouler et al., (1998a); accession no. U90222
AbiQ	pSRQ900	183	28	936 sp. (p2), c2-like sp. (p2)	Late	Emond et al. 1998

* AbiD, AbiD1 and AbiF are similar (26-47% identity)

ates in death of the infected cell because corruption of host functions has already been initiated prior to instigation of the defense activity. However, the 'suicidal' sacrifice of a single cell is insignificant as long as phage levels do not increase. Abi systems are phenotypically manifested by a reduction in EOP, a decrease in the ECOI (*E*fficiency of the *C*entre *O*f *I*nfection) and the development of smaller plaques due to decreased burst size. Although many Abi systems have been identified and characterised even to the DNA sequence level (Table 5), it is still the case that the exact mechanisms of action and the accompanying altruism remain unresolved in most cases. However, research efforts directed towards elucidating the modes of action of Abi systems are intensifying and valuable information has been gained from studies of phage mutants capable of overcoming the abortive mechanism (Bidnenko et al. 1995; Dinsmore & Klaenhammer 1997). In turn, efforts to determine the specific functionality of Abi systems at a molecular level are unveiling the specific events involved in the intracellular development of phage.

Seventeen lactococcal *abi* genes have been sequenced to date (Table 5) and at least two other genetically undefined systems have been recorded (Geis et al. 1992; Coakley et al. 1997), while in *S. thermophilus*, a single report on a possible Abi mechanism has been made (Larbi et al. 1992). All lactococcal Abi systems are characterised by an atypically low G+C content (24–31%) which is significantly below the average (37%) determined for other lactococcal genes, reflecting the possibility of horizontal gene transfer from another species of low G+C content (O'Connor et al. 1996). All predicted proteins appear to be of average hydrophobicity and suggest a cytosolic location; only AbiC (Durmaz et al. 1992) was reported to possess two putative transmembrane domains. The systems are commonly expressed by a single ORF, with the exceptions of AbiE (Garvey et al. 1995b), AbiG (O'Connor et al. 1996) and AbiL (Deng et al. 1999) which require two genes for expression. Of the 17 systems that have been sequenced only two are encoded by chromosomal DNA, the remainder are plasmid-encoded. In only three instances were independently isolated systems found to be identical, i.e. the pTR2030 and pCI829-derived *abiA* genes (Hill et al. 1990; Coffey et al. 1991) which, ironically, were the first two to be sequenced. In addition, the phage resistance plasmid pAJ2074 (a deletion derivative of pAJ1106; Jarvis 1988, 1992) was recently shown to harbour a copy of the pNP40-derived *abiF* gene (O. Cahalane, personal communication). Forde et al. (1999) also reported a plasmid-located homologue of the abortive infection gene *abiB*, which, when originally identified by Cluzel et al. (1991) had not been localised to either plasmid or chromosomal DNA.

The other significant homologies exist between the AbiD, AbiD1 and AbiF proteins (summarised by Garvey et al. 1995a; Allison & Klaenhammer 1998), which share 26–47% identity and display a number of highly conserved regions with 18% of the residues being common to all three. This genetic relatedness could be indicative of a family of Abi systems in *Lactococcus*. However, although they are effective against similar phage species, their individual modes of action and regulatory requirements appear to be quite different. Firstly, the product of *abiF* was shown to inhibit the rate of intracellular replication of ø712 and øc2 DNA implying an early acting mechanism and while AbiD confers resistance to øsk1, replication is not affected which is suggestive of a late-acting Abi mechanism. In recent studies, a ø712 mutant was derived which showed partial resistance to the action of AbiF but which remained sensitive to AbiD1, indicating that even if these proteins are similar, they probably do not act on the same target (A. Rince & G.F. Fitzgerald, unpublished results). There may also be differences in the way in which the expression of these Abi systems is determined. It has been demonstrated that the phenotype mediated by AbiF was only completely expressed in the presence of the chromosomally encoded *recA* gene (Garvey et al. 1995b, 1997). The *abiD* promoter region was shown to contain large inverted repeats which interestingly also exhibit high homology (71%) with the cos site of øc2 (McLandsborough et al. 1995), suggesting an evolutionary link between the two or indeed phage participation in gene expression. Despite this extensive homology, the AbiD system is marginally more effective against øsk1 than øc2 itself. For *abiD1* (formerly known as *abi105*), a strong consensus promoter was shown to drive the production of a small transcript of about 50 bases which stops at a terminator located upstream of the gene. Transcripts corresponding to *abiD1* were not detectable indicating that gene expression was low and tightly controlled. AbiD1 was actually found to be toxic in *L. lactis* and *E. coli* when expressed under the control of a strong T7 promoter (Anba et al. 1995; Bidnenko et al. 1995).

AbiA and AbiK (Emond et al. 1997), which confer temperature sensitive defences, are also predicted to be in the same protein family based on amino acid composition, molecular weight, content of bulky and small residues, average hydrophobicity and average charge. Indeed, phage DNA replication and structural protein synthesis was undetected in infected cells harbouring either Abi system, suggesting a similar role for both. An increase in the copy number of *abiA* significantly enhanced the efficacy of the phage resistance mechanism (Casey et al. 1992; Dinsmore & Klaenhammer 1994), and similar results were obtained following an alteration in the gene dosage of *abiK*. However, Emond and co-workers (1997) were unable to overexpress *abiK* constitutively in *Lactococcus* suggesting that it is tightly regulated and may be toxic to the cell, whereas *abiA* is not.

AbiJ (previously designated Abi-859; Deng et al. 1997) also displayed some relatedness to AbiA (exhibiting 22.7% identity and 48.7% similarity using the BESTFIT programme), but an alignment between the two proteins failed to identify any significant conserved regions. Both AbiJ and AbiD1 (Anba et al. 1995) were proposed to have helix-turn-helix motifs (two and one, respectively) implicating a possible mode of action involving direct interaction between the gene products and the phage DNA (or RNA in the case of AbiD1; Anba et al. 1995). Also noteworthy is the absence of cysteine residues in the products of *abiJ*, *abiC* and *abiEi*, but it is not known if this is coincidental or if it has a significance in terms of the functionality of these Abi proteins.

Mode of action of lactococcal Abi systems

According to Garvey et al. (1995a), Abi systems are segregated depending on whether they act prior to or at the level of DNA replication (early) or after replication has occurred (late). Based on these criteria, the previously mentioned AbiA, AbiF and AbiK systems fall into the category of 'early' acting mechanisms, and more recently the Abi-containing pMRC01 was also demonstrated to mediate a 5 min delay and an inhibition of øc2 DNA replication in cells harbouring the plasmid (Coakley et al. 1997). Other systems which have been examined and which have not been correlated with deficient phage DNA replication may be classified as 'late' acting mechanisms (Table 5).

Although a number of systems fall within the 'late' acting category, in only relatively few cases has the specific target or cellular effect been identified. This includes the as yet genetically undefined Abi system encoded by plasmid pBU1-8 from *L. lactis* ssp. *lactis* Bu2, which operates by completely blocking phage transcription (Geis et al. 1992). Recent investigations by O'Connor et al. (1999) demonstrated that øsk1 RNA synthesis was inhibited in cells containing AbiG, encoded by the lactococcal plasmid pCI750. Furthermore, evidence was also presented for the possible existence of a second abortive mechanism on pCI750. Parreira et al. (1996b) have shown that while the onset of RNA synthesis in *L. lactis* ssp. *lactis* IL1403 cells harbouring AbiB was normal, a dramatic degradation of sensitive øbIL70 transcripts was evident 10–15 min post-infection. Mapping of the 5' ends of the degradation products revealed endonucleolytic activity with cleavage occurring preferentially at U/U, A/U and U/A sites, resembling the cleavage specificity reported for panceatic RNase A and RNase M from *E. coli*. Expression of the 753 bp *abiB* gene was shown to be mediated by a promoter situated in an upstream iso-IS*S1* element and the RNase activity observed was proposed to be either induced or stimulated by an early phage product. In contrast, the *abiB* homologue recently located on the plasmid pCI642 from *L. lactis* ssp. *cremoris* HO2 was found to be poorly expressed and did not promote degradation of ø712 RNA transcripts (Forde et al. 1999). Defects in the MCP (ie. major capsid protein) synthesis of øu136 was observed in the case of AbiC-containing cells (Moineau et al. 1993b). Monoclonal antibodies raised against this protein were utilised in an ELISA test system for quantitative assessment of phage protein accumulation in these cells and was found to be reduced by 50% (Moineau et al. 1993b, c).

Researchers in the laboratories of T.R. Klaenhammer, M.-C. Chopin and G.F. Fitzgerald have provided important insights as to the mechanisms of action of the AbiA, AbiD1 and AbiF systems, respectively, through the study of Abi-resistant phages. Heteroduplex analysis of ø31 and its AbiA-insensitive derivative, ø31A, allowed Dinsmore & Klaenhammer (1997) to identify a point mutation (G to T) in a 738 bp open reading frame (ORF245) of the ø31A genome, which resulted in an arginine-to-leucine change at the amino acid level. The presence of the mutated ORF245 of ø31A reduced the sensitivity of ø31 to AbiA when provided *in trans*. Immediately upstream of this region two other ORFs, ORF169 and ORF71 were identified which were separated by an 84 bp intergenic region consisting of two inverted repeats, and sequence analysis of an independently isolated AbiA-insensitive mutant, ø31B, revealed a G to A substitution in one of the inverted repeats. A 118 bp fragment (termed L2R2) from the wild-type ø31, encompassing this 84 bp region, virtually eliminated AbiA activity against ø31 when present *in trans*, establishing a correlation between AbiA and this region of the ø31 genome. The authors speculated that the L2R2 fragment provided a false target for AbiA or alternatively, titrated away a phage regulatory factor resulting in an excess of the AbiA phage target. In addition, it was noted that ORFs 169 and 245 are not present in the prolate-headed øc2 genome even though AbiA does retard øc2 infections, albeit at a lower efficiency than the small isometric-headed phages. Based on this observation it was proposed that AbiA could function differently against different phage species. More recent investigations revealed that mutations in a predicted leucine zipper motif and an adjacent charged region of the AbiA protein affected phage resistance levels to varying degrees, demonstrating the necessity of these regions for complete activity of the mechanism (Dinsmore et al. 1998).

A middle-expressed phage operon (termed the M-operon) consisting of four ORFs was implicated in the action of AbiD1 (Bidnenko et al. 1995). In this instance, the operon, which was derived from øbIL66 phage mutants capable of overcoming the AbiD1 phage resistance, allowed the development of AbiD1-sensitive phages when present *in trans* in AbiD1 host cells. Sequence comparisons of a 72 bp region of *orf1* in the wild-type bIL66 and four independent bIL66 mutant phages revealed that a point mutation had occurred suggesting that this *orf* is necessary for sensitivity to AbiD1. The sensitivity was partly suppressed when *orf3* was overexpressed *in trans*. The specified protein of this *orf* exhibited a putative ATP-binding site and 43% identity with the putative gene product of an ORF upstream of the lysin genes of bIL67 and øvML3. It has been proposed that the gene product of *orf1* interacts with the *abiD1* gene product to prevent the translation of *orf3* RNA, the expression of which is critical for normal phage development. More recently, Bidnenko et al. (1998) revealed that the *orf3* product displayed amino acid homology with the *E. coli* RuvC Holliday junction resolvase, and observed endonucleolytic activity upon induction of the operon in an *E. coli* expression system. Analysis has shown that both *orf2* and *orf3* are essential for nucleolytic activity, leading the authors to speculate that these code for a structure-specific endonuclease (M-nuclease), which might be essential for phage multiplication. Thus it is possible that AbiD1 acts by limiting an essential phage-encoded endonuclease.

The procedure of applying heteroduplex analysis was successfully used to detect a mutation (G to A; valine to isoleucine) in the *l6* gene of a øc2 mutant which was resistant to the AbiF mechanism (A. Rince & G.F. Fitzgerald, unpublished results) implicating this region of the phage genome as the AbiF target. This group also isolated AbiF-resistant mutants of ø712 and identified a 1 kb region from øsk1 which could protect ø712 from the abortive mechanism. It was found that the AbiF-resistant mutants were sensitive to the action of AbiA, confirming that the mechanisms of operation of the two Abi's are different. As previously stated, there was also no relationship between AbiF- and AbiD1-resistant mutants despite the genetic similarities between the Abi proteins.

It is evident that considerable advances have been made in lactococcal Abi research and the list of novel unrelated Abi's is continuously expanding. Only within the last year sequence data have become available for five novel abortive mechanisms: AbiN (Prevots et al. 1998), AbiO (Prevots & Ritzenthaler 1998), AbiP (formerly Abi420) (Portugal et al. unpublished data; Schouler et al. 1998a; Genbank accession no. U90222), AbiQ (Emond et al. 1998) and AbiL (Deng et al. 1999). They exhibit no extensive homology to any of the other Abi systems, and beyond determining the extent of resistance that each confers, the specific mechanisms by which they halt infections have not yet been elucidated. They are likely to involve multiple targets (DNA, RNA or proteins) with expression of the genes being controlled by many factors including gene copy number (*abiN* and *abiO* are potentially toxic when cloned on high copy number plasmids), association with IS elements, temperature sensitivity and host background. Sequence comparisons of ORFs adjacent to the *abiN* gene have indicated a relationship with the temperate phage, ørlt (van Sinderen et al. 1996; Prevots et al. 1998) and Emond et al. (1998) speculated that the morphogenic pathway of c2- and 936-type phages is targeted by the AbiQ protein.

It appears that Abi systems have developed independently through different pathways and are directed against different phage types. They are frequently combined with other phage defence mechanisms within a single strain or indeed more than one Abi system can exist within a given strain. *L. lactis* ssp. *lactis* ME2 represents the prototype phage resistant strain, harbouring genes for adsorption inhibition, two R/M systems and two abortive infection mechanisms (Hill et al. 1989; Hill 1993; Klaenhammer & Fitzgerald 1994; Garvey et al. 1995a). A single conjugative plasmid, pNP40, codes for at least three distinct phage resistance mechanisms, two (AbiE and AbiF) of which mediate abortive infection and the third which inhibits phage DNA injection (Garvey et al. 1995b, 1996). Abortive infection presumably comes into play only in those instances where a phage infection could not be sufficiently hindered at previous steps, so that the incidences of cell death are reduced to a minimum (Josephsen & Neve 1998). While analysis of lactococcal Abi systems has proved rewarding in view of their genetic diversity, researchers are constantly challenged to answer many currently unresolved questions concerning the abortive response. The occurrence of Abi-resistant phages will continue to provide information as to the nature and activity of these varied mechanisms and the creative manner in which they are evaded by the phages themselves.

Engineered phage resistance

While nature itself has donated to bacterial hosts many of the mechanistic strategies to cope with phage attack, the development of cleverly contrived genetic engineering regimes has provided additional weapons towards combating phage proliferation. These 'non-natural' mechanisms suffer the weakness of being highly specific, and of course, industrial and consumer acceptance of recombinant DNA technology remains a grey area (Verrips & van den Berg 1996; Daly et al. 1998). Although restricted in this capacity, they do indicate the potential of modern biotechnological practices for the design and implementation of effective phage resistance systems. The method of gene displacement by which Garbutt et al. (1997) manipulated the Pip protein to engineer phage resistance was addressed in a previous section. Other systems exploit sequences derived from the phage genome as a means of inhibiting the parent or related phage types in a manner similar to abortive infection. They are based on the principle that the phage life cycle can be interrupted if, in the host, its gene products are present in a dysfunctional form, present in excess, or expressed at the wrong stage of development (Sanford & Johnston 1985).

One such system, known as Per (*P*hage *e*ncoded *r*esistance), is based on a detailed knowledge of the phage genome. Hill et al. (1990) first introduced a 1.4 kb fragment containing the origin of replication (*ori*) of ø50 into *L. lactis* ssp. *lactis* NCK203 and enhanced the host's insensitivity to ø50 infection, manifested by a reduction in the EOP and by smaller plaques. The Per construct provided false signals for phage replication and ultimately retarded phage development. A second Per mechanism, Per31, was later generated by cloning the ø31 *ori*, leading to a similar phenotypic response (O'Sullivan et al. 1993). In a subsequent study, both Per50 and Per31 were successfully utilised as part of a phage defence rotation strategy (PDRS) to help maximise the resistance against mutant and recombinant phages (Durmaz & Klaenhammer 1995). When the genetic region coding for the putative replisome organiser protein of the temperate lactococcal bacteriophage Tuc2009 was supplied on a high copy number plasmid vector, it was found to confer a Per phenotype on its host (McGrath et al. 1999). The Per effect has also recently been detected in *S. thermophilus* strains, whose phages are segregated into at least two distinct replication groups. A non-coding region, located just downstream of the putative primase of øSfi21 (Foley et al. 1998) and identical in the closely related øO1205 (Stanley et al. 1997), was identified as the replicating origin of these phages (replication group I). The cloned origins exhibited an inhibitory activity against infections by blocking the accumulation of phage DNA, while two putative origins from a group II phage, ø7201, provided more modest resistance. It was also notable that a cross-resistance, affecting both replication groups (albeit at a lower level) was provided by the ø1205 *ori* (E. Stanley, personal communication). Per systems are generally energy-efficient and target-specific, being effective primarily against the phage group from which the *ori* is derived. They are also very much dependent on the copy number of the plasmid carrying the *ori* and on the strain in which the construct is placed. An in depth description of Per-based insensitivity is presented by G. Venema et al. (1999).

An alternative means of generating engineered phage resistance was based on an antisense RNA technology (Kim & Batt 1991c; Batt et al. 1995). A lactococcal phage ORF of unknown function, termed *gp51C*, was cloned in an antisense orientation with respect to a strong lactococcal promoter. The construct provided phage resistance when introduced to a lactococcal recipient, although further constructs based on truncated versions of the gene were less effective. Antisense RNA directed against other genes, *gp18C* and *gp24C* of ø7-9 (Kim et al. 1992), and the major capsid protein of øF4-1 (Chung et al. 1992) provided a less impressive level of insensitivity. Recently, Polzin *et al.* (1996) reported no inhibition against four early and three late ORFs of øc2 using an antisense strategy, and suggested that the technology may not be as effective against prolate-headed phages. The efficiency of inhibition requires care in selecting the phage gene and depends on the level of transcription of the gene as well as the amount of antisense RNA produced. In general, the greater the ratio of antisense to sense RNA, the greater the inhibition. Batt and co-workers (1995) have also pursued the technology for rapid identification of conserved regions of phage genomes.

Another approach used to engineer phage-derived resistance was devised by Djordjevic et al. (1997) where they cloned the lethal *Lla*I three-gene restriction cassette downstream of a ø31 inducible promoter to trigger the expression of an explosive suicide system. This combination of a phage 'trap' and programmed cell death mimiced an abortive infection phenotype, which was more potent when the regulatory protein LlaIC (see R/M section) is supplied *in trans*. Another interesting application of the *Lla*I R/M system was its exploitation by Djordjevic & Klaenhammer (1996) for the development of a positive-selection cloning vector for Gram positive microorganisms.

Response of bacteriophages and prospects for culture improvement

Given that LAB possess several unrelated genes, gene systems and mechanisms (either natural or 'non-natural') devoted towards controlling phage attack, the apparent ease with which the viruses can adapt to and overcome them is quite disillusioning. The emergence of 'new' lytic phages, whether factory- or genetically-derived has been observed in a number of instances where exposure to resistance systems has contributed to the appearance of novel phages. This phenomenon was first described by Hill et al. (1991) who noted that ø50 was capable of overcoming the strong resistance encoded by *Lla*I on pTR2030. The acquisition of a functional truncated methylase gene (*Lla*PI) from pTR2030 was demonstrated to be responsible for the circumvention of host restriction activities on subsequent phage infections. Studies have also suggested that phages resistant to endonucleolytic activity may be due to the paucity or loss of restriction sites during the course of phage evolution (Moineau et al. 1993a). Resistances to AbiA, AbiC, AbiD1, AbiF, Per50 and Per31 have been shown to involve specific mutational responses and/or gross genetic rearrangements (O'Sullivan et al. 1993; Dinsmore & Klaenhammer 1994, 1997; Moineau et al. 1994b; Bidnenko et al. 1995; A. Rince & G.F. Fitzgerald, unpublished). To overcome the AbiC mechanism, øu136 actually mutated to become a morphologically distinct phage, øu137, by incorporating DNA fragments from the *L. lactis* ssp. *lactis* NCK203 host chromosome. This newly formed phage was also insensitive to Per50, and disruption of the particular chromosomal sequences in the starter culture prevented the selection of phage mutants corresponding to øu137 during subsequent infections with øu136 (Moineau et al. 1994b).

It has long been recognised that the conjugative nature of many phage resistance plasmids can be exploited to produce cultures tailored to exhibit more desirable insensitivity properties (Sanders et al. 1986; Sing & Klaenhammer 1986; Jarvis 1988; Coffey et al. 1989; Kelly et al. 1990; Klaenhammer 1991; Harrington & Hill 1991; Coakley et al. 1997). In a few instances the procedure has been facilitated by the concomitant presence on the plasmids of food-grade selectable markers such as nisin (Harrington & Hill 1991) and lacticin 3147 (Coakley et al. 1997). As stated previously, the use of isogenic single strain starter rotation schemes, in which complementary defences are rotated within one starter background (Sing & Klaenhammer 1993; Durmaz & Klaenhammer 1995), can limit the exposure of phages to particular resistance mechanisms. In turn, it increases the life span of the modified phage resistant strain since the selective pressure on phages to adapt and survive is reduced. O'Sullivan et al. (1998) successfully manipulated the protocol for the development of a phage resistant single-strain starter, *L. lactis* ssp. *cremoris* DPC5000, for use in Cheddar cheese manufacture. They sequentially stacked three plasmids encoding distinct phage resistance mechanisms (adsorption inhibition, R/M and Abi) followed by introduction of the lactose, proteinase plasmid pLP712 to generate a host with acceptable fermentation characteristics. It represents the first case where a stacking approach included an adsorption inhibition mechanism, and emphasises the need for the regular characterisation of new mechanisms which can be incorporated in rotation systems to help minimise the emergence of novel phages. The industrial applications of phage resistance plasmids are discussed in greater detail by Moineau in a separate chapter.

Concluding remarks

The resources and vigilant efforts which have been employed for the analysis of LAB phage-host interactions over the past two decades have been of paramount importance in both fundamental and applied research. Accompanying the substantial level of information which has been generated is the tremendous advances in the construction of customised phage resistant strains for industrial applications. It has paved the way for the continuous identification of novel phage defences and the development of 'non-natural' systems to achieve the ultimate goal of stringent, long term phage protection. The instances of 'non-natural' resistance which have already been described may signal the focus of future research. In addition, the pool of complete phage genomic sequences rapidly emerging is invaluable for the understanding and control of phage proliferation and evolution. Thus, considering the actual and anticipated benefits of these research efforts, the development of efficient starter cultures with in-built effective phage protection is now an expectation which is likely to be realised in the near future.

References

Ackermann H-W & DuBow MS (1987) Bacteriophage taxonomy. In: Viruses of prokaryotes, Vol. 1, (pp. 1–11). CRC Press, Boca Raton

Akcelik M & Tunail N (1992) A 30 kDa cell wall protein produced by plasmid DNA which encodes inhibition of phage adsorption in *Lactococcus lactis* ssp. *lactis* P25. Milchwissenschaft 47: 215–217

Allison GE & Klaenhammer TR (1998) Phage resistance mechanisms in lactic acid bacteria. Int. Dairy J. 8: 207–226

Anba J, Bidnenko E, Hillier A, Ehrlich SD & Chopin M-C (1995) Characterisation of the lactococcal *abiD1* gene coding for phage abortive infection. J. Bacteriol. 177: 3818–3823

Arendt EK, Daly C, Fitzgerald GF & van de Guchte M (1994) Molecular characterisation of lactococcal bacteriophage Tuc2009 and identification and analysis of genes encoding lysin, a putative holin and two structural proteins. Appl. Environ. Microbiol. 60: 1875–1883

Auad L, Peril MAA, Holgado AAPR & Raya RR (1998) Evidence of a restriction-modification system in *Lactobacillus delbrueckii* ssp. *lactis* CNRZ 326. Curr. Microbiol. 36: 271–273

Auvray F, Coddeville M, Ritzenthaler P & Dupont L (1997) Plasmid integration in a wide range of bacteria mediated by the integrase of *Lactobacillus delbrueckii* bacteriophage mv4. J. Bacteriol. 179: 1837–1845

Batt CA, Erlandson K & Bsat N (1995) Design and implementation of a strategy to reduce bacteriophage infection of dairy starter cultures. Int. Dairy J. 5: 949–962

Benbadis L, Faelen M, Slos P, Fazel A & Mercenier A (1990) Characterisation and comparison of virulent bacteriophages of *Streptococcus thermophilus* isolated form yoghurt. Biochimie: 72: 855–862

Benbadis L, Garel JR & Hartley DL (1991) Purification, properties and sequence specificity of *Ssl*I, a new type II restriction endonuclease from *Streptococcus salivarius* ssp. *thermophilus*. Appl. Environ. Microbiol. 57: 3677–3678

Beresford TPJ, Ward LJH & Jarvis AW (1993) Temporally regulated transcriptional expression of the genomes of lactococcal bacteriophages c2 and sk1. Appl. Environ. Microbiol. 59: 3708–3712

Bickle TA & Kruger DH (1993) Biology of DNA restriction. Microbiol. Rev. 57: 434–450

Bickle TA (1993) The ATP-dependent restriction enzymes. In: Linn SM, Lloyd RS & Roberts RJ (Eds) Nucleases, 2nd edn (pp 89–109). Cold Spring Harbor Laboratory Press, Cold Spring Harbor, New York

Bidnenko E, Ehrlich SD & Chopin M-C (1995) Phage operon involved in sensitivity to the *Lactococcus lactis* abortive infection mechanism AbiD1. J. Bacteriol. 177: 3824–3829

Bidnenko E, Ehrlich SD & Chopin MC (1998) *Lactococcus lactis* phage operon coding for an endonuclease homologous to RuvC. Mol. Microbiol. 28: 823–834

Birkeland N-D (1994) Cloning, molecular characterisation and expression of the genes encoding the lytic functions of lactococcal bacteriophage øLC3: a dual lysis system of modular design. Can. J. Microbiol. 40: 658–665

Botstein (1980) A theory of modular evolution for bacteriophages. Annals of the New York Academy of Science 354: 484–491

Boyce JD, Davidson BE & Hillier AJ (1995) Sequence analysis of the *Lactococcus lactis* bacteriophage BK5-T and demonstration that the phage DNA has cohesive ends. Appl. Environ Microbiol. 61: 4089–4098

Bradley DE (1967) Ultrastructure of bacteriophages and bacteriocins. Bacteriol. Rev. 31: 230–314

Brondsted L & Hammer K (1999) Use of the integration elements encoded by the temperate lactococcal bacteriophage TP901-1 to obtain chromosomal single-copy transcriptional fusions in *Lactococcus lactis*. Appl. Environ. Microbiol. 752–758

Brooker BE (1976) Cytochemical observations on the extracellular carbohydrate produced by *Streptococcus cremoris*. J. Dairy Res. 43: 283–290

Bruttin A, Desiere F, Lucchini S, Foley S & Brussow H (1997) Characterisation of the lysogeny DNA module from the temperate *Streptococcus thermophilus* bacteriophage øSfi21. Virology 233: 136–148

Budde-Niekiel A & Teuber M (1987) Electron microscopy of the adsorption of bacteriophages to lactic acid streptococci. Milchwissenschaft 42: 551–553

Callegari ML, Cocconcelli PS, Kok J, Venema G & Morelli L (1996) Heterologous expression of *Lactobacillus helveticus* CNRZ 892 S-layer gene in *Lactococcus lactis* mg1363. In: Proceedings of 5th Symposium on Lactic Acid Bacteria, Genetics, Metabolism and Applications, Veldhoven, the Netherlands. Abs D12

Callegari ML, Riboli B, Sanders JW, Cocconcelli PS, Kok J, Venema G, Morelli L (1998) The S-layer gene of *Lactobacillus helveticus* CNRZ 892: cloning sequence and heterologous expression. Microbiology 144: 719–726

Casey CN, Morgan E, Daly C & Fitzgerald GF (1993) Characterisation and classification of virulent lactococcal bacteriophages isolated from a Cheddar cheese plant. J. Appl. Bacteriol. 74: 268–275

Casey J, Daly C & Fitzgerald GF (1992) Controlled integration into the *Lactococcus* chromosome of the pCI829-encoded abortive infection gene from *Lactococcus lactis* ssp. *lactis* UC811. Appl. Environ. Microbiol. 58: 3283–3291

Chandry PS, Davidson BE & Hillier AJ (1994) Temporal transcription map of the *Lactococcus lactis* bacteriophage sk1. Gene 138: 123–126

Chandry PS, Moore SC, Boyce JD, Davidson BE & Hillier AJ (1997) Analysis of the DNA sequence, gene expression origin of replication and modular structure of the *Lactococcus lactis* lytic bacteriophage sk1. Mol. Microbiol. 26: 49–64

Chopin A, Chopin M-C, Mollio-Batt A & Langella P (1984) Two plasmid determined restriction and modification systems in *Streptococcus lactis*. Plasmid 11: 260–263

Christiansen B, Brondsted L, Vogensen FK & Hammer K (1996) A resolvase-like protein is required for the site-specific integration of the temperate lactococcal bacteriophage TP901-1. J. Bacteriol. 178: 5164–5173

Chung DK, Chung SK & Batt CA (1992) Antisense RNA directed against the major capsid protein from *Lactococcus lactis* ssp. *cremoris* bacteriophage F4-1 confers partial resistance to the host. Appl. Microbiol. Technol. 37: 79–83

Chung DK, Kim JH & Batt CA (1991) Cloning and nucleotide sequence of the major capsid protein from *Lactococcus lactis* ssp. *cremoris* bacteriophage F4-1. Gene 101: 121–125

Cluzel P-J, Chopin A, Ehrlich SD and Chopin M-C (1991) Phage abortive infection mechanism from *Lactococcus lactis* ssp. *lactis*, expression of which is mediated by an iso-ISS*1* element. Appl. Environ. Microbiol. 57: 3547–3551

Coakley M, Fitzgerald GF & Ross RP (1997) Application and evaluation of phage resistance- and bacteriocin-encoding plasmid pMRC01 for the improvement of dairy starter cultures. Appl. Environ. Microbiol. 63: 1434–1440

Coffey AG, Fitzgerald GF & Daly C (1989) Identification and characterisation of a plasmid encoding abortive infection from *Lactococcus lactis* ssp. *lactis* UC811. Neth. Milk. Dairy J. 43: 229–244

Coffey AG, Fitzgerald GF & Daly C (1991) Cloning and characterisation of the determinant for abortive infection from the lactococcal plasmid pCI829. J. Gen. Microbiol. 143: 1355–1362

Costello V (1988) Characterisation of bacteriophage-host interaction in *Streptococcus cremoris* UC503 and related streptococci. Ph.D. thesis, National University of Ireland

Crow VL & Gopal PK (1995) Cell surface differences in lactococcal strains. Int. Dairy J. 5: 45–68

Daly C, Fitzgerald GF & Davis R (1996) Biotechnology of lactic acid bacteria with special reference to bacteriophage resistance. Antonie van Leeuwenhoek 70: 99–110

Daly C, Fitzgerald GF, O'Connor L & Davis R (1998) Technological and health benefits of dairy starter cultures. Int. Dairy J. 8: 195–205

Davis R, van der Leile D, Mercenier A, Daly C & Fitzgerald GF (1993) *Scr*FI restriction-modification system of *Lactococcus lactis* ssp. *cremoris* UC503: cloning and characterisation of two *Scr*FI methylase genes. Appl. Environ. Microbiol. 59: 777–785

de Ruyter PGGA, Kuipers OP, Meijer WC & de Vos WM (1997) Food-grade controlled lysis of *Lactococcus lactis* for accelerated cheese ripening. Nat. Biotechnol. 15: 976–979 de Vos WM (1989) On the carrier state of bacteriophages in starter lactococci: an elementary explanation involving a bacteriophage resistance plasmid. Neth. Milk Dairy J. 43: 221–227

de Vos WM, Underwood HM & Davies FL (1984). Plasmid-encoded bacteriophage resistance in *Streptococcus cremoris* SK11. FEMS Microbiol. Lett. 23: 175–178

Deng Y-M, Harvey ML, Liu CQ & Dunn NW (1997) A novel plasmid-encoded phage abortive infection system for *Lactococcus lactis* biovar *diacetylactis*. FEMS Microbiol. Lett. 146: 149–154

Deng YM, Liu CQ & Dunn NW (1999) Genetic organisation and functional analysis of a novel abortive infection system, AbiL, from *Lactococcus lactis*. J. Biotechnol. 67: 135–149

Diersken KP, Ebel W, Marks J, Sandine WE & Trempy JE (1995) Polysaccharide expression in lactococci. In: Ferretti JJ, Gilmore MS, Klaenhammer TR & Brown F (Eds) Genetics of streptococi, enterococci and lactococci (pp 469-480). Dev. Biol. Stand. Basel, Karger

Dinsmore PK & Klaenhammer TR (1994) Phenotypic consequences of altering the copy number of *abiA*, a gene responsible for aborting bacteriophage infections in *Lactococcus lactis*. Appl. Environ. Microbiol. 60: 1129–1136

Dinsmore PK & Klaenhammer TR (1995) Bacteriophage resistance in *Lactococcus*. Mol. Biotechnol. 4: 297–314

Dinsmore PK & Klaenhammer TR (1997) Molecular characterisation of a genomic region in a *Lactococcus* bacteriophage that is involved in its sensitivity to the phage defence mechanism AbiA. J. Bacteriol. 179: 2949–2957

Dinsmore PK, O'Sullivan DJ & Klaenhammer TR (1998) A leucine repeat motif in AbiA is required for resistance of *Lactococcus lactis* to phages representing three species. Gene 212: 5–11

Djordjevic GM & Klaenhammer TR (1996) Positive selection, cloning vectors for Gram positive bacteria based on a restriction endonuclease cassette. Plasmid 35: 37–45

Djordjevic GM & Klaenhammer TR (1997) Genes and gene expression in *Lactococcus* bacteriophages. Int. Dairy J. 7: 489–508

Djordjevic GM, O' Sullivan DJ, Walker SA, Conkling MA & Klaenhammer TR (1997) Bacteriophage-triggered defence systems: phage adaptation and design improvements. Appl. Environ. Microbiol. 4370–4376

Dupont L, Boizet-Bonhoure B, Coddeville M, Auvray F & Ritzenthaler P (1995) Characterisation of genetic elements required for site-specific integration of *Lactobacillus delbrueckii* ssp. *bulgaricus* bacteriophage mv4 and construction of an integration-proficient vector for *Lactobacillus plantarum*. J. Bacteriol. 177: 586–595

Durmaz E & Klaenhammer TR (1995) A starter culture rotation strategy incorporating paired restriction-modification and abortive infection bacteriophage defences in a single *Lactococcus lactis* strain. Appl. Environ. Microbiol. 61: 1266–1273

Durmaz E, Higgins DL & Klaenhammer TR (1992) Molecular characterisation of a second abortive phage resistance gene present in *Lactococcus lactis* ssp. *lactis* ME2. J. Bacteriol. 174: 7463–7469

Dybvig K & Yu H (1994) Regulation of a restriction and modification system via DNA inversion in *Mycoplasma pulmonis*. Mol. Microbiol. 12: 387–395

Emond E, Dion E, Walker SA, Vedamuthu ER, Kondo JK & Moineau S (1998) AbiQ, an abortive infection mechanism from *Lactococcus lactis*. Appl. Environ. Microbiol. 64: 4748–4756

Emond E, Holler BJ, Boucher PA, Vandenbergh P, Vedamuthu ER, Kondo JK & Moineau S (1997) Phenotypic and genetic characterisation of the bacteriophage abortive infection mechanism AbiK from *Lactococcus lactis*. Appl. Environ. Microbiol. 63: 1274–1283

Engel G, Altermann E, Jurgen RK & Henrich B (1998) Structure of a genome region of the *Lactobacillus gasseri* temperate phage øadh covering a repressor gene and cognate promoters. Gene 210: 61–70

Fitzgerald GF, Daly C, Brown LR & Gingeras TR (1982) *Scr*FI: a new sequence-specific endonuclease from *Streptococcus cremoris*. Nuc. Acid Res. 10: 8171–8179

Foley S, Lucchini S, Zwahlen MC & Brussow H (1998) A short non-coding viral DNA element showing characteristics of a replication origin confers bacteriophage resistance to *Streptococcus thermophilus*. Virology 250: 377–387

Forde A, Daly C & Fitzgerald GF (1999) Identification of four phage resistance plasmids from *Lactococcus lactis* ssp. *cremoris* HO2. Appl. Environ. Microbiol. 65: 1540–1547

Forde A & Fitzgerald GF (1999) Analysis of exopolysaccharide (EPS) production mediated by the bacteriophage adsorption blocking plasmid, pCI658, isolated from Lactococcus lactis ssp. cremoris HO2. Int. Dairy J (In press)

Fremaux C, De Antoni GL, Raya RR & Klaenhammer TR (1993) Genetic organisation and sequence of the region encoding integrative functions from *Lactobacillus gasseri* temperate bacteriophage øadh. Gene 126: 61–66

Froseth BR, Harlander SK &McKay LL (1988) Plasmid-mediated reduced phage sensitivity in *Streptococcus lactis* KR5. J. Dairy Sci. 71: 275–284

Garbutt KC, Kraus J & Geller BL (1997) Bacteriophage resistance in *Lactococcus lactis* engineered by replacement of a gene for a bacteriophage receptor. J. Dairy Sci. 80: 1512–1519

Garvey P, van Sinderen D, Twomey DP, Hill C & Fitzgerald GF (1995a) Molecular genetics of bacteriophage and natural phage defence systems in the genus *Lactococcus*. In. Dairy J. 5: 905–947

Garvey P, Fitzgerald GF & Hill C (1995b) Cloning and DNA sequence analysis of two abortive infection phage resistance determinants from the lactococcal plasmid pNP40. Appl. Environ. Microbiol. 61: 4321–4328

Garvey P, Rince A, Hill C & Fitzgerald GF (1997) Identification of a RecA homolog ($RecA_{LP}$) on the conjugative lactococcal phage resistance plasmid pNP40: evidence of a role for chromosomally encoded $RecA_L$ in abortive infection. Appl. Environ. Microbiol. 63: 1244–1251

Garvey PA, Hill C & Fitzgerald GF (1996) The lactococcal plasmid pNP40 encodes a third bacteriophage resistance mechanism, one which affects phage DNA penetration. Appl. Environ. Microbiol. 62: 676–679

Geis A, Janzen T, Teuber M & Wirsching F (1992) Mechanism of plasmid-mediated bacteriophage resistance in lactococci. FEMS Microbiol. Lett. 94: 7–14

Geis A, Neve H & Teuber M (1987) Plasmid-dependent bacteriophage resistance in lactic acid streptococci: isolation, characterisation and cloning of plasmids. FEMS Microbiol. Rev. 46: Abs 16

Geller BL, Ivey RG, Trempy JE & Hettinger-Smith B (1993) Cloning of a chromosomal gene required for phage infection of *Lactococcus lactis* ssp. *lactis* C2. J. Bacteriol. 175: 5510–5519

Gopal PK & Crow VL (1993) Characterisation of loosely associated material from the cell surface of *Lactococcus lactis* ssp. *cremoris* E8 and its phage-resistant variant strain 398. Appl. Environ. Microbiol. 59: 3177–3182

Gopal PK & Reilly KI (1995) Molecular architecture of lactococcal cell surface as it relates to important industrial properties. Int. Dairy J. 5: 1095–1111

Guimont C, Henry P & Linden G (1993) Restriction-modification in *Streptococcus thermophilus*: isolation and characterisation of a type II restriction endonuclease *Sth*455I. Appl. Microbiol. Biotechnol. 39: 216–220

Harrington A & Hill C (1991) Construction of a bacteriophage resistance derivative of *Lactococcus lactis* ssp. *lactis* 425A by using the conjugal plasmid pNP40. Appl. Environ. Microbiol. 57: 3405–3409

Harrington A & Hill C (1992) Plasmid involvement in the formation of a spontaneous bacteriophage insensitive mutant *Lactococcus lactis*. FEMS Microbiol. Lett. 96: 132–142

Hill C (1993) Bacteriophage and bacteriophage resistance in lactic acid bacteria. FEMS Microbiol. Lett. 12: 87–108

Hill C, Miller LA & Klaenhammer TR (1990a) Cloning, expression, and sequence determination of a bacteriophage fragment encoding bacteriophage resistance in *Lactococcus lactis*. J. Bacteriol. 172: 6419–6426

Hill C, Miller LA & Klaenhammer TR (1990b) Nucleotide sequence and distribution of the pTR2030 resistance determinant (*hsp*) which aborts bacteriophage infection in lactococci. Appl. Environ. Microbiol. 56: 2255–2258

Hill C, Miller LA & Klaenhammer TR (1991) *In vivo* genetic exchange of a functional domain from a type II A methylase between lactococcal plasmid pTR2030 and a virulent bacteriophage. J. Bacteriol. 173: 4363–4370

Hill C, Pierce K & Klaenhammer TR (1989) The conjugative plasmid pTR2030 encodes two bacteriophage defence mechanisms in lactococci, restriction-modification (R^+/M^+) and abortive infection (Hsp^+). Appl. Environ. Microbiol. 55: 2416–2419

Jarvis AW & Klaenhammer TR (1986) Bacteriophage resistance conferred on lactic streptococci by the conjugative plasmid pTR2030: effects on small isometric-, large isometric- and prolate headed phages. Appl. Environ Microbiol. 51: 1272–1277

Jarvis AW (1988) Conjugal transfer in lactic streptococci of plasmid-encoded insensitivity to prolate- and small isometric-headed bacteriophages. Appl. Environ. Microbiol. 54: 777–784

Jarvis AW (1989) Bacteriophages of lactic acid bacteria. J. Dairy Sci. 72: 3406–3428

Jarvis AW (1992) Analysis of phage resistance mechanisms encoded by lactococcal plasmid pAJ2074. Can. J. Microbiol. 39: 252–258

Jarvis AW (1995) Relationships by DNA homology between lactococcal phages 7-9, P335 and New Zealand industrial lactococcal phages. Int . Dairy J. 5: 355–366

Jarvis AW, Fitzgerald GF, Mata M, Mercenier A, Neve H, Powell IB, Ronda C, Saxelin M & Teuber M (1991) Species and types of phages of lactococcal bacteriophages. Intervirol. 32: 2–9

Johnsen MG, Appel KF, Madsen PL, Vogensen FK, Hammer K & Arnau J (1996) A genomic region of lactocccal temperate bacteriophage TP901-1 encoding major virion proteins. Virology 218: 306–315

Johnsen MG, Neve H, Vogensen FK & Hammer K (1995) Virion positions and relationships of lactocccal temperate bacteriophage TP901-1 proteins. Virology 212: 595–606

Josephsen J & Klaenhammer TR (1990) Stacking of three different restriction and modification systems in *Lactococcus lactis* by co-transformation. Plasmid 23: 71–75

Josephsen J & Neve H (1998) Bacteriophages and Lactic Acid Bacteria. In: Salminen S & van Wright A (Eds) Lactic Acid Bacteria: Microbiology and Functional Aspects, 2nd edn (pp 385–436). Marcel Dekker Inc., New York

Josephsen J & Vogensen FK (1989) Identification of three different plasmid-encoded restriction-modification systems in *Streptococcus lactis* ssp. *cremoris* W56. FEMS Microbiol. Lett. 59: 161–166

Kannan P, Cowan GM, Daniel AS, Gann AAF & Murray NE (1989) Conservation and organisation in the specificity polypeptideds of two families of type I restriction enzymes. J. Mol. Biol. 209: 335–344

Kelly W, Dobson J, Jorck-Ramberg D, Fitzgerald GF & Daly C (1990) Introduction of bacteriophage resistance plasmids into commercial *Lactococcus* starter cultures. FEMS Microbiol. Rev. 87: Abst. C20

Keogh BP & Pettinghill g (1983) Adsorption of bacteriophage eb7 on *Streptococcus cremoris* EB7. Appl. Environ. Microbiol. 45: 1946–1948

Kim SG & Batt CA (1991a) Nucleotide sequence and deletion analysis of a gene coding for a structural protein of *Lactococcus lactis* bacteriophage F4-1. Food Microbiol. 8: 27–36

Kim SG & Batt CA (1991b) Identification of a nucleotide sequence conserved in *Lactococcus lactis* bacteriophages. Gene 98: 95–100

Kim SG & Batt CA (1991c) Antisense mRNA-mediated bacteriophage resistance in *Lactococcus lactis* ssp. *lactis*. Appl. Environ. Microbiol. 57: 1109–1113

Kim SG, Bor Y-C & Batt CA (1992) Bacteriophage resistance in *Lactococcus lactis* ssp. *lactis* using antisense ribonucleic acid. J. Dairy Sci. 75: 1761–1767

King WR, Collins EB & Barrett EL (1983) Frequencies of bacteriophage-resistant and slow-acid producing variants of *Streptococcus cremoris*. Appl. Environ. Microbiol. 45: 1481–1485

Klaenhammer TR & Fitzgerald GF (1994) Bacteriophage and bacteriophage resistance. In: Gasson MJ & de Vos WM (Eds) Genetics and Biotechnology of Lactic Acid Bacteria (pp 106–168). Blackie Academic and Professional, Glasgow

Klaenhammer TR (1991) Development of bacteriophage resistant strains of lactic acid bacteria. Biochemical Soc. Transactions 19: 675–681

Kneale GG (1994) A symmetrical model for the domain structure of type I DNA methylatransferases. J. Mol. Biol. 243: 1–5

Kodaira K-I, Oki M, Kakikawa M, Watanabe N, Hirakawa M, Yamada K & Taketo A (1997) Genome structure of the *Lactobacillus* temperate phage øgle: the whole genome sequence and the putative promoter/repressor system. Gene 187: 45–53

Kulakauskas S, Lubys A & Ehrlich SD (1995) DNA restriction-modification systems mediate plasmid-maintenance. J. Bacteriol. 177: 3451–3454

Langella P & Chopin A (1989) Effect of restriction-modification systems on transfer of foreign DNA into *Lactococcus lactis* ssp. *lactis*. FEMS Microbiol. Lett. 59: 301–306

Larbi D, Decaris B & Simonet J -M (1992) Different bacteriophage resistance mechanisms in *Streptococcus salivarius* ssp. *thermophilus*. J. Dairy Res 59: 349–357

Lawrence RC (1978) Action of bacteriophage on lactic acid bacteria: consequences and protection. N.Z. J. Dairy Sci. Technol. 11: 251–256

Lillehaug D & Birkeland N -K (1993) Characterisation of genetic elements required for site-specific integration of the temperate lactococcal bacteriophage øLC3 and construction of integration-negative øLC3 mutants. J. Bacteriol. 175: 1745–1755

Lillehaug D, Lindquist BH & Birkeland N -K (1991) Characterisation of øLC3, a *Lactococcus lactis* ssp. *cremoris* temperate bacteriophage with cohesive single-stranded DNA ends. Appl. Environ. Microbiol. 57: 3206–3211

Lillehaug D, Nes IF & Birkland N-K (1997) A highly efficient and stable system for site-specific integration of genes and plasmids into the phage øLC3 attachment (*attB*) of the *Lactococcus lactis* chromosome. Gene 188: 129–136

Limsowtin GKY & Terzaghi BE (1976) Phage resistant mutants: their selection and use in cheese factories. N.Z. J. Dairy Sci. 11: 251–256

Limsowtin, GKY, Powell IB & Parente, E (1996) Types of starters. In: Cogan TM and Accolas J-P (Eds) Dairy Starter Cultures (pp 101-129). VCH Publishers Inc., New York

Lubbers MW, Waterfield NR, Beresford TPJ, LePage RWF & Jarvis AW (1995) Sequencing and analysis of the prolate-headed lactococcal bacteriophage c2 genome and identification of the structural genes. Appl. Environ. Microbiol. 61: 4348–4356

Lucey M, Daly C & Fitzgerald GF (1992) Cell surface characteristics of *Lactococcus lactis* harbouring pCI528, a 46 kb plasmid encoding inhibition of bacteriophage adsorption. J. Gen. Microbiol. 138: 2137–2143

Madsen A & Josephsen J (1998a) Characterisation of *Lla*CI, a new restriction-modification system from *Lactococcus lactis* ssp. *cremoris* W15. Biol. Chem. 379: 443-449 Madsen A & Josephsen J (1998b) Cloning and characterisation of the lactococcal plasmid-encoded type II restriction-modification system, *Lla*DII. Appl. Environ. Microbiol. 64: 2424–2431

Madsen PL & Hammer K (1998) Temporal transcription of the lactococcal temperate phage TP901-1 and DNA sequence of the early promoter region. Microbiology 144: 2203–2215

Malone T, Blumenthal RM & Cheng X (1995) Structure-guided analysis reveals nine sequence motifs conserved among DNA-methyltransferases and suggests a catalytic mechanism for these enzymes. J. Mol. Biol. 253: 618–632

Marshall RJ & Berridge NJ (1976) Selection and some properties of phage resistant starters for cheese-making. J. Dairy Res. 43: 449–458

Mayo B, Hardisson C & Brana AF (1991) Nucleolytic activities in *Lactococcus lactis* ssp. *lactis* NCDO497. FEMS Microbiol. Lett. 79: 195–198

McGrath S, Seegers JF, Fitzgerald GF & van Sinderen D (1999) Molecular characterisation of a phage-encoded resistance system in *Lactococcus lactis*. Appl. Environ. Microbiol. 65: 1891–1899

McLandsborough LA, Kolaetis KM, Requena T & McKay LL (1995) Cloning and characterisation of the abortive infection genetic determinant *abiD* isolated from pBF61 of *Lactococcus lactis* ssp. *lactis* KR5. Appl. Environ. Microbiol. 61: 2023–2026

Mikkonen M, Raisanen L & Alatossava T (1996) The early gene region completes the nucleotide sequence of *Lactobacillus delbrueckii* ssp. *lactis* phage LL-II. Gene 175: 49–57

Moineau S, Fortier J, Ackermann HW & Pandian S (1992) Characterisation of lactococcal bacteriophages from Quebec cheese plants. Can. J. Microbiol. 38: 875–882

Moineau S, Pandian S & Klaenhammer TR (1993a) Restriction-modification systems and restriction endonucleases are more effective on lactococcal bacteriophages that have emerged recently in the dairy industry. Appl. Environ. Microbiol. 59: 197–202

Moineau S, Durmaz E, Pandian S & Klaenhammer TR (1993b) Differenciation of two abortive mechanisms by using monoclonal antibodies directed towards lactococcal bacteriophage capsid proteins. Appl. Environ. Microbiol. 59: 208–212

Moineau S, Bernier D, Jobin M, Herbert J, Klaenhammer TR & Pandian S (1993c) Production of monoclonal antibodies against the major capsid protein of the *Lactococcus* bacteriophage u136 and development of an enzyme-linked immunosorbent assay for direct phage detection in whey and milk. Appl. Environ. Microbiol. 59: 2034–2040

Moineau S, Borkaev M, Walker SA, Vedamuthu ER & Vandenbergh PA (1994a) Novel starter rotation system based on phage species sensitivity. J. Dairy Sci. 77: 18

Moineau S, Pandian S & Klaenhammer TR (1994b) Evolution of a lytic bacteriophage via DNA acquisition from the *Lactococcus lactis* chromosome. Appl. Environ. Microbiol. 60: 1832–1841

Moineau S, Walker SA, Vedamuthu ER & Vandenbergh PA (1995a) Cloning and sequencing of *Lla*II restriction-modification genes from *Lactococcus lactis* and relatedness of this system to the *Streptococcus pneumoniae Dpn*II system. Appl. Environ. Microbiol. 61: 2193–2202

Moineau S, Walker SA, Holler BJ, Vedamuthu ER & Vandenbergh PA (1995b) Expression of a*Lactococcus lactis* phage resistance mechanism by *Streptococcus thermophilus*. Appl. Environ. Microbiol. 61: 2461-2466

Monteville MR, Ardestani B & Geller BR (1994) Lactococcal phages require a host cell wall carbohydrate and a plasma membrane protein for adsorption and ejection of DNA. Appl. Environ. Microbiol. 60: 3204–3211

Murray NE, Daniel AS, Cowan GM & Sharp PM (1993) Conservation of motifs within the unusually variable polypeptide sequences of type I restriction and modification enzymes. Mol. Microbiol. 9: 133–143

Naito Y, Naito T & Kobayashi I (1998) Selfish restriction-modification genes: resistance of a resident R/M plasmid to displacement by an incompatible plasmid mediated by host killing. Biol. Chem. 379: 429–436

Nauta A, van Sinderen D, Karsens H, Smit E, Venema G & Kok J (1996) Inducible gene expression mediated by a repressor-operator system isolated from *Lactococcus lactis* bacteriophage r1t. Mol. Microbiol. 19: 1331–1341

Neve H (1996) Bacteriophage. In: Cogan TM and Accolas J-P (Eds) Dairy Starter Cultures (pp 157–189). VCH Publishers Inc., New York

Neve H, Zenz KI, Desiere F, Kock A, Heller K & Brussow H (1998) Comparison of the lysogeny modules from the temperate *Streptococcus thermophilus* bacteriophages TP-J34 and Sfi21: implications for the modular theory of evolution. Virology 241: 61–72

Nyengaard N, Vogensen FK & Josephsen J (1993) *Lla*AI and *Lla*BI, two type II restriction endonucleases from *Lactococcus lactis* ssp. *cremoris* W9 and W56 recognising, respectively, 5'-/GATC-3' and 5'-C/TRYAG-3'. Gene 136: 371–372

Nyengaard N, Vogensen FK & Josephsen J (1995) Restriction-modification systems in *Lactococcus lactis*. Gene: 157: 13–18

Nyengaard NR, Falkenberg Klok J & Josephsen J (1996) Cloning and analysis of the restriction-modification system *Lla*BI, a bacteriophage resistance system from *Lactococcus lactis* ssp. *cremoris* W56. Appl. Environ. Microbiol. 62: 3494–3498

O'Connor L, Coffey A, Daly C & Fitzgerald GF (1996) AbiG, a genotypically novel abortive infection mechanism encoded by plasmid pCI750 of *Lactococcus lactis* ssp. *cremoris* UC653. Appl. Environ. Microbiol. 62: 3075–3082

O'Connor L, Tangney M & Fitzgerald GF (1999) Expression, regulation and mode of action of the AbiG abortive infection system of *Lactococcus lactis* ssp. *cremoris* UC653. Appl. Environ. Microbiol. 65: 330–335

O'Sullivan D, Coffey A, Fitzgerald GF, Hill C & Ross RP (1998) Design of a phage-insensitive lactococcal dairy starter via sequential transfer of naturally occurring conjugative plasmids. Appl. Environ. Microbiol. 64: 4618–4622

O'Sullivan DJ & Klaenhammer TR (1998) Control of expression of *Lla*I restriction in *Lactococcus lactis*. Mol. Microbiol. 27: 1009–1020

O'Sullivan DJ, Hill C & Klaenhammer TR (1993) Effect of increasing the copy number of bacteriophage origins of replication, *in trans*, on incoming-phage proliferation. Appl. Environ. Microbiol. 59: 2449–2456

O'Sullivan DJ, Walker SA, West SG & Klaenhammer TR (1996) Development of an expression system using a lytic phage to trigger explosive plasmid amplification and gene expression. Biotechnology 14: 82–87

O'Sullivan DJ, Zagula K & Klaenhammer TR (1995) *In vivo* restriction by *LlaI* is encoded by three genes, arranged in an operon with *llaIM*, on the conjugative *Lactococcus* plasmid pTR2030. J. Bacteriol. 177: 134–143

O'Sullivan T, van Sinderen D & Fitzgerald GF (1999) Structural and functional analysis of pCI65st, a 6.5 kb plasmid from *Streptococcus thermophilus* NDI-6. Microbiology 145: 127–134

Oram, J D & Reiter B (1968) The adsorption of phage to group N streptococci. The specificity of adsorption and the location of phage receptor substances in cell wall and plasma membrane fractions. J. Gen. Virol. 3: 103–11

Parreira R, Valyasevi R, Lerayer LS, Ehrlich SD & Chopin M-C (1996a) Gene organisation and transcription of a late-expressed region of a *Lactococcus lactis* phage. J. Bacteriol. 178: 6158–6165

Parreira R, Ehrlich SD & Chopin M-C (1996b) Dramatic decay of phage transcripts in lactococcal cells carrying the abortive infection determinant AbiB. Mol. Microbiol. 19: 221–230

Platteeuw C & de Vos WM (1992) Location, characterisation and expression of lytic enzyme-encoding gene lytA, of *Lactococcus lactis* bacteriophage øUS3. Gene 118: 115–120

Poch MT, Somkuti GA & Solaiman DKY (1997) *Sth*132I, a novel class IIS restriction endonuclease of *Streptococcus thermophilus* ST132. Gene 195: 201–206

Polzin KM, Collins LJ, Lubbers MW & Jarvis AW (1996) Effect of various mRNAs on bacteriophage c2 replication. In: Proceedings of 5th Symposium on Lactic Acid Bacteria, Genetics, Metabolism and Applications, Veldhoven the Netherlands. Abs F2

Prevots F & Ritzenthaler P (1998) Complete sequence of the new lactococcal abortive phage resistance gene *abiO*. J. Dairy Sci. 81: 1483–1485

Prevots F, Daloyau M, Bonin O, Dumont X & Tolou S (1996) Cloning and sequencing of the novel abortive infection gene *abiH* of *Lactococcus lactis* ssp. *lactis* biovar. *diacetylactis* S94. FEMS Microbiol. Lett. 142: 295–299

Prevots F, Mata M & Ritzenthaler P (1990) Taxonomic differenciation of 101 lactococcal bacteriophages and characterisation of bacteriophages with unusually large genomes. Appl. Environ. Microbiol. 57: 1313–1318

Prevots F, Tolou S, Delpech B, Kaghad M & Daloyau M (1998) Nucleotide sequence and analysis of the new chromosomal abortive infection gene *abiN* of *Lactococcus lactis* ssp. *cremoris* S114. FEMS Microbiol. Lett. 159: 331–336

Raya RR, Fremaux C, De Antoni GL & Klaenhammer TR (1992) Site-specific integration of the temperate bacteriophage øadh into the *Lactobacillus gasseri* chromosome and molecular characterisation of the phage (*attP*) and bacterial (*attB*) attachment sites. J. Bacteriol. 174: 5584–5592

Reyes-Gavilan CG, Limsowtin GKY, Sechaud L, Veaux M & Acolas JP (1990) Evidence for a plasmid-linked restriction-modification system in *Lactobacillus helveticus*. Appl. Environ. Microbiol. 56: 3412–3419

Sanders ME & Klaenhammer TR (1983) Characterisation of phage-sensitive mutants from a phage-insensitive strain of *Streptococcus lactis*: evidence for a plasmid determinant that prevents phage adsorption. Appl. Environ. Microbiol. 46: 1125–1133

Sanders ME, Leonard PJ, Sing WD & Klaenhammer TR (1986) Conjugal strategy for the construction of fast-acid producing, bacteriophage-resistant lactic streptococci for use in dairy fermentations. Appl. Environ. Microbiol. 52: 1101–1107

Sanford JC & Johnston SA (1985) The concept of parasite-derived resistance- deriving resistance genes from the parasites own genome. J. Theor. Biol. 113: 395–405

Schafer A, Geis A, Neve H & Teuber M (1991) Bacteriophage receptors of *Lactococcus lactis* ssp. *diacetylactis* F7/2 and *Lactococcus lactis* ssp. *cremoris* Wg2-1. FEMS Microbiol. Lett. 78: 69–74

Schouler C, Bouet C, Ritzenthaler P, Drouet X & Mata M (1992) Characterisation of *Lactococcus lactis* phage antigens. Appl. Environ. Microbiol. 58: 2479–2484

Schouler C, Clier F, Lerayer AD Ehrlich SD & Chopin M -C (1998a) A type IC restriction-modification system in *Lactococcus lactis*. J. Bacteriol. 180: 407–411

Schouler C, Gautier S, Ehrlich SD & Chopin M -C (1998b) Combinational variation of restriction-modification specificities in *Lactococcus lactis*. Mol. Microbiol. 28: 169–178

Schouler C, Ehrlich SD and Chopin M-C (1994) Sequence and organisation of the lactococcal prolate-headed bIL67 phage genome. Microbiology 140: 3061–3069

Shearman CA, Jury K & Gasson MJ (1992) Autolytic *Lactococcus lactis* expressing a lactococcal bacteriophage lysin gene. Bio/Technology 10: 196–199

Shearman CA, Underwood H, Jury K & Gasson M (1989) Cloning and DNA sequence analysis of a *Lactococcus* bacteriophage lysin gene. Mol Gen. Genet. 218: 214–221

Sheehan MM, Garcia JL, Lopez R & Garcia P (1996) Analysis of the catalytic domain of the lysin of the lactococcal bacteriophage Tuc2009 by chimeric gene assembling. FEMS Microbiol. Lett. 140: 23–28

Sijtsma L, Hellingwerf KJ & Wouters JTM (1991) Composition and phage binding capacity of cell walls isolated from *Lactococcus lactis* ssp. *cremoris* SK110 and SK112. Neth. Milk Dairy J. 45: 81–95

Sijtsma L, Jansen N, Hazeleger WC, Wouters JTM & Hellingwerf KJ (1990) Cell surface characteristics of bacteriophage resistant *Lactococcus lactis* ssp. *cremoris* SK110 and its bacteriophage sensitive variant SK112. Appl. Environ.Microbiol. 56: 3230–3233

Sijtsma L, Sterkenburg A & Wouters JTM (1988) Properties of the cell walls of *Lactococcus lactis* ssp. *cremoris* SK110 and their relation to bacteriophage resistance. Appl. Environ. Microbiol. 54: 2808–2811

Sing WD & Klaenhammer TR (1986) Conjugal transfer of bacteriophage resistance determinants on pTR2030 into *Streptococcus cremoris* strains. Appl. Environ. Microbiol. 51: 1264–1271

Sing WD & Klaenhammer TR (1991) Characterization of restriction-modification plasmids from *Lactococcus lactis* ssp. *cremoris* and their effects when combined with pTR2030. J. Dairy Sci. 74: 1133–1144

Sing WD & Klaenhammer TR (1993) A strategy for rotation of different bacteriophage defences in a lactococcal single-strain starter culture system. Appl. Environ. Microbiol. 59: 365–372

Solaiman DKY & Somkuti GA (1990) Isolation and characterisation of a type II restriction endonuclease from *Streptococcus thermophilus* ST117. FEMS Microbiol. Lett. 80: 261–266

Solaiman DKY & Somkuti GA (1991) A type II restriction endonuclease from *Streptococcus thermophilus*. FEMS Microbiol. Lett. 67: 261–266

Stanley E, Fitzgerald GF, Le Marrec C, Fayard B & van Sinderen D (1997) Sequence analysis and characterisation of ø01205, a temperate bacteriophage infecting *Streptococcus thermophilus* CNRZ1205. Microbiology 143: 3417–3429

Steenson LR & Klaenhammer TR (1986) Plasmid heterogeneity in *Streptococcus cremoris* M12R: effects on proteolytic activity and host-dependent phage replication. J. Dairy Sci. 69: 2227–2236

Stingele F, Neeser JR & Mollet B (1996) Identification and characterisation of the *eps* (exopolysaccharide) gene cluster from *Streptococcus thermophilus* Sfi6. J. Bacteriol. 178: 1680–1690

Su P, Harvey M, Im HJ & Dunn NW (1997) Isolation, cloning and characterisation of the *abiI* gene from *Lactococcus lactis* ssp. *lactis* M138 encoding abortive phage infection. J. Biotechnol. 54: 95–104

Su P, Im H, Hsieh H, Kang' AS & Dunn NW (1999) *Lla*FI, a type III restriction and modification system in *Lactococcus lactis*. Appl. Environ. Microbiol. 65: 686–693

Takano T, Ochi A & Yamamoto N (1990) Restriction enzyme from *Lactobacillus fermentum*. FEMS Microbiol. Rev. 87: Abs C64

Timinkas A, Butkus V & Janulaitas A (1995) Sequence motifs characteristic for DNA [cytosine-N4] and DNA [adenine-N6] methyltransferases. Classification of all DNA methyltransferases. Gene 157: 3–11

Tortorello ML, Chang PK, Ledford RA & Dunny GM (1990) Plasmid associated antigens associated with resistance to phage adsorption in *Lactococcus lactis*. In Abstracts of the 3rd International ASM Conference on Streptococcal Genetics, Abs 50

Tremblay DM & Moineau S (1999) Complete genomic sequence of the lytic bacteriophage DT1 of *Streptococcus thermophilus*. Virology 255: 63–76

Twomey DP, Davis R, Daly C & Fitzgerald GF (1993) Sequence of the gene encoding a second *Scr*FI m^5C methyltransferase of *Lactococcus lactis*. Gene 136: 205–209

Twomey DP, Gabillet N, Daly C & Fitzgerald GF (1997) Molecular characterisation of the restriction endonuclease gene (*scrFIR*) associated with the *Scr*FI restriction-modification system from *Lactococcus lactis* ssp. *cremoris* UC503. Microbiology 143: 2277–2286

Twomey DP, McKay LL & O'Sullivan DJ (1998) Molecular characterisation of the *Lactococcus lactis Lla*KR2I restriction-modification system and effect of an IS*982* element positioned between the restriction and modification genes. J. Bacteriol. 180: 5844–5854

Valeyasevi R, Sandine WE & Geller BL (1990) Bacteriophage kh receptor of *Lactococcus lactis* ssp. *cremoris* KH is rhamnose of extracellular wall polysaccharide. Appl. Environ. Microbiol. 56: 1882–1889.

Valyasevi R, Sandine WE & Geller BL (1991) A membrane protein is required for bacteriophage c2 infection of *Lactococcus lactis* ssp. *lactis* C2. J. Bacteriol. 173: 6095–6100

Valeyasevi R, Sandine WE & Geller BL (1994) *Lactococcus lactis* ssp. *lactis* C2 bacteriophage sk1 receptor involves rhamnose and glucose moieties in the cell wall. J. Dairy Sci. 77: 1–6

van de Guchte M, Daly C, Fitzgerald GF & Arendt EK (1994a) Identification of the putative repressor-encoding gene cI of the temperate lactococcal bacteriophage Tuc2009. Gene 144: 93–95

van de Guchte M, Daly C, Fitzgerald GF & Arendt EK (1994b) Identification of *int* and *attP* on the genome of lactococcal bacteriophage Tuc2009 and their use for site-specific plasmid integration in the chromosome of Tuc2009-resistant *Lactococcus lactis* MG1363. Appl. Environ. Microbiol. 60: 2324–2329

van Kranenburg R, Marugg JD, van Swam II, Willem NJ & de Vos WM (1997) Molecular characterisation of the plasmid-encoded *eps* gene cluster essential for exopolysaccharide biosynthesis in *Lactococcus lactis*. Mol. Microbiol. 24: 387–397

van Sinderen D, Karsens H, Kok J, Terpstra P, Ruiters MHJ, Venema g & Nauta A (1996) Sequence analysis and molecular characterisation of the temperate lactococcal bacteriophage r1t. Mol. Micorbiol. 19: 1343–1355

Verrips CT & van den Berg DJC (1996) Barriers to application of genetically modified lactic acid bacteria. Antonie van Leeuwenhoek 70: 299–316

Vlegels PAP, Hazeleger WC, Helmerborst TH & Wouters JTM (1988) Phage resistance of *Streptococcus cremoris* due to low adsorption efficiency. Neth. Milk Dairy J. 42: 195–206

Walker SA, O'Sullivan DJ, West SG & Klaenhammer TR (1996) Molecular characterisation of a phage inducible middle promoter from the lactococcal bacteriophage ø31. In: Proceedings of 5th Symposium on Lactic Acid Bacteria, Genetics, Metabolism and Applications, Veldhoven, the Netherlands. Abs F25

Watanabe K, Ishibashi K, Nadashima Y & Sakurai T (1984) A phage-resistant mutant of *Lactobacillus casei* which permits phage adsorption but not genome injection. J. Gen. Virol. 65: 981–986

Whitehead HR & Cox GA (1935) The occurrence of bacteriophage in lactic streptococci. N.Z. J. Dairy Sci. Technol. 16: 319–320

Wilson GG & Murray NM (1991) Restriction and modification systems. Annu. Rev. Genet. 25: 585–627

Xu G, Willert J, Kapfer W & Trautner TA (1995) *Bsu*CI, a type I restriction-modification system in *Bacillus subtilis*. Gene 157: 59

Antonie van Leeuwenhoek **76:** 115–137, 1999.

Acquired antibiotic resistance in lactic acid bacteria from food *

Michael Teuber*, Leo Meile & Franziska Schwarz
*Laboratory of Food Microbiology, Swiss Federal Institute of Technology, ETH-Zentrum, CH-8092 Zurich, Switzerland. (*Author for correspondence)*

Key words: antibiotic resistance, Lactobacillus, Lactococcus, Entrococcus, food

Abstract

Acquired antibiotic resistance, i.e. resistance genes located on conjugative or mobilizable plasmids and transposons can be found in species living in habitats (e.g. human and animal intestines) which are regularly challenged with antibiotics. Most data are available for enterococci and enteric lactobacilli. Raw material from animals (milk and meat) which are inadvertantly contaminated with fecal matters during production will carry antibiotic resistant lactic acid bacteria into the final fermented products such as raw milk cheeses and raw sausages. The discovered conjugative genetic elements of LAB isolated from animals and food are very similar to elements studied previously in pathogenic streptococci and enterococci, e.g. θ-type replicating plasmids of the pAMβ1, pIP501-family, and transposons of the Tn*916*-type. Observed resistance genes include known genes like *tetM, ermAM, cat, sat* and *vanA*. A composite 29'871 bp resistance plasmid detected in *Lactococcus lactis* subsp. *lactis* isolated from a raw milk soft cheese contains *tetS* previously described in *Listeria monocytogenes*, *cat* and *str* from *Staphylococcus aureus*. Three out of five IS elements on the plasmid are almost or completely identical to IS*1216* present in the *vanA* resistance transposon Tn*1546*. These data support the view that in antibiotic challenged habitats lactic acid bacteria like other bacteria participate in the communication systems which transfer resistance traits over species and genus borders. The prevalence of such bacteria with acquired resistances like enterococci is high in animals (and humans) which are regularly treated with antibiotics. The transfer of antibiotic resistant bacteria from animals into fermented and other food can be avoided if the raw substrate milk or meat is pasteurized or heat treated. Antibiotic resistance traits as selectable markers in genetic modification of lactic acid bacteria for different purposes are presently being replaced, e.g. by metabolic traits to generate food-grade vectors.

1. Definition of lactic acid bacteria (LAB) and their use in food

Lactic acid bacteria (LAB) are a group of Gram-positive anaerobic bacteria which excrete lactic acid as a main fermentation product into the medium. As a consequence, this biochemical definition associates lactic acid bacteria of different phylogenetic branches of bacterial evolution: the 'low GC' taxa, e.g. *Enterococcus, Lactobacillus, Lactococcus, Leuconostoc, Pediococcus* and *Streptococcus* and the 'high GC' genus*Bifidobacterium* (Schleifer & Ludwig 1995). LAB are important microorganisms in parts of the human and animal body (intestine, nasopharyngal and vaginal mucosa), and in environments (mainly plants) where spontaneous fermentations of carbohydrate containing substrates occur. These spontaneous events developed into important food fermentations of plant material (sour dough, sauerkraut, etc.) and animal products (sausages, cheeses and sour milk from meat and milk, respectively) (Teuber 1993). Such food has been and still is an important vehicle of living bacteria from animal and plant sources into the human body. At cell densities of 10^8 to 10^9 viable bacteria per gram, one portion (100 g) of fermented product delivers 10^{10} to 10^{11} viable microorganisms. Indeed, some LAB strains from animal and human intestinal microflora have been adopted as 'probiotic' food supplements including *Enterococcus faecium, Lactobacillus plantarum, Lb. acidophilus, Lb. casei* subsp. *rhamnosus*, and several *Bifidobacterium* and *Propionibacterium* species (Tannock 1998).

* Dedicated to KH Schleifer, a pioneer in the molecular taxonomy of lactic acid bacteria, on the occasion of his 60th birthday.

2. Definitions of antibiotic resistance phenomena, transfer mechanisms and their significance

Extensive recent reviews of the application of antibiotics in human and veterinary medicine (Levy 1997; WHO 1997), agriculture (National Research Council & Insitute of Medicine US 1998; MAFF 1998; Falkiner 1998) and aquaculture (Reilly & Käferstein 1997) have documented the evolution and enrichment of antibiotic resistant bacteria: the phenomenon is regularly observed upon the introduction of a new antibiotic (Levy 1997). The evolution of antibiotic resistance in microbial communities is enhanced by horizontal transfer of resistance genes over species and genus borders by conjugative plasmids, transposons, the possession of integrons and insertion elements, as well as lytic and temperate bacteriophages (Davies 1994). During 50 years of large scale antibiotic application several mechanisms of resistance against a particular antibiotic have evolved including enzymatic inactivation of the antibiotic (e.g. by β-lactamases, aminoglycoside acetyl-, nucleotidyl- and phosphoryltransferases), restricted import of antibiotics (e.g. penicillin binding proteins), active export of antibiotics (e.g. by membrane inserted ATP-dependent efflux systems) or target modification (e.g. methylation of 23S rRNA, mutation of aminoacid sequence of topoisomerase) (Davies 1997). For certain pathogenic and potentially pathogenic bacteria like staphylococci and enterococci, the evolution of highly resistant clones has inaugurated a crisis in antibiotic resistance (Neu 1992). In the case of some vancomycin resistant enterococci, there remains no clinically useful antibiotic for a successful treatment, (Jett et al. 1994; McDonald et al. 1997).

The basic hypothesis governing antibiotic resistance evolution of bacteria have been accentuated by Levy (1997) in a number of drug resistance equations and rules:

- Antibiotic application plus resistance traits/genes generate a resistance problem.
- The severity and speed of resistance development are dependent on the selection densitiy, i.e. the amount of antibiotic per treated individual (human or animal) per geographic area.
- Reversal of resistance levels in a biotope is more dependent on the survival of susceptible bacteria, and a re-colonization by those than on the loss of resistance traits.
- Resistance reservoirs are provided by pathogenic and commensal bacteria which have come into contact with antibiotics for a prolonged time.
- Low level (chromosomal) resistance acquired by spontaneous mutation enhances the evolution of high level resistance by horizontal gene transfer (Baquero et al. 1997).

It is the scope of this report to review the distribution and origin of antibiotic resistance in LAB, the potential mechanisms of transfer and their possible role in the dissemination of antibiotic resistances in animal and human populations. The application of resistance genes as selection markers in the molecular biology and genetic modification of LAB is briefly discussed.

3. Antibiotic resistance profiles of LAB

The knowledge of intrinsic chromosomally coded resistance of LAB to common antibiotics is necessary to recognise acquired resistance traits.

The resistance spectrum of *Bifidobacterium* was recently described by Charteris et al. (1998). The investigated (probiotic) bifidobacteria are susceptible to ampicillin, penicillin G, cephalosporin, bacitracin, chloramphenicol, erythromycin, clindamycin, nitrofurantoin and tetracycline. Resistances – some being most likely intrinsic – exist towards vancomycin, gentamicin, kanamycin, streptomycin, fusidic acid, trimethoprim, norfloxacin, nalidixic acid, metronidazol, polymyxin B and colistin. The mechanisms of resistances are unknown.

The resistance profiles of enterococci *(E. faecalis* and *E. faecium)* isolated from food or patients are widely varying containing many acquired traits (see below). A possible baseline is provided by *Enterococcus faecium* strain 68 which is used as probiotic for man and animal, and as silage inoculant. This strain was isolated around 1920 in the pre-antibiotic area. It is susceptible to erythromycin (15 μg disc), framycetin (100 μg), streptomycin/penicillin (streptopen 35 μg), gentamicin (10 μg), penicillin G, tetracycline (30 μg) and chloramphenicol (30 μ). It is intrinsically resistant to kanamycin (30 μg), streptomycin (10 μg) and oxacillin (5 μg) (Teuber, unpublished data for an *E. faecium* strain # 68 isolated from a pharmaceutical preparation 1985, 1997).

Investigated strains of *Lactococcus lactis* were sensitive to amikacin, ampicillin, 1st generation cephalosporine, chloramphenicol, erythromycin, gentamicin, imipenem, oxacillin, penicillin, piperacillin, sulfonamide, tetracycline, trimethoprim/sulfomethoxazol, and vancomycin (de Fabrizio et al. 1994). A slighty lowered susceptibility was observed towards carbenicillin, ciprofloxacin, dicloxacillin, and norfloxacin. (Intrinsic?) resistances were recorded towards colistin, fosfomycin, pipemidic acid and rifamycin. Recently, multiple drug efflux proteins were discovered in *L. lactis* subsp. *lactis* MG1363 (van Veen & Konings 1998), one being an ABC transporter (*lmrA*), the other a proton motive force-dependent drug transporter (*lmrP*). Both are responsible for a resistance to high concentrations of ethidium bromide. The natural substrates are not known. A similar transporter (*horA*) described in *Lactobacillus brevis* ABBC45 and encoded on a high copy number plasmid is obviously responsible for a resistance to the bitter-tasting hop compounds humulone (acylphloroglucine substituted with two isoprenoid side chains) and isohumulone, and a low level resistance towards ethidium bromide (Sami et al 1997). The membrane permeability disturbing hop compounds (Teuber & Schmalreck 1973) have antimicrobial activities against Gram-positive bacteria (Schmalreck et al. 1975). A resistance to them induces a serious beer spoilage potential of lactobacilli (Sami et al. 1997). Indeed, it should be investigated whether the detected efflux proteins have a natural function in providing resistance to phenolic compounds in bacteria which live in and on plant tissues like many LAB including *Lactococcus lactis* (Teuber 1995).

Twenty-six strains of *L. lactis* subsp. *cremoris* and subsp. *lactis* were all resistant to trimethoprim and almost all to sulfathiazole. Resistances to gentamicin, kanamycin, lincomycin, nafcillin, neomycin, nisin, rifampin and streptomycin varied (Orberg & Sandine 1985).

The industrially important lactobacilli for the food industry include the thermophilic *Lb. delbrueckii* subsp. *bulgaricus* and subsp. *lactis*, *Lb. helveticus* and *Lb. acidophilus* (as probiotic) the mesophilic *Lb. plantarum*, *Lb. casei* subsp. *rhamnosus*, *Lb. sake*, *Lb. curvatus*, and the heterofermentative *Lb. brevis*, *Lb. kefir*, *Lb. fermentum* and *Lb. sanfrancisco* (Teuber 1993). Data on antibiotic resistance in these species are scarce.

Thirty-one strains of *Lb. delbrueckii* subsp. *bulgaricus* as components of yoghurt cultures showed intrinsic resistance towards mycostatin, nalidixic acid, neomycin, polymyxin B, trimethoprim, colimycin, sulfamethoxazol, and sulfonamides. Susceptibilities to cloxacillin, dihydrostreptomycin, doxycyclin, furadantin, novobiocin, oleandomycin, oxacillin, and streptomycin were prominent, but kanamycin and streptomycin susceptibilities varied (Sozzi & Smiley 1980). Many strains of *Lb. plantarum, Lb. casei, Lb. salivarius, Lb. leichmannii, Lb. acidophilus* carry intrinsic resistance towards vancomycin which is due to the presence of D-alanine: D-alanine ligase-related enzymes (Elisha & Courvalin 1995). Intrinsic vancomycin resistance of *Lactobacillus, Leuconostoc* and *Pediococcus* species has been used to separate them from other Gram-positive bacteria on vancomycin-supplemented selective media (Simpson et al. 1988). An important observation is the high chromosomal mutation rate (10^{-4}) to nitrofurazone, kanamycin and streptomycin resistance as compared to erythromycin and tetracycline resistance (10^{-9}) (Curragh & Collins 1991). Fifteen strains of *Streptococcus thermophilus* from yoghurt cultures showed varying resistance levels to colimycin, gentamicin, kanamycin, mycostatin, nalidixic acid, neomycin, polymyxin B., trimethoprim/sulfamethoxazol, streptomycin and sulfonamides (Sozzi & Smiley 1980).

4. Natural transfer tools of LAB

The main mechanisms of horizontal gene transfer in bacteria in natural environments (i.e. outside the microbiological laboratory) are conjugation (Clewell 1995) and most probably bacteriophages (e.g. Schmieger & Schicklmaier 1999).

Conjugal gene transfer between enterobacteria was discovered due to detection of high level multiple antibiotic resistance in *Shigella* and *Escherichia coli* isolates during a dysentery epidemic in Japan (Watanabe 1963). The identification and exploitation of conjugative plasmids in these bacteria have revolutionized genetic work and finally enabled genetic engineering. From the very beginning antibiotic resistance genes were important tools which served as selective markers in genetically modified organisms. Later on, it became evident that conjugative plasmids were also common in Gram-positive bacteria like *Staphyococcus aureus*, *Enterococcus* and *Streptococcus* species. Again, conjugative plasmids contributed to the dissemination of antibotic resistance traits in natural biotopes (humans, animals, plants and their environments). In addition, like in the genetic enineering of Gram-negative bacteria, some of the conjugative antibiotic resistance plasmids from Gram-positive bacteria became the basis of the construction of vectors bearing selectable resitance traits (see below). Conjugative transposons are characterized by transposase genes which induce excisson of itself and – after conjugative transfer – incorporation into the genomes of recipient Gram-negative and Gram-positive bacteria. A well known example is the tetracycline resistance transposon Tn*916* from *Enterococcus faecalis* which has been fully sequenced (Flannagan et al. 1994).

A prerequisite to acquire antibiotic resistance genes from other bacteria is the potential of LAB to communicate actively and passively with these other bacteria with the aid of conjugative plasmids and transposons. Indeed, plasmids are common in enterococci, lactococci, leuconostoc, pediococci, and present in some strains of *S. thermophilus*, lactobacilli and bifidobacteria (Janzen et al. 1992; Mercenier 1994; Dellaglio et al. 1995; Devriese & Pot 1995; Sgorbati et al. 1995; Simpson and Taguchi 1995; Teuber 1995). Conjugative transposons (broad and narrow host range) have been described in enterococci, lactococci and streptococci (Clewell 1993).

Transfer of antibiotic resistance genes within LAB by bacteriophages and prophages seems theoretically possible, but has not been studied to any extent in LAB. In addition, the host range of such transfers would be rather limited to closely related strains within one species.

4.1 Plasmids as conjugative genetic elements in LAB with regard to antibiotic resistance transfer

E. faecalis and *E. faecium* clearly represent the most promiscuous bacterial species discovered within the Gram-positive bacteria. Enterococci are part of the human and animal intestinal microflora. They have developed into important human pathogens especially in immuno compromised patients (Jett et al. 1994). Infection is from the patients own microflora or from other patients, hospital personnel or possibly food. In food, enterococci are common in raw meat and milk (due to faecal contamination during production) and products made from these raw material without pasteurization (raw milk cheeses and raw fermented sausages and ham) (Teuber et al. 1996; Giraffa et al. 1997).

A well characterized broad host range conjugative plasmid is pAMβ1 which was isolated from *E. faecalis*. It carries a constitutive MLS resistance (to macrolides, lincosamides and streptogramin B). Self-transfer at frequencies of about 10^{-4} per donor (and lower) has been observed in filter mating experiments to enterococci, staphylococci, *Clostridium, Lactobacillus, Lactococcus* and *Bacillus* species (Macrina & Archer 1993). Elaborate studies have shown that pAMβ1 could move from *E. faecalis* into the plasmid-free strain Bu2-60 of *L. lactis* subsp. *lactis* biovar *diacetylactis* during filter mating. From there it conjugated to 8 out of 18 commercial isolates of *S. thermophilus* with frequencies of 3.5×10^{-5} to 7.6×10^{-9} per recipient: *S. thermophilus* transconjugants conjugated pAMβ1 into other *S. thermophilus* strains at 10^{-4} to 10^{-7} per recipient. Back-transfer into *Lactococcus* Bu2-60 occurred at higher frequencies of about 10^{-4} (Kleinschmidt et al. 1993). pAMβ1 was quite stably maintained in *S. thermophilus* at 37 °C, but lost within 90 generations of growth at 42 °C.

Out of 13 *Lb. debrueckii* subsp. *bulgaricus*, 23 *Lb. delbrueckii* subsp. *lactis*, 44 *Lb. acidophilus*, 1 *Lb. helveticus*, 1 *Lb. brevis*, 5 *Lb. plantarum*, 6 *Lb. casei* subsp. *casei*, 4 *Lb. casei* subsp. *rhamnosus*, 1 *Lb. fermentum* and 1 *Lb. salivarius* subsp. *salicinius* strains only two strains accepted pAMβ1 during filter mating at about 10^{-7} frequency: *Lb. brevis* 3030/62 and *Lb. helveticus* 3048b. In skimmed milk at 37 °C, pAMβ1 was lost from *Lb. brevis* within 10 generations, at 30 °C within 100 generations (Soeding et al. 1993). In cabbage juice or tomato juice, pAMβ1 stability was 80% within 77 generations at 30 °C.

Obviously many lactobacilli in contrast to lactococci (Gasson & Davies 1980) are resistant to conjugative transfer of the broad host range resistance plasmid pAMβ1. Another broad host range plasmid, pIP501, originally isolated from *Streptococcus agalactiae* can be conjugated into *Streptococcus, Staphylococcus, Clostridium, Listeria* and *Pediococcus* species. Like pAMβ1, pIP501 replicates by the θ-type mechanism, carries a MLS resistance gene, a region responsible for conjugative transfer, and a gene for a chloramphenicol acetyltransferase (Macrina & Archer 1993). Although the conjugal transfer of pIP501 to other LAB has not been studied in greater detail, the replicon and resistance genes of this plasmid and some derivative have been extensively used for the construction of cloning vectors for lactococci (de Vos and Simons 1994). Indigenous conjugation systems in lacotococci are very common, reflecting the abundance of plasmid DNA (Neve et al. 1987; Gasson & Fitzgerald 1994). Transferrable antibiotic resistance plasmids, however, have normally not been observed in starter cultures most of which stem from the pre-antibiotic aera. In contrast, reports of native conjugation systems in *Lactobacillus* are few: one plasmid-associated transfer of lactose fermenting ability in *Lb. casei* and bacteriocin production and immunity in *Lb. johnsonii* (Kullen & Klaenhammer 1998). Likewise, conjugative plasmids have not been discovered in *Bifidobacterium* (Kullen and Klaenhammer, 1998). A similar situation is evident in *Leuconostoc* and *Pediococcus* which, however, can accept broad host range antibiotic resistance plasmids like pAMβ1, pVA797::Tn*917*, and pIP501 by high-frequency-conjugation from *Lactococcus* harbouring these plasmids (Dessart & Steenson 1991). Another efficient transfer system (10^{-2} in broth) is the sex pheromone induced conjugation by rolling-circle-replicating large plasmids of the pAD1 type (Clewell 1993). Recently, it was demonstrated that clumping of donor and recipient cells (a prerequisite for conjugation) is induced by subinhibitory levels of chloramphenicol, erythromycin or tetracyline even in the absence of pheromone production in the recipient (Wu et al. 1999).

4.2 Conjugative transposons in LAB

Conjugative transposons are a main type of vehicle regarding antibiotic resistance transport in Gram-positive bacteria. They have been discovered in *E. faecalis* (Tn*916*, Tn*918*, Tn*920*, Tn*925*, Tn*2702*), *E. faecium* (Tn*5233*), *S. pyogenes* (Tn*3701*), *S. agalactiae* (Tn*93951*), and *Lactococcus lactis* (Tn*5276*, Tn*5301*). In enterococci and streptococci, they determine resistances to tetracycline (*tetM*), erythromycin (*ermAM, erm*), chloramphenicol (*cat*) and kanamycin (*aphA-3*). In lactococci, they code for nisin (*nis*) production and sucrose fermentation (*sac*). They vary in size between 16 and 70 kb. They may be inserted into plasmids or the chromosome in one or multiple copies (inserts). Immunity and entry exclusion are absent. They may mobilize plasmids or chromosomal genes. The most remarquable observation, however, is the extreme host range which is a property of the Tn*916*/Tn*1545* family. It includes Gram-positive and Gram-negative bacteria of the following genera: *Acetobacterium, Acholeplasma, Alcaligenes, Bacillus, Butyrivibrio, Citrobater, Clostridium, Enterococcus, Escherichia, Eubacterium, Fusobacterium, Haemophilus, Lactobacillus, Lactococcus, Leuconostoc, Listeria, Mycoplasma, Neisseria, Peptostreptococcus, Staphylococcus, Streptococcus, Thermus* and *Veillonella*, and probably more which have not yet been investigated (Clewell & Flannagan 1993). Tn*916* has been completey sequenced (Flannagan et al. 1994). Resistance transfer rates are in the order of 10^{-9} to 10^{-6} per donor in filter matings. More complex transposons like Tn*5253* (65.5 kb) may have an internal Tn*916*-like region which is capable of independent conjugative transposition (Rice 1998).

Another important though not conjugative transposon is Tn*1546* which carries the *vanA* gene cluster responsible for vancomycin resistance in enterococci. This transposon is easily mobilized, however, by conjugative plasmids (Arthur & Courvallin 1993).

5. LAB with acquired and transferable antibiotic resistances in food

5.1 Enterococci

Enterococci are commonly found in the intestines of farm animals and humans (Teuber 1993). In food microbiology, they have been – like *E. coli* – regarded as indicators of fecal contamination. When it became known that enterococci are widely dispersed with feces and sewage in the environment and are able to multiply in a variety of organic materials (milk, meat, vegetables) or in fermented food made from meat and milk (e.g. sausages and cheeses) they have been looked at as index organisms for unhygienic food processing if pasteurized substrates were used (Devriese & Pot 1995). Food items consisting of raw meat or milk or made from these substrates by fermentation are almost always contaminated with enterococci. For some artisanal products, enterococci isolated from traditional raw milk cheeses are used as starter cultures (Giraffa et al 1997). Certain *E. faecium* strains are applied as probiotics for farm animals and humans, and may be a component of silage inoculants (Elmer et al 1996). Although enterococci were for a long time considered to be medically unimportant, within the last 20 years they have developed into serious pathogens specifically in nosocomial infections where they cause abdominal abscesses, urinary tract infections, endocarditis and bacteremia (Jett et al 1994).

The extremely high level of antibiotic resistance observed in these bacteria (Table 1) has made them feared infectious agents in intensive care wards. Possible pathogenicity factors like hemolysins and cytolysins have been described. The most important species are *E. faecalis* and *E. faecium*, the first being more common in human illnesses, the second one (though less common in human infections) may pose a larger resistance threat (Huycke et al 1998).

Table 1. Antibiotic resistance levels (%) of enterococcal isolates from meat, animals, humans and other sources

	Europe 1995 32 strains animals, human	Germany VRE 1998 38 strains beef, pork	Germany enterococci 1998 209 strains beef, pork	Denmark enterococci 1995 57 strains human clinical	Denmark E. faecium 1995: 54 strains broiler	Denmark E. faecium 1996 71 strains broiler	Denmark E. faecium 1995 12 strains cattle	Denmark E. faecium 1995 40 strains cattle	Denmark E. faecalis 1995 225 strains pigs	Denmark E. faecalis 1996 38 strains pigs	Denmark E. faecium 1995 58 strains pigs	Denmark E. faecium 1996 23 strains pigs	Denmark enterococci 1998 88 strains chicken meat	Sweden enterococci 1998 55 strains chicken meat	Switzerland E. faecium 1998 38 strains broiler intestine	Switzerland E. faecalis 1998 27 strains broiler intestine	Switzerland enterococci 1998 78 strains chicken meat	Finland E. faecium & faecalis 1997 533 strains broiler	USA entero-cocci 1993 50 strains pork	USA entero-cocci 1993 50 strains human clinical	USA entero-cocci 1993 49 strains water	row
ampicillin	47	0	0	14	0		0		0	0	0	0	0	2	0	0	0	1	0	4	0	1
penicillin	100	26.3	9.5	18	33	1	0	0	1	0	52	0	28	11				15	2	4	0	2
methicillin		100	99.5																			3
amoxillin-clavulanic acid		0	0																68	94	98	4
ceftriaxone	75	40.5	65		80	55	92	80	53	55	84	74							62	76	92	5
cephalotin		100	92																32	38	57	6
imipenem		50	18																6	4	10	7
vancomy-cin	100	100	0.5	0	56	18		0	0	0	24	0	0	24	0	5	3	7	0	0	2	8
teicoplanin		79	0		17	15	0	0	1	0	7	0										9
LY333328		0	0																			10
gentamicin		0	0	21	0	0	0	0	0	8	0	0			0	0	5		36	74	88	11
strepto-mycin	28	34	1		0	8	8	0	29		40	4			26	3	20					12
erythro-mycin	91	100	69	42	65		38		91		91		9	38	37	45	86	14	38	62	35	13
tetra-cyclines	94	94	20	60	19	41	8	10	87	13	69	9	63	31	41	76	71	63	88	56	37	14
cipro-floxacin		26	36	68	50	28	38	53	10	8	43	39	3	0	22	0	87		6	62	29	15
chloramphe-nicol	9	29	2		11		15		24		7		0	0	7	8	87	1	0	9	0	16
clinda-mycin		100	24																86	92	92	17
rifampin		18	24																20	26	49	18
trimetho-prim/sulfa-methoxa-zole		21	0	12	70	4	100	0	44	3	97	0	0	0			97	1	0	36	22	19
avoparcin		100	0		59	17	0	0	0	0	29	0										20
virginia-mycin		100	3		43	54	8	10			47	22										21
tylosin		100	7		59	48	8	3	91	8	90	4										22
spiramycin					93	86	85	63	97	97	97	65										23
lincomycin					91	82	77		100		100											24
neomycin					9	4	15	3	42	34	34	4										25
pristina-mycin					37	42	15	8			53	13										26
bacitracin						28		0		0		0			100	100	99					27
prevalence of resistant isolates %	38 vanco mycin	8 vanco-mycin	nc	nc	100	nc	100	nc	100	nc	100	nc	55	73	100	100	50	53	~ 100	100	100	28
MDR (%)	100	100	ng	ng	93	ng	92	ng	63	ng	98	ng	ng	ng	ng	ng	87	42	> 88	> 98	< 96	29
conjugative transfer	nd	nd	nd	nd	nd	nd	nd	nd	nd	nd	44%	nd	nd	4 out of 9 VRE	nd	nd	nd	nd	nd	nd	nd	30
reference	Klare et al 1995a	Klein et al 1998	Klein et al 1998	Aarestrup et al 1998	Aarestrup et al 1998	Aarestrup et al 1998	Aarestrup et al 1998	Aarestrup et al 1998	Aarestrup et al 1998	Aarestrup et al 1998	Aarestrup et al 1998	Aarestrup et al 1998	Quednau et al 1998	Quednau et al 1998	ETHZ	ETHZ	ETHZ	Minist.of Agriculture Finland 1997	Knudtson & Hartman 1993	Knudtson & Hartman 1993	Knudtson & Hartman 1993	

MDR: multiple drug resistance; nd: not determined; ng: not given; nc: not calculated; ETHZ: Goldenberger, Frei, Loeffel and Teuber, ETH Zurich, manuscript in preparation.

In both species, the evolutionary development of resistance has been attributed to the possession of broad host range and extremely mobile genetic elements like conjugative plasmids and transposons (e.g. pAMβ1 or Tn*916*, Clewel et al. 1993). The molecular details of the structures and functions of these elements are fairly well studied and becoming understood (Clewell et al. 1995; Marra & Scott 1999). It is noteworthy that transcription of the transfer functions of Tn*916* requiring excission of the element is dramatically increased in the presence of tetracycline (Celli & Trieu-Cuot 1998).

Enterococci in farm animals and meat. It has been shown that raw meat, raw milk and products made from them without proper heat treatment frequently contain resistant enterococci (Rollins et al. 1985; Batish & Ranganathan 1986; Knutson & Hartman 1993; Teuber et al. 1996; Giraffa et al 1997; Aarestrup et al 1998; Klein et al. 1998; Quednau et al. 1998). Some recent data for meat (chicken, turkey, pigs, beef) are presented in Table 1, which also contains recent data on human strains and surface water for comparison. Not only are the tested isolates resistant to antimicrobial feed additives used at the time of testing (avoparcin, virginiamycin, tylosin, and bacitracin, rows 20 to 27) but also to a battery of medically important antibiotics like penicillins, cephalosporins, vancomycin, streptomycin, erythromycin, fluoroquinolones, chloramphenicol, tetracyclines, neomycine, sulfonamides and trimethoprim, rifampicin, and clindamycin. The prevalence of multiple drug resistant strains is high (Klare et al. 1995; Klein et al. 1998).

Table 1 also exemplifies the extensive nature of the data on antibiotic resistance presented in different publications.

In epidemiological terms, however, this information provides important facts regarding the antibiotic resistance situation in enterococci from animals and their meat compared to clinical strains from humans. Resistance levels to tetracyclines (20 to 97%) and erythromycin (9 to 100%) are highly similar to recently investigated clinical isolates, e.g. from the UK, Holland and Switzerland (Guiney & Urwin 1993; Liassine et al. 1998; van den Braak et al. 1998; Wuest et al. 1998). In contrast, ampicillin resistance is low, but penicillin and methicillin resistances are high, as is resistance to fluoroquinolones (row 15). Trimethoprim and sulfonamides range in antibiotic efficiency between zero and 100%, indicating the dependence on the use of these antibiotics in these animals. Clear evidence that feeding antibiotics to chicken induces the selection of resistant enterococci, is provided in the case of bacitracin, a substance legally available in Europe as feed additive. If bacitracin is used (see columns 6, 15 to 17) resistance levels are between 28 and 100%. In enterococci from cattle and pigs where bacitracin has not been employed, resistance levels are zero. Likewise, if tylosin is not used for feeding chicken or pigs, cross resistance to erythromycin is low (see data from Denmark and Finland, rows 13 and 18 in Table 1). Although the incidence of resistance to gentamicin is low in Europe, some strains isolated in our laboratory from chicken meat exported from China to Europe showed a high level gentamicin resistance (MIC >1024 mg/l) coupled with a high level streptomycin resistance (MIC>1024 mg/l). Gentamicin resistant enterococci have also been reported in veterinary samples (e.g. from horses) and chicken meat in the USA (Thal et al. 1995). The gentamicin resistant *E. faecalis* investigated in our laboratory contained the 6′-aminoglycoside acetyltransferase – 2″-aminoglycoside phosphotransferase gene (AAC-APH) as detected by hybridization with a specific nucleotide probe derived from the nucleotide sequence of the AAC-APH from human isolates. This gene was not seen in the American isolates from chicken [Goldenberger, Frei, Löffel and Teuber, ETH Zurich, unpublished].

When determined, the prevalence of antibiotic resistant enterococci in farm animals and their meat was always high, and multiple drug resistance (i.e. to more than 1 antibiotic) was between 63 and 100% of the resistant isolates. Although Knutson & Hartman (1993) stated that the human, pork and water isolates investigated did not show multiple resistance to antibiotics such as gentamicin, ampicillin, streptomycin, and vancomycin, this statement is not supported by their data (see columns 19 to 21 in Table 1) and remains questionable.

A frightening observation comes from the reports that vancomycin resistant enterococci isolated from chicken meat and pork after enrichment in the presence of vancomycin reveal 100% multiple drug resistance (see columns 1 and 2 in Table 1), for example an *E. faecium* isolate from minced pork was resistant to vancomycin, teicoplanin, avoparcin, penicillin, ampicillin (intermediate resistance), streptomycin, erythromycin, tetracycline, chloramphenicol, and ciprofloxacin (Klare et al. 1995a, 1995b; Klein et al. 1998).

The problem of vancomycin resistance in enterococci (VRE) has been reviewed in great detail (McDonald et al 1997). If VRE from farm animals and food are to be a threat to the human population, colonization of a predisposed person has to precede infection. A wound infection of a chicken farm worker reported from the UK was obviously caused by a vancomycin resistant enterococcus (Bates et al 1993). The ban of avoparcin as a feed additive in Europe has resulted in a sharp drop of vancomycin resistant enterococci between 1995 and 1996 in farm animals as shown in the Danish surveillance study (columns 5 and 11 versus columns 6 and 12 in Table 1; Aarestrup et al 1998). A preliminary report from Germany has described a similar phenomenon in enterococci from human intestinal microflora obtained from healthy persons of the community (Witte 1998). VRE have been reported in chicken meat from Sweden (Quedneau 1998), a country where antibiotic feeds have been banned since 1986 (Swedish Ministry of Agriculture 1997). Vancomycin resitance was transferrable by conjugation to other enterococci *in vitro*. Whether Tn*1546* was involved has not yet been investigated.

The ban of avoparcin use as being a contributor to vancomycin resistance was only politically achievable because molecular methods proved unequivocally that the resistance genes involved (*vanA*) were identical with the genes characterized in human clinical isolates (Klare et al. 1995a, b). In addition, avoparcin-induced vancomycin resistance was shown in the laboratory to be capable of transfer by conjugation to other enterococci and staphylococci (Noble et al. 1992; Klare et al. 1995a). A comparison of VRE from poultry (79% of 305 isolates including 58% high-level-vancomycin reistant *E. faecium*) with VRE from humans by genome-typing with pulsed-field gel electrophoresis did not reveal genetic overlap. Nucleotide-sequencing of the intergenic *vanC-vanY* region, however, showed full sequence conservation in more than 50% of the strains suggesting that dissemination (by conjugation) of the resistance genes carried on transposable elements may be of greater importance than clonal dissemination of resistant strains (van den Braak et al. 1998).

Enterococci in fermented food. The numbers of viable enterococci in contaminated raw milk, poultry, pork and beef are usually in the range of 10^2 to 10^4 per gram (Teuber et al 1996). If these products are properly heat treated during cooking or processing (calculating with a D_{65} value of 2 min and a z-value of 7 to 8) viable enterococci will be eliminated. If fermented products are made from the raw substrates meat and milk like sausages and cheeses, the contaminating enterococci (in addition to staphylococci, and eventually enterobacteria) may multiply to high numbers, e.g. more than 10^7 CFU per gram in soft cheeses (Teuber et al. 1996). Based on the high prevalence of antibiotic resistant enterococci in meat and milk (see Table 1), antibiotic resistant enterococci should be expected to occur in fermented products from raw milk and meat.

In 1980, a preliminary study of 67 European cheeses made from raw and pasteurized milk yielded 27 (40%) samples containing enterococci (10^3 and 10^7 CFU per gram) (Sievers et al 1993). In 15 (55%) of these cheeses,*E. faecalis*, *E. faecium*, and *E. durans* strains were isolated showing resistance to one or more of the following antibiotics: penicillin, cefalotin, furadoin, fucidin, erythromycin, tetracycline and chloramphenicol. The multiple drug resistance level was 68% (Teuber et al 1996); 25% of pasteurized and 67% of raw milk cheeses contained enterococci. The presence of enterococci in pasteurized cheeses is due to recontamination after pasteurization. A follow-up study in 1991 and 1995 investigating 53 hard, semi-hard and soft cheeses presumably made from raw milk yielded 39 antibiotic resistant *E. faecalis*, 12 *E. faecium*, 3 *E. casseliflavius*, 2 *E. durans*, and 3 undefined enterococcal strains. Resistance to penicillin (18%), erythromycin (48%), gentamicin (80%), tetracycline (59%), rifampicin (7%), chloramphenicol (32%), fusidic acid (14%) and vancomycin (4%) was detected. Prevalence of multiple drug resistance was 51%. A common multiple drug resistance type included resistance to tetracycline, chloramphenicol, erythromycin and gentamicin. The vancomycin resistant isolates have not yet been characterized.

Seventeen samples of minced meat, raw meat sausages and ham yielded 8 *E. faecalis*, 6 *E. faecium*, 3 *E. casseliflavius*, 3 *E. durans* and 26 undefined enterococcal strains (between 10^2 and 10^5 CFU per gram) which were resistant to one or more antibiotics including streptomycin, erythromycin, penicillin, tetracycline, rifampicin, gentamicin, and chloramphenicol. The incidence was 100% for streptomycin, 26% each for tetracycline and chloramphenicol, 13% for gentamicin, 8% for rifampicin, 6% for chloramphenicol, and 4% for penicillin.Using erythromycin and tetracycline as markers, 15% of the enterococcal isolates from cheese and sausage could transfer resistances to plasmid-free *E. faecalis* JH2-2 or *L. lactis*

Table 2. Conjugal transfer of tetracycline, erythromycin and chloramphenicol resistances from *Enterococcus* strains isolated from food (Teuber et al. 1996) to plasmid-free *Enterococcus faecalis* JH2-2 and *Lactococcus lactis* subsp. *lactis* biovar. *diacetylactis* Bu2-60 during filter mating

Product	Donor	Resistances	Selection	Recipients with transfer frequency		Resistance gene (hybridization)	Localization in the donor strains
				E. faecalis JH2-2	*L. lactis lactis* Bu2-60		
Cheese							
Bösbachli	*E. faecalis* FO1	Tet Lm Erm Cm Gm	Tet5	10^{-5}	10^{-6}	*tetM*	Chromosome Plasmid (ca 30 kb)
			Erm5 Cm20	10^{-8}	–	ermAM (*cat*pIP501)	Plasmid (ca. 27 kb)
Bösbachli	*E. faecalis* FO5	Tet Lm Erm Cm Gm	Tet5	ne	10^{-6}	*tetM*	Chromosome Plasmid (ca 30 kb)
			Erm5 cm20	10^{-8}	–	*ermAM* (*cat*pIP501)	Plasmid (ca. 30 kb)
Crottin	*E. faecalis* FE1	Tet Lm Gm	Tet5	ne	10^{-8}	*tetM*	Chromosome
Bric de Meaux	*E. faecalis* FF4	Lm Erm Gm Pen	Erm5	10^{-2}	–	*ermAM*	Plasmid (ca. 30 kb)
Appenzeller 2	*E. faecalis* AP21	Tet Lm Erm Cm Gm	Erm5	10^{-5}	–	*ermAM*	Plasmid (ca. 30 kb)
Appenzeller 5	*E. faecalis* AP51	Tet Lm Erm Gm	Erm5	10^{-6}	–	*ermAM*	Plasmid (ca. 27 kb)
	E. faecalis AP53	Tet Cm Gm Pen	Tet5	10^{-5}	–	*tetM*	Chromosome
Appenzeller 6	*E. faecalis* AP62	Tet Lm Erm Cm Gm	Erm5	10^{-5}	–	*ermAM*	Plasmid (ca. 30 kb)
Tête de Moine 1	*E. faecalis* TM11	Tet Lm Erm Cm Gm	Erm5	10^{-6}	–	*ermAM*	Plasmid (ca. 27 kb)
San Carlo TI	*E. faecalis* 216	Lm Erm Cm Gm	Erm5	10^{-5}	–	*ermAM*	Plasmid (> 50 kb)
Appenzeller 3	*E. faecium* AP34	Tet Gm	Tet5	10^{-5}	–	nh	Plasmid (> 50 kb)
Meat products							
Salami 6	*Enterococcus* sp. RE45	Lm Tet	Tet5	ne	10^{-9}	*tetM*	Chromosome Plasmid (> 50 kb)
Salami 3	*E. faecium* RE14	Tet Lm Erm	Tet5	ne	10^{-9}	*tetM*	Chromosome Plasmid (> 50 kb)
			Erm5	10^{-6}	–	*ermAM*	Plasmid (ca. 30 kb)
Salami 2	*E. faecium* RE1	Lm Erm Pen	Erm5	10^{-5}	–	*ermAM*	Plasmid (ca. 30 kb)
Landjaeger 3	*E. hirae* RE22	Lm Erm	Erm5	10^{-7}	–	*ermAM*	Plasmid (ca. 30 kb)
Landjaeger 5	*E. faecalis* RE25	Tet Lm Erm	Erm5	10^{-5}	10^{-6}	*ermAM*	Plasmid (49 kb)
		Cm	Cm20	10^{-5}	10^{-6}	(*cat*pIP501)	
Landjaeger 5	*E. faecium* RE27	Lm Erm	Erm5	10^{-6}	–	**ermAM**	Plasmid (ca. 30 kb)
Salami 5	*E. casseliflavus* RE37	Tet Lm Pen		10^{-6}	–	*ermAM*	Plasmid (ca. 30 kb)
Minced meat	*Enterococcus* sp. RE39		Erm5	10^{-4}	10^{-3}	**ermAM**	Plasmid (26.5 kb)
Pancetta	*E. faecalis* RE52	Tet Lm Erm	Erm5	10^{-4}	–	**ermAM**	Plasmid (>50 kb)

Erm5, erythromycin [5μg/ml]; cm20, chloramphenicol [20 μg/ml]; Tet5, tetracycline [5μg/ml]; Gm, gentamicin; Lm, lincomycin; Pen, penicillin; ne, not examined; -, no transfer; nh, no hybridization with *tetM* gene probe. Conjugative transfer and allocation of of resistance genes to transferred plasmids or the chromosome was investigated by Southern hybridization of total DNA extracts of resistant transconjugants after gel electrophoresis according to Perreten et al. (1997a).

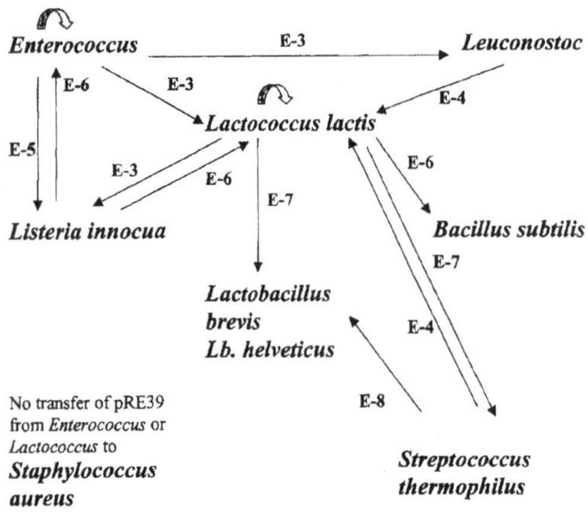

Figure 1. Conjugative transfer of plasmid pRE39 between *Enterococcus* sp. RE39, *E. faecalis* JH 2-2, *Leuconostoc mesenteroides*, *Listeria innocua*, *Lactococcus lactis*, some lactobacilli and *Bacillus subtilis* during filter matings. Plasmid pRE39 does not conjugate into *Staphylococcus aureus* (modified from Teuber et al. 1996). In addition, conjugative transfer of the closely related plasmid pAMβ1 from *Streptococcus thermophilus* to *Lactococcus lactis* and *Lb. brevis* and *Lb. helveticus* is shown (modified from Soeding et al. 1993, and Kleinschmidt et al. 1993).

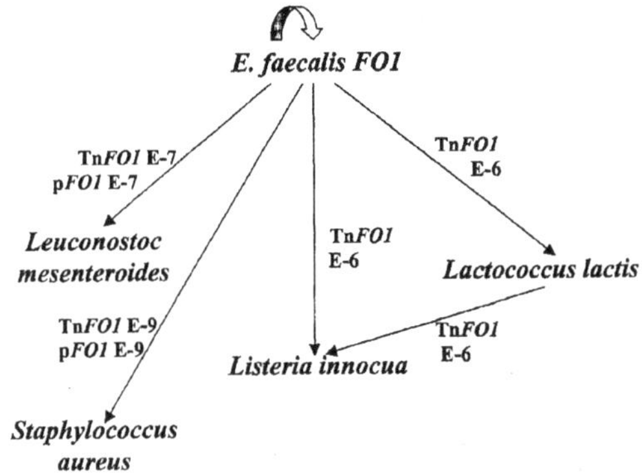

Figure 2. Conjugative transfer of transposon Tn*FO1* from *Enterococcus faecalis* to *Staphylococcus aureus* 80CR5, *Listeria innocua* L17, *Leuconostoc mesenteroides* M7-1, *Lactococcus lactis* subsp. *lactis* Bu-60 and *E. faecalis* JH2-2 during filter mating. *E. faecalis* FO1 was isolated from a raw milk mountain cheese. Transposon Tn*FO1* is found on the chromosome and on a 30-kbp plasmid p*FO1* (modified from Perreten et al. 1997a).

Bu2-60 by conjugation in filter mating experiments (see Table 2). Probably, conjugative plasmids like pRE39 and transposons like Tn*FO1* are involved since resistance genes can be located to chromosomal and/or plasmid DNA (see below).

Molecular characterization of resistance of food borne enterococci

The first studies of food borne enterococci on a molecular level (Rollins et al 1985) have clearly shown that the responsible erythromycin resistance genes were very similar in enterococci isolated from humans, chicken and pigs. In addition, they were encoded on Tn*917*-like conjugative transposons. Recently, 88% of the enterococci isolated from broilers, cattle, pigs and humans in the Danish surveillance studies (Aarestrup et al. 1998) were reported to carry erythromycin resistance genes identified by specific nucleotide probes as *ermB* (Jensen et al. 1999).

Following a similar line of experimentation, 3 of the conjugative enterococcal strains from our investigations were further characterized on a molecular level (see Table 2).

Enterococcus sp. RE39 from minced beef contained a conjugative 26-kb plasmid pRE39 similar to pAMβ1; however, the MLS resistance was inducible. pRE39 encoded a rRNA methylase with an amino acid-sequence identical to the enzyme encoded on pAMβ1 (Teuber et al. 1996). This gene was identified in additional 18 of the enterococcal strains investigated from sausages and cheese by hybridization with a specific nucleotide probe and PCR, thus corroborating the Danish observations (Jensen et al. 1999). The host range of conjugative transfer by filter mating of pRE39 includes *E. faecalis* JH2-2, *L. lactis* Bu2-60, *Listeria innocua, Leuconostoc mesenteroides* and *Bacillus subtilis*. Transfer frequencies on a receptor basis were high being 10^{-3} to 10^{-6} (see Figure 1).

The chloramphenicol and erythromycin resistances in *E. faecalis* RE25 are encoded on a 49-kb conjugative plasmid pRE25 which transfers these resistances to *Listeria innocua*, *E. faecalis* JH2-2, and *Lactococcus lactis* BU2-60 at frequencies of 10^{-5} to 10^{-6}. The nucleotide sequence of the cloned chloramphenicol acetyltransferase is 100% identical with that of the *Streptococcus agalactiae* pIP501 gene (Perreten, Moschetti &Teuber GeneBank Acc.No. X92945). The erythromycin resistance region of this plasmid hybridized with the specific *erm* nucleotide probe from pAMβ1. The complete nucleotide sequence of this plasmid is currently investigated.

Transfer of the tetracycline resistance gene (at the nucleotide level 99.8% identical to *tetM* of transposon Tn*916*) from E. *faecalis* FO1 isolated from a raw milk mountain cheese (about 10^4 CFU per gram), to *Staphylococcus aureus* (10^{-9}), *E. faecalis* JH2-2, *Listeria innocua* (10^{-6}), *Leuconostoc mesenteroides* (10^{-7}), and *L. lactis* Bu2-60 (10^{-6}) was achieved by filter mating (see Figure 2). The mobile genetic element was identified as the Tn*916*-type transposon Tn*FO1*. It resides on the chromosome and on a 30-kb plasmid in this strain (Perreten et al. 1997a).

It must be concluded that antibiotic resistant enterococci are present in some of our daily food supply. This explains their presence in the intestinal microflora of people having not been treated with antibiotics but providing a community reservoir (Corpet 1988; Bates et al. 1993; Liassine et al. 1998; van den Braak 1998). The consumption of raw sausages or cheeses containing 10^5 to 10^7 multiple drug resistant enterococci per gram leads to the ingestion of 10^9 resistant bacteria in a meal containing 100 g of product. When these enterococci reach the intestine, at least a transient colonization is most likely to occur as has been shown in the case of intestinal 'probiotic' bacteria including lactobacilli, bifidobacteria, and enterococci (Elmer et al. 1996; Tannock 1998). Transfer of resistance genes into the endogenous intestinal microflora of the consumer is highly probable (Corpet 1988; van den Braak 1998; Witte 1998). If the *Enterococcus*-contaminated food is carefully cooked or heat treated, the spread of resistance would be blocked. However, as stated recently (MacDonald et al. 1998), a paucity of information exists on enterococcal colonization of the intestinal tract and the development of infection in humans. Nonetheless, colonization of germ-free animals and conjugative transfer of plasmids and transposons in this biotope has been reported in several instances (Doucet-Populaire et al 1991; Brockmann et al 1992). The clinical relevance of vancomycin-resistant enterococci in Europe seems fortunately still to be low (Grüneberg & Hryniewicz 1998).

5.2. Lactobacillus

For bacteria to survive in an antibiotic containing habitat they must be expected to pick up antibiotic resistance genes from other bacteria. This has been observed (albeit in a small number of studies) for lactobacilli in habitats such as the bovine udder surface, the intestinal, vaginal and nasopharyngal mucosae of man and animals (see Table 3). Strains of lactobacilli resistant to vancomycin (intrinsic), cefazolin, penicillin, tetracyclin, trimethoprim-sulfamethoxazole and ciprofloxacin have been occasionally identified in cases of human *Lactobacillus*-bacteremia (Antony et al 1992). As with food borne pathogens, lactic acid bacteria contaminate milk and meat obtained from animals treated with antibiotics and carrying populations of resistant bacteria.

Indeed, tetracycline resistant *Lb. plantarum* has been found in several raw milk soft cheeses (G. Dasen, ETH Zurich, unpublished). Plasmid-borne erythromycin resistance (MLS type) was identified in a poultry isolate of *Lb. reuteri* (Tannock et al. 1994). Another poultry isolate of *L. reuteri* contained a 7-kb plasmid pTC82 harboring a gene for a chloramphenicol acetyltransferase homologous to that on the *Staphylococcus aureus* plasmid pC194 (Lin et al 1996). A *Lb. fermentum* isolated from pig feces ($>10^8$ CFU/gram) carried a 5.7-kb plasmid with an *erm* gene conferring high level erythromycin resistance (MIC >1 mg/ml) which was 98.2% identical to the gene of the enterococcal conjugative transposon Tn*1545* (Fons et al. 1997). As mentioned previously, the conjugative plasmid pAMβ1 can move from *Lactococcus* and *Enterococcus* into *Lb. reuteri* and back again (Tannock 1987). Clearly, enteric lactobacilli participate in promiscuous gene transfer in their natural biotopes and in the laboratory. It is not surprising in the light of the experiences with enterococci that lactoacilli (e.g. *Lb. plantarum*, *Lb. acidophlus*, *Lb. brevis*, *Lb. casei*) isolated from 'home-made' Spanish cheese (Serena, Gamonedo, Cabrales) did show resistances towards penicillin G, cloxacillin, streptomycin, gentamicin, tetracycline, erythromycin and chloramphenicol (Herrero et al. 1996). Also, antibiotic resistant lactococci and leuconostocs were isolated from the same material.

Of 14 *Lactobacillus* strains isolated from the crops of broiler chicken (*Lb. acidophilus*, *Lb. fermentum*, *Lb. crispatus*, *Lb. casei* subsp. *casei* and *Lb. salivarius*) 12 strains were resistant to ciprofloxacin, 11 to gentamicin, 11 to streptomycin, 14 to trimethoprim/sulfamethoxazole, 8 to tetracycline, 2 to erythromycin, 5 to vancomycin (intrinsic), 14 to metronidazole (intrinsic), and all 14 to bacitracin, an antibiotic used as feed additive for the parent animals (see also data on bacitracin resistance of enterococci from the same animals in Table 1, row 27 in columns 15, 16 and 17; Goldenberger et al., manuscript submitted), none to chloramphenicol, ampicillin, penicillin G and rifampicin. All strains showed multiple drug resistance (3 to 6 fold), but the question of intrinsic reistances has not been finally cleared. However, intrinsic resistances to tetracyline and erythromycine have not been reported for these bacteria in the literature.

Table 3. Antibiotic resistances and resistance plasmids in enteric *Lactobacillus* species from animals

Species	Source	Plasmid name and size	Observed resistances or resistance genes	Reference
Lb. acidophilus strain PA3	pig		tetracycline erythromycin oleandomycin kanamycin neomycin gentamicin	Klaenhammer & Sutherland 1980
Lb. acidophilus (20 strains)	pig and calf feces	several	penicillin (1 strain) ampicillin (1) cloxacillin (7) aminoglycosides (20) tetracycline (15) erythromycin (10) chloramphenicol (1) bacitracin (6)	Vescovo et al. 1982
Lb. reuteri (16 strains)	pig and calf feces	several	cloxacillin (16 strains) aminglycosides (16) tetracycline (16) erythromycin (13) chloramphenicol (3) rifamycin (3) bacitracin (10)	Vescovo et al. 1982
Lb. reuteri	poultry	pET 633 9.8 kb	*ermGT* (similar to *ermC* of pE 194)	Tannock et al. 1989
Lb. reuteri	poultry	pLUL631	*erm* (similar to *ermB* of pIP501)	Connell & Tannock 1989
Lb. reuteri G4	chicken	pTC82	*catTC* (81% identity to *catPC* 194)	Lin et al. 1996
Lb. fermentum	pig	pLEM3 5.7 kb	*erm* (98.2% identity to Tn*1545 erm*)	Fons et al 1997
Lb. plantarum	raw pork	pCAT 8.5 kb	*cat*	Ahn et al. 1992

5.3. Lactococcus

It is known for some time that lactococci can accept antibiotic resitance genes like pAMβ1 by conjugation (Gasson & Davies 1980). However they should be sensitive to most clinically important antibiotics as summarized above in paragraph 3. Antibiotic resistant lactococci might develop in biotopes where they come into constant contact with antibiotics. Such a biotope is the dairy cow which is commonly treated with antibiotics to cure or prevent mastitis. From raw milk the bacteria are transferred into raw milk cheeses. We have isolated a multiply drug resistant *L. lactis* subsp. *lactis* pK214 from a raw milk soft cheese. The strain carries a resistance plasmid which was completely sequenced. The sequence is deposited at Genbank accession no. x92946.

Genomics of the 29871-bp multiresistance plasmid pK214 from Lactococcus lactis

Plasmid pK214 is resident in *L. lactis* strain K214 which is streptomycin-, tetracycline- and chloramphenicol-resistant (Perreten et al. 1997b). Genes encoding these three antibiotic resistances (*str*, *tet* and *cat)* are found on this plasmid (see Table 4 and Figure 4). Since we have no cured strains of the original host *L. lactis* K214, further genes encoding the three resistances could be present in its genome or the other 3 plasmids.

Organisation of pK214. A map describing the genetic organisation of pK214 is shown in Figure 3. The 29871-nucleotide sequence contains 29 open reading frames (ORFs) which may represent functionally translated or truncated genes. The deduced products and the most homologous proteins from the databanks are listed in Table 4. The average G + C content of pK214 is 32.4 mol% which is significantly lower than that of the *L. lactis* genomic DNA (36–38%; Kilpper-Bälz et al. 1982). It supports the idea that horizontal DNA transfer from other organisms with a lower G + C content has occurred. Comparative sequence analysis of the entire plasmid DNA and its putative ORF's demonstrates that the plasmid is composed of distinct DNA segments which originate in species like *Staphyloccoccus* (32–36 mol% G + C), *Listeria* (37–39 mol% G + C) and *Enteroccoccus* (34–39 mol% G + C).

The DNA segment containing nucleotides 1 to 9000 is assumed to be the basic backbone of the *Lactococcus* plasmid into which DNA segments from other organisms have been integrated. Part of this *Lactococcus* segment is a *repB* gene encoding a RepB protein with high homology (around 80%) to Rep proteins typically found in *Lactococcus* plasmids like pWV02, pCI305 pCI528 pSL2 and pVS40. The RepB protein of pVS40 (von Wright and Räty 1993) from *L. lactis* subsp. *lactis* biovar *diacetylatis* has a homology of 82.3% and that of pMRC01, a 60-kb conjugative plasmid from *L. lactis* DPC3147 strain, of 80.0% to pK214 (Dougherty et al. 1998). Interestingly, the pK214 region from ORF04 to ORF08 (positions 1824 to 6185) has an overall identity of 91.3% at the nucleotide level with a segment of pMRC01 and up to 99.4% identity to selected ORFs (Table 4). As in the proven theta-type replicon of plasmid pCI305, the putative *repB* gene in pK214 is preceded by an AT-rich sequence (iteron) of 22-bp (5′-CCGAAAAAAACTGTGTATATG-3′) which is tandemly repeated three and one-half times (Foley et al. 1996). Therefore, in analogy to plasmid pCIP305, it is concluded that the region between position 3970 and 4220 has to be regarded to contain the origin of replication (*oriV*) of pK214.

Other putative functions within the designated *Lactococcus* moiety could be assigned to another replication-associated protein (ORF02), a resolvase (ORF01) and a DNA nickase (ORF09). We detected 97.3% identity of this hypothetical protein to the *traA* protein of pMRC01. The nickase gene of pK214, named *nel*, is not preceded by a nicking site (*oriT*) typically found upstream of nickase genes in conjugative plasmids of both Gram-positive and Gram-negative bacteria (Climo et al. 1996). Furthermore, other conjugative transfer functions seem to be absent from pK214. It would also explain that we failed to demonstrate conjugation-mediated plasmid transfer but succeeded in electroporation (Perreten et al., 1997b). Surprisingly, the G + C mol% values of the potential genes in the '*Lactococcus*' segment are lower than those of *Lactococcus* genomes.

Due to high sequence homologies of predicted gene products, we propose that segments from *Staphylococcus, Listeria* and *Enteroccoccus* have been inserted into pK214 during its evolution. Since pK214 harbours 5 complete insertion sequence (IS) elements (3 of them almost identical to the IS*1216* elements from *Enterococcus faecium* (Table 5; Arthur & Courvalin 1993), the heterologous segment acquisition may have been driven by these elements.

Insertion sequence elements of pK214: Two IS*904* elements (Dodd et al 1990) obviously from *L. lac-*

Table 4. Genomics of the 29'871 bp plasmid pK214 from *Lactococcus lactis* subsp. *lactis* isolated from a raw milk soft cheese. Open reading frames (ORFs) are deduced from the nucleotide sequence and compared to known amino acid-sequences from databanks

ORF No.	5′end pos. #	3′end pos. #	Known protein with highest homology {organism}	% aa-Identity	% G/C-content	Accession No.
ORF01	21	593	DNA invertase of pMRC01 {*Lactococcus lactis*}	71.0	29.1	AE001272
ORF02	681	1448	*soj* protein {*Mycoplasma pneumoniae*}	26.7	27.9	U34816
ORF03	1451	1750	Hypothetical protein	–	27.7	–
ORF04	1824	2189	orf59 gene product of pMRC01 {*Lactococcus lactis*}	96.4	28.7	AE001272
ORF05	2204	2584	orf60 gene product of pMRC01 {*Lactococcus lactis*}	99.0	30.4	AE001272
ORF06	2610	3554	orf61 gene product of pMRC01 {*Lactococcus lactis*}	99.4	28.9	AE001272
ORF07	4236	5390	*repB* protein of pVS40 {*Lactococcus lactis*}	82.3	35.5	L02920
ORF08	5868	6185	orf63 gene product of pMRC01 {*Lactococcus lactis*}	95.9	31.1	AE001272
ORF09	7231	8925	*nel* protein (*traA* protein in *S. aureus*) of pMRC01 {*Lactococcus lactis*}	97.3	29.6	AE001272
ORF10	9592	10353	IS*904* transposase {*Lactococcus lactis*}	99.,6	39.1	D00696
ORF11	10534	11790	*mef214* protein (*mefA*) {*Streptococcus pyogenes*}	34.6	41.9U70055	
ORF12	11879	12331	*phnB* protein {*Eschericia coli*}	29.9	37.7	J05260
ORF13	12458	12991	*p35* protein {*Eschericia coli*}	38.0	44.4	X13270
ORF14	13008	13382	IS*904* truncated transposase {*Lactococcus lactis*}	100.0	37.9	D00696
ORF15	14256	13438	IS*861* transposase {*Streptococcus agalactiae*}	43.0	34.6	M22449
ORF16	15678	15127	*binL* protein (resolvase) {*Staphylococcus aureus*}	79.2	37.7	X52734
ORF17	16998	16153	*rob* protein {*Eschericia coli*}	32.7	37.9	M97495
ORF18	17155	17841	Transposase of Tn*1546* {*Enterococcus faecium*}	99.1	37.4	U49512
ORF19	18908	18060	Streptomycin adenylase of pS194 {*Staphylococcus aureus*}	98.8	24.7	X06627
ORF20	19755	18916	*repD* protein of pC221 {*Staphylococcus aureus*}	89.1	31.2	X02529
ORF21	21663	20452	*mob* protein (*repC*) {*Lactobacillus plantarum*}	46.3	34.8	M33531
ORF22	22563	21916	Chloramphenicol acetyltransferase of pSBK203 {*Staphylococcus aureus*}	99.5	24.1	M90091
ORF23	23036	22698	truncated replication initator {*Listeria monocytogenes*}	95.1	24.8	X68412
ORF24	23201	23887	Transposase of Tn*1546* {*Enterococcus faecium*}	99.1	37.4	U49512

Table 4. Continued

ORF No.	5′end pos. #	3′end pos. #	Known protein with highest homology {organism}	% aa-Identity	% G/C-content	Accession No.
ORF25	24429	24199	orf7 gene product of Tn*916* {*Enterococcus faecalis*}	56.2	37.7	L15633
ORF26	24906	25265	orf9 gene product of Tn*916* {*Enterococcus faecalis*}	53.6	36.1	L15633
ORF27	25946	25779	Hypothetical protein of pIP811 {*Listeria monocytogenes*}	100	35.1	L09756
ORF28	27974	26034	*tetS* protein of pIP811 {*Listeria monocytogenes*}	99.8	33.3	L09756
ORF29	29084	29770	Transposase of Tn*1546* {*Enterococcus faecium*}	100.0	37.4	U49512

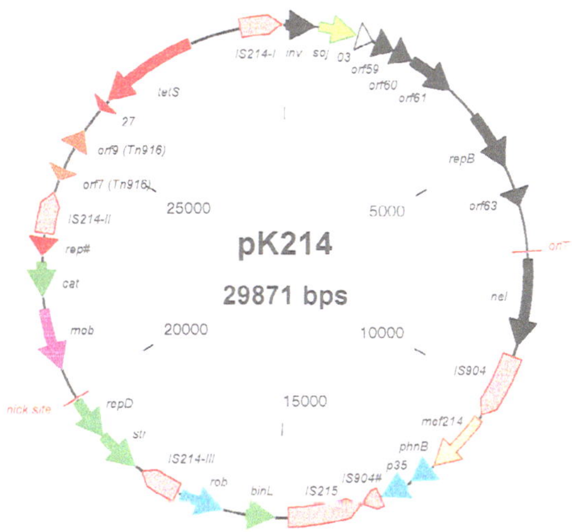

Figure 3. Open reading frames and assigned functions of plasmid pK214. For the identification of the ORF's see Table 4. Red: IS-elements; black: genes with highest identities to *Lactococcus lactis* genes; green: *Staphylococcus*-derived genes.

tis were found adjacent to the *Lactococcus* segment (described above) on the physical map of pK214 (Figure 3 and Table 5). One of these elements, however, is truncated. It seems to be interrupted by another IS element, named IS*215*. This new IS element encodes two hypothetical proteins, and contains two perfect inverted repeats of 14 nucleotides and two direct repeats of 5 bases. It belongs to the IS*3* family since its hypothetical transposase sequence has the characteristic D(35)E motif (Khan et al. 1991). Both invariant residues (Asp [D] at position 13722-13720, and Glu [E] at position 13614-13612, respectively) are surrounded by highly conserved residues of the motif.

Distinct segments resembling sequences from organisms other than *Lactococcus* are flanked by *Enterococcus faecium* IS-like elements (Table 5). Three IS*1216*-like elements (IS*214*-I, IS*214*-II, IS*214*-III) are unidirectionally dispersed on pK214 (Figure 4). The observed high sequence homologies of these elements prove that they are originating from *Enteroccus faecium* where they have been shown to be be involved in high-level vancomycin resistance-elements like Tn*1546* (Handwerger & Skoble 1995; Jensen 1998; Rice & Carias 1998). Such mobile units (e.g. Tn*5385* or Tn*1546*-like transposons) were found to be intersected by one or several IS*1216*–like IS elements. Their roles in transposition of these huge mobile ele-

Table 5. Structures of the insertion elements (IS) found on plasmid pK214 of *Lactococcus lactis* subsp. *lactis* K214

IS-Element (ORF No. Table 4)	**Position** 5′end	3′end	ORF/ inverted and direct repeats	Identity to	% Identity	Acc. No. GenBank
IS*904* (ORF10)	9149	10436		Nucleotide sequence of IS*904* {*Lactococcus lactis*}	99.5 (ntl)	D00696
	9149	9188	IV10-L			
	9206	9496	ORF10-1	ORF1 of IS*904* {*Lactococcus lactis*}	100.0 (aal)	D00696
	9592	10353	TnpA	ORF2 of IS*904* {*Lactococcus lactis*}	99.6 (aal)	D00696
	10393	10355	IV10-Ra			
	10436	10398	IV10-Rb			
IS*904* (ORF14)	13008	13382	ORF14	ORF2 of IS*904* (truncated) {*Lactococcus lactis*}	100.0 (ntl)	D00696
IS*215* (ORF15)	14854	13407		Nucleotide sequence of IS*1296* {*Mycoplasma mycoides*}	52.7 (ntl)	X84021
	13407	13411	DR15-R			
	13412	13425	IV15-R			
	14256	13438	TnpA	ORF2 of IS*861* {*Streptococcus agalactiae*}	43.0 (aal)	M22449
	14810	14253	ORFA	ORFA of IS*1223* {*Lactobacillus johnsonii*}	29.9 (aal)	U09558
	14847	14834	IV15-L			
	14854	14850	DR15-L			
IS *214-III* (ORF18)	17087	17895		Nucleotide sequence of Tn*1546* {*Enterococcus faecium*}	98.5 (ntl)	U49512
	17087	17109	IV18-L			
	17155	17841	TnpA	Transposase of Tn*1546* {*Enterococcus faecium*}	99.1 (aal)	U49512
	17895	17873	IV18-R			
IS*214-II* (ORF24)	23133	23941		Nucleotide sequence of Tn*1546* {*Enterococcus faecium*}	98.5 (ntl)	U49512
	23133	23155	IV-24L			
	23201	23887	TnpA	Transposase of Tn*1546* {*Enterococcus faecium*}	99.1 (aal)	U49512
	23941	23919	IV24-R			
IS*214-I* (ORF29)	29016	29824		Nucleotide sequence of Tn*1546* {*Enterococcus faecium*}	99.6 (ntl)	U49512
	29016	29038	IV29-L			
	29084	29770	TnpA	Transposase of Tn*1546* {*Enterococcus faecium*}	100.0 (aal)	U49512
	29824	29802	IV29-R			

IV = inverted repeat; L = left; R = right; DR = direct repeat; ntl = nucleotide level; aal = amino acid level

```
L. lactis pK214                 1 MTFNIINLETWDRKEYFNHYFNQQTTYSVTKEFDITLLKSMIKNKGYELYPALIYTIVNI
L. monocytogenes ΔpWDB100         ------------------------------------------------------------
S. aureus pSBK203                 ------------------------------------------------------------
S. aureus plasmid                 ------------------------------------------------------------

L. lactis pK214                61 INQNKVFRTGINSEGNLGYWEKLNPLYTVFNKETEKFSNIWTESNVSFNSFYNSYKSDLL
L. monocytogenes ΔpWDB100         --------------------D---------------------------------------
S. aureus pSBK203                 --------------------D---------------------------------------
S. aureus plasmid                 -------------S----F-D-------CLIN-----L---I------S-----------

L. lactis pK214               121 EYKDKNEMFPKKPIPENTVPISMIPWIDFSSFNLNIGNNSRFLLPIITIGKFYSKNNKIY
L. monocytogenes ΔpWDB100         ------------------------------------------------------------
S. aureus pSBK203                 ------------------------------------------------------------
S. aureus plasmid                 ----T-------------G-----------------------------------------

L. lactis pK214               181 LPVSLQVHHAVCDGYHVSLFMSEFQNIVDSVNEWI----
L. monocytogenes ΔpWDB100         ---------------------------------------     99.5% Id
S. aureus pSBK203                 ---------------------------------------     99.5% Id
S. aureus plasmid                 ----------------------------------DLNLK     94.4% Id
```

Figure 4. Alignment of the assigned chloramphenicol-acetyltransferase (CAT) encoded by pK214 with closely related Cat proteins. Id = GenBank accession no. are for *L. monocytogenes* LM2 ΔpWDB100, X68412; *S. aureus* R-plasmid pSBK203, M90091; *S. aureus* plasmid, M58515.

ments are still not clearly defined. In pK214, the pair IS*214*-I/IS*214*-II contains a segment probably originating from *Listeria* whereas the pair IS*214*-II/IS*214*-III could represent the frame for a *Staphylococcus aureus*-derived segment (see below).

Antibiotic resistance genes of pK214: The segment containing ORF's 16, 19, 20 and 22 seems to be derived from *Staphylococcus* since these ORFs show high homologies (78.8 to 99.5% identity) to proteins encoded on plasmids found in *Staphylococcus aureus*. Two hypothetical proteins represent antibiotic resistance determinants, namly ORF19 with 98.8% identity to a streptomycin-inactivating adenylase from the *Staphylococcus aureus* plasmid pS194 and ORF22 with 99.5% identity to the chloramphenicol acetyltransferase (Cat) from the *Staphylococcus aureus* plasmid pSBK203 (Table 4). Moreover, a homologous Cat protein was also identified on the multiresistent plasmid pWDB100 from *Listeria monocytogenes* (Hadorn et al. 1994). An alignment of the *cat* gene products belonging to the cat_{pC223} class (Trieu-Cuot et al. 1993) convincingly demonstrates their close evolutionary relationship (Figure 4). In addition, the regulatory region preceding the *cat* gene of pK214 is 100% identical to that of pWDB100 and pS194, respectively. A third antibiotic resistance gene, named *tetS* corresponding to ORF28 resides on a pK214 segment probably derived from *Listeria monocytogenes* (Figure 3). Its encoded gene product has a 99.8% identity with the tetracyline resistance protein found on plasmid pIP811 of *Listeria monocytogenes* BM4210 (Charpentier et al. 1993). A fourth antibiotic resistance protein is probably encoded by *mef214* (ORF 11) which is expressed and confers resistance to eryhromycin in *E. coli*, but not in the original host *Lactococcus lactis*. It has the molecular characteristic of an efflux protein with 12 membrane spanning domains (Schwarz et al. 1999). It is related to proton drug antiporters MefA and MefE of *Streptococcus pyogenes* and *S. pneumoniae*, respectively (Ainsa et al. 1998). No homology to the previously characterized lactococcal proton motive force-dependent drug transporter LmrP (Bolhuis et al. 1995) exists.

It is noteworthy that only 1 out of 29 ORF's (ORF3) of plasmid pK214 does not shown any significant homology to other sequences in the public gene banks. In the 60 kb plasmid pMRC01 from *Lactococcus lactis* this rate is 25 out of 64 putative ORF's (Dougherty et al. 1998).

6. Acquired antibiotic resistances in the genetic modification of lactic acid bacteria

Molecular biology of LAB started when Gasson & Davies (1980) demonstrated conjugal transfer of the drug resistance plasmid pAMβ into lactic streptococci and when McKay's and other laboratories discovered a wealth of metabolic plasmids in commercial starter cultures of *Lactococcus lactis* and its subspecies (Pechmann & Teuber 1980; McKay 1983).

Vectors designed for pathogenic streptococci were the first to be applied in lactococcal gene cloning experments: derivatives of the 4.2 kb cryptic plasmid VA 380-I from *Streptococcus ferus*, of the 30 kb plamsid pIP501 from *Streptococcus agalactiae*, and of the 26.5 kb plasmid pAMβ1 from *E. faecalis* (review see de Vos & Simons 1994) together with their original resistance genes if present. Lactococcal vectors were constructed from small cryptic plasmids such as pWV01 and pSH71. Antibiotic resistance genes as selectable markers were chosen on the basis of known experiences in other Gram-positive and Gram-negative bacteria: the chloramphenicol acetyltransferase gene from pC194, the erythromycin resistance gene from pE194, and the kanamycin nucleotidyl transferase (*knt-110*) gene from pUB110, all originating from *Staphylococcus aureus*. Very clearly, molecular biologists have exploited the natural principles of gene transfer and antibiotic resistance genes found in humans, animals and food which are regarded as acquired antibiotic resistances. Recently, food grade markers and vectors have been developed for LAB to avoid release of antibiotic resistant genetically modified LAB with food if they should be applied one day (for further literature see de Vos & Simons 1994; Leenhouts et al. 1998).

7. Discussion and conclusions

Horizontal gene transfer

From a human point of view LAB have several significant functions (Teuber 1993):

- they are important human and animal pathogens (e.g. *Streptococcus pneumoniae, S. pyogenes, S. agalactiae* and others)
- they may be opportunistic pathogens in immunocompromised persons (eg *Enterococcus faecalis, E. faecium*, certain lactobacilli)
- they are commensal bacteria in human and animal habitats (mucosae, intestine) including 'low and high GC' species (e.g. bifidobacteria, lactobacilli, lactococci, streptococci, enterococci)
- they are responsible for economically important spontaneous fermentations of plant subtsrates (e.g. lactobacilli, leuconostocs, pediococci, oenococci for sour dough, sauerkraut, silage, soj sauce and wine)
- they are applied as probiotic cultures to produce food with health claims (eg bifidobacteria, lactobacilli, enterococci)
- they are main components of dairy starter cultures for the fermentation worldwide of more than 150 million tons of milk per year into jogurt, sour milk and cheese (eg lactococci, lactobacilli, leuconostocs and *S. thermophilus*)
- the LAB in some fermented products are consumed in life form by man and animals in high numbers (up to more than 10^9 bacteria per gram of product)

Since all these functions are executed by LAB when living in association with other microbes in heavily crowded environments (between 10^9 and 10^{10} bacteria per cm^3), the resulting close contact with other bacteria (e.g. in the intestine or on the surface of a mucosa) is an excellent precondition for horizontal gene transfer with the aid of conjugative transposons and plasmids.

It has been proposed that the excission and conjugative transfer of the broad host range transposon Tn*916* is triggered by physical cell to cell contact (Scott 1992). As described in paragraph 4.2, Tn*916* first discovered in *E. faecalis* has been shown to conjugate into many other species of the human intestinal microflora. This behaviour is even more understandable because (for still unknown reasons) Tn*916* is not restricted in a new host, and a new host cell can accept more than one copy in different locations on the chromosome or on a plasmid. Since excission and integration are often mutagenic leading to base substitutions or, sometimes, to shifts in the reading frame, conjugative transposons may be important in creating and maintaining genome plasticity (Scott 1992). Transposons are common in many LAB and are a preferred tool to distribute antibiotic resistance genes (Clewell et al. 1995). Tn*916* contains the *tetM* gene. A similar transposon has been detected in *E. faecalis* FO1 in a raw milk mountain cheese. Tn*FO1*, in contrast to Tn*916*, is able to excise from the *Lactococcus* chromosome and conjugate, e.g. into *Listeria innocua* (Perreten et al. 1997a). All other tetracycline resistant enterococci from cheeses and sausages able to donate the tetracycline resistance to other bacteria by conjugation (see Table 2) carry *tetM* in a chromosomal location in the donor and recipients. Although we have not proven the tranposon nature of these processes it is tempting to propose and investigate this hypothesis.

The occurrence of insertion sequence (IS) elements necessary for DNA excission and transposition has been amply demonstrated in LAB (Schaefer et al. 1991; Gasson & Fitzgerald 1994). The detection of IS*1181* in *Streptococcus thermophilus, S. salivarius, Lactococcus lactis* and *Lb. delbrueckii* subsp. *bulgaricus* has been taken as evidence for a horizontal gene transfer between these bacteria during dairy fermentations (Guédon et al. 1995). Horizontal gene transfer between pathogenic (*L. monocytogenes*), potentially pathogenic (*E. faecalis*) and starter culture bacteria (*Lactococcus lactis*) is supported by the occurrence of identical *tetS* genes in these species (Poyart-Salmeron et al. 1988; François et al. 1997; Perreten et al. 1997b).

Other tools for horizontal gene transfer are the many described conjugative and mobilizable plasmids in LAB (Gasson & Fitzgerald 1994; Teuber 1995). In lactococci, conjugative plasmids are carriers for the genetic informations for vital technological functions including lactose transport and metabolism, citrate transport, casein degradation, oligopeptide transport, bacteriophage resistance, extracellular polysaccharide synthesis and bacteriocin production. But metabolic functions have also been described in transposons, e.g. sucrose fermentation on the *Lactococcus lactis* conjugative nisin transposons Tn*5276*, Tn*5301*, Tn*5306* and Tn5307 (Gasson & Fitzgerald 1994). The complete nucleotide sequence of the conjugative bacteriocin-producing plasmid pMRC01 from *Lactococcus lactis* DPC3147 demonstrates a transfer region which is highly homologous with the corresponding regions of the staphylococcal plasmids pSK41 and pGO1, and the streptococcal plamid pIP501 (Dougherty et al 1998). On the other side, the lactococcal replication segment is 79 to 100% homologous with that of the multiresistance plasmid pK214 discussed in detail in this review (see paragraph 5.3).

Horizontal gene transfer is obviously an essential mechanism for LAB to adapt and therefore survive in their specific environments. If necessary, antibiotic resistance genes are using the horizontal gene transfer routes available in LAB for metabolic and regulatory functions. Publication of the complete genome sequences, e.g. of *E. faecalis, Lactococcus lactis* and *Lactobacillus acidophilus*, hopefully in the near future will automatically provide the information on the incidence of transposon-like sequences and IS elements in these genomes.

Conclusions

There have been few systematic studies to investigate acquired antibiotic resistance in lactic acid bacteria in food. Most data exist for the potentially and opportunistic pathogenic as well as hygiene indicator bacteria *E. faecalis* and *E. faecium* (e.g. Teuber et al. 1996; Aarestrup et al. 1998). The number of reports on lactococci and lactobacilli is limited. However, it is recently expanding due to increased interest in probiotic lactic acid bacteria (Kullen & Klaenhammer 1998) and genetic modification of LAB for different purposes (see de Vos & Simons 1994).

Examination of these data yields the following general observations:

There may be intrinsic resistance traits, e.g. vancomycin resistance in *Leuconostoc*, certain *Lactobacillus* and others, or resistance to nalidixic acid.

Distinction between intrinsic and acquired resistance is difficult as long as it is not possible to trace an investigated strain into the pre-antibiotic aera. This, however, is possible, e.g. for *E. faecium* strains 68 and many starter cultures isolated at the turn of the last century by starter culture producers (e.g. Ch. Hansen, Labor Wiesby, Marshall Miles and others).

If LAB live in a biotope challenged regularly with antibiotics (human intestine, animal intestine, bovine udder) acquired antibiotic resistance is found in LAB from such habitats including *Enterococcus, Lactococcus*, and *Lactobacillus* species.

The dependence of acquired resistance on the application of antibiotics can be clearly demonstrated with vancomycin resistant enterococci in chicken fed with avoparcin (Aarestrup 1995; van den Braak et al. 1998; Witte 1998), virginiamycin (Hammerun et al. 1998), or bacitracin (Goldenberger et al., ETH Zürich, unpublished). Likewise, fermented food made from raw milk and meat may contain to a certain percentage LAB from the originating animals including enterococci, lactobacilli and lactococci.

There is no barrier between pathogenic (e.g. streptococci), potentially pathogenic (e.g. enterococci) and commensal LAB (e.g.enteric lactobacilli, lactococci) regarding acquired resistances. Identical genes responsible for resistance are found, e.g. for tetracycline (e.g. *tetM*), erythromycin (*ermAM*), chloramphenicol (*cat*), streptomyin (*str*), streptogramin (*sat*) in all three cohorts.

Natural resitsance transfer mechanisms identified in these LAB are also very similar and center around conjugative plasmids (e.g. pAMβ1, pAD1 or pIP501 type) and conjugative transposons (e.g. Tn*916* type). By such mechanisms LAB communicate in their environments with neighbourig bacteria. This statement has been supported by *in vitro* and *in vivo* experiments (Doucet-Populaire et al. 1991; Brockmann et al 1996;

van den Braak et al. 1998; Witte 1998) and the detection of conjugative and nonconjugative antibiotic resistance plasmids in commensal LAB (Teuber et al. 1996; Perreten et al. 1997b, Kullen & Klaenhammer 1998).

For the human population, recent work in Holland has shown that it is not the enterococcal clones coming from farm animals that invade permanently into the human intestinal microflora. It is rather a dispersal of conjugative antibiotic (e.g. vancomycin) resitance genes placed on trasnposons (e.g. Tn*1546*) or plasmids (van den Braak et al. 1998).

Molecular biology has used antibiotic resistance genes found in pathogenic bacteria (*cat, ery, nt* from *S. aureus*) to construct cloning vectors for LAB. These vectors are now being replaced by food-grade cloning and expression systems (see Leenhouts et al 1998 for further references).

Like all other bacteria, LAB are prone to gene exchange to enhance their survival in antibiotic containing environments. For the food microbiologist, there is no doubt that it is necessary to avoid the distribution of bacteria with mobilizable antibiotic resistances (see WHO 1997). Measures include the use of proper starter cultures and of proper substrates (e.g. pasteurized milk) for food fermentations. In addition, generation of antibiotic resistant bacteria in food animals and plant has to be minimized by a prudent use of antibiotics. This includes the ban of antibiotics with clinical application as growth promotors in animal husbandry as recently established politically in the European Union and Switzerland: antibiotics with cross resistance to compounds used in human medicine (tylosine, virginiamycin) or used as such in human medicine like bacitracin, will be banned as of July 1, 1999, in the EU. All antibiotics for feeding purposes have been banned in Switzerland since January 1, 1999. In addition, Swiss farmers will have to keep track of therapeutic and prophylactic applications of antibiotics in their animals.

Acknowledgement

We thank D. Goldenberger, G. Dasen, F. Frei, B. Löffel, U. Schuler-Schmid and V. Perreten for permission to make reference to their work prior to publication. The data on entero cocci were compiled during a sabbatical leave of the first author in the laboratory of Julian Davies (University of British Columbia, Vancouver, Canada) whose hospitality is gratefully acknowledged.

References

Ahn C, Collins-Thompson D, Duncan C & Stiles ME (1992) Mobilization and location of the genetic determinant of chloramphenicol resistance from *Lactobacillus plantarum* caTC2R. Plasmid 27: 169–176

Ainsa JA, Blokpoel MCJ, Otal I, Young DB, de Smet KAL & Martin C (1998) Molecular cloing and characterization of tap, a putative multidrug efflux pump present in *Mycobacterium fortuitum* and *Mycobacterium tuberculosis*. J. Bacteriol. 180: 5836–5843

Antony SJ, Stratton CW & Dummer JS (1996) *Lactobacillus* bacteremia: description of the clinical course in adult patients without endocarditis. Clin. Infect. Dis. 23: 773–778

Aarestrup FM, Bager F, Jensen NE, Madsen M, Meyling A & Wegener HC (1998) Resistance to antimicrobial agents used for animal therapy in pathogenic-, zoonotic- and indicator bacteria isolated from different food animals in Denmark: a baseline study for the Danish Integrated Antimicrobial Resistance Monitoring Programme (DANMAP). APMIS 106: 745–770

Arthur M & Courvalin P (1994) Genetics and mechanisms of glycopeptide resistance in enterococci. Antimicrob. Agents Chemother. 37: 1563–1571

Baquero F, Negir MC, Morosini MI & Blazquez J (1997) The antibiotic selection process: concentration-specific amplification of low-level resistant populations. *In:* Antibiotic Resistance. Origins, evolution, selection and spread (pp. 87–105). Chadwick DJ & Goode J (Eds.) John Wiley & Sons, Chichester

Bates J, Jordens Z & Selkon JB (1993) Evidence for an animal origin of vancomycin-resistant enterococci. The Lancet 342: 490–49

Batish VK & Ranganathan B (1986) Antibiotic susceptibility of deoxyribonuclease-positive enterococci isolated from milk and milk products and their epidemiological significance. Int. J. Food Microbiol. 16: 203–206

Berg T, Firth N, Apisiridej S, Hettiaratchi A, Leelaporn A & Skurray R A (1998) Complete nucleotide sequence of pSK41: Evolution of staphylococcal conjugative multiresistance plasmids. J. Bacteriol. 180: 4350–4359

Bolhuis H, Poelarends G, van Veen HW, Poolman B, Driessen AJM & Konings WN (1995) The lactococcal lmrP gene encodes a proton motive force-dependent drug transporter. J. Biol. Chem. 270: 26092–26098

Brockmann E, Jacobsen BL, Hertel C, Ludwig W, Schleifer KH (1998) Monitoring of genetically modified *Lactococcus lactis* in gnotobiotic and conventional rats by using antibiotic resistance markers and specific probe or primer based methods. System. Appl. Microbiol. 19: 203–212

Celli J & Trieu-Cuot P (1998) Circularization of Tn*916* is required for expression of the transposon-encoded transfer functions: characterization of long tetracycline-inducible transcripts reading through the attachment site. Mol. Microbiol. 28: 103–117

Charpentier E, Gerbaud G & Courvalin P (1994) Presence of the *Listeria* tetracycline resistance gene *tet(S)* in *Enterococcus faecalis*. Antimicrob. Agents Chemother. 38: 2330–2335

Charpentier E, Gerbaud G, Jacquet C, Rocourt J & Courvalin P (1995) Incidence of antibiotic resistance in *Listeria* species. J. Infect. Dis. 172: 277–281

Charteris WP, Kelly PM, Morelli L & Collins JK (1998) Antibiotic susceptibility of potentially probiotic *Bifidobacterium* isolates from the human gastrointestinal tract. Lett. Appl. Microbiol. 26: 333–337

Clewell DB (1994) Bacterial Conjugation (DB Clewell, Ed.). Plenum Press, New York

Clewell DB, Flannagan SE & Jaworsky DD (1995) Unconstrained bacterial promiscuity: the Tn*916*-Tn*1545* family of conjugative transposons. Trends Microbiol. 3: 229–236

Climo M, Sharma VK & Archer GL (1996) Identification and characterization of the origin of conjugative transfer (*ori*T) and gene (*nes*) encoding a single-stranded endonuclease on the staphylococcal plasmid pGO1. J. Bacteriol. 178: 4975–4983

Corpet EE (1988) Antibiotic resistance from food (letter). New England J. Med. 318: 1206–1067

Curragh HJ & Collins MA (1992) High levels of spontaneous drug resistance in *Lactobacillus*. J. Appl. Bacteriol. 73: 31–36

Davies JE (1994) Inactivation of antibiotics and the dissemination of resistance genes. Science 264: 375–382

Davies JE (1997) Origins, acquisition and dissemination of antibiotic resistance determinants. *In:* Chadwick DJ & Goode J (Eds.) Antibiotic Resistance. Origins, evolution, selection and spread (pp. 15–27). John Wiley & Sons, Chichester

De Fabrizio SV, Parada JL & Ledford RA (1994) Antibiotic resistance of *Lactococcus lactis* – an approach of genetic determinants location through a model system. Microbiologie–Aliments–Nutritions 12: 307–315

Dellaglio F, Dicks LMT & Torriani S (1995) The genus *Leuconostoc*. In: Wood BJB & Holzapfel WH (Eds.) The Genera of Lactic Acid Bacteria (pp. 235–278). Blackie Academic & Professional, London

Dessart SR & Steenson LR (1991) High frequency intergeneric and intrageneric conjugal transfer of drug resistance plasmids in *Leuconostoc mesenteroides* ssp. *cremoris*. J. Dairy Sci. 74: 2912–2919

Devriese LA & Pot B (1995) The genus *Enterococcus*. In: Wood BJB and Holzapfel WH (Eds.) The genera of lactic acid bacteria (pp. 327–367). Blackie Academic & Professional, London

De Vos WM & Simons GFM (1994) Gene cloning and expression systems in Lactococci. In: Gasson MJ & de Vos WM (Eds.) Genetics and Biotechnology of Lactic Acid Bacteria (pp. 52–105). Blackie Acadmic & Professional, London

Dodd HM, Horn N, Gasson MJ (1990) Analysis of the genetic determinant for production of the peptide antibiotic nisin. J. Gen. Microbiol. 136: 555 566

Doucet-Populaire F, Trieu-Cuot P, Dosbaa I, Andremont A & Courvalin P (1991) Inducible transfer of conjugative Transposon Tn*1545* from*Enterococcus faecalis* to *Listeria monocytogenes* in the digestive tracts of gnotobiotic mice. Antimicrob. Agents & Chemother. 35: 185–187

Dougherty BA, Hill C, Weidman JF, Richardson DR, Venter JC & Ross RP (1998) Sequence and analysis of the 60 kb conjugative, bacteriocin-producing plasmid pMRC01 from *Lactococcus lactis* DPC3147. Mol. Microbiol. 29: 1029–1038

Elisha BG & Courvalin P (1995) Analysis of genes encoding D-alanine: D-alanine ligase-related enzymes in *Leuconostoc mesenteroides* and *Lactobacillus* spp. Gene 152: 79–83

Elmer GW, Surawics ChM & McFarland LV (1996) Biotherapeutic agents. A neglected modality for the treatment and prevention of selected intestinal and vaginal infections. J. Am. Med. Assoc. 275: 870–876

Falkiner FR (1998) The consequences of antibiotic use in horticulture. J. Antimicrob. Chemother. 41: 429–431

Flannagan SE, Zitzow LA, Su YA & Clewell DB (1994) Complete nucleotide sequence of the 18-kb conjugative transposon Tn*916* from*Enterococcus faecalis*. Plasmid 32: 350–354

Foley S, Bron S, Venema G, Daly C & Fitzgerald GF (1996) Molecular analysis of the replication origin of the *Lactococcus lactis* plasmid pCI305. Plasmid 36: 125–141

Fons M, Hégé T, Ladiré M, Raibaud P, Ducluzeau R & Maguin E (1997) Isolation and characterization of a plasmid from *Lactobacillus fermentum* conferring erythromycin resistance. Plasmid 37: 199–203

François B, Charles M & Courvallin P (1997) Conjugative transfer of tet(S) between strains of Enterococcus faecalis is associated with the exchange of large fragments of chromosomal DNA. Microbiology 143: 2145–2154

Gasson MJ & Davies FL (1980) Conjugal transfer of the drug resistance plasmid pAMβ1 in the lactic streptococci. FEMS Microbiol. Lett. 7: 51–53

Gasson MJ & Fitzgerald GF (1994) Gene transfer systems and transposition. In: Gasson MJ & de Vos WM (Eds.) Genetics and Biotechnology of Lactic Acid Bacteria (pp. 1–51). Blackie Academic & Professional, London

Giraffa G, Carminati D & Neviani E (1997) Enterococci isolated from dairy products: a review of risks and potential technological use. J. Food Protect. 60: 732–738

Grüneberg RN & Hryniewicz W (1998) Clinical relevance of a European collaborative study on comparative susceptibility of Gram-positive clinical isolates to teichoplanin and vancomycin. Intern. J. Antimicrob. Ag. 10: 271–277

Guédon G, Bourgoin F, Pébay M, Roussel Y, Colmin C, Simonet JM & Decaris B (1995) Characterization and distribution of two insertion sequences, IS*1191* and iso-IS*981*, in *Streptococcus thermophilus*: does intergeneric transfer of insertion sequences occur in lactic acid bacteria co-cultures? Mol. Microbiol. 16: 69–78

Guiney M & Urwin G (1993) Frequency and antimicrobial susceptibility of clinical isolates of enterococci. Eur. J. Clin. Microbiol. Infect. Dis. 12: 362–366

Hadorn K, Kayser FH & Hächler H (1994) Miniplasmid derived from *Listeria monocytogenes* multiresistance plasmid pWDB100 upon conjugal transfer into *Staphylococcus epidermidis* carries chloramphenicol resistance gene identical with staphylococcal gene. System. Appl. Microbiol. 17: 492–500

Hadorn K, Hächler H, Schaffner A & Kayser FH (1993) Genetic characterization of plasmid-encoded multiple antibiotic resistance in a strain of *Listeria monocytogenes* causing endocarditis. Eur. J. Clin. Microbiol. Infect. Dis. 12: 928–937

Hammerum AM, Jensen LB, Aarestrup FM (1998) Detectin of the *satA* gene and transferability of virginiamycin resistance in *Enterococcus faecium* from food-animals. FEMS Microbiol. Lett. 168: 145–151

Handwerger S & Skoble J (1995) Identification of chromosomal mobile element conferring high-level vancomycin resistance in *Enterococcus faecium*. Antimicrob. Agents Chemother. 39: 2446–2453

Herrero M, Mayo B, Gonzales B & Suarez JE (1996) Evaluation of technologically important traits in lactic acid bacteria isolated from spontaneous fermentations. J. Appl. Bacteriol. 81: 565–570

Huycke MM, Sahm DF & Gilmore MS (1998) Multiple-drug resistant enterococci: The nature of the problem and agenda for the future. Emerging Infect. Dis. 4: 239–249

Janzen T, Kleinschmidt J, Neve H & Geis A (1992) Sequencing and characterization of pST1, a cryptic plasmid from *Streptococcus thermophilus*. FEMS Microbiol. Lett. 95: 175–180

Jensen LB, Fridmodt-Moller N & Aarestrup FM (1999) Presence of *erm* gene classes in Gram-positive bacteria of animal and human origin in Denmark. FEMS Microbiol. Lett. 170: 151–158

Jensen LB (1998) Internal variations in Tn*1546*-like elements due to the presence of IS*1216V*. FEMS Microbiol. Lett. 169: 349–354

Jett BD, Huycke M & Gilmore MS (1994) Virulence of enterococci. Clin. Microbiol. Rev. 7: 462–478

Khan E, Mack JPG, Katz RA, Kulkosky J & Skalka AM (1990) Retroviral integrase domain: DNA binding and the recognition of LTR sequences. Nucleic Acids Research 19: 851–861

Kilpper-Bälz R, Fischer G & Schleifer K-H (1982) Nucleic acid hybridization of group D streptococci. Curr. Microbiol. 7: 245–250

Klaenhammer TR & Sutherland SM (1980) Detection of plasmid deoxyribonucleic acid in an isolate of *Lactobacillus acidophilus*. Appl. Environ. Microbiol. 39: 671–674

Klare I, Heier H, Claus H, Böhme G, Marin S, Seltmann G, Hakenbeck R, Altanassova V & Witte W (1995) *Enterococcus faecium* strains with vanA-mediated high-level glycopeptide resistance isolated from animal foodstuffs and fecal samples of humans in the community. Microb. Drug Resistance 1: 265–272

Klare I, Heier H, Claus H, Reissbrodt R & Witte W (1995) *vanA*-mediated high-level glycopeptide resistance in *Enterococcus faecium* from animal husbandry. FEMS Microbiol. Lett. 125: 165–172

Klein G, Pack A & Reuter G (1998) Antibiotic resistance patterns of enterococci and occurrence of vancomycin-resistant enterococci in raw minced beef and pork in Germany. Appl. Environ. Microbiol. 64: 1825–1830

Kleinschmidt J, Soeding B, Teuber M & Neve H (1993) Evaluation of horizontal and vertical gene transfer and stability of heterologous DNA in *Streptococcus thermophilus* isolated from yogurt and yogurt starter cultures. System. Appl. Microbiol. 16: 287–295

Knudtson LM & Hartman PA (1993) Antibiotic resistance among enterococci isolates from environmental and clinical sources. J. Food Protect. 56: 489–492

Kullen MJ & Klaenhammer TR (1999) Genetic Modification of Intestinal Lactobacilli and Bifidobacteria. In: Tannock GW (Ed.) Probiotics, a critical review (pp. 65–83). Horizon Scientific Press, Wymondham

Leenhouts K, Bolhuis A, Venema G, Kok J (1998) Construction of a food-grade multiple-copy integration system for *Lactococcus lactis*. Appl. Microbiol. Biotechnol. 49: 417–423

Levy SB (1997) Antibiotic resistance: An ecological imbalance. In: Chadwick DJ & Goode J (Eds.) Antibiotic Resistance. Origins, evolution, selection and spread (pp. 1–14). John Wiley & Sons, Chichester

Liassine N, Frei R, Jan I & Auckenthaler R (1998) Characterization of glycopeptide-resistant enterococci from a Swiss hospital. J. Clin. Microbiol. 36: 1853–1858

Lin ChF, Fung ZF, Wu ChL & Chung TCh (1996) Molecular characterization of a plasmid-borne (pTC82) chloramphenicol resistance determinant (*cat*-TC) from *Lactobacillus reuteri* G4. Plasmid 36: 116–124

Marra D & Scott JR (1999) Regulation of excision of the conjugative transposon Tn916. Mol. Microbiol. 31: 609–621

Macrina FL & Archer GL (1993) Conjugation and broad host range plasmids in streptococci and staphylococci. In: Clewell DB, (Ed.) Bacterial Conjugation (pp. 313–368). Plenum Press, New York

McDonald LC, Kuehnert MJ, Tenover FC & Jarvis WR (1997) Vancomycin-resistant enterococci outside the health-care setting: prevalence, sources, and public health implications. Emerging Infect. Dis. 3: 311–317

McKay L Funcional properties of plasmids in lactic streptococci. Antonie van Leeuwenhoek 49: 259–274

Mercenier A, Pouwels PH & Chassy BM (1994) Genetic engineering of lactobacilli, leuconostocs and *Streptococcus thermophilus*. In: Gasson MJ & de Vos WM (Eds.) Genetics and Biotechnology of Lactic Acid Bacteria (pp. 252–293). Blackie Academic & Professional, London

Ministry of Agriculture and Forestry (Finland) (1997) Tylosin and spiramycin as feed additives – Influence on the efficacy of therapeutic macrolides, Report from the Republic of Finland according to article 29 of the treaty between Member States of the European Union and the Republic of Finland on the scientific arguments for the adaptation to prohibit the use of tylosin and spiramycin as feed additives in animal nutrition by the National Veterinary and Food Research Institut (EELA) (E. Tast, T Honkanen-Buzalski & P Mannerkop, Eds.). Publications of Ministry of Agriculture and Forestry 5/1997, Helsinki

Ministry of Agriculture, Fisheries and Food (UK) (1998) A review of antimicrobial resistance in the food chain, July 1998. A technical report for MAFF. MAFF Publications, London

Ministry of Agriculture (Sweden), Government Official Reports 132 (1997) Antimicrobial feed additives. Report from the Commission on Antimicrobial Feed Additives. Norsteedts Tryckeri AG., Stockholm

National Research Council & Institute of Medicine (1998) The use of drugs in food animals: benefits and risks. National Academy Press, Washington

Neu HC (1992) The crisis in antibiotic resistance. Science 257: 1064–1973

Neve H, Geis A & Teuber M (1987) Conjugation, a common plasmid transfer mechanism in lactic acid streptococci of dairy starter cultures. System. Appl. Microbiol. 9: 151–157

Noble WC, Virani Z & Cree RGA (1992) Co-transfer of vancomycin and other resistance genes from *Enterococcus faecalis* NCTC 12201 to *Staphylococcus aureus*. FEMS Microbiol. Lett. 93: 195–198

Orberg PK & Sandine WE (1985) Survey of antimicrobial resistance in lactic streptococci. Appl. Environ. Microbiol. 49: 538–542

Pechmann H & Teuber M (1980) Plasmid pattern of Group N (lactic) Streptococci. Zbl. Bakt. I. Orig. C1: 133–136

Perreten V, Kollöffel B & Teuber M (1997a) Conjugal transfer of the Tn*916*-like transposon Tn*FO*1 from *Enterococcus faecalis* isolated from cheese to other Gram-positive bacteria. System. Appl. Microbiol. 20: 27–38

Perreten V, Schwarz F, Cresta L, Boeglin M, Dasen G & Teuber M (1997b) Antibiotic resistance spread in food. Nature 389: 801–802

Poyart-Salmeron C, Carlier C, Trieu-Cuot P, Courtieu AL & Courvalin P (1990) Transferable plasmid-mediated antibiotic resistance in *Listeria monocytogenes*. The Lancet 335: 1422–1426

Quednau M, Ahrne S, Petersson AC & Molin G (1998) Antibiotic-resistant strains of *Enterococcus* isolated from Swedish and Danish retailed chicken and pork. J. Appl. Microbiol. 84: 1163–1170

Reilly A & Käferstein F (1997) Food safety hazards and the application of the principles of the hazard analysis and critical control point (HACCP) system for their control in aquaculture production. Aquaculture Res. 28: 735–752

Rice LB & Carias LL (1997) Transfer of Tn*5385*, a composite, multiresistance chromosomal element from *Enterococcus faecalis*. J. Bacteriol. 180: 714–721

Rice LB (1998) Tn*916* family conjugative transposons and dissemination of antimicrobial resistance determinants. Antimicrob. Agents Chemother. 42: 1871–1877

Rinckel LA & Savage DC (1990) Characterization of plasmids and plasmid-borne macrolide resistance from *Lactobacillus* sp. strains. Plasmid 23: 100–133

Rollins LR, Lee LN & LeBlanc DJ (1985) Evidence for a disseminated erythromycin resistance determinant mediated by Tn*917*-like sequences among group D streptococci isolated from pigs, chickens, and humans. Antimicrob. Agents Chemother. 27: 439–444

Sami M, Yamashita H, Hirono T, Kadokura H, Kitamoto K, Yoda K & Yamasaki M (1997) Hop-resistant *Lactobacillus brevis* contains novel plasmid harboring a multidrug resistance-like gene. J. Fermentation & Bioengineering 84: 1–6

Schäfer A, Jahns A, Geis A & Teuber M (1991) Distribution of the IS elements IS*S1* and IS*904* in lactococci. FEMS Microbiology Letters 80: 311–318

Schleifer KH & Ludwig W (1995) Phylogenetic relationships of lactic acid bacteria. In: Wood BJB & Holzapfel WH (Eds.) The Genera of Lactic Acid Bacteria (pp. 7–18). Blackie Academic & Professional, London

Schmalreck AF, Teuber M, Reininger W & Hartl A (1975) Structural features determining the antibiotic potencies of natural and synthetic hop bitter resins, their precursors and derivatives. Canad. J. Microbiol. 21: 205–212

Schmieger H & Schicklmaier P (1999) Transduction of multiple drug resistance of *Salmonella enterica* serovar *thyphimurium* DT104. FEMS Microbiol. Lett. 170: 251–256

Schwarz F, Perreten V & Teuber M (1999) Mef214, a putative efflux protein, of *Lactococcus lactis* subsp. *lactis* K214 conferring increased macrolide resistance in *Escherichia coli.* (paper submitted).

Scott JR (1992) Sex and the single circle: conjugative transposition. J. Bacteriol. 174: 6005–6010

Sgorbati B, Biavati B & Palenzona D (1995) The genus *Bifidobacterium.* In: Wood BJB & Holzapfel WH (Eds.). The Genera of Lactic Acid Bacteria (pp. 279–306). Blackie Academic & Professional, London

Sievers M, Teuber M, Wirsching F, Perreten V, Eisenring R, Fähndrich P, Schlaepfer S, Krusch U, Simm Ch & Löhmer A (1993) Antibiotic resistance properties of enterococci and staphylococci from fermented food. FoodMicro '93, Book of Abstracts (pp. 191)

Simpson WJ, Hammond JRM & Miller RB (1988) Avoparcin and vancomycin: Useful antibiotics for the isolation of brewery lactic acid bacteria. J. Appl. Bacteriol. 64: 299–310

Simpson WJ & Taguchi H (1995) The genus *Pediococcus,* with notes on the genera *Tetratogenococcus* and *Aerococcus.* In: The Genera of Lactic Acid Bacteria (pp. 125–172). Blackie Academic & Professional, London

Soeding B, Kleinschmidt J, Teuber M & Neve H (1993) Assessment of abilities of conjugal transfer and stability of pAMβ1 in dairy lactobacilli with emphasis on thermophilic and nonstarter lactobacilli. System. Appl. Microbiol. 16: 296–302

Sozzi T & Smiley MB (1980) Antibiotic resistances of yogurt starter cultures *Streptococcus thermophilus* and *Lactobacillus bulgaricus.* Appl. Environ. Microbiol. 40: 862–865

Tannock GW (1987) Conjugal transfer of plasmid pAMβ1 in *Lactobacillus reuteri* and between lactobacilli and *Enterococcus faecalis.* Appl. Environ. Microbiol. 53: 2693–2695

Tannock GW (1998) Probiotics. A critical review. Horizon Scientific Press, Wymondham

Tannock GW, Luchansky JB, Miller L, Connell H, Thode-Andersen S, Mercer AA & Klaenhammer TR (1994) Molecular characterization of a plasmid-borne (pGT633) erythromycin resistance determinant (*ermGT*) from *Lactobacilllus reuteri* 100–163. Plasmid 31: 60–71

Teuber M and Schmalreck AF (1973) Membrane leakage in *Bacillus subtilis* 168 induced by the hop constituents lupulone, humulone, isohumulone and humulinic acid. Arch. Mikrobiol. 94: 159–171

Teuber M (1993) Lactic acid bacteria. In: J Rehm, Reed G, Pühler A & Stadler P (Eds.) 'Biotechnology' Vol. 1, 2nd edition (pp. 326–365). Verlag Chemie, Weinheim

Teuber M (1995) The genus Lactococcus. In Wood BJB & Holzapfel WH (Eds.) The Genera of Lactic Acid Bacteria (pp. 173–234). Blackie Academic & Professional, London

Teuber M, Perreten V & Wirsching F (1996) Antibiotikumresistente Bakterien: eine neue Dimension in der Lebensmittelmikrobiologie. Lebensmittel-Technologie 29: 182–199

Thal LA, Chow JW, Mahayni R, Bonilla H, Perri MB, Donabedia SA, Silverman J, Taber S & Zervos MJ (1995) Characterization of antimicrobial resistance in enterococci of animal origin. Antimicrob. Agents Chemother. 39: 2112–2115

Trieu-Cuot P, de Cespédès G, Bentorcha F, Delbos F, Gaspar E & Horaud T (1993) Study of heterogeneity of chloramphenicol acetyltransferase (CAT) genes in streptococci and enterococci by polymerase chain reaction: Characterization of a new CAT determinant. Antimicrob. Agents Chemother. 37: 2593–2598

Van den Braak N, van Belkum A, van Keulen M, Vliegenthart J, Verbrugh HA & Endtz HP (1998) Molecular characterization of vancomycin-resistant enterococci from hospitalized patients and poultry products in the Netherlands. J. Clin. Microbiol. 36: 1927–1932

Van Veen HW & Konings WN (1998) The ABC family of multidrug transporters in microorganisms. Biochim. Biophys. Acta 1365, 31–36

Vescovo M, Morelli L & Bottazzi V (1982) Drug resistance plasmids *in Lactobacillus acidophilus* and *Lactobacillus reuteri.* Appl. Environ. Microbiol. 43: 50–56

Von Wright A & Räty K (1993) The nucleotide sequence for the replication region of pVS40, a lactococcal food grade cloning vector. Lett. Appl. Microbiol. 17: 25–28

Watanabe T (1963) Infective heredity of multiple drug resistance in bacteria. Bacteriol. Rev. 27: 87–115

WHO (1997) The medical impact of the use of antimicrobials in food animals. Report of a WHO Meeting, Berlin, Germany, 13–17 October 1997. WHO, Geneva

Witte W (1998) Medical consequences of antibiotic use in agriculture. Science 279: 996–997

Wu K, An FY & Clewell DB (1999) *Enterococcus faecalis* Pheromone-responding plasmid pAD1 gives rise to an aggregation (clumping) response when cells are exposed to subinhibitory concentrations of chloramphenicol, erythromycin or tetracycline. Plasmid 41: 82–88

Wüst J, Zbinden R & Kayser FH (1998) Die Empfindlichkeit von Bakterien gegen Chemotherapeutika (Zürich 1996). Praxis 87: 403–412

Antonie van Leeuwenhoek **76:** 139–155, 1999.

Multi-domain, cell-envelope proteinases of lactic acid bacteria

Roland J. Siezen
NIZO food research, PO Box 20, 6710BA Ede, The Netherlands, and Wageningen Centre for Food Sciences, PO Box 557, 6700AN Wageningen, The Netherlands; E-mail: siezen@nizo.nl

Key words: cell envelope, cell surface, domains, lactic acid bacteria, proteinase, subtilase

Abstract

The multi-domain, cell-envelope proteinases encoded by the genes *prtB* of *Lactobacillus delbrueckii* subsp. *bulgaricus, prtH* of *Lactobacillus helveticus, prtP* of *Lactococcus lactis, scpA* of *Streptococcus pyogenes* and *csp* of *Streptococcus agalactiae* have been compared using multiple sequence alignment, secondary structure prediction and database homology searching methods. This comparative analysis has led to the prediction of a number of different domains in these cell-envelope proteinases, and their homology, characteristics and putative function are described. These domains include, starting from the N-terminus, a pre-pro-domain for secretion and activation, a serine protease domain (with a smaller inserted domain), two large middle domains A and B of unknown but possibly regulatory function, a helical spacer domain, a hydrophilic cell-wall spacer or attachment domain, and a cell-wall anchor domain. Not all domains are present in each cell-envelope proteinase, suggesting that these multi-domain proteins are the result of gene shuffling and domain swapping during evolution.

Introduction

Comparative sequence analysis has become one of the cornerstones of modern molecular biology. It has played an essential role in studies of gene family evolution, functional annotation of sequenced genomes, protein structure prediction and modelling, and prediction of biochemical function and/or active sites.

The *domain structure* of a protein is one of the types of information that can be predicted from both multiple-sequence alignments and motif or pattern matches. Here we use comparative sequence analysis to analyze the multi-domain structure of extracellular, cell-envelope proteinases (generically abbreviated here as CEP) of lactic acid bacteria, also referred to as cell-surface proteinases, cell wall-bound proteinases or lactocepins. The presently known proteinase gene sequences are from *Lactococcus lactis, Lactobacillus paracasei, Lactobacillus helveticus, Lactobacillus delbrueckii, Streptococcus pyogenes* and *Streptococcus agalactiae*. Our objective is to predict (i) which domains could be present in these large extracellular proteins, (ii) what the structure and function of each domain could be, (iii) which domains are common to CEPs or unique, and (iv) which domains may be essential, disposable or interchangeable.

Multi-domain cell-surface proteins

Many proteins consist of domains and/or modules, particularly the extracellular proteins of eukaryotes, but they are also observed in prokaryotes, albeit less frequently (Baron et al. 1991; Bork 1991; Coggins 1991; Doolittle & Bork 1993; Campbell & Downing, 1994). Domains are generally compact folded structures of several hundred residues with a well-defined function, while modules are mostly smaller than 100 residues and often found as (imperfect) repeated units. Many modules correspond to single exons that can be exchanged between proteins during evolution through a process called exon shuffling. Hundreds of different modules have now been identified, and similar domains and modules were found to be widespread among seemingly unrelated extracellular proteins. Sequence similarity (homology) searches in databases allows related domains and modules to be identified and classified into certain families, each with their own characteristic 3-dimensionally folded structure. Once the structure of *one* domain/module of a family has been determined, the 3-dimensional structure of homologous units can be predicted by homology modelling. Determination of the 3-dimensional structure of intact multi-domain proteins has proven to be extremely difficult, because they crystallize poorly and are often membrane-associated. However, in recent years the expression of individual domains and modules has allowed the structure determination of important functional regions (Baron et al. 1991; Campbell & Downing 1994). Many modules are found to contain predominantly ß-sheet structure, and the conserved residues that make up the consensus sequence of a module generally form part of a hydrophobic core.

Nearly all extracellular proteins are known to be involved in a variety of protein-protein and other binding interactions. Examples are proteins associated with clotting, fibrinolysis, complement, extracellular matrix, cell adhesion, cell-surface receptors, etc. The various functions of domains and modules can be divided into (i) *recognition and binding*, such as lectin and immunoglobulin domains, Ca-binding domains, cofactor-binding domains and receptors for signal transduction, (ii) *enzymes*, particularly proteases, (iii) *regulation*, of binding or enzyme activity, (iv) *linkers and spacers*, often with unusual amino acid sequences, allowing formation of (triple) helix coiled structures, or highly extended flexible structures, (v) *trafficking and sorting*, such as transmembrane and cell-sorting sequences. Many of these multi-domain proteins also have cytoplasmic domains, particularly enzymes and signal transduction domains. Unfortunately, the functions of many domains and modules have not been elucidated yet.

Protease domains are frequently linked to other domains. Examples are the intracellular, calcium-activated neutral protease (also called calpain) of eukaryotes that has a cysteine protease domain, two calmodulin-like Ca-binding domains which regulate protease activity and a Gly-rich membrane-binding domain (Suzuki 1987; Molinari & Carafoli 1997). Various proteins of the complement, blood-clotting and fibrinolysis systems have a C-terminal serine protease domain of the thrombin/plasmin/trypsin family, and these are linked to a variety of repeated regulatory modules (Bork 1991). Eukaryotic proprotein convertases (PCs; or pro-hormone convertases, PHCs) are involved in activation of peptide hormones, growth factors, viral proteins, plasma proteins, receptors, etc. (Barr 1991; Van de Ven et al. 1993; Nakayama 1997). These PHCs have an N-terminal subtilisin-like serine protease domain, linked to an essential (possibly regulatory) middle domain; some also have repeated Cys-rich modules, a transmembrane domain and a cytoplasmic, cell-sorting domain.

Gram-positive bacteria can also have multi-domain extracellular proteins. Well-known are the large surface proteins of streptococci and staphylococci that often play a role in evading host defence (Fischetti et al. 1991; Yother and Briles 1992; Goward et al. 1993). These include fibronectin-binding proteins, a plasminogen receptor, a complement-inactivating C5a peptidase (see below), immunoglobulin-binding proteins and the M protein, a virulence factor of *Streptococcus*. The M protein for instance consists of four different regions of repeated modules, together forming a long α-helical coiled-coil structure extending away from the cell-surface; in addition it has a Pro/Gly-rich domain spanning the peptidoglycan cell wall and a hydrophobic membrane anchor (Fischetti 1991).

The multi-domain cell-envelope proteinases of lactic acid bacteria

Virulent streptococci

Streptococcus pyogenes (synonymous with Group A Streptococcus or GAS) is an important human pathogen, causing throat and skin infections (Hardey and Whiley 1995). GAS possess various multi-domain surface proteins, most of which are virulence factors involved in attachment, colonization, and invasion of the human host. Among these is the complement-inactivating C5a endopeptidase ScpA. This *S. pyogenes* ScpA specifically cleaves human chemotaxin C5a at its Lys68-Asp69 bond (Wexler et al. 1985), destroying its ability to serve as a chemoattractant. The *scpA* gene is chromosally located in a virulence regulon (Simpson et al. 1990). The *scp* genes have been sequenced from GAS serotype M12 strain CS24 (Chen and Cleary 1990), serotype M49 strain CS101 (Podbielski et al. 1995) and GBS strain 78-471 (Chmouryguina et al. 1996), and found to encode proteins of 1150–1167 residues with greater than 98% sequence identity. *Streptococcus agalactiae* (synonymous with Group B Streptococcus or GBS) is an important cause of mastitis in cattle, and it is the principal cause of neonatal meningitis, sepsis and pneumonia in humans.

The cell-surface proteinase gene (*csp*) sequenced is from strain COH1, a clinical isolate originally obtained from the blood of a septic neonate (TO Harris and CE Rubens, personal communication). The function of this Csp proteinase of 1571 residues is as yet unknown but it may also be a virulence factor; this strain also has the *scp* gene for C5a peptidase (TO Harris and CE Rubens, personal communication).

Dairy lactic acid bacteria

Lactic acid bacteria are used in the production of a wide range of fermented dairy products such as cheese, yoghurt and quarg. Most dairy lactic acid bacteria need to take up various essential amino acids from the medium since they lack the ability to synthesize them. In milk the caseins are the main source of amino acids. Therefore, dairy lactic acid bacteria depend for growth in milk on a proteolytic system for degradation of caseins. The peptides and amino acids generated also play a key role in flavour development. Proteolysis is a cascade process involving numerous components, such as (i) the extracellular proteinase for initial degradation of casein into peptides, (ii) transport systems for uptake of peptides and amino acids, (iii) various intracellular peptidases for degradation into amino acids, and (iv) different enzymes for converting amino acids into volatile flavour components. Several excellent reviews of this proteolytic system have appeared in recent years (Kok and de Vos 1994; Kunji et al. 1996; Mierau et al. 1997).

A single, cell wall-bound extracellular proteinase (CEP) is generally considered to be responsible for the initial breakdown of caseins by dairy lactic acid bacteria (Smid et al. 1991; Tan et al. 1993; Pritchard & Coolbear 1993; Kok & De Vos 1994; De Vos & Siezen 1994). Gene deletion and modification studies have demonstrated that strains cannot grow in milk in the absence of a functional, extracellular proteinase (for reviews see Kok 1990; Pritchard and Coolbear 1993; Kok and de Vos 1994). CEPs of several dairy lactic acid bacteria have been identified and characterized biochemically, and include predominantly strains of *Lactococcus lactis* subsp. *cremoris*, *Lactobacillus casei* subsp. *casei*, *Lactobacillus delbruckii* subsp. *bulgaricus* and *Lactobacillus helveticus* (reviewed by Kunji et al. 1996). Some *Lb. helveticus* (Gilbert et al. 1997; JA Pederson & JE Steele, personal communication) and *Lb. bulgaricus* strains (Stefanitsi et al. 1995) have been reported to contain two different CEPs. Recently, an additional function of the CEP of *Lactococcus lactis* in the control of autolysis by degradation of the autolysin AcmA has been described (Buist et al. 1998).

Three distinctly different types of genes encoding CEPs have been cloned and sequenced from dairy lactic acid bacteria, which we will refer to as *prtB*, *prtH* and *prtP*, following nomenclature previously proposed (Gilbert et al. 1997).

The *prtP* gene has been cloned and sequenced from the *L. lactis* strains Wg2 (Kok et al. 1988), SK11 (Vos et al. 1989a), NCDO763 (Kiwaki et al. 1989), and from *Lactobacillus. paracasei* NCDO151 (Holck & Naes 1992) and found to encode PrtP proteins with greater than 95% sequence identity. This *prtP* gene can be either plasmid or chromosomally encoded (Kok & Venema 1988; Kok & de Vos 1994). Each PrtP consists of 1902 amino acid residues, except for PrtP of *L. lactis* SK11, which has 1962 residues due to an imperfect duplication of 60 residues near the C-terminus.

In thermophilic lactobacilli different *prt* genes have been found. The chromosomally encoded *prtB* gene has been sequenced from *Lb. delbrückii* subsp. *bulgaricus* strain NCDO1489 and encodes a protein of 1946 residues (Gilbert et al. 1996), while the *prtH* gene from *Lb. helveticus* strain CNRZ32 presumably encodes a protein of 1849 residues (JA Pederson and JE Steele, personal communication).

General properties of proteinases

Localization, isolation and characterization

The extracellular, cell-surface location of CEPs is supported by various kinds of data, mostly obtained for the enzyme from *Lactococcus lactis*. Firstly, a C-terminally truncated but active PrtP can be liberated from the cell wall by treating cells with Ca-free buffer (Mills and Thomas 1981; reviewed by Exterkate 1995). The released form of PrtP has a M_w of about 135–145 kDa, implying that about 500 C-terminal residues have been removed. This release is believed to result from an intramolecular autoproteolytic event, although the cleavage site(s) has not been identified (Laan and Konings 1989). Secondly, C-terminal deletion analysis of the *prtP* gene has demonstrated that large C-terminal segments can be deleted without loss of PrtP activity; all truncated forms lacking at least the last 130 residues are secreted into the medium and no longer remain attached to the cell-envelope (Kok et al. 1988; Bruinenberg et al. unpublished results). These simple methods have al-

Table 1. Number (and numbering) of residues per domain

	Ll PrtP		Lh PrtH		Ld PrtB		Sa Csp		Sp ScpA	
	residues	numbering	residues	numbering	residues	numbering	residues	numbering	residues	numbering
PP	187	1–187	176	1–176	192	1–192	143	1–143	101	1–101
PR	512	188–699	511	177–687	497	193–689	495	144–638	482	102–583
A	410	700–1109	414	688–1101	386	690–1075	438	639–1076	449	584–1032
B	476	1110–1585	494	1102–1595	488 (+169)	1076–1563 (1564–1732)				
H	210	1586–1795	72	1596–1667			459	1077–1535		
W	131	1796–1926	182	1668–1849	214	1733–1946			94	1033–1126
AN	36	1927–1962					36	1536–1571	41	1127–1167

lowed purification and characterization of PrtP from various *L. lactis* and *Lb. paracasei* strains. Treatment of whole *L. lactis* cells with lysozyme to disrupt the cell wall liberates a protein of 180 kDa, which is close to the predicted size of the mature PrtP deduced from the primary sequence (Coolbear et al. 1992). Thirdly, electron microscopy of immuno-gold labelled PrtP has confirmed a localization in the cell wall (Hugenholtz et al. 1987). Finally, all genes encoding CEPs specify a typical N-terminal signal sequence (see below), which is required for translocation across the cell membrane (Kok & De Vos 1994; De Vos & Siezen 1994).

CEPs from thermophilic lactobacilli are much more resistant to release in Ca-free buffer, and more rigourous treatments have been applied such as lysozyme treatment, osmotic shock and membrane solubilization: proteinases of 170-180 kDa have been purified from cell walls of *Lb. delbrueckii* subsp. *bulgaricus* CNRZ397 (Laloi et al. 1991) and from the cell-membrane fraction of *Lb. helveticus* L89 (Martin-Hernandez et al. 1989). However, it is not known whether these proteinases correspond to the products of *prtB*, *prtH* or *prtP* genes since no sequencing was performed.

The biochemical characterization of various CEPs from lactococci and lactobacilli has been reviewed previously (Kok & de Vos 1994; Kunji et al. 1996). The *scpA* gene of *S. pyogenes* was expressed in *E.coli* and the extracted ScpA of 140 kDa compared with that extracted from *S. pyogenes* strain CS101 by phage-associated lysin, purified by chromatography and characterized biochemically (Chen and Cleary 1989, 1990). The purified protein was functionally active in its ability to inactivate serum chemotaxins.

Domain analysis

The complete amino acid sequences of the cell-envelope proteinases *Lb. delbrueckii* PrtB (Gilbert et al. 1996; EMBL code L48487), *Lb. helveticus* PrtH (JA Pederson and JE Steele, personal communication), *L. lactis* PrtP (Vos et al. 1989a; L04962), *S. pyogenes* ScpA (Chen & Cleary 1990; J05229), and *S. agalactiae* Csp (TO Harris & CE Rubens, personal communication) were analyzed and compared by multiple sequence alignment using PILEUP (Devereux et al. 1984), multiple alignment secondary structure prediction using PHD (Rost & Sander 1993, 1994) and DSC (Ross & King 1999) and database homology searching using BLAST (Altschul et al. 1990, 1997) and (T)FASTA (Pearson & Lipman 1988).

This comparative analysis has led to the prediction of a number of different domains in these CEPs (Table 1, Figure 1), and their homology, characteristics and putative function are described below and summarized in Tables 2 and 3.

Pre-pro domain (PP-domain)

Most subtilisin-like serine proteases are synthesized as pre-pro-enzymes, subsequently translocated over a cell membrane via the signal peptide-dependent pathway, and finally activated by cleavage of the pro-peptide. A detailed comparison of the pre-pro sequences and the putative processing sites of these enzymes has identified two main types of pro-peptide (Siezen et al. 1995).

The N-terminal sequence of each of the five encoded CEPs of lactic acid bacteria begins with a typical signal sequence of 31–39 residues for Sec-dependent translocation. This is followed by a prosequence, which for PrtP and PrtB is known to

Table 2. Domain properties

Domain	*Lc. lactis* PrtP	*Lb. helveticus* PrtH	*Lb. delbruecki* PrtB	*S. agalactiae* Csp	*S. pyogenes* ScpA	function (putative)
PP					'propeptide' stays attached	secretion, activation
PR	homology to subtilases	homology to subtilases	homology to subtilases	homology to subtilases	homology to subtilases	serine protease
A	β-sheet	β-sheet	β-sheet	β-sheet	β-sheet	regulation ?
B	β-sheet high T (16%)	β-sheet high T (11%), K (12%) no C, M	β-sheet repeat (54 res): *b1-b3* high K (12%)			stability ?
H	helices (73% of residues) high A (22%), L (13%), T (12%) no C, M	helices (75% of residues) high A (21%), D (17%), K (22%), L (13%) no C, H, I, M, P, R, W		helices (40% of residues) high L (13%) no C, W		stalk, spacer ?
W	repeat (60 res): *a1-a2* high G (22%), T (21%), D (11%) no H, C, F, W, Y, M, V	homology to S-layer proteins high K (15%) no C, W	repeats (41 res): *c1-c2*; (4 res): *d1-d21* high K (32%), S (11%) no C, W, M		repeat (17 res): *e1-e4* high K (17%), P(14%), Q (11%) no H, C, F, W, Y, V, M, I	cell wall attachment, spacer
AN	LPxTG sorting motif high G (14%), L (14%), E (11%), V (11%) no D, N, H, C, Y, W			LPxTG sorting motif high I (11%), K (14%), L (19%), S (11%), T (14%) no C, H, Q, V, W	LPxTx sorting motif high K (17%), T (17%), L (15%)	cell wall anchor

Table 3. Domain homology (% identical amino acids)

Lh PrtH	51%	43%	PrePro	B				
	65%	48%	Protease	A				
Ld PrtB	34%	29%	32%	28%	PrePro	B		
	49%	33%	46%	34%	Protease	A		
Sa Csp	ns	—	ns	—	ns	—	PrePro	B
	36%	ns	38%	26%	38%	26%	Protease	A
Sp ScpA	ns	—	ns	—	ns	—	ns	—
	35%	27%	34%	25%	34%	28%	41%	32%
	Ll PrtP		Lh PrtH		Ld PrtB		Sa Csp	

ns=not significant (i.e. < 25%).

be removed by autocatalytic processing (Kok et al. 1988; Vos et al. 1989a; Kiwaki et al. 1989; Holck & Naes 1992; Gilbert et al. 1996). The propeptides of PrtP, PrtH and PrtB contain about 150 residues and are clearly homologous (Table 3), also suggesting that they are processed at a common sequence KVY(Y/H)(P/A)(T/N) ↓D.

The predicted 70-residue propeptide of ScpA before the protease domain is not removed (Chen & Cleary 1990), while a 99-residue propeptide of Csp is predicted to be processed at SKETPQ↓N. The streptococcal pre-propeptides are not homologous to each other or to those of PrtP, PrtH or PrtB. It may be possible to make homology models of the CEP propeptides, based on the 3-dimensional structure of the propeptide of subtilisin complexed to its protease domain (Gallagher et al. 1995; Jain et al. 1998).

Catalytic or protease domain (PR-domain)

The catalytic domain of CEPs belongs to the superfamily of subtilisin-like serine proteases, also abbreviated *subtilases* (Siezen et al. 1991; Siezen and Leunissen 1997). These subtilases are generally extracellular and occur in archaea, bacteria, fungi, yeasts and higher eukaryotes. In many cases, an N-terminal catalytic domain is coupled to C-terminal extensions which display additional properties such as sequence repeats, Cys-rich domains or transmembrane segments. The catalytic domain is defined as the segment with sequence homology to subtilisins and thermitase (Siezen et al. 1991). From several known crystal structures and a multiple alignment of almost 200 amino acid sequences, a core 3-dimensional structure was predicted for the catalytic domain of all subtilases (Siezen & Leunissen 1997). Predictions can be made for subtilases of unknown three-dimensional structure concerning essential conserved residues, allowable insertions and deletions, disulfide bonds, Ca-binding sites, substrate-binding site residues, ionic and aromatic interactions and surface loops.

The predicted structure of the PrtP catalytic domain from *L. lactis* SK11 is shown in Figure 2 (Siezen et al. 1993; de Vos and Siezen 1994). Based on this structure, strategies for protein engineering were developed and implemented, aimed at modulating either stability, catalytic activity or substrate specificity (Siezen et al. 1991, 1993, 1994; Bruinenberg et al. 1994a, 1994b).

The sequence alignment of the CEP catalytic domains is shown in Figure 3. The catalytic domains show a higher degree of similarity than the other domains (Table 3); the PrtH and PrtP are by far the most related, i.e. they share 65% identical residues. Also, the two streptococcal PR-domains are more related to each other (41%) than to PR-domains of lactococci and lactobacilli (34–38%). In a family tree analysis of all known subtilases (Siezen and Leunissen 1997), these five CEPs would form a separate cluster, and the next most-related enzyme is Vpr from *Bacillus subtilis*, also shown in Figure 3 for comparison. The conserved catalytic residues Asp32, His64 and Ser221 and the oxyanion-hole residue Asn155 are highlighted. Conserved core elements (black bars) and secondary structure are indicated (Siezen et al. 1991, Heringa et al. 1995; Siezen & Leunissen 1997). The five CEPs all have a few large inserts outside this conserved core structure, where the sequence homology is lowest (Figure 3); these correspond to inserts in surface loops (Figure 2).

I(nsert)-domain

An unusually large insert of 134–156 residues is seen in all five CEPs between residues 180–190 of the protease domain (Figure 2). Its size warrants a classification as a separate internal domain, the I-domain. The

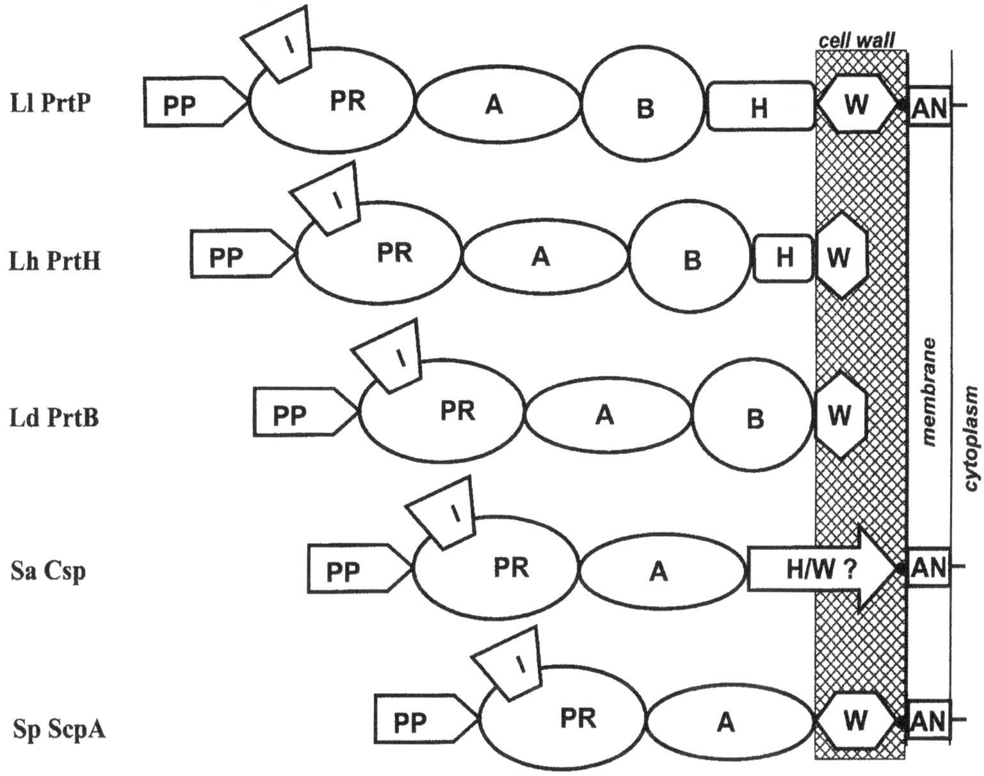

Figure 1. Schematic representation of the predicted domains in cell-envelope proteinases of different lactic acid bacteria. PP = pre-pro domain; PR = protease domain; I = insert domain; A = A-domain; B = B-domain, H = helical domain; W = cell-wall domain; AN = anchor domain (black dot indicates LP×TG motif).

sequence identity is significantly lower (at least 10%) in the I-domain than in the rest of the PR-domain. Deletion analysis of the entire I-domain of PrtP from *L. lactis* SK11 showed that this internal domain is not essential for folding or activity of the protease domain, but it does influence substrate specificity (Bruinenberg et al. 1994a).

Interestingly, a homologous I-domain also occurs in *B. subtilis* Vpr protease domain, and in all plant subtilases known to date, but in no other subtilases (Siezen & Leunissen 1997).

Substrate specificity of proteinases

The structural basis of substrate specificity of subtilases has been reviewed by Perona and Craik (1995). The substrate-binding cleft in subtilases is mainly between the sequence segments 96–107 and 125–130 (Figure 6 in Siezen & Leunissen 1997). The main residues involved in substrate binding in subtilisins and thermitase are marked by an arrow in Figure 3; the marked residues between 150–170 line the P1-binding pocket. Genetic variants of PrtP differ in their specificity towards degradation of α_{s1}-, β- and κ-caseins, and towards chromogenic peptide substrates (Visser et al. 1986; Exterkate 1990; Exterkate et al. 1991). The specificity of genetic variants was analysed using the enzymes isolated from 16 different *L. lactis* strains and fragment 1–23 of α_{s1}-casein as substrate (Exterkate et al. 1993). This led to a classification into seven groups, which displayed a whole range of different specificities. The genetic basis for these differences were first analyzed by constructing hybrids of the enzymes from *L. lactis* subsp. *cremoris* strains Wg2 and SK11 (Vos et al. 1991). Large segments of the PrtP-Wg2 and PrtP-SK11 enzymes were swapped to analyse which of the 44 amino acid differences in 1902 residues were responsible for the marked difference in substrate specificity. These studies limited the differences in the proteinase specificities to just a few amino acid substitutions, located mostly in the segments 96–107 and 125–130 (see Figure 3). Subsequent protein engineering of single or multiple residues in these

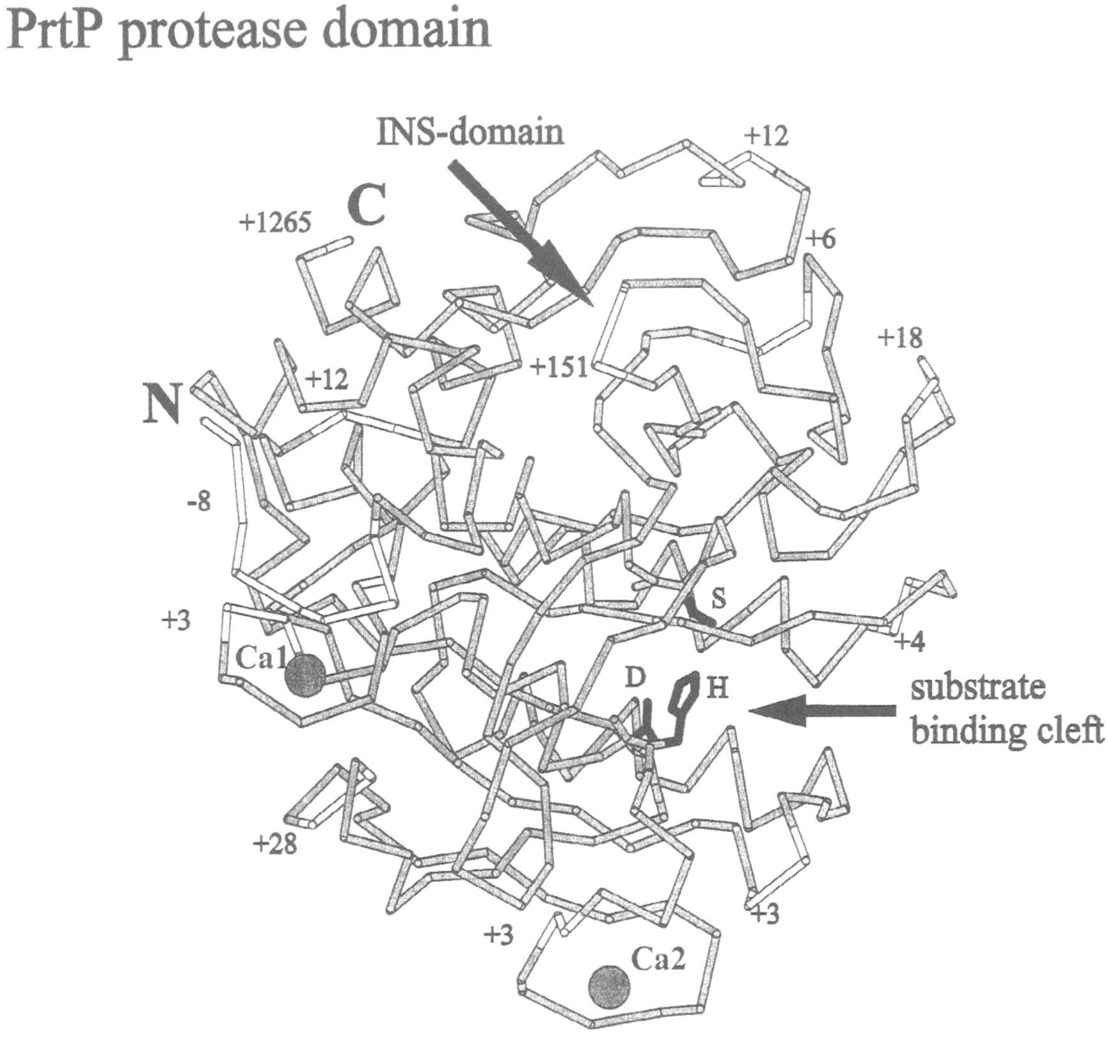

Figure 2. α-Carbon backbone model of the three-dimensional structure of the protease domain of PrtP of *Lactococcus lactis*. The grey backbone represents the conserved core of subtilases (Siezen and Leunissen 1997); white backbone indicates the approximate position where extra residues (and their number) are predicted to be inserted or deleted (only N-terminus). Side chains of the catalytic residues Asp, His and Ser are shown in black. Spheres show the position of predicted bound calcium ions. N- and C-termini of the protease domain are shown: the additional 1265 residues correspond to the other C-terminal domains of PrtP. The substrate binding cleft has been modelled in detail (Siezen et al. 1993).

segments by site-directed mutagenesis confirmed that major changes in specificity do occur as a result of subtle variations in the substrate-binding region of PrtP (Siezen et al. 1993; Bruinenberg et al. 1994b).

By comparing the corresponding predicted substrate-binding residues in the five CEPs shown in Figure 3, we see only minor differences in the segments 125–130 and 150–170, but major variations in the segment 96–107. As yet, there is very little known about substrate specificity of PrtH, PrtB, Csp and ScpA, but any differences between these CEPs may therefore be due to the variations in segment 96–107. Homology modelling of their PR-domain, as previously done for PrtP (Figure 2) could provide more insight into substrate specificity.

A-domain

Based on sequence alignment (Figure 4) a novel A-domain of about 400 residues can now be predicted,

```
                              10         20         30          40                                    50
                               |          |          |           |                                     |
basbpn              AQSVPYGVSQIKAPALHSQGYTGSNVKVAVIDSGIDSSHPDL--------------------------------KVAGGASMVPSETNP
blscar              AQTVPYGIPLIKADKVQAQGFKGANVKVAVLDTGIQASHPDL--------------------------------NVVGGASFVAGE-AY
tvther       YTPNDPYFSSRQYGPQKIQAPQAWD-IAEGSGAKIAIVDTGVQSNHPDLAG------------------------------KVVGGWDFVDNDSTP
bsvpr                  MDDSAPYIGANDAWDLGYTGKGIKVAIIDTGVEYNHPDLKK----------------------------NFGQYKGYDFVDNDYDP
llprtp               DAKANSMANVQAVWSN-YKYKGEGTVVSVIDSGIDPTHKDMRLSDDKDVKL-TKSDVEKFTDTVK--HGRYFNSKVPYGFNYADNNDTI
lhprth              <DSSADNMANVSTVWNN-YKYKGEGTVVSIIDTGIDPNHKDLRLSDDSKVKL-TKDKVNAFTKESG--YGRYFTDKVPYGHNYSDNNDNI
ldprtb              <DESADQMAQVQDVWQE-QKLKGEGMVISIIDTGIDSSHQDLKLDSGVSTAL-SKSEVESDKSKLG--HGKYYTEKVPYGYNYADKNDQI
sacsp               <NIDSNKIITVPKVWNT--GYKGEGTVVAIIDSGLDTNHDALQLNDSTKAKYQNEQQMNAAKAKAGINYGKWYNNKVIFGHNYVDVNTEL
spscpa   (59)TADTPATSKATIRDLNDPSQVKTLQEK-AGKGAGTVVAVIDAGFDKNHEAWRLTDKTKARYQSKEDLEKAKKEHGITYGEWVNDKVAYYHDYSKDGKT-
sec.st                   hhhhhtthhhhh tt   tt eeeeeeb   tt tt                                eeeeee  tt
core
binding                 C                           ↑        C                                       C   C C

          60         70               80         90          100        110        120        130          140
           |          |                |          |            |          |          |          |            |
basbpn   F---------QDNNSHGTHVAGTVAAL-----NNSIGVLGVSPSASLYAVKVLGA---DGSGQYSWIINGIEWAIANNMDVINMSLGGPSG-----SAALKAAVD
blscar   N---------TDGNGHGTHVAGTVAAL-----DNTTGVLGVAPSVSLYAVKVLNS---SGSGTYSGIVSGIEWATTNGMDVINMSLGGPSG-----STAMKQAVD
tvther   ----------QNGNGHGTHCAGIAAAVT----NNSTGIAGTAPKASILAVRVLDN---SGSGTWTAVANGITYAADQGAKVISLSLGGTVG-----NSGLQQAVN
bsvpr    KETPTGDPRGEATDHGTHVAGTVAANGT---------IKGVAPDATLLAYRVLGP---GGSGTTENVIAGVERAVQDGADVMNLSLGNSLN---NPDWATSTALD
llprtp   TDD-------KVDEQHGMHVAGIIGANGTG-DDPAKSVVGVAPEAQLLAMKVFSNSDTSAKTGSATVVSAIEDSAKIGADVLNMSLGSNSG-NQTLEDPELAAVQ
lhprth   TDD-------NPSEQHGMHVAGIVAANGTA--DSVNSVVGVAPEAQLLAMKAFSNSDSSASTDSTSIIGAIDDSAKLGADVLNMSLGSVSG-EQTEDDPEVAAVE
ldprtb   VDN-------GCGEMHGQHVAGIAGANGQ---------VKGVAPDAQLLAMKVFSNNAKNSGAYDDDIISAIEDSVKLGADVINMSLGSVSS-DVGPSDPQQQAVA
sacsp    KE--------VKSTSHGMHVTSIATANPSK-KDTNELIYGVAPEAQVMFMRVFSD--EKRGTGPALYVKAIEDAVKLGADSINLSLGGANGSLVNADDRLIKALE
spscpa   ----------AVDQEHGTHVSGILSGNAPSETKEPYRLEGAMPEAQLLLMRVEIV--NGLADYARNYAQAIRDAVNLGAKVINMSFGNAALAYANLPDETKKAFD
sec.st             bbbhhhhhhhhhhh      bbbbb   bbtt eeeeee    t   tb   hhhhhhhhhhhhhhtt beeee     bb      hhhhhhhh
core
binding              ↑                 C                 C ↑      ↑↑↑↑↑  ↑                ↑↑↑↑↑↑          ↑

              150        160                     170        180                     190              200        210
                |          |                       |          |                       |                |          |
basbpn   KAVASGVVVAAAAGNEGTSGSSS-----------------TVGYPGKYPSVIAVGAVDS---------------SNQRASFSSVGPE-------LDVMAPGVSIQSTLP
blscar   NAYARGVVVVAAAGNSGSSGNTN-----------------TIGYPAKYDSVIAVGAVDS---------------NSNRASFSSVGAE-------LEVMAPGAGVYSTYP
tvther   YAWNKGSVVVAAAGNAGNT---------------------APNYPAYYSNAIAVASTDQ---------------NDNKSSFSTYGSW-------VDVAAPGSSIYSTYP
bsvpr    WAMSEGVVAVTSNGNSGPNGW-------------------TVGSPGTSREAISVGATQLPLN-(128)-KALGEQVADFSSRGPVMDTWMIKPDISAPGVNIVSTIP
llprtp   NANESGTAAVISAGNSGTSGSATEGVNKDYYGLQDNEMVGSPGTSRGATTVASAENTDV-(145)-KYTEDKMSDFTSYGPVSNLS-FKPDITAPGGNIWSTQN
lhprth   RATKKGTAAVISAGNSGTSNSEIEGVNKAYYGNPDMETLGNPGTARSATTVASAENTKA-(145)-ARQTDLMSDFTSYGPTSSLA-FKPDISAPGGHIWSTQN
ldprtb   KASEAGVINVISAGNSGVAGSTADGNPVNNTGTSELSTVGTPGVTPDALTVASAENSKV-(150)-NSRAGKMSDFTSWGPTPELD-FKPEITAPGGKIYSLAN
sacsp    MARLAGVSVVIAAGNDGTFGSGASKPSALY---PDYGLVGSPSTAREAISVASYNNTTL-(139)-NPNAGVLSDFSSWGLTADGQ-LKPDLSAPGGSIYAAIN
spscpa   YAKSKGVSIVTSAGNDSSFGGKTRLPLADH---PDYGVVGTPAAADSTLTVASYSPDKQ-(128)-TASGTKLSRFSSWGLTADGN-IKPDIAAPGQDILSSVA
sec.st   hhhhtt eeeee                           tttbtt eeeeee t               tb    tt    btt          eeee bbeeeeet
core                                                                INS
binding              ↑ ↑↑↑                      ↑  ↑                                    ↑               C           ↑

                      220        230                      240        250                   260        270
                        |          |                        |          |                     |          |
basbpn   ------GNKYGAYNGTSMASPHVAGAAALILSKHP---------------NWTNTQVRSSLENTTTKLG---------DSFYYGKGLINVQAAAQ
blscar   ------TSTYATLNGTSMASPHVAGAAALILSKHP---------------NLSASQVRNRLSSTATYLG---------SSFYYGKGLINVEAAAQ
tvther   ------TSTYASLSGTSMATPHVAGVAGLLASQGR-----------------SASNIRAAIENTADKISG---------TGTYWAKGRVNAYKAVQY
bsvpr    THDPDHPYGYGSKQGTSMASPHIAGAVAVIKQAKP---------------KWSVEQIKAAIMNTAVTLKDSD--GEVYPHNAQGAGSARIMNAIKAD(209)
llprtp   ------NNGYTNMSGTSMASPFIAGSQALLKQALNNKNNPFYAYYKQLKGTALTDFLKTVEMNTAQPINDINYNNVIVSPRRQGAGLVDVKAAIDA(1263)
lhprth   ------NNGYTNMSGTSMASPFIAGTQALVSQTMNDKNGAFYATYQKMSAEERTPFIKTLEMNTASIQPDISHDNVIVSPRRQGAGFINANATIQAL(1162)
ldprtb   ------DNKYQQMSGTSMASPFVAGSEALILQGIK-------KQGLNLSGEELVQFAKNSAMNTSHPVYDTEHTKEIISPRRQGSGEINVKDAINN(1257)
sacsp    ------DNEYDMMSGTSMASPHVAGATALVKQYLL-------KEHPELKKGDIERTVKYLLMSTAKAHLNKD-TCAYTEPRQQGAGIIDVAAAVQT(933)
spscpa   ------NNKYAKLSGTSMSAPLVAGIMGLLQKQYE-------TQYPDMTPSERLDLAKKVLMSSATALYDED-EKAYFSPRQQGAGAVDAKKASAAT(582)
sec.st         tteeeee bhhhhhhhhhhhhhhhhh                   hhhhhhhhhh                        hhhh
core
binding           ↑↑↑↑↑
```

Figure 3. Multiple sequence alignment of the protease domains. Alignment was initially performed using the PILEUP program (Devereux et al. 1984), and improvements were made manually by taking into account the structure-based alignment (Siezen et al. 1991; Heringa et al. 1995). Inserts were judged to occur most likely in turns in external loops. Residue numbering at the top corresponds to that of mature subtilisin BPN' (basbpn). Catalytic residues Asp32, His64 and Ser221 are highlighted, as is the oxyanion-hole residue Asn155. Structurally conserved regions of the core are shown as solid bars; common secondary structure elements are shown as: h = helix, e = extended ß-sheet, b = bend and t = ß-turn. INS indicates the position and number of additional residues in the large insert in the catalytic domain. ↑, residues involved in substrate binding; *C*, residues with side chain ligands to calcium ions (Siezen et al. 1991; Perona and Craik 1995). Each sequence begins at the mature N-terminus; an N-terminus based on the predicted pro-peptide cleavage site is indicated as (<). basbpn, *Bacillus amyloliquefaciens* subtilisin BPN'; blscar, *Bacillus licheniformis* subtilisin Carlsberg; tvther, *Thermoactinomyces vulgaris* thermitase; bsvpr, *Bacillus subtilis* Vpr; llprtp, *Lactococcus lactis* PrtP; lhprth, *Lactobacillus helveticus* PrtH; ldprtb, *Lactobacillus delbrueckii* subsp. *bulgaricus* PrtB; sacsp, *Streptococcus agalactiae* Csp; spscpa, *Streptococcus pyogenes* ScpA.

that is present in all five CEPs and located directly after the PR-domain (Figure 1; Tables 1, 2). The overall sequence identity (25–48%) is lower than for the PR-domain, but still quite significant. This is even more apparent in the last 300 residues which show several short segments of much higher sequence identity (Figure 4). The predicted secondary structure of the A-domain is largely β-strand, and these conserved segments may therefore represent the β-sheet hydrophobic core of a globular domain.

A domain

```
Ll PrtP    -(106)-FVEGFLNF-(   )-LPYMGFFGDW-(   )-AFSSDKN-
Lh PrtH    -(105)-YVEGFLTF-(   )-LPYMGFFGDW-(   )-AISNAKN-
Ld PrtB    - (99)-FVEGYVGF-(   )-LPYMGFFGSY-(   )-AISPNGD-
Sa Csp     -(101)-FLEGYVRF-(   )-IPYVGFKGEF-(   )-AISPNDD-
Sp ScpA    -(101)-FLEGFVRF-(   )-IPYIGFRGDF-(   )-AISPNGD-
Consensus         fvEGfv.F      lPYmGFfGd.       AiSpn.d
sec.st            eeeeeeee      eeeeeeee         eee

Ll PrtP    -(   )-LLRNISN-(   )-KTYYN-(   )-WDGTYYDQ-
Lh PrtH    -(   )-LYRNANN-(   )-KTYYN-(   )-WDGTYFDQ-
Ld PrtB    -(   )-FDRNYKE-(   )-KDYYS-(   )-WDGTDADG-
Sa Csp     -(   )-FLRNFNN-(   )-KNYYS-(   )-WKGTTTDG-
Sp ScpA    -(   )-FLRNAKN-(   )-KNYNN-(   )-WDGKDKDG-
Consensus         flRN..n       K.Yyn       WdGt..Dg
sec.st            eee           eeeee

Ll PrtP    -(   )-DGSYTYRI-(   )-PFKLDSKAPTV-(   )-LTDNASN-(14)-
Lh PrtH    -(   )-DGNYTYRI-(   )-PVKVDSVAPVV-(   )-LFDNAAN-(14)-
Ld PrtB    -(   )-DGQYIYKV-(   )-PVKVDSQAPEV-(   )-LADYAGN-(14)-
Sa Csp     -(   )-DGKYKYVL-(   )-DITLDRQAPTL-(   )-VEDYAGN-(14)-
Sp ScpA    -(   )-NGTYTYRV-(   )-DVIVDNTTPEV-(   )-VEDMAGN-(14)-
Consensus         dG.YtYr.      pvkvDs.aP.v       l.D.AgN
sec.st               eeeee      eeee     e        ee
```

B domain

```
Ll PrtP    -(13)-KISSTTTGY-(   )-TYTFSGTY-(   )-FTASMPV-(   )-LLKHFD-
Lh PrtH    -(13)-KISSQTKGY-(   )-TYVFSGTY-(   )-FVAKLPL-(   )-LLKKQD-
Ld PrtB    -(13)-KITSQSPAY-(   )-TYTVTGTY-(   )-FEVKLPV-(   )-LIKDGD-
Consensus        KIsSqt.gY       TYtfsGTY       F.aklPv       LlK..D
sec.st           eeeeeeeee       eeeeeeee       eeeee

Ll PrtP    -(   )-FSVDVPVNYGDN-(   )-TATDKDGN-(   )-TITGKVKHPTTTLQVDG-
Lh PrtH    -(   )-FSTELPVSFGEN-(   )-VATDSNGN-(   )-TLTGKVKRPTTTLQIGG-
Ld PrtB    -(   )-FSQKVDLNYGDN-(   )-TATDADGN-(   )-TVTGKLAYPVARFQLIG-
Consensus         FS..vpvnyGdN       tATD.dGN       T.TGKvk.PtttlQ..G
sec.st            eeeeeeee           eee            eeeeee   eeeee

Ll PrtP    -(   )-LDAVAPTLSLD-(   )-VYTNDPNF-(   )-VAVTDSEDN-(   )-LAAPTV-
Lh PrtH    -(   )-VDAEAPTLNLD-(   )-VYTNKDKF-(   )-VKVTDIQGN-(   )-LAEPSV-
Ld PrtB    -(   )-VDLAAPSLQLE-(   )-VYTNKDVY-(   )-VTLTDGSDN-(   )-LEAPEV-
Consensus         vDa.APtL.Ld       VYTNkd.f       V.vTD..dN       LaaP.V
sec.st            ee   eee          eee   ee       eeeeee          eeeeee

Ll PrtP    -(   )-NGTFKFKSTDLYGNES-(17)-
Lh PrtH    -(   )-NGTVQFKAVDKYGNES-(17)-
Ld PrtB    -(   )-NGKVEFKYQDVYGNES-(17)-
Consensus         NGtv.FK..D.YGNES
sec.st              eeeeeeee
```

Figure 4. Multiple sequence alignment of most conserved segments in domains A and B. Consensus residues in bold: uppercase for totally conserved, lower case for highly conserved. e = predicted ß-sheet secondary structure. Numbers of residues at begin and end of each domain are shown in brackets.

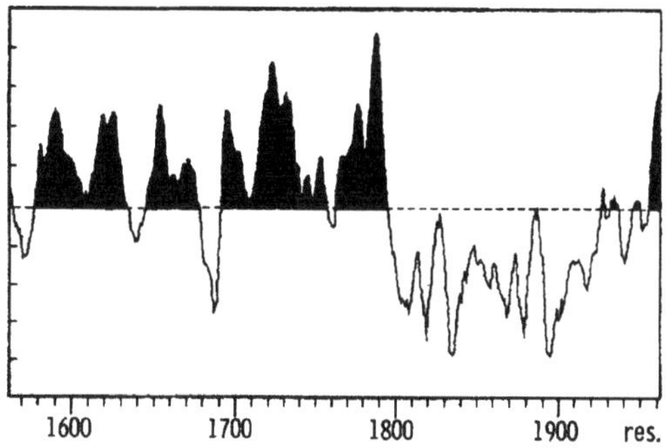

Figure 5. α-helix secondary structure prediction in C-terminal residues 1562–1962 of PrtP of *Lactococcus lactis* SK11. Black areas indicate highest probability helix regions.

The function of this A-domain is as yet unknown, but we speculate that it is essential for proteolytic activity and that it is involved in regulating proteolytic activity and/or specificity of the PR-domain. Firstly, protein engineering studies of PrtP have shown that residues 747 and 748 in this A-domain (located just preceding the conserved K.Yyn segment in Figure 4) play a role in substrate specificity (Vos et al. 1991; Siezen et al. 1993). Secondly, the subtilisin-like, multi-domain extracellular proprotein convertases of eukaryotes, belonging to another subfamily of the subtilases (Siezen & Leunissen 1997), also have a highly conserved middle domain, that is essential for activity and plays a regulatory role (Nakayama 1997).

Database searches show that this A-domain sequence is unique for these CEPs and is not found in any other sequence.

B-domain

A novel B-domain of about 500 residues can also be predicted, that is located directly after the putative A-domain and found only in the CEPs of lactococci and lactobacilli, i.e. PrtB, PrtH, and PrtP (Tables 1, 2). The overall sequence identity (Table 3: 28–43%) is lower than for either the PR-domain (46–65% identity) or A-domain (33–48% identity), but again there are numerous segments of much higher sequence identity (Figure 4). Like the A-domain, this putative B-domain has a high prediction of β-sheet secondary structure and the conserved segments may form the hydrophobic core of the domain. High levels of Lys and Thr residues are present (Table 2).

An unusual property of the B-domain of *Lb. delbrueckii* PrtB is an extension of 169 residues due to a three-fold imperfect repeat of the C-terminal residues. Such repeats are more common in the cell-wall domain of the other CEPs and represent repeated modular units (see below).

No function can be ascribed to this B-domain yet, since it does not have any homologue in current databases. It is not required for proteolytic activity since it is absent in the streptococcal CEPs.

However, in PrtP there are two lines of evidence that this putative B-domain may play a role in stabilizing the protease domain. Progressive C-terminal deletion analysis of PrtP has shown that all of the sequence beyond the A-domain can be deleted without loss of activity or specificity, but the remaining enzyme becomes progressively unstable with truncation within the B-domain, leading to extensive autoproteolysis (Kok et al. 1988; Bruinenberg et al. unpublished results). Secondly, the release of truncated PrtP from the cell in Ca-free medium is the indirect result of the loss of weakly bound Ca ions, leading to exposure of a site susceptible to autoproteolysis (Exterkate & Alting 1999). This Ca-binding site is probably in the B-domain, since autoproteolysis most likely occurs in the B-domain (Bruinenberg et al. unpublished results).

Helix domain (H-domain)

Both PrtP and PrtH have an additional domain with a very high predicted content (about 75% with highest probability) of α-helix secondary structure (Tables 1, 2). While their lengths are unequal (210 vs 72 residues) and their sequences are not directly homologous, they both have high contents of typical α-helix forming residues, i.e. Ala, Leu, Thr, Lys, Asp.

Therefore, the structure of this putative helical domain could be elongated if long helices exist that form a stalk-like structure, as also found e.g. in streptococcal M protein (Fischetti 1991), pneumococcal surface protein (Yother & Briles 1992), type I macrophage receptor (Kodama et al. 1990) and influenza hemagglutinin (Bullough et al. 1994). A 210-residue helix of the PrtP H-domain could form a stalk of over 30 nm length, but it could also consist of shorter helices that fold back into a shorter stalk. In fact, the secondary structure predictions suggest at least six separate helices (Figure 5). Several helices can form a coiled-coil structure (Lupas 1996a), but the prediction algorithms Paircoil (Berger et al. 1995) and COILS (Lupas 1996b) only predict 40 residues as coiled coil in this PrtP H-domain.

The function of this putative H-domain may therefore be as a long single-helix spacer to position the

```
La SlpA      -(288)-ATSNTNGKS-( )-IMHNAYYYDKD-( )-NGKTYYQVVE-
Lh SlpH      -(282)-ATSNKNGKS-( )-IMHNAYFYDKD-( )-NGISYYEVIE-
Lc CbsB      -(296)-ATSDVNGKS-( )-IMHNAYYYKED-( )-GDKNFYEVIK-
Lh PrtH     -(1682)-AVSNxNTSS-( )-VMFKSVLYTKD-( )-KGHYFYKVVD-
Consensus           AtSn.Ngks      iMhnay.Y.kD      .g...Y.V..

La SlpA     -( )-YINAANIDGTKRTLKHNAYVYASSKKRANKVVLKKGEVVTTYG-
Lh SlpH     -( )-YINADNIDGTKRTLKHNAYVYKTSKKRANKVVLKKGTEVTTYG-
Lc CbsB     -( )-YINADNIDGTKRTLKHNAYVYKTSKKRANKVVLKKGEEVTTYG-
Lh PrtH     -( )-YVRVRNVTGTKATLKRNSFVYQSNGKKASRKLLKKGTTITVYG-
Consensus        Yina.NidGTKrTLKhNayVY..skKrAnkvvLKKG..vTtYG

La SlpA   -( )-FKNGQKY.YKIGD-( )-YVKVANF-(1)
Lh SlpH   -( )-FKNGQRY.YKIGA-( )-YVKVANF-(1)
Lc CbsB   -( )-FKNGKQY.YKIGN-( )-YVKASNF
Lh PrtH   -( )-LKHYKKYAYRIGE-( )-YIKSVNV-(7)
Consensus      fKng..Y.YkIG.       YvK..Nf
```

Figure 6. Comparison of most homologous segments of the C-terminal W-domain of *L. helveticus* PrtH (Lh PrtH) to the C-terminal 150-160 residues of S-layer proteins of lactobacilli: La SlpA, *Lactobacillus acidophilus* SlpA; Lh SlpH, *Lactobacillus helveticus* SlpH; Lc CbsB, *Lactobacillus crispatus* CbsB. Consensus residues in bold: uppercase for totally conserved, lower case for highly conserved. Numbers of residues preceding or following each domain are shown in brackets.

```
                                  ↓
S. pyogenes M6               LPTTGET-ANPFFTAAALTVMATAGVAAVV-KRKEEN
G streptococci protein G     LPETGEG-SNPFFTAAALAVMAGAGALAVASKRKED
S.aureus protein A           LPETGE--ENPLIGTTVFGGLSLAGAALLAGRRREL
S.mutans WapA                LPSTGEQ-AGLLLTTVGLVIVAVAGVYFY--RTRR
S.faecalis Asc10             LPKTGEK-QN-VLLTVVGSLAAMLGLAGLGFKRRKETK
Lc. lactis NisP              LPVTGDG-ED-FLPALGIVCISILGIL----KRKTKN
Lc. lactis PrtP              LPKTGETTERPAFGFLGVIVVSLMGVLGL--KRKQREE
S.agalactiae Csp             LPKTGESTPKITYTILLFSLSMDLGLATI--KLKSIKRE
S.pyogenes ScpA              LPTTNDKDTNRLHLLKLVMTTFFFGLVAHIFKTKRQKETKK

consensus                    LP.Tge..................G......krk.
                                        ←----hydrophobic--→←charged→
```

Figure 7. Comparison of C-terminal anchor sequence of *Lactococcus lactis* PrtP, *Streptococcus agalactiae* Csp,and *Streptococcus pyogenes* ScpA with some other streptococcal and staphylococcal surface proteins. The consensus LPxTG sequence (highlighted), hydrophobic membrane-spanning sequence (boxed), and cytoplasmic charged tail are shown. A vertical arrow shows the cleavage site between Thr and Gly residues.

protease and associated A- and B-domains away from the cell surface.

A large C-terminal segment (459 residues) of *S. agalactiae* Csp fits neither into the above profile of a helix domain, nor that of a cell-wall domain, as described below. The predicted secondary structure is 40% helix, and little sheet structure, which is not enough to form long helices. Except for a high content of Leu (13%) the amino acid composition is rather balanced, and not at all like the unusual cell-wall domain compositions of the other CEPs (Table 2). More likely, this segment may represent a globular domain, or even more than one domain. Its structure and function remain unclear, since it is unique and has no homologue in sequence databases.

Cell-wall spacer (W-domain)

Common to many surface proteins of gram-positive bacteria is a cell-wall domain, with an unusual amino acid composition, often high in Pro/Gly or Ser/Thr, that precedes the membrane anchor (Fischetti et al. 1991). Hence, this domain is likely to be extended and span the peptidoglycan layer, functioning as a cell-wall spacer.

Such a cell-wall domain appears to be present in 4 of the 5 CEPs compared here, albeit that they are not homologous to each other and each have their own unusual characteristics. The W-domains of PrtP and ScpA have several features in common: they (i) precede a typical AN-domain (see below), (ii) are quite hydrophilic, (iii) lack most hydrophobic and aromatic residues (C, F, W, Y, M, V, H), (iv) are rich in certain residues, i.e. PrtP is Gly/Thr/Asp-rich (54%) and ScpA is Lys/Pro/Gln-rich (42%), (v) contain imperfect modular repeats, i.e. 2×60 residues in PrtP (although only in *L. lactis* SK11 PrtP) and 4×17 residues in ScpA. In an fully extended form this W-domain could span a cell wall of at least 25 nm.

PrtH and PrtB from lactobacilli lack the AN-domain (see below) and appear to be attached to the cell wall by other mechanisms, presumably through interactions of their W-domain with the cell wall. The W-domain of PrtB is very basic, mainly due to an abundance (32%) of Lys residues which are spaced quasi-periodically; it also has an imperfect 41-residue repeat (Gilbert et al. 1996). Binding to the cell wall may therefore be through electrostatic interactions. Such very basic C-terminal domains with sequence repeats are also attached to some extracellular subtilases from *Bacillus subtilis*, and it was speculated that these may also function in binding to the cell wall (Sloma et al. 1988; Margot & Karamata 1996).

Most interesting is the recent discovery that the W-domain of PrtH is homologous to the family of S(urface)-layer proteins of lactobacilli: the C-terminal 150–160 residues of these S-layer proteins are up to 33% identical to the W-domain of PrtH (JA Pederson and JE Steele, personal communication); an alignment of the most homologous segments is shown in Figure 6. This conserved C-terminal domain of S-layer proteins of lactobacilli is thought to be responsible for cell-wall attachment (Boot et al. 1995). These S-layer proteins are known to cover the cell surface in regular arrays (Sleytr and Sara 1997). Hence the W-domain of PrtH may use the same mechanism of binding to the cell wall as S-layer proteins. One extracellular subtilase from *Bacillus* may also use this mechanism, since it has a C-terminal extension with homology to S-layer proteins from *Bacillus* and *Clostridium* (Schmidt et al. 1995). An extracellular pullulanase from *Clostridium* also has a C-terminal domain with homology to S-layer proteins, that is proposed to function in attachment to the cell surface (Matuschek et al. 1994).

Cell-wall anchor (AN-domain)

The most C-terminal part is conserved in many surface proteins of Gram-positive bacteria and carries a sorting signal, typically LP×TG, followed by a hydrophobic putative membrane-spanning α-helix and a small charged tail (Navarre and Schneewind, 1994) (see Figure 7). After translocation, the LPxT↓G sequence is cleaved (at the position indicated by the arrow) and the carboxyl group of threonine is probably covalently linked to an N-terminal glycine that is part of the cross-bridges in the peptidoglycan layer (Navarre & Schneewind 1994).

This AN-domain is clearly present in *L. lactis* PrtP and in *S. agalactiae* Csp (Figure 7), suggesting that these proteinases become covalently linked to the cell wall. The C-terminal sequence in *S. pyogenes* ScpA has most of the characteristics of this AN-domain, but differs slightly in that it has Asn instead of Gly in the sorting signal, and more charged residues preceding the hydrophobic membrane-spanning sequence. The AN-domain is absent in *Lb. delbruecki* PrtB and *Lb. helveticus* PrtH.

Maturation protein

A maturation protein PrtM is required for the activation of PrtP-like proteinases. The *prtM* gene, located directly upstream of the *prtP* gene and oppositely transcribed, has been identified in *L. lactis* Wg2, SK11, NCDO763, BC101 and *Lb. paracasei* NCDO 151 (Haandrikman et al. 1989; Vos et al. 1989b; Holck & Naes 1992; Nissen-Meyer et al. 1992). The *prtM* gene encodes a protein of 299 amino acids with an N-terminal sequence typical of a prokaryotic pro-lipoprotein (Haandrikman et al. 1989; Vos et al. 1989b; Haandrikman et al. 1991; Holck & Naes 1992; Sankaran & Wu 1994). PrtM is homologous to the PrsA protein of *B. subtilis* (Jacobs et al. 1993), that acts as a general chaperone for different extracellular proteins. PrtM and PrsA appear to belong to a family of homologous proteins, one of which is the peptidyl-prolyl cis/trans isomerase PpiC of *E.coli* (Rahfeld et al. 1994; Rudd et al. 1995). Although it has not been directly shown yet that PrtM is a also a peptidyl-prolyl isomerase, the unique requirement for PrtM in maturation of PrtP-like proteinases may therefore be linked to specific proline residues not found in the other cell-envelope proteinases. Possible candidates are clustered proline residues in the W-domain just preceding the anchoring LPxTG sorting motif.

No homologue of the *prtM* gene has been identified in *Lb. delbrückii* subsp. *bulgaricus* (Gilbert et al. 1996) or in *Lb. helveticus* (JA Pederson and JE Steele, personal communication) in regions directly flanking the proteinase gene. In *S. pyogenes* the *scpA* gene is part of a virulence regulon, but the other genes of this cluster do not encode a PrtM-like protein (Podbielski et al. 1995). In *S. agalactiae* there is an ORF upstream of the *csp* gene encoding a protein of 231 residues with a lipoprotein motif; it has no apparent homology to PrtM or PrsA but it could be a maturation protein (TO Harris and CE Rubens, personal communication). The absence of *prtM*-like genes in the near vicinity does not necessarily imply that no chaperone is required for maturation. In fact, a PrtM-like chaperone could be encoded elsewhere on the chromosome, as in the case of PrsA of *B. subtilis* (Jacobs et al. 1993).

Discussion

The availability and comparison of gene sequences encoding cell-envelope proteinases from different lactic acid bacteria has provided new insight in the multi-domain structure of these large, extracellular proteins. While each domain presumably has a different function, not all functions are clear yet, in particular that of the newly predicted middle domains A and B. The putative A-domain appears to be essential since it is always present, and it may play a regulatory role in catalytic activity. The putative B-domain is only found in *Lactococcus* and *Lactobacillus* CEPs, and is therefore not essential for proteolytic activity, but it may be necessary for stabilization of the PR-domain. Since these dairy lactic acid bacteria, in contrast to the virulent streptococci, use their CEP to generate peptides required for growth, an alternative function for the B-domain may be in transport of the peptides generated by the protease domain towards the cell membrane, where other dedicated peptide-transport systems (reviewed by Kunji et al. 1996) facilitate translocation into the cell. The putative H(elix)-domain is novel for dairy lactic acid bacteria, but similar helical stalk or spacer domains are common in virulent streptococci in association with immunoglobulin-binding domains, e.g. as in streptococcal M proteins (Fischetti 1991). Domains for three different mechanisms of attachment to the cell surface can be distinguished, each of which has also been found in other gram-positive bacteria. First, and most common, the LP×TG sorting motif of the AN-domain, which ultimately leads to covalent binding of the Thr residue to the peptidoglycan of the cell wall (Navarre & Schneewind 1994). Secondly, the highly hydrophilic or positively charged W-domain that binds non-covalently through electrostatic interactions and hydrogen bonds. Thirdly, the novel S-layer type domain only found in PrtH to date (JA Pederson and JE Steele, personal communication), that may incorporate into the regular arrays of S-layer proteins covering the cell surface (Sleytr & Sara 1997).

The acquirement and coupling of multiple domains in proteins is generally ascribed to evolutionary events such as gene duplication, gene fusion and rearrangement. In eukaryotes, this rearrangement is frequently the result of exon duplication and shuffling. In prokaryotes, such a transfer of cassettes, encoding various types of domains, would give maximum speed of evolution and diversity of similar, but not necessarily identical, functions in cell-surface proteins of different species. The degree of sequence conservation is found to vary for the different domains in the order PR>PP ~ A>B (Table 3). Since the entire *prtP* gene is found both in lactococci and lactobacilli, the horizontal exchange between species of the complete *prtP* gene also occurs. Furthermore, the *prtH* gene of *Lb. helveticus* is much more homologous to *prtP* genes of lactococci than to the *prtB* gene of *Lb. delbrueckii* (Table 3). Hence, the genes for CEPs are certainly not appropriate markers for determining phylogenetic relationships between lactic acid bacteria.

Future directions

To answer some of the remaining questions about the functions and/or necessity of these domains one could construct genes encoding CEPs lacking one or more domains ("domain knock-outs"). A longer, more tedious way would be to try to isolate and crystallize individual domains and determine their three-dimensional structure; comparison with 3D structure databases might then provide insight in functionality.

The present knowledge of domain structure and function in CEPs could be exploited in several ways. Novel hybrid enzymes could be made by recombination of the domains from different LAB species, a process similar to natural evolution. Such hybrid enzymes, but less directed toward domain exchange, were already made between two highly homologous

PrtP variants and provided insight in residues involved in determining substrate specificity of the protease (Vos et al. 1991). Hybrids formed from more distantly related CEPs could lead to improved stability or cell-wall attachment, and presumably also to altered substrate specificity. Furthermore, the target region in PrtP for the maturation protein PrtM could be identified in this way and lead to a strategy for PrtM-independent maturation of PrtP. Entirely new properties could be added to the surface of lactic acid bacteria by coupling other proteins to (parts of) these CEPs. A successful example in *Staphylococcus carnosus* is described for the secreted lipase that was tethered to the cell surface by coupling to the cell-wall anchor of *S. aureus* fibronectin-binding protein (Strauss & Gotz 1996). In *Lactococcus lactis* attempts have been made to introduce vaccine antigens into the protease domain of PrtP (Wells et al. 1996). In a similar way, one could add domains with properties such as enzymatic activities, adhesion properties or bacteriocidal activity.

Acknowledgement

I am greatly indebted to Theresa Harris, Craig Rubens, Jeff Pederson and Jim Steele for communicating their sequence data prior to publication, to Bernadette Renckens for performing database searches and sequence comparisons, and to Willem de Vos, Michiel Kleerebezem and Richard van Kranenburg for stimulating discussions. Use of the services and facilities of the Dutch National NWO/SURF Expertise Center CAOS/CAMM is gratefully acknowledged. Supported by the EU-BIOTECH programme (contract B104 CT960016).

References

Altschul SF, Gish W, Miller W, Myers EW & Lipman DJ (1990) Basic local alignment search tool. J Mol Biol 215: 403–410
Altschul SF, Madden TL, Schaffer AA, Zhang J, Zhang Z, Miller W & Lipman DJ (1997) Gapped BLAST and PSI-BLAST: a new generation of protein database search programs. Nucl Acids Res 25: 3389–3402
Baron M, Norman DG & Campbell ID (1991) Protein modules. Trends Biochem. Sci. 16: 13–17
Barr PJ (1991) Mammalian subtilisins: The long-sought dibasic processing endoproteases. Cell 66: 1–3
Berger B, Wilson DB, Wolf E, Tonchev T, Milla M & Kim PS (1995) Predicting coiled coils by use of pairwise residue correlations. Proc Natl Acad Sci 92: 8259–8263
Boot HJ, Kolen CPAM & Pouwels PH (1995) Identification, cloning and nucleotide sequence of a silent S-layer protein gene of *Lactobacillus acidophilus* ATCC 4356 which has extensive similarity with the S-layer protein gene of this species. J. Bacteriol. 177: 7222–7230
Bork P (1991) Shuffled domains in extracellular proteins. FEBS Letters 286: 47–54
Bruinenberg PG, Doesburg P, Alting AC, Exterkate FA, deVos WM & Siezen RJ (1994a) Evidence for a large dispensable segment in the subtilisin-like catalytic domain of the *Lactococcus lactis* cell-envelope proteinase. Protein Eng. 7: 991–996
Bruinenberg PG, deVos WM & Siezen RJ (1994b) Prevention of C-terminal autoprocessing of *Lactococcus lactis* SK11 cell-envelope proteinase by engineering of an essential surface loop. Biochem. J. 302: 957–963
Buist G, Venema G & Kok J (1998) Autolysis of *Lactococcus lactis* is influenced by proteolysis. J. Bacteriol. 180, 5947–5953
Bullough PA, Hughson FM, Skehel JJ & Wiley JJ (1994) Structure of influenza heamagglutinin at the pH of membrane fusion. Nature 371: 37–43
Campbell ID & Downing AK (1994) Building protein structure and function from modular units. Trends Biotech. 12: 168–172
Chen CC & Cleary PP (1989. Cloning and expression of the streptococcal C5a peptidase gene in *Escherichia coli*: linkage to the type 12 m protein gene. Infect. and Immun. 57: 1740–1745
Chen CC & Cleary PP (1990) Complete nucleotide sequence of the streptococcal C5a peptidase gene of *Streptococcus pyogenes*. J. Biol. Chem. 265: 3161–3167
Chmouryguina I, Suvorov A, Ferrieri P & Cleary PP (1996) Conservation of the C5a peptidase genes in group A and B streptococci. Infect. and Immun. 64: 2387–2390
Coggins JR (1991) Deletions, fusions and domain rearrangements. Current Opinion Biotech. 2: 576–581
Coolbear T, Reid JR & Pritchard GG (1992) Stability and specificity of the cell-wall associated proteinase from *Lactococcus lactis* subsp. H2 released by treatment with lysozyme in the presence of calcium ions. Appl. Environm. Microbiol. 58: 3263–3270
Devereux J, Haeberli P & Smithies O (1984) A comprehensive set of sequence analysis programs for the VAX. Nucleic. Acid Res. 12: 387–395
De Vos WM & Siezen RJ (1994) Engineering pivotal proteins for lactococcal proteolysis. In: Andrews AT & Varley J (Eds) Biochemistry of Milk Products, Royal Society of Chemistry, Cambridge, UK, pp. 56–71
Doolittle RF & Bork P (1993) Evolutionary mobile modules in proteins. Sci. American, Oct.: 34–40
Exterkate FA (1990) Differences in short peptide-substrate cleavage by two cell-envelope-located serine proteinases of *Lactococcus lactis* subsp. *cremoris* are related to secondary binding specificity. Appl. Microbiol. Biotechnol. 33: 401–406
Exterkate FA, Alting AC & Slangen CJ (1991) Specificity of two genetically related cell-envelope proteinases of *Lactococcus lactis* subsp. *cremoris* towards α_{s1}-casein-(1-23)-fragment. Biochem. J. 273: 135–139
Exterkate FA, Alting AC & Bruinenberg PG (1993) Diversity of cell envelope proteinase specificity among strains of *Lactococcus lactis* and its relationship to charge characteristics of the substrate-binding region. Appl. Environm. Microbiol. 59: 3640–3647
Exterkate FA (1995) The lactococcal cell envelope proteinases: differences, calcium-binding effects and role in cheese ripening. Int. Dairy Journal 5: 995–1018
Exterkate FA & Alting AC (1999) The role of calcium in the activity and stability of the *Lactococcus lactis* cell-envelope proteinase. Appl. Environm. Microbiol., in press
Fischetti VA (1991) Streptococcal m protein. Sci. American, June: 32–39

Fischetti VA, Pancholi V & Schneewind O (1991) Common characteristics of the surface proteins from gram-positive cocci. In: Dunny GM, Cleary PP & McKay LL (Eds) Genetics and Molecular Biology of Streptococci, Lactococci, and Enterococci, American Society for Microbiol., pp. 290–294

Gallagher T, Gilliland G, Wang L & Bryan P (1995) The prosegment-subtilisin BPN' complex: crystal structure of a specific 'foldase'. Structure 3: 907–914

Gilbert C, Atlan D, Portalier R, Germond GJ, Lapierre L & Mollet B (1996) A new cell surface proteinase: sequencing and analysis of the prtB gene from *Lactobacillus delbrueckii* subsp. *bulgaricus*. J. Bacteriol.78: 3059–3065

Gilbert C, Blanc B, Frot-Coutaz J, Portalier R & Atlan D (1997) Comparison of cell surface proteinase activities within the *Lactobacillus* genus. J. Dairy Res. 64: 561–571

Goward CR, Scawen MD, Murphy JP & Atkinson T (1993) Molecular evolution of bacterial cell-surface proteins. Trends Biochem. Sci. 18: 136–140

Haandrikman AJ, Kok J, Laan H, Soemitro S, Ledeboer AM, Konings WN & Venema G (1989) Identification of a gene required for maturation of an extracellular lactococcal serine proteinase. J. Bacteriol. 171: 2789–2794

Haandrikman AJ, Kok J & Venema G (1991) Lactococcal proteinase maturation protein PrtM is a lipoprotein. J. Bacteriol. 173: 4517–4525

Hardie JM & Wiley RA (1994) The genus *Streptococcus*. In: Wood BJB & Holzapfel WH (Eds) The Lactic Acid Bacteria vol. 2, The Genera of Lactic Acid Bacteria, Blackie Academic and Professional, London, pp. 55–124

Heringa J, Argos P, Egmond MR & de Vlieg J (1995) Increasing thermal stability of subtilisin from mutations suggested by strongly interacting side-chain clusters. Protein Eng. 8: 21–30

Holck A & Naes H (1992) Cloning, sequencing and expression of the gene encoding the cell-envelope-associated proteinase from *Lactobacillus paracasei* subsp. *paracasei* NCDO 151. J. Gen. Microbiol. 138: 1353–1364

Hugenholtz J, van Sinderen D, Kok J & Konings WN (1987) Cell wall-associated proteases of *Streptococcus cremoris* Wg2. Appl. Environm. Microbiol. 53: 853–859

Jain SC, Shinde U, Yuyan L, Inouye M & Berman H. The crystal structure of an autoprocessed Ser221Cys-subtilisin E-propeptide complex at 2.0 Å resolution. J.Mol. Biol. 284: 137–144

Jacobs M, Andersen JB, Kontinen V & Sarvas M (1993) *Bacillus subtilis* PrsA is required in vivo as an extracytoplasmic chaperone for secretion of active enzymes synthesized either with or without prosequences. Mol. Microbiol. 8: 957–966

King RD & Sternberg MJE (1999) Identification and application of concepts important for accurate and reliable protein secondary structure prediction. Protein Science, in press

Kiwaki M, Ikemura H, Shimizu-Kadota M & Hirashima A (1989) Molecular characterization of a cell wall-associated proteinase gene from *Streptococcus lactis* NCDO763. Molec. Microbiol. 3: 359–369

Kodama T, Freeman M, Rohrer L, Zabrecky J, Matsudaira P & Krieger M (1990) Type I macrophage scavenger receptor contains α-helical and collagen-like coiled coils. Nature 343: 531–535

Kok J (1990) Genetics of the proteolytic system of lactic acid bacteria. FEMS Microbiol. Rev. 87: 15–42

Kok J & Venema G (1988) Genetics of proteinases in lactic acid bacteria. Biochimie 70: 475–488

Kok J & de Vos WM (1994) The proteolytic system of lactic acid bacteria. In: Gasson MJ & de Vos WM (Eds) Genetics and Biotechnology of Lactic Acid Bacteria Blackie and Professional, London, pp. 169–210

Kok J, Leenhouts KJ, Haandikman AJ, Ledeboer AM & Venema G (1988) Nucleotide sequence of the cell wall proteinase gene of *Streptococcus cremoris* Wg2. Appl. and Environm. Microbiol. 54: 231–238

Kok J, Hill D, Haandrikman AJ, de Reuver MJB, Laan H & Venema G (1988) Deletion analysis of the proteinase gene of *Streptococcus cremoris* Wg2. Appl Environm Microbiol 54: 239–244

Kunji ERS, Mierau I, Hagting A, Poolman B & Konings WN (1996) The proteolytic systems of lactic acid bacteria. Ant. van Leeuwenhoek 70: 187–221

Laan H & Konings WN (1989) The mechanism of protein release from *Lactococcus lactis* subspecies *cremoris* Wg2. Appl. Environm. Microbiol. 55: 3103–3106

Laloi P, Atlan D, Blanc B, Gilbert C & Portalier R (1991) Cell-wall associated proteinase of *Lactobacillus delbrueckii* subsp. *bulgaricus* CNRZ397: differential extraction, purification and properties of the enzyme. Appl. Environm. Microbiol. 36: 196–204

Lupas A (1996a) Coiled coils: new structures and new functions. Trends Biochem. Sci. 21: 375–382

Lupas A (1996b) Prediction and analysis of coiled-coil structures. Meth Enzymology 266: 513–525

Margot P & Karamata D (1996) The wprA gene of *Bacillus subtilis* 168, expressed during exponential growth, encodes a cell-wall-associated protease. Microbiol. 142: 3437–3444

Martín-Hernández MC, Alting AC & Exterkate FA (1994) Purification and characterization of the mature, membrane-associated cell-envelope proteinase of *Lactobacillus helveticus* L89. Appl. Microbiol. Biotechnol. 40: 828–834

Matuschek M, Burchhardt G, Sahm K & Bahl H (1994) Pullalanase of *Thermoanaerobacterium thermosulphurigenes* EM1 (*Clostridium thermosulphurogenes*): molecular analysis of the gene, composite structure of the enzyme, and a common model for its attachment to the cell surface. J. Bacteriol. 176: 3295–3302

Mierau I, Kunji ERS, Venema G & Kok J (1997) Casein and peptide degradation in lactic acid bacteria. Biotechn. Gen. Engin. Rev. 14: 279–301

Mills OE & Thomas TD (1981) Nitrogen sources for growth of lactic streptococci in milk. N.Z. J. Dairy Sci. Technol. 16: 43–55

Molinari M & Carafoli E (1997) Calpain: A cytosolic proteinase active at the membranes. J. Membrane Biol. 156: 1–8

Nakayama K (1997) Furin: a mammalian subtilisin/Kex2p-like endoprotease involved in processing of a wide variety of precursor proteins. Biochem. J. 327: 625–635

Navarre WW & Schneewind O (1994) Proteolytic cleavage and cell wall anchoring at the LPXTG motif of surface proteins in Gram-positive bacteria. Mol. Micobiol. 14: 115–121

Nissen-Meyer J, Lillehaug D & Nes IF (1992) The plasmid-encoded lactococcal envelope-associated proteinase is encoded by a chromosomal gene in *Lactococcus lactis* subsp. *cremoris* BC 101. Appl. and Environm. Microbiol. 58: 750–753

Pearson WR & Lipman DJ (1988) Improved tools for biological sequence comparison. Proc. Natl. Acad. Sci. USA 85: 2444–2448

Perona JJ & Craik CS (1995) Structural basis of substrate specificity in the serine proteases. Protein Science 4: 337–360

Podbielski A, Flosdorff A & Weber-Heynemann J (1995) The group A streptococcal *virR49* gene controls expression of four structural *vir* regulon genes. Infect. and Immun. 63: 9–20

Pritchard GG & Coolbear T (1993) The physiology and biochemistry of the proteolytic system in lactic acid bacteria. FEMS Microbiol. Rev. 12: 179–206

Rahfeld JU, Rucknagel KR, Schelbert B, Ludwig B, Hacker J, Mann K & Fischer G (1994) Confirmation of the existence of a third family among peptidyl-prolyl *cis/trans* isomerases. FEBS Letters 352: 180–184

Rost B & Sander C (1993) Prediction of protein secondary structure at better than 70% accuracy. J. Mol. Biol. 232: 584–599

Rost B & Sander C (1994) Combining evolutionary information and neural networks to predict protein secondary structure. Proteins 19: 55–77

Rudd KE, Sofia JH, Koonin EV, Plunkett G III, Lazar S & Rouviere PE (1995) A new family of peptidyl-prolyl isomerases. Trends Biochem. Sci. 20: 12–14

Sankaran K & Wu HC (1994) Lipid modification of bacterial lipoprotein: transfer of diacylglyceryl moiety from phosphatidylglycerol. J. Biol. Chem. 269: 19701-19706

Schmidt BF, Woodhouse L, Adams RM, Ward T, Mainzer SE & Lad PJ (1995) Alkalophilic *Bacillus* sp. strain LG12 has a series of serine protease genes. Appl. Environm. Microbiol. 61: 4490–4493

Siezen RJ, de Vos WM, Leunissen JAM & Dijkstra BW (1991) Homology modelling and protein engineering strategy of subtilases, the family of subtilisin-like serine proteases. Protein Engin. 4: 719–737

Siezen RJ, Bruinenberg PG, Vos P, van Alen-Boerrigter IJ, Nijhuis M, Alting AC, Exterkate FA, de Vos WM (1993) Engineering of the substrate binding region of the subtilisin-like, cell-envelope proteinase of *Lactococcus lactis*. Protein Engin. 6: 927–937

Siezen RJ, Leunissen JAM & Shinde U (1995) Homology analysis of the propeptides of subtilisin-like serine proteases (subtilases). In: Shinde, U, (Ed) Intramolecular Chaperones and Folding, R.G. Landes Company, pp 231–253

Siezen RJ & Leunissen JAM (1997) Subtilases: the superfamily of subtilisin-like serine proteases. Prot. Science 6: 501–523

Simpson WJ, LaPenta D, Chen C & Cleary PP (1990) Coregulation of type 12m protein and streptococcal C5a peptidase genes in group A streptococci: evidence for a virulence regulon controlled by the *virR* Locus. J. Bacteriol. 172: 696–700

Sleytr UB & Sára M (1997) Bacterial and arcaeal S-layer proteins: structure-function relationships and their biotechnological applications. Trends Biotech. 15: 20–26

Sloma A, Ally A, Ally D & Pero J (1988) Gene encoding a minor extracellular protease of*Bacillus subtilis*. J Bacteriol. 170: 5557–5563

Smid EJ, Poolman B & Konings WN (1991) Casein utilization by lactococci. Appl. Environm. Microbiol. 57: 2447– 2452

Stefanitsi D, Sakellaris G & Garel J-R (1995) The presence of two proteinases associated with the cell wall of Lactobacillus bulgaricus. FEMS Microbiol. Lett. 128: 53–58

Strauss A & Götz F (1996) *In vivo* immobilization of enzymatically active polypeptides on the cell surface of *Staphylococcus carnosus*. Mol. Microbiol. 21: 491–500

Suzuki K (1987) Calcium-activated neutral protease: domain structure and activity regulation. Trends Biochem. Sci. 12: 103–105

Tan PST, Poolman B & Konings WN (1993) Proteolytic enzymes of *Lactococcus lactis*. J. Dairy Res. 60: 269–286

Ven WJM van de, Roebroek AJM, van Duijnhoven HLP (1993) Structure and function of eukaryotic proprotein processing enzymes of the subtilisin family of serine proteases. Crit. Rev. Oncogenesis 4: 115–136

Visser S, Exterkate FA, Slangen CJ & de Veer GJCM (1986) Comparative study of action of cell wall proteinase from various strains of *Streptococcus cremoris* on bovine α_{s1}-, β- and κ-casein. Appl. Environ. Microbiol. 52: 1162–1166

Vos P, Simons G, Siezen RJ & deVos WM (1989a) Primary structure and organization of the gene for a prokaryotic cell envelope-located serine proteinase. J. Biol. Chem. 264: 13579–13585

Vos P, van Asseldonk M, van Jeveren F, Siezen RJ, Simons G & deVos WM (1989b) A maturation protein is essential for the production of active forms of *Lactococcus lactis* SK11 serine proteinase located in or secreted from the cell envelope. J. Bacteriol. 171: 2795–2802

Vos P, Boerrigter IJ, Buist G, Haandrikman AJ, Nijhuis M, de Reuver MB, Siezen RJ, Venema G, de Vos WM & Kok J (1991) Engineering of the *Lactococcus lactis* serine proteinase by construction of hybrid enzymes. Protein Eng. 4: 479–484

Wells JM, Chamberlain RLM, Schofield KM & Le Page RWF (1996) Lactic acid bacteria as vaccine delivery vehicles. Ant. Van Leeuwenhoek 70: 317–330

Wexler DE, Chenoweth DE & Cleary PP (1985) Mechanism of action of group A streptococcal C5a inactivator. Proc.Natl. Acad. Sci. USA 82: 8144–8148

Yother J & Briles DE (1992) Structural properties and evolutionary relationships of PspA, a surface protein of *Streptococcus pneumoniae*, as revealed by sequence analysis. J. Bacteriol. 174: 601–609

METABOLISM

Antonie van Leeuwenhoek 76: 159–184, 1999.

The biosynthesis and functionality of the cell-wall of lactic acid bacteria

Jean Delcour*, Thierry Ferain, Marie Deghorain, Emmanuelle Palumbo & Pascal Hols
Université Catholique de Louvain, Unité de Génétique, Croix du Sud 5, B-1348 Louvain-la-Neuve, Belgium.
*(*Author for correspondence; E-mail: delcour@gene.ucl.ac.be)*

Key words: autolysin, cell wall, D-alanine, D-serine, lactic acid bacteria, peptidoglycan, polysaccharide, S-layer, teichoic acid, vancomycin

Abstract

The cell wall of lactic acid bacteria has the typical Gram-positive structure made of a thick, multilayered peptidoglycan sacculus decorated with proteins, teichoic acids and polysaccharides, and surrounded in some species by an outer shell of proteins packed in a paracrystalline layer (S-layer). Specific biochemical or genetic data on the biosynthesis pathways of the cell wall constituents are scarce in lactic acid bacteria, but together with genomics information they indicate close similarities with those described in *Escherichia coli* and *Bacillus subtilis*, with one notable exception regarding the peptidoglycan precursor. In several species or strains of enterococci and lactobacilli, the terminal D-alanine residue of the muramyl pentapeptide is replaced by D-lactate or D-serine, which entails resistance to the glycopeptide antibiotic vancomycin. Diverse physiological functions may be assigned to the cell wall, which contribute to the technological and health-related attributes of lactic acid bacteria. For instance, phage receptor activity relates to the presence of specific substituents on teichoic acids and polysaccharides; resistance to stress (UV radiation, acidic pH) depends on genes involved in peptidoglycan and teichoic acid biosynthesis; autolysis is controlled by the degree of esterification of teichoic acids with D-alanine; mucosal immunostimulation may result from interactions between epithelial cells and peptidoglycan or teichoic acids.

Introduction

Already in 1675, Antonie van Leeuwenhoek had noticed the distinctive coccal, rod-like and spiral shapes of 'clear globules', as he described bacteria, 'without being able to discern any film that hold them together, or contained them' (quoted in Salton 1994). Three centuries later, Chapman and Hillier (1953) visualized the bacterial cell wall for the first time in a thin section of *Bacillus cereus* under the electron microscope. Today, fifty more years have elapsed, and a tremendous amount of information has been gained, allowing futurist applications such as using crystalline S-layer proteins for the construction of molecular nanodevices to come closer to reality (Pum & Sleytr 1999). Being specifically focused on lactic acid bacteria (LAB), this review will nevertheless address more general questions on the structure and function of the four components of the Gram-positive cell wall: peptidoglycan (PG, also called murein), teichoic acids, S-layer, and polysaccharides (Figure 1).

The sacculus

The bacterial bag-shaped PG enveloppe (sacculus) is the structural equivalent of the exoskeleton of insects. Like chitin, PG is basically a polymer of N-acetylglucosamine, but it is much more resistant to cracking and shearing than the insect cuticle, due to the covalent nature of the crosslinking of individual chains, as opposed to weak bonds holding together the chitin glycan units. The best image for describing a PG layer is that of a fisherman's net containing the bacterial cellular contents (Figure 2). The mesh of the net is made of two segments of parallel, rather inextensible glycan threads, held together by two small elastic peptide crosslinks allowing the net to expand or shrink. Thus, the whole surface is made of a single gigantic covalently closed macromolecule with remarkable elasticity and tensile strength. Single-layered PG is proper to Gram-negative bacteria, which can stand turgor pressures of the order of 2-5 atm (more than that of a car tyre!). Gram-positives have a much thicker

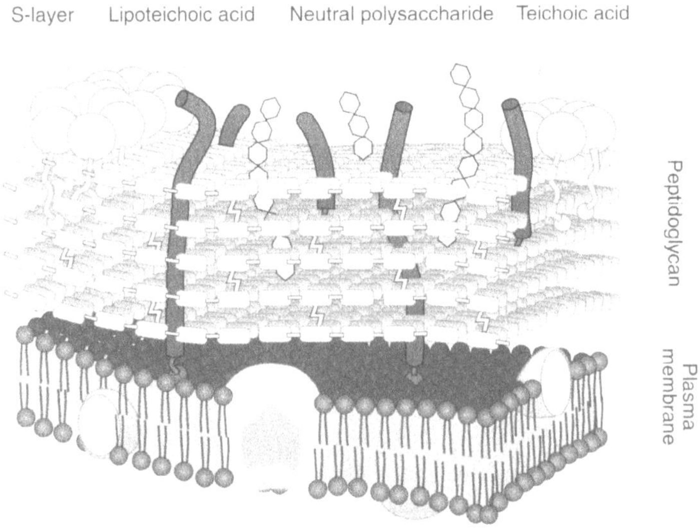

Figure 1. Artist's view of the cell wall of a Gram-positive bacterium. The bilipidic plasma membrane with embedded proteins is covered by a mutilayered peptidoglycan shell decorated with neutral polysaccharides, lipoteichoic acids and teichoic acids, surrounded by an outer envelope of S-layer proteins; for the sake of clarity, cell wall-associated proteins are not depicted.

Figure 2. Schematic representation of the peptidoglycan network structure of LAB. Expanded (above) and shrunken (below) configurations are drawn to scale. M = MurNAc, G = GlcNAc. Bars represent the donor (short) and acceptor (long) crosslinking peptides, respectively; for the sake of clarity, higher-order crosslinks are omitted. The view shown corresponds to either a radial or a tangential cut of the PG layer.

cell-wall (typically 50 nm), which allows them to stand turgor pressures of the order of 20 atm. In this case, several concentric PG layers are covalently held together by additional radial crosslinks which create transverse meshes re-inforcing the whole structure. Diverse 2- and 3-dimensional PG architectures are made possible because each glycan thread has potential half-crosslinks available every other monomer, and these are pointing perpendicular to the glycan axis at right angles of each other (Figure 2). How PG biosynthesis is controlled in time and space so as to ensure faithful shaping and division of bacteria with so many diverse morphologies remains a largely open question which lies beyond the scope of this paper (for recent reviews on cell growth and division, see Ayala et al. 1994; Koch 1995; Rothfield & Zhao 1996; Rothfield & Justice 1997; Giesbrecht et al. 1998; Höltje 1998; Nanninga 1998).

The fisherman's net organization of the peptidoglycan sheets not only provides the sacculus with a container function, but makes it also behave as a sieve (Dijkstra & Keck 1996). Recent studies on isolated sacculi from *E. coli* and *B. subtilis* (Demchick & Koch 1996) indicate that in both cases the mean diameter of the meshes is approximately 4 nm, which corresponds to a cut-off value of about 25 kDa for globular proteins. Taking into account the increase in mesh size resulting from the turgor pressure imposed on the net *in vivo*, these authors conclude that bacterial cells should exclude molecules larger than 55 kDa. This sieving effect imposes restrictions to the secretion of large proteins, as that of levansucrase in *B. subtilis* which has been shown to be unable to freely diffuse through the cell wall and can only rely on the 'inside-to-outside' cell wall expansion process (see below) to be dragged towards the outer face of the thick PG enveloppe and ultimately be released passively into the medium (Kemper et al. 1993). The sieving properties of the cell wall has recently been exploited for the development of an assay for cell wall integrity using fluorescent *in situ* hybridization (Bidnenko et al. 1998).

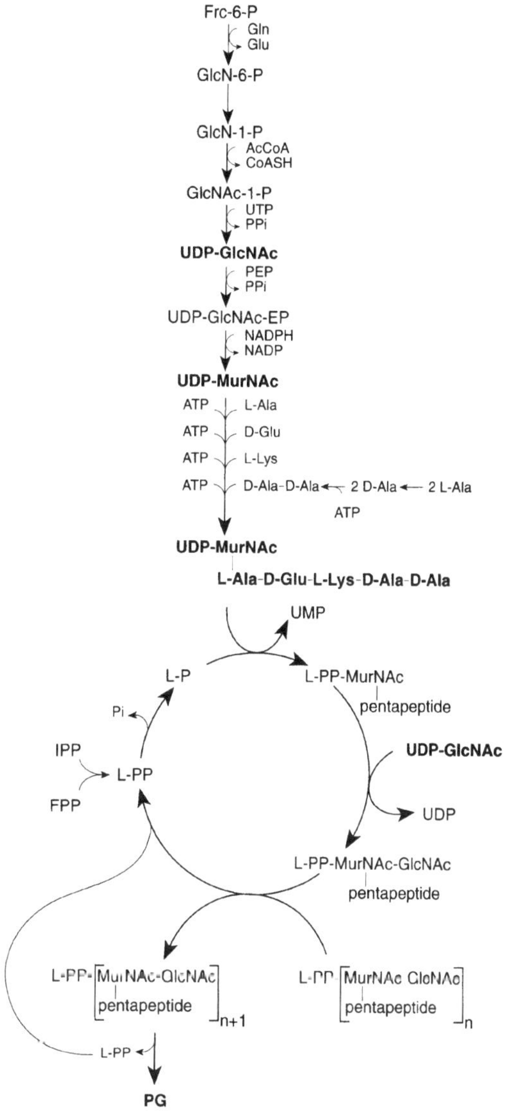

Figure 3. Pathway of peptidoglycan synthesis. L = undecaprenol (see text for explanations).

N-acetyl-glucosamine (GlcNAc) and N-acetyl-muramic acid (MurNAc), the basic components of the PG glycan thread

In LAB, like in all eubacteria, the PG glycan thread is a polymer of the disaccharide N-acetyl-glucosamine-$\beta(1\rightarrow4)$-N-acetyl-muramic acid. UDP-derivatives of the constituent hexoses are used as activated precursors in PG synthesis (Figure 3). Following polymerization and incorporation into the cell wall PG (see below), GlcNAC and/or MurNAC may undergo three different modifications. First, acetyl groups can be removed from cell wall PG through the action of N-deacetylase, as has been described for some bacilli (Araki et al. 1980; Zipperle et al. 1984) and *Lactobacillus fermentum* (Logardt & Neujahr 1975). This can have an incidence on the physiology of the cell wall *in vivo* (for instance, increased sensitivity to autolysis) as well as on the behaviour of the cells *in vitro* when subjected to common experimental lysis procedures, as glycosidic bonds adjacent to non-acetylated MurNAC residues are refractory to most muramidases (which cleave the $\beta1\rightarrow4$ linkages between disaccharide units), including the phage T4 and egg-white lysozymes (Amano et al. 1977; Kawagishi et al. 1980; Kleppe et al. 1981). Second, the cell wall MurNAC residues can be 6-O-Acetylated. This has been observed in 11 species of Gram-negative and Gram-positive bacteria, including the pathogen *Staphylococcus aureus* as well as several species of LAB such as *Enterococcus faecalis, Lactobacillus casei, Lactobacillus acidophilus*, and *L. fermentum* (for a review, see Clarke & Dupont 1992). The extent of naturally occuring O-acetylated residues is quite variable in different species, and amounts to 60% in *L. acidophilus* (Coyette & Ghuysen 1970) and *L. casei* (Billot-Klein et al. 1997). The contribution of O-acetylation to the physiology of bacteria is unknown, but again it is noteworthy that most muramidases (including egg-white lysozyme) are inactive against glycosidic bonds adjacent to O-acetylated MurNAc residues, with the notable exception of mutanolysin (Hamada et al. 1978; for a review on bacterial lysozymes, see Höltje 1996). Finally, C6 of MurNAc can be substituted by cell wall teichoic and teichuronic acids, as will be discussed later.

The pentapeptide, the canonical crosslinking agent

The next stage in the process of PG biosynthesis is the attachment on the D-lactyl carboxyl group of MurNAc of a pentapeptide classically made of alternating L- and D-aminoacids (Figure 4; for a review on the biosynthesis of the PG unit, see van Heijenoort 1994). The first three aminoacids are attached successively on MurNAc, while the last two residues are first united by a D-alanine/D-alanine ligase to form a dipeptide which is then added to the tripeptide to yield the complete MurNAC-pentapeptide (Figure 3). In LAB, the consensus sequence (from the D-lactyl residue) is L-alanine/D-glutamate/*meso*-diaminopimelic acid or L-lysine/D-alanine/D-alanine. In *L. fermentum* as well as in some species of the genus *Bifidus*, L-ornithine is found at the central position, and in a few others (see below), D-serine or D-lactate replace D-ala at the C-terminus. Most residues are derived from normal metabolism: D-lactate is the end-product of fermentation, L-alanine and L-lysine are common aminoacids,

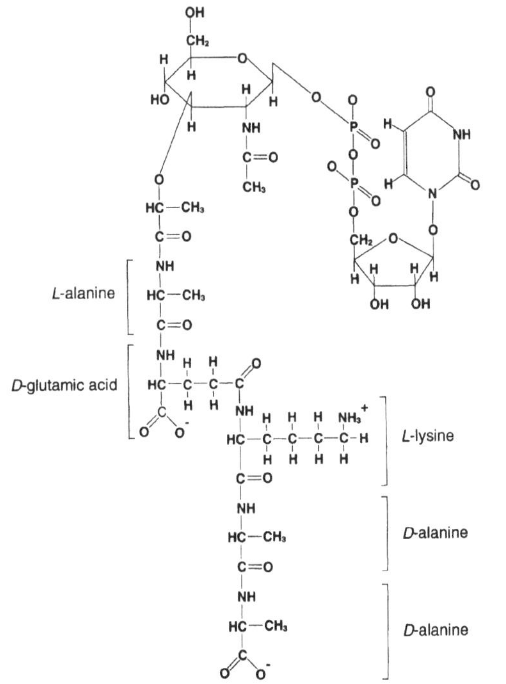

Figure 4. Structure of the disaccharide unit with its attached pentapeptide. Note the peculiar peptide linkage between the γ-carboxyl group of D-glu and the α-amino group of L-lys.

meso-diaminopimelic acid is the precursor of L-lysine and L-ornithine is an intermediate in L-arginine biosynthesis. However, the synthesis of D-aminoacids requires specific transamination or racemisation steps (Hols et al. 1997; Fotheringham et al. 1998).

The pentadepsipeptide, a peculiarity of some LAB

In early 1992, Arthur et al. provided evidence that in a vancomycin-resistant strain of *Enterococcus faecium* (a species which normally produces L-lactate), the C-terminal D-alanine residue of the MurNAc-pentapeptide was substituted by D-lactate (hence the name 'pentadepsipeptide'). This was experimentally confirmed a few months later by three different groups (Allen et al. 1992; Handwerger et al. 1992; Messer et al. 1992). In the strains studied, the resistance phenotype was shown to result from the presence of a large transposon (Tn*1546*). In addition to the transposase and resolvase genes needed for its mobility, Tn*1546* was shown to encode 7 *van* genes among which *vanH*, a D-lactate dehydrogenase, and *vanA*, a D-alanine/D-lactate ligase (Arthur et al., 1993). The strains were shown to be inducibly resistant to high doses of vancomycin and teichoplanin (a related glycopeptide). This vancomycin resistance phenotype was defined as VanA. Two other related phenotypes of vancomycin resistance were soon reported: VanB (inducibly resistant to high doses of vancomycin only) and VanC (constitutively resistant to low doses of vancomycin). VanB strains were found to also produce D-lactate-ending depsipeptides, whereas in the VanC group D-serine was found to replace D-lactate (for a recent review, see Hakenbeck 1994; Arthur et al. 1996). At first, substitution of the terminal D-alanine seemed to be restricted to vancomycin-resistant strains of enterococci, but a number of other LAB species known as intrinsically resistant to vancomycin were later shown to also use D-lactate-ending UDP-MurNAc-pentadepsipeptides as PG precursors: *L. casei, Lactobacillus plantarum, Leuconostoc mesenteroides,* and *Pediococcus pentosaceus* (Billot-Klein et al. 1994; Handwerger et al. 1994; Ferain et al. 1996).

In order to understand why LAB endowed with depsipeptide-bearing PG precursors are resistant to vancomycin, one must refer to the mechanism of action of this glycopeptide antibiotic. Vancomycin comes into contact with the PG precursors on the cell wall side of the cytoplasmic membrane and binds to the D-ala/D-ala terminus of the pentapeptide, preventing polymerization of PG precursors. In the case of VanA and VanB resistance, the affinity of vancomycin for the depsipeptide is reduced about 1000-fold due to the loss of a single hydrogen bond (out of 5) between an oxygen from vancomycin and the NH group of the amide D-ala-D-ala linkage (Bugg et al. 1991). As to VanC, the hydrogen-binding network is preserved, but the bulkier side chain of serine sterically hinders the interaction (Healy et al. 1998).

The production and assembly of muramyl-pentadepsipeptide precursors ending in D-lac first requires that D-lactate is available. As explained above, VanA-type enterococci (L-lactate producers) may acquire this ability horizontally through the uptake of the Tn *1546* transposon. Likewise, the VanB phenotype of a clinical vancomycin-resistant *E. faecium* isolate was recently shown to result from the presence of another transposon, Tn*5382* (Carias et al. 1998). D/L-lactate producers such as *L. plantarum, L. mesenteroides* or *P. pentosaceous* pose no problems, but the source of D-lactate in *L. casei* (a L-lactate producer) is not known; the source of D-serine in VanC resistance is not known either. Second, the ligase must be able to accept D-lac or D-ser instead of D-ala. We have seen that Tn*1546* encodes the VanA ligase which meets this requirement. In the case of LAB intrinsically resistant to vancomycin, the resident ligase most probably acquired this property through evolution of the ancestral enzyme. The advantage of using D-lactate (a waste end-product of metabolism) rather than alanine (a precious aminoacid) is obvious, but the production and use of D-serine makes less sense. Recent reports clearly show that the *E. coli* D-ala-D-ala ligase can easily be changed into a D-ala-D-lac or D-ala-D-ser enzyme through a limited number of directed aminoacid substitutions (Park et al. 1996; Healy et al. 1998). In our laboratory, we have recently shown that a double *ldhL/ldhD* knockout mutant of *L. plantarum* produces about 40% of its muramyl precursors with D-ala C-termini as compared to 100% D-lac in the wild-type (Ferain et al. 1996). In this case, further investigations are needed to establish whether the resident ligase is either ambivalent, mutated or replaced by an alternative enzyme.

Lipid II, the membrane carrier

Lipid II is composed of the lipid-soluble undecaprenol (a C-55 isoprenoid compound also known as bactoprenol) attached through a pyrophosphate linkage to the C-1 of the GlcNAC-$\beta(1\rightarrow4)$-MurNAC-penta(depsi)peptide disaccharide unit (Figure 5). As will become clear throughout this review, the undecaprenyl moiety of lipid II is used as lipid carrier in several macromolecular syntheses related to the biosynthesis of cell wall components. Undecaprenyl pyrophosphate synthetase (UPPS), the enzyme responsible for the last step in undecaprenyl pyrophosphate synthesis, has been characterized in several species, and most extensively studied in *L. plantarum* (Allen 1985); yet, despite the interest in the biochemistry of this enzyme, its gene has only been identified very recently by Apfel et al. (1999). These authors have cloned and characterized the wanted *uppS* gene from *E. coli* using a genomics strategy, and found 27 putative *uppS* genes in genomic databases from a variety of species, including *B. subtilis* and *E. faecalis*; they provide evidence that the gene is essential in *E. coli*.

The main role of lipid II in PG biosynthesis is to convey the pentapeptide disaccharide building blocks pre-fabricated in the cytoplasm up to their site of polymerization and crosslinking on the external face of the cytoplasmic membrane (Figure 5). The physical and

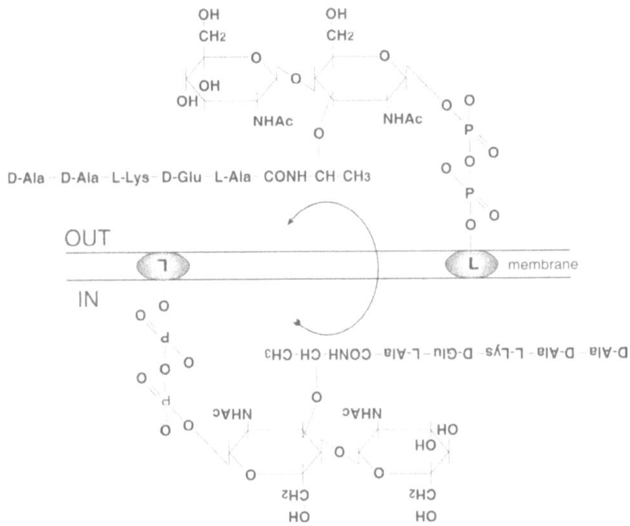

Figure 5. Structure of Lipid II and representation of its crossing through the membrane (L = undecaprenol).

chemical properties of the penta(depsi)peptide made of alternating L- and D-isomers are best suited for allowing an easy passage through the membrane, as energy minimization calculations predict a compact folding with a hydrophobic surface (Labichinski et al. 1985; Koch 1991). The mechanisms allowing lipid II to cross the membrane remain largely unknown. The physical properties and the size of undecaprenol are compatible with a passive process allowing the precursor to go across and curl back through the phospholipid membrane (Koch 1995), but the need for an active mechanism requiring a 'flippase' is often invoked in view of the high rate of translocation required to match the rate of PG synthesis (Nanninga 1998).

In some species, before crossing the cell membrane, lipid II may undergo additional processing. As will be discussed later, cross-linking of the glycan threads takes place in LAB between the third and fourth residues of facing pentapeptides. Depending on the species, the crosslink may either be direct, or require a 'cross-bridge' of one (frequently D-asp in LAB) or several aminoacids hooked on lipid II through (consecutive) peptide linkage(s) to the ω-amino group of the central diaminoacid. Since in some instances this process requires aminoacyl-tRNAs and ATP (Matsuhashi et al. 1967; Green & Vold 1993), it can only take place on lipid II still facing the cytoplasm. In addition, the α-position of D-glu, the ϵ-position of *meso*-diaminopimelic acid or the α-position of the cross-bridge D-asp may be amidated (Linnet & Strominger 1974).

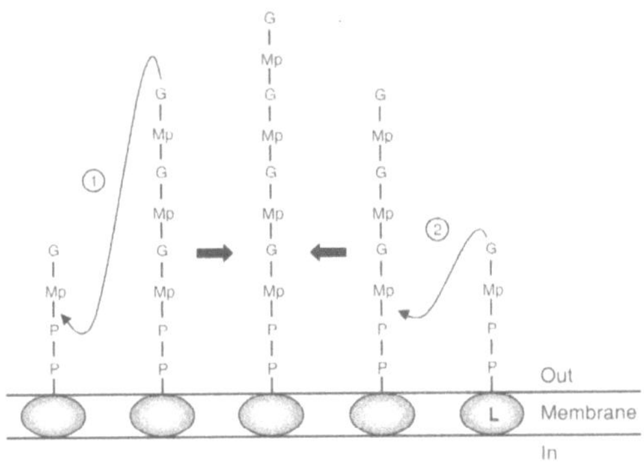

Figure 6. The two modes of transglycosylation during elongation of the peptidoglycan chain: (1) = elongation from the reducing end; (2) = elongation from the non-reducing end. g = GlcNAc; Mp = MurNAc-pentapeptide; L = undecaprenol.

Assembly of nascent PG chains

This step of PG synthesis is the target of several classes of antibiotics, in particular penicillin and its analogs in the β-lactam family (see below), or vancomycin and related glycopeptides (as already discussed). Polymerization of lipid II-bound disaccharides takes place through transglycosylation on the cell wall side of the cytoplasmic membrane. Two opposite directions of chain elongation are known. Either the terminal GlcNAC 4-OH of the nascent chain splits the energetic pyrophosphate bond linking the incoming disaccharide to its undecaprenol carrier ($\beta 4 \rightarrow 1$

elongation from the reducing end of the nascent chain; reaction 1, Figure 6). Or else, the GlcNAC 4-OH of the incoming disaccharide splits the energetic pyrophosphate linkage attaching the nascent chain to its own undecaprenol carrier ($\beta 1 \rightarrow$ elongation from the non-reducing end of the nascent chain; reaction 2, Figure 6). No information is available on the direction of elongation in LAB.

The main actors of PG assembly are membrane-bound enzymes known as 'penicillin-binding proteins' (PBPs) (for a review, see Ghuysen 1991). PBPs can be divided in two groups: the multimodular, high molecular weight PBPs (>60 kDa, further clustered into the A and B subclasses), and the monofunctional, low molecular weight PBPs. They all possess a penicillin-binding module (PB) with an acyl serine transferase activity homologous to that of β-lactamases responsible for inactivation of β-lactam antibiotics such as penicillin. This module provides PBPs with the ability to break D-alanyl-D-alanine bonds in two ways: the D,D-transpeptidase activity of high molecular weight PBPs is responsible for cross-linking, whereas the D,D-carboxypeptidase activity of low molecular weight PBPs is involved in PG maturation (see below). In addition to acyl serine transferase activity, high molecular weight PBPs are provided with another, non-penicillin binding module (nPB) responsible for the polymerization of disaccharides through transglycosylation in Class A PBPs, and for the shaping or the division of the cells in class B PBPs (Matsuhashi 1994; Ayala et al. 1994; for a recent review on high molecular weight PBPs, see Goffin & Ghuysen 1998). Exceptionally, the major transglycosylase activity in *Streptococcus pneumoniae* does not belong to the PBP family (Park et al. 1985). Whether this would hold true for lactic streptococci is not known.

In LAB, PBPs have been mostly studied in enterococci (El Kharroubi et al. 1991; Piras et al. 1993; Signoretto et al. 1994; Zorzi et al. 1996; Pucci et al. 1997; Duez et al. 1998). Recent work in two laboratories have provided information on PBPs in dairy LAB. In *Streptococcus thermophilus*, a mutant with a *Tn916*-insertion in the *pbp2b* gene encoding a class B high molecular weight PBP was obtained (Stingele & Mollet 1996). The mutant grows slowly in curled and twisted chains, and the cells appear asymmetrical and nearly wedge-shaped, in agreement with a specific role of PBP2-type penicillin-binding proteins in shaping the cell morphology. Interestingly, the mutant is also defective in the production of extracellular polysaccharides, probably because of an expansion of the pool of vacant lipid II (as PG assembly is reduced in the mutant), resulting in a shortage of undecaprenyl-phosphate carriers also needed for EPS production (Stingele & Mollet 1996). In *L. lactis*, IS*S1*-insertion mutants were obtained in three distinct genes encoding ORFs closely related to PBP2A and PBP2B from *S. pneumoniae*, and PBP1A from *B. subtilis*, respectively (F. Rallu & E. Maguin, personal communication). These *pbp* mutations were selected as extragenic suppressors of IS*S1*-insertion mutations responsible for resistance to acid stress: the first two restored acid stress sensitivity in a mutant with an IS*S1* insertion into an ORF closely related to the *relA* gene encoding ppGpp synthetase in *Streptococcus equisimilis* (Mechold et al. 1993), while the third one suppressed acid stress resistance in another mutant with an IS*S1* insertion into an ORF related to the Acr regulator of *E. coli* controlling the expression of an ion efflux pump (Ma et al. 1996). The physiological mechanisms underlying these genetic interactions remain to be elucidated.

As indicated above, the D,D-transpeptidase activity of high molecular weight PBPs is responsible for the cross-linking of the nascent polysaccharide threads. In LAB, the ω-amino group of the central diaminoacid of one penta(depsi)peptide (or the N-terminus of its cross-bridge peptide) is esterified to the α-carboxyl group of the fourth residue (D-alanine) freed by cleavage of the D-ala-D-ala or D-ala-D-lac (depsi)peptide bond belonging to an opposite penta(depsi)peptide from a parallel nascent oligosaccharide chain (Figure 7) . Cross-linking is only possible with a complete penta(depsi)peptide as donor, as the reaction takes place outside the cytoplasm and can only rely on the energy stored in the D-ala-D-ala or D-ala-D-lac ester bond.

The level of cross-linking in the cell wall PG is usually about 20% in single-layered Gram-negative bacteria, but may go close to 100% in multi-layered Gram-positives (Labischinski & Maidhof 1994). This depends on many factors. The 2- or 3-dimensional structure of PG is of importance, due to the helical orientation of the successive peptides along the polysaccharide thread (Figure 3). The length, folding and flexibility of the cross-links also play a role in making contacts between the partner peptides more or less easy. Cross-linking is also dependent on the D,D-carboxypeptidase activity of the low-molecular weight PBPs which clip off the C-terminal D-alanine in the pentapeptide or the D-lactate in the pentadepsipeptide, making them inapt as transpeptidation donors. Further

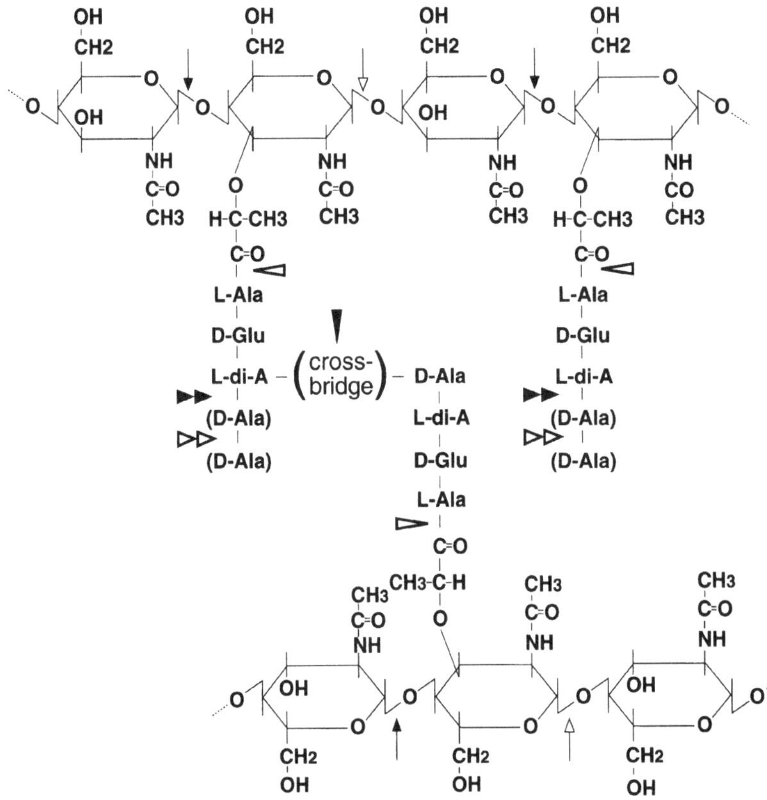

Figure 7. Modes of action of peptidoglycan hydrolases. Open arrows: *N*-acetylmuramidase; filled arrows: *N*-acetylglucosaminidase; open single arrowheads: *N*-acetylmuramoyl-L-alanine amidase; filled single arrowheads: endopeptidase; double open arrowheads: D,D-carboxypeptidase; double filled arrowheads: L,D-carboxypeptidase.

trimming or cutting of the peptides, either vacant or engaged into crosslinks, is performed by a variety of peptidases involved in the maturation and expansion of PG, and possibly also in determining whether a cell will elongate or divide (Begg et al. 1990). As a consequence, several dozens of different fragments are obtained when PG is cleaved with muramidase, as examplified by the complexity of the pattern obtained by Billot-Klein et al. (1997) with *L. casei*.

In cell wall PG, the ω-amino group of the central diaminoacid of vacant pentapeptides (or the N-terminus of its attached cross-bridge) may undergo covalent bonding with a class of cell wall-bound proteins sharing the LPXTG consensus (Navarre & Schneewind 1994, 1999). This anchoring system has been exploited for the cell wall display of various heterologous proteins (for a review, see Stahl & Uhlen 1997), and in particular for the cell wall presentation of antigens or epitopes in LAB (Pozzi et al. 1992; Piard et al. 1997; Steidler et al. 1998; Navarre & Schneewind 1999; see also Leenhouts, this symposium).

The structural diversity of PG cross-links is an important tool in systematic bacteriology (Ghuysen 1968; Sneath et al. 1986). In eubacteria, there are two main groups of cross-linkages (A and B) depending on the ranking of the acceptor residue on the penta(depsi)peptide (third or second, respectively) (Schleifer & Kandler 1972). All LAB have type A crosslinks. We have indicated earlier that besides differences in aminoacid sequence, cross-links also differ by secondary modifications consisting in various degrees of amidation of their free carboxyl groups. The resulting changes in the overall charge of the peptides and of their corresponding cross-links are likely to have an incidence on their folding, flexibility, or sensitivity to cleavage, with consequences on global charge, elasticity and stability of the whole PG envelope.

PG expansion during cell growth

The bacterial cell can be viewed as a container under pressure, due to the high osmolarity of the cytoplasm as compared to most external media to which the cells are exposed. The tensile strength of PG is the only defense against bursting, and the structural integrity of the net has therefore to be maintained in all circumstances. How this is compatible with the need for PG expansion during cell growth, which requires cleaving the meshes to covalently insert new material, remains one of the most challenging areas of bacterial physiology. Bacterial shapes are so diverse and PG architectures so different (more than 100 different PG structures are known in eubacteria), that no universal mechanism can be envisioned. Two general principles seem to prevail in Gram-positives: 'make before break' and 'inside to outside'. The first rule implies that incorporation of new PG material takes place through its cross-linking into a pre-existing, covalently closed PG sheat, and precedes cleavage of the cross-links at the site of insertion, thus making sure that shears be anticipatively repaired. The second rule states that new, un-stressed material be first assembled at the internal side of the wall in close contact with the cell membrane, and then be pushed out radially by more recent PG material.

In some way, PG expansion can be compared to molting during the larval stages of insects, where a shrunken, un-stressed envelope assembles underneath the stressed cuticle. In PG, the glycan threads are rather inextensible, and elasticity is mainly due to the cross-links which can stretch considerably (up to four-fold), from an (un-stressed) globular folding to a (stressed) extended configuration, allowing the meshes to enlarge (Koch 1991; Koch & Woeste 1992). The 'surface stress theory' of PG expansion (for a review, see Koch, 1985, 1995) states that PG layers closer to their site of synthesis on the cytoplasmic membrane have meshes with compact and unstressed cross-links, and bear appendant, 'made-before-break' glycans. While the layers are being pushed towards more peripheral zones and in reaction to the turgor pressure, their meshes first enlarge through the stretching of the cross-links. Then, the cross-links start to be stressed and undergo facilitated enzymatic cleavage by endopeptidase (see below), and the 'made-before-break' cross-linked polysaccharides automatically fill the cracks. The most peripheral layers are those under the strongest stress, and when all the 'made-before-break' chains have been used, PG shears without further repair and autolyses, giving a fibrous appearance to the cell wall periphery, as clearly seen on electron-microscopic thin sections prepared by freeze-drying (Graham & Beveridge 1990).

LAB belong to three morphological groups: bacilli, cocci, and bifidi. As each specific shape is maintained through successive generations, PG expansion must take place in a well-controlled way preventing for instance the diameter from widening or the length from increasing. Bifidobacteria have no model system to refer to with respect to the mode of cell growth. As to the bacilli and cocci, reports on *B. subtilis* or *Enterococcus hirae*, respectively, could enlight our understanding of how the PG fabric expands. The growth of *B. subtilis* has been approximated to that of an expanding cylinder with two inert hemispherical caps, and rules governing the extension of a cylindrical soap bubble have been shown to apply, implying that numerous foci of PG expansion are scattered along the whole surface of the cylinder so as to make it behave as a fluid surface (for a review, see Koch 1985). How the diameter of the cylinder is kept constant during growth is still a matter of discussion. In one of the models proposed for *B. subtilis,* the polysaccharide threads in each concentrical layer are laid in the hoop direction, so that incorporation of new chains through cross-linking will only lengthen the cylinder (Labischinski et al. 1985). In *B. subtilis*, the average PG chain is approximately 100 disaccharides long, and it has been estimated that it would take about 16 chains to circle the cylinder (Ward 1973). The rules in lactobacilli are not known, but it seems doubtful that lactobacilli with such variable morphologies as for instance *L. delbrueckii* ssp. *bulgaricus* (Botazzi 1988) would follow a simple mode of PG expansion. As to *E. hirae,* it has the shape of two elliptic poles sealed through a central annulus and cell wall expansion in this case is zonal rather than dispersive. Following the splitting of the annulus, the old poles remain almost inert while new poles grow in between the split annuli concomitantly to the formation of an inward-growing annular septum. Here too, the mode of lateral growth can be mimicked by that of two soap bubbles hold on each side and fused through a central septum (Koch 1995). Again, whether a given lactic coccus will follow the rules prevailing in *E. hirae* is hardly predictable.

PG hydrolases

Introduction of new PG material into the murein net requires that cross-links be hydrolyzed, and the inside-to-outside mode of cell wall growth of Gram-positive bacteria involves the constant removal of the most peripheral murein layers. A variety of cellular hydrolases are known to be able to cleave covalent bonds in PG (Figure 7). Murein hydrolases can be grouped into two different categories. The first group is made of enzymes which do not weaken the 2- or 3-dimensional meshing of the net, such as D,D-carboxypeptidases and L,D-carboxypeptidases. As indicated previously, D,D-carboxypeptidase activity is contributed by low-molecular weight PBPs. The existence of L,D-carboxypeptidase activity is attested in some species by the existence of crosslinks with tripeptide acceptors, notably in *E. faecium* where the ratio of murotripeptides over murotetrapeptides is very high (Billot-Klein et al. 1996), but the responsible enzyme has so far remained elusive. The second group comprises murein hydrolases (known as autolysins) which split glycan bonds and cross-links, and therefore unknit the PG fabric. Autolysins have been described in a variety of Gram-positive and Gram-negative bacteria (for general reviews, see Höltje & Tuomanen 1991; Shockman & Höltje 1994; Shockman et al. 1996; Höltje 1996; for a review on *B. subtilis* autolysins, see Doyle & Koch 1987; for reviews on LAB autolysins, see Chapot-Chartier 1996; Gasson 1996). Several bands of autolytic activity are usually resolved on zymograms following SDS-PAGE, indicating the complexity of the cell autolytic machinery (for LAB, see examples in Valence & Lortal 1995; Ostlie et al. 1995; Lepeuple et al. 1998; Blackman et al. 1998). Strikingly, the *B. subtilis* complete genome sequence reveals the presence of 16 genes encoding murein hydrolases, among which 11 *N*-acetylmuramoyl-L-alanine amidases (Kunst et al. 1997).

Autolysins are primarily needed for the separation of daughter cells following septation, as shown by the filamentous growth of some autolysin-defective mutants (as an example, see Buist et al. 1995). Interestingly, *B. subtilis* wild-type cells grow as filaments when cultured at 48° under high aeration, and these can be resolved into single cells by the addition of exogenous muramidase such as egg-white lysozyme (Fan 1970). Conflicting reports have been published on *B. subtilis* lacking both glucosaminidase (LytD) and the major amidase (LytC): according to Margot et al. (1994) , the mutant showed no detectable morphological changes and grew and divided normally, whereas a recent paper by Blackman et al. (1998) reports notable filamentation changes compared to the parent strain. Differences in experimental conditions probably explain this difference in phenotype.

Another role must be specifically devoted to autolysins in Gram-positive bacteria, whose cell wall is not only increasing in surface during growth, but also in thickness. Whereas surface enlargement is taken care of by cell division, the only way to compensate for wall thickening is to continuously shed the peripheral PG layers, as in skin desquamation. Thus, a delicate balance between PG synthesis and degradation is needed to maintain cell wall integrity, and it is therefore not too surprising that inhibitors of PG synthesis, in particular β-lactam antibiotics, trigger autolysis in growing cells (Pisabarro et al. 1990). Many conditions have been reported to induce autolysis, especially in Gram-positive bacteria. For instance, a number of strains of *L. lactis* autolyse upon prolonged incubation in culture medium, but to a different extent according to the carbon source or to the temperature (Sandholm & Sarimo 1981; Vegarud et al. 1983; Buist et al. 1997). A variety of LAB strains autolyse when incubated in buffer (Mou et al. 1976; Lortal et al. 1989, 1991), but to a much lower extent at acidic pH (Riepe et al. 1997). Autolysis of *B. subtilis* has been shown to depend on the state of energization of the cellular membrane (Joliffe et al. 1981) and can be triggered by cold-shock (Yamanaka et al. 1997).

Our knowledge of the biochemical and genetical mechanisms responsible for the induction or prevention of autolysis remains disappointingly scarce. Interestingly, stationary phase-induced autolysis in *B. subtilis* has been shown to be drastically enhanced in mutants defective for one or several proteins belonging to the cold-shock family (Graumann et al. 1997). In the same species, the LytC amidase has been reported to be transcriptionally down-regulated by *lytR*, and its activity was shown to switch from random to processive under the control of its modifier LytB (Herbold & Glaser 1975a, b; Lazarevic et al. 1992; Margot & Karamata 1992). In *E. hirae*, genetic evidence suggests that the export of muramidase-2 into the medium is under the control of the *arpU* gene (Lleo et al. 1995). In the same species, the sensitivity of exponential and stationary phase cells to hen egg white lysozyme as well as autolysis in buffer or during stationary phase (Massidda et al. 1996), seem to be controlled by the *psr* gene (a negative regulator of *pbp5* expression, see Ligozzi et al. 1993).

A few reports concern autolysin genes in LAB. Three muramidase genes have been cloned in *E. faecalis* (Beliveau et al. 1991), *E. hirae* (Chu et al. 1992), and *Lactococcus lactis* (*acmA* gene, Buist et al. 1995), respectively. The encoded proteins comprise several repeats in their C-terminal sequence which are thought to direct cell wall binding of the enzyme (Joris et al. 1992). The function of the *L. lactis* muramidase has been studied by examining the phenotype of knockout mutants. The AcmA muramidase is the only autolysin present in the strain MG1363 of *L. lactis* subsp. *cremoris*, as the zymogram is completely negative in the knockout mutant (Buist et al. 1995). It must be recalled that the second major muramidase found in the AM2 strain of *L. lactis* subsp. *cremoris* has recently been shown to be of prophage origin (Lepeuple et al. 1998). Knockout of *acmA* prevents autolysis upon prolonged incubation of the MG1363 strain (Buist et al. 1995), or upon suspension of the CO strain in buffer (Riepe et al. 1997). The role of AcmA is mainly related to cell separation, as AcmA-deficient cells grow in long chains (Buist et al. 1995; Riepe et al. 1997). Interestingly, the separation defect in the MG1363 AcmA mutant can be reverted by the addition of culture supernatant from the wild-type strain (Buist et al. 1995). A recent report also shows that the AcmA muramidase is sensitive to the proteolytic action of the membrane-bound proteinase PrtP (Buist et al. 1998). Moreover, induced overproduction of AcmA has been achieved in MG1363, leading to increased autolysis (Buist et al. 1997). Altogether, these observations open interesting perpectives for the design of food-grade lysis systems aiming at a better control of cheese ripening (Crow et al. 1995).

It must be recalled that LAB, together with numerous other groups of bacteria, are found in close contact with, and sometimes adhere to, the mucosal surfaces of the gastro-intestinal, urogenital and broncho-nasal tracts. It is estimated, for instance, that the normal healthy microflora of the female urogenital tract is dominated by lactobacilli (Redondo-Lopez et al. 1990). Some LAB species or strains found in these ecological niches are thought to be endowed with 'probiotic' properties contributing to the maintenance of health, including the ability to stimulate the mucosal immune system so as to enhance the non-specific (immunostimulation) or specific (immunoadjuvanticity) responses (Hamann et al. 1998; for a review on probiotics, see Tannock, 1999). Interestingly, PG fragments released at these sites through autolysis or as a result of exogenous lysozyme attack could contribute to this protective action, as muramylpeptides derived from *L. plantarum* ATCC8014 for instance are known to display immunoadjuvant activity (Kotani et al. 1975; for a review on the immunological properties of muramylpeptides, see Adam et al. 1981). In contrast, there are indications in the literature that cell wall fragments from members of the normal enteric flora (including *E. faecium*) may have the potential to induce a wide range of inflammatory responses, from transient acute to chronic erosive joint disease (Stimpson et al. 1986; Schwab 1993). Attention should be paid, when selecting LAB for probiotic use, to these potentially beneficial or harmful effects of cell wall (auto)lysis on human health.

Teichoic acids

Besides PG, the cell wall of most Gram-positive bacteria also comprises teichoic acids (*sensu lato*) which may account for more than half of the weight of the wall. LAB deserve a special mention when considering these components of paramount importance for the functionality of the cell wall, as their discovery sprung forty years ago (Armstrong et al. 1958) from investigations conducted by Baddiley and his colleagues in search of a function for glycerol- and ribitol-cytidine diphosphates serendipitously found as 'contaminants' in bacterial extracts from *L. arabinosus* (now *L. plantarum*) prepared for the study of coenzyme A biosynthesis (for a recollection, see Baddiley 1989). Teichoic acids are quite diverse in structure and abundance, depending on the species or strain, stage or rate of growth, pH of the medium, carbon source, availability of phosphate, etc. (for general reviews on teichoic acids, see Archibald & Baddiley 1986; Fischer 1988, 1990, 1994; Pooley & Karamata 1994; for a review on anionic polymers from bacilli, see Archibald et al. 1993). The structures of some teichoic acids found in LAB are given in Figure 8. As will be discussed below, they are synthesized on the outer face of the cytoplasmic membrane, and their size and physicochemical properties are compatible with an extended configuration across the cell wall. By virtue of their lipid anchor, lipoteichoic acids (LTA) and lipoglycans (LG) may remain attached to the cytoplasmic membrane, but a fraction of them are found free in the cell wall or even released into the medium; teichoic acids (TA) and teichuronic acids (TUA), on the other

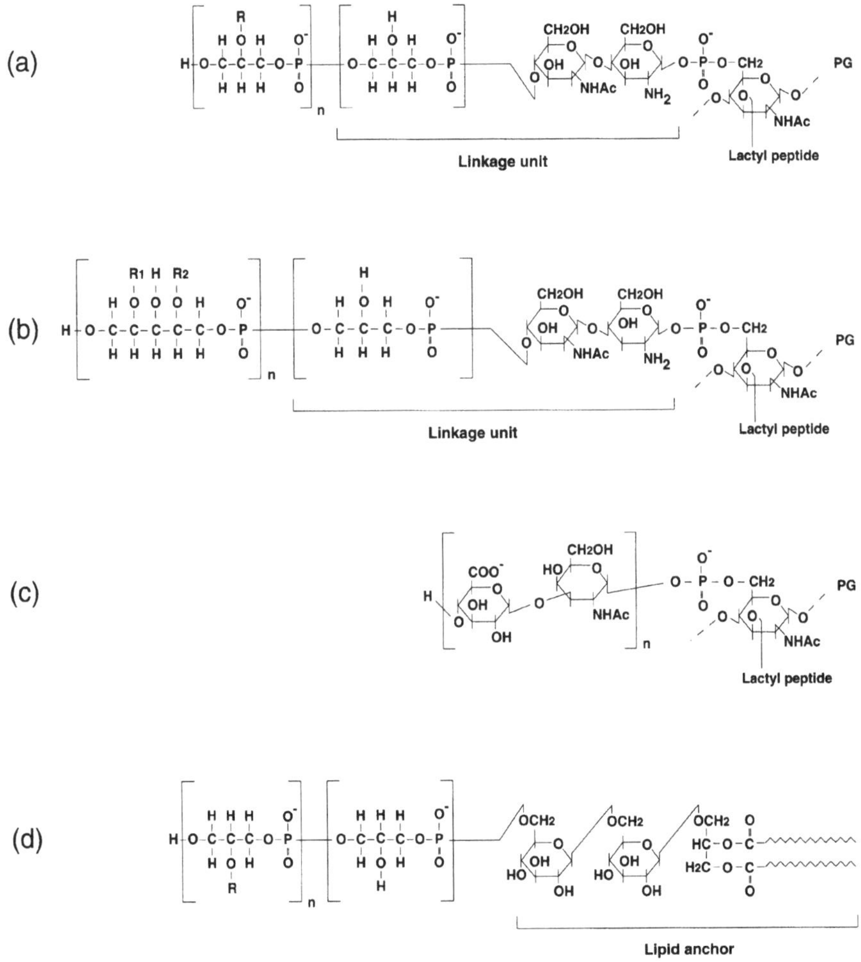

Figure 8. Examples of structures of teichoic acids *sensu lato.* (a): poly(glycerol phosphate) teichoic acid (*Bacillus subtilis*); (b): poly(ribitol phosphate) teichoic acid (*Lactobacillus plantarum*); (c): teichuronic acid (*Bacillus subtilis*); (d): lipoteichoic acid (*Bacillus subtilis*); – PG: peptidoglycan chain; R, R_1, R_2 = substituent (glucose, D-alanine, GlcNAc...) (see text for explanations).

hand, are covalently bound to PG and are sometimes generically called 'wall teichoic acids' (WTA).

Teichoic acids contribute in many respects to the functionality of the cell wall. Those which are made of repeating units of glycerol- or ribitol-phosphate constitute a reservoir of phosphate which can be mobilized when phosphate becomes scarce (Grant 1979). As anionic polymers, they play a role as scavenger of cations, in particular Mg^{++} (Hughes et al. 1970; Lambert et al. 1975b), and may create a pH gradient across the wall by sequestering the protons expelled through the cytoplasmic membrane as a result of metabolism (Urrutia et al. 1992; Kemper et al. 1993). Teichoic acids are also known to influence the activity of autolysins (Höltje & Tomasz 1975; Yamada et al. 1975; Bierbaum & Sahl 1987) and are involved in phage adsorption (Archibald 1976; Briehl et al. 1989). LTA provide the main component of the hydrophobicity of the cell enveloppe (Miörner et al. 1983), and may contribute in this way to its adhesiveness (for a review on hydrophobic interactions in relation to bacterial adhesion, see Rosenberg & Kjelleberg 1986). In this respect, it was shown recently that the adhesion of the *Lactobacillus johnsonii* La1 strain to Caco-2 human intestinal cells could be inhibited in a concentration-dependent way by LTA purified from that strain, as well as by *L. johnsonii* La1 culture supernatant containing LTA (Granato et al. 1999). Last but not least, teichoic acids are potent immunogens and can be regarded as the gram-positive equivalent of the gram-negative lipopolysaccharides (Knox & Wicken 1973; Wicken & Knox 1975, 1980). LTA is able to weakly stimulate cytokine synthesis, and this effect is lost upon removal of the D-alanine substituents (see below) (Bhakdi et al. 1991; Tsutsui et al. 1991; for a general review on bacterial modulins, see Henderson et al. 1996).

Biosynthesis of wall teichoic acids

As shown in Figure 8, both poly(glycerol phosphate) and poly(ribitol phosphate) teichoic acids have the same overall structure made of a chain of phosphodiester-bound glycerol or ribitol residues attached to a terminal 'linkage unit' allowing their covalent fixation on PG. The structure of the linkage unit of the *L. plantarum* poly(ribitol phosphate) TA is shown in Figure 8b as an example in LAB (Kojima et al. 1985a). It is made of the disaccharide *N*-acetylmannosaminyl$\beta(1\rightarrow4)$glucosamine followed by glycerol phosphate. This type of structure seems to be well conserved despite the wide variety of TA; differences may occur in the nature of the sugars and the number of glycerol phosphate residues (Archibald et al. 1993). The two teichoic acids may be found together in the same species (but attached to distinct PG glycan chains); however, in *L. casei* (Neuhaus 1985) and in *L. lactis* subsp. *cremoris* (Valyasevi et al. 1990), no wall teichoic acids are present and the anionic character of the cell envelope is exclusively due to LTA. The biosynthesis of both types of TA follows a similar scheme (Figure 9). TA and PG biosyntheses share the same lipid carrier (undecaprenyl phosphate), and the formation of the TA linkage unit starts with the same precursor as that used in priming PG synthesis (UDP-GlcNAc); both PG and main chain TA precursors are nucleotide-activated monomers synthesized in the cytoplasm, although CDP-glycerol or CDP-ribitol are used in TA biosynthesis instead of UDP-bound sugars in PG. Biosynthesis of the polymer takes place most probably on the outer side of the cytoplasmic membrane. It starts with the assembly of the linkage unit on the lipid carrier, followed by the cyclic addition of a few dozens of glycerol or ribitol phoshate units. Monomers are added either at the end remote from the anchor lipid (as in *B. subtilis*, Archibald et al. 1993), or at the end adjoining the linkage unit, depending on the species. In the course of polymerization, the C-2 hydroxyl of glycerol can be variably substituted with α-D-glucose or D-alanine ester (TA and LTA substitution will be discussed below); in the case of poly(ribitol phosphate) TA, the C-2 and C-4 hydroxyls can also be substituted in a similar way. The completed TA is finally hooked through its linkage unit on the C-6 of a MurNAc residue of a growing PG glycan chain.

Most of the genes involved in TA biosynthesis have been identified in *B. subtilis,* except those responsible for the synthesis of the linkage unit (Mauël et al. 1991; Pooley & Karamata 1994; Lazarevic et al. 1995). Insertional mutagenesis studies indicate that TA is essential in this species (Mauël et al. 1989), and conditional-lethal (thermosensitive) mutations could only be obtained, in particular in the *tagD* gene encoding glycerol-3-phosphate cytidylyltransferase (Pooley et al. 1991). Thermosensitive Tag$^-$ mutants display grossly deformed cell shapes when incubated at the restrictive temperature, demonstrating complex metabolic interactions between TA and PG syntheses (Briehl et al., 1989). Two genes from the *B. subtilis tag* divergon (*tagGH*) have been reported to encode a two-component ABC transporter likely to be responsible for the translocation through the membrane of poly(glycerol phosphate) as well as a minor poly(glucose-galactosamine phosphate) TA, or their precursors (Lazarevic & Karamata 1995). Genes related to TA biosynthesis have not yet been described in LAB.

Biosynthesis of teichuronic acids

The phosphate-free teichuronic acids have only been described so far in a few species (mainly bacilli), but the phylogenic relatedness of the genus *Bacillus* to LAB makes it possible that they would also exist in that group. The structure of TUA differs from that of TA in two respects: the main chain is made of sugar monomers directly linked by glysosidic bonds, and no linkage unit is present (Figure 8c). The nature of the sugar monomers varies in the different species studied. In *B. subtilis* (strain W23) it is a disaccharide composed of N-acetyl-galactosamine linked to glucuronic acid, but more complex tri- or tetra-saccharide may be found in other bacilli (Ward 1981). The negative charges in TUA are provided by the carboxyl groups of glucuronic acid. In *B. megaterium*, TUA are the only anionic polymers present, but in *B. subtilis* TUA replace TA under phosphate starvation conditions (Lang

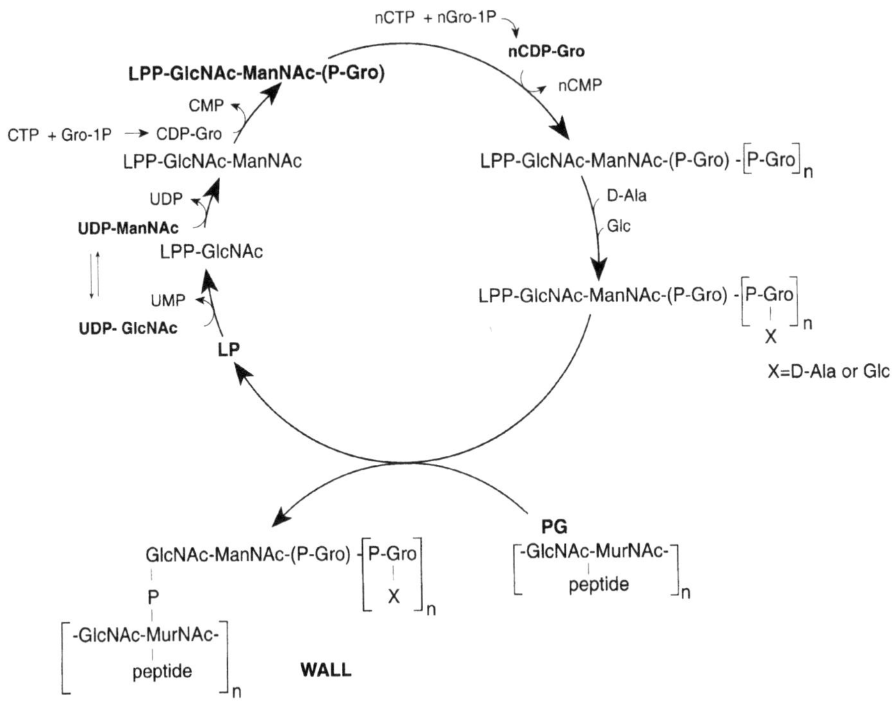

Figure 9. Pathway of teichoic acid synthesis. Gro = glycerol; L = undecaprenol; ManNAc = *N*-acetyl mannosamine (see text for explanations).

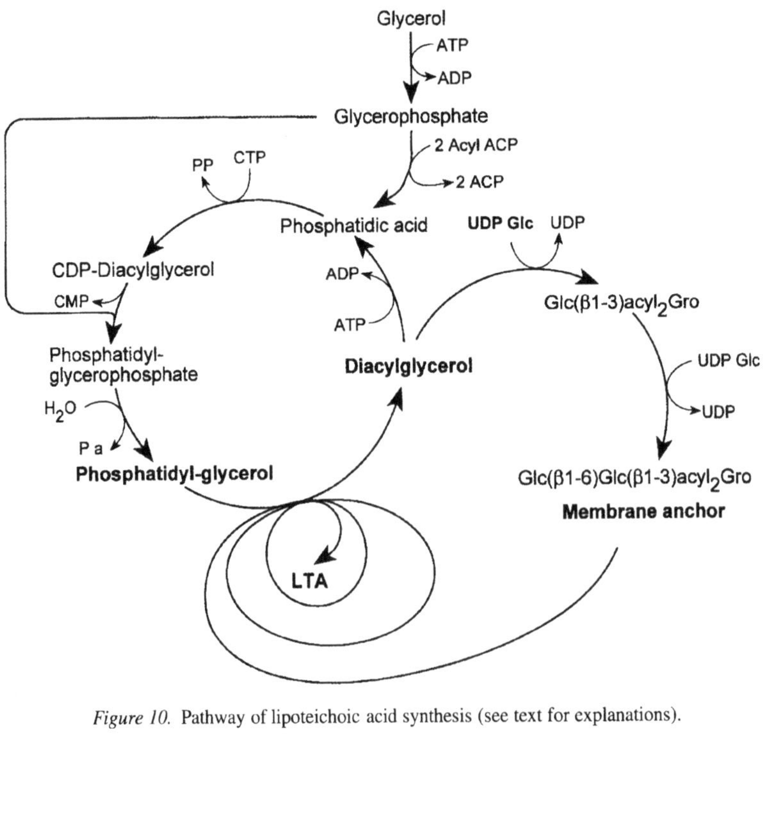

Figure 10. Pathway of lipoteichoic acid synthesis (see text for explanations).

et al. 1982). Recent genetic studies clearly show in this species that transcriptional repression of the *tag* divergon (responsible for TA synthesis) together with transcriptional induction of the *tuaA* promoter (needed for TUA synthesis, see below) is governed by the two-component PhoP-PhoR regulation system in response to phosphate concentration (Liu et al. 1998; Qi & Hulett 1998). Interestingly, genetic evidence indicates that TUA contributes to the ability of an alkalophilic *Bacillus* sp. to grow at alkaline pH (Aono & Ohtani 1990; for a review, see Krulwich et al. 1997).

Biosynthesis of TUA follows the same logics as that of TA. Undecaprenyl phosphate serves as lipid anchor for the growing chain, and nucleotide-activated sugars are used as precursors (UDP-GalNAc and UDP-GlcA in *B. subtilis* W23). The disaccharide monomers are cyclically inserted between the lipid carrier and the nascent TUA chain (elongation from the reducing end), and the completed polymer is finally cleaved from its undecaprenylphosphate anchor (together with one phosphate) to be attached through a phosphodiester bond to the C-6 of MurNAC on a growing PG chain. Recently, the 8-cistron *tua* operon encoding the genes needed for TUA biosynthesis has been characterized in *B. subtilis* strain 168 (Soldo et al. 1999). Comparison of GC contents of *tua* and *tag* loci suggests that TUA is an ancestral bacillar anionic polymer, while operons involved in the synthesis of TA were acquired by horizontal transfer from low-GC content species. This strengthens the possibility that *tua* genes could also be found in some LAB genomes.

Biosynthesis of lipoteichoic acids

LTA are widely distributed in LAB and have been found in enterococci, lactobacilli, lactococci, leuconostocs, and streptococci (Fischer 1994). LTA with a poly(glycerol phosphate) main chain (Figure 8d) represents the most common type of membrane-anchored anionic polymer. Interestingly, it was first discovered in the lactic acid bacterium *L. fermentum* (Wicken & Knox 1970). In *L. lactis*, it has been estimated that the glycolipid anchor (see below) of poly(glycerol phosphate) LTA constitutes every fifth lipid molecule of the outer leaflet of the cytoplasmic membrane (Fischer 1981). Other types of LTA are known, as for example in *Lactococcus garvieae* where the repeating units are made of (α1-6)-digalactosyl-glycerophosphate with no D-alanyl substituent (Koch & Fischer 1978). Several species of bifidobacteria have lipoglycans of various structures instead of LTA, and these are substituted with L-alanine instead of the usual D-isomer (Iwasaki et al. 1990; Sutcliffe & Shaw 1991). The main chain of poly(glycerol phosphate) LTA is very similar to that of poly(glycerol phosphate) TA except that the chiralities of glycerol are opposite, reflecting differences in their respective modes of biosynthesis (see below). The reducing end of the main chain is linked by a phosphodiester bond to the O-6 of the non-reducing hexosyl terminus of a glyceroglycolipid anchor whose precise structure may vary according to the genus or species, but obeys the same general pattern, i.e. diacylglycerol bound to a di- or tri-saccharide (itself bearing sometimes an O-6-acyl or -phosphatidyl substituent). Membrane anchoring is ensured by the hydrocarbon chains of the fatty acids. As in TA, the C-2 hydroxyls of the main chain LTA glycerolphosphate residues carry glucosyl (mostly D-glucose, D-galactose or GlcNAc) or D-alanyl substituents (to be discussed below).

The pathway of biosynthesis of poly(glycerol phosphate) LTA in bacilli is outlined in Figure 10. It starts with the recruitment of a free membrane glycolipid to which the *sn*-glycerol-1-phosphate moiety of phosphatidylglycerol is esterified, together with the release of diacylglycerol. Elongation proceeds by further addition of glycerophosphate residues to the glycerol-terminal end of the growing chain, using again phosphatidylglycerol precursors (and not CDP-glycerol as in TA, hence the difference in chirality). Substitution with glycosyl or alanyl residues takes place concurrently on the same chain during elongation, but may also continue on completed chains. The majority of the released diacylglycerol is recycled through phosphorylation to phosphatidic acid, itself a precursor of phosphatidylglycerol, while the remainder is used for membrane glycolipid synthesis. LTA chains in a given strain are not only variable in size, but also in the extent of substitution and in the identity of the lipid anchor fatty acids, as illustrated by the polydisperse pattern obtained from *E. faecalis* LTA (Fischer 1994). Although LTA are born anchored to the cytoplasmic membrane and may remain attached when completed, a fraction of the chains may be released through deacylation, while others still holding their glyceroglycolipid anchor may become loose, possibly through dragging by the inside-to-outside PG expansion mechanism. Finally, the heterogeneous 'secreted' fraction of LTA will assemble into various kinds of micelles (Wicken et al. 1986) and contribute to the glycocalyx or be released into the medium (for a general review on bacterial cell surface amphiphiles, see Wicken & Knox 1980).

TA and LTA substitution

The main chains of TA and LTA are decorated with glycosyl or D-alanyl substituents which contribute greatly to the functionality of these anionic polymers, as evidenced by the phenotypes of strains with genetic changes altering the substitution process. It was shown more than 30 years ago that glucosylation of TA was required for the adsorption of certain phages on *B. subtilis* (Young 1967) and phage-resistant mutants deficient in glucosylation were mapped in three distinct loci named *gtaA, B*, and *C* (Young et al. 1969). The corresponding genes were later shown to encode the three genes needed in turn for glucosylation: phosphoglucomutase (*gtaC*) ensuring the production of glucose-1-P from the glycolytic intermediate glucose-6-P (Pooley et al. 1987), UDPglucose pyrophosphorylase (*gtaB*, also known as *galU*) responsible for the production of UDPglucose from glucose-1-P (Pooley et al. 1987), and UDP poly(glycerol phosphate) α-glucosyltransferase (*gtaA*, also known as *tagE* or *rodC*) allowing the transfer of the glucose moiety from UDPglucose onto TA (Honeyman & Stewart 1989). UDPglucose is brought to the membrane-bound GtaA protein following coupling to undecaprenyl phosphate, thereby relating the TA glucosylation process to other membrane-bound macromolecular syntheses requiring the same lipid carrier (PG, EPS). Recently, a mutant (*gtcA*) deficient in poly(ribitol phosphate) TA galactosylation has been obtained in *Listeria monocytogenes* (Promadej et al. 1999). Curiously, a *gtcA*-related gene is present in the genomic sequence of *B. subtilis* (Glaser et al. 1993), a species not described as having galactose-decorated TA. Little information is available on glycosylation in LAB. An early report showed that the loss of glucose substitution in poly(ribitol phosphate) TA confers phage resistance to a mutant of *L. plantarum* (Douglas & Wolin 1971); likewise, the plasmid-encoded resistance of the *L. lactis* subsp. *cremoris* strain SK110 to phage sk11G (de Vos et al. 1984) results from the acquisition of galactosylated LTA (Sijtsma et al. 1988, 1990). In neither case, did the changes in the glycosylated status of anionic polymers bring any phenotypic traits other than phage resistance. In contrast, knockout of the *galE* gene in *L. lactis*, which encodes the UDP-glucose 4-epimerase allowing UDP-glucose to be converted into UDP-galactose during growth on glucose, was shown recently to impact on the separation of cells following division, leading to the formation of long chains of cells connected with tiny peptidoglycan filaments. This peculiar mode of growth in long chains is reminiscent of the phenotypes of the AcmA (Buist et al. 1995) and DltD (Duwat et al. 1997; see below) mutants, and it can be speculated that the lack of UDP-galactose, by preventing substitution of LTA with galactosyl residues, would lead to an increased inhibition of autolytic activity by LTA (B. Grossiord, personal communication).

D-alanylation takes place during the assembly of the LTA main chain, but may also occurs postsynthetically through transesterification from one site to another within LTA (Childs III et al. 1985). Whether D-alanylation of TA can also be accomplished by the same enzymatic machinery as that used in LTA substitution is still unclear (Perego et al. 1995). Pulse-chase experiments strongly suggest that in *Staphilococcus aureus* D-alanine substituents can be transesterified from LTA to TA (Koch et al. 1985), but this does not exclude that primary D-alanyl substitution of TA could also take place through a direct (common or distinct) process.

Altering the process of D-alanylation usually has a dramatic incidence on the functionality of the cell wall (for a review, see Heaton and Neuhaus 1993). This can easily be understood by reference to the variety of functions devoted to anionic polymers in connexion with their acidic nature, keeping in mind that D-alanine is esterified to TA and LTA, and therefore brings its positive charge as a counterion to those of phosphate or carboxyl. It is known that a reduction in D-alanine substitution increases the binding capacity of LTA for magnesium ions (Heptinstall et al. 1970; Lambert et al. 1975a). Early *in vitro* studies also showed that the inhibitory effect of LTA on *S. aureus* autolysin activity was negatively correlated to the level of D-alanylation (Fischer et al. 1981). This conclusion was recently supported *in vivo* by the reduced autolytic activity of alanine racemase mutants in the same species (Kullik et al. 1998), although in this case D-alanine deprivation may also affect PG synthesis and/or cross-linking. More convincingly, *L. casei* mutants defective in LTA D-alanyl esters show aberrant C-shaped morphologies and defects in cell separation indicative of reduced autolytic activity (Ntamere et al. 1987). Opposite evidence supporting a repressor role of D-alanine substitution on autolysis is provided by the increased autolysis of alanine racemase mutants in *B. subtilis* (Ferrari et al. 1985;

Heaton et al. 1988), and *L. plantarum* or *L. lactis* (Hols et al. in preparation), but again these mutations may have pleiotropic effects. A stronger argument is provided by the enhanced buffer- and β-lactam-induced lysis of *B. subtilis* mutants solely defective in D-alanylation (Wecke et al. 1996, 1997); no changes in growth or morphology were observed in this case. It seems difficult at first sight to reconcile these apparent conflicting results, although species-related diversity in autolysins and anionic polymers is expected to bring differences in the control of autolysis.

Recently, in a search for UV-sensitive *ISS1* insertion mutants in *L. lactis*, Duwat et al. (1997) have found a hit in the *dltD* gene involved in LTA D-alanylation (see below). Besides being UV-sensitive, the mutant was reported to grow slowly in long chains, to be less efficient in conjugation, and not to require glycine for becoming electrocompetent, all indicative of alterations in the functionality of the cell wall. Using a similar *ISS1* insertion approach, F. Rallu & E. Maguin (personal communication) have selected two distinct mutations in the *dltD* gene, which respectively suppress the acid stress resistance of the RelA and AcrR mutants already mentioned (see above). The molecular mechanisms accounting for these genetic interactions are unknown.

The biochemistry and molecular biology of D-alanylation have been studied in *B. subtilis* and in *L. casei* (for a review, see Neuhaus et al., 1996). In *B. subtilis*, 5 genes responsible for D-alanine decoration of both LTA and TA have been found clustered (*dltA*→*E*,in that order) in an operon driven by three consecutive promoters (σ^A-σ^D-σ^A); the same type of organization has been observed in *L. casei*, although in this case a single, vegetative promoter was detected. The encoded proteins are similar in both species, and the following common mechanism can be envisioned. D-alanine is first activated by the DltA enzyme (also called Dcl in *L. casei*) in the presence of ATP and Mg^{++} to yield a D-alanyl adenylate. The same enzyme transfers its D-alanyl moiety to the DltC (Dcp) carrier protein (an analog of acyl-carrier proteins involved in fatty acid synthesis). The next steps are still a matter of debate. It is thought (Perego et al. 1995; Neuhaus et al. 1996) that the membrane-bound DltB protein (for which no enzymatic activity has been demonstrated so far) transfers D-alanine to the lipid carrier undecaprenyl phosphate, and that DltD is finally responsible for D-alanyl transesterification of LTA (and possibly TA, see above) on the outer face of the cytoplasmic membrane. No functions have been allocated to the *dltE* gene yet; this gene, unlike the four others, is dispensable for the process.

S-Layer

The bacterial envelope may comprise one (or two) superimposed surface layer(s) (S-layer) made of protein subunits, sometimes glycosylated, packed in a paracrystalline hexagonal or tetragonal monolayer. S-layers have been found in more than 400 species belonging to about 100 eubacterial and archaebacterial genera, and more than 30 different S-layer protein genes have already been cloned. As the subject has been reviewed extensively over the last three years (Boot & Pouwels 1996; Bahl et al. 1997; Beveridge et al. 1997; Messner et al. 1997; Rachel et al. 1997; Sleytr 1997; Sleytr & Sara 1997; Sleytr et al. 1997; Pum & Sleytr 1999), this chapter will only address a few points of specific interest to lactic acid bacteria.

In LAB, S-layers have been found so far only in the genus *Lactobacillus* (Masuda & Kawata 1983; Messner & Sleytr 1992; Lortal 1993). S-layer proteins have been studied in some details in *L. helveticus* (Lortal et al. 1992; Mozes & Lortal 1995), *L. brevis* (Vidgren et al. 1992; Yasui et al. 1995), *L. acidophilus* (Boot et al. 1993), and *L. crispatus* (Toba et al. 1995). Genes have been cloned and characterized in *L. brevis* (Vidgren et al. 1992), *L. acidophilus* (Boot et al. 1993), and *L. helveticus* (Callegari et al. 1998). In *L. acidophilus*, two *slp* genes are present in opposite orientation in the same locus; one (*slpA*) is higly expressed in the majority of the cells, while the other (*slpB*) is silent except in a small fraction of the population where it has been exchanged with *slpA* by inversional homologous recombination. This situation is reminiscent of the phenomenon of antigenic variation of S-protein described in several pathogens (Boot et al. 1995, 1996a).

The S-layer lattice is strongly (yet, non-covalently) anchored to the underlying PG through a repetitive modular element (SLH motif) that occurs at the amino- or carboxy-terminal end of the S-layers monomers and also of many other cell-wall associated proteins (Lupas et al. 1994; Lupas 1996; Sara et al. 1998). This anchoring system offers interesting perspectives of applications for the surface-display of heterologous proteins, including antigens in live recombinant vaccine vectors (Mesnage et al. 1999; for a review on LAB as live mucosal vaccine vectors, see Mercenier 1999).

Knockout of the *L. acidophilus slpA* gene has been attempted in two laboratories with no success, suggesting that the gene is essential (Boot et al. 1996a). This is not too surprising, considering the enormous burden imposed on the cell economy by the production of up to 500 S-layer subunits per second (Sleytr et al. 1993), amounting to 15–20% of the total cellular protein content (Yasui et al. 1995). This very high rate of production is due to a conjunction of strong transcription/translation/secretion signals and remarkably stable mRNA (reported half-lives: 15 min; see Boot et al. 1996b; Kahala et al. 1997). Taking advantage of these features, Savijoki et al. (1997) have developed an expression vector allowing the production of the β-lactamase reporter in various lactobacilli at a rate of up to 5×10^5molecules/cell $\times$ hr, yielding 80 mg/l of secreted protein. These are probably amongst the highest rates of heterologous expression reported so far in lactobacilli.

Despite intensive research efforts over the last 30 years, the contribution of the S-layer(s) to the functionality of the cell wall remains elusive. Several putative functions have been invoked: virulence factor in pathogenic species, sieve allowing retention of useful macromolecules or exclusion of harmful hydrolases, masking of phage receptor, adhesion factor (for a review, see Sleytr et al. 1993). In *L. crispatus*, the S-layer has been suggested to act as an adhesin, as it was reported to promote binding to components of the extracellular matrix of target cells *in vitro*, particularly collagen (Toba et al., 1995), in contrast with earlier work indicating no participation of the S-layer to adhesion to Caco-2 enterocytes *in vivo* (Greene & Klaenhammer 1994).

Cell wall polysaccharides

Cell wall polysaccharides (WPS) are ubiquitous components of the cell envelope, and form the basis of many serological typing systems, as examplified in LAB by the Lancefield classification of streptococci (Lancefield, 1933). They are sometimes called neutral polysaccharides to distinguish them from the anionic polysaccharide teichoic acids, although some may also be acidic as they may be branched with anionic substituents, in particular glycerol phosphate (Coyette & Ghuysen 1970; Kojima et al. 1985b, 1986). Conceptually, a distinction is made between the capsular polysaccharides (CPS), which form a thick outermost shell intimately associated with, and often covalently bound to, the cell wall, and WPS *sensu stricto*, which decorate the cell enveloppe and may again be either covalently bound to the PG (as an example, see Coyette & Ghuysen 1970) or loosely associated with it (as an example, see Gopal & Crow 1993). Extracellular polysaccharides (EPS) are another category of polysaccharides which in contrast to capsules do not remain attached to the cells but are released into the medium (for a review on EPS, see Whitfield 1988). Operationally, differentiation between these various classes is often difficult, as for instance EPS transiting through the thick Gram-positive cell wall may be recovered as loosely associated material, and as sample handling may remove most of the capsular material and leave only residual PG-bound polysaccharides mistakenly taken as WPS.

Capsules have been extensively studied as they contribute to the virulence of many pathogens, including pathogenic streptococci such as *S. pneumoniae* where more than 90 different serotypes have been described, each producing its own, specific CPS (Henrichsen 1995; Garcia & Lopez 1997; for a general review on CPS, see Roberts 1996). This diversity is also observed in EPS, in particular in 'ropy' dairy LAB which produce a range of different texturing polysaccharides of high technological interest (for a review on EPS in LAB, see Cerning 1990). These two classes of polysaccharides lie beyond the focus of this review and will not be covered here.

Like polysaccharides in general, WPS are indeed complicated structures which differ not only in the nature of sugar monomers, but also in their mode of linkage, branching, and substitution. A unified nomenclature for polysaccharide synthesis and gene designation has recently been proposed (Reeves et al. 1996). Further complications arise, as different WPS may co-exist in the same strain, as their structure may be strain-dependent, and as their composition may also vary with the growth conditions (Wicken et al. 1983). Quite often, the only information available is the identity of the monomers, sometimes together with their molar ratios. Rhamnose is very often found as a constituent of WPS in LAB, in particular in lactobacilli (Hall & Knox 1965; Coyette & Ghuysen 1970; Douglas & Wolin 1971; Wicken et al. 1983), and in streptococci (Schleifer & Kilpper-Bälz, 1987). The precise structures of WPS acting as co-aggregation receptors for the formation of the dental plaque have been elucidated in several oral streptococci (for a review on oral bacteria adhesion mechanisms, see Kolenbrander & London 1993; Whittaker et al. 1996), as well as in a few non-pathogenic LAB species, in particular *L. casei* (Nagaoka et al. 1990), *Bifidobacterium longum* (Nagaoka et al. 1995), and *Bifidobacterium catenulatum* (Nagaoka et al. 1996). In the latter cases, as the polysaccharide fractions analyzed had been released from the cell wall following *N*-acetylmuramidase treatment, they are to be considered as authentic WPS covalently bound to the PG, most probably through a phosphodiester linkage between the reducing end of the WPS and the C-6 of MurNAc (Murazumi et al. 1986).

Information about WPS biosynthesis pathways is scarce. Two reports available in the litterature concern the biosynthesis of the acidic and neutral WPS in *B. subtilis* (Kojima et al. 1986; Murazumi et al. 1986). In both cases, undecaprenyl phosphate serves as lipid carrier, and nucleotide sugars are used as precursors: glucose, galactose, mannose, GlcNAc and ManNAc precursors are used as UDP derivatives, whereas the rhamnose precursor is a dTDP-derivative. This overall scheme is similar to that established for PG and WTA biosyntheses (see above), as well as to those of CPS (Roberts 1996) and EPS (Stingele et al. 1996; van Kraenenburg et al. 1997) biosyntheses. It is also reminiscent of the mode of biosynthesis of the O-antigen of the Gram-negative lipopolysaccharides (for a review on LPS, see Raetz 1996).

Some of the genes responsible for the biosynthesis of the *Streptococcus mutans* serotype c-specific WPS have recently been cloned. This polysaccharide is made of a backbone of 1,2- and 1,3-linked rhamnosyl units decorated with glucose side chains. Using a random genomic sequencing approach, Tsukioka et al. (1997a, b) have identified four genes (*rmlA, B, C* and *D*) respectively similar to the gram-negative *rbfA, B, D* and *C* genes responsible for the conversion of glucose-1-phosphate into the dTDP-rhamnose precursor used for the synthesis of the LPS O-antigen (Raetz 1996). Southern blotting analysis revealed the presence of genes hybridizing strongly to the three *rml* probes in various *Streptococcus* species (including *S. salivarius*); weaker hybridization signals were also found with *L. casei*. Two recent papers from the same group report the cloning of 7 additional genes required for the conversion of glucose-1-phosphate into the UDP-glucose precursor of side chain glucose units (*gluA*; Yamashita et al. 1998a), and for the cell wall localization and side chain formation of the antigen (*rgpA*→*F*; Yamashita et al. 1998b). Interestingly, the RgpC and D proteins are homologous to components of the ATP-binding cassette transporter system and may be involved in polysaccharide export, while RgpE has features typical of glycosyltransferases. Nothing is known about the enzyme(s) or the gene(s) required for WPS anchoring to the cell wall, and we ignore how this process is coordinated with PG elongation and WTA decoration (the latter taking place concomitantly on the same MurNAc C-6 site).

Our understanding of the contribution of WPS to the functionality of the cell wall is far from being clear, in particular in non-pathogenic LAB. Interestingly, knockout of the *gluA* gene in *S. mutans*, which prevents glucose decoration of the c-antigen WPS without affecting the synthesis of its rhamnose backbone, brings about an increased sensitivity to acidic pH (Yamashita et al. 1998a). The physiological explanation for this peculiar phenotype is unknown. Probably the main beneficial function of WPS in general is to mediate interactions with components of the environment. WPS may play a role in non-specific interactions with inert surfaces by contributing to the wall surface physical chemistry (Rosenberg & Kjelleberg 1986; Pelletier et al. 1997). They may also behave as specific lectin receptors mediating adhesion to other microorganisms (as for example in the case of co-aggregation of oral microorganisms in the dental plaque; Mcintire et al. 1988; Cassels & London 1989) or to eukaryotic cells (as for instance in the case of colonizing lactobacilli in the gut; Greene & Klaenhammer 1994; Adlerberth et al. 1996). Adsorption of phages has also been shown to rely on specific interactions with WPS in *L. plantarum* (Douglas & Wolin 1971), *L. casei* (Ishibashi et al., 1982), and *L. lactis* (Valyasevi et al. 1990; Monteville et al. 1994).

Concluding remarks

Over the last two decades, LAB research has witnessed fantastic progress in many different fields ranging from molecular taxonomy to metabolic engineering. Over the years, the five preceding symposia in this series have offered numerous reviews on a wide variety of subjects, of both fundamental and applied interest. Strikingly, the topic of LAB cell wall structure and function had never been addressed before in this context, for the simple reason that little was known – and still is – in this area of 'LAB genetics, metabolism and applications', as the title of the symposia series goes. Yet, we have seen that the cell wall plays an essential role in many aspects of LAB physiology, not only as an exoskeleton 'holding together' the cellular contents, but also as a surface interacting with the environment in many ways. Truly, the physiology and molecular biology of the cell wall may not seem easily approachable, as it is made of many different polymeric components with complicated biosyntesis pathways. Fortunately, we are now entering the genomics era, as the complete genome sequences of several LAB species will shortly be available (see Ehrlich, this symposium). This will allow identification *in silico* of many genes involved in LAB cell wall biosynthesis, and offer tremendous opportunities for tackling problems of cell wall structure and function in LAB through the creation of isogenic pairs of strains differing by only a single genetic lesion.

Acknowledgements

We are grateful to Sylvie Lortal and Christophe Boonaert for the gift of personal documentation on some of the topics covered in this review. We acknowledge Emmanuelle Maguin and Benoît Grossiord for allowing us to report their unpublished results. We are indebted to Bernard Hallet for the drawing of Figure 1. We thank Marie-Paule Delcour for her kind assistance in the preparation of the manuscript. Research on the molecular biology of lactic acid bacteria has been funded in our laboratory by EC contracts (AGRE-CT90-0041, BIOT-CT94-3055, BIO4-CT96-0439, BIO4-CT96-0542, BIO4-CT96-0498 and BIO4-CT98-0118) and FNRS grants (3.4578.70 and 2.4594.96). MD and EP hold a doctoral bursary from FRIA.

References

Adlerbeth I, Ahrné S, Johansson ML, Molin G, Hanson LA & Wold AE (1996) A mannose-specific adherence mechanism in *Lactobacillus plantarum* conferring binding to the human colonic cell line HT-29. Appl. Environ. Microbiol. 62: 2244–2251

Allen CM (1985) Purification and characterization of undecaprenylpyrophosphate synthetase. Methods Enzymol. 110: 281–299

Adam A, Petit JF, Lefrancier P & Lederer E (1981) Muramyl peptides. Mol. Cell. Biochem. 41: 27–47

Allen NE, Hobbs JN Jr, Richardson JM & Riggin RM (1992) Biosynthesis of modified peptidoglycan precursors by vancomycin-resistant *Enterococcus faecium*. FEMS Microbiol. Lett. 98: 109–116

Amano K, Hayashi H, Araki Y & Ito E (1977) The action of lysozyme on peptidoglycan with N-unsubstituted glucosamine residues. Isolation of glycan fragments and their susceptibility to lysozyme. Eur. J. Biochem. 76: 299–307

Aono R and Ohtani M (1990) Loss of alkalophily in cell-wall-component-defective mutants derived from alkalophilic Bacillus C-125. Isolation and partial characterization of the mutants. Biochem. J. 15: 933–936

Apfel CM, Takacs B, Fountoulakis M, Stieger M & Keck W (1999) Use of genomics to identify bacterial undecaprenyl pyrophosphate synthetase: cloning, expression and characterization of the essential *uppS* gene. J. Bacteriol. 181: 483–492

Araki Y, Oba S and Ito E (1980) Enzymatic deacetylation of N-acetylglucosamine residues in cell wall peptidoglycan. J. Biochem. Tokyo 88: 469–479

Archibald AR (1976) Cell wall assembly in *Bacillus subtilis:* development of bacteriophage-binding properties as a result of the pulsed incorporation of teichoic acid. J. Bacteriol. 127: 956–960

Archibald AR and Baddiley J (1986) The teichoic acids. In: Wolfrom ML & Timpson RS, (Eds) Advances in Carbohydrate Chemistry. (pp 325–375). Academic Press, New York

Archibald AR, Hancock IC & Harwood CR (1993) Cell wall structure, synthesis and turnover. In: Hoch JA & Losick R (Eds). (pp 381–410) *Bacillus subtilis* and other Gram-Positive Bacteria American Society for Microbiology, Washington, DC

Armstrong JJ, Baddiley J, Buchanan JG, Carss B & Greenberg GR (1958) Isolation and structure of ribitol phosphate derivatives (teichoic acids) from bacterial cell walls. J. Chem. Soc. 4344–4354

Arthur M, Molinas C, Bugg TDH, Wright GD, Walsh CT & Courvalin P (1992) Evidence for *in vivo* incorporation of D-lactate into peptidoglycan precursors of vancomycin-resistant enterococci. Antimicrob. Agents Chemother. 36: 867–869

Arthur M, Molinas C, Depardieu F & Courvalin P (1993) Characterization of TN*1546*, a Tn*3*-related transposon conferring glycopeptide resistance by synthesis of depsipeptide peptidoglycan precursors in *Enterococcus faecium* BM4147. J. Bacteriol. 175: 117–127

Arthur M, Reynolds P & Courvalin P (1996) Glycopeptide resistance in enterococci. Trends Microbiol 4: 401–407

Ayala JA, Garrido T, de Pedro MA & Vicente M (1994) Molecular biology of bacterial septation. In: Ghuysen JM & Hakenbeck R, (Eds). Bacterial Cell Wall. Elsevier, Amsterdam, pp 73–101

Baddiley J 1989. Bacterial cell walls and membranes. Discovery of the teichoic acids. BioEssays 10: 207–210

Bahl H, Scholz H, Bayan N, Chami M, Leblon G & Gulik-Krzywicki T et al. (1997) Molecular biology of S-layers. FEMS Microbiol. Rev. 20: 47–98

Begg KJ, Tagasuga A, Edwards DH, Dewar SJ, Spratt BG, Adachi H, Ohta T, Matsuzawa H & Donachie WD (1990) The balance between different peptidoglycan precursors determines whether *Escherichia coli* cells will elongate or divide. J. Bacteriol. 172: 6697–6703

Béliveau C, Potvin C, Trudel J, Asselin A & Bellemare G (1991) Cloning, sequencing and expression in *Escherichia coli* of a *Streptococcus faecalis* autolysin. J. Bacteriol. 173: 5619–5623

Beveridge T, Pouwels PH, Sara M, Kotiranta A, Lounatmaa K, Kari K et al. (1997) Functions of S-layers. FEMS Microbiol. Rev. 20: 99–149

Bhakdi S, Klonisch P, Nuber P & Fischer W (1991) Stimulation of monokine production by lipoteichoic acids. Infect. Immun. 59: 4614–4620

Bidnenko E, Mercier C, Tremblay J, Tailliez P & Kulakauskas S (1998) Estimation of the state of the bacterial cell wall by fluorescent in situ hybridization. Appl. Environ. Microbiol. 64: 3059–3062

Bierbaum G and Sahl HG (1987) Autolytic system of *Staphylococcus simulans* 22: influence of cationic peptides on activity of *N*-acetylmuramoyl-L-alanine amidase. J. Bacteriol. 169: 5452–5458

Billot-Klein D, Gutmann L, Sable S, Guittet E & van Heijenoort J (1994) Modification of peptidoglycan precursor is a common feature of the low-level vancomycin-resistant VanB-type *Enterococcus* D366 and of the naturally glycopeptide-resistant species *Lactobacillus casei*, *Pediococcus pentosaceus*, *Leucon-*

ostoc mesenteroides, and *Enterococcus gallinarum*. J. Bacteriol. 176: 2398–2405

Billot-Klein D, Shlaes D, Bryant D, van Heijenoort J & Gutmann L (1996) Peptidoglycan structure of *Enterococcus faecium* expressing vancomycin resistance of the VanB type. Biochem. J. 313: 711–715

Billot-Klein D, Legrand R, Schoot B, van Heijenoort J & Gutmann L (1997) Peptidoglycan structure of *Lactobacillus casei*, a species highly resistant to glycopeptide antibiotics. J. Bacteriol. 179: 6208–6212

Blackman SA, Smith TJ & Foster SJ (1998) The role of autolysins during vegetative groth of *Bacillus subtilis* 168. Microbiology 144: 73–82

Boot HJ and Pouwels PH (1996) Expression, secretion and antigenic variation of bacterial S-layer proteins. Mol. Microbiol. 21: 1117–1123

Boot HJ, Kolen CPAM, van Noort JM & Pouwels PH (1993) S-layer protein of *Lactobacillus acidophilus* ATCC 4356: purification, expression in *Escherichia coli*, and nucleotide sequence of the corresponding gene. J. Bacteriol. 175: 6089–6096

Boot HJ, Kolen CPAM & Pouwels PH (1995) Identification, cloning, and nucleotide sequence of a silent S-layer protein gene of *Lactobacillus acidophilus* ATCC 4356 which has extensive similarity with the S-layer protein gene of this species. J. Bacteriol. 177: 7222–7230

Boot HJ, Kolen CPAM & Pouwels PH (1996a) Interchange of the active and silent S-layer protein genes of *Lactobacillus acidophilus* by inversion of the chromosomal *slp* segment. Mol. Microbiol. 21: 799–809

Boot HJ, Kolen CPAM, Andreadaki FJ, Leer RJ & Pouwels PH (1996b) The *Lactobacillus acidophilus* S-layer protein gene expression site comprises two consensus promoter sequences, one of which directs transcription of stable mRNA. J. Bacteriol. 178: 5388–5394

Bottazzi V (1988) An introduction to rod-shaped lactic-acid bacteria. Biochimie 70: 303–315

Briehl M, Pooley HM & Karamata D (1989) Mutants of *Bacillus subtilis* 168 thermosensitive for growth and wall teichoic acid synthesis. J. Gen. Microbiol. 135: 1325–1334

Bugg DH, Wright GD, Dutka-Malen S, Arthur M, Courvalin P & Walsh CT (1991) Molecular basis for vancomycin resistance in *Enterococcus faecium* BM4147: biosynthesis of a depsipeptide peptidoglycan precursor by vancomycin resistance protein VanH and VanA. Biochemistry 30: 10408–10415

Buist G, Kok J, Leenhouts KJ, Dabrowska M, Venema G and Haandrikman AJ (1995) Molecular cloning and nucleotide sequence of the gene encoding the major peptidoglycan hydrolase of *Lactococcus lactis*, a muramidase needed for cell separation. J. Bacteriol. 177: 1554–1563

Buist G, Karsens H, Nauta A, van Sinderen D, Venema G and Kok J (1997) Autolysis of *Lactococcus lactis* caused by induced overproduction of its major autolysin, AcmA. Appl. Environ. Microbiol. 63: 2722–2728

Buist G, Venema G & Kok J (1998) Autolysis of *Lactococcus lactis* is influenced by proteolysis. J. Bacteriol. 180: 5947–5953

Callegari ML, Riboli B, Sanders JW, Cocconcelli PS, Kok J, Venema G & Morelli L (1998) The S-layer gene of *Lactobacillus helveticus* CNRZ 892: cloning, sequence and heterologous expression. Microbiology 144: 719–726

Carias LL, Rudin SD, Donskey CJ & Rice LB (1998) Genetic linkage and cotransfer of a novel, *vanB*-containing transposon (Tn*5382*) and a low-affinity penicillin-binding protein 5 gene in a clinical vancomycin-resistant *Enteroccoccus faecium* isolate. J. Bacteriol. 180: 4426–4434

Cassels FJ and London J (1989) Isolation of a coaggregation-inhibiting cell wall polysaccharide from *Streptococcus sanguis* H1. J. Bacteriol. 171: 4019–4025

Cerning J (1990) Exocellular polysaccharides produced by lactic acid bacteria. FEMS Microbiol. Rev. 87: 113-130

Chapman G and Hillier J (1953) Electron microscopy of ultra-thin sections of bacteria. I. Cellular division in *Bacillus cereus*. J. Bacteriol. 66: 362–373

Chapot-Chartier MP (1996) Les autolysines des bactéries lactiques. Lait 76: 91–109

Childs WC III, Taron DJ & Neuhaus FC (1985) Biosynthesis of D-alanyl-lipoteichoic acid by *Lactobacillus casei:* interchain transacylation of D-alanyl ester residues. J. Bacteriol. 162: 1191–1195

Chu CP, Kariyama R, Daneo-Moore L & Shockman GD (1992) Cloning and sequence analysis of the muramidase-2 gene from *Enterococcus hirae*. J. Bacteriol. 174: 1619–1625

Clarke AJ & Dupont C (1992) *O*-acetylated peptidoglycan: its occurrence, pathobiological significance, and biosynthesis. Can. J. Microbiol. 38: 85–91

Coyette J and Ghuysen JM (1970) Structure of the walls of *Lactobacillus acidophilus* strain 63 AM Gasser. Biochemistry 9: 2935–2943

Crow VL, Coolbear T, Gopal PK, Martley FG, McKay LL & Riepe H (1995) The role of autolysis of lactic acid bacteria in the ripening of cheese. Int. Dairy J. 5: 855–875

de Vos WM, Underwood HM & Davies FL (1984) Plasmid-encoded bacteriophage resistance in *Streptococcus cremoris* SK11. FEMS Microbiol. Lett. 23: 175–179

Demchick P & Koch AL (1996) The permeability of the wall fabric of *Escherichia coli* and *Bacillus subtilis*. J. Bacteriol. 178: 768–773

Dijkstra A & Keck W (1996) Peptidoglycan as a barrier to transenvelope transport. J. Bacteriol. 178: 5555–5562

Douglas LJ & Wolin MJ (1971) Cell wall polymers and phage lysis of *Lactobacillus plantarum*. Biochemistry 10: 1551–1555

Doyle RJ & Koch AL (1987) The functions of autolysins in the growth and division of *Bacillus subtilis*. CRC Crit. Rev. Microbiol. 15: 169–222

Duez C, Thamm I, Sapunaric F, Coyette J & Ghuysen JM (1998) The division and cell wall gene cluster of *Enterococcus hirae* S185. DNA Sequence 9: 149–161

Duwat P, Cochu A, Ehrlich SD & Gruss A (1997) Characterization of *Lactococcus lactis* UV-sensitive mutants obtained by IS*S1* transposition. J. Bacteriol. 179: 4473–4479

El Kharroubi A, Jacques P, Piras G, Van Beeumen J, Coyette J & Ghuysen JM (1991) The *Enterococcus hirae* R40 penicillin-binding protein 5 and the methicillin-resistant *Staphylococcus aureus* penicillin-binding protein 2' are similar. Biochem. J. 280: 463–469

Fan DP (1970) Autolysin(s) of *Bacillus subtilis* as dechaining enzyme. J. Bacteriol. 103: 494–499

Ferain T, Hobbs JN, Richardson J, Bernard N, Garmyn D, Hols P, Allen NE & Delcour J (1996) Knockout of the two *ldh* genes has a major impact on peptidoglycan precursor synthesis in *Lactobacillus plantarum*. J. Bacteriol. 178: 5431–5437

Ferrari E, Henner DJ & Yang MY (1985) Isolation of an alanine racemase gene from *Bacillus subtilis* and its use for plasmid maintenance in *B. subtilis*. Biotechnology 3: 1003–1007

Fischer W (1981) Glycerophosphoglycolipids: presumptive biosynthetic precursors of lipoteichoic acids. *In* Shockman GD & Wacken AJ, (Eds). Chemistry and Biological Activities of Bacterial Surface Amphiphiles (pp 209–228). Academic Press, New York

Fischer W (1988) Physiology of lipoteichoic acids in bacteria. Adv. Microb. Physiol. 29: 233–302

Fischer W (1990) Bacterial phosphoglycolipids and lipoteichoic acids. *In* Kates M (Ed.) Handbook of lipid research: glycolipids, phospholipids and sulfoglycolipids (pp 123–234). Plenum Press, New York

Fischer W (1994) Lipoteichoic acids and lipoglycans. *In*: Ghuysen JM & Hakenbeck R (Eds) Bacterial Cell Wall. (pp 199–215). Elsevier, Amsterdam

Fischer W, Rösel P & Koch HU (1981) Effect of alanine ester substitution and other structural features of lipoteichoic acids on their inhibitory activity against autolysins of *Staphylococcus aureus*. J. Bacteriol. 146: 467–475

Fotheringham IG, Bledig SA & Taylor PP (1998) Characterization of the genes encoding D-amino acid transaminase and glutamate racemase, two D-glutamate biosynthetic enzymes of *Bacillus sphaericus* ATCC 10208. J. Bacteriol. 180: 4319–4323

Garcia E and Lopez R (1997) Molecular biology of the capsular genes of *Streptococcus pneumoniae*. FEMS Microbiol. Lett. 149: 1–10

Gasson MJ (1996) Lytic systems in lactic acid bacteria and their bacteriophages. Antonie van Leeuwenhoek 70: 147–159

Ghuysen JM (1968) Use of bacteriolytic enzymes in determination of wall structure and their role in cell metabolism. Bacteriol. Rev. 32: 425–464

Ghuysen JM (1991) Serine β-lactamases and penicillin-binding proteins. Annu. Rev. Microbiol. 45: 37–667

Giesbrecht P, Kersten T, Maidhof H & Wecke J (1998) Staphylococcal cell wall: morphogenesis and fatal variations in the presence of penicillin. Microbiol. Mol. Biol. Rev. 62: 1371–1414

Glaser P, Kunst F, Arnaud M, Coudart MP, Gonzales W, Hullo MF et al. (1993) *Bacillus subtilis* genome project: cloning and sequencing of the 97 kb region from 325° to 333°. Mol. Microbiol. 10: 371–384

Goffin C & Ghuysen JM (1998) Multimodular penicillin-binding proteins: an enigmatic family of orthologs and paralogs. Microbiol. Mol. Biol. Rev. 62: 1079–1093

Gopal PK & Crow VL (1993) Characterization of loosely associated material from the cell surface of *Lactococcus lactis* subsp. *cremoris* E8 and its phage-resistant variant strain 398. Appl. Environ. Microbiol. 59: 3177–3182

Graham LL & Beveridge TJ (1990) Evaluation of freeze-substitution and conventional embedding protocols for routine electron microscopic processing of eubacteria. J. Bacteriol. 172: 2141–2149

Granato D, Perotti F, Masserey I, Rouvet M, Golliard M, Servin A & Brassart D (1999) Cell surface-associated lipoteichoic acid acts as an adhesion factor for attachment of *Lactobacillus johnsonii* La1 to human enterocyte-like Caco-2 cells. Appl. Environ. Microbiol. 65: 1071–1077

Grant WD (1979) Cell wall teichoic acid as a reserve phosphate source in *Bacillus subtilis*. J. Bacteriol. 137: 35–43

Graumann P, Wendrich TM, Weber MHW, Schröder K & Marahiel MA (1997) A family of cold shcock proteins in *Bacillus subtilis* is essential for cellular growth and for efficient protein synthesis at optimal and low temperatures. Mol. Microbiol. 25: 741–756

Green CJ & Vold BS (1993) *Staphylococcus aureus* has clustered tRNA genes. J. Bacteriol. 175: 5091–5096

Greene JD & Klaenhammer TR (1994) Factors involved in adherence of lactobacilli to human Caco-2 cells. Appl. Environ. Microbiol. 60: 4487–4494

Hakenbeck R (1994) Resistance to glycopeptide antibiotics. *In* Ghuysen JM & Hakenbeck R (Eds). Bacterial Cell Wall. (pp 535–558). Elsevier, Amsterdam

Hall EA & Knox KW (1965) Properties of the polysaccharide and mucopeptide components of the cell wall of *Lactobacillus casei*. Biochem. J. 96: 310–318

Hamada S, Torii M & Kotani S et al. (1978) Lysis of *Streptococcus mutans* cells with mutanolysin, a lytic enzyme prepared from a culture liquor of *Streptomyces globisporus* 1829. Arch. Oral Biol. 23: 543–549

Hamann L, El-Samalouti V, Ulme AJ, Flad HD & Rietschel ET (1998) Components of gut bacteria as immunomodulators. Int. J. Food Microbiol. 41: 141–154

Handwerger S, Pucci MJ, Volk KJ, Liu J & Lee MS (1992) The cytoplasmic peptidoglycan precursor of vancomycin-resistant *Enterococcus faecalis* terminates in lactate. J. Bacteriol. 174: 5982–5984

Handwerger S, Pucci MJ, Volk KJ, Liu J & Lee MS (1994) Vancomycin-resistant *Leuconostoc mesenteroides* and *Lactobacillus casei* synthesize cytoplasmic peptidoglycan precursors that terminate in lactate. J. Bacteriol. 176: 260–264

Healy VL, Park IS & Walsh CT (1998) Active-site mutants of the VanC2 D-alanyl-D-serine ligase, characteristic of one vancomycin-resistant bacterial phenotype, revert towards wild-type D-alanyl-D-alanine ligases. Chem. Biol. 5: 197–207

Heaton MP & Neuhaus FC (1993) The significance of secondary cell wall polymers in gram-positive organisms: *Lactobacillus casei* as a model system for the study of D-alanyl-lipoteichoic acid biosynthesis and function. *In*: Foo EL et al. (Eds) The Lactic Acid Bacteria (pp 89–98). Horizon Scientific Press, Norfolk

Heaton MP, Johnston RB & Thompson TL (1988) Controlled lysis of bacterial cells utilizing mutants with defective synthesis of D-alanine. Can. J. Microbiol. 34: 256–261

Henderson B, Poole S & Wilson M (1996) Bacterial modulins: a novel class of virulence factors which cause host tissue pathology by inducing cytokine synthesis. Microbiol. Rev. 60: 316–341

Henrichsen J (1995) Six new recognized types of *Streptococcus pneumoniae*. J. Clin. Microbiol. 33: 2759–2762

Heptinstall S, Archibald AR & Baddiley J (1970) Teichoic acids and membrane function in bacteria. Nature 225: 519–521

Herbold DR & Glaser L (1975a) *Bacillus subtilis* N-acetylmuramic acid L-alanine amidase. J. Biol. Chem. 250: 1676–1682

Herbold DR & Glaser L (1975b) interaction of *N*-acetylmuramic acid L-alanine amidase with cell wall polymers. J. Biol. Chem. 250: 7231–7238

Hols P, Defrenne C, Ferain T, Derzelle S, Delplace B & Delcour J (1997) The alanine racemase gene is essential for growth of *Lactobacillus plantarum*. J. Bacteriol. 179: 3039–3042

Höltje JV (1996) Bacterial lysozymes. *In*: Joliès P (Ed). Lysozymes: model enzymes in biochemistry and biology (pp 65–74). Birkhäuser Verlag, Basel

Höltje JV (1998) Growth of the stress-bearing and shape-maintaining murein sacculus of *Escherichia coli*. Microbiol. Mol. Biol. Rev. 62: 181–203

Höltje JV & Tomasz A (1975) Lipoteichoic acid: a specific inhibitor of autolysin activity in*Pneumococcus*. Proc. Natl. Acad. Sci. USA 72: 1690–1694

Höltje JV & Tuomanen EI (1991) The murein hydrolases of *Escherichia coli*: properties, functions and impact on the course of infections in vivo. J. Gen. Microbiol. 137: 441–454

Honeyman AL & Stewart GC (1989) The nucleotide sequence of he *rodC* operon of *Bacillus subtilis*. Mol. Microbiol. 3: 1257–1268

Hughes AH, Hancock IC & Baddiley J (1970) Teichoic acids in cation control in bacterial membranes. Biochem. J. 132: 83–93

Ishibashi K, Takesue S, Watanabe K & Oishi K (1982) Use of lectins to characterize the receptor sites for bacteriophage PL-1 of *Lactobacillus casei*. J. Gen. Microbiol. 128: 2251–2259

Iwasaki H, Araki Y, Ito E, Nagaoka M & Yokokura T (1990) Structure of macroamphiphiles from several *Bifidobacterium* strains. J. Bacteriol. 171: 845–852

Jolliffe LK, Doyle RJ & Streips UN (1981) The energized membrane and cellular autolysis in *Bacillus subtilis*. Cell 25: 753–763

Joris B, Englebert S, Chu CP, Kariyama R, Daneo-Moore L, Shockman GD & Ghuysen JM (1992) Modular design of the *Enterococcus hirae* muramidase-2 and *Streptococcus faecalis* autolysin. FEMS Microbiol. Lett. 91: 257–264

Kahala M, Savijoki K & Palva A (1997) In vivo expression of the *Lactobacillus brevis* S-layer gene. J. Bacteriol. 179: 284–286

Kawagishi S, Araki Y & Ito E (1980) *Bacillus cereus* autolytic endoglucosaminidase active on cell wall peptidoglycan with N-unsubstituted glucosamine residues. J. Bacteriol. 141: 137–143

Kemper MA, Urrutia MM, Beveridge TJ, Koch AL & Doyle RJ (1993) Proton motive force may regulate cell wall-associated enzymes of *Bacillus subtilis*. J. Bacteriol. 175: 5690–5696

Kleppe G, Vasstrand E & Jensen HB (1981) The specificity requirements of bacteriophage T4 lysozyme. Involvement of N-acetamido groups. Eur. J. Biochem. 119: 589–593

Knox KW & Wicken AJ (1973) Immunological properties of teichoic acids. Bacteriol. Rev. 37: 215–257

Koch AL (1985) How bacteria grow and divide in spite of internal hydrostatic pressure. Can. J. Microbiol. 31: 1071–1084

Koch AL (1991) Effective growth by the simplest means: the bacterial way. ASM News 57: 633–637

Koch AL (1995) Bacterial growth and form. Chapman & Hall, New York, 423 pp

Koch AL & Woeste S (1992) Elasticity of the sacculus of *Escherichia coli*. J. Bacteriol. 174: 4811–4819

Koch HU & Fischer W (1978) Acyldiglucosyldiacylglycerol-containing lipoteichoic acid with a poly(3-*O*-galabiosyl-2-*O*-galactosyl-*sn*-glycero-1-phosphate) chain from *Streptococcus lactis* Kiel 42172. Biochemistry 17: 5275–5281

Koch HU, Doker R & Fischer W (1985) Maintenance of D-alanine ester substituion of lipoteichoic acid by re-esterification in *Staphylococcus aureus*. J. Bacteriol. 164: 1211–1217

Kojima N, Araki Y & Ito E (1985a) Structural studies on the linkage unit of ribitol teichoic acid of *Lactobacillus plantarum*. Eur. J. Biochem. 148: 29–34

Kojima N, Araki Y & Ito E (1985b) Structural studies on the acidic polysaccharide of *Bacillus cereus* AHU 1356 cell walls. Eur. J. Biochem. 148: 479–484

Kojima N, Araki Y & Ito E (1986) Biosynthesis of the wall acidic polysaccharide in *Bacillus cereus* AHU 1356. Eur. J. Biochem. 155: 513–519

Kolenbrander PE & London J (1993) Adhere today, here tomorrow: oral bacterial adherence. J. Bacteriol. 175: 3247–3252

Kotani S, Watanabe Y, Shimono T, Kinoshita F, Narita T, Kato K, Stewart-Tull DES & Morisaki I (1975) Immunoadjuvant activities of peptidoglycan subunits from the cell walls of *Staphyloccus aureus* and *Lactobacillus plantarum*. Biken J. 18: 93–103

Krulwich TA, Ito M, Gilmour R & Guffanti AA (1997) Mechanisms of cytoplasmic pH regulation in alkalophilic strains of *Bacillus*. Extremophiles 1: 163–169

Kullik I, Jenni R & Berger-Bächi B (1998) Sequence of putative alanine racemase operon in *Staphylococcus aureus:* insertional interruption of this operon reduces D-alanine substitution of lipoteichoic acid and autolysis. Gene 219: 9–17

Kunst F, Ogasawara N, Moszer I, Albertini AM, Alloni G, Azevedo V et al. (1997) The complete genome sequence of the Gram-positive bacterium *Bacillus subtilis*. Nature 390: 249–256

Labischinski H & Maidhof H (1994) Bacterial peptidoglycan: overview and evolving concepts. *In*: Ghuysen JM & Hakenbeck R, (Eds). Bacterial Cell Wall (pp 23–38). Elsevier, Amsterdam

Labischinski HR, Barnickel D, Naumann D & Keller P (1985) Conformational and topological aspects of the three-dimensional architecture of bacterial peptidoglycan. Ann. Inst. Pasteur Microbiol. 136A: 45–50

Lambert PA, Hancock IC & Baddiley J (1975a) Influence of alanyl ester residues on the binding of magnesium ions to teichoic acids. Biochem. J. 151: 671–676

Lambert PA, Hancock IC & Baddiley J (1975b) The interaction of magnesium ions with teichoic acids. J. Biochem. 149: 519–524

Lancefield RC (1933) A serological differentiation of human and other groups of hemolytic streptococci. J. Exp. Med. 59: 571-591

Lang WK, Glassey K & Archibald AR (1982) Influence of phosphate supply on teichoic acid and teichuronic acid content of *Bacillus subtilis* cell walls. J. Bacteriol. 151: 367–375

Lazarevic V & Karamata D (1995) The *tagGH* operon of *Bacillus subtilis* 168 encodes a two-component ABC transporter involved in the metabolism of two wall teichoic acids. Mol. Microbiol. 16: 345–355

Lazarevic V, Margot P, Soldo B & Karamata D (1992) Sequencing and analysis of the *Bacillus subtilis lytRABC* divergon: a regulatory unit encompassing the structural genes of the *N*-acetylmuramoyl-L-alanine amidase and its modifier. J. Gen. Microbiol. 138: 1949–1961

Lazarevic V, Mauël C, Soldo B, Freymond PP, Margot P & Karamata D (1995) Sequence analysis of the 308° to 311° segment of the *Bacillus subtilis* 168 chromosome, a region devoted to cell wall metabolism, containing non-coding grey holes which reveal chromosomal rearrangements. Microbiology 141: 329–335

Lepeuple AS, Van Gemert E & Chapot-Chartier MP 1998) Analysis of the bacteriolytic enzymes of the autolytic *Lactococcus lactis* subsp. *cremoris* strain AM2 by renaturing polyacrylamide gel electrophoresis: identification of a prophage-encoded enzyme. Appl. Environ. Microbiol. 64: 4142–4148

Ligozzi M, Pittaluga F & Fontana R (1993) Identification of a genetic element (*psr*) which negatively controls expression of *Enterococcus hirae* penicillin-binding protein 5. J. Bacteriol. 175: 2046–2051

Linnett PE & Strominger JL (1974) Amidation and cross-linking of the enzymatically synthesised peptidoglycan of *Bacillus stearothermophilus*. J. Biol. Chem. 249: 2489–2496

Liu W, Eder S & Hulett FM (1998) Analysis of *Bacillus subtilis tagAB* and *tagDEF* expression during phosphate starvation identifies a repressor role for PhoP–P. J. Bacteriol. 180: 753–758

Lleo MM, Fontana R & Solioz M (1995) Identification of a gene (*arpU*) controlling muramidase-2 export in *Enterococcus hirae*. J. Bacteriol. 177: 5912–5917

Logardt IM & Neujahr HY (1975) Lysis of modified walls from *Lactobacillus fermentum*. J. Bacteriol. 124: 73–77

Lortal S (1993) Crystalline surface-layers of the genus *Lactobacillus*. *In*: Beveridge TJ & Koval SF, (Eds). Advances in Bacterial Paracrystalline Surface Layers (pp 57–65). Plenum Press, New York

Lortal S, Boyaval P & van Heijenoort J (1989) Influence de plusieurs facteurs sur l'autolyse de *Lactobacillus helveticus* CNRZ414. Lait 69: 223–231

Lortal S, Rousseau M, Boyaval P & van Heijenoort J (1991) Cell wall and autolytic system of *Lactobacillus helveticus* ATCC 12046. J. Gen. Microbiol. 137: 549–559

Lortal S, van Heijenoort J, Gruber K & Sleytr UB (1992) S-layer of *Lactobacillus helveticus* ATCC 12046: isolation, chemical characterization and re-formation after extraction with lithium chloride. J. Gen. Microbiol. 138: 611–618

Lupas A (1996) A circular permutation event in the evolution of the SLH domain? Mol. Microbiol. 20: 897–898

Lupas A, Engelhardt H, Peters J, Santarius U, Volker S & Baumeister W (1994) Domain structure of the *Acetogenium kivui* surface layer revealed by electron crystallography and sequence analysis. J. Bacteriol. 176: 1224–1233

Ma D, Alberti M, Lynch C, Nikaido H & Hearst JE (1996) The local repressor AcrR plays a modulating role in the regulation of *acrAB* genes of *Escherichia coli* by global stress signals. Mol. Microbiol. 19: 101–112

Margot P & Karamata D (1992) Identification of the structural genes for *N*-acetylmuramyl-L-alanine amidase and its modifier in *Bacillus subtilis* 168: inactivation of these genes by insertional mutagenesis has no effect on growth or cell separation. Mol. Gen. Genet. 232: 359–366

Margot P, Mauël C & Karamata D (1994) The gene of the *N*-acetylglucosaminidase, a *Bacillus subtilis* 168 cell wall hydrolase not involved in vegetative cell autolysis. Mol. Microbiol. 12: 535–545

Massidda O, Kariyama R, Daneo-Moore L & Shockman GD (1996) Evidence that the PBP 5 synthesis repressor (*psr*) of *Enterococcus hirae* is also involved in the regulation of cell wall composition and other cell wall-related properties. J. Bacteriol. 178: 5272–5278

Masuda K & Kawata T (1983) Distribution and chemical characterization of regular arrays in the cell walls of strains of the genus *Lactobacillus*. FEMS Microbiol. Lett. 20: 145–150

Matsuhashi M (1994) Utilization of lipid-linked precursors and the formation of peptidoglycan in the process of cell growth and division: membrane enzymes involved in the final steps of peptidoglycan synthesis and the mechanism of their regulation. *In*: Ghuysen JM & Hakenbeck R, (Eds). Bacterial Cell Wall. (pp 55–71). Elsevier, Amsterdam

Matsuhashi M, Dietrich CP & Strominger JL (1967) The role of soluble RNA and lipid intermediates in glycine incorporation in *Staphylococcus aureus*. J. Biol. Chem. 242: 3191–3206

Mauël C, Young M, Margot P & Karamata D (1989) The essential nature of teichoic acids in *Bacillus subtilis* as revealed by insertional mutagenesis. Mol. Gen. Genet. 215: 388–394

Mauël C, Young M & Karamata D (1991) Genes concerned with synthesis of poly(glycerol phosphate), the essential teichoic acid in *Bacillus subtilis* strain 168, are organized in two divergent transcription uints. J. Gen. Microbiol. 137: 929–941

McIntire FC, Crosby LK, Vatter AE, Cisar JO, McNeil MR, Bush CA, Tjoa SS & Fennessey PV (1988) A polysaccharide from *Streptococcus sanguis* 34 that inhibits coaggregation of *S. sanguis* 34 with *Actinomyces viscosus* T14V. J. Bacteriol. 170: 2229–2235

Mechold U, Steiner K, Vetterman S & Malke H (1993) Genetic organization of the streptokinase region of the *Streptococcus equisimilis* H46A chromosome. Mol. Gen. Genet. 241: 129–140

Mercenier A (1999) Lactic acid bacteria as live vaccines. *In*: Tannock G (Ed). Probiotics: A critical review Horizon (pp 113–127). Scientific Press, Wymondham, U.K.

Mesnage S, Tosi-Couture E & Fouet A (1999) Production and cell surface anchoring of functional fusions beteen the SLH motifs of the *Bacillus anthracis* S-layer proteins and the *Bacillus subtilis* levansucrase. Mol. Microbiol. 31: 927–936

Messer J & Reynolds PE (1992) Modified peptidoglycan precursors produced by glycopeptide-resistant enterococci. FEMS Microbiol. Lett. 94: 195–200

Messner P & Sleytr UB (1992) Crystalline bacterial cell-surface layers. Adv. Microb. Physiol. 33: 213–275

Messner P, Allmaier G, Schäffer C, Wugeditsch T, Lortal S, König H, Niemetz R, Dorner M (1997) Biochemistry of layers. FEMS Microbiol. Rev. 20: 25–46

Miörner H, Johansson G & Kronvall G (1983) Lipoteichoic acid is the major cell wall component responsible for surface hydrophobicity of group A streptococci. Infect. Immun. 39: 336–343

Monteville MR, Ardestani B & Geller BL (1994) Lactococcal bacteriophages require a host cell wall carbohydrate and a plasma membrane protein for adsorption and ejection of DNA. Appl. Environ. Microbiol. 60: 3204–3211

Mou L, Sullivan JJ & Rao GR (1976) Autolysis of *Streptococcus cremoris*. J. Dairy Res. 43: 275–282

Mozes N & Lortal S (1995) X-ray photoelectron spectroscopy and biochemical analysis of the surface of *Lactobacillus helveticus* ATCC 12046. Microbiology 141: 11–19

Murazumi N, Araki Y & Ito E (1986) Biosynthesis of the wall neutral polysaccharide in *Bacillus cereus* AHU 1356. Eur. J. Biochem. 161: 51–59

Nagaoka M, Muto M, Nomoto K, Matuzaki T, Watanabe T & Yokokura T (1990) Structure of polysaccharide-peptidoglycan complex from the cell wall of *Lactobacillus casei* YIT9018. J. Biochem. 108: 568–571

Nagaoka M, Shibata H, Kimura I, Hashimoto S, Kimura K, Sawada H & Yokokura T (1995) Structural studies on a cell wall polysaccharide from *Bifidobacterium longum* YIT4028. Carbohyd. Res. 274: 245–249

Nagaoka M, Hashimoto S, Shibata H, Kimura I, Kimura K, Sawada H & Yokokura T (1996) Structure of a galactan from cell walls of *Bifidobacterium catenulatum* YIT4016. Carbohyd. Res. 281: 285–291

Nanninga N (1998) Morphogenesis of *Escherichia coli*. Microbiol. Mol. Biol. Rev. 62: 110–129

Navarre WW & Schneewind O (1994) Proteolytic cleavage and cell wall anchoring at the LPXTG motif of surface proteins in gram-positive bacteria. Mol. Microbiol. 14: 115–121

Navarre WW & Schneewind O (1999) Surface proteins of gram-positive bacteria and mechanisms of their targeting to the cell wall envelope. Microbiol. Mol. Biol. Rev. 63: 174–229

Neuhaus FC (1985) Inter-chain transacylatiobn of D-alanine ester residues of lipoteichoic acid: a unique mechanism of membrane communication. Biochem. Soc. Trans. 13: 987–990

Neuhaus FC, Heaton MP, Debarov DV & Zhang Q (1996) The *dlt* operon in the biosynthesis of D-alanyl-lipoteichoic acid in *Lactobacillus casei*. Microbial Drug Resist. 2: 77–84

Ntamere A, Taron DJ & Neuhaus FC (1987) Assembly of D-alanyl-lipoteichoic acid in *Lactobacillus casei:* mutants deficient in the D-alanyl ester content of this amphiphile. J. Bacteriol. 169: 1702–1711

Ostlie HM, Vegarud G & Langsrud T (1995) Autolysis of lactococci: detection of lytic enzymes by polyacrylamide gel electrophoresis and characterization in buffer systems. Appl. Environ. Microbiol. 61: 3598–3603

Park W, Seto H, Hakenbeck R & Matsuhashi M (1985) Major peptidoglycan transglycosylase activity in *Streptococcus pneumoniae* that is not a penicillin binding protein. FEMS Microbiol. Lett. 27: 45–48

Park IS, Lin CH & Walsh CT (1996) Gain of D-alanyl-D-lactate or D-lactyl-D-alanine synthetase activities in three active-site mutants of the *Escherichia coli* D-alanyl-D-alanine ligase B. Biochemistry 35: 10464–10471

Pelletier C, Bouley C, Cayuela C, Bouttier S, Bourlioux P & Bellon-Fontaine MN (1997) Cell surface characteristics of *Lactobacillus casei* subsp. *casei*, *Lactobacillus paracasei* subsp. *paracasei*, and *Lactobacillus rhamnosus* strains. Appl. Environ. Microbiol. 63: 1725–1731

Perego M, Glaser P, Minutello A, Strauch MA, Leopold K & Fischer W (1995) Incorporation of D-alanine into lipoteichoic acid and wall teichoic acid in *Bacillus subtilis*. J. Biol. Chem. 270: 15598–15606

Piard JC, Hautefort I, Fischetti VA, Ehrlich SD, Fons M & Gruss A (1997) Cell wall anchoring of the *Streptococcus pyogenes* M6 protein in various lactic acid bacteria. J. Bacteriol. 179: 3068–3072

Piras G, Raze D, El Kharroubi A, Hastir D, Englebert S, Coyette J & Ghuysen JM (1993) Cloning and sequencing of the low-affinity penicillin-binding protein 3^r-encoding gene of *Enterococcus hirae* S185: modular design and structural organization of the protein. J. Bacteriol. 175: 2844–2852

Pisabarro AG, De Pedro MA & Ishiguro EE (1990) Dissociation of the ampicillin-induced lysis of amino acid-deprived *Escherichia coli* into two stages. J. Bacteriol. 172: 2187–2190

Pooley HM & Karamata D (1994) Teichoic acid synthesis in *Bacillus subtilis:* genetic organization and biological roles. *In*: Ghuysen JM & Hakenbeck R, (Eds). Bacterial Cell Wall (pp 187–198). Elsevier, Amsterdam

Pooley HM, Paschoud D & Karamata D (1987) The *gtaB* marker in *Bacillus subtilis* 168 is associated with a deficiency in UDPglucose pyrophosphorylase. J. Gen. Microbiol. 133: 3481–3493

Pooley HM, Abellan FX & Karamata D (1991) A conditional-lethal mutant of *Bacillus subtilis* 168 with a thermosensitive glycerol-3-phosphate cytidylyltransferase, an enzyme specific for the synthesis of the major wall teichoic acid. J. Gen. Microbiol. 137: 921–928

Pozzi G, Contorni M, Oggioni MR, Manganelli R, Tommasino M, Cavalieri F & Fischetti VA (1992) Delivery and expression of a heterologous antigen on the surface of streptococci. Infect. Immun. 60: 1902–1907

Promadej N, Fiedler F, Cossart P, Dramsi S & Kathariou S (1999) Cell wall teichoic acid glycosylation in *Listeria monocytogenes* serotype 4b rrequires *gtcA*, a novel, serogroup-specific gene. J. Bacteriol. 181: 418–425

Pucci MJ, Thanassi JA, Discotto LF, Kessler RE & Dougherty TJ (1997) Identification and characterization of cell wall-cell division gene clusters in pathogenic gram-positive cocci. J. Bacteriol. 179: 5632–5635

Pum D & Sleytr UB (1999) The application of bacterial S-layers in molecular nanotechnology. Trends Biotechnol. 17: 8–12

Qi Y & Hulett FM (1998) Role of PhoP–P in transcriptional regulation of genes involved in cell wall anionic polymer biosynthesis in *Bacillus subtilis*. J. Bacteriol. 180: 4007–4010

Rachel R, Pum D, Smarda J, Smajs D, Komrska J, Krzyzanek V, Rieger G & Stetter KO (1997) Fine structure of S-layers. FEMS Microbiol. Rev. 20: 13–23

Raetz CRH (1996) Bacterial lipopolysaccharides: a remarkable family of bioactive macroamphiphiles. *In*: Neidhardt FC, (Ed.) *Escherichia coli* and *Salmonella*, cellular and molecular biology. 2nd ed. (pp 1035–1063). ASM Press, Washington DC (USA)

Redondo-Lopez V, Cook RL & Sobel JD (1990) Emerging role of lactobacilli in the control and maintenance of the vaginal bacterial micro-flora. Rev. Infect. Dis. 12: 856–872

Reeves PR, Hobbs M, Valvano MA, Skurnik M, Whitfield C, Coplin D, Kido N, Klena J, Maskell D, Raets CRH & Rick PD (1996) Bacterial polysaccharide synthesis and gene nomenclature. Trends Microbiol. 4: 495–503

Riepe HR, Pillidge CJ, Gopal PK & McKay LL (1997) Characterization of the highly autolytic *Lactococcus lactis* subsp. *cremoris* strains CO and 2250. Appl. Environ. Microbiol. 63: 3757–3763

Roberts IS (1996) The biochemistry and genetics of capsular polysaccharide production in bacteria. Annu. Rev. Microbiol. 50: 285–315

Rosenberg M & Kjelleberg S (1986) Hydrophobic interactions: role in bacterial adhesion. Microbiol. Ecol. 9: 353–393

Rothfield LI & Zhao CR (1996) How do bacteria decide where to divide? Cell 84: 183–186

Rothfield LI & Justice SS (1997) Bacterial cell division: the cycle of the ring. Cell 88: 581–584

Salton MRJ (1994) The bacterial cell envelope – a historical perspective. *In*: Ghuysen JM & Hakenbeck R (Eds) Bacterial Cell Wall (pp 1–22). Elsevier, Amsterdam

Sandholm E & Sarimo SS (1981) Autolysis of *Streptococcus thermophilus*. FEMS Microbiol. Lett. 11: 125–129

Sara M, Egelseer EM, Dekitsch C & Sleytr UB (1998) Identification of two binding domains, one for peptidoglycan and another for a secondary cell wall polymer, on the N-terminal part of the S-layer protein SbsB from *Bacillus stearothermophilus* PV72/p2. J. Bacteriol. 180: 6780–6783

Savijoki K, Kahala M & Palva A (1997) High level heterologous protein production in *Lactococcus* and *Lactobacillus* using a new secretion system based on the *Lactobacillus brevis* S-layer signals. Gene 186: 255–262

Schleifer KH & Kandler O (1972) Peptidoglycan types of bacterial cell walls and their taxonomic implications. Bacteriol. Rev. 36: 407–477

Schleifer KH & Kilpper-Bälz R (1987) Molecular and chemotaxonomic approaches to the classification of streptococci, enterococci and lactococci: a review. Syst. Appl. Microbiol. 10: 1–19

Schwab JH (1993) Phlogistic properties of peptidoglycan-polysaccharide polymers from cell walls of pathogenic and normal-flora bacteria which colonize humans. Infect. Immun. 61: 4535–4539

Shockman GD & Höltje JV (1994) Microbial peptidoglycan (murein) hydrolases. *In*: Ghuysen JM & Hakenbeck R (Eds) Bacterial Cell Wall (pp 131–166). Elsevier, Amsterdam

Shockman GD, Daneo-Moore L, Kariyama R & Massidda O (1996) Bacterial walls, peptidoglycan hydrolases, autolysins, and autolysis. Microb. Drug Resist. 2: 95–98

Signoretto C, Boaretti M & Canepari C (1994) Cloning, sequencing and expression in *Escherichia coli* of the low affinity penicillin-binding protein of *Enterococcus faecalis*. FEMS Microbiol. Lett. 123: 99–106

Sijtsma L, Sterkenburg A & Wouters JT (1988) Properties of the cell walls of *Lactococcus lactis* subsp. *Cremoris* SK110 and SK112 and their relation to bacteriophage resistance. Appl. Environ. Microbiol. 54: 2808–2811

Sijtsma L, Wouters JTM & Hellingwerf KJ (1990) Isolation and characterization of lipoteichoic acid, a cell envelope component involved in preventing phage adsorption, from *Lactococcus lactis* subsp. *cremoris* SK110. J. Bacteriol. 172: 7126–7130

Sleytr UB (1997) Basic and applied S-layer research: an overview. FEMS Microbiol. Rev. 20: 5–12

Sleytr UB & Sara M (1997) Bacterial and archaeal S-layer proteins: structure-function relationships and their biotechnological applications. Trends Biotechnol. 15: 20–26

Sleytr UB, Messner P, Pum D & Sara M (1993) Crystalline bacterial cell surface layers. Mol. Microbriol. 10: 911–916

Sleytr UB, Bayley H, Sara M, Breitwieser A, Küpcü S, Mader et al. C. (1997) Applications of S-layers. FEMS Microbiol. Rev. 20: 151–175

Sneath PHA, Mair NS, Sharpe ME & Holt JG (Eds.) (1986) Bergey's manual of systematic bacteriology (Vol. 2). Williams and Wilkins, Baltimore

Soldo B, Lazarevic V, Pagni M & Karamata D (1999) Teichuronic acid operon of *Bacillus subtilis* 168. Mol. Microbiol. 31: 795–805

Stahl S & Uhlen M (1997) Bacterial surface display: trends and progress. Trends Biotechnol. 15: 185–192

Steidler L, Viaene J, Fiers W & Remaut E (1998) Functional display of a heterologous protein on the surface of *Lactococcus lactis* by means of the cell wall anchor of *Staphylococcus aureus* Protein A. Appl. Environ. Microbiol. 64: 342–345

Stimpson SA, Brown RR, Anderle SK, Klapper DG, Clark RL, Cromartie WJ & Schwab JH (1986) Arthropathic properties of cell wall polymers from normal flora bacteria. Infect. Immun. 51: 240–249

Stingele F & Mollet B (1996) Disruption of the gene encoding penicillin-binding protein 2b (*pbp2b*) causes altered cell morphology and cease in exopolysaccharide production in *Streptococcus thermophilus* Sfi6. Mol. Microbiol. 22: 357–366

Stingele F, Neeser JR & Mollet B (1996) Identification and Characterization of the *eps* (Exopolysaccharide) Gene Cluster from *Streptococcus thermophilus* Sfi6. J. Bacteriol. 178: 1680–1690

Sutcliffe IC & Shaw N (1991) Atypical lipoteichoic acids of gram-positive bacteria. J. Bacteriol. 173: 7065–7069

Tannock G (Ed.) (1999) Probiotics: A critical review. Horizon Scientific Press, Wymondham, U.K.

Toba T, Virkola R, Westerlund B, Björkman Y, Sillanpää J, Vartio T, Kalkkinen N & Korhonen TK (1995) A collagen-binding S-layer protein in *Lactobacillus crispatus*. Appl. Environ. Microbiol. 61: 2467–2471

Tsukioka Y, Yamashita Y, Oho T, Nakano Y & Koga T (1997a) Biological function of the dTDP-rhamnose synthesis pathway in *Streptococcus mutans*. J. Bacteriol. 179: 1126–1134

Tsukioka Y, Yamashita Y, Nakano Y, Oho T & Koga T (1997*b*) Identification of a fourth gene involved in dTDP-rhamnose synthesis in *Streptococcus mutans*. J. Bacteriol. 179: 4411–4414

Tsutsui O, Kikeguchi S, Matsumara K & Kato K (1991) Relationship of the chemical structure and immunobiological activities of lipoteichoic acid from *Streptococcus faecalis* (*Enterococcus hirae*) ATCC 9790. FEMS Microbiol. Immunol. 76: 211–218

Urrutia MM, Kemper M, Doyle R & Beveridge TJ (1992) The membrane-induced proton motive force influences the metal binding ability of *Bacillus subtilis* cell walls. Appl. Environ. Microbiol. 58: 3837–3844

Valence F & Lortal S (1995) Zymogram and preliminary characterization of *Lactobacillus helveticus* autolysins. Appl. Environ. Microbiol. 61: 3391–3399

Valyasevi R, Sandine WE & Geller BL (1990) The bacteriophage kh receptor of *Lactococcus lactis* subsp. *cremoris* KH is the rhamnose of the extracellular wall polysaccharide. Appl. Environ. Microbiol. 56: 1882–1889

van Heijenoort J (1994) Biosynthesis of the bacterial peptidoglycan unit. *In*: Ghuysen JM & Hakenbeck R (Eds) Bacterial Cell Wall (pp 39–54). Elsevier, Amsterdam

Van Kranenburg R, Marugg JD, van Swan II, Willem NJ & de Vos WM (1997) Molecular characterization of the plasmid-encoded *eps* gene cluster essential for exopolysaccharide biosynthesis in *Lactococcus lactis*. Mol. Microbiol. 24: 387–397

Vegarud G, Castberg HB & Langsrud T (1983) Autolysis of group N Streptococci. Effects of media composition modifications and temperature. J. Dairy Sci. 66: 2294–2302

Vidgren G, Palva I, Pakkanen R, Lounatmaa K & Palva A (1992) S-layer protein gene of *Lactobacillus brevis:* cloning by polymerase chain reaction and determination of the nucleotide sequence. J. Bacteriol. 174: 7419–7427

Ward JB (1973) The chain length of glycans in bacterial cell walls. Biochem. J. 133: 395–398

Ward JB (1981) Teichoic and teichuronic acids: biosynthesis, assembly and location. Microbiol. Rev. 45: 211–243

Wecke J, Perego M & Fischer W (1996) D-alanine deprivation of *Bacillus subtilis* teichoic acid is without effect on cell growth and morphology but affects the autolytic activity. Microb. Drug Resist. 2: 123–129

Wecke J, Madela K & Fischer W (1997) The absence of D-alanine from lipoteichoic acid and wall teichoic acid alters surface charge, enhances autolysis and increases susceptibility to methicillin in *Bacillus subtilis*. Microbiology 143: 2953–2960

Whitfield C (1988) Bacterial extracellular polysaccharides. Can. J. Microbiol. 34: 415–420

Whittaker CJ, Klier CM & Kolenbrander PE (1996) Mechanisms of adhesion by oral bacteria. Annu. Rev. Microbiol. 50: 513–552

Wicken AJ & Knox KW (1970) Studies in the group F antigen of lactobacilli: isolation of a teichoic acid-lipid complex from *Lactobacillus fermenti* NCTC6991. J. Gen. Microbiol. 60: 293–301

Wicken AJ & Knox KW (1975) Lipoteichoic acids: a new class of bacterial antigen. Science 187: 1161–1167

Wicken AJ & Knox KW (1980) Bacterial cell surface amphiphiles. Biochim. Biophys. Acta 604: 1–26

Wicken AJ, Ayres A, Campbell LK & Knox KW (1983) Effect of growth conditions on production of rhamnose-containing cell wall and capsular polysaccharides by strains of *Lactobacillus casei* subsp. *rhamnosus*. J. Bacteriol. 153: 84–92

Wicken AJ, Evans JD & Knox KW (1986) Critical micelle concentrations of lipoteichoic acids. J. Bacteriol. 166: 72–77

Yamada M, Hirose A & Matsuashi M (1975) Association of lack of cell wall teichuronic acid with formation of cell packets of *Micrococcus lysodeikticus (luteus)* mutants. J. Bacteriol. 123: 678–686

Yamanaka K, Araki J, Takano M & Sekiguchi J (1997) Characterization of *Bacillus subtilis* mutants resistant to cold shock-induced autolysis. FEMS Microbiol. Lett. 150: 269–275

Yamashita Y, Tsukioka Y, Nakano Y, Tomihisa K, Oho T & Koga T (1998*a*) Biological functions of UDP-glucose synthesis in *Streptococcus mutans*. Microbiology 144: 1235–1245

Yamashita Y, Tsukioka Y, Tomihisa K, Nakano Y & Koga T (1998*b*) Genes involved in cell wall localization and side chain formation of rhamnose-glucose polysaccharide in *Streptococcus mutans*. J. Bacteriol. 180: 5803–5807

Yasui T, Yoda K & Kamiya T (1995) Analysis of S-layer proteins of *Lactobacillus brevis*. FEMS Microbiol. Lett. 133: 181–186

Young FE (1967) Requirement of glycosylated teichoic acid for adsorption of phage in *Bacillus subtilis* 168. Proc. Natl. Acad. Sci. USA. 58: 2377–2384

Young FE, Smith C & Reilly BE (1969) Chromosomal location of genes regulating resistance to bacteriophage in *Bacillus subtilis*. J. Bacteriol. 98: 1087–1097

Zipperle GF Jr, Ezzell JW Jr & Doyle RJ (1984) Glucosamine substitution and muramidase susceptibility in *Bacillus anthracis*. Can. J. Microbiol. 30: 553–559

Zorzi W, Zhou XY, Dardenne O, Lamotte J, Raze D, Pierre J, Gutmann L & Coyette J (1996) Structure of the low-affinity penicillin-binding protein 5 PBP5fm in wild-type and highly penicillin-resistant strains of *Enterococcus faecium*. J. Bacteriol. 178: 4948–4957

Antonie van Leeuwenhoek **76:** 185–198, 1999.

Bacteriocins: mechanism of membrane insertion and pore formation

Gert N. Moll, Wil N. Konings & Arnold J.M. Driessen*
*Department of Microbiology and the Groningen Biomolecular Sciences and Biotechnology Institute, University of Groningen, Kerklaan 30, 9751 NN Haren, The Netherlands. (*Author for correspondence; E-mail: a.j.m.driessen@biol.rug.nl)*

Key words: lactococcin, lantibiotic, lipid, peptide, pore, proton motive force

Abstract

Lactic acid bacteria produce several types of pore forming peptides. Class I bacteriocins are lantibiotics that contain (methyl)lanthionine residues that may form intramolecular thioether rings. These peptides generally have a broad spectrum of activity and form unstable pores. Class II bacteriocins are small, heat stable peptides mostly with a narrow spectrum of activity. Most bacteriocins interact with anionic lipids that are abundantly present in the membranes of Gram-positive bacteria. 'Docking molecules' may enhance the conductivity and stability of lantibiotic pores, while 'receptors' in the target membrane may determine specificity of class II bacteriocins. Insertion into the membrane of many bacteriocins is proton motive force driven. Lantibiotics may form pores according to a 'wedge-like' model, while class II bacteriocins may enhance membrane permeability either by the formation of a 'barrel stave' pore or by a 'carpet' mechanism.

Introduction

In the struggle for a niche and nutrients bacteria produce peptides that serve as weapons to kill competing bacteria. One of these peptides, nisin, is applied as food preservative in more than 50 countries (Delves-Broughton et al. 1996). In view of the fall of the efficiency of classical antibiotics due to a spreading and increase in resistance mechanisms, antibacterial peptides are gaining further interest as potential antibiotics. Many of these peptides show an impressive activity against bacteria that are resistant to conventional antibiotics. The best antibacterial peptides kill a broad range of susceptible bacteria *in vitro* at concentrations ranging from 0.25 to 4 μg ml^{-1}. Bacteriocins are a subgroup of the antimicrobial peptides that were originally defined as proteinaceous compounds that kill closely related bacteria (Tagg et al. 1976). Bacteriocins can be classified into three groups (Nes et al. 1996). At present the following classes can be discerned: (i) Lantibiotics, small (<5 kDa) peptides that owe their name to special amino acids with intramolecular thioether rings such as lanthionine and β-methyl-lanthionine. Type 'A' lantibiotics are elongated, cationic and pore forming peptides some of which consist of two components (Gilmore et al. 1994; Hill et al. 1998; Navaratna et al. 1998). Some examples of type A lantibiotics are shown in Figure 1 and listed in Table 1. Type 'B' lantibiotics have a rather globular structure, are enzyme inhibitors and immunologically active peptides (Jung 1991); (ii) Small (<10 kDa) heat stable peptides that do not contain lanthionines. Four subgroups can be defined: IIA: *Lysteria*-active peptides which have an N-terminal consensus sequence (Tyr-Gly-Asn-Gly-Val-Xaa-Cys), IIB: Two peptide bacteriocins, IIC: Sec dependent bacteriocins (Leer et al. 1995; Worobo et al. 1995; Cintas et al. 1997), IID: Class II bacteriocins that do not belong to the other subgroups. Some class II bacteriocins are listed in Table 2; and (iii) Large heat labile proteins (>30 kDa).

A wide range of bactericidal activities (Moll et al. 1996) have been for bacteriocins described ranging from enzyme activity modulation, inhibition of outgrowth of spores, anion carrier activity to the formation of selective or non-selective pores. These peptides may have a broad or narrow target spectrum. In the latter case, their activity is often presumed to be associated with a receptor-like molecule at the surface of the target cell. Here we will describe the mechan-

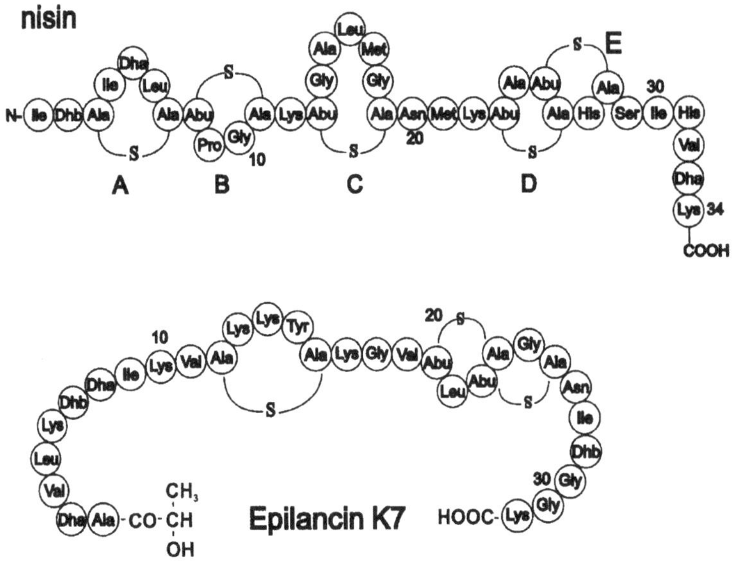

Figure 1. Schematic representation of two lantibiotics: Nisin A, Epilancin K7.

isms of membrane insertion and of pore formation and selectivity of class I and II bacteriocins.

Structure of bacteriocins

Bacteriocins are in general cationic (i.e., they contain an excess of lysyl and arginyl residues) amphipathic molecules composed of 12 to 45 amino acid residues. They are usually unstructured in aqueous solution, but have the propensity to form an α-helical structure when exposed to structure promoting solvents such as trifluoroethanol or when mixed with anionic phospholipid membranes. Some peptides form loop structures owing to a disulphide bridge or a covalent bond. In particular, the presence of intramolecular ring structures is a characteristic feature of lantibiotics (Class I). The structure of several lantibiotics has been elucidated by means of nuclear magnetic spin resonance (NMR) (Freund et al. 1991a,b; Van de Kamp et al. 1995a,b; Van den Hooven 1995). Nisin A has five ring structures termed A, B, C, D and E, of which the last two are intertwined (Figure 1). These ring structures are rigid but interconnected by flexible hinge regions. One of the hinge regions connects ring B and C while the other connects ring C and D. Due to the presence of these hinges, part of the molecule may insert into the membrane, while the other part remains bound to the membrane surface. In rings A, B and C the hydrophobic part, composed of the residues Ile4, Leu6, Pro9, Leu16 and Met17 are located opposite the Lys12 and the thioether bonds. Also the region from residue 21–28 is amphipathic and α-helical, especially in the presence of trifluoroethanol. Nisin A is amphipathic in two ways: First, the majority of the residues in the N-terminal half is hydrophobic and only a single charged residue is present, Lys12, whereas the C-terminal part contains most charged and hydrophilic residues. Secondly, both the first domain, containing ring A, B and C as well as the second domain, containing ring D and E, have a hydrophobic and a hydrophilic side (van den Hooven 1995).

In contrast to the lantibiotics, the three-dimensional structure of only one class IIa bacteriocin, leucocin A of 37 residues, has been solved (Fregeau Gallagher et al. 1997). In water or aqueous DMSO solution, leucocin A exists as a random coil. In lipophilic media two domains can be identified. First, a three stranded anti-parallel β-sheet domain of residue 2–16, linked by a disulphide bridge, and second an amphiphilic α-helix of residue 17–31. The α-helical part has been suggested to be responsible for the target specificity, whereas the β-sheet exerts the antimicrobial activity (Fregeau Gallagher et al. 1997). Interestingly, in analogy to the class I bacteriocins, also class II bacteriocins are amphipathic in two respects. The hydrophobicity of several class II bacteriocins increases gradually from the N-terminal to the C-terminal side of the molecule(s), while the hydrophobicity of the utmost C-terminal part

Table 1. Amino acid sequences of non-modified pore-forming (type A) lantibiotics (Class I). The amino-terminal pre-sequence is not indicated

Nisin A	----ITS--ISLCTPGC-KTG--ALMGCNMKTATCH----CSIHVSK-	Gross and Morell 1971
Nisin Z	----ITS--ISLCTPGC-KTG--ALMGCNMKTATCN----CSIHVSK-	Mulders 1991
Subtilin	----WKS--ESLCTPGC-VTG--ALQTCFLQTLTCN----CKISK---	Gross 1973
Gallidermin	----IAS--KFLCTPGCAKTG--SFN------SYC-----C-------	Kellner 1991
Epidermin	----IAS--KFICTPGCAKTG--SFN------SYC-----C-------	Allgaier 1986
Epilancin K7	------X--ASVLKTSIKVSK--KYCKG--VTLTCG----CNITGGK-	Van de Kamp 1995ab
Pep5	----TAGPAIRASVKQCQKT---LKATR-LFTVSCKGKNGCK------	Kellner 1991
Lactocin S	-----STPVLASVAVSMELLP--TASVLYSDVAGCF----KYSAKHHC	Mørtvedt 1991
Salivaricin A	---KRGSGWIATITDDCPNS-----------VFVC-----C-------	Ross 1993
Lacticin 481	---KGGSGVIHTISHECNMNS--WQF-----VFTC-----CS------	Piard 1993
Streptococcin A-FF22	----GKNGVFKXIXHEXHLNX---------WAFLA------------	Jack 1994
Cytolysin LL	TTVCAVAATAAASSAACGWVGGGIFTGVTVVVSLKH----C-------	Gillmore 1994
Cytolysin LS	-----------TTPACFTIG--LGVGALFSAKFC------------	Gillmore 1994

slightly decreases (Figure 2). This orientation is inverted in comparison to that suggested for nisin (Van den Hooven 1995). Secondly, most class II bacteriocins have a hydrophilic and a hydrophobic side perpendicularly to the N-terminal-to-C-terminal axis. Figure 3 shows an α-helical wheel presentation of the α and β peptide of lactococcin G (Nissen-Meyer et al. 1992). Such peptides often contain a helix-breaker such as a proline or glycine in the α-helical domain. The exact role of these residues is not known, but they may facilitate membrane insertion or wedge formation.

Membrane interaction

Gram-positive bacteria are characterised by a high content of anionic lipids in the membrane. Owing to the cationic character of bacteriocins, the role of anionic lipids in membrane binding has been emphasised. Nisin binds rapidly to anionic liposomes but not to zwitter-ionic liposomes (Driessen et al. 1995). This interaction is strong since nisin is able to diffuse only slowly to other liposomes. Fragments of nisin have been used to identify the regions that are involved in membrane interaction. An amino-terminal nisin fragment Nis^{1-12} only loosely interacts with the membrane (Moll et al. 1997). Mutagenesis of the charged amino acids in nisin has revealed that the positive charges in the carboxyl-terminal region are important for membrane binding (Demel et al. 1996; Breukink et al. 1997; Van Kraaij et al. 1997).

Bacteriocins of class II also depend on anionic phospholipids for the initial membrane interaction. Chen et al. (1998) showed specific binding of pediocin PA-1 (class IIa) to anionic phospholipids. The class IIa consensus motif, Tyr-Gly-Asn-Gly-Val-Xaa-Cys, neither determines the ability of pediocin PA-1 to bind to the membrane (Chen et al. 1997) nor does it specify the ability to kill *Listeria* species (Fleury et al. 1996). The capacity to inhibit growth of *L. ivanovii* by analogues of mesentericin Y105 has been analysed. Various regions of the mesentericin Y105 molecule appear essential for activity: the amino terminal Lys-Tyr-Tyr sequence, the presence of a disulphide loop, a helical segment in the sequence and the carboxyl-terminal tryptophan residue (Fleury et al. 1996).

Table 2 shows an alignment of the amino acid sequences of various class II peptides. Obviously, class IIa bacteriocins resemble each other. Classes IIb, c and d bacteriocins are only remotely homologous. The C-terminal region of some class II bacteriocins is rich in positively charged residues, e.g., lactococcin Gα (RKKKH), lactococcin A (HH), carnobactericin B2 (RRP), lactacin F (KIRK), pediocin PA-1 (HKC) and plantaricin J (RR). An analysis of chimers that consist of pediocin-like molecules, i.e, pediocin PA-1, sakacin P, curvacin A and leucocin A, indicates that the C-terminal part of these bacteriocins is responsible for the target specificity (Fimland et al. 1996). A C-terminal fragment of pediocin PA-1 specifically inhibits pediocin PA-1 activity (Fimland et al. 1998) suggesting that this fragment competes with the intact molecule for the same binding sites at the target membrane, or alternatively, associates with the intact molecule to yield an inactive oligomer. The C-

Table 2. Amino acid sequences of mature class II bacteriocins. Class IIA: bacteriocins with the consensus sequence Tyr-Gly-Asn-Gly-Val-Xaa-Cys and anti-Listeria activity. Class IIB: two-component class 2 bacteriocins. Class IIC: Sec-dependent bacteriocins. Class IID: other class II bacteriocins

CLASS IIA		
Bavaricin A	-------KYYGNGVHXGKHSXTVDWG--------TAIGNIGNNAAANXATGXNAGG------------------	Larsen 1993
Sakacin P	-------KYYGNGVHCGKHSCTVDWG--------TAIGNIGNNAAANXATGXNA--------------------	Tichaczek 1994
Sakacin 674	-------KYYGNGVHCGKHSCTVDWG--------TAIGNIGNNAAANWATGGNAGNK-----------------	Holck 1994
Piscicolin 126	-------KYYGNGVSCHKHGCTVDWS--------KAIGIIGNNAAANLTTGGAAGWNKG---------------	Jack 1996
Pediocin PA-1	-------KYYGNGVTCGKHSCSVDWG--------KATTCIINNGAMAWATGGHQGNHKC---------------	Marugg 1992
Leucocin A	-------KYYGNGVHCTKSGCSVNWG--------EAF----SAGVHRLANG---GNGFW----------------	Hastings 1991
Mesentericin Y105	-------KYYGNGVHCTKSGCSVNWG--------EAA----SAGIHRLANG---GNGFW----------------	Fleury 1996
Carnobactericin B2	-------VNYGNGVSCSKTKCSVNWG--------QAFQERYTAGINSFVSGVASGAGSIGRRP-----------	Quadri 1994
Sakacin A	------ARSYGNGVVCNNKKCWVNRG--------EAT----QSIIGGMISGWASGLAGM---------------	Holck 1992
Curvaticin A	------ACSYGNGVYCNNKKSWVNRG--------EAT----QXXIX---XG-----------------------	Tichaczek 1993
Bavaricin MN	------TKYYGNGVVX-NSKXWVDWG--------QAAGGIGQTVVXGWLGGAIPGK------------------	Kaiser 1996
CLASS IIB		
Thermophilin 1	-------YSGKDCLKD-MGGYALAGAGSGALWGAPAGG---VGALPGAFVGAHVGAIAGGFACMGGMIGNKFN-	Marciset 1997
Thermophilin 2	------QINWGSVVGHCIGGAIIGGA--------FSGG---AAAGVGCLVGSGKAIINGL--------------	
Acidocin J1132a	-------------NPK-VAHCASQIG---------------RSTAWGAVSGA----------------------	Tahara 1996
Acidocin J1132b	------------GNPK-VAHCASQIG---------------RSTAWGAVSGA----------------------	
Plantaricin J	------------------GAWKNFWS----------------SLRKGFYDGEAGRAIRR---------------	Diep 1996
Plantaricin K	-------------RRSRKNGIGYAIG------------Y--AFGAVERAVLGGSRDYNK---------------	
Lactococcin Ga1	--------GTWDDIGQGIGRVAYWVG--------KAMG---NMSDVNQASRINRKKKH----------------	Nissen-M. 1992
Lactococcin Gb	------KKWGWLAWVDPAYEFIKGFG---------------KGAIKEGNKDKWKNI------------------	
Lactobin A	-----NRWTNAYSAALGCAVPGVKYG---------KKLGGVWGAVIGGVGGAAVCGLAGYVRKG----------	Contreras 1997
Lactacin X	-------NRWGDTVLSAASGAGTGIK---------ACKSFGPWGMICGVGGAAIGGYFGYTHN-----------	Fremaux 1993/
Lactacin F	------RNNWQTNVGGAVGSAMIGAT---------VGG---TICGPACAVAGAHYLPILWTGVTAATGGFGKIRK-	Allison 1993
Plantaricin Sa	-------KKKKQSWYAAAGDAIVSFG---------------EGFLNAW--------------------------	Jiménez-D. 1995/
Plantaricin Sb	-----------------RNKLAYNMG---------------HYAGKATIFGLAAWALLA---------------	Stephens 1998
Plantaricin E	----------FNRGGYNFGKSVRHVD---------------AIGSVAGIRGILKSIR-----------------	Diep 1996
Plantaricin F	-------VFHAYSARGVRNNYKSAVG--------------PADWVISAVRGFIHG-------------------	

Table 2. Continued.

CLASS IIC		
Divergicin A	--------AAPKITQKQKMCVNGQLG---------GMLAGALGGPGGVVLGGIGGAIAGGCFN-----------	Worobo 1995
Enterocin P	-----ATRSYGNGVYCNNSKCWVNWG--------EAK----ENIAGIVISGWASGLAGMGH-------------	Cintas 1997
CLASS IID		
Carnobactericin A	-------DQMSDGVNYGKGSSLSKGG---------AKC---GLGIVGGLATIPSGPLGWLAGAAGVINSCMK--	Worobo 1994
Enterocin B	----ENDHRMPNELNR--PNNLSKGG---------AKC---GAAIAGGLFGIPKGPLAWAAGLANVYSKCN---	Casaus 1997
Curvaticin FS47	-------YTAKQCLQA-IGSCGIATG---------AGA---AGGPAGAFVGAXVXI------------------	Garver 1994
Lactococcin A	KLTFIQSTAAGDLYYNNTWTHKVYQQT------QNAFGAAANTIVNGWMGGAAGGFGLHH--------------	Holo 1991
Acidocin 8912	-----------------KTHYPTNAX---------------KSLRKGFXESLRXTD------------------	Kanatani
Lactococcin B	------SLQYVMSAGPYTWYKDTRTG---------KTICKQTIDTASYTFGVMAEGWGKTFH------------	van Belkum 1992
Plantaricin A	----------------KSSAYSLQMG---------------ATAIKQVKKLFKKWGW-----------------	Nissen-M. 1993

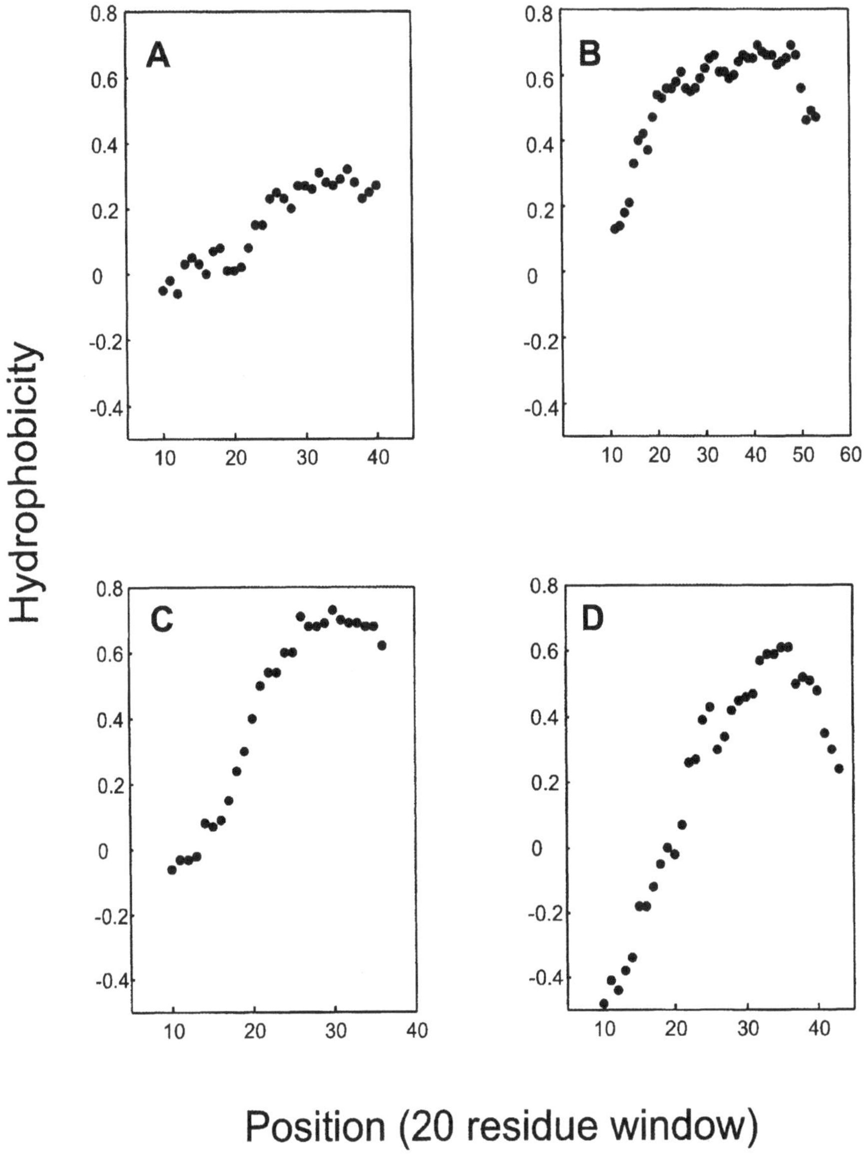

Figure 2. Hydrophobicity analysis of bacteriocins. (A): Averaged hydrophobicity profile of a group of aligned class IIA bacteriocins. (B): thermophilin A (class IIB). (C): divergicin A (class IIC). (D): enterocin B (class IID). The hydrophobicity was calculated with a 20 residue window using the hydrophobicity scale of Eisenberg et al.(1984).

terminus of bavaricin MN is suspected to be essential for membrane interaction (Kaiser & Montville 1996).

A strong membrane interaction was also observed for the two-peptide class IIb bacteriocin lactococcin G. This bacteriocin consists of two amphipathic α-helical peptides, α and β. The individual peptides are capable of binding independently to the membrane, but both peptides need to be present for activity. It also appears that there is no diffusion from cell to cell, once the peptides are bound by the membrane surface of the target cells. Lactococcin G interacts with liposomes (Hauge et al. 1998b) but forms pores only with intact cells (Moll et al. 1996a).

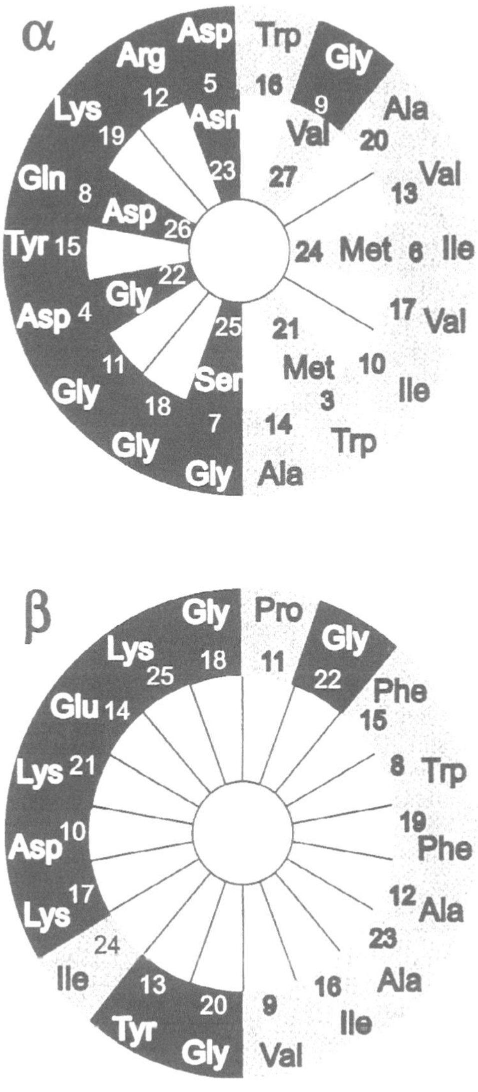

Figure 3. Helical wheel presentation of the α and β peptides of lactococcin G (Nissen-Meyer et al. 1992).

Membrane insertion

Insertion of lantibiotics into lipid monolayers has been observed. Demel et al. (1996) demonstrated that the ability of nisin to interact with anionic lipids correlates with its antimicrobial activity. Amino-terminally extended nisin Z, ITPQ-nisin Z, shows a 3-5 fold reduced monolayer insertion capacity that is paralleled by a lowered antimicrobial activity. The replacement of the amino-terminal isoleucine by a bulky tryptophan residue decreases the ability of nisin to penetrate into the membrane. On the basis of these experiments, it appears that the amino-terminal part of the nisin molecule inserts into monolayers. Membrane insertion and activity is also reduced when the flexibility of the hinge region is reduced that connects ring C and D (see Figure 1). Nisin Z insertion is maximal in monolayers composed of 70% of the anionic phosphatidylglycerol. The lantibiotic lacticin 481 also inserts more efficiently into a monolayer of anionic phospholipids than of zwitter-ionic phospholipids, but the phospholipid preference appears less dramatic than found for nisin Z.

Cell wall passage and receptors

Type A lantibiotics are able to form ion-conducting pores in black lipid membranes. The lantibiotics nisin and epidermin, but not Pep 5 or epilancin K7, interact with the peptidoglycan precursor, lipid II. In black lipid membranes, the presence of lipid II enhances the ability of nisin to depolarise the transmembrane electrical potential, $\Delta\psi$. The lifetime of pores increases from milliseconds to about 6 seconds in the presence of lipid II. Similarly, the apparent size of the pore increased from about 1 to 2–2.5 nm (Wiedeman et al. 1998). Cells that were depleted of lipid II by preincubation with the lipopeptide ramoplanin, are less sensitive to nisin and epidermin (Brötz et al. 1998). These data suggest that lipid II acts as a kind of docking molecule, or as nucleus of pore formation. Similarly, carnocin UI49 might require a 'docking molecule' for activity. The high sensitivity of nisin producing strains to carnocin UI49 was suggested to be related to a protein, NisP (Stoffels et al. 1994). NisP is the nisin leader peptidase (Van der Meer et al. 1993), but it is unclear how it would enhance the sensitivity to carnocin UI49.

The cell wall of Gram-positive bacteria allows passage of bacteriocins. Anionic cell surface polymers like teichoic acid and lipoteichoic acid have been suggested to play a role in the initial interaction with cationic bacteriocins (Jack et al. 1995). Various lines of evidence suggest that such interactions may be important for the majority of class II bacteriocins, for which 'receptors' have been implicated to explain their narrow target specificity (Van Belkum et al. 1991). Many of these bacteriocins are unable to form pores in liposomes (Zajdel et al. 1985; Moll et al. 1996a; Moll et al. 1999). For instance, lactostrepcin 5 is inactive against protoplasts, and the activity is reduced 10-fold when the cells are pre-treated with typsin. This suggests the involvement of a proteinaceous factor at the cell surface (Zajdel et al. 1985). Involvement of a receptor-like factor has also been implicated for pediocin PA-1 (Chikindas et al. 1993) and leucocin A (Fregeau Gallagher et al. 1997).

Membrane insertion and the proton motive force

The lantibiotic nisin disturbs the lipid bilayer organisation when it binds to the membrane (Driessen et al. 1995). In equilibrium, nisin orients parallel to the surface of the membrane (Breukink et al. 1998). Membrane insertion is promoted by a $\Delta\psi$, trans negative, or by the transmembrane pH gradient (ΔpH), cis acid (Moll et al. 1996b, 1997). The imposition of a $\Delta\psi$, promotes the penetration of epilancin K7 into the hydrophobic carbon region of the lipid bilayer (Driessen et al. 1995). Also in cells, the $\Delta\psi$ (Jung & Sahl 1991; Moll et al. 1996b) and ΔpH (Moll et al. 1997) stimulate membrane insertion and pore formation. However, even in the absence of a $\Delta\psi$, nisin efficiently inhibits cell growth (Moll 1998a). At nisin concentrations below 400 nM, ΔpH induced pore formation depends on the magnitude of the ΔpH (Moll et al. 1997). Membrane insertion of lacticin 3147, a two-component lantibiotic appears to be enhanced by the $\Delta\psi$ (McAuliffe et al. 1998).

Less information is available on the role of an energised membrane on the insertion of IIa bacteriocins. Even in the absence of $\Delta\psi$, pediocin PA-1 (Chikindas et al. 1993) and lactococcin A (Van Belkum et al. 1991) are able to form pores in the membrane. Pediocin PA-1 (Chen et al. 1997b) and bavaricin MN (Kaiser et al. 1996) elicit carboxyfluorescein efflux from liposomes, and this process is stimulated by the $\Delta\psi$. Several class IIb and IIc bacteriocins do not need a proton motive force for activity, e.g. lactococcin G (Moll et al. 1996a), lactacin F (Bruno & Montville 1993) and lactococcin B (Venema et al. 1993).

Models for pore formation

Various models for pore-formation have been proposed during the years. For the lantibiotic nisin, pore-formation is thought to occur through a series of distinct steps. In equilibrium, nisin seems to orient parallel to the surface of the membrane (Breukink et al. 1998). Both the $\Delta\psi$ (trans negative) and the ΔpH (cis acid) are able to induce pore formation, possibly by driving the membrane insertion of nisin domains. The C-terminus inserts deeply into the membrane (Martin et al. 1996; Winkowsky et al. 1996) and it has even been suggested that the entire nisin translocates across the membrane (Van Kraaij et al. 1998). Nisin induces the transmembrane movement of fluorescent phospholipid (Moll et al. 1998c) suggesting that the membrane insertion of the C-terminus of nisin causes intermonolayer contact of phospholipids. A wedge-like model for nisin-induced pore formation (Driessen et al. 1995) may involve a proton motive force driven co-insertion of lipids and nisin domains (Figure 4A). The hinge(s) in the nisin molecule might allow bending of the C-terminal part and thus its insertion into the membrane. Multiple inserted nisin molecules may give rise to a large local disturbance of the lipid bilayer organisation causing formation of transient lipid-protein pores. Such structures are intrinsically unstable due to the hydrophobic forces that will drive the re-arrangements of the lipids into their original bilayer organisation.

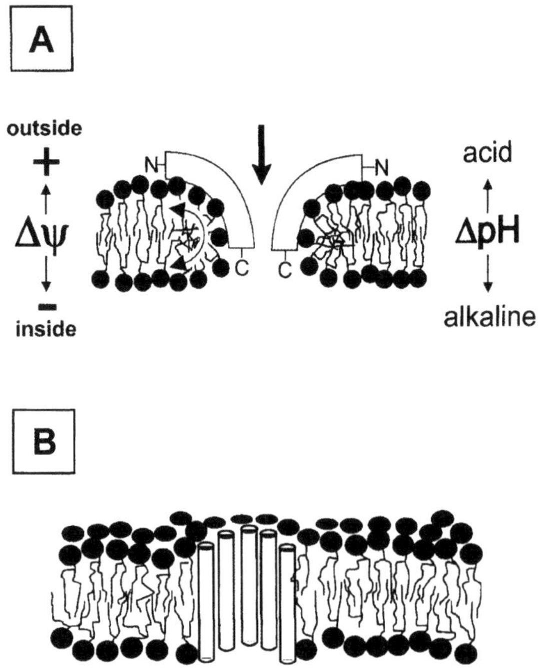

Figure 4. Models of pore formation. (A): Wedge-like pore. (B): Barrel-stave pore.

Upon membrane insertion, the class II bacteriocins are thought to form a bundle of α-helical peptides much akin a barrel-stave like pore (Figure 4B). The presence of a helix-breaking amino acid residue in the middle of their sequence may facilitate the insertion of the peptide into the membrane from an initial surface bound state. According to the barrel-stave model, the hydrophilic faces of a bundle of amphipathic α-helical peptides form the inner wall of the water-filled pore. The outer, hydrophobic side of these helical bundles will face the fatty acyl chains of the membrane lipids. Alternatively, a 'carpet-like' model could explain peptide-induced pore formation. According to this model, single peptide molecules might be oriented parallel to the membrane surface and interfere with the membrane bilayer organisation without forming a peptide aggregate. Once there are sufficient peptides in each other's vicinity, the membrane will temporarily collapse due to a strong phospholipid mobilising activity that co-operatively results in a local and transient permeability (Bechinger 1997). The negative charge of the membrane lipids has been suggested to confer cation selectivity to such pores (Homblé et al. 1998).

A pore model has been proposed for thermophilin 13 (class 2b, Table 2). It was suggested that thermophilin A would form two transmembrane α-helices. By contrast three β-turns were predicted for thermophilin B at distances that would allow the sequences in between to span the membrane as antiparallel β-sheets. Thermophilin A may form a transmembrane pore that is stabilised – in a one to one ratio – by the perpendicularly oriented transmembrane and β-sheet structured portion of thermophilin B. In this respect, thermophilin A is active on its own, but its activity is enhanced by thermophilin B (Marciset el al. 1997). Other two-component class IIB bacteriocins, for instance lactococcin G, plantaricin EF and plantaricin JK differ from thermophilin both structurally and functionally. These peptides are α-helical in the presence of membrane-mimicking solvents (Hauge et al. 1998b, 1999). For the pore-forming activity, both components are strictly required. These pores show a strong ion-selectivity that seems different for the various bacteriocins.

Pore specificity

Lantibiotics are commonly described as to form large non-specific pores. They mediate efflux of ions, amino acids, and ATP from cells (Sahl et al. 1995), and allow efflux of ions and carboxyfluorescein from (proteo)liposomes (Moll et al. 1996b). Carboxyfluorescein efflux is faster than potassium efflux suggesting that there may be some anion preference (Breukink et al. 1997). In black lipid membranes the estimated pore size is around 1 nm at low (0.1–0.5 μM) concentrations of nisin and Pep5, while subtilin may form a pore with a diameter of 2 nm (Benz et al. 1991). These calculated pore sizes, however depend on both the lantibiotic concentration and on the pH. Black lipid membrane studies reveal little ion-selectivity. The pore lifetime of most lantibiotics is usually in the order of milliseconds, whereas gallidermin may form pores with a lifetime of up to 30 s. Little is known about pore specificity of two-component lantibiotics. Interestingly, the two-component lantibiotic lacticin 3147 causes an increase in the ΔpH, but dissipates the $\Delta\psi$, which indicates a strong ion selectivity that excludes protons as permeating species (Mc Auliffe et al. 1988).

The capacity of nisin to dissipate the $\Delta\psi$ in the absence of a ΔpH dramatically increases when the extracellular pH is lowered from 8 to 6 (Moll et al. 1997). Also the ability of nisin to cause efflux of carboxyfluorescein from liposomes strongly diminishes when the pH is lowered (Winlowsky et al. 1996). In contrast, nisin is equally effective in dissipating the ΔpH in the absence of a $\Delta\psi$, within the pH range of pH 6 to 7 (Moll et al. 1997). Taken together, these data point at a change in pore specificity as a function of pH and suggest that under physiological conditions the ΔpH might be an important target for nisin action.

Class IIa bacteriocins dissipate the proton motive force (e.g. Bruno & Montville 1993; Hauge et al. 1998a), while pediocin PA-1 at high concentration permits efflux of relatively large molecules (Chikindas et al. 1993). Two peptide bacteriocins (class IIb) seem to form relatively specific pores. All two component bacteriocins are known to dissipate the $\Delta\psi$. On the basis of their ion-selectivity, these peptides can be classified in two subgroups: (1) monovalent cation conducting bacteriocins as lactococcin G (Moll et al. 1998b) and plantaricin EF (Moll et al. 1999). In contrast to plantaricin EF, whose action causes the dissipation of the ΔpH (Moll et al. 1999), lactococcin G does not conduct proton-movement; and (2) anion conducting bacteriocins like plantaricin JK (Moll et al. 1999) and possibly acidocin J1132 (Tahara et al. 1996). Plantaricin JK permits efflux of anions like glutamate, but it does not elicit the release of potassium ions from the cell. Acidocin J1132 also causes glutamate efflux. In contrast to the above bacteriocins, lactacin F seems induce efflux of both potassium ions and phosphate (Abee et al. 1994).

Concomitantly with a dissipation of the $\Delta\psi$, lactococcin G and plantaricin EF cause the ΔpH to increase. Subsequently, ΔpH is either rapidly dissipated (plantaricin EF) or not affected at all (lactococcin G). Plantaricin JK, acidocin J1132 and thermophilin 13 (Marciset et al. 1997) immediately dissipate the ΔpH without an initial transient increase. This suggests that the former group of bacteriocins does not conduct the transmembrane movement of protons or hydroxy ions, and that their ability to collapse the entire proton motive force stems from indirect effects.

Lactococcin B (class IIc) dissipates the proton motive force in sensitive cells, but about 150-fold higher concentrations are required for efflux of glutamate (Venema et al. 1993). This indicates that lactococcin B mediates ion-movements, but that it does not form large and non-specific pores.

How do pores cause growth inhibition?

How does the bacteriocin mediated transmembrane ion flow lead to growth inhibition? Often the dissipation of the $\Delta\psi$ is taken to explain growth inhibition. However, when the $\Delta\psi$ is dissipated by the addition of both potassium ions and the ionophore valinomycin to the growth medium, only little growth inhibition is observed. A proper interpretation of the latter result is complicated by the fact that multidrug resistance proteins may excrete ionophores thereby rendering them inactive (Bolhuis et al. 1996). However, dissipation of the ΔpH by nigericin results in a rapid and complete growth inhibition of *Lactococcus lactis* and other lactic acid bacteria (Moll et al. 1998a). Strong cytotoxic effects presumably result from proton influx. This leads to a drop of the intracellular pH and consequently many enzymatic processes will become inhibited. Growth inhibition may also be the consequence of influx of cytotoxic sodium ions (Cheng et al. 1997) and be accelerated by an osmotic imbalance of the turgor pressure of the cell. Pores in the cytoplasmic membrane clearly affect the energetic status of the cell, i.e., dissipation of the proton motive force causes an arrest of ΔpH- and $\Delta\psi$ dependent (transport) processes. Some bacteriocins directly affect the cellular energy by causing efflux of ATP (Zajdel et al. 1985). Dissipation of ion gradients may lead to futile cycles and concomitantly to the depletion of ATP. The acceleration of the hydrolysis of ATP may lead to an exhaustion of high energy phosphorylated intermediates. Most interestingly, in the case of nisin indirect ATP depletion by the induction of hydrolysis appears more prominent than ATP efflux (Winkowski et al. 1994).

Resistance and immunity

Bacteriocin producing cells are protected against their own bacteriocin as long as they are present in the – inactive – precursor form. The leader sequences of bacteriocins that direct their excretion often consist of 18 to 30 amino acid residues. The final step of bacteriocin biosynthesis is secretion of the precursor. Often after secretion, helix breaking residues at the positions −2 and/or −1 of the cleavage site may serve as recognition sites for the leader peptidase that cleaves this bond to release the mature and active peptide in the medium (Jack et al. 1996). Bacteriocin secretion systems are thought to contribute to self protection or immunity of the bacteriocin producing strain. For instance, cells expressing the *nisEFG* genes that encode a putative ABC transporter that is involved in the secretion of the nisin, exhibit an increased nisin-resistance (Koponen et al. 1998). Such a self protecting mechanism has also been observed for epidermin producing cells (Otto et al. 1998). These ABC transporters use ATP to secrete the bacteriocin across the membrane. Interestingly, colicin V is secreted by an export system, which has been termed 'MDR like' on the basis of one of the encoding nucleotide sequences (Gilson et al. 1990). It is not known to what extent such bacteriocin export systems confer resistance to a broader range of toxic substances except for the exported bacteriocin. Some of the ABC transporters excreting the bacteriocins harbour a covalently linked cytosolic domain that functions as the leader peptidase (Håverstein et al. 1995). The proteolytic domain resides in the N-terminal part of these proteins that carry out a dual function, i.e., the removal of the leader peptide from its substrate, and the subsequent translocation of the bacteriocin across the cytoplasmic membrane. Thereby they efficiently prevent that the mature and active bacteriocin remains in the cytosol.

Resistance to nisin has also been correlated to an altered fatty acid and phospholipid composition (Abee et al. 1994; Mazzota & Montville 1997; Verheul et al. 1997). Crandell & Montville (1998) found a decrease in the anionic lipid content of a *Listeria* strain that showed an elevated resistance to nisin (class Ia) and pediocin PA-1 (class IIa). In yeast, a specific cell wall protein was found to confer an increased resistance against high nisin concentrations (Dielbandhoesing et al. 1998). The bacteriocin encoding operons usually also contain a gene coding for an immunity protein. These proteins confer the producing cells with a high resistance to the bacteriocins that are secreted by the cells. For instance, nisin producing bacteria protect themselves against nisin not only by the *nisEFG* transport system but also by the product of the *nisI* gene (Kuipers et al. 1993). NisI is a lipoprotein that is both present intracellularly and on the outer surface of the cytoplasmic membrane. NisI is most likely a nisin binding protein. Interestingly, NisI stimulates nisin activity on sensitive cells suggesting a dual function as an immunity and an enhancing factor (Koponen et al. 1998). Immunity genes are known that are not directly associated with the bacteriocin gene cluster (Brurberg et al. 1997; Eijsink et al. 1998). The mechanism of action of immunity proteins is not understood. They could shield a receptor, prevent pore formation by binding the bacteriocins or even block the pores. In conjunction with cellular proteinases, they may also facilitate degradation of the bacteriocin to prevent it from acting against the producer cells. Class IIa immunity proteins resemble each other to some extent. The immunity protein of a sakacin P producer confers resistance to pediocin PA-1, curvacin A, sakacin P and enterocin A (Eijsink et al. 1998). Interestingly, the lactococcin A immunity factor (Nissen Meyer et al. 1993) has been suggested to interact directly with the lactococcin A receptor (Venema et al. 1994).

Prospects

The observation that the activity of nisin can be enhanced by the peptidoglycan precursor lipid II, indicates that the target specificity of bacteriocins may be defined by more complicated factors than originally anticipated. The dramatic enhancement of pore lifetime in the presence of lipid II opens the possibility to stabilise the pore with the prospect of detailed structural and functional analysis. For class II bacteriocins, the target cell specificity seems to depend on both structural domains within the bacteriocins and on receptors-like molecules in the target cell. The nature of such 'receptors' remains to be elucidated. A detailed analyses of these specificity determinants may permit the rational design of bacteriocins with desired target spectrum and/or ion specificity. The combinatorial treatment of target cells with bacteriocins of complementary ion specificity provides new leads to realise rapid cell death.

Acknowledgement

Dr. J. Lolkema is acknowledged for the use of the MEM Gen program for hydrophobicity analyses.

References

Abee T, Klaenhammer T-R & Letellier L (1994) Kinetic studies of the action of lactacin F, a bacteriocin produced by *Lactobacillus johnsonii* that forms poration complexes in the cytoplasmic membrane. Appl. Environ. Microbiol. 60: 1006–1013

Abee T, Rombouts FM, Hugenholz J, Guihard G & Letellier L (1994) Mode of action of nisin Z against *Listeria monocytogenes* Scott A grown at high and low temperatures. Appl. Environ. Microbiol. 60: 962–1978

Allgaer H, Jung G, Werner RG Schneider U & Zähner H (1986) Epidermin: sequencing a heterodetic tetracyclic 21-peptide amide antibiotic. Eur. J. Biochem. 160: 9–22

Allison GE, Fremaux C & Klaenhammer TR (1994) Expansion of bacteriocin activity and host range complementation of two peptides encoded within the lactacin F operon. J. Bacteriol. 176: 2235–2241

Bechinger B (1997) Structure and function of channel-forming peptides: magainins, cecropins, melittin and alamethicin. J. Membr. Biol. 156: 197–211

Benz R, Jung G & Sahl H-G (1991) Mechanism of channel formation by lantibiotics in black lipid membranes. In: Jung G & Sahl H-G (Eds) Nisin and novel lantibiotics (pp 359–372). Escom, Leiden

Bolhuis H, Molenaar D, van Veen HW, Poolman B, Driessen AJM & Konings WN (1996) Multidrug resistance in *Lactococcus lactis*: evidence for ATP-dependent drug extrusion from the inner leaflet of the cytoplasmic membrane. EMBO J. 15: 4239–4245

Breukink E, van Kraaij C, Demel RA, Siezen RJ, de Kruijff B & Kuipers OP (1998) The orientation of nisin in membranes. Biochemistry 37: 8153–8162

Breukink E, van Kraaaij C, Demel RA, Siezen RJ, Kuipers OP & de Kruijff B (1997) The C-terminal region of nisin is responsible for the initial interaction of nisin with the target membrane. Biochemistry 36: 6968–6976

Brötz H, Josten M, Wiedmann I, Schneider U, Gotz F, Bierbaum G & Sahl H-G (1998) Role of lipid-bound peptidoglycan precursors in the formation of pores by nisin, epidermin and other lantibiotics. Mol. Microbiol. 30: 317–327

Bruno, MEC & Montville TJ (1993) Common mechanistic action of bacteriocins form lactic acid bacteria. Appl. Environ. Microbiol. 1993: 3003–3010

Bruno MEC, Kaiser A & Montville TJ (1992) Depletion of proton motive force by nisin in *Listeria monocytogenes* cells. Appl. Environ. Microbiol. 58: 2255–2259

Brurberg MB, Nes IF & Eijsink VGH (1997) Pheromone-induced production of antimicrobial peptides in *Lactobacillus*. Mol. Microbiol. 26: 347–360

Casaus P, Nilsen T, Cintas LM, Nes IF, Hernandez PE & Holo H (1997) Enterocin B, a new bacteriocin from *Enterococcus faecium* T136 which can act synergistically with enterocin A. Microbiology 143: 2287–2294

Chen Y, Ludescher RD & Montville TJ (1998) Influence of lipid composition on pediocin PA-1 binding to phospholipid vesicles. Appl. Environ. Microbiol. 64: 3530–3532

Chen Y, Ludescher RD & Montville (1997a) Electrostatic interactions but not the YGNGV consensus motif, govern the binding of pediocin PA-1 and its fragments to phospholipid vesicles. Appl. Environ. Microbiol. 63: 4770–4777

Chen Y, Shapira R, Eisenstein M & Montville TJ (1997b) Functional characterization of pediocin PA-1 binding to liposomes in the absence of a protein receptor and its relationship to a predicted tertiairy structure. Appl. Environ. Microbiol. 63: 524–531

Cheng J, Guffanti AA & Krulwich TA (1997) A two-gene ABC-type transport system that extrudes Na^+ in *Bacillus subtilis* is induced by ethanol or protonophore. Mol. Microbiol. 23: 1107–1120

Chikindas ML, García-Garcerá ML, Driessen AJM, Ledeboer AM, Nissen-Meyer J, Nes IF, Abee T, Konings WN & Venema G (1993) Pediocin PA-1, a bacteriocin from *Pediococcus acidilactici* PAC1.0, forms hydrophilic pores in the cytoplasmic membrane of target cells. Appl. Environ. Microbiol. 59: 3577–3584

Cintas LM, Casaus LS, Håverstein LS, Hernandez PE & Nes IF (1997) Biochemical and genetic characterization of enterocin P, a novel Sec-dependent bacteriocin from *Enterococcus faecium* P13 with a broad antimicrobial spectrum. Appl. Environ. Microbiol. 63: 4321–4330

Contreras BGL, de Vuyst L, Devreese B, Busanyova K, Raymaekers J, Bosman F, Sablon E & Vandamme EJ (1997) Isolation, purification and amino acid sequence of lactobin A, one of the two bacteriocins produced by *Lactobacillus amylovorus* LMG P-13139. Appl. Environ. Microbiol. 63: 13–20

Crandell D & Montville TJ (1998) Nisin resistance in *Listeria monocytogenes* ATCC 700302 is a complex phenotype. Appl. Environ. Microbiol. 64: 231–237

Delves-Broughton J, Blackburn P, Evans RJ & Hugenholtz J (1996) Applications of the bacteriocin, nisin. Antonie van Leeuwenhoek 69: 193–202

Demel RA, Peelen T, Siezen RJ, de Kruijff B & Kuipers OP (1996) Nisin Z, mutant nisin Z and lacticin 481 interactions with anionic lipids correlate with antimicrobial activity. A monolayer study. Eur. J. Biochem. 235: 267–274

Dielbandhoesing SK, Zhang H, Caro LHP, van der Vaart JM, Klis FM, Verrips CT & Brul S (1998) Specific cell wall proteins confer resistance to nisin upon yeast cells. Appl. Environ. Microbiol. 64: 4047–4052

Diep DB, Håverstein LS & Nes IF (1996) Characterization of the locus responsible for the bacteriocin production in *Lactobacillus plantarum* C11. J. Bacteriol. 178: 4472–4483

Driessen AJM, van den Hooven HW, Kuiper W, van de Kamp M, Sahl H-G, Konings RNH & Konings WN (1995) Mechanistic studies of lantibiotic-induced permeabilization of phospholipid vesicles. Biochemistry 34: 1606–1614

Eijsink VGH, Skeie M, Middelhove H, Brurberg MB & Nes IF (1998) Comparative studies of class IIa bacteriocins of lactic acid bacteria. Appl. Environ. Microbiol. 64: 3275–3281

Eisenberg D (1984) Three-dimensional structure of membrane and surface proteins. Ann. Rev. Biochem. 53: 595–623

Fimland G, Blingsmo OR, Sletten K, Jung G, Nes IF & Nissen-Meyer J (1996) New biologically active hybrid bacteriocins constructed by combining regions from various pediocin-like bacteriocins: the C-terminal region is important for determining specificity. Appl. Environ. Microbiol. 62: 3313–3318

Fimland G, Jack R, Jung G, Nes IF & Nissen-Meyer J (1998) The bactericidal activity of pediocin PA-1 is specifically inhibited by a 15-mer fragment that spans the bacteriocin from the center towards the C-terminus. Appl. Environ. Microbiol. 64: 5057–5060

Fleury Y, Dayem MA, Montagne JJ, Chaboisseau E, Lecaer JP, Nicolas P & Delfour A (1996) Covalent structure, synthesis, and structure-function studies of mesentericin Y 105^{37}, a defensive peptide from Gram-positive bacteria *Leuconostoc mesentoroides*. J. Biol. Chem 271: 14421–14429

Fregeau Gallagher NL, Sailer M, Niemczura WP, Nakashima TT, Stiles ME & Vederas JC (1997) Three-dimensional structure of leucocin A in trifluoroethanol and dodecylphosphocholine micelles: spatial location of residues critical for biological activity in type IIa bacteriocins from lactic acid bacteria. Biochemistry 36: 15062–15072

Fremaux C, Ahn C & Klaenhammer TR (1993) Molecular analysis of the lactacin F operon. Appl. Environ. Microbiol. 59: 3906–3915

Freund S, Jung G, Gutbrod O, Folkers G & Gibbons WA (1991a) The three dimensional solution structure of gallidermin determined by NMR-based molecular graphics. In: Jung G & Sahl H-G (Eds) Nisin and novel lantibiotics (pp 91–103). Escom publishers, Leiden, The Netherlands

Freund S, Jung G, Gibbons WA & H-G-Sahl (1991b) NMR and circular dichroism studies on Pep5. In: Jung G & Sahl H-G (Eds) Nisin and novel lantibiotics (pp 103–113). Escom publishers, Leiden, The Netherlands

Garver KI & Muriana PM (1994) Purification and partial amino acid sequence of Curvaticin FS47, a heat-stable bacteriocin produced by *Lactobacillus curvatus* FS47. Appl. Environ. Microbiol. 60: 2191–2195

Gillmore MS, Segarra RA, Booth MC, Bogie CP, Hall LR & Clewell DB (1994) Genetic structure of the *Enterococcus faecalis* plasmid pAS1-encoded cytolitic toxin system and its relationship to lantibiotic determinants. J. Bacteriol. 176: 7355–7344

Gilson L, Mahanty HK & Kolter R (1990) Genetic analysis of an MDR-like export system: the secretion of colicin V. EMBO J. 9: 3875–3884

Gross E & Morell JL (1971) The structure of nisin. J. Am. Chem. Soc. 93: 4634–4635

Gross E, Klitz HH & Nebelin E (1973) Die structur des subtilins. Hoppe Seyler's Z. Physiol. Chem. 354: 810–812

Hastings JW, Sailor M, Johnson K, Roy L, Vederas JC, Stiles ME (1991) Characterization of leucocin A-UAL 187 and cloning of the bacteriocin gene from *Leuconostoc gelidum*. J. Bacteriol. 173: 7491–7500

Hauge HH, Mantzilas D, Moll GN, Konings WN, Driessen AJM, Eijsink VGH & Nissen-Meyer J (1998a) Plantaricin A is an amphiphilic α-helical bacteriocin-like pheromone which exerts antimicrobial and pheromone activities through different mechanisms. Biochemistry 37: 16026–16032

Hauge HH, Nissen-Meyer J, Nes IF & Eijsink VGH (1998b) Amphiphilic α-helices are important structural motifs in the α and β peptides that constitute the bacteriocin lactococcin G. Eur. J. Biochem. 251: 565–572

Hauge HH, Mantzilas D, Eijsink VGH & Nissen-Meyer J (1999) membrane-mimicking entities induce structuring of the two-peptide bacteriocins plantaricin E/F and plantaricin J/K. J. Bacteriol. 181: 740–747

Håverstein LS, Diep DB & Nes IF (1995) A family of bacteriocin ABC transporters carries out proteolytic processing of their substrates concomitant with export. Mol Microbiol 16: 229–240

Hill C, Meaney B & Ross RP (1998) Exploitation of the novel two-component lantibiotic, lacticin 3147: control of mastitis & food uses. Third international workshop on lantibiotics. Blaubeuren, Germany, April 5–8, p 39

Holck A, Axelsson L, Birkeland SE, Aukrust T & Blom H (1992) Purification and amino acid sequence of sakacin A, a bacterocin from *Lactobacillus sake* Lb706. J. Gen. Microbiol. 138: 2715–272

Holck AL, Axelsson L, Huhne K & Krockel L (1994) Purification and cloning of sakacin 674, a bacteriocin from *Lactobacillus sake* Lb674. FEMS Microbiol. Lett. 115: 143–149

Holo H, Nilssen O & Nes IF (1991) Lactococcin A, a new bacteriocin from *Lactococcus lactis* subsp. *cremoris*: isolation and characterization of the protein and its gene. J. Bacteriol. 173: 3879–3887

Homblé F, Cabiaux V & Ruyssschaert J-M (1998) Channel or channel-like activities associated with pore forming proteins or peptides? Mol. Microbiol. 27: 1261–1263

Jack RW, Carne A, Metzger J, Stefanovitc S, Sahl H-G, Jung G & Tagg JR (1994) Elucidation of the structure of SA-FF22, a lanthionine-containing antibacterial peptide produced by *Streptococcus pyogenes* strain FF22. Eur. J. Biochem. 220: 455–462

Jack RW, Tagg JR & Ray B (1995) Bacteriocins of Gram-positive bacteria. Microbiol. Rev. 59: 171–200

Jack RW, Wan J, Gordon J, Harmark K, Davidson BE, Hillier AJ, Wettenhall REH, Hickey MW & Coventry MJ (1996) Characterization of the chemical and antimicrobial properties of piscicolin 126, a bacteriocin produced by *carnobacterium piscicola* JG 126. Appl. Environ. Microbiol. 62: 2897–2903.

Jiménez-Díaz R, Ruiz-Barba JL, Cathcart DP, Holo H, Nes IF, Sletten KH & Warner PJ (1995) Purification and partial amino acid sequence of plantaricin S, a bacteriocin produced by *Lactobacillus plantarum* LPCO10, the activity of which depends on the complementary action of two peptides. Appl. Environ. Microbiol. 61: 4459–4463

Jung G & Sahl H-G (1991) Nisin and novel lantibiotics. Escom Leiden, The Netherlands

Kaiser AL & Montville TJ (1996) Purification of the bacteriocin bavaricin MN and characterization of its mode of action against *Listeria monocytogenes* Scott A cells and lipid vesicles. Appl. Environ. Microbiol. 62: 4529–4535

Kanatani K, Tahara T, Oshimura M, Sano K & Umezawa C (1995) Cloning and nucleotide sequence of the gene for acidocin 8912, a bacteriocin from *Lactobacillus acidophilus* TK 8912. Lett. Appl. Microbiol. 21: 384–386

Kellner R, Jung G, Horner T, Zahner H, Schnell N, Entian K-D & Gotz F (1988) Gallidermin: a new lanthionine-containing polypeptide antibiotic. Eur. J. Biochem. 177: 53–59

Kellner R, Jung G & Sahl H-G (1991) Structure elucidation of the tricyclic lantibiotic Pep5 containing eight positively charged amino acids. In: Jung G & Sahl H-G (Eds) Nisin and novel lantibiotics (pp 141–1581). Escom publishers, Leiden, The Netherlands

Klaenhammer TR (1993) Genetics of bacteriocins produced by lactic acid bacteria. FEMS Microbiol. Rev. 12: 39–86

Koponen O, Takala TM, Kilpeläinen N & Saris PE (1998) Dual function of NisI, an immunity lipoprotein at the membrane surface and a factor stimulating nisin activity in the soluble form. Third International Workshop on Lantibiotics, Blaubeuren, p 29

Kuipers OP, Beerthuyzen MM, Siezen RJ & de Vos WM (1993) Characterization of the nisin gene cluster *nisABTCIPR* of *Lactococcus lactis*. Requirement of expression of the *nisA* and *nisI* genes for the development of immunity. Eur. J. Biochem. 216: 281–291

Larsen AG & Norrung B (1993) Inhibition of *Listeria monocytogenes* by Bavaricin A, a bacteriocin produced by *Lactobacillus bavaricus* MI401. Lett. Appl. Microbiol. 17: 132–134

Leer RJ, van der Vossen JMBM, van der Giezen M, van Noort JM & Pouwels PH (1995) Genetic analysis of acidocin B, a novel bacteriocin produced by *Lactobacillus acidophilus*. Microbiology 141: 1629–1635

Marciset O, Jeronimus-Stratingh MC, Mollet B & Poolman B (1997) Thermophilin 13, a nontypical antilisterial poration complex bacteriocin that functions without a receptor. J. Biol. Chem. 22: 14277–14284

Martin LJM, Ruysschaert JM, Sanders D & Giffard C (1996) Interaction of the lantibiotic nisin with membranes revealed by fluorescence quenching of an introduced typtophan. Eur. J. Biochem. 239: 156–164

Marugg JD, Gonzalez CF, Kunka BS, Ledeboer AM, Pucci MJ, Toonen MY, Walker SA, Zoetmulder LCM & Vandenberg PA (1992) Cloning, expression, and nucleotide sequence of genes involved in production of pediocin PA-1, a bacteriocin from *pediococcus acidilactici* PAC 1.0. Appl. Environ. Microbiol. 58: 2360–2367

Mazzota A & Montville TJ (1997) Nisin induces changes in the membrane fatty acid composition of *Listeria monocytogenes* nisin-resistant strains at 10 °C and 30 °C. J. Appl. Microbiol. 82: 32–38

McAuliffe O, Ryan PM, Ross PR, Hill C, Breeuwer P & Abee T (1998) Lacticin 3147, a broad-spectrum bacteriocin which selectively dissipates the membrane potential. Appl. Environ. Microbiol. 64: 439–445

Moll GN (1998a) unpublished data

Moll GN, Ubbink-Kok T, Hauge HH, Nissen-Meyer J, Nes IF, Konings WN & Driessen AJM (1996a) Lactococcin G is a potassium ion-conducting, two component bacteriocin. J. Bacteriol. 178: 600–605

Moll GN, Hauge HH, Nissen-Meyer J, Nes IF, Konings WN & Driessen AJM (1998b) Mechanistic properties of the two component bacteriocin lactococcin G. J. Bacteriol. 180: 96–99

Moll GN, Roberts GCK, Konings WN & Driessen AJM (1996b) Mechanism of lantibiotic-induced pore-formation. Antonie van Leeuwenhoek 69: 185–191

Moll GN, Konings WN & Driessen AJM (1998c) The lantibiotic nisin induces transmembrane movement of a fluorescent phospholipid. J. Bacteriol. 180: 6565–6570

Moll GN, Clark J, Chan WC, Bycroft BW, Roberts GCK, Konings WN & Driessen AJM (1997) Role of transmembrane pH gradient and membrane binding in nisin pore formation. J. Bacteriol 179: 135–140

Moll GN, van den Akker E, Hauge HH, Nes IF, Nissen-Meyer J, Konings WN & Driessen AJM (1999) Complementary and overlapping ion-selectivity of plantaricin EF and JK. Submitted

Mørtvedt CI, Nissen-Meyer J, Sletten K & Nes IF (1991) Purification and amino acid sequence of lactocin S, a bacteriocin produced by *Lactobacillus sake* L45. Appl. Environ. Microbiol. 57: 1829–1834

Mulders JWM, Boerrigter IJ, Rollema HS, Siezen RJ & de Vos WM (1991) Identification and characterization of the lantibiotic nisin Z, a natural nisin variant. Eur. J. Biochem. 201: 581–584

Navaratna MADB, Sahl H-G & Tagg JR (1998) Two-component anti-*Staphylococcus aureus* lantibiotic activity produced by *Staphylococcus aureus* C55. Appl. Environ. Microbiol. 64: 4803–4808

Nes IF, Diep DB, Håverstein LS, Brurberg MB, Eijsink V & Holo H (1996) Biosynthesis of bacteriocins in lactic acid bacteria. Antonie van Leeuwenhoek 70: 113–128

Nissen-Meyer J, Håverstein LS, Holgo H, Sletten K & Nes IF (1993) Association of the lactococcin A immunity factor with the cell membrane: purification and characterization of the immunity factor. J. General Microbiol. 139: 1503–1509

Nissen-Meyer J, Holo H, Haverstein LS, Sletten K & Nes IF (1992) A novel lactococcal bacteriocin whose activity depends on the complementary action of two peptides. J. Bacteriol. 174: 5686–5692

Okereke A & Montville TJ (1992) Nisin dissipates the proton motive force of the obligate anaerobe *Clostridium sporogenes* PA 3679. Appl. Environ. Microbiol. 58: 2463–2467

Otto M, Peschel A & Götz F (1998) Producer self-protection against the lantibiotic epidermin by the ABC transporter ApiFEG of *Staphylococcus epidermis* Tü3298. FEMS Microbiol. Lett. 166: 203–211

Piard J-C, Kuipers OP, Rollema HS, Desmazeaud MJ & de Vos WM (1993a) Structure, organization and expression of the *lct* gene for lacticin 481, a novel lantibiotic produced by *Lactococcus lactis*. J. Biol. Chem. 268: 16361–16368

Quadri LEN, Sailor M, Roy KL, Vederas JC and Stiles ME (1994) Chemical and genetic characterization of bacteriocins produced by *Carnobacterium piscicola* LV17B. J. Biol. Chem. 269: 12204–12211

Ross KF, Ronson CW & Tagg JR (1993) Isolation and characterization of the lantibiotic salivaricin A and its structural gene *salA* from *Streptococcus salivarius* 20P3. Appl. Environ. Microbiol. 59: 2014–2021

Sahl H-G, Jack RW & Bierbaum G (1995) Biosynthesis and biological activities of lantibiotics with unique post-translational modifications. Eur. J. Biochem. 230: 827–853

Stephens SK, Floriano B, Cathcart DP, Bayley SA, Witt VF, Jiménez Díaz R, Warner PJ & Ruiz-Barba JL (1998) Molecular Analysis of the locus responsible for the production of plantaricin S, a two-component bacteriocin produced by *Lactobacillus plantarum* LPCO10. Appl. Environ. Microbiol. 64: 1871–1877

Stoffels G, Guðmundsdóttir Á & Abee T (1994) Membrane-associated proteins encoded by the nisin cluster may function as a receptor for the lantibiotic carnocin U149. Microbiology 140: 1443–1450

Tagg JR, Dadjani AS & Wannamaker LW (1976) Bacteriocins of Gram-positive bacteria. Microbiol. Rev. 40: 722–756

Tahara T, Oshimura M, Umezawa C & Kanatani K (1996) Isolation, partial characterisation, and mode of action of acidocin J1132, a two component bacteriocin produced by *Lactobacillus acidophilus* JCM 1132. Appl. Environ. Microbiol. 62: 892–897

Tichaczek PS, Vogel RF & Hammes WP (1993) Cloning and sequencing of *curA* encoding curvacin A, the bacteriocin produced by *Lactobacillus* LTH1174. Arch. Microbiol. 160: 279–283

Tichaczek PS, Vogel RF & Hammes WP (1994) Cloning and sequencing of *sakP* encoding sakacin P, the bacteriocin produced by *Lactobacillus sake* LTH673. Microbiology 140: 361–367

Van Belkum M, Kok J, Venema G, Holo H, Nes IF, Konings WN & Abee T (1991) The bacteriocin lactococcin A specifically increases permeability of lactococcal cytoplasmic membranes in a voltage-independent, protein mediated manner. J. Bacteriol. 173: 7934–7941

Van Belkum MJ, Kok J & Venema G (1992) Cloning, sequencing and expression in *Escherichia coli* of *lcnB*, a third bacteriocin determinant from the lactococcal bacteriocin plasmid p9B4-6. Appl. Environ. Microbiol. 58: 572–577

Van de Kamp M, van den Hooven HW, Konings RNH, Bierbaum G, Sahl H-G, Kuipers OP, Siezen RJ, de Vos WM, Hilbers CW & van de Ven FJM (1995) Elucidation of the primary structure of the lantibiotic epilancin K7 from *Staphylococcus epidermis* K7. Cloning and characterization of the epilancin-K7-encoding gene and NMR analysis. Eur. J. Biochem. 230: 587–600

Van de Kamp M, Horstink LM, van den Hooven HW, Konings RNH, Hilbers CW, Frey A, Sahl H-G, Metzger JW & van de Ven FJM (1995) Sequence analysis by NMR spectroscopy of the peptide lantibiotic epilancin K7 from *Staphylococcus epidermis* K7. Eur. J. Biochem. 227: 757–771

Van der Meer JR, Polman J, Beerthuizen MM, Siezen RJ, Kuipers OP & de Vos WM (1993) Characterization of the *Lactococcus lactis* nisin A operon genes *NisP*, encoding a subtilin-like serine protease involved in precursor processing and *nisR*, encoding a regulatory protein involved in nisin biosynthesis. J. Bacteriol. 175: 2578–2588

Van de Ven FJM, van den Hooven HW, Konings RNH & Hilbers CW (1991) The spatial structure of nisin in aqueous solution. In: Jung G & Sahl H-G (Eds) Nisin and novel lantibiotics (pp 35–44). Escom publishers, Leiden, The Netherlands

Van den Hooven (1995) Structure elucidation of the lantibiotic nisin in aqueous solution and in membrane-like environments. Ph.D. Thesis, University of Nijmegen, The Netherlands

Van Kraaij C, Breukink E, Noordermeer MA, Demel R, Siezen RJ, Kuipers OP & de Kruijff B (1998) Pore formation by nisin involves translocation of its C-terminal part across the membrane. Biochemistry 37: 16033–16040

Van Kraaij C, Breukink E, Rollema HS, Siezen R, Demel, de Kruijff B & Kuipers OP (1997) Influence of charge differences in the C-terminal part of nisin on antimicrobial activity and signalling capacity. Eur. J. Biochem. 247: 114–120

Venema K, Abee T, Haandrikman AJ, Leenhouts KJ, Kok J, Konings WN and Venema G (1993) Mode of action of lactococcin B, a thiol-activated bacteriocin from *Lactococcus lactis*. Appl. Environ. Microbiol. 59: 1041–1048

Venema K, Dost MHR, Venema G & Kok J (1996) Mutational analysis and chemical modification of Cys 24 of lactococcin B, a bacteriocin produced by *Lactococcus lactis*. Microbiology 142: 2825–2830

Venema K, Haverkort RE, Abee T, Haandrikman AJ, Leenhouts KJ, de Ley L, Venema G & Kok J (1994) Mode of action of LciA, the lactococcin A immunity protein. Mol. Microbiol. 14: 521–532

Verheul, A, Russell NJ, Van 't Hof R, Rombouts FM & Abee T (1997) Modifications of membrane phospholipid composition in nisin-resistant *Listeria monocytogenes* Scott A. Appl. Environ. Microbiol. 63: 3451–3457

Wiedeman I, Benz R & Sahl H-G (1998) Pore formation by the peptide antibiotic nisin in the presence of the bacterial peptidoglycan lipid II: a black lipid membrane study. Third International Workshop on Lantibiotics, Blaubeuren, p 72

Winkowsky K, Bruno MEC & Montville TJ (1994) Correlation of bioenergetic parameters with cell death in *Listeria monocytogenes* cells exposed to nisin. Appl. Environ. Microbiol. 60: 4186–4188

Winkowsky KD, Ludescher RD & Montville TJ (1996) Physicochemical characterization of the nisin-membrane interaction with liposomes derived from *Listeria monocytogenes*. Appl. Environ. Microbiol. 62: 323–327

Worobo RW, van Belkum MJ, Sailer M, Roy KL, Vederas JC & Stiles ME (1995) A signal peptide secretion-dependent bacteriocin from *Carnobacterium divergens*. J. Bacteriol. 177: 3143–3149

Worobo RW, Henkel T, Sailer M, Roy KL, Vederas JC & Stiles ME (1994) Characteristics and genetic determinant of a hydrophobic peptide bacteriocin, carnobacteriocin A, produced by *Carnobacterium piscicola* LV17A. Microbiology 140: 517–526

Zajdel JK, Ceglowsky P & Dobrzanski WT (1985) Mechanism of action of lactostrepcin 5, a bacteriocin produced by *Streptococcus cremoris*. Appl. Environ. Microbiol. 49: 969–974

Antonie van Leeuwenhoek **76**: 199–205, 1999.

Intestinal microflora and the interaction with immunocompetent cells

Stephanie Blum, Susana Alvarez, Dirk Haller, Pablo Perez & Eduardo J. Schiffrin*
*Nestlé Research Center, Nestec LTD, Vers-chez-les-Blanc, 1000 Lausanne 26, Switzerland. (*Author for correspondence; E-mail: eduardo.schiffrin@rdls.nestle.com)*

Key words: infections, inflammation, intestinal microflora, lactic acid bacteria, mucosal immunology

Abstract

The intestinal mucosal surface is colonised by the comensal microflora that attains very high numbers of bacterial cells in the distal intestine, more specifically in the colon. At the same time these extensive areas are the interface with the external environment, through which most pathogens initiate infectious processes in mammals. Intestinal mechanisms of defense need to discriminate accurately between comensal, symbiotic microflora, and exogenous pathogens. Today we do not fully understand the essence of the mechanism of discrimination but, probably, innate as well as adaptive immune responses participate in this process. We have explored , in *in vitro* models, the capacity of mucosal immunocompetent cells to discriminate amongst signals delivered by different types of bacteria. We have found at least two different patterns of innate response to gram-negative and gram-positive bacteria, and within this last group big differences are observed between species. We have only worked with non-pathogenic bacteria in what may represent the modulation of the physiological host status. The understanding of these modulatory functions could render a unique possibility for the use of food-borne bacteria to prevent or correct intestinal problems associated with food allergy, inflammatory bowel disease, and autoimmunity.

Abbreviations: GIT – gastrointestinal tract; IEC – intestinal epithelial cells; IEL – intra-epithelial lymphocytes; IgA – Immunoglobulin A; PBMC – Peripheral blood mononuclear cells; PRR – Pattern recognition receptor; GALT – Gut associated lymphoid tissue

Introduction

Mucosal surfaces are large areas covered by epithelial cells offering a potential portal of entry to pathogens and represent the main site of initiation of human infectious processes. In physiological situations, mucosal surfaces can be sterile, such as the distal respiratory track, or colonised, as the gastrointestinal tract (GIT). A major feature of the protective mechanisms localized at colonised mucosal surfaces, such as the intestine, is the capacity to discriminate between non-pathogenic luminal elements (foods, commensal bacteria) and antigens derived from enteropathogens (Underdown & Schiff 1986). The lack of a specific response to soluble food antigens following ingestion is a process called oral tolerance, for which a number of different mechanisms have been proposed (Garside, Mowat & Khoruts 1999). The intestinal immune system remains non-reactive to the resident microflora, which has also been explained on the basis of immunological tolerance. This phenomenon is of major importance for preservation of gut integrity. It is sug gested that loss of immunological tolerance towards commensal microflora may be one mechanism in inflammatory bowel disease (IBD) (Duchmann, Kaiser & Hermann 1995).

The function of the immune system depends on two types of recognition systems: innate (non-clonal) and adaptive (clonal) host defence. The receptors of the innate immune system, called pattern recognition receptors (PRR), are coded in the germ line and recognize conserved molecular structures shared by a large variety of pathogens (Lipford, Heeg & Wagner 1998; Stahl & Ezekowictz 1998). On the other hand, T and B cell antigen-recognition receptors are highly specific, as they undergo gene rearrangements and somatic mutations generating an adapted specificity for the antigen. However, innate and adaptive immunity are not two independent mechanisms of defense. It has been shown that receptors of the innate immune system can contribute to signaling that activates the adaptive immune system. Moreover, this activation strongly depends on bacterial (or bacterial product) interactions with PRR on mammalian cells (Kirschning Wesche, Ayres & Rothe 1998). These mechanisms are of major importance for host defense and may also play a role in the homeostasis of a tissue under permanent exposure to a changing external environment. In particular, the host intestinal mucosa is the interface to a variable bacterial environment containing the commensal microflora with some potential pathogenic components, food microbes and, occasionally, true pathogens. The changing luminal environment requires fine tuned host responses to maintain a permanent low level of immune activation that prevents inflammatory tissue damage on one hand, and permits energetic responses upon the arrival of a harmful pathogen, on the other. In the orchestration of this discriminative response, endogenous mediators as well as luminal factors may play important roles (Klapproth, Donnenberg, Abraham & Mobley, 1995). Recent work conducted in our group was aimed at understanding how commensal and non-pathogenic food microbes can contribute to the modulation of intestinal reactions against pathogens and facilitate the preservation of gut integrity.

Mucosal defenses

(a) The mucosal secretory immune system

The mucosal secretory immune response is probably the most extensively studied mechanism of defense against enteropathogens. The existing paradigm recognizes a compartmentalisation of inductive and effector sites. The induction takes place at lymphoid aggregates, i.e. Peyer's patches (Abreu-Martin & Targan 1996). They are covered by a specialised epithelium that contains the M cell, adapted for sampling the intestinal content (Neutra, Frey & Kraehenbuhl 1996). The underlying lymphocytes are arranged in follicles that contain T and B cell compartments. B cells in the gut associated lymphoid tissue (GALT) are preferentially committed to immunoglobulin A (IgA) production. B cell immunoglobulin isotype switching occurs upon interactions with T cells. $CD4^+$ T cells (T helper lymphocytes) expressing CD40 ligand and producing IL-4 , IL-10, IL-5, co-localise with germinal center centrocytes (B cells in maturation) that are undergoing isotype switching, and participate actively in this process (Liu & Arpin 1997). Antigen-specific B and T lymphocytes leave the inductive sites, travel through lymph and the blood stream and home back into the mucosa. The specific homing of mucosal primed lymphocytes is directed, in part, by the expression of $\alpha4$ $\beta7$ integrin, that recognizes the mucosal addressin cell surface molecule (MAdCAM-1) expressed on mucosal blood vessels (Berlin, Berg, Briskin & Butcher 1993).

B cells complete their differentiation into plasmacytes in the lamina propria where they secrete dimeric IgA. The IgA is then transported through the epithelial layer towards the intestinal lumen by the polymeric Ig receptor produced by the intestinal epithelial cells (Underdown & Schiff 1986). Secretory IgA (sIgA) reaches the intestinal lumen and reacts with specific antigens so as to prevent their attachment to the mucosal surface. This process is called immune exclusion and does not imply activation of inflammatory processes. Production and secretion of IgA are further regulated at the Lamina Propria (LP) by (i) endogenous environmental mediators, such as TGF-β and IL-5, mainly produced by regulatory T cells (Coffman, Lebman & Shrader 1989; Lebman, Lee & Coffman 1990; Stavnezer & Shocket 1991) and (ii) intestinal bacterial colonisation (Kett, Baklein, Kral, Fausa & Brandtzaeg 1995). Lamina propria helper T cells ($CD4^+$) are predominantly memory cells with high expression of activation markers and a pattern of cytokine production that corresponds to the TH2 type (IL-4, IL-5). This cytokine environment promotes IgA production by lamina propria B cells (Smart, Trejdosiewicz, Badr-el-Din & Heatley 1988).

(b) Epithelial compartment

Only in recent years the epithelial compartment has been recognized as a central player in mucosal defenses and tissue homeostasis (Eckmann Jung, Schürer-Maly, Panja & Kagnoff, 1993; Jung, Eckmann, Yang, Panja & Kagnoff 1995). It involves at least two major cellular components: the intraepithelial lymphocytes (IEL) and the intestinal epithelial cells (IEC) . IEL are located within the epithelial compartment in contact with the basolateral membrane of the IEC. Upon stimulation, IEL are able to modify the IEC immune-phenotype and its immunological or defensive functions (Cerf-Benussan Quaroni, Kurnick & Bhan 1984). These effects could be mediated by cell-to-cell interactions such as αE β7/E-cadherin interactions, and also through secreted cytokines, like IFN-γ. Most of the IELs are mature T cells (CD3$^+$) of the CD8$^+$ (suppressor/cytotoxic) phenotype, which express the homodimeric form of the CD8 molecule (CD8$\alpha\alpha$), in evidence of their extra-thymic origin. In contrast, peripheral blood CD8+ T cells express the CD8 $\alpha\beta$ heterodimer (Guy-Grand, Cerf-Benussan, Malissen & Vasalli 1991). Moreover, human and murine IELs are enriched in T cells that express $\gamma\delta$ TCR. In humans, IEL express a limited array of TCR $\alpha\beta$ and to a lesser extent, TCR $\gamma\delta$, indicating that they recognize a restricted range of antigens in the context of MHC class I – related molecules (Blumberg, Terhorst, Bleicher & Balk 1991; Panja, Blumberg, Balk & Mayer 1993; Tanaka, Morita, Nieves, Brenner & Bloom 1995). It has been reported that IEL homing into the epithelial compartment depends on the expression of αE β7 integrin, found on all IELs. Its ligand, E-cadherin, is expressed on the basolateral membrane of the enterocyte (Cepek, Shaw, Parker & Brenner 1994). These molecules can function as accessory molecules in TCR-mediated responses (Sarnacki, Begue, Buc, Le Deist & Cerf-Benussan 1992; Lefrançois, Barret, Havran & Puddington 1994). αE β7 expression is up-regulated by TGF-β (Rocha, Guy-Grand & Vasalli 1995), predominantly derived from IEC. Therefore, nutrients with the capacity to modulate TGFβ production by IEC could have an impact on mucosal mechanisms of defense. IEL show high cytolytic (CTL) activity *in vitro*, suggesting that one of their main functions *in vivo* could be the elimination of infected or highly-damaged cells. It has been shown that a fraction of IELs contains intracytoplasmic perforin and granzyme-containing granules, which exert immediate cytolytic activity upon TCR stimulation. Furthermore, IELs express Fas-ligand and are able to kill Fas-bearing targets. To date it is not clear which cytolytic pathway, Fas or perforin-mediated, is preferentially used by IEL (Gelfanov, Lai & Liao 1996).

IEC are considered to be a constitutive component of the mucosal immune system. They are the first host cell in contact with luminal antigens and microorganisms, both commensal or pathogens, and were proven to be antigen presenting cells (Bland & Warren 1986). Upon stimulation, IEC can change their immune-phenotype (Barclay & Mason 1982; Hughes, Block, Bhan & Harmatz 1991) and produce a wide range of chemokines and immunomodulatory cytokines (Eckmann,Jung, Schürer-Maly, Panja & Kagnoff 1993; Jung, Eckmann, Yang, Panja & Kagnoff 1995). These changes render IEC more active in the recruitment of immune cells to control invading pathogens. They become accessory antigen presenting cells, or effector cells in anti-infectious processes. This last possibility remains speculative. However, IEC reactivitiy to luminal stimuli occurs mainly from bacterial origin and is part of the innate host defenses. Little is known about the modulation of epithelial immunocompetence by non-pathogenic bacteria and their role in the physiology of mucosal defense. The use of specific bacterial strains in foods (functional nutrients), which provide the consumer with health benefits, particularly for the control of gastrointestinal infections, makes this a field of high interest (Brassart & Schiffrin 1997).

Indigenous microflora and mechanisms of defence

The so-called 'barrier effect' of the microflora or colonisation resistance proposes that the resident microflora prevents colonisation of the gut by pathogens. This effect can be achieved through different mechanisms: (i) the competition for mucosal-specific niches or (ii) nutrients; and (iii) the production of bactericidal or bacteriostatic products. The endogenous microflora seems to have a capacity to modulate cellular events related to host defenses and tissular homeostasis. We have studied, in a series of reductionist models, the interaction between non-pathogenic bacterial strains and immunocompetent cells that are present in the mucosal tissues.

Non-pathogenic intestinal microflora and the epithelial compartment

Bacterial interaction with intestinal monolayers in vitro

Our group has emphasized the role of the enterocyte in the reaction of mucosal tissues to bacterial stimulation. Non-pathogenic bacteria do not invade the host. Consequently, to deliver an immunomodulatory signal they need to induce functional modifications in epithelial cells that, in turn, may transduce the signal to the immune cells in the mucosal environment. Based on this hypothesis, it was demonstrated that lactic acid bacteria and Gram-negative bacteria, both components of the resident microflora, exert a different pattern of IEC phenotye modification (Delneste,

Table 1. Bacterial modulation of HLA-DR, CD54 and CD95 Expression on HT-29 cells

Stimulus	HLA -DR	CD54	CD95
No stimulus	0	5	5
L. johnsonii La1	0	5	4
E. coli	0	26**	5
IFNγ	9	20**	10
LPS	0	6	5
La1 + IFNγ	15*	27*	14*
E.coli + IFNγ	26*	80*	16*
LPS + IFNγ	19*	60*	16*

HT-29 cells (60-80% confluence) were cultured for 24 h with bacteria (50 bacteria/HT-29 cell) with or without 25 U/ml IFNg. HLA-DR, CD54 (ICAM-1) and CD95 (Fas) expression was determined by flowcytometry. Results are presented in mean fluorescence intensity (MFI) substrated by the MFI of the isotype control.
*Significant change ($p<$ 0.05) compared to IFNγ single treatment. ** Significant change ($p<$ 0.05) compared to basal expression.

Donnet-Hughes & Schiffrin 1998). Relevant IEC immune markers can be grouped in molecules involved in (i) antigen presentation (MHC class II, Cd1), an early event of the immune response; (ii) cross-talk between IEC and lymphocytes (ICAM-1, Fas, IFN-γ receptor), crucial for homeostatic regulation, and, finally (iii) soluble mediators such as cytokines and chemokines (IL-8, TNF-α, MCP-1), promoting the recruitment and activation of immune cells in the different intestinal micro-environments. Treatment of the human intestinal cell line HT-29 *in vitro* with a non-pathogenic *E. coli* increased the basal expression of ICAM-1 (CD54) and IFN-γR (CD119). Furthermore, when *E. coli* was combined with IFN-γ, a product of IEL, a synergistic effect was detected for the expression of MHC class II molecules (HLA-DR), ICAM-1 and Fas (Table 1). In addition, pro-inflammatory cytokines such as TNF-α, IL-8, MCP-1 and GM-CSF were induced by the Gram-negative bacteria and the induction was increased significantly when *E. coli* was combined with IFN-γ. In contrast, lactic acid bacteria (LAB) did not exert any agonistic effect with respect to the induction of pro-inflammatory cytokines. However, the combination of LAB and IFN-γ increased the expression of IFN-γ-R. The fact that LAB alone or in combination with IFN-γ did not induce any of the pro-inflammatory cytokines suggests that LAB could participate in tissue protection against the deleterious effect of an ongoing inflammatory process. Bacterial-epithelial cell contact was a pre-requisite for the induction of IEC phenotypic changes. The underlying physical interaction between procaryote and eukaryotic cells has not been elucidated, but these observations could imply that either specific receptors or physical surface properties such as hydrophobicity (Perez, Minaard, Disalvo & De Antoni 1998), are important to cell-to-cell interactions.

Non-pathogenic intestinal microflora and professional immune cells

Bacteria interaction with blood leucocytes

Although commensal bacteria do not massively invade the host, it is possible that in defined mucosal environments, like the Peyer's patches, they may interact directly with immune cells and induce their activation. Fluorescent-labelled bacteria (*Lactobacillus* spp., *Bifidobacteria* spp.) were used to assess the interaction with human blood leucocyte subpopulations using flow cytometry (Perez et al., submitted). Of all strains tested, *Lactobacillus johnsonii* La1 showed the strongest interactions with lymphocytes, monocytes and granulocytes (Table 2). *L. johnsonii* La1 is also highly adherent to IEC. A general correlation between adhesion to epithelium and level of interaction with blood leucocytes was not observed for the other bacterial strains investigated. Assuming that bacteria/leucocyte cell-to-cell interactions could lead to the stimulation of immune cells, an assessment of cell proliferation and cytokine production was performed. ^{3}H-thymidine incorporation by proliferating leucocytes was determined at 3 and 5 days after incubation with bacterial cells. Although all bacteria induced proliferation at day 5,*L. johnsonii* La1 was significantly stronger. The analysis of expression of activation antigens on lymphocyte subsets identified natural killer cells (NK, CD3^{-}/56^{+}) as the subpopulation preferentially stimulated by bacteria. In addition, lactobacilli induced IL-12 and IFN-γ secretion. This could not be observed with Gram-negative *E. coli* (Haller, Blum, Bode, Hammes, & Schiffrin submitted).

Table 2. Percentage of leucocyte subpopulations showing association of fluorescent labeled bacteria

Strain	Lymphocytes	Monocytes	Granulocytes
Bifidobacterium bifidum CIDCA 536	1.48 ± 0.04	6.43 ± 0.01	13.8 ± 0.64
Bifidobacterium infantis CIDCA 538	0.90 ± 0.01	9.33 ± 0.32	32.1 ± 0.76
Lactobacillus johnsonii La1	4.64 ± 0.05	25.05 ± 2.14	46.89 ± 1.99
Lactobacillus acidophilus La10	2.41 ± 0.67	11.59 ± 1.48	27.83 ± 0.56

Human unstimulated PBMC were incubated with BCEF-AM fluorescent labeled bacteria in a ratio 20:1 (bacteria:leukocytes), washed three times and analysed by flowcytometry. The interaction of BCEF-AM-bacteria with leukocyte subpopulations is demonstrated as the percentage of fluorescence positve cells within the gated population.

Non-pathogenic intestinal microflora and mixed mucosal cell populations

Bacterial interaction with mucosal mixed cell populations

It is highly possible that bacterial signaling to the host needs to be 'processed' by a network of mucosal immune cells, as an integrated response from a larger cellular network that dictates the host reaction to a constantly changing microbial intestinal environment.

To investigate such interactions, an *in vitro* model was established by co-culturing CaCO-2 cells with peripheral blood mononuclear cells (PBMC) using insert culture technique (Haller, Bode, Hammes, Pfeifer, Schiffrin & Blum, submitted). The immune response to non-pathogenic bacteria was assessed by determination of cytokine expression at the mRNA and protein levels. The pro-inflammatory cytokines TNF-α, IL-8 and MCP-1 were induced in Caco-2 cells upon stimulation with non-pathogenic *E. coli* K12 and also with the Gram-positive non-pathogenic *Lactobacilllus sakei* 681. *L. johnsonii* La1 did not stimulate the expression of these cytokines but, rather, up-regulated TGF-β mRNA levels. Cytokine production by Caco-2 cells in response to bacterial challenge was depended on the presence of PBMC in the basolateral compartment of the *in vitro* system. PBMC located in the basolateral chamber responded with a distinct reaction to bacterial stimulation, even though the bacteria had no direct access to this compartment. In fact, TNF-α and GM-CSF were detected only in response to *E. coli and L. sakei* applied on the apical surface of Caco-2 cells, whereas no stimulation of these cytokines occurred after treatment with *L. johnsonii* La1. Epithelial cell-immunocyte cross talk appears responsible for the innate reaction that can discriminate amongst different intestinal non-pathogenic bacteria. This discriminative response, occurs in both compartments of the *in vitro* model, and is probably coordinated by cell secretory products that remain to be identified. The differential modulation of defense mechanisms by non-pathogenic bacteria may also play a role in the homeostasis of the mucosal tissue.

Discussion and conclusions

IEC permanently interact with the luminal content, including the comensal microflora, and the endogenous cellular network of professional immune cells. Work performed with the *in vitro* models emphasized the different type of responses triggered by non-patogenic bacteria on IEC or PBMC. This remains to be the case when using more complex models that combine an epithelial layer with underlying immune cells in order to mimic the intestinal lamina propria. With regard to specific molecules involved in cell-to-cell interactions, modulation of the immune phenotype and cytokine secretion, suggests a role for IEC in the regulation of the cellular environment at the intra-epithelial compartment and at the lamina propria. One appropriate function of IEC is to adapt the physiological reactivity of the host tissues to a highly changing intestinal content. A dysfunction at this interphase could promote a discordance between luminal signals and response initiation. This could cause pathological conditions because of exaggerated responses to non-dangerous signals (like food antigens), resulting in food allergy or in chronic inflammation, e.g. inflammatory bowel disease, IBD. Both conditions could result from the breakdown of oral tolerance. One of the underlying mechanisms proposed for oral tolerance is the secretion of inhibitory cytokines such as TGF-β. It is notable to emphasize that the probiotic strain *L. johnsonii* La1 selectively induced the production of TGF-β. A weak communication of a harmful luminal signal could limit the appropriate response for controlling luminal pathogens, as has been observed in some cases of intestinal immunological disorders (Kotler, Giang, Thiim & Orenstein 1995; Wicox, Schwartz, Cotsonis & Thompson 1996). The fine tuning of this dynamic interphase is not only dependent on IEC function but also on an intimate cell-to-cell crosstalk, where IEL and LP immune cells may participate. This has been partially demonstrated with the *in vitro* co-culture model described above. However, it is strongly suggested that the IEC play a central role in delivering early signals to the neighboring cells. Knowledge about the downstream cascade of events will give us unique possibilities to modify gut homeostasis by nutritional interventions.

References

Abreu-Martin MT & Targan SR (1996) Regulation of immune responses of the intestinal mucosa. Crit. Rev. Immunol. 16: 277–309

Berlin C, Berg EL, Briskin MJ, Andrew DP, Kilshaw PJ, Holzmann B, Weismann IL, Hamann A & Butcher EC (1993) Alpha-4 beta-7 intefrin mediates lymphocyte binding to the mucosal addressin MAdCAM-1. Cell 74: 158–195

Blumberg RS, Terhorst C, Bleicher P, McDermott FV, Allan CH, Landau SB, Trier JS & Balk S. (1991) Expression of a nonpolymorphic MHC class I-like molecule, CD1d, by human intestinal epithelial cells. J. Immunol. 147: 2518–2524

Bland PW & Warren LG (1986) Antigen presentation by epithelial cells of the rat small intestine. Immunology 58: 1-7. Barclay AN, Mason DW. (1982) Induction of Ia antigen in rat epidermal cells and gut epithelium by immunological stimuli. J. Exp. Med. 156: 1665–1669

Brassart D & Schiffrin EJ (1997) The use of probiotics to reinforce mucosal defence mechanisms. Trends Food Sci. Technol. 8: 321–326

Coffman RL, Lebman DA & Shrader B (1989) Transforming growth factor beta specifically enhances IgA production by liposaccharide-stimulated murine B-lymphocytes. J. Exp. Med. 170: 1039–1044

Cerf-Bensussan N, Quaroni A, Kurnick JT & Bhan AK (1984) Intraepithelial lymphocytes modulate Ia expression by intestinal epithelial cells. J. Immunol. 132: 2244–2252

Cepek KL, Shaw SK, Parker CM, Russell GJ, Morrow JS, Rimm DL & Brenner MB (1994) Adhesion between epithelial cells and lymphocytes is mediated by E-cadherin and the alpha E beta-7 integrin. Nature 372: 190–193

Delneste Y, Donnet-Hughes A & Schiffrin EJ (1998) Functional Foods: mechanisms of action on immunocompetent cells. Nutrition Reviews 56: S93–S98

Duchmann R, Kaiser I & Hermann E (1995) Tolerance exists towards resident intestinal flora but is broken in active inflammatory bowel disease (IBD). Clin. Exp. Immunol. 102: 448–455

Eckmann L, Jung HC, Schürer-Maly C, Panja A, Morzycka-Wrobleska E & Kagnoff MK (1993) Differential cytokine expression by human intestinal epithelial cell lines: regulated expression of Interleukin 8. Gastroenterology 105: 1689–1697

Garside P, Mowat A McI & Khoruts A (1999) Oral tolerance and disease. Gut. 44: 137–142

Guy-Grand D, Cerf-Benussan N, Malissen B, Malasssis-Seris M, Briottet C & Vasalli P (1991) Two gut intraepithelial CD8+ lymphocyte populations with different T cell receptors: a role for the gut epithelium in T cell differentiation. J. Exp. Med. 173: 471–478

Gelfanov V, Lai YG & Liao NS (1996) Activated $\alpha\beta$-CD8$^+$ but not $\alpha\alpha$-CD8$^+$, TCR-$\alpha\beta^+$ murine intestinal intraepithelial lymphocytes can mediate perforin-based cytotoxicity whereas both subsets are active in Fas-based cytotoxicity. J. Immunol. 157: 35–41

Haller D, Bode Ch, Hammes WP, Pfeifer AMA, Schiffrin EJ & Blum S (1998) Bacterial modulation of immune response in intestinal epithelial cells after co-cultivation with PBMCs *in vitro*. International Congress of Immunology, New Dehli. The Immunologist, Supplement 1: 548

Hughes A, Block KJ, Bhan AK, Gillen D, Giovino VC & Harmatz PR (1991) Expression of MHC class II (Ia) antigen by the neonatal enterocyte: the effect of treatment with IFNγ. Immunology 72: 491–496

Jung HC, Eckmann L, Yang SK, Panja A, Fierer J, Morzycka-Wrobleska E & Kagnoff MK (1995) A distinct array of pro-inflammatory cytokines is expressed in human colon eoithelial cells in response to bacterial invasion. J. Clin. Invest. 95: 55–65

Kett K, Baklein K, Bakken A, Kral JG, Fausa O & Brandtzaeg P (1995) Intestinal B-cell isotype response in relation to bacterial load: evidence for immunoglobulin A subclass adaptation. Gastroenterology 109: 819–825

Klapproth JM, Donnenberg MS, Abraham JM, Mobley HLT & James SP (1995) Products of enteropathogenic *Escherichia coli* inhibit lymphocyte activation and lymphokine production. Infection and Immunity 63: 2248–2254

Kirschning CJ, Wesche H, Ayres TM & Rothe M (1998) Human toll-like receptor 2 confers responsiveness to bacterial lipopolysaccharide. J. Exp. Med. 188: 2091–2097

Kotler DP, Giang TT, Thiim M, Nataro JP, Sordillo EM & Orenstein JM (1995) Chronic bacterial enteropathy in patients with AIDS. Infectious Diseases 171: 552–558

Lipford GB, Heeg K & Wagner H (1998) Bacterial DNA as immune activator. Trends Microbiol. 6: 496–500

Liu YJ & Arpin C (1997) Germinal centre development. Immunol. Rev. 156: 11–126

Lebman DA, Lee FD & Coffman RL (1990) Mechanism for transforming growth factor beta and IL-2 enhancement for IgA expression in liposaccharide-stimulated B cell cultures. J.Immunol. 144: 952–959

Neutra MR, Frey A & Kraehenbuhl JP (1996) Gateways for mucosal infection and immunization. Cell 86: 345–348

Panja A, Blumberg RS, Balk S & Mayer L (1993) Cd1d is involved in T cell-intestinal epithelial cell interactions. J. Exp. Med. 178: 1115–1119

Perez PF, Minaard J, Disalvo EA & De Antoni GL (1998) Surface properties of bifidobacterial strains of human origin. Appl. Environ. Microbiol. 64: 21–26

Rocha B, Guy-Grand D & Vasalli P (1995) Extrathymic T cell differentiation. Current Opinion in Immunology 7: 235–242

Sarnacki S, Begue B, Buc H, Le Deist F & Cerf-Benussan N (1992) Enhancement of CD3 induced activation of human intestinal intraepithelial lymphocytes by stimulation of the $\beta 7$-containing integrin defined by HLM-1 monoclonal antibody. Eur. J. Immunol. 22: 2887–2892

Smart CJ, Trejdosiewicz LK, Badr-el-Din S & Heatley RV (1988) T lymphocytes of the human colonic mucosa: functional and phenotypic analysis. Clin. Exp. Immunol. 73: 63–67

Stavnezer J & Shocket P (1991) Effects of cytokines on switching to IgA and alpha germline transcripts in the B loymphoma 1.29. Transforming growth factor-beta activates transcription of the unrearranded C alpha gene. J. Immunol. 147: 4374–4383

Stahl PD & Ezekowitz AB (1998) The mannose receptor is a pattern recognition receptor involved in host defense. Current Opinion in Immunology 10: 50–55

Tanaka Y, Morita CT, Nieves E, Brenner MB & Bloom BR (1995) Natural and synthetic non peptide antigens recognized by human $\gamma\delta$ T cells. Nature 375: 155–158

Underdown BJ & Schiff JM (1986) Strategic defense initiative at the mucosal surfaces. Annu Rev Immunol 4: 389–417

Wicox CM, Schwartz DA, Cotsonis G, Thompson SE. (1996) Chronic unexplained diarrhea in human immunodeficiency cirus infection: determination of the best diagnostic approach. Gastroenterology 110: 30–37

Antonie van Leeuwenhoek **76:** 207–215, 1999.

Bioactive peptides encrypted in milk proteins: proteolytic activation and thropho-functional properties

Hans Meisel[1,*] & Wilhelm Bockelmann[2]
Federal Dairy Research Centre,
[1]*Institute for Chemistry and Physics,*
[2]*Institute for Microbiology, P.O. Box 6069, D-24121 Kiel, Germany. (*Author for correspondence)*

Key words: bioactive peptides, functional foods, lactic acid bacteria, milk proteins

Abstract

The bioactivities of peptides encrypted in major milk proteins are latent until released and activated by enzymatic proteolysis, e.g. during gastrointestinal digestion or food processing. The proteolytic system of lactic acid bacteria can contribute to the liberation of bioactive peptides. *In vitro*, the purified cell wall proteinase of *Lactococcus lactis* was shown to liberate oligopeptides from β- and α-caseins which contain amino acid sequences present in casomorphins, casokinines, and immunopeptides. The further degradation of these peptides by endopeptidases and exopeptidases of lactic acid bacteria could lead to the liberation of bioactive peptides in fermented milk products. However, the sequences of practically all known biologically active peptides can also be cleaved by peptidases from lactic acid bacteria. Activated peptides are potential modulators of various regulatory processes in the body: Opioid peptides are opioid receptor ligands which can modulate absorption processes in the intestinal tract, angiotensin-I-converting enzyme (ACE)-inhibitory peptides are hemodynamic regulators and exert an antihypertensive effect, immunomodulating casein peptides stimulate the activities of cells of the immune system, antimicrobial peptides kill sensitive microorganisms, antithrombotic peptides inhibit aggregation of platelets and caseinophosphopeptides may function as carriers for different minerals, especially calcium. Bioactive peptides can interact with target sites at the luminal side of the intestinal tract. Furthermore, they can be absorbed and then reach peripheral organs. Food-derived bioactive peptides are claimed to be health enhancing components which can be used for functional food and pharmaceutical preparations.

Introduction

In the last two decades a number of studies have been performed on bioactive peptides that are present in the amino acid sequence of milk proteins. Although other animal as well as plant proteins contain potential bioactive sequences, milk proteins are currently the main source of a range of biologically active peptides (Table 1). The structures of biologically active sequences were obtained from *in vitro* enzymatic and/or by *in vivo* gastrointestinal digests of the appropriate precursor proteins; chemical synthesis has been carried out to confirm the sequence of potential bioactive peptides.

Milk-protein derived bioactive peptides are inactive within the sequence of the parent protein and can be released and activated by enzymatic proteolysis for example by the action proteinases and peptidases from lactic acid bacteria. Lactic acid bacteria are nutritionally fastidious and need amino acids for nitrogen nutrition. In milk, caseins are the major source for amino acids (Juillard et al. 1995). The proteolytic system of lactic acid bacteria consists of a cell wall bound proteinase and several intracellular peptidases (Pelissier 1984; Monnet et al. 1993; Bockelmann 1995; Poolman et al. 1995; Law & Haandrikman 1997). Transport systems specific for amino acids, di- and tripeptides and oligopeptides (up to 18 amino acids in length) are present in lactic acid bacteria for nitrogen uptake (Konings et al. 1997). Longer oligopeptides, not transported into the cells can be a source for the liberation of bioactive peptides in fermented milk products when further degraded, for example by in-

Table 1. Bioactive peptides derived from milk proteins

Bioactive peptide	Protein precursor	Bioactivity
Casomorphins	α-, β-Casein	Opioid agonist
α-Lactorphin	α-Lactalbumin	Opioid agonist
β-Lactorphin	β-Lactoglobulin	Opioid agonist
Lactoferroxins	Lactoferrin	Opioid antagonist
Casoxins	κ-Casein	Opioid antagonist
Casokinins	α- β-Casein	ACE-inhibitory
Lactokinins	α-Lactalbumin, β-Lactoglobulin, Serum albumin	ACE-inhibitory
Immunopeptides	α- β-Casein	Immunomodulatory
Lactoferricin	Lactoferrin	Antimicrobial
Casocidin	α_{S2}-Casein	Antimicrobial
Isracidin	α_{S1}-Casein	Antimicrobial
Casoplatelins	κ-Casein	Antithrombotic
Phosphopeptides	α-, β-Casein	Mineral binding

tracellular peptidases of lactic acid bacteria after cell lysis (Law & Haandrikman 1997). In the human gastrointestinal tract, digestive enzymes will contribute to the further breakdown of long casein-derived oligopeptides which may also lead to release of bioactive peptides. Once they are liberated in the body, bioactive peptides may act as regulatory compounds with hormone-like activity. The following overview is based on recent reviews on the specific biochemical properties and possible dietary and pharmaceutical applications of peptides derived from milk proteins (Meisel & Schlimme 1996; Meisel 1997a, b; Meisel 1998).

Phenomenon of multifunctional bioactive peptides

Many milk-derived peptides reveal multifunctional properties: i.e. specific peptide sequences having two or more different biological activities have been reported (Table 2; peptides are indicated in bold numbers as listed in the table). Some regions in the primary structure of caseins contain overlapping peptide sequences that exert different biological effects. These regions have been considered as 'strategic zones' which are partially protected from proteolytic breakdown (Fiat & Jollès 1989; Schanbacher et al. 1997; Meisel 1998). Three strategies have generally been used in the identification and characterisation of such peptides, i.e. isolation of inhibitory peptides from *in vitro* enzymatic or from *in vivo* gastrointestinal digests of precursor proteins and chemical synthesis of peptides or peptide fragments having similar structures to those known to be bioactive.

Opioid peptides, i.e. opioid receptor (μ-, δ- or κ-type) ligands with agonistic activity, originate from different milk proteins and exert naloxone-inhibitable opioid activities in both receptor studies and during bioassays (Brantl et al. 1981; Chiba and Yoshikawa 1986; Loukas et al. 1993). The atypical opioid peptides **1-7** derived from milk proteins have N-terminal sequences different from that of the 'typical' endogenous opioid peptides (Teschemacher et al. 1997). The major exogenous opioid peptides, β-casomorphins, are fragments of the β-casein sequence **1** (Meisel 1986) and have been characterized as μ-type ligands (Teschemacher & Brantl 1994). β-caso-morphins were also found in analogous positions in sheep, water buffalo and human β-casein (for review: Fiat and Jollès 1989). Five opioid peptides were isolated from wheat gluten (Fukudome & Yoshikawa 1994). These gluten exorphins were found at 15 sites in the primary structure of glutenin and were highly selective for δ-receptors.

Opioid antagonists have been found in bovine and human κ-casein (casoxins **8**, **9**) and in human α_{S1}-casein (Chiba et al. 1989; Yoshikawa et al. 1994). Furthermore the opioid antagonist lactoferroxin has been found in human lactoferrin (Yoshikawa et al. 1988). Various casoxins which are opioid receptor ligands of the μ-type were isolated as C-terminally methoxylated peptides, e.g. the derivatives corresponding to the κ-casein sequences **8** (Chiba & Yoshikawa 1986). The

Table 2. Examples for multifunctional bioactive peptides encrypted in bovine milk proteins[1]

No.	Peptide sequence [2]	Fragment	Name	Preparation	Biological activity: Opioid IC_{50}^{3}	ACE-inhibitory IC_{50}^{4}	Immunomodulatory[5]	Ca^{2+} binding K_{app} $(l \cdot mol^{-1})^{6}$
1	YPFPGPIPNSL	β-CN(f60-70)	β-Casomorphin-11	Intest. Chyme	10			
2	YPFPGPI	β-CN(f60-66)	β-Casomorphin-7	Trypsin	14	500	−21/+26	
3	YPFPG	β-CN(f60-64)	β-Casomorphin-5	Trypsin	1.1	0		
4	RYLGYLE	α_{S1}-CN(f90-96)	α-Casein Exorphin	Pepsin	1.2			
5	YGLF	α-LA(f50-53)	α-Lactorphin	Synthesis	300	733		
6	YLLF	β-LG(f102-105)	β-Lactorphin	Synthesis	160	172		
7	YGFQNA	SA(f399-404)	Serorphin	Pepsin	85			
8	SRYPSY$\cdot$OCH$_3$	κ-CN(f33-38)	Casoxin 6	Pepsin	15 (↓)			
9	YIPIQYVLSR	κ-CN(f25-34)	Casoxin C	Trypsin	50 (↓)			
10	YL	β-LG(f102-103)	Lactokinin	Synthesis		122		
11	ALPMHIR	β-LG(f142-148)	Lactokinin	Trypsin		43		
12	VPP	β-CN (f84-86)	β-Casokinin	Sour Milk		9		
13	IPP	β-CN (f74-76)	β-Casokinin	Sour Milk		5		
14	AVPYPQR	β-CN(f177-183)	β-Casokinin	Trypsin		15		
15	FFVAPFPEVFGK	α_{S1}-CN(f23-34)	α_{S1}-Casokinin	Trypsin		77		
16	FFVAP	α_{S1}-CN(f23-27)	α_{S1}-Casokinin	Tryp.+Peptidase		6		
17	VAP	α_{S1}-CN(f25-27)	α_{S1}-Casokinin	Synthesis		2		
18	YQQPVLGPVR	β-CN(f193-202)	β-Casokinin-10	Synthesis		300	−28/+14	
19	TTMPLW	α_{S1}-CN(f194-199)	α_{S1}-Immunocasokinin	Trypsin		16	+162	
20	PGPIPN	β-CN(f63-68)	Immunopeptide	Synthesis			+139	
21	LLY	β-CN(f191-193)	Immunopeptide	Synthesis			+148	
22	YG	α-LA(f50-51)(f18-19),	κ-CN(f38-39)	Synthesis		>1000	+101	
23	YGG	α-LA(f18-20)	Immunopeptide	Synthesis			+35	
24	RELEELNVPGEIVES*LS*S*S*EESITR	β-CN(f1-25)4P	Caseinophosphopeptide	Trypsin				629
25	DIGS*ES*TEDQAMEDIM	α_{S1}-CN(f43-58)2P	Caseinophosphopeptide	Trypsin				328
26	QMEAES*IS*S*S*EEIVPNS*VEQK	α_{S1}-CN(f59-79)5P	Caseinophosphopeptide	Trypsin				841
27	FKCRRWQWRMKKLGAPSITCVRRAF	LF(f17-41)	Lactoferricin	Pepsin	antimicrobial: disruption of bacterial cell membranes			
28	MAIPPKKNQDK	κ-CN(f106-116)	Casoplatelin	Trypsin	antithrombotic: inhibition of platelet aggregation			

[1]Details are available within the references used to generate this Table (adapted from Meisel 1997): Bellamy et al. 1992; Brantl et al. 1981; Chiba & Yoshikawa 1986; Chiba et al. 1998; FitzGerald & Meisel 1999; Jollès et al. 1986; Kayser & Meisel 1996; Loukas et al. 1983; Meisel 1986; Meisel & Schlimme 1994; Migliore-samoure et al. 1989; Mullally et al. 1996, 1997; Nakamura et al. 1995; Teschemacher et al. 1997.
[2]The one-letter amino acid codes were used; Phosphoserin=S*.
[3]IC_{50} values (μmol/l) are given for peptide concentrations inhibiting [^{3}H]-ligand binding by 50%; (↓) indicates antagonistic activity.
[4]IC_{50} values (μmol/l) are given for peptide concentrations inhibiting the activity of angiotensin-converting-enzyme (ACE) by 50%.
[5]Figures indicate the maximum% stimulation (+) and/or inhibition (−), respectively, in relation to control (=100).
[6]Apparent calcium binding constants (K_{app}) were determined with samples obtained from tryptic fractions containing the indicated peptide as the main component.

chemically modified casoxins were more active than the non-methoxylated fragments.

Casein derived immunopeptides including fragments of α_{S1}-casein **19** and β-casein **18**, **20**, **21** stimulate phagocytosis of sheep red blood cells by murine peritoneal macrophages and exert a protective effect against *Klebsiella pneumoniae* infection in mice after intravenous administration of peptides (Migliore-Samour et al. 1989). The C-terminal β-casein sequence 193-209 containing β-casokinin-10 **18** induced a significant proliferative response in rat lymphocytes (Coste et al. 1992). Depending on peptide concentration β-casokinin-10 **18** and β-casomorphin-7 **2** showed a suppression as well as a stimulation of lymphocyte proliferation (Table 2). β-Casomorphin-7 **2** inhibits the proliferation of human colonic lamina propria lymphocytes where the anti-proliferative effect was reversed by the opiate receptor antagonist naloxone (Elitsur & Luk, 1991). Recently, Kayser and Meisel (1996) reported that the peptides Tyr-Gly **22** and Tyr-Gly-Gly **23** significantly enhanced the proliferation of human peripheral blood lymphocytes *in vitro*.

Antimicrobial peptides have been derived from the whey protein lactoferrin (Tomita et al. 1991). Lactoferrin is an iron-binding glycoprotein present in most biological fluids of mammals including milk. The existence of an antimicrobial sequence named lactoferricin **27** near the N-terminus of lactoferrin in a region distinct from its iron-binding sites has been reported (Bellamy et al. 1992). The antimicrobial activity of lactoferricin and synthetic analogs seems to be correlated with the net positive charge of the peptides. These cationic peptides kill sensitive microorganisms by increasing cell membrane permeability (Bellamy et al. 1993). Dionysius & Milne (1998) reported on the antibacterial activity of lactoferricins towards enterotoxigenic *Escherichia coli* and *Listeria monocytogenes*. The α_{S2}-casein fragment casocidin-I (position 165-203) containing a high proportion (10 of 39) of basic amino acid residues was found to inhibit the growth of *Esherichia coli* and *Staphylococcus carnosus* (Zucht et al. 1995, 1996). Another antibacterial α_{S1}-casein fragment (position 1-23) was significantly effective against lethal infection of mice by *Staphylococcus aureus* (Lahov & Regelson 1996).

Antithrombotic peptides, casoplatelins, are derived from the C-terminal part (caseinoglycomacropeptide) of bovine κ-casein. The main antithrombotic peptides of κ-casein are the sequence **28** and smaller fragments thereof (Bouhallab et al. 1992).

Several food protein sources including fish gelatin and maize protein contain peptides which are inhibitors of the Angiotensin-Converting-Enzyme and thus have an antihypertensive effect (for review: Ariyoshi 1993; Meisel 1993; Yamamoto 1997). ACE-inhibitors derived from milk protein represent different fragments of caseins **12-18** (casokinins; Meisel & Schlimme 1994) or whey proteins **10**, **11** (lactokinins; FitzGerald & Meisel 1999).

Mineral binding caseinophosphopeptides correspond to different phosphorylated regions of α_{S1}-, α_{S2}- and β-casein **24**, **25**, **26** (for review: FitzGerald 1998). The negatively charged side chains of phosphoseryl residues represent the anionic binding sites for minerals. Binding and solubilisation of calcium in the presence of phosphate has many potential beneficial consequences, e.g. inhibition of the precipitation of insoluble calcium phosphate.

Proteolytic activation of encrypted bioactive peptides

Relatively high amounts of bioactive peptides could potentially be produced during proteolysis of 1 g of each of the major casein and whey protein components (Meisel 1998). For chemical characterization and identification, most of the yet known bioactive peptides were proteolytically released from their parent milk protein by hydrolysis with pancreatic proteinases preferably with trypsin (Table 2). However, other enzyme activities and combinations of endoproteinases including chymotrypsin, pepsin, thermolysin and proline-specific endopeptidase were also used to generate bioactive peptides (for review: Meisel & Schlimme 1996, and cited ref.). Proteinase preparations of bacterial, fungal, plant and animal origin have recently been used to produce caseinophosphopeptides (McDonagh & FitzGerald 1998).

The release of bioactive peptides from their immediate precursor sequence is a prerequisite for any functional role in the living system. In addition to the possible liberation of bioactive peptides during intestinal proteolysis such peptides may already be generated during manufacture of several milk products and thus be ingested as food components. For example, partially hydrolyzed milk proteins for hypoallergenic infant formulae and for clinical applications in enteral nutrition consist exclusively of peptides.

A number of bacterial species used in food production or which occur in the human intestine may produce bioactive peptides. Cheese contains phosphopeptides as natural constituents and secondary proteolysis during cheese ripening leads to formation of various ACE-inhibitory peptides (Meisel et al. 1997). Data obtained from 'non-starter' lactic acid bacteria (*Lactobacillus* ssp.) present in fermented milk products and in the human intestine indicate that their proteolytic system is comparable to *Lactococcus lactis*. In *Lactococcus lactis*, cell wall-bound proteinases catalyze the first steps of casein degradation (Kunji et al. 1996). Two types of proteinases (PI, PIII) were detected which are distinguished by their specificity, PI with a preference for β-casein, PIII cleaving α- and β-casein at the same rate (Law & Haandrikman 1997). The specificities of these proteinases towards caseins have been determined by several authors. Early papers indicated that only small parts of caseins were degraded, and that relatively large fragments (chain lengths >10 amino acids) were formed (Visser et al. 1988; Monnet et al. 1989; Reid et al. 1991; Pritchard & Coolbear 1993). In a more recent study, Juillard et al. (1995) analysed 95% of the peptides formed by a PI-type proteinase by HPLC and mass spectrometry. The authors found that hydrolysis of β-casein proceeded much further than reported previously, more than 40% of the peptide bonds were cleaved resulting in the formation of more than 100 different oligopeptides. Many of the peptides had chain lengths of less than 10 amino acids and thus were substrates for the oligopeptide transport system. A further intracellular breakdown is catalyzed by a number of peptidases (Mierau et al. 1997). In spite of the presence of a di-/tripeptide transport system in *Lactococcus* the liberation of di- and tripeptides was not detected. Casein-derived peptides with chain lengths of more than 10 amino acids are abundant and contain various amino acid sequences representing bioactive peptides within a longer peptide (Juillard et al. 1995). Two degradation products of β-casein with biological activity were detected, β-CN(f60-68) and β-CN(f190-193) which are part of the sequence of β-casomorphin-11 **1** and immunopeptide **21**, respectively. The presence of the tetrapeptide can be concluded indirectly from the presence of two flanking peptides β-CN(f183-189) and β-CN(f194-209) in proteolytic digests of β-casein (Juillard et al. 1995).

Several casokinins derived from β-casein were liberated by a cell-wall-associated serine-type proteinase isolated from *Lactobacillus helveticus* CP790 (Yamamoto et al. 1994) and fermentation of milk with starter cultures containing *Lactobacillus helveticus* CP790 and *Saccharomyces cerevisiae* produced the β-casokinins **12** and **13**. An antihypertensive peptide, β-CN(f169-175), was identified from casein hydrolysate produced by the purified extracellular proteinase from *Lactobacillus helveticus* (Maeno et al. 1994). This peptide did not have a strong ACE-inhibitory activity (IC_{50}>1000 μmol/l). However, the corresponding hexapeptide, Lys-Val-Leu-Pro-Val-Pro, which was obtained after liberation of the C-terminal Gln residue by pancreatic digestion *in vitro*, had a strong ACE-inhibitory activity (IC_{50}=5 μmol/l) as well as a remarkable antihypertensive effect *in vivo* (Maeno et al. 1994). Liberation of α_{S1}-casokinins **15**, **16** and **19** by the lactococcal proteinase seems unlikely since no appropriate cleavage sites were found (Reid et al. 1991). Cleavage sites in α_{S1}-casein were often within the sequences of bioactive peptides (α_{S1}-CN f23/24, 33/34, 37/38, 74/75). Studies on the degradation of κ-casein showed that proteolytic cleavage sites were within the sequences of bioactive peptides (κ-CN f32/33, 106/107). One peptide, κ-CN(f33-41), was isolated that contained casoxin-6 **8** (Reid et al. 1994).

The formation of bioactive peptides by lactic acid bacteria in fermented milk products seems to be a rare event. Various long oligopeptides are liberated by degradation of caseins which could be precursors of peptides with biological activity when cleaved by other enzymes. In fermented milk products intracellular peptidases of lactic acid bacteria will most likely contribute to further degradation after cell lysis (Poolman et al. 1995; Kunji et al. 1996; Mierau et al. 1997). The specificities of the known peptidases suggest that practically all peptide bonds in caseins can be cleaved (Law & Haandrikman 1997). The formation of casomorphins in fermented milk products is particularly unlikely since all lactic acid bacteria (*Lactococcus* ssp. and *Lactobacillus* ssp.) are equipped with a X-prolyl-dipeptidyl-aminopeptidase (Law & Haandrikman 1997). The alternating X-Pro sequence of the casomorphins is the ideal substrate for this peptidase (Bockelmann et al. 1991). It was found that enzymatic degradation of β-casomorphins with bacterial cell lysate of *Lactococcus lactis* subsp. *cremoris* was influenced by the combination of pH and salt concentrations and proceeded over 15 weeks at cheese storage temperatures (Muehlenkamp & Warthesen 1998); β-casomorphins were more resistant to proteolysis at lower pH and higher salt concentrations (e.g. pH 5.0, 5% salt). In bovine milk samples incubated with caseolytic bacterial species such as *Pseudomonas aeruginosa* and *Bacillus cereus*, varying amounts of β-

casomorphin immunoreactive materials were detected (Hamel et al. 1985). With regard to casokinins and lactokinins, Pihlanto-Leppälä et al. (1998) have been shown that commercial lactic acid starters were not able to produce ACE-inhibitory activity from casein or whey *in vitro*, but further proteolysis with pepsin and trypsin was needed to liberate ACE-inhibitory peptides. No data are available on the action of the proteolytic enzymes from lactic acid bacteria in the human gut after ingestion of fermented milk products.

Tropho-functional properties

Milk-protein derived bioactive peptides may function as exogenous regulatory substances which may act on different intestinal and peripheral target sites of the mammalian organism. Orally ingested peptides or intestinal proteolytic products may produce local effects at the luminal side of the gastrointestinal tract. Moreover, regulatory peptides may be absorbed via carrier-mediated transport or via paracellular transport and then reach target tissues. The paracellular permeability is regulated by intestinal tight junctions, and modulation of tight-junction structure by other food substances facilitates paracellular transport (Shimizu 1999).

Bioactive peptides that have been produced by limited proteolysis during processing and/or intestinal digestion of milk proteins could be further digested by intestinal proteinases or brush-border peptidases. However, evidence for the liberation of β-casomorphins (Svedberg et al. 1985; Meisel 1986; Meisel & Frister 1989) and casein phosphopeptides (Meisel & Frister 1988; Kasai et al. 1992) from casein into the intestinal lumen of mammals after intake of milk or a casein-containing diet, i.e. under *in vivo* conditions, has already been obtained. Furthermore, evidence for *in vivo* production of antibacterial peptides has been provided by Tomita et al. (1994) who isolated lactoferricin from gastrointestinal contents of rats fed a diet containing bovine lactoferrin.

Opioid peptides derived from milk proteins appear to have physiological significance in the female organism (liberation of casomorphin in the mammary gland) and in neonates, and they appear to participate in the control of gastrointestinal functions in adults (Teschemacher et al. 1997). Opioid receptors are located in the nervous endocrine and immune systems as well as in the intestinal tract of the mammalian organism and can interact with their endogenous ligands as well as with exogenous opioids and opioid antagonists (Teschemacher & Brantl 1994). Orally given opioid peptides derived from milk proteins are able to modulate absorption processes in the gut. The enhancement of net water and electrolyte absorption by β-casomorphins in the small and large intestine is a major component of their anti-diarrhoeal action (Daniel et al. 1990a, b). This effect could be mediated via subepithelial opioid receptors or specific luminal binding sites at the brush border membrane (Tomé et al. 1987; Brandsch et al. 1994). β-Casomorphins **1**, **2**, **3** are claimed to be rapidly degraded once they enter the blood stream. However, the presence of β-casomorphin-7 immunoreactive material has been demonstrated in the plasma of newborn calves after their first milk intake (Umbach et al. 1985). This material revealed a greater molar mass than β-casomorphin-7 **1** and thus has been considered as a β-casomorphin precursor. Such pre-casomorphins could reach any potential site of action in the organism to elicit effects after liberation of the protected active sequence from the precursor molecule. Opioid casein fragments have not been detected in the plasma of adult mammals (Umbach et al. 1985; Teschemacher et al. 1986). Thus, only the neonatal intestine appears to be permeable to (pre-)casomorphins. A variety of opioid activities in further test systems have been described which may be of pharmacological but not of physiological significance (for review: Teschemacher et al. 1997).

The immunopotentiatory peptides Tyr-Gly **22** and Tyr-Gly-Gly **23** were found to be the active components in a dialysed leukocyte extract from normal donors that was recently used in a large multicenter trial to inhibit the development of infections in patients with pre-AIDS (Hadden 1991). It was suggested that casein peptides are implicated in the stimulation of the newborn's immune system (Migliore-Samour et al. 1989). The immunostimulating activities may also have a direct effect on the resistance to bacterial and viral infections of adult humans.

Bactericidal peptides may assist in protecting against a microbial challenge, especially in the neonatal intestinal tract, and thus support the non-immune defence of the gut (Schanbacher et al. 1997). It is not yet clear how antibacterial effects of food-derived peptides are of physiological importance.

Casoplatelins are inhibitors of the aggregation of ADP-activated platelets as well as binding of human fibrinogen γ-chain to a specific receptor site on the platelet surface (Jollès et al. 1986). Data on the physiological significance of food-derived peptides with antithrombotic activity are not yet available.

Inhibition of the multifunctional enzyme ACE which is located in different tissues (e.g. plasma, lung, kidney, heart, skeletal muscle, pancreas, brain, mammary arteries, testes, uterus, intestine) may influence different regulatory systems of the organism (Ondetti & Cushman 1982; Bruneval et al. 1986; Johnston 1992; Meisel 1993). ACE has been classically associated with the renin-angiotensin system regulating peripheral blood pressure where ACE-inhibition results in an antihypertensive effect. When ACE-inhibitory peptides were orally given to rats blood pressure was reduced in a dose-dependent manner (Suetsuna & Osajima 1989). It is known that small peptides, such as di- and tripeptides, are easily absorbed in the intestine without being decomposed by digestive enzymes. Accordingly, small ACE-inhibitory peptides can directly pass across the intestine to reach peripheral target sites (for review: Yamamoto 1997). In addition, ACE-inhibitory peptides having Pro residues, especially at the C-terminal end, are highly resistant to the degradation by digestive enzymes. Recently a placebo-controlled study showed that the daily intake of 95 ml of *Calpis* sour milk results in a significant decrease of blood pressure in hypertensive subjects (Hata et al. 1996). The sour milk drink, which was prepared by fermenting reconstituted skim milk powder with a starter culture of *Lactobacillus helveticus* CP790 and *Saccharomyces cerevisiae*, contains two ACE-inhibitory tripeptides Val-Pro-Pro **12** and Ile-Pro-Pro **13** (Nakamura et al. 1995) that were not present in the placebo drink. It is worth noting that the ingested dose of these peptides was in the range of only 1.2 to 1.6 mg.

The formation of soluble complexes with minerals and the proteolytic resistance of caseinophosphopeptides (CPPs) in the intestinal lumen is a prerequisite for their function as mineral carriers (for review: FitzGerald 1998). Under normal physiological conditions the paracellular transport of calcium plays a major role independent of vitamin D in calcium absorption from the distal small intestine (for review: Kitts & Yuan 1992). Hence CPPs might exert an influence on the absorption of calcium or other minerals and trace elements in the intestine. A recent human feeding trial reported an increase in calcium and zinc absorption following CPP incorporation into a rice-based infant food (Hansen 1997). Furthermore, it has been shown that calcium binding CPPs can have an anticariogenic effect in that they inhibit caries lesions through recalcification of the dental enamel (Reynolds 1987).

Potential uses in functional foods and pharmaceuticals

Although a number of studies indicate the regulatory potential of food-derived biologically active peptides, their role as food hormones or 'formones' needs further clarification. Even if some bioactive peptides are not released under physiological conditions *in vivo*, they could be produced commercially and used as ingredients in functional foods. A physiologically functional food can be defined as a food derived from naturally occurring substances that can and should be consumed as part of the daily diet, and thus provides health benefits in reducing the risk of developing a disease. Functional foods should not be confused with medical food products that are designed to supply missing nutrients or to be used for prevention and treatment of disease where pharmacologically active compounds (drugs) are needed. The question of what kinds of bioactive peptides are beneficial and desirable as food constituents or as drugs should be always carefully examined. However, the possibilities for the design of new dietary products and drugs to help reduce or control diet-related chronic diseases look promising. According to the present state of knowledge ACE-inhibitory peptides, caseinophosphopeptides and immunomodulating peptides are the most favourite bioactive peptides for application to foodstuffs formulated to provide specific health benefits.

Casein-derived peptides, which can be manufactured on industrial scale, have already been considered for interesting applications both as dietary supplements and as pharmaceutical preparations (for review: Meisel 1998). The production of bioactive peptides during food processing, for example by use of specific bacterial enzymes or genetically transformed microorganisms, is of interest for future research work.

Acknowledgements

Prof. E. Schlimme is gratefully acknowledged for the generous and stimulating support of the work on bioactive peptides and for helpful discussions.

References

Ariyoshi Y (1993) Angiotensin-converting enzyme inhibitors derived from food proteins. Trends Food Sci. Technol. 4: 139–144

Bellamy W, Takase M, Yamauchi K, Kawase K, Shimamura S & Tomita M (1992) Identification of the bactericidial domain of lactoferrin. Biochim. Biophys. Acta 1121: 130–136

Bellamy W, Wakabayashi H, Takase M, Kawase K, Shimamura S & Tomota M (1993) Role of cell-binding in the antibacterial mechanism of lactoferricin B. J. Appl. Bacteriol. 75: 478–484

Bockelmann W (1995) The proteolytic system of starter and non-starter bacteria: components and their importance for cheese ripening. Int. Dairy J. 5: 977–994

Bockelmann W, Fobker M. & Teuber M (1991) Purification and characterization of the X-prolyl- dipeptidyl- aminopeptidase from *Lactobacillus delbrueckii* subsp. *bulgaricus* and *Lactobacillus acidophilus*. Int. Dairy J. 1: 51–66

Bouhallab S, Mollé D & Léonil J (1992) Tryptic hydrolysis of caseinomacropeptide in membrane reactor: preparation of bioactive peptides. Biotechnol. Lett. 14: 805–810

Brandsch M, Brust P,Neubert K & Ermisch A (1994). β-Casomorphins – chemical signals of intestinal transport systems. In: Brantl V & Teschemacher H (Eds) β-Casomorphins and Related Peptides: Recent Developments (pp 207–219). VCH, Weinheim

Brantl V, Teschemacher H, Bläsig J, Henschen A & Lottspeich F (1981) Opioid activities of β-casomorphins. Life Sci. 28: 1903–1909

Bruneval P, Hinglais N & Alhenc-Gelas F (1986) Angiotensin I converting enzyme in human intestine and kidney. Ultrastructural immunohistochemical localization. Histochemistry 86: 73–80

Chiba H, Tani F & Yoshikawa M (1989) Opioid antagonist peptides derived from κ-casein. J. Dairy Res. 56: 363–366

Chiba H & Yoshikawa M (1986) Biologically functional peptides from food proteins: New opioid peptides from milk proteins. In: Feeney RE & Whitaker JR (Eds) Protein Tailoring for Food and Medical Uses (pp 123–153). Marcel Dekker Inc., New York

Daniel H, Vohwinkel M & Rehner G (1990a) Effect of casein and β-casomorphins on gastrointestinal motility in rats. J. Nutr. 120: 252–257

Daniel H, Wessendorf A, Vohwinkel M & Brantl V (1990b) Effect of D-Ala^{2,4}Tyr^{5}-β-casomorphin-5-amide on gastrointestinal functions. In: Nyberg F & Brantl V (Eds) β-Casomorphins and Related Peptides (pp 95–104). Fyris-Tryck AB, Uppsala

Dionysius DA & Milne JM (1998) Antibacterial peptides of bovine lactoferrin: purification and characterizaton. J. Dairy Sci. 80: 667–674

Elitsur Y & Luk GD (1991) β-casomorphin (BCM) and human colonic lamina propria lymphocyte proliferation. Clin. Experiment. Immunol. 85: 493–497

Fiat AM & Jollès P (1989) Caseins of various origins and biologically active casein peptides and oligosaccharides: structural and physiological aspects. Mol. Cell. Biochem. 87: 5–30

FitzGerald RJ (1998) Potential uses of caseinophosphopeptides. Int. Dairy J. 8: 451–457

FitzGerald RJ & Meisel H (1999) Lactokinins: Whey Protein-derived ACE Inhibitory Peptides. Nahrung/Food 431: 165–167

Fukudome S-I & Yoshikawa M (1994) Isolation and characterizaton of opioid peptides derived from wheat gluten. In: V. Brantl & H. Teschemacher (Eds) β-Casomorphins and related peptides: recent developments (pp 27–33). VCH, Weinheim

Hadden JW (1991) Immunotherapy of human immunodeficiency virus infection. Trends Pharmaceutical Sci. 12: 107–111

Hamel U, Kielwein G & Teschemacher H (1985) β-casomorphin immunoreactive materials in cow's milk incubated with various bacterial species. J. Dairy Res. 52: 139–148

Hansen M, Sandstöm B, Jensen M & Sörensen SS (1997) Casein phosphopeptides improve zinc and calcium absorption from rice-based but not from whole grain infant cereal. J. Pediatr. Gastroenterol. Nutr. 24: 56–62

Johnston CI (1992) Renin-angiotensin system: a dual tissue and hormonal system for cardiovascular control. J. Hypertension 10: S13–S26

Jollès P, Lévy-Toledano S, Fiat AM, Soria C, Gillessen D, Thomaidis A, Dunn FW & Caen JB (1986) Analogy between fibrinogen and casein. Eur. J. Biochem. 158: 379–384

Juillard V, Laan H, Kunji ERS, Jeronimus-Stratingh CM, Bruins AP & Konings WN (1995) The extracellular PI-type proteinase of *Lactococcus lactis* hydrolyzes β-casein into more than one hundred different oligopeptides. J. Bacteriol. 177: 3472–3478

Kasai T, Honda T & Kiriyama S (1992) Caseinophosphopeptides (CPP) in feces of rats fed casein diet. Biosci. Biotechnol. Biochem. 56: 1150–1151

Kayser H & Meisel H (1996) Stimulation of human peripheral blood lymphocytes by bioactive peptides derived from bovine milk proteins. FEBS Lett. 383: 18–20

Kitts DD & Yuan YV (1992) Caseinophosphopeptides and calcium bioavailability. Trends in Food Sci. & Technol. 3: 31–35

Konings WN, Lolkema JS, Bolhuis H, van Veen HW, Poolman B. & Driessen AJM (1997) The role of transport processes in survival of lactic acid bacteria. Antonie van Leeuwenhoek 71: 117–128

Kunji ERS, Mierau I, Hagting A, Poolman B & Konings WN (1996) The proteolytic systems of lactic acid bacteria. Antonie van Leeuwenhoek 70: 187–221

Lahov E & Regelson W (1996) Antibacterial and immunostimulating casein-derived substances from milk: casesidin isracidin peptides. Fd. Chem. Toxic. 34: 131–145

Law J & Haandrikman A (1997) Proteolytic enzymes of lactic acid bacteria. Int. Dairy J. 7: 1–11

Loukas S, Varoucha D, Zioudrou C, Streaty RA & Klee WA (1983) Opioid activities and structures of α-casein-dervied exorphins. Biochemistry 22: 4567–4573

Maeno M, Yamamoto Y & Takano T (1998) Identification of an antihypertensive peptide from casein hydrolysate produced by a proteinase from *Lactobacillus helveticus* CP790. J. Dairy Sci. 79: 1316–1321

McDonagh D & FitzGerald RJ (1998) Production of caseinophosphopeptides (CPPs) from sodium caseinate using a range of commercial protease preparations. Int. Dairy J. 8: 39–45

Meisel H (1986) Chemical characterization and opioid activity of an exorphin isolated from *in vivo* digests of casein. FEBS Lett. 196: 223–227

Meisel H (1993) Casokinins as inhibitors of Angiotensin-Converting-Enzyme. In: Sawatzki G & Renner B (Eds) New Perspectives in Infant Nutrition (pp 153–159). Thieme, Stuttgart, New York

Meisel H (1997a) Biochemical properties of bioactive peptides derived from milk proteins: potential nutraceuticals for food and pharmacological applications. Liv. Prod. Sci. 50: 125–138

Meisel H (1997b) Biochemical properties of regulatory peptides derived from milk proteins. Biopoly. 43: 119–128

Meisel H (1998) Overview on milk protein-derived peptides. Int. Dairy J. 8: 363–373

Meisel H & Frister H (1988) Chemical characterization of a caseinophosphopeptide isolated from *in vivo* digests of a casein diet. Biol. Chem. Hoppe-Seyler 369: 1275–1279

Meisel H & Frister H (1989) Chemical characterization of bioactive peptides from *in vivo* digests of casein. J. Dairy Res. 56: 343–349

Meisel H & Schlimme E (1994) Inhibitors of Angiotensin-Converting-Enzyme derived from bovine Casein (Casokinins). In: Brantl V & Teschemacher H (Eds) β-Casomorphins and related peptides: recent developments (pp 27–33). VCH , Weinheim

Meisel H & Schlimme E (1996) Bioactive peptides derived from milk proteins: Ingredients for functional foods? Kieler Milchwirtschaftl. Forschungsber. 48: 343–357

Meisel H, Goepfert A & Günther S (1997) Occurence of ACE inhibitory peptides in milk products. Milchwissenschaft 52 307–311

Mierau I, Kunji ERS, Venema G & Kok J (1997) Casein and peptide degradation in lactic acid bacteria. Biotech. Genetic Engineering Rev. 14: 279–301

Migliore-Samour D, Floc'h F & Jollès P. (1989) Biologically active casein peptides implicated in immunomodulation. J. Dairy Res. 56: 357–362

Monnet V, Bockelmann W, Gripon JC & Teuber M (1989) Comparison of cell wall proteinases from *Lactococcus lactis* subsp. *cremoris* AC1 and *Lactococcus lactis* subsp. *lactis* NCDO 763. II. specificity towards bovine b-casein. Appl. Microbiol. Biotechnol. 31: 112–118

Monnet V, Chapot-Chartier MP & Gripon JC (1993) Lactococcal peptidases. Lait 73: 97–108

Muehlenkamp MR & Warthesen JJ (1996) β-Casomorphins: analysis in cheese and susceptibility to proteolytic enzymes from *Lactobacillus lactis* ssp. *cremoris*. J. Dairy Sci. 79: 20–26

Mullally MM Meisel H & FitzGerald RJ (1996) Synthetic peptides corresponding to α-lactalbumin and β-lactoglobulin sequences with angiotensin-I-converting enzyme inhibitory activity. Biol. Chem. Hoppe-Seyler 377: 259–260

Mullally MM, Meisel H & FitzGerald RJ (1997) Identification of novel angiotensin-I-converting enzyme inhibitory peptide corresponding to a tryptic fragment of β-lactoglobulin. FEBS Lett. 402: 99–101

Nakamura Y, Yamamoto N, Sakai K, Okubo A, Yamazaki S & Takano T (1995) Purification and characterization of angiotensin I-converting enzyme inhibitors from a sour milk. J. Dairy Sci. 78: 777–783

Ondetti MA & Cushman DW (1982) Enzymes of the renin-angiotensin system and their inhibitors. Ann. Rev. Biochem. 51: 283–308

Pelissier JP (1984) Proteolysis of caseins. Sciences des Aliments 4: 1–35

Pihlanto-Leppälä A, Rokka T & Korhonen H (1998) Angiotensin I converting enzyme inhibitory peptides from bovine milk proteins. Int. Dairy J. 8: 325–331

Poolman B, Kunji, ERS, Hagting A, Juillard V & Konings WN (1995) The proteolytic pathway of *Lactococcus lactis*. J. Appl. Bacteriol. Symp. Supp. 79: 65S–75S

Pritchard GG & Coolbear T (1993) The physiology and biochemistry of the proteolytic system in lactic acid bacteria. FEMS Microbiol. Rev. 12: 179–206

Reynolds E (1987) Phosphopeptides. PCT Int. Patent Application WO 87/07615 A1

Reid JR, Coolbear T, Pillidge CJ & Pritchard GG (1994) Specificity of hydrolysis of bovine κ-casein by cell envelope-associated proteinases from *Lactococcus lactis* strains. Appl. Environ. Microbiol. 60: 801–806

Reid JR, Moore CH, Midwinter GG & Pritchard GG (1991) Action of cell wall proteinase from *Lactococcus lactis* subsp. *cremoris* SK11 on bovine α_{S1}-casein. Appl. Microbiol. Biotechnol. 35: 222–227

Sato R, Naguchi T & Naito H (1986) Casein phosphopeptide (CPP) enhance calcium absorption from the ligated segment of rat small intestine. J. Nutri. Sci. Vitaminol. 32: 67–76

Schanbacher FL, Talhouk RS & Murray FA (1997) Biology and origin of bioactive peptides in milk. Liv. Prod. Sci. 50: 105–123

Shimizu M (1999) Modulation of intestinal functions by food substances. Nahrung/Food 143: 154–158

Suetsuna K & Osajima K (1989) Blood pressure reduction and vasodilatory effects *in vivo* of peptides originating from sardine muscle (in Japan). J. Jp. Soc. Nutr. Food Sci. 42: 47–54

Svedberg J, de Haas J, Leimenstoll G, Paul F & Teschemacher H (1985) Demonstration of β-casomorphin immunoreactive materials in *in vitro* digests of bovine milk and in small intestine contents after bovine milk ingestion in adult humans. Peptides 6: 825–830

Teschemacher H & Brantl V (1994) Milk protein derived atypical opioid peptides and related compounds with opioid antagonist activity. In: Brantl V & Teschemacher H (Eds) β-Casomorphins and Related Peptides: Recent Developments (pp 3–17). VCH, Weinheim

Teschemacher H, Umbach M, Hamel U, Praetorius K, Ahnert-Hilger G, Brantl V, Lottspeich F & Henschen A (1986) No evidence for the presence of β-casomorphins in human plasma after ingestion of cows' milk or milk products. J. Dairy Res. 53: 135–138

Teschemacher H, Koch G & Brantl V (1997) Milk protein-derived opioid receptor ligands. Biopoly. 43: 99–117

Tomé D, Dumontier AM, Hautefeuille M & Desjeux JF (1987) Opiate activity and transepithelial passage of intact β-casomorphins in rabbit ileum. Am. J. Physiol. 253: G737–G744

Tomita M, Bellamy W, Takase M, Yamauchi K, Wakabayashi H & Kawase K (1991) Potent antibacterial peptides generated by pepsin digestion of bovine lactoferrin. J. Dairy Sci. 74: 4137–4142

Tomita M, Takase M, Bellamy W & Shimamura S (1994) A review: the active peptide of lactoferrin. Acta Pediatr. Jpn. 36: 585–591

Umbach M, Teschemacher H, Praetorius K, Hirschhäuser R & Bostedt H (1985) Demonstration of a β-casomorphin immunoreactive material in the plasma of newborn calves after milk intake. Regulatory Peptides 12: 223–230

Visser S, Slangen J, Exterkate FA & de Veer GJCM (1988) Action of a cell wall proteinase (PI) from streptococcus cremoris HP on bovine β-casein. Appl. Microbiol. Biotechnol. 29: 61–66

Yamamoto N (1997) Antihypertensive peptides derived from food proteins. Biopoly. 43: 129–134

Yamamoto N, Akino A & Takano T (1994) Antihypertensive effect of the peptides derived from casein by an extracellular proteinase from *Lactobacillus helveticus* CP790. J. Dairy Sci. 77: 917–922

Yoshikawa M, Tani F & Chiba H (1988) Structure-activity relationship of opioid antagonist peptides derived from milk proteins. In: Shiba T (Ed) Peptide Chemistry (pp 473–476). Protein Research Foundation, Osaka

Zucht HD, Raida M, Andermann K, Mägert H-J & Forssman WG (1995) Casocidin-I: a casein-α_{S2} derived peptide exhibits antibacterial activity. FEBS Lett. 372 185–188

Antonie van Leeuwenhoek **76:** 217–246, 1999.

Peptidases and amino acid catabolism in lactic acid bacteria

Jeffrey E. Christensen[2], Edward G. Dudley[2], Jeffrey A. Pederson[1] & James L. Steele[1,*]
[1]*Department of Food Science,* [2]*Department of Bacteriology, University of Wisconsin-Madison, Madison, WI 53706-1565, USA*
*(*Author for correspondence; E-mail: jlsteele@facstaff.wisc.edu)*

Key words: amino acid catabolism, lactic acid bacteria, peptidases, physiology

Abstract

The conversion of peptides to free amino acids and their subsequent utilization is a central metabolic activity in prokaryotes. At least 16 peptidases from lactic acid bacteria (LAB) have been characterized biochemically and/or genetically. Among LAB, the peptidase systems of *Lactobacillus helveticus* and *Lactococcus lactis* have been examined in greatest detail. While there are homologous enzymes common to both systems, significant differences exist in the peptidase complement of these organisms. The characterization of single and multiple peptidase mutants indicate that these strains generally exhibit reduced specific growth rates in milk compared to the parental strains. LAB can also catabolize amino acids produced by peptide hydrolysis. While the catabolism of amino acids such as Arg, Thr, and His is well understood, few other amino acid catabolic pathways from lactic acid bacteria have been characterized in significant detail. Increasing research attention is being directed toward elucidating these pathways as well as characterizing their physiological and industrial significance.

Introduction

Lactic acid bacteria (LAB) are commonly defined as Gram-positive, nonsporulating, catalase negative, aerotolerant, acid tolerant, nutritionally fastidious, strictly fermentative organisms that lack cytochromes and produce lactic acid as the major end-product of carbohydrate metabolism (Axelsson 1998). The genera typically included are *Aerococcus*, *Carnobacterium*, *Enterococcus*, *Lactobacillus*, *Lactococcus*, *Leuconostoc*, *Oenococcus*, *Pediococcus*, *Streptococcus*, *Tetragenococcus*, *Vagococcus*, and *Weissella* (Axelsson 1998).

The conversion of peptides to free amino acids and the subsequent utilization of these amino acids is a central metabolic activity in LAB. The peptidase system is involved in the hydrolysis of exogenous peptides to obtain essential amino acids for growth and hydrolysis of peptides formed by housekeeping proteinases. The amino acids formed by this system can be utilized for processes such as protein synthesis, generation of metabolic energy, and recycling of reduced cofactors. This review will cover our current knowledge of the enzymes and metabolic pathways involved in peptide and amino acid catabolism in LAB, with a focus on LAB that are associated with food fermentations.

Peptidases of lactic acid bacteria

The lactococci and lactobacilli associated with dairy environments generally have between six and fourteen amino acid auxotrophies. The ability of LAB to grow to high cell density in milk is dependent on a proteolytic system that can liberate essential amino acids from casein derived peptides. Although this review will not elaborate on the proteinase and peptide transport activities discussed in previous reviews (Kunji et al. 1996; Law & Haandrikman 1997), the relevance of recent articles concerning this function of the proteolytic system will be addressed.

The PrtP hydrolysis and Opp transport of peptides derived from purified β-casein by *Lc. lactis* has been investigated by subtractive evaluation of the remaining media peptides and accumulation of intracellular peptides and amino acids (Kunji et al. 1998). The results indicate that most of the essential amino acids can be obtained from transported peptides derived from the C-terminal end (residues 161–191) of purified β-casein. The exception was His, which was previously shown to be a required supplement for growth on purified β-casein (Kunji et al. 1995). Opp was shown to transport from ten to fourteen of the β-casein derived peptides ranging in size from five to ten residues. However, subsequent kinetic analysis of Opp indicates that peptides from four to at least eighteen residues can be transported with little specificity for particular side chains (Detmers et al. 1998). These studies begin to define the potential peptidase substrates available to *Lc. lactis* during growth in milk.

Peptidase specificity

The available substrate hydrolysis information for peptidases was tabulated (see Table 1 references) in order to obtain a more comprehensive view of the specificity of a given peptidase. The cumulative data is comprised of a variety of amino acid ρ-nitroanilide (−ρNA) and β-naphthilamide (−βNAP) substrates, ∼180 dipeptides, ∼100 tripeptides, and ∼40 oligopeptides (data not shown). The amino acids of the natural (non-chromogenic) substrates were classified according to their characteristics (i.e. hydrophilicity, charge, side chain grouping). The data was sorted with regard to the order and class of the amino acids, resulting in the grouping of peptides with common features. For example, Ala-Phe and Leu-Tyr were grouped together (hydrophobic-aromatic), as were Lys-Pro and His-Pro (basic-imino). The substrate hydrolysis generalizations that were notable are included with the descriptions of individual peptidases.

While the identification and classification of peptidases according to the hydrolysis of chromogenic substrates or small peptides is convenient, this information does little to elucidate the specific role an enzyme may have in hydrolysis of exogenous peptides or general cellular physiology.

Aminopeptidase C

The broad specificity aminopeptidase C (PepC) has been purified and characterized from strains of *Lb. delbrueckii, Lb. helveticus, Lc. lactis* and *S. thermophilus* (see Table 1) (Neviani et al. 1989; Wohlrab & Bockelmann 1993; Chapot-Chartier et al. 1994; Mistou et al. 1994; Fernández de Palencia 1999). Significant activity of PepC on AA-βNAP substrates is reported for residues that are basic (Arg, His, and Lys), acidic (Glu and Asp), hydrophobic/uncharged (Ala and Leu), and aromatic (Phe). A similar level of activity is found for the corresponding AA-ρNA substrates, including Met- and Gly-ρNA. No PepC activity is detectable for Pro-ρNA, Pro-βNAP, Xaa-Pro-ρNA, or Xaa-Pro-βNAP substrates. Hydrolysis is also reported for a variety of di- and tripeptides with uncharged or basic residues in the amino terminal position.

A systematic evaluation of *Lc. lactis* PepC catalytic properties indicates the enzyme contains an extended binding site (Mistou & Gripon 1998). PepC was determined to have preference for a hydrophobic residue in the P′1 (Gly-Xaa) and P′2 (Gly-Gly-Xaa) position of the respective substrates. Analysis of various di-, tri-, and tetrapeptides revealed an increasing affinity (decreasing K_m) with increasing substrate amino acid residues. Additionally, a comparison of enzymatic parameters measured for polyglycine substrates (Gly_n where n=2 through 5) demonstrated the dependence of catalytic activity on substrate length. The highest substrate affinity and hydrolysis were measured with the tetrapeptides Gly-Gly-Phe-Leu and Gly-Gly-Gly-Ala.

Structural modeling of the yeast bleomycin hydrolase and *Lc. lactis* PepC suggest that the carboxy-terminal residues are inserted into the active site toward the catalytic residues of these enzymes, permitting interaction between the enzyme α-carboxy group and the substrate α-amino group (Joshua-Tor et al. 1995; Mata et al. 1997). Characterization of a site-directed mutant enzyme, missing the C-terminal residue (PepC-ΔAla^{435}), confirmed dependence on this residue for strict aminopeptidase activity (Mata et al. 1997).

Northern analysis with a *pepC* (∼1.3-kb ORF) specific probe identified two transcript sizes, ∼1.7 and ∼2.7-kb, from *Lb. helveticus* grown in MRS broth under pH controlled conditions (5.85) (Vesanto et al. 1994). Both *pepC*-hybridized transcripts were present at late exponential phase, but below detectable levels at stationary phase. Nucleotide sequence analysis of the regions proximal to *pepC* identified a downstream ORF (840-bp) that encodes a putative 30 kDa transmembrane protein with 40.4% amino acid identity to a *Staphylococcus xylosus* glucose uptake protein (Brückner 1998). Northern analysis with a probe specific to the downstream ORF revealed co-transcription with *pepC* during late exponential phase, while at stationary phase the ORF is transcribed from its own promoter resulting in a 1.0-kb monocistronic tran-

Table 1. Peptidases of lactic acid bacteria

Peptidase		Strain	kDa[a]	Structure[b]	Class[c]	Motifs[d]	Reference
Aminopeptidase	PepC	*Lb. delbrueckii*	54[H]	tetramer	C		Wohlrab & Bockelmann (1993)
		Lb. delbrueckii	51[N]		C	PS00139[e]	Klein et al. (1994a)
		Lb. helveticus	49[N]		C	PS00139, PS00639	Fernández et al. (1994)
		Lb. helveticus	51[N]		C	PS00139, PS00639	Vesanto et al. (1994)
		Lb. helveticus	50[S]	tetramer	C		Fernández de Palencia (1999)
		Lc. lactis	50[H]	hexamer	C		Neviani et al. (1989)
		Lc. lactis	50[N]		C	PS00139, PS00639	Chapot-Chartier et al. (1993)
		Lc. lactis	50	hexamer[X]			Mistou et al. (1994)
		S. thermophilus	50[N]	hexamer	C	PS00139, PS00639	Chapot-Chartier et al. (1994)
Aminopeptidase	PepN	*Lb. casei*	95[S]	monomer	M		Fernández de Palencia et al. (1997)
		Lb. casei	87[S]	monomer	M		Arora & Lee (1992)
		Lb. rhamnosus	89[S]	monomer	M		Arora & Lee (1994)
		Lb. delbrueckii	98[S]	monomer	M		Tsakalidou et al. (1993)
		Lb. delbrueckii	95[S]		M		Bockelmann et al. (1992)
		Lb. delbrueckii	95[N]	monomer	M	PS00142	Klein et al. (1993)
		Lb. helveticus	97[S]	monomer	M		Khalid & Marth (1990a)
		Lb. helveticus	96[N]		M	PS00142	Christensen et al. (1995)
		Lb. helveticus	97[S]	monomer	M		Blanc et al. (1993)
		Lb. helveticus	92[S]	monomer	M		Miyakawa et al. (1992)
		Lb. helveticus	95[S]	monomer	M		Sasaki et al. (1996)
		Lb. helveticus	96[N]		M	PS00142	Varmanen et al. (1994)
		Lb. lactis	78[S]	monomer	M		Eggimann & Bachmann (1980)
		Lb. sanfrancisco	75[S]	monomer	M		Gobbetti et al. (1996)
		Lc. lactis	95[H]		M		Exterkate et al. (1992)
		Lc. lactis	95[S]	monomer			Van Alen-Boerrigter et al. (1991)
		Lc. lactis	95[N]		M	PS00142	Strøman (1992)
		Lc. lactis	95[S]	monomer	M		Tan & Konings (1990)
		Lc. lactis	95[N]		M	PS00142	Tan et al. (1992)
		Lc. lactis	85[H]		M		Desmazeaud & Zevaco (1979)
		S. thermophilus	92[S]	monomer	M		Tsakalidou & Kalantzopoulos (1992)
		S. thermophilus	97[S]	monomer	M		Rul et al. (1994)
		S. thermophilus	98[S]	monomer	M		Midwinter & Pritchard (1994)
		S. thermophilus	89[S]	monomer	M		Tsakalidou et al. (1993)
Aminopeptidase	Unclassified	*Lb. acidophilus*	38[S]	tetramer	M		Machuga & Ives (1984)
		Lb. delbrueckii	32[S]	octamer	M		Wohlrab & Bockelmann (1994)
		Lb. casei	30[S]	tetramer	S		Fernández de Palencia et al. (1997)
		Lc. lactis	36[S]		M		Geis et al. (1985)
	PepA	*Lc. lactis*	40[S]	hexamer	M		Bacon et al. (1994)
		Lc. lactis	43[S]	trimer	M		Exterkate & De Veer (1987)
		Lc. lactis	38[N]				I'Anson et al. (1995)
		Lc. lactis	41[S]	hexamer	M		Niven (1991)
		S. thermophilus	45[S]	octamer	M		Rul et al. (1995)
XPDAP	PepX	*Lb. acidophilus*	95[S]	dimer	S		Bockelmann et al. (1991)
		Lb. casei	79[S]	monomer	S		Habibi-Najafi & Lee (1994)
		Lb. delbrueckii	95[S]	dimer	S		Bockelmann et al. (1991)
		Lb. delbrueckii	82[S]		S		Atlan et al. (1990)
		Lb. delbrueckii	90[S]	trimer	S		Miyakawa et al. (1991)

Table 1. Continued

Peptidase		Strain	kDa[a]	Structure[b]	Class[c]	Motifs[d]	Reference
		Lb. delbrueckii	88[N]	monomer	S	361-GRSYLG	Meyer-Barton et al. (1993)
		Lb. helveticus	72[S]	monomer	S		Khalid & Marth (1990b)
		Lb. helveticus	90[N]		S	361-GRSYLG	Yüksel & Steele (1996)
		Lb. helveticus	87[S]	monomer	S		Miyakawa et al. (1994)
		Lb. helveticus	91[N]	dimer	S	361-GRSYLG	Vesanto et al. (1995)
		Lb. lactis	90[S]	dimer	S		Meyer & Jordi (1987)
		Lc. lactis	117[H]		S		Booth et al. (1990b)
		Lc. lactis	88[S]	dimer[X]			Chich et al. (1995)
		Lc. lactis	88[N]			346-GKSYLG	Nardi et al. (1991)
		Lc. lactis	85[S]	dimer	S		Zevaco et al. (1990)
		Lc. lactis	88[S]	dimer	S		Yan et al. (1991)
		Lc. lactis	90[S]	dimer	S		Kiefer-Partsch et al. (1989)
		Lc. lactis	88[N]			346-GKSYLG	Mayo et al. (1991)
		Lc. lactis	82[S]	dimer	S		Lloyd & Pritchard (1991)
		S. thermophilus	80[S]	dimer	S		Tsakalidou et al. (1998)
		S. thermophilus	80[S]	dimer	S		Meyer & Jordi (1987)
Dipeptidase	PepD	*Lb. helveticus*	53[N]				Dudley et al. (1996)
		Lb. helveticus	53[N]	octamer	C		Vesanto et al. (1996)
Dipeptidase	PepV	*Lb. casei*	46[S]	monomer	M		Fernández-Esplá & Martín-Hernández (1997)
		Lb. delbrueckii	51[S]	monomer	M		Wohlrab & Bockelmann (1992)
		Lb. delbrueckii	52[N]	monomer	M	PS00758, PS00759	Vongerichten et al. (1994)
		Lb. helveticus	51[N]			PS00758, PS00759	Yüksel & Steele (1997b)
		Lb. helveticus	50[S]		M		Tan et al. (1995)
		Lb. sake	50[S]	monomer	M		Montel et al. (1995)
		Lb. sanfrancisco	65[S]	monomer	M		Gobbetti et al. (1996)
		Lc. lactis	100[H]		M		Hwang et al. (1981)
		Lc. lactis	51[N]		M	PS00758, PS00759	Hellendoorn et al. (1997)
		Lc. lactis	49[S]	monomer	M		Van Boven et al. (1988)
		Lc. lactis	50[H]		M		Desmazeaud & Zevaco (1977)
Pro-iminopeptidase	PepI	*Lb. delbrueckii*	33[N]		S	105-GQSWGG	Atlan et al. (1994)
		Lb. delbrueckii	34[S]	trimer	S		Gilbert et al. (1994)
		Lb. delbrueckii	33[N]		S	104-GQSWGG	Klein et al. (1994b)
		Lb. helveticus	34[N]	dimer	S	103-GQSWGG	Varmanen et al. (1996a)
Pro-iminopeptidase		*Lc. lactis*	50[S]	dimer	M		Baankreis & Exterkate (1991)
Prolidase	PepQ	*Lb. casei*	41[S]	monomer	M		Fernández-Esplá et al. (1997)
		Lb. delbrueckii	41[N]	dimer		PS00491	Morel et al. (1999b)
		Lb. delbrueckii	41[N]		M	PS00491	Rantanen & Palva (1997)
		Lb. delbrueckii	41[N]		M	PS00491	Stucky et al. (1995)
		Lb. helveticus	41[N]			f	Yüksel & Steele (1997a)
		Lc. lactis	42[H,D]		M		Booth et al. (1990a)
		Lc. lactis	43[H,D]		M		Kaminogawa et al. (1984)
Prolinase	PepR	*Lb. helveticus*	35[N]			109-GQSWGG	Dudley & Steele (1994)
		Lb. helveticus	33[S]	tetramer	S		Shao et al. (1997)
		Lb. helveticus	35[N]			109-GQSWGG	Varmanen et al. (1996b)
		Lb. rhamnosus	34[N]			107-GQSWGG	Varmanen et al. (1998)
		Lb. curvatus	32[S]	dimer			Magboul & McSweeney (1998)

Table 1. Continued

Peptidase		Strain	kDa[a]	Structure[b]	Class[c]	Motifs[d]	Reference
	PepL	*Lb. delbrueckii*	35[N]		S	105-GHSWGG	Klein et al. (1995)
	PepP	*Lc. lactis*	43[S]	monomer	M		Mars & Monnet (1995)
		Lc. lactis	46[N]		M	PS00491	Matos et al. (1998)
Tripeptidase	PepT	*Lc. lactis*	55[S]	dimer	M		Bacon et al. (1993)
		Lc. lactis	46[N]			PS00758, PS00759	Mierau et al. (1994)
		Lc. lactis	52[S]	dimer	M		Bosman et al. (1990)
Tripeptidase	Unclassified	*Lb. delbrueckii*	29[S]	trimer	M		Bockelmann et al. (1995)
		Lb. delbrueckii	38[S]	dimer	M		Bockelmann et al. (1997)
		Lb. sake	55[S]	monomer	M		Sanz et al. (1998)
		Lc. lactis	23[S]	trimer			Sahlstrøm et al. (1993)
		Lc. lactis	75[H]		M		Desmazeaud & Zevaco (1979)
		P. pentosaceus	45[S]	dimer	M		Simitsopoulou et al. (1997)
Endopeptidase	PepE	*Lb. helveticus*	52[N]		C	PS00139, PS00639	Fenster et al. (1997)
	PepG	*Lb. delbrueckii*	50[N]		C	PS00139, PS00639	Klein et al. (1997)
Endopeptidase	PepO	*Lb. helveticus*	71[N]		M	PS00142	Chen & Steele (1998)
		Lc. lactis	71[N]		M	PS00142	Lian et al. (1996)
		Lc. lactis	71[N]		M	PS00142	Mierau et al. (1993)
		Lc. lactis	70[S]	monomer	M		Pritchard et al. (1994)
		Lc. lactis	71[N]			PS00142	Tynkkynen et al. (1993)
Endopeptidase	PepF1	*Lc. lactis*	70[N]	monomer	M	[g]	Monnet et al. (1994)
	PepF1	*Lc. lactis*	70[N]			[g]	Nardi et al. (1997)
	PepF2	*Lc. lactis*	70[N]			PS00142	Nardi et al. (1997)
Endopeptidase	Unclassified	*Lb. delbrueckii*	68[S]	monomer	M		Bockelmann et al. (1996)
		Lb. paracasei	30[S]	multimer	M		Tobiassen et al. (1997)
		Lc. lactis	98[S]	monomer	M		Yan et al. (1987b)
		Lc. lactis	180	multimer	M		Baankreis et al. (1995)
		Lc. lactis	40[S]	dimer	M		Yan et al. (1987a)
		Lc. lactis	70[S]	monomer	M		Baankreis et al. (1995)
		Lc. lactis	70[S]	monomer	M		Stepaniak & Fox (1995)
		Lc. lactis	52[S]	multimer	M		Stepaniak et al. (1998)
		Lc. lactis	93[H]		M		Muset et al. (1989)
		Lc. lactis	70[S]	monomer	M		Tan et al. (1991)
		Lc. lactis	140[H]		M		Ohmiya & Sato (1975)
		Lc. lactis	50[H]		M		Desmazeaud & Zevaco (1976)

[a]Molecular weight of a subunit calculated from the derived amino acid sequence (N) when nucleotide sequence was available; in other cases individual subunit size is reported from SDS-PAGE (S) or gel filtration (H), with preference given to SDS-PAGE values when both are reported.

[b]Quaternary structure predicted from comparison of gel filtration value with either derived amino acid sequence or SDS-PAGE values; X-ray crystal analysis (X).

[c]Catalytic class of enzyme determined biochemically or by sequence motif identification: C, cysteine-peptidase; M, metallo-peptidase; or S, serine-peptidase.

[d]Active site motif for prolyl oligopeptidase family (see Rawlings et al. 1991) (residue position-GxSxxG) or PROSITE pattern identification determination (Hofmann et al. 1999): PS00139, Thiol Protease Cys; PS00639, Thiol Protease His; PS00142, Neutral zinc metallopeptidases, zinc-binding region signature; PS00758, ArgE/*dapE*/ACY1/CPG2/*yscS* family signature 1; PS00759, ArgE/*dapE*/ACY1/CPG2/*yscS* family signature 2; and PS00491, Aminopeptidase P and proline dipeptidase signature.

[e]The sequence of *Lb. delbrueckii* contains a single residue mismatch to the current PROSITE Thiol Protease His signature.

[f]The sequence of *Lb. helveticus* contains a one residue mismatch to the current PROSITE Aminopeptidase P and proline dipeptidase signature.

[g]The sequence encoded by *pepF1* contains two adjacent residue mismatches to the current PROSITE Neutral zinc metallopeptidases (zinc-binding region) signature. However, the same region is coincidently overlapped by PROSITE Signal peptidases I serine active site signature.

script. Currently, there is no information available on the transcriptional regulation of *pepC* for LAB grown in milk.

The physiological role of PepC in *Lc. lactis* has been investigated for deletion mutant derivatives as a function of specific growth rate (Mierau et al. 1996a). The growth rate of the PepC deletion mutant was reported to be the same as for the wild type strain in complex media (GM17 broth) while in milk a subtle decrease in growth rate is observed (μ_{max} 0.60 vs 0.65 h^{-1}), corresponding to a 10% longer generation time (Mierau et al. 1996a).

Aminopeptidase N

The broad specificity aminopeptidase N (PepN) has been purified and characterized from strains of *Lb. casei*, *Lb. delbrueckii*, *Lb. helveticus*, *Lc. lactis* and *S. thermophilus* (see Table 1). In general, the activity of PepN on AA-ρNA substrates indicates the highest specificity for the basic amino acids Lys and Arg, followed by the hydrophobic/uncharged residues Leu and Ala. Significant activity is also observed for Met and Phe-ρNA, while in general low to undetectable activity is reported for Asp-, Glu-, and Gly-ρNA. Niven et al. (1995) reported a general increase in activity with increase in the hydrophobicity of the carboxy-terminal residue of an Arg-Xaa dipeptide. Also, a trend of increasing K_m and V_{max} with peptide chain length was observed for Lys-Phe-$(Gly)_n$ substrates, with the highest specific activity for the Lys-Phe-$(Gly)_4$ hexapeptide. Several tryptic digest derived oligopeptides (7–16 residues) of β-casein were at least partially hydrolyzed by purified PepN, demonstrating the capacity to liberate residues from even larger substrates (Tan et al. 1993). Additionally, a study using cell-free extracts prepared from peptidase mutants of *Lb. helveticus* has shown that PepN is predominant in the liberation of amino-terminal Tyr residue from β-casein f193-209 (Christensen & Steele 1998a). These results accentuate the range of peptide size and conditional specificity of PepN beyond the amino terminal residue of the substrate.

Northern analyses with *pepN* specific probes derived from *Lb. helveticus* and *Lc. lactis* suggest that the transcripts from the genes are monocistronic (Strøman 1992; Varmanen et al. 1994). In *Lb. helveticus*, grown in MRS broth under pH controlled conditions (5.85), the level of transcription of *pepN* is reported to remain constant through exponential and stationary phase (Varmanen et al. 1994). However, evaluation of the data provided in the publication figure indicates the ratio between *pepN* mRNA (dpm) and cell density nearly doubles in the two generations preceding the onset of stationary phase. Similarly, the ratio of total aminopeptidase activity to message level increases nearly six-fold in the same time frame. The fact that PepN activity remains constant 12 h into stationary phase while *pepN* transcript levels decrease suggests the enzyme is relatively stable and/or there is an elongated balance of turnover and expression under pH controlled conditions. In contrast, a study of PepN activity from MRS broth grown *Lb. helveticus* without pH control indicates a decrease in PepN activity from end of exponential phase (pH ~4.8) and reduction by four-fold during transition to stationary phase (Christensen & Steele 1998b). The activity of *Lb. helveticus* PepN accounts for greater than 99% of the hydrolysis of the substrates (Lys-, Leu-, Met-, and Ala-ρNA) used to measure total aminopeptidase activity in the previously mentioned studies (Christensen & Steele 1996).

The regulation of PepN expression in *Lc. lactis* has been demonstrated to be strain dependent (Meijer et al. 1996). Total aminopeptidase activity is equivalent or higher in cultures grown in milk compared to complex amino acid or peptide-containing media (Meijer et al. 1996). Additionally, the dipeptide Pro-Leu is reported to decrease PepN and PrtP expression in *Lc. lactis* MG1363 (Marugg et al. 1995). Currently, there is no information available on the transcriptional regulation of *pep*N for LAB grown in milk.

The physiological role of PepN in *Lb. helveticus* and *Lc. lactis* has been investigated for deletion mutant derivatives as a function of specific growth rate. The growth rate and final cell densities of *Lb. helveticus* PepN deletion mutants are the same as for the wild type strain in complex media (MRS broth) and complete amino acid defined media (Christensen & Steele 1996). The growth rate in milk was slightly slower for the PepN deletion mutant compared to wild type (μ_{max} 0.56 vs 0.60 h^{-1}), corresponding to a 7% longer generation time, but the final cell densities were similar. Therefore, despite the *Lb. helveticus* auxotrophy for fourteen amino acids, the remaining peptidases are capable of liberating essential amino acids from milk derived peptides at a velocity that nearly maintains the parental growth rate.

The growth rate of a *Lc. lactis* PepN deletion mutant was reported to be the same as for the wild type strain in complex media (GM17 broth) (Mierau et al. 1996a). However, the *Lc. lactis* PepN mutant displays a decreased growth rate in milk corresponding to a 25% longer generation time compared to wild type (μ_{max} 0.52 vs 0.65 h^{-1}) (Mierau et al. 1996a; Mierau et al. 1996b). These results suggest that PepN plays a more important role in hydrolysis of milk derived peptides in *Lc. lactis* than in *Lb. helveticus*.

Aminopeptidase A

Aminopeptidase A (PepA), also referred to as glutamyl aminopeptidase, has been purified and characterized from several strains of *Lc. lactis* (Exterkate & De Veer 1987; Niven 1991; Bacon et al. 1994) and *S. thermophilus* (Rul et al. 1995). PepA activity is essentially defined by the characteristic hydrolysis of N-terminal acidic amino acid residues. The enzyme hydrolyses Glu- and Asp-ρNA, but low activity was reported for Glu- and Asp-βNAP with PepA of *S. thermophilus* (Rul et al. 1995). Hydrolysis of Glu-Xaa and Asp-Xaa dipeptides is observed when the C-terminus amino acid is basic (-Lys), uncharged (-Gly), hydrophobic/uncharged (-Ala, -Leu), polar/uncharged (-Ser), or aromatic (-Phe, -Tyr). The dipeptides Ser-Xaa are also hydrolyzed when the C-terminus amino acid is hydrophobic/uncharged (-Ala, -Leu) or aromatic (-Phe, -Tyr). PepA is also capable of releasing acidic residues from peptides as large as a decapeptide (Bacon et al. 1994). Significant hydrolysis is not observed for di- and tripeptides with N-terminal amino acids from any other class.

The physiological role of PepA in *Lc. lactis* has been investigated for insertionally inactivated *pepA* deficient derivatives as a function of growth in milk (l'Anson et al. 1995). The hydrolytic activity for Glu-ρNA from cell extracts of three *pepA* derivatives was reported to be deficient relative to wild type. The insertion derivative strains displayed a statistically significant increase in lag phase, equivalent to a half-generation, but grew to the same final cell densities as wild type. However, the cell density and milk acidification data indicates that the *pepA* derivatives had the same growth and acidification rates as the parental strain.

X-prolyl dipeptidyl aminopeptidase

X-prolyl dipeptidyl aminopeptidase, PepX, has been purified and characterized from strains of *Lb. acidophilus*, *Lb. delbrueckii*, *Lb. casei*, *Lb. helveticus*, *Lc. lactis* and *S. thermophilus* (see Table 1). PepX liberates Xaa-Pro dipeptides from the N-terminus of peptides. In addition to peptidase activity, PepX has been demonstrated to execute amidase and esterase reactions (Houbart et al. 1995; Yoshpe-Besançon et al. 1994). The highest activities reported for PepX on Xaa-Pro-ρNA substrates are when the N-terminal residues are uncharged (Ala-, Gly-) or basic (Arg-). While liberation of amino acids from dipeptides does not occur, PepX releases Xaa-Pro dipeptides from substrates containing from three to seven amino acid residues. No kinetics data is reported for comparison with respect to substrate size (Kiefer-Partsch et al. 1989; Zevaco et al. 1990; Booth et al. 1990b; Lloyd & Pritchard 1991; Tsakalidou et al. 1998). Liberated Xaa-Pro dipeptides contain residues that are basic (Arg-, His-, Lys-), aromatic (Phe-, Tyr-), and hydrophobic/uncharged (Ala-, Ile-, Val-, Gly-). Three Xaa-Pro dipeptides are sequentially liberated from β-casomorphin f60-66 (Tyr-Pro-Phe-Pro-Gly-Pro-Ile) and β-casein f176-182 (Lys-Ala-Val-Pro-Tyr-Pro-Gln) (Kiefer-Partsch et al. 1989; Zevaco et al. 1990; Lloyd & Pritchard 1991). The hydrolysis of the latter substrate confirms the additional specificity for Xaa-Ala-(Xaa)$_n$ first identified by activity with Ala-Ala-ρNA (Meyer & Jordi 1987; Zevaco et al. 1990; Khalid & Marth 1990b). PepX is also capable of hydrolyzing Pro-Pro-(Xaa)$_n$ substrates, but little or no hydrolysis is observed for Xaa-Pro-Pro (including when Xaa is Pro) (Booth et al. 1990b; Miyakawa et al. 1991; Miyakawa et al. 1994). This is corroborated by the lack of hydrolysis of a bradykinin (Arg-Pro-Pro-Gly-Phe-Ser-Pro-Phe-Arg) and Lys-bradykinin (Lys-Pro-Pro-Gly-Phe-Ser-Pro-Phe-Arg), although the possibility of substrate size limitation of these nanopeptides has not been investigated (Lloyd & Pritchard 1991).

Northern analysis with a *pepX* (~2.4-kb ORF) specific probe identified a single transcript of ~2.6-kb from *Lb. helveticus* grown in MRS broth under pH controlled conditions (5.85) (Vesanto et al. 1995). The monocistronic *pepX* transcript level was constant into late exponential phase and is then slowly degraded in stationary phase. Under the same conditions, the PepX activity increases four-fold in the last two generations prior to stationary phase and remains constant for at least 12 more hours. A similar PepX expression profile was reported for *Lb. delbrueckii* grown in MRS without pH control, but ~50% of the activity is lost after 12 h in stationary phase (Atlan et al. 1990). The regulation of expression of PepX in *Lc. lactis* has been demonstrated to be strain dependent, with the activity being equivalent or higher in cultures grown in milk than from complex amino acid or peptide-containing media (Meijer et al. 1996). Currently, there is no information available on the transcriptional regulation of *pepX* for LAB grown in milk.

The physiological role of PepX in *Lb. helveticus*, *Lb. delbrueckii*, and *Lc. lactis* has been investigated for deletion mutants (Mayo et al. 1993; Mierau et al. 1996a; Yüksel & Steele 1996; Yüksel & Steele 1999), gene disrupted mutants (Matos et al. 1998), and chemically derived mutants (Atlan et al. 1990; Nardi et al. 1991) as a function of specific growth rate. Chemical mutagen derivatives of *Lb. delbrueckii* expressing no detectable PepX activity displayed elongated lag phases and lower total biomass in MRS medium, but the specific growth rate was the same as the parental strain (Atlan et al. 1990). Chemical mutagen derivatives of *Lc. lactis* expressing no detectable PepX activity displayed decreased growth rates in milk media corresponding to a 60% longer generation time compared to wild type (μ_{max} 1.0 vs 1.6 h^{-1}) (Nardi et al. 1991). Gene disrupted derivatives of *pepX* in *Lc. lactis* expressed no detectable PepX activity and displayed decreased growth rates in milk media corresponding to a 31% longer generation time compared to wild type (μ_{max} 0.70 vs 0.93 h^{-1}) (Matos et al. 1998). Complementation data was not reported as confirmation of physiological role for the chemically derived or gene disrupted mutants.

Deletion derivatives of *pepX*, devoid of PepX activity, were constructed by two step recombination procedure in *Lb. helveticus* (Yüksel & Steele 1996). The resulting *pepX* deletion derivatives displayed equivalent growth profiles and final cell densities in MRS broth. However, the growth rate of the PepX negative strains in milk was decreased, corresponding to approximately two-fold longer generation times compared to wild type (μ_{max} 0.61 vs 0.28 h^{-1}) (Yüksel & Steele 1999). However, as with growth in MRS, the final cell densities were similar for parental and deletion derivative strains. Introduction of a plasmid-encoded copy of *pepX* into a deletion derivative strain resulted in the reversion to wild type PepX activity and growth rate in milk.

Mutant derivatives of *pepX* have also been constructed in *Lc. lactis* (Mayo et al. 1993; Mierau et al. 1996a). The resulting *pepX* deficient derivatives were reported to have acidification or growth rates similar to the wild type strains. However, the decrease in milk growth rate for one *pepX* deletion derivative corresponds to a 15% longer generation time (μ_{max} 0.56 vs 0.65 h^{-1}), though the final cell densities were equivalent to the parental strain (Mierau et al. 1996a).

PepD

The dipeptidase, PepD, has been purified and characterized from *Lb. helveticus* (Vesanto et al. 1996). The information on activity of PepD for dipeptide substrates is limited but similar to that reported for the broad specificity PepV (Dudley et al. 1996; Vesanto et al. 1996). Possible exceptions include the inability of PepD to hydrolyze the Val-Xaa (-Arg, -Gly, -Leu) and Ile-Xaa (-Gln, -Val) dipeptides, though other dipeptides with hydrophobic/uncharged N-terminal residues are hydrolyzed. As with PepV, no hydrolysis of AA-ρNA substrates and no significant hydrolysis of Pro containing peptides is reported.

Northern analyses with *pepD* (~1.4-kb ORF) specific probes derived from *Lb. helveticus* and indicate that the transcript encoding PepD is monocistronic (1.5-kb) (Dudley et al. 1996; Vesanto et al. 1996). The level of *pepD* transcript was determined to be highest in late exponential phase for cells grown in MRS broth under pH controlled conditions. Total dipeptidase activity was measured under the same growth conditions and remained proportional to cell density. However, analysis of a PepD mutant of *Lb. helveticus* suggests that this enzyme does not contribute significantly to the total dipeptidase activity (Dudley et al. 1996).

The physiological role of PepD in *Lb. helveticus* has been investigated for deletion mutant derivatives as a function of specific growth rate (Dudley et al. 1996). The growth rate and final cell densities of *Lb. helveticus* PepD deletion mutants are the same as for the wild type strain in complex media (MRS broth), complete amino acid defined media, and milk. Additionally, total dipeptidase activity was the same for *Lb. helveticus* wild type and the *pepD* deletion derivative with several dipeptides previously identified as PepD substrates from the original *E. coli* subclones. The growth rate and activity results suggest that PepD does not play a significant role in the liberation of essential amino acids from milk derived peptides.

PepV

The dipeptidase, PepV, has been purified and characterized from strains of *Lb. casei*, *Lb. delbrueckii*, *Lb. helveticus*, *Lb. sake*, *Lb. sanfrancisco* and *Lc. lactis* (Desmazeaud & Zevaco 1977; Hwang et al. 1981; Van Boven et al. 1988; Wohlrab & Bockelmann 1992; Vongerichten et al. 1994; Montel et al. 1995; Tan et al. 1995; Gobbetti et al. 1996; Fernández-Esplá & Martín-Hernández 1997). PepV is considered a broad specificity dipeptidase. As with PepD, no hydrolysis of AA-ρNA substrates and no significant hydrolysis of Pro containing peptides is reported. A wide range of dipeptides are hydrolyzed containing basic (Arg-, His-, Lys-), hydrophobic/uncharged (Ala-, Ile-, Leu-, Val-), aromatic (Phe-, Tyr-), and Met residues in the N-terminal position. In contrast, PepV generally does not hydrolyze dipeptides containing Gly as the N-terminal residue. PepV also hydrolyses several β-Ala-Xaa and Xaa-β-Ala substrates, including carnosine (β-Ala-His) (Vongerichten et al., 1994).

Northern analyses with *pepV* (~1.6-kb ORF) specific probes derived from *Lc. lactis* suggest that the transcript from the gene is monocistronic (1.7-kb transcript) (Hellendoorn et al. 1997). However, no further analysis or explanation was given for a second, lighter ~3.6-kb hybridization band. While the Northern hybridization data suggests it is possible that the *pepV* is also encoded by a multicistronic transcript, the reported sequence data adjacent to *pepV* is insufficient to allow analysis for ORFs that could account for the larger transcript size.

The physiological role of PepV in *Lc. lactis* has been investigated for insertionally inactivated derivatives as a function of specific growth rate (Hellendoorn et al. 1997). The growth rate and final cell densities of the PepV deficient strain are the same as for the wild type in complex media (M17 broth). However, the PepV deficient strain displays a decreased growth rate in milk corresponding to a 22% longer generation time compared to wild type (μ_{max} 0.58 vs 0.71 h^{-1}).

Proline iminopeptidases

Proline iminopeptidase activity is defined by specificity for N-terminal proline of peptides. PepI has been purified and characterized from *Lb. delbrueckii* and *Lb. helveticus* (Gilbert et al. 1994; Varmanen et al. 1996a). The *Lactobacillus* PepI deduced sequence contains the motif GQSWGG, which conforms to the active site consensus sequence of the prolyl oligopeptidase family (GxSxGG) (Rawlings et al. 1991) (see also description of PepR). Chemical and site-directed mutation analysis the *Lb. delbrueckii* PepI has determined that the consensus region Ser-107, Asp-246 and His-273 constitute the catalytic triad of this enzyme (Morel et al. 1999a). These three amino acid residues are also conserved in PepI of *Lb. delbrueckii* and *Lb. helveticus*.

Hydrolytic activity has been reported for Pro-Xaa substrates when the C-terminal residue is hydrophobic/uncharged (-Ala, -Gly, -Ile, -Leu, -Val), acidic (-Glu), and aromatic (-Phe, -Tyr). The PepI of *Lb. delbrueckii* hydrolyses a variety of other dipeptides at low rates, the majority of which contain at least one hydrophobic/uncharged residue (Gilbert et al. 1994). In addition, this enzyme liberates proline from the N-terminus of a tripeptide, but not from a limited number of the tetra- and pentapeptides analyzed. The iminopeptidase of *Lc. lactis* displays relatively high activity for liberation of proline from several tripeptides and Pro-Phe-Gly-Lys, but not the pentapeptides tested (Baankreis & Exterkate 1991). The *Lactobacillus* PepI hydrolyze Pro-Pro and show no significant activity for AA-ρNA derivatives except Pro-ρNA. In contrast, no activity was detected for these two substrates with the *Lc. lactis* proline iminopeptidase (Baankreis & Exterkate 1991). Differences in hydrolytic capabilities with several additional substrates, together with the metallopeptidase classification, suggests the *Lc. lactis* enzyme is not closely related to PepI of *Lactobacillus*.

Transcript containing *pepI* was not detectable by Northern blot from pH controlled (5.6 minimum) whey broth grown culture (Varmanen et al. 1996b). However, RT-PCR using primers to *pepI* and the flanking ORFs resulted in identification of products consistent with the cotranscription of *pepI* and either of the adjacent ORFs. This evidence suggests a multicistronic transcript with a minimum size of ~2.9-kb encoding PepI and two other ORFs.

The physiological role of PepI in *Lb. helveticus* has been investigated for deletion mutant derivatives as a function of specific growth rate. The growth rate and final cell densities of *Lb. helveticus* PepI deletion mutants are the same as for the wild type strain in complex media (MRS broth) and complete amino acid defined media (Yüksel & Steele 1999). However, the growth rate in milk was slower for the PepI deletion mutant compared to wild type (μ_{max} 0.56 vs 0.61 h^{-1}), corresponding to a 9% longer generation time, but the final cell densities were similar.

Prolidase (PepQ)

PepQ has been purified and characterized from *Lb. casei, Lb. delbrueckii, Lb. helveticus*, and *Lc. lactis* (Kaminogawa et al. 1984; Booth et al. 1990c; Fernández-Esplá et al. 1997; Yüksel & Steele 1999; Morel et al. 1999b). Prolidase activity is defined by specificity for Xaa-Pro dipeptides. However, not all prolidases are exclusively Xaa-Pro dipeptidases, nor do they hydrolyze all Xaa-Pro substrates. The predominant activity of PepQ from LAB for Xaa-Pro substrates includes N-terminal residues that are hydrophobic/uncharged (Ala-, Ile-, Leu-, Val), basic (His-), aromatic (Phe-, Tyr), and sulfur containing (Met-). Conflicting results have been obtained for approximately twenty non-proline-containing dipeptides. Not all of the characterized enzymes are able to hydrolyze Pro-Pro, but none are reported to hydrolyze Gly-Pro. Additionally, some of these enzymes appear to have activity for Pro-Ala, Pro-Pro, and Pro-Val substrates.

Northern analyses with *pepQ* (1.1-kb ORF) specific probes derived from *Lb. delbrueckii* indicate that the transcript from the gene is monocistronic (~1.1-kb transcript) (Rantanen & Palva 1997). The *pepQ* transcript levels reach a maximum at the transition between exponential and stationary phase during growth under pH controlled conditions (6.00) in MRS broth. No data was reported for expression of total prolidase activity for these growth conditions.

The physiological role of PepQ in *Lb. helveticus* has been investigated for deletion mutant derivatives as a function of specific growth rate. The growth rate and final cell densities of *Lb. helveticus* PepQ deletion mutants are the same as for the wild type strain in complex media (MRS broth) and complete amino acid defined media (Yüksel & Steele 1999). However, the growth rate in milk was slower for the PepQ deletion mutant compared to wild type (μ_{max} 0.54 vs 0.61 h^{-1}), corresponding to a 13% longer generation time, but the final cell densities were similar. Comparison of peptidase activity from wild type and PepQ deficient strains indicated that PepQ accounts for greater than 99% of the hydrolysis of Met-Pro, Leu-Pro, and Phe-Pro. No significant differences were observed for the hydrolysis of Met-Ala or Pro-Phe, which are known PepR and/or PepI substrates.

Prolinase (PepR)

A dipeptidase with prolinase activity, PepR, has been purified and characterized from *Lb. helveticus* (Shao et al. 1997). Prolinase activity is defined by specificity for Pro-Xaa dipeptides. The *Lactobacillus* PepR deduced sequences contain the motif GQSWGG (Dudley & Steele 1994; Varmanen et al. 1996b; Varmanen et al. 1998), which conforms to the active site consensus sequence of the prolyl oligopeptidase family (G×S×GG) (Rawlings et al. 1991). Replacement of consensus region serine residue (Ser-111 to Ala) resulted in the loss of detectable activity, confirming the predicted active site (Shao et al. 1997).

The *Lactobacillus* PepR displays a relatively broad dipeptidase specificity as well as limited activity for larger peptides and Pro-(ρNA, βNAP) substrates. Hydrolysis of Pro-Xaa substrates has been reported for C-terminal residues that are hydrophobic/uncharged (-Ala, -Ile, -Leu, -Val), aromatic (-Phe), and sulfur containing (-Met). Relatively high activity for several non-proline dipeptides is reported for PepR of *Lb. helveticus*, including Met-(Ala, Leu, Phe), Leu-(Arg, Ser), Ser-Phe, and Thr-Leu (Shao et al. 1997). Hydrolysis of Xaa-Pro dipeptides has not been reported.

Northern analyses with a *pepR* (~0.9-kb ORF) specific probe derived from *Lb. helveticus* suggest that the transcript is monocistronic (1.25-kb) (Varmanen et al. 1996b). Transcript level and prolinase activity was determined to be highest at late exponential phase for cells cultured in complex media (MRS broth) under pH controlled conditions.

Northern analyses with a *pepR* (~0.9-kb ORF) specific probe derived from *Lb. rhamnosus* identified two transcript sizes (1.0 and 1.5-kb) (Varmanen et al. 1998). The larger transcript was consistent with the expected size of a product encoding *pepR* and a downstream ORF of unknown function. Both transcript sizes were at their maximum level at the exponential to stationary transition phase, although no later time points were reported.

The physiological role of PepR in *Lb. helveticus* and *Lb. rhamnosus* has been investigated for deletion mutant derivatives as a function of specific growth rate. The growth rate and final cell densities of *Lb. helveticus* and *Lb. rhamnosus* PepR deletion mutants are the same as for the respective wild type strains in complex media (MRS broth) and milk (Shao et al. 1997; Varmanen et al. 1998). The total hydrolytic activity from cell extracts of *Lb. helveticus* (grown in MRS broth) and *Lb. rhamnosus* (grown in whey medium) for the substrate Pro-Leu was determined to be reduced by ~95% in the PepR deletion mutants. Additionally, PepR of *Lb. helveticus* accounts for significant activity from cell extracts toward Pro-Met (~90%), Thr-Leu (~90%), Pro-Phe (~80%), Gly-Leu (~55%), and Met-Ala (30%) (Shao et al. 1997).

PepL

A peptidase, PepL, has been cloned and characterized from *Lb. delbrueckii* that displays high specificity for

Leu-ρNA and Ala-ρNA (Klein et al. 1995). Growth tests with a Leu / Pro auxotrophic *E. coli* strain expressing recombinant PepL suggest hydrolysis occurs with several di- and tripeptides with N-terminal Leu residues. Growth was also obtained with Pro-Pro, Pro-Ser, and Gln-Pro as the sole source of proline.

Northern analyses was conducted with a *pepL* (~0.9-kb ORF) specific probe and a probe derived from a partially sequenced upstream ORF with limited C-terminal deduced sequence identity to several peptide transport related proteins, including *dciAE* of *Bacillus subtilis* (Klein et al. 1995). A 1.5-kb transcript was detected with both probes using mRNA isolated from *Lb. delbrueckii* grown in complex media (MRS broth). The transcript is proposed to be a product of *pepL* and an upstream gene (Klein et al. 1995). However, a 1.5-kb transcript (and three smaller) is also detected with the *pepL* probe from mRNA of *E. coli* CM89 with pJUK20, which contains a 6.6-kb insert with *pepL* and only ~200-bp of upstream sequence, and is therefore inadequate to provide the full length transcript. Additionally, the 1.5-kb transcript is not large enough to include *pepL* and a homologue of the peptide transporters of *B. subtilis*, since the ORFs that encode the latter proteins are >1.6-kb (Mathiopoulos et al. 1991; Perego et al. 1991).

PepP

The aminopeptidase, PepP, has been purified and characterized from *Lc. lactis* (Mars & Monnet 1995). PepP liberates the N-terminal amino acid from peptides with general specificity for Xaa-Pro-Pro-$(Yaa)_n$ sequences (Mars & Monnet 1995). The rate of hydrolysis was highest for the pentapeptides Arg-Pro-Pro-Gly-Phe (bradykinin f1-5) and Leu-Pro-Pro-Ser-Arg. Relatively high activity was observed with peptides ranging from three to nine residues, but the dipeptides tested were not hydrolyzed. Specificity was observed for the N-terminal (Xaa-) residues Arg, Met, Lys, Leu and Tyr. Increased hydrolysis was observed with the esterified derivatives of Xaa-Pro-Pro-$(Yaa)_n$ peptides containing Glu residues at the P′3 and P′5 position (*n*=4) or P′7 and P′9 position (*n*=8). These results suggest that PepP contains an extended binding site and hydrolysis may be prevented by the presence of negatively charged substrate residues beyond the P′2 position (Mars & Monnet 1995).

The physiological role of PepP in *Lc. lactis* has been investigated for a gene-disrupted mutant as a function of specific growth rate (Matos et al. 1998). The growth rate of *Lc. lactis* PepP gene disrupted mutants are the same as for the wild type strain in complex media (M17 broth) and only slightly slower than wild type when grown in milk (μ_{max} 0.89 vs 0.93 h^{-1}). However, the assertion that PepP is not involved in hydrolysis of casein-derived prolyl peptides during growth in milk is premature without analysis of the intracellular peptide composition.

PepT and other tripeptidases

The tripeptidase, PepT, has been purified and characterized from *Lc. lactis* (Bosman et al., 1990; Bacon et al. 1993). PepT hydrolyses tripeptides with a wide range amino acids including substrates comprised of hydrophobic/uncharged, aromatic, basic, acidic, and sulfur-containing residues. Hydrolysis is observed for Pro-Gly-Gly and Leu-Xaa-Pro (-Gly-, -Ala-), but not for any Xaa-Pro-Yaa substrates tested. In accordance with the strict substrate size specificity of PepT, no activity has been reported for any di-, tetra-, or larger oligopeptides.

Several other tripeptidases have been purified and characterized from *Lb. delbrueckii*, *Lb. sake*, *Lc. lactis*, and *Pediococcus pentosaceus* (Desmazeaud & Zevaco 1979; Sahlstrøm et al. 1993; Bockelmann et al. 1995; Bockelmann et al. 1997; Simitsopoulou et al. 1997; Sanz et al. 1998). Like PepT, these enzymes are generally metallopeptidases and have high activity for tripeptides comprised of hydrophobic/uncharged and aromatic residues.

The physiological role of PepT in *Lc. lactis* has been investigated for deletion mutant derivatives as a function of specific growth rate (Mierau et al. 1996a). The growth rate and final cell densities of the PepT deficient strain are the same as for the wild type in complex media (M17 broth). However, the PepT deficient strain displays a decreased growth rate in milk corresponding to a 10% longer generation time compared to wild type (μ_{max} 0.60 vs 0.65 h^{-1}). Comparison of wild type and PepT deficient total activity with tripeptide substrates was not reported.

PepE

The endopeptidase, PepE, has been purified and characterized from *Lb. helveticus* (Fenster et al. 1997). PepE was initially identified on the basis of hydrolysis of the N-terminal and C-terminal blocked substrate *N*-benzoyl-Phe-Val-Arg-ρNA (Fenster et al. 1997). Purified PepE was also capable of hydrolyzing Met-enkephalin (Tyr-Gly-Gly-Phe-Met) at the Gly_3-Phe_4

bond and Bradykinin (Arg-Pro-Pro-Gly-Phe-Ser-Pro-Phe-Arg) at the Gly_4-Phe_5 bond. No activity was reported with β-casomorphin, *N*-benzoyl-Val-Gly-Arg-ρNA, or with Arg-, Phe-, Pro-, Lys- Gly-, or Val-ρNA.

PepG

PepG has been purified and characterized from *Lb. delbrueckii* (Klein et al. 1997). Comparison of PepG with the deduced amino acid sequence of *Lb. helveticus* PepE and *Lb. delbrueckii* PepC reveals 71.2% and 44.6% identity, respectively. Activity for several AA-, di-, and tripeptide-βNAP substrates was compared from *E. coli* CM89 expressing recombinant PepG and PepC from *Lb. delbrueckii* (Klein et al. 1997). Nearly all of the twenty-one substrates tested were hydrolyzed by PepC with the exception of Pro- and Gly-Pro-βNAP. In contrast, PepG hydrolyzed Gly-Phe-βNAP and Leu-Gly-βNAP, but not the smaller counterparts Phe-βNAP and Gly-βNAP, suggesting the enzyme may have dipeptidyl specificity. However, PepG also hydrolyzed Ala- and not Ala-Ala-βNAP, but displayed a lower rate of hydrolysis with Ala-ρNA compared to Ala-Ala-ρNA (Klein et al. 1997). Additionally, PepG had no detectable activity for several other AA-ρNA substrates that were hydrolyzed by PepC (Gly-, Glu-, Leu-, Lys-, Phe-, and Tyr-ρNA). Growth tests with a Leu/Pro auxotroph of *Salmonella typhimerium* expressing recombinant PepG suggest hydrolysis of di-/tripeptides and a tetrapeptide containing Leu, but no growth was obtained with any peptides containing Pro as the sole essential amino acid.

PepO

The endopeptidase, PepO, has been purified and characterized from *Lc. lactis* (Tan et al. 1991; Pritchard et al. 1994; Stepaniak & Fox 1995; Lian et al. 1996). PepO is reported to hydrolyze oligopeptides ranging in length from five (Met- and Leu-enkephalin) to thirty-five residues (α_{S1}-casein f165-199) (Tan et al. 1991; Pritchard et al. 1994; Stepaniak & Fox 1995; Lian et al. 1996). Hydrolysis also occurs with bradykinin, angiotensin, neurotensin, and insulin β-chain. Similar to thermolysin, several points of hydrolysis by PepO occur at peptide bonds in which Leu or Phe occupy the P1N site (Pritchard et al. 1994; Stepaniak & Fox 1995). In general, the affinity and rate of hydrolysis increases for YGGF$(X)_n$ substrates as the values increase for n=(1 to 13) (Lian et al. 1996). Although PepO has activity with several large casein derived fragments, the native caseins are not detectably hydrolyzed (Tan et al. 1991; Pritchard et al. 1994; Stepaniak & Fox 1995).

Northern analysis with a *pepO* (~1.9-kb ORF) specific probe identified two transcript sizes, ~1.5 and ~2.2-kb, from *Lb. helveticus* grown in MRS broth (Chen & Steele 1998). Northern analysis of total mRNA from the *pepO* deletion mutant also revealed two transcripts, ~1.5 and ~1.8-kb. The size of the *pepO* probe (1.5-kb) and the position of the deletion (~1.2-kb from transcriptional start site) indicate that the 1.5-kb transcript detected in the wild type and deletion mutant strain is not a degradation product of the 2.2-kb *pepO* transcript.

The physiological role of PepO in *Lb. helveticus* and *Lc. lactis* has been investigated for insertion disrupted and deletion mutant derivatives as a function of specific growth rate. The growth rates, acidification rates, and final cell densities of the insertion disrupted *pepO* mutants of *Lc. lactis* were reported to be indistinguishable from the parental or wild type strains when grown in milk (Mierau et al. 1993; Tynkkynen et al. 1993). The growth rate and final cell densities of *Lb. helveticus* PepO deletion mutants are the same as for the wild type strain in complex media (MRS broth), amino acid defined media, and milk (Chen & Steele 1998). Similar results were obtained for growth and acidification rates of wild type and PepO deletion mutants of *Lc. lactis* in complex media (GM17 broth) (Mierau et al. 1996a). However, the *Lc. lactis* PepO deletion mutant displays a decreased growth rate in milk corresponding to a 9% longer generation time compared to wild type (μ_{max} 0.60 vs 0.65 h^{-1}) (Mierau et al. 1996a).

PepF

The endopeptidase, PepF, has been purified and characterized from *Lc. lactis* (Monnet et al. 1994). The original report of PepF was from *Lc. lactis* ssp. *lactis* in which *pepF1* is located chromosomally (Monnet et al. 1994). A PepF homologue (encoded by *pepF2*) is expressed in *Lc. lactis* ssp. *cremoris* that has 84% identity to the deduced amino acid sequence of the enzyme encoded by *pepF1* located on the lactose-proteinase plasmid of this strain (Nardi et al. 1997).

PepF is reported to hydrolyze oligopeptides ranging in length from seven (β-casomorphin f60-66) to seventeen residues (adrenorticotropic hormone f1-17) (Monnet et al. 1994). However, a clear preference is revealed for substrates of about eight or nine residues as indicated by the rates of hydrolysis. The enzyme was shown to hydrolyze ACTH f1-17 at three positions, liberating peptides of three to seven residues. No hydrolysis was observed for ACTH f1-24, glucagon (29 residues), or insulin B chain (30 residues), suggesting the upper size limit for substrates is less than twenty-four residues. Additionally, the native caseins are not detectably hydrolyzed.

Northern analyses with *pepF2* (~1.8-kb ORF) specific probes derived from *Lc. lactis* suggest that the transcripts from the chromosomally located *pepF* genes are polycistronic and include two adjacent ORFs (~3.7-kb transcript) (Nardi et al. 1997). However, comparison of luciferase activity from *lux* fusions with the promoter corresponding to the 3.7-kb transcript and a second potential promoter upstream of *pepF2* clearly indicated the greater strength of the latter promoter, which would result in a 2.6-kb bicistronic transcript.

The physiological role of PepF in *Lc. lactis* has been investigated for deletion mutant derivatives as a function of specific growth rate. The growth rate of *Lc. lactis* PepF1 or PepF2 deletion mutants are the same as for the wild type strain in complex media (M17 broth) (Nardi et al. 1997). However, loss of either copy of PepF from *Lc. lactis* ssp. *cremoris* resulted in a decrease in growth rate in a minimal medium. Similar results were obtained with a double mutant, devoid of PepF, resulting in decrease growth rates in minimal media corresponding to a 16% increase in generation time compared to wild type (μ_{max} 0.12 vs 0.14 h^{-1}). These results suggest a potential role for PepF in protein turnover under nitrogen limiting conditions.

Location of peptidases

Several lines of evidence strongly suggest the intracellular location of all the peptidases identified in LAB to date. Included are the absence of recognized signal sequences or membrane anchors in any of the peptidases for which sequence is available. In addition, more recent evidence for the ability of Opp to transport several large casein derived peptides negates the previous assertions that peptidase secretion would be required for liberation of essential amino acids (Detmers et al. 1998; Kunji et al. 1998). A thorough evaluation of evidence for localization of peptidases can be found in a previous review (Kunji et al. 1996).

Physiological role of peptidases

The primary focus and evidence for physiological role of peptidases of LAB is derived from the evaluation of mutant growth rate differences in milk and complex or defined media. In addition, a few studies have evaluated the effect of multiple peptidase mutations (Christensen & Steele 1996; Mierau et al. 1996a; Matos et al. 1998; Yüksel & Steele 1999). Single and multiple peptidase mutants have been constructed in *Lb. helveticus* (*pepC, pepD, pepI, pepN, pepO, pepQ, pepR, pepX*) and *Lc. lactis* (*pepA, pepC, pepF, pepN, pepO, pepT, pepX, pepV*) (see respective sections). In general, no differences are reported for the growth parameters of single and multiple mutants compared to the wild type strain when grown in complex media (MRS or M17 broth) or complete amino acid defined media. These results suggest the mutation(s) do not effect the normal protein turnover or protein maturation requirements of the peptidase mutant strain to the degree that is detectable as a growth rate deficiency.

The reduction of growth rates in milk exhibited by LAB peptidase mutants follows the generalization of greater deficiency with greater numbers of mutations (Mierau et al. 1996a; Yüksel & Steele 1999). This trend can be interpreted as reduction in rate of liberation of essential amino acids from the exogenous peptide supply, thereby limiting the availability of amino acids for new protein synthesis. While this interpretation is direct, there is also evidence for regulation of peptidases as a function of the class of intracellular nitrogen accumulated. The regulation of PepN and CEP in *Lc. lactis* are affected by the exogenous supply of amino acids or peptides, with specific cases reported for dipeptides such as Pro-Leu which result in decreased expression of these enzymes (Marugg et al. 1995; Meijer et al. 1996). These results suggest the pool of accumulated peptides in a peptidase mutant may regulate the activity of remaining peptidases, further reducing the hydrolytic capabilities, and indirectly causing a reduction in growth rate. Evaluation of remaining peptidase activities of single and multiple peptidase mutants indicate only minor fluctuations in levels of expression (Christensen & Steele 1996; Mierau et al. 1996a). However, these comparisons were made from cultures in complex media, conditions where no effect on growth rates were observed, and may not be representative of the expression levels during growth in milk or under conditions where intracellular peptides accumulate.

The observation that a peptidase is non-essential for growth in milk is clearly different from what is essential for survival in a population should the mutation arise naturally. Differences in the specific growth rates of single peptidase mutants relative to wild type are often described as insignificant, even when the deficiency corresponds to a 10% increase in the generation time. In all cases where the specific growth rates in milk were reported for peptidase mutants the mean values indicate a reduction in growth rate relative to wild type. These results indicate that the difference in growth rate is not simply a function of assay error, although the precision of some of these measurements may be low. Furthermore, while the difference may not be considered statistically significant due to a small number of replicates and inherent assay error, any measurable reduction in growth rate is clearly significant to the survival of a genotype in a population. For example, a 10% increase in generation time for a mutant strain would result in progressive reduction to 0.1% of the total biomass after competitive growth for ~110 wild type generations starting from equivalent inoculations. Additionally, this estimation only accounts for differences observed in exponential growth and little is known about the effect of specific LAB peptidase mutations on transitional phases and survival at stationary phase.

It may be considered that the evaluation of specific growth rates in milk is not the most accurate or informative means by which the importance of a given peptidase mutation can be expressed. A more sensitive analysis may include the experimental determination of the rate of loss of a genotype from a parental population in competitive growth conditions (as described above). Additionally, alternate roles may be revealed by evaluating the survival of a mutant strain under conditions that LAB are known to encounter, including carbohydrate depleted and acidic environments. Analysis such as these may in part elucidate the capacity of peptidase mutant (or over expressing) strains for survival in cheese, especially since the pathways for amino acid catabolism are clearly dependent on the availability of substrate.

The physiological role of peptidases, in functions not related to the hydrolysis of exogenous peptides, may be more appropriately evaluated in minimal media as was reported for PepF mutants of *Lc. lactis* (see PepF section) (Nardi et al. 1997). The impact of peptidase mutations are likely to be most evident when protein turnover is required under conditions of starvation and when new protein synthesis occurs at transitional phases (Gottesman & Maurizi 1992; Miller 1987). The precedence for this type of peptidase dependence was established in studies with *E. coli* and *S. typhimurium* multiple peptidase mutants (Yen et al. 1980a; Yen et al. 1980b; Miller & Green 1981; Miller & Green 1983).

Amino acid catabolism by LAB

The catabolism of amino acids by LAB has implications with regard to the quality and safety of fermented foods. Amino acid catabolism by anaerobic/fermentative microorganisms is also believed to have an important role in their ability to obtain energy in nutrient-limited environments. For example, Arg and His are known to provide energy via substrate level phosphorylation and precursor/product exchange coupled to amino acid decarboxylation, respectively. However, the catabolic pathways for many amino acids remain unknown or only partially characterized in LAB. This section of the review will summarize our current knowledge of amino acid catabolic pathways identified in LAB. The relationships between these pathways and cellular physiology, development of cheese flavor, and production of compounds with human health implications will also be discussed when possible. The catabolism of sulphur containing amino acids is not reviewed as this topic is covered elsewhere in these proceedings.

Arginine catabolism

Two pathways for the catabolism of arginine have been described. Strains of *Lb. fermentum* have been demonstrated to produce nitric oxide from arginine. Morita et al. (1997) postulated that this reaction was catalyzed by a nitric oxide synthase that could also produce citrulline as a coproduct. A more common pathway for the catabolism of arginine in LAB is via the arginine deiminase (ADI) pathway (Manca de Nadra et al. 1982; Cunin et al. 1986; Konings et al. 1995). The enzymes involved in the arginine deiminase pathway and corresponding arginine/ornithine antiporter are shown in Figure 1. This pathway results in the conversion of one mole of arginine into two moles of ammonia and one mole each of ornithine, carbon dioxide, and ATP. This is the only known LAB amino acid catabolic pathway that results in the production of ATP via substrate level phosphorylation.

A number of studies have examined the occurrence of ADI pathway enzymes in LAB (Crow & Thomas

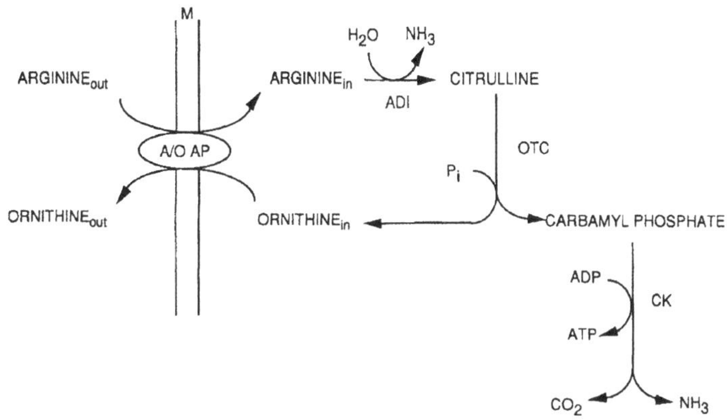

Figure 1. The arginine deiminase pathway and arginine/ornithine antiporter (A/O AP). Abbreviations include: M, cytoplasmic membrane; ADI, arginine deiminase; OTC, ornithine transcarboxylase; CK, carbamate kinase.

1982; Manca de Nadra et al. 1982; Liu et al. 1995). Collectively, these studies indicate that the pathway is typically present in heterofermentative lactobacilli, while it is a variable trait in *Lc. lactis*, *Oenococcus oenos*, and the *Thermobacterium* subgenera of *Lactobacillus*. Additionally, numerous LAB have been described that have an incomplete ADI pathway (Crow & Thomas 1982; Manca de Nadra et al. 1982). The gene cluster encoding the ADI pathway from *Lb. sake* has been characterized in detail (Zúñiga et al. 1998). The genes contained within this cluster and their order were determined to be: *arcA*, encoding ADI; *arcB*, encoding ornithine transcarboxylase; *arcC*, encoding carbamate kinase; *arcT*, encoding a noncharacterized transaminase; *arcD*, encoding an arginine-ornithine antiporter. The regulation of this pathway has been examined in *Lb. buchneri, Lb. leichmannii, Lb. sake, Lc. lactis*, and *O. oenos*. In general, arginine induces the expression of the ADI pathway enzymes and some carbohydrates repress their synthesis. A gene encoding *ahrC*, a regulator of arginine biosynthesis and catabolism, has been identified in *Lc. lactis* (Rallu et al. 1996); it is possible that this protein is involved in the induction of the ADI pathway enzymes by arginine. Repression by carbohydrates is dependent both on the organism examined and the carbohydrate utilized, suggesting that catabolite repression can be involved in the regulation of this pathway. A recent study by Zúñiga et al. (1998) identified two *cre* sequences present in the promoter region upstream of *arcA*; identification of these sequences strongly supports the involvement of the regulatory factor CcpA, and hence catabolite repression as a mechanism of regulation of this pathway in *Lb. sake*.

Both the production of ATP and ammonia from arginine have physiological implications. It has been demonstrated in *Lb. buchneri* (Liu & Pilone 1998), *Lc. lactis* (Crow & Thomas 1982), and *O. oenus* (Liu & Pilone 1998) that the energy derived from arginine catabolism can be coupled to growth. Additionally, survival at low pH is enhanced by the activity of the ADI pathway, most likely the result of an increase in culture pH resulting from ammonia production (Marquis et al. 1987).

Arginine catabolism by LAB during wine production can result in the production of ethyl carbonate precursors. This has human health implications, as ethyl carbonate is a known animal carcinogen (Liu & Pilone 1998).

Aspartate catabolism

In LAB, three different Asp catabolic pathways have been identified, however the details and distribution of these pathways are still not well understood. First, an Asp aminotransferase has been purified from *Lb. murinus* by Rollan et al. (1988) and cloned from *Lc. lactis* (Dudley & Steele 1999). Secondly, an Asp decarboxylase has been reported in a nonspeciated *Lactobacillus* strain by Abe et al. (1996). Lastly, an aspartase, which catalyzes the reversible reaction Asp to fumarate and NH_3^+, has been purified from *Lb. murinus* (Rollan et al. 1985).

The gene encoding an Asp aminotransferase (*aspC*) has recently been characterized from *Lc. lactis* LM0230 (Dudley & Steele 1999). The deduced amino acid sequence indicates that the lactococcal AspC is highly related to other pyridoxal-5′-phosphate dependent Asp aminotransferases from prokaryotes.

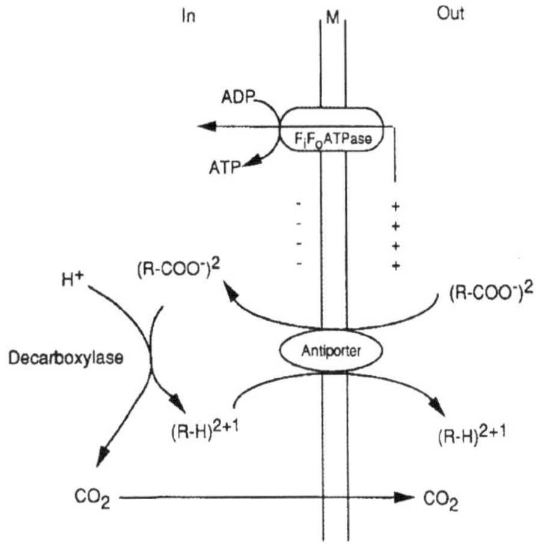

Figure 2. A generalized schematic of the generation of metabolic energy and regulation of intracellular pH by decarboxylation and electrogenic antiport. Abbreviations include: M, cytoplasmic membrane; R-COO$^-$, an amino acid; R-H, the corresponding amine.

Also, it was demonstrated that cell-free extracts of *E. coli* strains overexpressing the lactococcal *aspC* could catalyze Asp transamination with α-ketoglutarate, but no detectable activity with the other 19 natural amino acids was observed (Dudley & Steele 1999). The lactobacilli and lactococcal AspC, like their counterparts in other organisms, are likely involved in the last step of aspartate biosynthesis (Paulus 1993). This is supported by the observation that *Lc. lactis* LM0230(Δ*aspC*) cannot grow in defined medium lacking both Asp and Asn (Dudley & Steele 1999). Since Asp is believed to serve as a precursor for other cellular amino acids and nucleotides in LAB, this enzyme likely has a central role in LAB physiology. However, it is not known whether the AspC of LAB are involved in catabolism Asp *in vivo*.

Other than the purification of the aspartase from *Lb. murinus*, we are not aware of any other studies concerning this enzyme in LAB. Therefore, the physiological role of this enzyme is currently unknown.

Decarboxylation of Asp to Ala may be used to generate metabolic energy and regulate intracellular pH in *Lactobacillus* sp. (Abe et al. 1996; Konings et al. 1995) via a general mechanism common to a variety of decarboxylases. These two mechanisms are diagrammed in Figure 2. In the cytoplasm, the decarboxylation of an amino acid results in the consumption of one proton. The transport of the corresponding amine outside of the cell leads to 'indirect proton extrusion' (Maloney et al. 1992), which may be used to regulate intracellular pH. In addition, amine transport results in an electrochemical gradient that can be used by the cell to drive energy-consuming reactions such as the transport of nutrients or to generate ATP via a F_1F_0 ATPase (Konings et al. 1997).

Glutamate catabolism

The identified pathways of Glu catabolism in LAB are initiated by an aminotransferase, a dehydrogenase, or a decarboxylase. The first two enzymes result in the formation of α-ketoglutarate from Glu, while γ-aminobutyrate (GABA) is the product of decarboxylation. As discussed below, it appears unlikely that these products are catabolized further.

All purified and cloned aminotransferases identified to date from LAB are involved in catabolism of Glu, as this amino acid serves as the amino donor in these reactions. While a number of prokaryotes encode an α-ketoglutarate dehydrogenase, which converts α-ketoglutarate to succinyl-CoA, this enzyme has not been found in LAB (Morishita & Yajima 1995; Lapujade et al. 1998). Therefore, based upon our current knowledge, it appears catabolism of Glu by this pathway is restricted to the cycling of Glu and α-ketoglutarate by aminotransferases.

Glu decarboxylase has been purified from *Lactobacillus brevis* (Ueno et al. 1997) and the genes encoding the decarboxylase (*gadB*), putative Glu/γ-aminobutyrate antiporter (*gadC*), and a transcriptional activator (*gadR*) of these genes have been characterized from *Lc. lactis* (Sanders et al. 1998). The decarboxylase and antiporter are organized in an operon, and *gadR* is essential for *gadCB* transcription. Additionally, *gadCB* transcription is highest in complex medium around the onset of stationary phase and is positively regulated by NaCl, Glu and low pH.

Glu dehydrogenase activity has been detected in cell-free extracts of *Lb. fermentum* and *Lc. lactis* (Misono 1985), however a second report was unable to detect Glu dehydrogenase in *Lc. lactis* (Lapujade et al. 1998). Therefore, it is not clear how widely distributed this activity is in lactococci.

Higuchi et al. (1997) have demonstrated whole cells of *Lactobacillus* sp. incubated in the presence of Glu are capable of synthesizing ATP in a ΔpH and F_1F_0 ATPase dependent manner as depicted in Figure 2. Sanders et al. (1998) demonstrated that *Lc. lactis* cells preincubated in a defined medium with 0.3 M NaCl and 1 mM Glu had an approximately 2500–5000 fold increase in viability after a lactic acid challenge (pH 3.5) when compared to cells that had been pre-incubated in defined medium with no additions, NaCl alone or Glu alone. This effect was dependent upon a functional *gadR* and *gadB*. The same acid sensitivity was seen with a *gadC* mutant, however this may be the result of a polar effect on the downstream *gadB*. The results of these two studies clearly implicate Glu decarboxylation in the generation of metabolic energy and regulation of intracellular pH via the mechanisms summarized in Figure 2.

GABA production is known to occur in cheese, however we are not familiar with any reports suggesting that GABA has a direct or indirect impact on cheese flavor. Zoon & Allersma (1996) have correlated increased production of CO_2 and GABA in cheese with an increase in the number of eyes. The authors suggest Glu decarboxylase activity of *S. thermophilus* and/or *Lb. helveticus* was responsible for the development of eyes. However, the identification of the culture(s) responsible and the construction of isogenic strains lacking Glu decarboxylase activity are needed to further support this conclusion.

Histidine catabolism

Histidine decarboxylases, which catalyze the reaction histidine→histamine+CO_2, are broadly distributed in *Escherichia*, *Salmonella*, *Clostridia*, *Bacillus*, and *Lactobacillus* (Voight and Eitenmiller 1977). Most decarboxylases, including many histidine decarboxylases, require pyridoxal-5′-phosphate as a coenzyme (Boecker & Snell 1972; Guirard & Snell 1964). However, there is a large subclass of amino acid decarboxylases that utilize covalently bound pyruvate rather than pyridoxal-5'-phosphate (Chang & Snell 1968; Huynh et al. 1984; Recsei et al. 1983). The best studied of these enzymes is histidine decarboxylase of *Lactobacillus* 30a, which is synthesized as a proenzyme containing 310 amino acids (Parks et al. 1985; Hackert et al. 1981). The proenzyme is activated via an autocatalytic event with cleavage between residues Ser_{81} and Ser_{82} (Recsei and Snell 1981). The gene encoding histidine decarboxylase (*hdcA*) exists in an operon with an uncharacterized ORF, designated *hdcB*. Transcription of this operon is induced 3-fold in the presence of histidine (Vanderslice et al. 1986; Copeland et al. 1989).

The likely physiological roles for histidine decarboxylation include the regulation of intracellular pH and the generation of metabolic energy (Molenaar et al. 1993). The generalized scheme is described in the Asp catabolism section and summarized in Figure 2.

Histamine production in fermented foods is of importance for health and food safety concerns. Histamine is a biogenic amine that can result in food poisoning (tenBrink et al. 1990; Stratton et al. 1991; Santos et al. 1996). Several outbreaks of histamine poisoning have occurred subsequent to the consumption of cheese containing high levels of histamine (Doeglas et al. 1967; Kahana & Todd 1981; Taylor et al. 1982).

Threonine catabolism

Threonine catabolism by LAB is initiated by threonine aldolase, an enzyme widely distributed in LAB. Threonine aldolase converts Thr to acetaldehyde and Gly (Lees & Jago 1976; Marshall & Cole 1983). The enzyme has been partially purified from *Lb. delbrueckii* ssp. *bulgaricus* YOP_{12} and determined to have a molecular weight of 190 kDa (Manca de Nadra et al. 1987). Threonine aldolase activity in *Lb. delbrueckii* ssp. *bulgaricus* and *S. thermophilus* is stimulated by Thr and inhibited by Gly and Cys (Lees & Jago 1976; Manca de Nadra et al. 1987; Marranzini et al. 1989). Some LAB convert acetaldehyde to ethanol via an alcohol dehydrogenase (Marshall & Cole, 1983). Recently, a gene encoding an alcohol dehydrogenase (*adhE*) has been cloned and characterized from *Lc. lactis* (Arnau et al. 1998).

It is generally believed that the physiological role of threonine aldolase is the production of Gly. The observation that *Lc. lactis* Z8, which lacks threonine aldolase activity, requires Gly for growth supports this hypothesis (Reiter & Oram 1962).

Acetaldehyde contributes to the development of typical yogurt flavor (Law 1981; Sandine & Elliker 1970; Tamine & Seeth 1980). In addition to Thr catabolism, it is known that carbohydrate fermentation can result in the formation of acetaldehyde (Shankar 1977; Wilkins et al. 1986). Further research is required to clarify the relative contribution of carbohydrates and Thr to the production of acetaldehyde by cultures used in yogurt production.

Aromatic amino acid catabolism

In most LAB, aromatic amino acid catabolism is initiated by an aminotransferase (Gao et al. 1997; Yvon et al. 1997; Gummalla & Broadbent 1998). The aminotransferase initiated pathways are illustrated in Figure 3 and described below. Additionally, some LAB

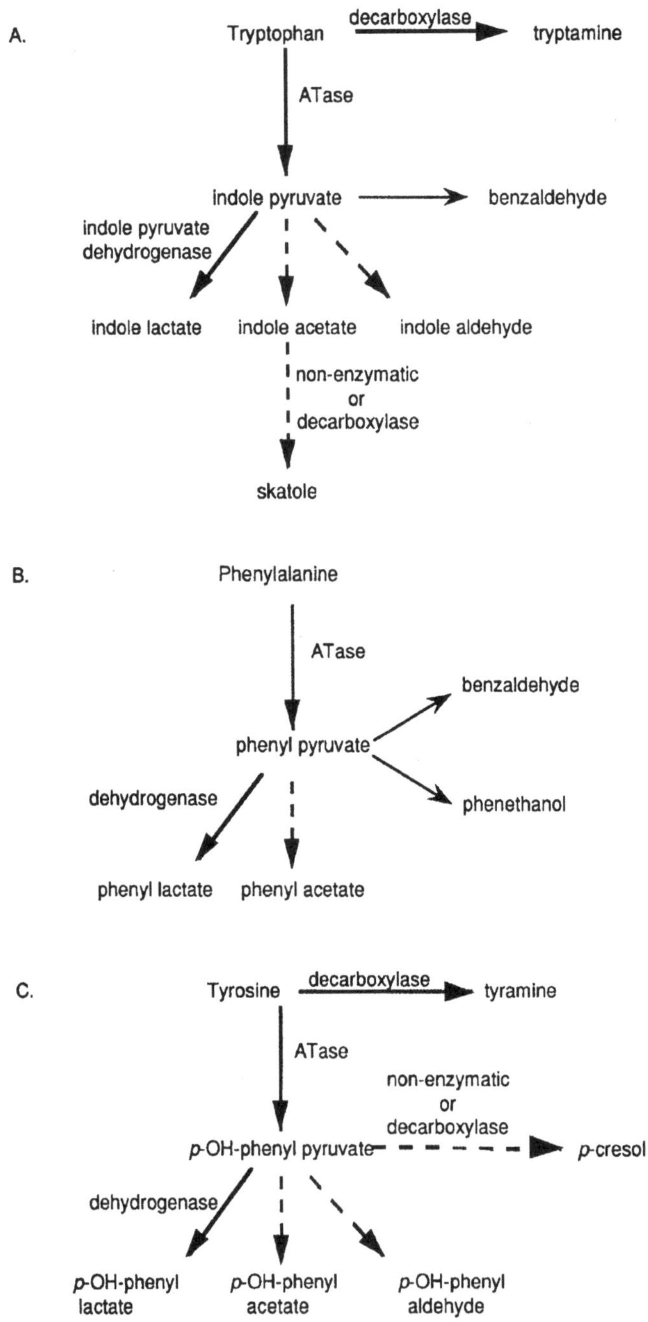

Figure 3. Pathways for the catabolism of tryptophan (A), phenylalanine (B), and tyrosine (C) by starter, adjunct and nonstarter bacteria. Thick arrows denote enzymatic reactions. Thin arrows indicate nonenzymatic chemical reactions. Dashed arrows indicate that the mechanism of conversion is unknown. Enzymes: ATase, aminotransferase.

express decarboxylases that have activity on aromatic amino acids.

Indole pyruvate (IpyA), the product of aminotransferase activity on Trp, can be catabolized further to indole lactic acid (ILA), indole acetic acid (IAA), indole aldehyde (IAld) and benzaldehyde (Nakazuma et al. 1977; Hummel et al. 1984; Gao et al. 1997). In certain strains of lactobacilli, IpyA is converted to ILA via IpyA dehydrogenase (Hummel et al. 1984, 1986; Gummalla & Broadbent 1998). IAA and IAld may also be produced enzymatically from IpyA. Gao et al. (1997) demonstrated that the accumulation of IAA and IAld are strain-specific in lactococci. This observation is significant because it suggests that differences between strains of lactococci may result in the accumulation of distinct Trp metabolic products. Formation of IAA is of particular importance as it can be non-enzymatically converted to skatole (Urbach 1995). In addition, Honeyfield & Carlson (1990) have demonstrated that a non-speciated strain of lactobacilli is capable of converting IAA to skatole via a decarboxylase.

Aminotransferase activity on Phe results in the formation of phenylpyruvate (PPA). Under the conditions employed by Gao et al. (1997), PPA was stable and further catabolism of PPA by lactococci was not detected. However, PPA, phenyl lactate, and phenyl acetate were detected as metabolites of lactococcal Phe catabolism by Yvon et al. (1997). These differences can be attributed to the different reaction conditions employed in these two studies. Also, non-enzymatic breakdown of PPA to benzaldehyde and phenethanol has been described (Kong et al. 1996; Groot & Bont 1998).

Aminotransferase activity on Tyr results in the formation of *p*-OH-phenyl pyruvate (HPPA). Gao et al. (1997) demonstrated that lactococci convert HPPA to 4-hydroxyl-benzaldehyde and *p*-OH-phenyl acetate. Certain strains of lactobacilli can decarboxylate HPPA to produce *p*-cresol (Yokoyama & Carlson 1981).

Aromatic amino acid aminotransferases (Ar-AT) have been purified and characterized from *Lc. lactis* NCDO763 and S3 (Yvon et al. 1997, Gao et al. 1998). The Ar-AT from NCDO763 is an 86 kDa pyridoxal-5′-phosphate-dependent enzyme that has activity on aromatic amino acids, Leu and Met. In *Lc. lactis* S3, two oligomeric species of Ar-AT have been identified (Gao et al. 1998). The more abundant species, Ar-AT1, exists as a homodimer with a molecular mass of 84kDa. Ar-AT2 is a homotetrameric enzyme composed of the same subunits as Ar-AT1. The Ar-AT purified from *Lc. lactis* NCDO763 is similar to Ar-AT1 with regard to optimum pH, subunit size, native molecular mass, and effect of NaCl and substrate specificity. However, the relative activity with aromatic amino acids for Ar-AT1 and the Ar-AT from NCDO763 are Trp>Tyr>Phe and Phe>Tyr>Trp, respectively.

It remains unclear which dehydrogenase(s) is responsible for the conversion of PPA and HPPA to phenyl lactate and *p*-OH-phenyl lactate, respectively. However, L-hydroxyisocaproate dehydrogenase (L-Hic) has broad substrate specificity, including activity towards PPA (Feil et al. 1994, 1997). Additionally, some evidence suggests that the reduction of PPA is catalyzed by an NAD(H)-dependent 2-hydroxyacid dehydrogenase (D-Hic) (Hummel & Kula 1989). These enzymes are described in greater detail in the branched-chain amino acid catabolism section.

The decarboxylation of Tyr and Trp results in the formation of tyramine and tryptamine, respectively. Organisms known to produce tyramine include *Lb. brevis, Lb. curvatus, Lb. delbrueckii, Lb. paracasei*, and *Leuconostoc lactis* (Straub et al. 1995; Masson et al. 1996; Roig-sagues et al. 1997). Recently, one of nine strains of *Lb. casei* examined was also demonstrated to produce tyramine from Tyr (Timpone & Steele, unpublished observations). Tyr decarboxylase activity has been detected in *Lb. brevis* (Joosten & Stadhouders 1987; Joosten & Northolt 1989) and in non-cheese related strains of *Lc. lactis* subsp. *cremoris* (Roig-Sagués et al. 1996). Trp decarboxylase activity has received less research attention. Gummalla & Broadbent (1998) reported the presence of Trp decarboxylase activity in cell-free extracts prepared from three strains of *Lb. casei* and two strains of *Lb. helveticus*. However, tryptamine was not detected in the supernatants of these cultures. The likely physiological roles for Tyr and Trp decarboxylation include the regulation of intracellular pH and the generation of metabolic energy. The generalized scheme is described in the Asp catabolism section and summarized in Figure 2.

A major concern of the dairy industry is the occurrence of objectionable (unclean) flavors. Compounds believed to be formed via aromatic amino acid catabolism that have been associated with objectionable flavors include *p*-cresol, phenethanol, phenylacetaldehyde, indole, and skatole (Dunn & Lindsay 1985; Vangtal & Hammond 1986; Wijesundera & Urbach 1993; Milo & Reineccius 1997).

Biogenic amines, such as tyramine, are frequently detected in cheese (Voight et al. 1974). Tyramine is of particular concern as it is the most common cause of monoamine intoxication (MAI). MAI is characterized by an increase in blood pressure that results in palpitations, severe headaches, hypertension, nausea, vomiting, and prostration. The majority (80%) of MAIs have been associated with consumption of cheese (McCabe 1986). It is important to understand the factors that influence tyramine accumulation in cheese, so that approaches can be developed to reduce its accumulation.

Branched-chain amino acid catabolism

Catabolism of branched-chain amino acids (BCAAs) is most likely initiated in *Lc. lactis* by an aminotransferase, converting Leu, Ile and Val to α-ketoisocaproate (KIC), α-keto-β-methylvalerate (KMV) and α-ketoisovalerate (KIV), respectively. Subsequently, enzymes have been identified in *Lc. lactis* and other LAB that convert these α-keto acids to aldehydes, alcohols or fatty acids. The details of these pathways are described below.

Aminotransferase activity on BCAAs was first reported in a set of studies interested in identifying the origin of a malt-flavor defect in milk (Jackson & Morgan 1954). Recently, a number of aminotransferases from *Lc. lactis* with activity on BCAAs have been described (Alting & Engels 1996; Yvon 1997; Gao & Steele 1998; Roudot-Algaron & Yvon 1998; Atiles et al. 1999). The Ar-ATs described by Yvon et al. (1997) and Gao & Steele (1998) differ significantly from the enzymes described by Alting & Engels (1996) and Atiles et al. (1999), as these latter two enzymes do not have activity with Tyr or Trp. The deduced amino acid sequence from a gene encoding a BCAA aminotransferase (*ilvE*) in *Lc. lactis* LM0230 has high identity to other BCAA aminotransferases (Atiles et al. 1999). A derivative of *Lc. lactis* LM0230 lacking IlvE activity [LM0230(Δ*ilvE*)] was constructed and determined to differ from LM0230 in that it was unable to grow in defined medium when either KIV or KMV was substituted for Val or Ile, respectively. However, LM0230(Δ*ilvE*) was able to grow when KIC was substituted for Leu (Atiles et al. 1999). This supports the biochemical evidence that lactococci have at least two aminotransferases with activity on Leu. Therefore, while it appears likely that lactococci initiate catabolism of BCAAs via aminotransferase activity, it is unclear how many lactococcal enzymes catalyze this reaction.

The conversion of the BCAA-derived α-ketoacids to their corresponding aldehydes is catalyzed by a decarboxylase. Tucker & Morgan (1967) identified a decarboxylase in *Lc. lactis* var. *maltigenes* strains that converts KIC to 3-methylbutanal and catalyzes the decarboxylation of KIV and KMV. Additionally, one strain of *Lc. lactis,* which is not part of the *maltigenes* biovariant, has been demonstrated to produce limited quantities of aldehydes from Leu and Val (MacLeod & Morgan 1958), suggesting this enzyme may be present in lactococci outside of the *maltigenes* biovariant cluster. A branched-chain α-ketoacid decarboxylase has also been identified in *Lb. casei* (Hickey et al. 1983). Additionally, production of the aldehydes expected from catabolism of Leu and Val by *Lactobacillus maltaromicus* (later classified by Collins et al. 1991, as *Carnobacterium*) has been described (Miller et al. 1974).

The most thoroughly characterized enzymes from LAB that catalyze the reduction of BCAA-derived α-keto acids are L-2-hydroxyisocaproate dehydrogenase (L-Hic) (Schütte et al. 1984) and D-2-hydroxyisocaproate dehydrogenase (D-Hic) (Hummel et al. 1985). In a screen of five strains of *Bifidobacterium* and 45 strains of *Lactobacillus* and *Leuconostoc*, L-Hic activity was only observed in *Lactobacillus confusus* (Schütte et al. 1984). The purified enzyme was NADH dependent and exhibited highest activity towards C_5 and C_6 acids with a preference towards branched-chain α-keto acids. The deduced amino acid sequence of the cloned gene is 30% identical to the L-lactate dehydrogenase (L-LDH) gene from *Lb. casei* (Lerch et al. 1989). The coenzyme-binding residues and the three substrate binding amino acids conserved in all L-LDHs are also present in L-Hic (Lerch et al. 1989). These results indicate that L-Hic is structurally related to L-LDHs.

In comparison to L-Hic, D-Hic activity appears to be more widely distributed among LAB. D-Hic activity has been identified in three strains of *Leuconostoc*, three strains of *Lb. casei* (Hummel et al. 1985) and in *Lb. delbrueckii* subsp. *bulgaricus* (Bernard et al. 1994). As with L-Hic, this enzyme is active on a variety of α-keto acids (Hummel et al. 1985; Kallwass 1992), and is NADH dependent. Comparison of deduced amino acid sequences of D-Hic and D-lactate dehydrogenase suggests the enzymes are related. Similar analysis suggests that D-Hic and L-Hic are genetically unrelated proteins (Lerch et al. 1989; Kochhar et al. 1992). In addition to L-Hic and D-Hic, at least two other LAB dehydrogenases with activity on BCAA-derived α-keto acids are known: a D-(-)-mandelic acid dehydrogenase from *Lactobacillus curvatus* (Hummel et al. 1988); and an alcohol dehydrogenase from *Lc. lactis* var. *maltigenes* (Morgan et al. 1965; Harrison et al. 1969).

The production of branched-chain fatty acids (BCFAs) from branched-chain amino acids by LAB does not appear to be common. However, one strain of *Lc. lactis* subsp. *lactis* var. *diacetylactis* and one non-speciated strain of *Lactobacillus* have been shown to produce isobutyric acid (i-C_4) from Val, and isovaleric acid (i-C_5) or another C_5-acid from Leu and Ile (Nakae & Elliott 1965). An aminotransferase may initiate this pathway, but the other enzymes required for i-C_4 and i-C_5 formation remain unknown.

The physiological significance of BCAA catabolism is poorly defined in LAB. In other bacteria, IlvE executes the last step of BCAA biosynthesis (Umbarger 1996). However, as dairy LAB are typically auxotrophic for BCAAs (Chopin 1993), *ilvE* likely has a different role in these organisms. The detection of an enzyme in *Lc. lactis* that catalyzes the reduction of KIC suggests BCAA catabolism may serve to regenerate NAD^+ in lactococci (Gao and Steele, unpublished results). Additionally, IlvE might be involved in the recycling of Glu from α-ketoglutarate, which can then be used to catalyze the production of other amino acids synthesized by aminotransferase reactions. However, the growth of LM0230(Δ*ilvE*) in milk is indistinguishable from the growth of isogenic *ilvE*$^+$ strains (Atiles et al. 1999). Therefore, a physiological role for IlvE during growth of LAB in milk is not yet apparent. Physiological roles of BCAA catabolism in other organisms include the generation of metabolic energy and carbon (Massey et al. 1976), the production of BCFAs for cell membrane biosynthesis (de Mendoza et al. 1993), and the production of α-keto acids and α-hydroxyacids that serve as siderophores in low iron environments (Kingsley et al. 1996). However, data supporting similar roles in LAB is lacking.

In the production of fermented dairy products, BCAA catabolism is typically associated with the generation of compounds having a detrimental effect on product aroma or flavor. The aroma of the three aldehyde derivatives of BCAA catabolism have each been described as moderately malty (Griffith & Hammond 1989). The corresponding alcohols have similar aromas, but with approximately 50-fold higher flavor thresholds (Sheldon et al. 1971). The conversion of Leu to 3-methylbutanal by *Lc. lactis* var. *maltigenes* has been implicated in the development of malty off-flavors in fermented milks (Jackson & Morgan 1954), as well as 2-methylpropanal and the alcohol derivatives of these two aldehydes (Morgan et al. 1965; Sheldon et al. 1971). Although 2-methylbutanal has a similar odor characteristic and flavor threshold to 3-methylbutanal, it has not been detected in significant concentrations in fermented milks. The aldehyde and alcohol products of BCAA catabolism are also known to cause defects in Cheddar cheese. Cheeses containing 18–90 ppm 3-methylbutanal and 9–45 ppm 3-methylbutanol were graded lower for flavor quality and were maltier than the control cheeses (Braun & Olson 1986). Additionally, Dunn & Lindsay (1985) correlated flavors described as harsh, dull and suppressed with elevated concentrations (>200 ppb) of 3-methylbutanal and 2-methylpropanal. In Emmentaler, 3-methylbutanal can be present at levels 6–16 fold over that of the nasal or retronasal odor thresholds of this compound in sunflower oil (Preininger & Grosch 1994). However, Preininger et al. (1996) suggested that 3-methylbutanal is not one of the defining flavor or odor compounds of Emmentaler, but may play a role in suppressing the unpleasant, sweaty odor of butyric acid.

The fatty acids derived from BCAAs are described as sweaty, estery, rotten fruit-like and fatty acid-like at concentrations around 5 ppm, to sweet, fruity, waxy and apple-like at higher concentrations (Brennand et al. 1989). While these acids have been detected in Cheddar (Ha & Lindsay 1991a; Yang & Min 1993), they appear to be more prevalent in other cheeses such as Swiss, Italian and blue-veined varieties (Patton 1964; Ha & Lindsay 1991a,b; Bosset et al. 1993). In Swiss, BCFAs have been correlated with flavors described as afterburn, amine and fruity (Vangtal & Hammond 1986). The study of Biede (1979) clearly implicates *Propionibacterium shermanii* as the primary culture involved in the production of i-C_5 in Swiss cheese, however another report (Paulsen et al. 1980) suggests *P. shermanii* has little impact on i-C_5 accumulation. Therefore, more research is required to elucidate how i-C_5 is formed in Swiss cheese.

Concluding remarks

A significant body of information is now available concerning many of the enzymes that comprise the peptidase enzyme systems of lactobacilli and lactococci. A complement of enzymes with comparable specificities is present in each of the systems studied.

Table 2. Specific growth rates of peptidase mutants in milk

Peptidase Mutant	Strain	Mutation Type	Mutant vs. WT μ_{max} in Milk	% Increase in Generation Time	Reference
PepC	*Lc. lactis*	deletion	0.60/0.65	10	Mierau et al. (1996)
PepN	*Lb. helveticus*	deletion	0.56/0.60	7	Christensen et al. (1996)
PepN	*Lc. lactis*	deletion	0.52/0.65	25	Mierau et al. (1996)
PepA	*Lc. lactis*	disrupted	No difference		l'Anson et al. (1995)
PepX	*Lb. helveticus*	deletion	0.28/0.61	120	Yüksel & Steele (1999)
PepX	*Lc. lactis*	deletion	0.56/0.65	15	Mierau et al. (1996)
PepX	*Lc. lactis*	deletion	No difference		Mayo et al. (1993)
PepX	*Lc. lactis*	chemical	1.0/1.6	60	Nardi et al. (1991)
PepX	*Lc. lactis*	disrupted	0.70/0.93	31	Matos et al. (1998)
PepD	*Lb. helveticus*	deletion	No difference		Dudley et al. (1996)
PepI	*Lb. helveticus*	deletion	0.56/0.61	9	Yüksel & Steele (1999)
PepQ	*Lb. helveticus*	deletion	0.54/0.61	13	Yüksel & Steele (1999)
PepR	*Lb. helveticus*	deletion	No difference		Shao et al. (1997)
PepR	*Lb. rhamnosus*	deletion	No difference		Varmanen et al. (1998)
PepP	*Lc. lactis*	disrupted	0.89/0.93		Matos et al. (1998)
PepT	*Lc. lactis*	deletion	0.60/0.65	10	Mierau et al. (1996)
PepO	*Lb. helveticus*	deletion	No difference		Chen & Steele (1998)
PepO	*Lc. lactis*	disrupted	No difference		Mierau et al. (1993)
PepO	*Lc. lactis*	deletion	0.60/0.65	9	Mierau et al. (1996)
PepO	*Lc. lactis*	disrupted	No difference		Tynkkynen et al. (1993)

However, there are peptidases in lactobacilli (PepD, PepE, PepI, PepL, PepR) and lactococci (PepA, PepF, PepP) for which a genetic homologue has not been reported in the converse systems. Not yet known is whether this reflects differences in the peptidase genotypic set or simply peptidase genes that have yet to be identified and sequenced. In addition, while the regulation of peptidase expression has been examined in complex media, only a few studies have examined the expression of these peptidases during growth in milk or under the conditions these organisms encounter in ripening cheese. The potential differences in lactobacilli and lactococci pertaining to the pathway of hydrolysis of casein derived peptides for growth in milk, or as they impact the flavor and texture of fermented foods, remain to be determined.

The physiological role of individual peptidases have been investigated from numerous single and multiple peptidase mutants constructed in *Lb. helveticus* and *Lc. lactis*. The growth rate of mutants have been evaluated in milk and a general decrease in growth rates has been observed (see Table 2). However, the mechanism for this reduction in growth rate has not been determined. Information on peptide accumulation in peptidase mutants grown in milk may assist in elucidating the role of specific peptidases. Additionally, little information is available regarding the potential role of these peptidases in cellular protein turnover and post-translational modification. Peptidase mutant strains may be valuable for subsequent screening of activities not identified in the presence of enzymes in the wild type strains with higher and/or overlapping specificities. This may be important for identification of peptidases that have important physiological roles but have not or can not be identified using small peptides or amino acid chromogens as substrates.

Relatively little information is available concerning amino acid catabolism in LAB. In the past few years there has been an increased interest in elucidating the pathways of amino acid catabolism by LAB and the production of sensory compounds and biogenic amines. Recent studies have implicated amino acid catabolism in the regulation of intracellular pH and the generation of metabolic energy. However, we do not know the significance of these pathways for cell survival in a carbohydrate-starved environment, as in ripening cheese. As more genes involved in these pathways are characterized, it will be possible to examine their regulation and construct isogenic strains lacking a particular enzyme activity. Ideally, a more thorough understanding of the peptidolytic/catabolic pathways present in LAB would focus engineering or selection for strains with controlled production of amino acid catabolites. This may include reduction of biogenic amine accumulation, a human health consideration, as well as controlled production of specific flavor precursors or compounds in fermented foods.

Acknowledgments

The authors would like to thank our colleagues Jeff Broadbent and Carmen Peláez for communicating results prior to publication. The work on peptidases and amino acid catabolism conducted in our laboratory was supported in part by the College of Agricultural and Life Sciences at the University of Wisconsin-Madison and the Center for Dairy Research through funding from Dairy Management Inc.

References

Abe K, Hayashi H & Maloney PC (1996) Exchange of aspartate and alanine. Mechanism for development of a proton-motive force in bacteria. J. Biol. Chem. 271: 3079–3084

Alting AC & Engels WJM (1996) Conversion of methionine by enzymes from *Lactococcus lactis* subsp. *cremoris* B78 during cheese ripening. poster K3, Federation of European Microbiological Societies Fifth Symposium on Lactic Acid Bacteria, 8–12 September, Veldhoven, The Netherlands

Arnau J, Jorgensen F, Madsen SM, Vrang A & Israelsen H (1998) Cloning of the *Lactococcus lactis adhE* gene, encoding a multifunctional alcohol dehydrogenase, by complementation of a fermentative mutant of *Escherichia coli*. J. Bacteriol. 180: 3049–3055

Arora G & Lee BH (1992) Purification and characterization of aminopeptidase from *Lactobacillus casei* ssp. *casei* LLG. J. Dairy Sci. 75: 700–710

Arora G & Lee BH (1994) Purification and characterization of an aminopeptidase from *Lactobacillus casei* subsp. *rhamnosus* S93. Biotech. Appl. Biochem. 19: 179–192

Atlan D, Gilbert C, Blanc B & Portalier R (1994) Cloning, sequencing and characterization of the *pep*IP gene encoding a proline iminopeptidase from *Lactobacillus delbrueckii* subsp. *bulgaricus* CNRZ 397. Microbiol. 140: 527–535

Atlan D, Laloi P & Portalier R (1990) X-Prolyldipeptidyl aminopeptidase of *Lactobacillus delbrueckii* ssp. *bulgaricus*: Characterization of the enzyme and isolation of deficient mutants. Appl. Environ. Microbiol. 56: 2174–2179

Atiles MW, Dudley EG & Steele JL (1999) Characterization and inactivation of the branched-chain aminotransferase gene from *Lactococcus lactis* LM0230. Manuscript in preparation

Axelsson L (1998) Lactic acid bacteria: classification and physiology. In: Salminen S & von Wright A (Eds) Lactic acid bacteria. Microbiology and functional aspects, (pp 1–72). Marcel Dekker, Inc., New York

Baankreis R & Exterkate FA (1991) Characterisation of a peptidase from *Lactococcus lactis* ssp. *cremoris* HP that hydrolyses di- and tripeptides containing proline or hydrophobic residues as the amino-terminal amino acid. Syst. Appl. Microbiol. 14: 317–323

Baankreis R, Vanschalkwijk S, Alting AC & Exterkate FA (1995) The occurrence of two intracellular oligoendopeptidases in *Lactococcus lactis* and their significance for peptide conversion in cheese. Appl. Microbiol. Biotech. 44: 386–392

Bacon CL, Jennings PV, Fhaolain IN & O'Cuinn G (1994) Purification and characterisation of an aminopeptidase A from cytoplasm of *Lactococcus lactis* subsp. *cremoris* AM2. Int. Dairy. J. 4: 503–519

Bacon CL, Wilkinson M, Jennings PV, Fhaolain IN & O'Cuinn G (1993) Purification and characterization of an aminotripeptidase from cytoplasm of *Lactococcus lactis* subsp. *cremoris* AM2. Int. Dairy. J. 3: 163–177

Bernard N, Johnsen K, Ferain T, Garmyn D, Hols P, Holbrook JJ & Delcour J (1994) NAD^+-dependent D-2-hydroxyisocaproate dehydrogenase of *Lactobacillus delbrueckii* subsp. *bulgaricus*. Gene cloning and enzyme characterization. Eur. J. Biochem. 224: 439–446

Biede SL, Paulsen PV, Hammond EG & Glatz BA (1979) The flavor of Swiss cheese. In: Underkofler LA (Ed) Developments in Industrial Microbiology, (pp 203–210). Society for Industrial Microbiology, Arlington, VA

Blanc B, Laloi P, Atlan D, Gilbet C & Portalier R (1993) Two cell-wall-associated aminopeptidases from *Lactobacillus helveticus* and the purification and characterization of APII from strain ITGL1. J. Gen. Microbiol. 139: 1441–1448

Bockelmann W, Beuck HP, Lick S & Heller K (1995) Purification and characterization of a new tripeptidase from *Lactobacillus delbrueckii* ssp. *bulgaricus* B14. Int. Dairy. J. 5: 493–502

Bockelmann W, Fobker M & Teuber M (1991) Purification and characterization of the X-prolyl-dipeptidyl-aminopeptidase from *Lactobacillus delbrückii* subsp. *bulgaricus* and *Lactobacillus acidophilus*. Int. Dairy. J. 1: 51–66

Bockelmann W, Gollan V & Heller KJ (1997) Purification of a second tripeptidase from *Lactobacillus delbrueckii* subsp. *bulgaricus* B14. Milchwissenschaft 52: 500–503

Bockelmann W, Hoppeseyler T & Heller KJ (1996) Purification and characterization of an endopeptidase from *Lactobacillus delbrueckii* subsp. *bulgaricus* B14. Int. Dairy. J. 6: 1167–1180

Bockelmann W, Shulz Y & Teuber M (1992) Purification and characterization of an aminopeptidase from *Lactobacillus delbrückii* subsp. *bulgaricus*. Int. Dairy. J. 2: 95–107

Boeker EA & Snell EE (1972) Amino acid decarboxylases. In PD Boyer (Ed.) The Enzymes. 6: 217–253

Booth M, Donnelly WJ, Fhaoláin IN, Jennings PV & O'Cuinn G (1990a) Proline-specific peptidases of *Streptococcus cremoris* AM2. J. Dairy Res. 57: 79–88

Booth M, Fhaoláin IN, Jennings PV & O'Cuinn G (1990b) Purification and characterization of a post-proline dipeptidyl aminopeptidase from *Streptococcus cremoris* AM2. J. Dairy Res. 57: 89–100

Booth M, Jennings V, Ni Fhaoláin I & O'Cuinn G (1990c) Prolidase activity of *Lactococcus lactis* ssp. *cremoris* AM2: Partial purification and characterization. J. Dairy Res. 57: 245–254

Bosman BW, Tan PST & Konings WN (1990) Purification and characterization of a tripeptidase from *Lactococcus lactis* ssp. *cremoris* Wg2. Appl. Environ. Microbiol. 56: 1839–1843

Bosset JO, Collomb M & Sieber R (1993) The aroma composition of Swiss Gruyère cheese IV. The acidic volatile components and their changes in content during ripening. Lebensm. Wissen. Technol. 26: 581–592

Braun SD & Olson NF (1986) Microencapsulation of cell-free extracts to demonstrate the feasibility of heterogeneous enzyme systems and cofactor recycling for development of flavor in cheese. J. Dairy Sci. 69: 1202–1208

Brennand CP, Ha JK & Lindsay RC (1989) Aroma properties and thresholds of some branched-chain and other minor volatile fatty acids occurring in milkfat and meat lipids. J. Sensory Studies 4: 105–120

Brückner R (1998) Direct sequence submission to GenBank for *glt*A (glucose uptake protein) from *Staphylococcus xylosus* (Accession Y14043). Unpublished

Chang GW & Snell EE (1968) Histidine decarboxylase of *Lactobacillus* 30a. II. Purification, substrate specificity, and stereospecificity. Biochemistry 7: 2005–2012

Chapot-Chartier MP, Nardi M, Chopin MC, Chopin A & Gripon JC (1993) Cloning and sequencing of *pep*C, a cysteine aminopeptidase gene from *Lactococcus lactis* ssp. *cremoris* AM2. Appl. Environ. Microbiol. 59: 330–333

Chapot-Chartier MP, Rul F, Nardi M & Gripon JC (1994) Gene cloning and characterization of PepC, a cysteine aminopeptidase from *Streptococcus thermophilus*, with sequence similarity to the eukaryotic bleomycin hydrolase. Eur. J. Biochem. 224: 497–506

Chen YS & Steele JL (1998) Genetic characterization and physiological role of endopeptidase O from *Lactobacillus helveticus* CNRZ32. Appl. Environ. Microbiol. 64: 3411–3415

Chich J-F, Rigolet P, Nardi M, Gripon J-C, Ribadeau-Dumas B & Brunie S (1995) Purification, crystallization, and preliminary X-ray analysis of PepX, an X-Prolyl dipeptidyl aminopeptidase from *Lactococcus lactis*. Prot. Struct. Funct. Genet. 23: 278–281

Chopin A (1993) Organization and regulation of genes for amino acid biosynthesis in lactic acid bacteria. FEMS Microbiol. Rev. 12: 21–38

Christensen JE, Lin DL, Palva A & Steele JL (1995) Sequence analysis, distribution and expression of an aminopeptidase N-encoding gene from *Lactobacillus helveticus* CNRZ32. Gene 155: 89–93

Christensen JE & Steele JL (1996) Characterization of peptidase-deficient *Lactobacillus helveticus* CNRZ32 derivatives. *Fifth Symposium on Lactic Acid Bacteria*, Veldhoven, The Netherlands

Christensen JE & Steele JL (1998a) Hydrolysis of casein derived peptides by peptidase-deficient *Lactobacillus helveticus* CNRZ32 derivatives. *Joint Meeting of the American Dairy Science Association and the American Society of Animal Science*. Journal of Dairy Science, Denver, Colorado, USA.

Christensen JE & Steele JL (1998b) Unpublished results

Collins MD, Rodrigues U, Ash C, Aguirre M, Farrow JAE, Martinez-Murcia A, Phillips BA, Williams AM & Wallbanks S (1991) Phylogenetic analysis of the genus *Lactobacillus* and related lactic acid bacteria as determined by reverse transcriptase sequencing of 16S rRNA. FEMS Microbiol. Lett. 77: 5–12

Copeland WC, Domena JD & Robertus JD (1989) The molecular cloning, sequence and expression of the *hdcB* gene from *Lactobacillus* 30A. Gene 85: 259–265

Crow VL & Thomas TD (1982) Arginine metabolism in lactic streptococci. J. Bacteriol. 150: 1024–1032

Cunin R, Glansdorff N, Piérard A & Stalon V (1986) Biosynthesis and metabolism of arginine in bacteria. Microbiol. Rev. 50: 314–352

de Mendoza D, Grau R & Cronan JE Jr (1993) Biosynthesis and function of membrane lipids. In: Sonenshein AL, Hock JA & Losick R (Eds) *Bacillus subtilis* and other gram-positive bacteria: biochemistry, physiology, and molecular genetics, (pp 411–421). American Society for Microbiology, Washington, DC

Desmazeaud MJ & Zevaco C (1976) General properties and substrate specificity of an intracellular neutral protease from *Streptococcus diacetilactis*. Ann. Biol. Anim. Biochim. Biophys. 16: 851–868

Desmazeaud MJ & Zevaco C (1977) General properties and substrate specificity of an intracellular soluble dipeptidase from *Streptococcus diacetilactis*. Ann. Biol. Anim. Biochim. Biophys. 17: 723–736

Desmazeaud MJ & Zevaco C (1979) Isolation and general properties of two intracellular amino peptidases of *Streptococcus diacetylactis*. Milchwissenschaft 34: 606–610

Detmers FJM, Kunji ERS, Lanfermeijer FC, Poolman B & Konings WN (1998) Kinetics and specificity of peptide uptake by the oligopeptide transport system of *Lactococcus lactis*. Biochem. 37: 16671–16679

Doeglas HM, Huisman J & Nater JP (1967) Histamine intoxication after cheese. Lancet 2: 1361–1372

Dudley EG, Husgen AC, He W & Steele JL (1996) Sequencing, distribution, and inactivation of the dipeptidase A gene (*pep*DA) from *Lactobacillus helveticus* CNRZ32. J. Bacteriol. 178: 701–704

Dudley EG & Steele JL (1994) Nucleotide sequence and distribution of the *pep*PN gene from *Lactobacillus helveticus* CNRZ32. FEMS Microbiol. Lett. 119: 41–45

Dudley EG & Steele JL (1999) Characterization and inactivation of the aspartate aminotransferase gene from *Lactococcus lactis* LM0230. Manuscript in preparation

Dunn HC & Lindsay RC (1985) Evaluation of the role of microbial Strecker-derived aroma compounds in unclean-type flavors of Cheddar cheese. J. Dairy Sci. 68: 2859–2874

Eggimann B & Bachmann M (1980) Purification and partial characterization of an aminopeptidase from *Lactobacillus lactis*. Appl. Environ. Microbiol. 40: 876–882

Exterkate FA, De Jong M, De Veer GJCM & Baankreis R (1992) Location and characterization of aminopeptidase N in *Lacto-*

coccus lactis ssp. cremoris HP. Appl. Microbiol. Biotech. 37: 46–54

Exterkate FA & De Veer GJCM (1987) Purification and some properties of a membrane-bound aminopeptidase A from *Streptococcus cremoris*. Appl. Environ. Microbiol. 53: 577–583

Feil IK, Hendle J & Schomburg D (1997) Modified substrate specificity of L-hydroxyisocaproate dehydrogenase derived from structure-based protein engineering. Protein Eng 10: 255–262

Feil IK, Lerch HP & Schomburg D (1994) Deletion variants of L-hydroxyisocaproate dehydrogenase. Probing substrate specificity. Eur. J. Biochem. 223: 857–863

Fenster KM, Parkin KL & Steele JL (1997) Characterization of a thiol-dependent endopeptidase from *Lactobacillus helveticus* CNRZ32. J. Bacteriol. 179: 2529–2533

Fernández de Palencia P, López de Felipe F, Requena T & Peláez C (1999) The aminopeptidase C (PepC) from *Lactobacillus helveticus* CNRZ32. A comparative study of PepC from dairy lactic acid bacteria. Submitted

Fernández de Palencia P, Pelaez C & Martín-Hernández MC (1997) Characterization of the aminopeptidase system from *Lactobacillus casei* subsp. *casei* IFPL 731. J. Agric. Food Chem. 45: 3778–3781

Fernández L, Bhowmik T & Steele JL (1994) Characterization of the *Lactobacillus helveticus* CNRZ32 *pepC* gene. Appl. Environ. Microbiol. 60: 333–336

Fernández-Esplá MD & Martín-Hernández MC (1997) Purification and characterization of a dipeptidase from *Lactobacillus casei* ssp. *casei* IFPL 731 isolated from goat cheese made from raw milk. J. Dairy Sci. 80: 1497–1504

Fernández-Esplá MD, Martín-Hernández MC & Fox PF (1997) Purification and characterization of a prolidase from *Lactobacillus casei* subsp. *casei* IFPL 731. Appl. Environ. Microbiol. 63: 314–316

Gao S, Mooberry ES & Steele JL (1998) Use of 13C nuclear magnetic resonance and gas chromatography to examine methionine catabolism by lactococci. Appl. Environ. Microbiol. 64: 4670–4675

Gao S, Oh DH, Broadbent JR, Johnson ME, Weimer BC & Steele JL (1997) Aromatic amino acid catabolism by lactococci. Lait 77: 371–381

Gao S & Steele JL (1998) Purification and characterization of oligomeric species of an aromatic amino acid aminotransferase from *Lactococcus lactis* subsp. *lactis* S3. J. Food Biochem. 22: 197–211

Geis A, Bockelmann W & Teuber M (1985) Simultaneous extraction and purification of a cell wall-associated peptidase and beta-casein specific protease from *Streptococcus cremoris* AC1. Appl. Microbiol. Biotech. 23: 79–84

Gilbert C, Atlan D, Blanc B & Portalier R (1994) Proline iminopeptidase from *Lactobacillus delbrueckii* subsp. *bulgaricus* CNRZ 397: Purification and characterization. Microbiol. 140: 537–542

Gobbetti M, Smacchi E & Corsetti A (1996) The proteolytic system of *Lactobacillus sanfrancisco* CB1: Purification and characterization of a proteinase, a dipeptidase, and an aminopeptidase. Appl. Environ. Microbiol. 62: 3220–3226

Gottesman S & Maurizi MR (1992) Regulation by proteolysis: energy-dependent proteases and their targets. Microbiol Rev 56: 592–621

Griffith R & Hammond EG (1989) Generation of Swiss cheese flavor components by the reaction of amino acids with carbonyl compounds. J. Dairy Sci. 72: 604–613

Guirard BM & Snell EE (1964) Nutritional requirements of *Lactobacillus* 30a for growth and histidine decarboxylase production. J. Bacteriol. 87: 370–376

Gummalla S (1998) Tryptophan catabolism by *Lactobacillus* spp.: Biochemistry and implications on flavor development in reduced-fat cheddar cheese. Master of Science thesis, Utah State University

Gummalla S & Broadbent JR (1996) Indole production by *Lactobacillus* ssp. in cheese: a possible role for tryptophanase. J. Dairy Sci. 79(Suppl. 1): 101

Guthrie BD (1993) Influence of cheese-related microflora on the production of unclean-flavored aromatic amino acid metabolites in Cheddar cheese. Ph.D thesis, University of Wisconsin-Madison

Ha KJ & Lindsay RC (1991a) Contributions of cow, sheep, and goat milks to characterizing branched-chain fatty acid and phenolic flavors in varietal cheeses. J. Dairy Sci. 74: 3267–3274

Ha KJ & Lindsay RC (1991b) Volatile branched-chain fatty acids and phenolic compounds in aged Italian cheese flavors. J. Food Sci. 56: 1241–1247, 1250

Habibi-Najafi MB & Lee BH (1994) Purification and characterization of X-prolyl dipeptidyl peptidase from *Lactobacillus casei* subsp. *casei* LLG. Appl. Microbiol. Biotech. 42: 280–286

Hackert ML, Meador WE, Oliver RM, Salmon JB, Recsei PA & Snell EE (1981) Crystallization and subunit structure of histidine decarboxylase from *Lactobacillus* 30a. J. Biol. Chem. 256: 687–690

Harrison GA, Khairallah EA & Morgan ME (1969) Alcohol dehydrogenase in *Streptococcus lactis* var. *maltigenes*: In vivo relationship to glycolysis, in vitro kinetics. poster P109. American Society for Microbiology 69th Annual Meeting, 4–9 May, Miami Beach, Fla

Hellendoorn MA, Franke-Fayard BMD, Mierau I, Venema G & Kok J (1997) Cloning and analysis of the *pepV* dipeptidase gene of *Lactococcus lactis* MG1363. J. Bacteriol. 179: 3410–3415

Hickey MW, Hillier AJ & Jago GR (1983) Enzymic activities associated with lactobacilli in dairy products. Aust. J. Dairy Technol. 38: 154–157

Higuchi T, Hayashi H & Abe K (1997) Exchange of glutamate and γ-aminobutyrate in a *Lactobacillus* strain. J. Bacteriol. 179: 3362–3364

Hoffman K, Bucher P, Falquet L & Bairoch A (1999) The PROSITE database, its status in 1999. Nuc. Acids Res. 27: 215–219

Honeyfield DC & Carlson JR (1990) Assay for the enzymatic conversion of indoleacetic acid to 3-methylindole in a ruminal *Lactobacillus* species. Appl. Environ. Microbiol. 56: 724–729

Houbart V, Ribadeau-Dumas B & Chich JF (1995) Synthesis of enterostatin-amide by the Xaa-prolyl dipeptidyl aminopeptidase from *Lactococcus lactis* subsp. *lactis* NCDO 763. Biotech. Appl. Biochem. 21: 149–159

Hummel W & Kula MR (1989) Dehydrogenases for the synthesis of chiral compounds. Eur. J. Biochem. 184: 1–13

Hummel W, Schmidt E, Wandrey C & Kula MR (1986) L-phenylalanine dehydrogenase from *Brevibacterium* sp. for the production of L-phenylalaine by reductive amination of phenylpyruvate. Appl. Microbiol. Biotechnol. 25: 175–185

Hummel W, Schütte H & Kula M-R (1985) D-2-hydroxyisocaproate dehydrogenase from *Lactobacillus casei*. Appl. Microbiol. Biotechnol. 21: 7–15

Hummel W, Schütte H & Kula M-R (1988) D-(-)-mandelic acid dehydrogenase from *Lactobacillus curvatus*. Appl. Microbiol. Biotechnol. 28: 433–439

Hummel W, Weiss N & Kula MR (1984) Isolation and characterization of a bacterium possessing L-phenylalanine dehydrogenase activity. Arch. Microbiol. 137: 47–52

Huynh QK, Recsei PA, Vaaler GL & Snell EE (1984) Histidine decarboxylase of *Lactobacillus* 30a. Sequences of the overlapping peptides, the complete alpha chain, and prohistidine decarboxylase. J. Biol. Chem. 259: 2833–2839

Hwang IK, Kaminogawa S & Yamauchi K (1981) Purification and properties of a dipeptidase from *Streptococcus cremoris*. Agric. Biol. Chem. 45: 159–166

Jackson HW & Morgan ME (1954) Identity and origin of the malty aroma substance from milk cultures of *Streptococcus lactis* var. *maltigenes*. J. Dairy Sci. 37: 1316–1324

Joosten HM & Northolt MD (1989) Detection, growth, and amine-producing capacity of lactobacilli in cheese. Appl. Environ. Microbiol. 55: 2356–2359

Joosten HM & Stadhouders J (1987) Conditions allowing the formation of biogenic amines in cheese. 1. Decarboxylative properties of starter bacteria. Neth. Milk Dairy J. 41: 247–258

Joshua-Tor L, Xu HE, Johnston SA & Rees DC (1995) Crystal structure of a conserved protease that binds DNA: the bleomycin hydrolase, Gal6. Science 269: 945–950

Kahana LM & Todd E (1981) Histamine poisoning and reaction to cheese. Ann. Intern. Med. 88: 520–521

Kallwass HKW (1992) Potential of *R*-2-hydroxyisocaproate dehydrogenase from *Lactobacillus casei* for stereospecific reductions. Enzyme Microb. Technol. 14: 28–35

Kaminogawa S, Azuma N, Hwang IK, Suzuki Y & Yamauchi K (1984) Isolation and characterization of a prolidase from *Streptococcus cremoris* H61. Agric. Biol. Chem. 48: 3035–3040

Khalid NM & Marth EH (1990a) Partial purification and characterization of an aminopeptidase from *Lactobacillus helveticus* CNRZ 32. Syst. Appl. Microbiol. 13: 311–319

Khalid NM & Marth EH (1990b) Purification and partial characterization of a prolyl-dipeptidyl aminopeptidase from *Lactobacillus helveticus* CNRZ 32. Appl. Environ. Microbiol. 56: 381–388

Kiefer-Partsch B, Bockelmann W, Geis A & Teuber M (1989) Purification of an X-prolyldipeptidyl aminopeptidase from the cell wall proteolytic system of *Lactococcus lactis* ssp. *cremoris*. Appl. Microbiol. Biotech. 31: 75–78

Kingsley R, Rabsch W, Roberts M, Reissbrodt R & Williams PH (1996) TonB-dependent iron supply in *Salmonella* by α-ketoacids and α-hydroxyacids. FEMS Microbiol. Lett. 140: 65–70

Klein JR, Dick A, Schick J, Matern HT, Henrich B & Plapp R (1995) Molecular cloning and DNA sequence analysis of *pepL*, a leucyl aminopeptidase gene from *Lactobacillus delbrueckii* subsp. *lactis* DSM7290. Eur. J. Biochem. 228: 570–578

Klein JR, Henrich B & Plapp R (1994a) Cloning and nucleotide sequence analysis of the *Lactobacillus delbrueckii* ssp. *lactis* DSM7290 cysteine aminopeptidase gene *pep*C. FEMS Microbiol. Lett. 124: 291–299

Klein JR, Klein U, Schad M & Plapp R (1993) Cloning, DNA sequence analysis and partial characterization of *pep*N, a lysyl aminopeptidase from *Lactobacillus delbrueckii* ssp. *lactis* DSM7290. Eur. J. Biochem. 217: 105–114

Klein JR, Schick J, Henrich B & Plapp R (1997) *Lactobacillus delbrueckii* subsp. *lactis* DSM7290 *pep*G gene encodes a novel cysteine aminopeptidase. Microbiol. 143: 527–537

Klein JR, Schmidt U & Plapp R (1994b) Cloning, heterologous expression, and sequencing of a novel proline iminopeptidase gene, *pep*I, from *Lactobacillus delbrueckii* subsp. *lactis* DSM 7290. Microbiol. 140: 1133–1139

Kochhar S, Hunziker PE, Leong-Morgenthaler P & Hottinger H (1992) Evolutionary relationship of NAD^+-dependent D-lactate dehydrogenase: comparison of primary structure of 2-hydroxy acid dehydrogenases. Biochem. Biophys. Res. Comm. 184: 60–66

Kong Y, Strickland M & Broadbent JR (1996) Tyrosine and phenylalanine catabolism by *Lactobacillus casei* flavor adjuncts: biochemistry and implications in cheese flavor. J. Dairy Sci. 79(Suppl. 1): 101

Konings WN, Lolkema JS, Bolhuis H, van Veen HW, Poolman B & Driessen AJ (1997) The role of transport processes in survival of lactic acid bacteria. Energy transduction and multidrug resistance. Antonie Van Leeuwenhoek 71: 117–128

Konings WN, Lolkema JS & Poolman B (1995) The generation of metabolic energy by solute transport. Arch. Microbiol. 164: 235–242

Kunji ERS, Fang G, Jeronimus-Stratingh CM, Bruins AP, Poolman B & Konings WN (1998) Reconstruction of the proteolytic pathway for use of beta-casein by *Lactococcus lactis*. Mol. Microbiol. 27: 1107–1118

Kunji ERS, Hagting A, De Vries CJ, Juillard V, Haandrikman AJ, Poolman B & Konings WN (1995) Transport of beta-casein derived peptides by the oligopeptide transport system is a crucial step in the proteolytic pathway of *Lactococcus lactis*. J. Biol. Chem. 270: 1569–1574

Kunji ERS, Mierau I, Hagting A, Poolman B & Konings WN (1996) The proteolytic systems of lactic acid bacteria. Antonie van Leeuwenhoek 70: 187–221

l'Anson KJA, Movahedi S, Griffin HG, Gasson MJ & Mulholland F (1995) A non-essential glutamyl aminopeptidase is required for optimal growth of *Lactococcus lactis* MG1363 in milk. Microbiol. 141: 2873–2881

Lapujade P, Cocaign-Bousquet M & Loubiere P (1998) Glutamate biosynthesis in *Lactococcus lactis* subsp. *lactis* NCDO 2118. Appl. Environ. Microbiol. 64: 2485–2489

Law BA (1981) The formation of aroma and flavor compounds in fermented dairy products. Dairy Sci. Abstr. 43: 143

Law J & Haandrikman A (1997) Proteolytic enzymes of lactic acid bacteria. Int. Dairy. J. 7: 1–11

Lees GJ & Jago GR (1976) Formation of acetaldehyde from threonine by lactic acid bacteria. J. Dairy Res. 43: 75–83

Lerch H-P, Frank R & Collins J (1989) Cloning, sequencing and expression of the L-2-hydroxyisocaproate dehydrogenase-encoding gene of *Lactobacillus confusus* in *Escherichia coli*. Gene 83: 263–270

Lian W, Wu D, Konings WN, Mierau I & Hersh LB (1996) Heterologous expression and characterization of recombinant *Lactococcus lactis* neutral endopeptidase (Neprilysin). Arch. Biochem. Biophys. 333: 121–126

Liu S-Q & Pilone GJ (1998) A review: arginine metabolism in wine lactic acid bacteria and its practical significance. J. Appl. Microbiol. 84: 315–327

Liu S-Q, Pritchard GG, Hardman MJ & Pilone GJ (1995) Occurance of arginine deiminase pathway enzymes in arginine catabolism by wine lactic acid bacteria. Appl. Environ. Microbiol. 61: 310–316

Lloyd RJ & Pritchard GG (1991) Characterization of X-prolyldipeptidylaminopeptidase from *Lactococcus lactis* ssp. *lactis*. J. Gen. Microbiol. 137: 49–56

Machuga EJ & Ives DH (1984) Isolation and characterization of an aminopeptidase (EC 3.4.11.1) from *Lactobacillus acidophilus* R-26. Biochim. Biophys. Acta 789: 26–36

MacLeod P & Morgan ME (1958) Differences in the ability of lactic streptococci to form aldehydes from certain amino acids. J. Dairy Sci. 41: 908–913

Magboul AAA & McSweeney PLH (1998) Purification and characterisation of an aminopeptidase from *Lactobacillus curvatus* DPC 2024. Aust. J. Dairy Tech. 53: 114

Manca de Nadra MC, Pese de Ruiz Holgado AA & Oliver G (1982) Arginine dihydrolase activity in lactic acid bacteria. Milchwissenschaft 37: 669–670

Manca De Nadra MC, Raya RR, Pesce De Ruiz Holgado A & Oliver G (1987) Isolation and properties of threonine aldolase of *Lactobacillus bulgaricus* YOP_{12}. Milchwissenschaft 42: 92–94

Marquis RE, Bender GR, Murray DR & Wong A (1987) Arginine deiminase system and bacterial adaptation to acid environments. Appl. Environ. Microbiol. 53: 198–200

Marranzini RM, Schmidt RH, Shireman RB, Marshall MR & Cornell JA (1989) Effect of threonine and glycine concentrations on threonine aldolase activity of yogurt microorganisms during growth in a modified milk prepared by ultrafiltration. J. Dairy Sci. 72: 1142–1148

Mars I & Monnet V (1995) An aminopeptidase P from *Lactococcus lactis* with original specificity. Biochim. Biophys. Acta 1243: 209–215

Marshall VM & Cole WM (1983) Threonine aldolase and alcohol dehydrogenase activities in *Lactobacillus bulgaricus* and *Lactobacillus acidophilus* and their contribution to flavour production in fermented milks. J. Dairy Res. 50: 375–379

Marugg JD, Meijer W, van Kranenburg R, Laverman P, Bruinenberg PG & de Vos WM (1995) Medium-dependent regulation of proteinase gene expression in *Lactococcus lactis*: control of transcription initiation by specific dipeptides. J. Bacteriol. 177: 2982–2989

Massey LK, Sokatch JR & Conrad RS (1976) Branched-chain amino acid catabolism in bacteria. Bacteriol. Rev. 40: 42–54

Masson F, Eclache L, Compte T, Talon R & Montel MC (1996) Screening of microbial strains producing amines and isolated from meat products. In Proceedings of 42nd International Congress of meat science and technology. pp 546–547, Sept. 1–6, Lillehammer, Norway

Mata L, Erra-Pujada M, Gripon J-C & Mistou M-Y (1997) Experimental evidence for the essential role of the C-terminal residue in the strict aminopeptidase activity of the thiol aminopeptidase PepC, a bacterial bleomycin hydrolase. Biochem. J. 328: 343–347

Mathiopoulos C, Mueller JP, Slack FJ, Murphy CG, Patankar S, Bukusoglu G & Sonenshein AL (1991) A *Bacillus subtilis* dipeptide transport system expressed early during sporulation. Mol Microbiol 5: 1903–1913

Matos J, Nardi M, Kumura H & Monnet V (1998) Genetic characterization of PepP, which encodes an aminopeptidase P whose deficiency does not affect *Lactococcus lactis* growth in milk, unlike deficiency of the X-Prolyl dipeptidyl aminopeptidase. Appl. Environ. Microbiol. 64: 4591–4595

Mayo B, Kok J, Bockelmann W, Haandrikman A, Leenhouts KJ & Venema G (1993) Effect of X-prolyl dipeptidyl aminopeptidase deficiency on *Lactococcus lactis*. Appl. Environ. Microbiol. 59: 2049–2055

Mayo B, Kok J, Venema K, Bockelmann W, Teuber M, Reinke H & Venema G (1991) Molecular cloning and sequence analysis of the X-prolyldipeptidyl aminopeptidase gene from *Lactococcus lactis* ssp. *cremoris*. Appl. Environ. Microbiol. 57: 38–44

McCabe BJ (1986) Dietary tyramine and other pressor amines in MAOI regimens: A review. J. Amer. Diet. Assoc. 86: 1059–1064

Meijer W, Marugg JD & Hugenholtz J (1996) Regulation of proteolytic enzyme activity in *Lactococcus lactis*. Appl. Environ. Microbiol. 62: 156–161

Meyer J & Jordi R (1987) Purification and characterization of X-prolyldipeptidylaminopeptidase from *Lactobacillus lactis* and from *Streptococcus thermophilus*. J. Dairy Sci. 70: 738–745

Meyer-Barton EC, Klein JR, Imam M & Plapp R (1993) Cloning and sequence analysis of the X-prolyl-dipeptidyl-aminopeptidase gene (*pep*X) from *Lactobacillus delbrueckii* ssp. *lactis* DSM7290. Appl. Microbiol. Biotech. 40: 82–89

Midwinter RG & Pritchard GG (1994) Aminopeptidase N from *Streptococcus salivarius* subsp. *thermophilus* NCDO 573: Purification and properties. J. Appl. Bacteriol. 77: 288–295

Mierau I, Haandrikman AJ, Velterop O, Tan PST, Leenhouts KL, Konings WN, Venema G & Kok J (1994) Tripeptidase gene (*pep*T) of *Lactococcus lactis*: Molecular cloning and nucleotide sequencing of *pep*T and construction of a chromosomal deletion mutant. J. Bacteriol. 176: 2854–2861

Mierau I, Kunji ERS, Leenhouts KJ, Hellendoorn MA, Haandrikman AJ, Poolman B, Konings WN, Venema G & Kok J (1996a) Multiple-peptidase mutants of *Lactococcus lactis* are severely impaired in their ability to grow in milk. J. Bacteriol. 178: 2794–2803

Mierau I, Kunji ERS, Venema G, Poolman B & Kok J (1996b) Peptidases and growth of *Lactococcus lactis* in milk. Lait 76: 25–32

Mierau I, Tan PST, Haandrikman AJ, Mayo B, Kok J, Leenhouts KJ, Konings WN & Venema G (1993) Cloning and sequencing of the gene for a lactococcal endopeptidase, an enzyme with sequence similarity to mammalian enkephalinase. J. Bacteriol. 175: 2087–2096

Miller A, Morgan ME & Libbey LM (1974) *Lactobacillus maltaromicus*, a new species producing a malty aroma. Int. J. System. Bacteriol. 24: 346–354

Miller CG (1987) Protein degradation and proteolytic modification. Pages 680-691 in FC Neidhardt JLI, KB Low, B Magasanik, M Schaechter & HE Umbarger (Ed.) *Escherichia coli and Salmonella typhimurium: Cellular and Molecular Biology*. American Society for Microbiology, Washington, DC

Miller CG & Green L (1981) Degradation of abnormal proteins in peptidase-deficient mutants of *Salmonella typhimurium*. J. Bacteriol. 147: 925–930

Miller CG & Green L (1983) Degradation of proline peptides in peptidase-deficient strains of *Salmonella typhimurium*. J. Bacteriol. 153: 350–356

Milo C & Reineccius GA (1997) Identification and quantification of potent odorants in regular-fat and low-fat mild Cheddar cheese. J. Agric. Food Chem. 45: 3590–3594

Misono H, Goto N & Nagasaki S (1985) Purification, crystallization and properties of $NADP^+$-specific glutamate dehydrogenase from *Lactobacillus fermentum*. Agric. Biol. Chem. 49: 117–123

Mistou M-Y & Gripon J-C (1998) Catalytic properties of the cysteine aminopeptidase PepC, a bacterial bleomycin hydrolase. Biochim. Biophys. Acta 3: 63–70

Mistou M-Y, Rigolet P, Chapot-Chartier MP, Nardi M, Gripon J-C & Brunie S (1994) Crystallization and preliminary X-ray analysis of PepC, a thiol aminopeptidase from *Lactococcus lactis* homologous to bleomycin hydrolase. J. Mol. Biol. 237: 160–162

Miyakawa H, Hashimoto I, Nakamura T, Ishibashi N, Shimamura S & Igoshi K (1994) Purification and characterization of an X-prolyl dipeptidyl aminopeptidase from *Lactobacillus helveticus* LHE-511. Milchwissenschaft 49: 670–673

Miyakawa H, Kobayashi S, Shimamura S & Tomita M (1991) Purification and characterization of an X-prolyldipeptidyl aminopeptidase from *Lactobacillus delbrueckii* ssp. *bulgaricus* LBU-147. J. Dairy Sci. 74: 2375–2381

Miyakawa H, Kobayashi S, Shimamura S & Tomita M (1992) Purification and characterization of an aminopeptidase from *Lactobacillus helveticus* LHE-511. J. Dairy Sci. 75: 27–35

Molenaar D, Bosscher JS, Ten Brink B, Driessen AJM & Konings WN (1993) Generation of a proton motive force by histidine decarboxylation and electrogenic histidine/histamine antiport in *Lactobacillus buchneri*. J. Bacteriol. 175: 2864–2870

Monnet V, Nardi M, Chopin A, Chopin M-C & Gripon J-C (1994) Biochemical and genetic characterization of PepF, an oligopeptidase from *Lactococcus lactis*. J. Biol. Chem. 269: 32070–32076

Montel MC, Seronie MP, Talon R & Hebraud M (1995) Purification and characterization of a dipeptidase from *Lactobacillus sake*. Appl. Environ. Microbiol. 61: 837–839

Morel F, Frot-Coutaz J, Aubel D, Portalier R & Atlan D (1997) Direct sequence submission to GenBank for a prolidase from *Lactobacillus debrueckii* ssp. *bulgaricus* CNRZ397 (Accession 3821250). Unpublished

Morel F, Gilbert C, Geourjon C, Frot-Coutaz J, Portalier R & Atlan D (1999a) The prolyl aminopeptidase from *Lactobacillus delbrueckii* subsp. *bulgaricus* belongs to the alpha/beta hydrolase fold family. Biochim. Biophys. Acta-Protein Struct. Molec. Enzym. 1429: 501–505

Morel F, Frot-Coutaz J, Aubel D, Portalier R & Atlan D (1999b) Characterization of a prolidase from *Lactobacillus delbrueckii* subsp. *bulgaricus* CNRZ 397 with an unusual regulation of biosynthesis. Microbiol. 145:437–446

Morgan ME, Lindsay RC, Libbey LM & Pereira RL (1965) Identity of additional aroma constituents in milk cultures of *Streptococcus lactis* var. *maltigenes*. J. Dairy Sci. 49: 15–18

Morishita T & Yajima M (1995) Incomplete operation of biosynthetic and bioenergetic functions of the citric acid cycle in multiple auxotrophic lactobacilli. Biosci. Biotech. Biochem. 59: 251–255

Morita H, Yoshikawa H, Sakata R, Nagata Y & Tanaka H (1997) Synthesis of nitric oxide from the two equivalent guanidino nitrogens of L-arginine by *Lactobacillus fermentum*. J. Bacteriol. 179: 7812–7815

Muset G, Monnet V & Gripon JC (1989) Intracellular proteinase of *Lactococcus lactis* ssp. *lactis* NCDO 763. J. Dairy Res. 56: 765–778

Nakae T & Elliott JA (1965) Production of volatile fatty acids by some lactic acid bacteria. II. Selective formation of volatile fatty acids by degradation of amino acids. J. Dairy Sci. 48: 293–299

Nardi M, Chopin MC, Chopin A, Cals MM & Gripon JC (1991) Cloning and DNA sequence analysis of an X-prolyldipeptidyl aminopeptidase gene from *Lactococcus lactis* ssp. *lactis* NCDO 763. Appl. Environ. Microbiol. 57: 45–50

Nardi M, Renault P & Monnet V (1997) Duplication of the *pep*F gene and shuffling of DNA fragments on the lactose plasmid of *Lactococcus lactis*. J. Bacteriol. 179: 4164–4171

Neviani E, Boquien CY, Monnet V, Thanh LP & Gripon JC (1989) Purification and characterization of an aminopeptidase from *Lactococcus lactis* ssp. *cremoris* AM2. Appl. Environ. Microbiol. 55: 2308–2314

Nierop Groot MN & de Bont JAM (1998) Conversion of phenylalanine to benzaldehyde initiated by an aminotransferase in *Lactobacillus plantarum*. Appl. Environ. Microbiol. 64: 3009–3013

Niven GW (1991) Purification and characterization of aminopeptidase A from *Lactococcus lactis* ssp. *lactis* NCDO 712. J. Gen. Microbiol. 137: 1207–1212

Ohmiya K & Sato Y (1975) Purification and properties of intracellular proteinase from *Streptococcus cremoris*. Applied Microbiology 30: 738–745

Parks EH, Ernst SR, Hamlin R, Xuong NH & Hackert ML (1985) Structure determination of histidine decarboxylase from *Lactobacillus* 30a at 3.0 A resolution. J. Mol. Biol. 182: 455–465

Patton S (1964) Volatile acids of Swiss cheese. J. Dairy Sci. 47: 817–818

Paulsen PV, Kowalewska J, Hammond EG & Glatz BA (1980) Role of microflora in production of free fatty acids and flavor in Swiss cheese. J. Dairy Sci. 63: 912–918

Paulus H (1993) Biosynthesis of the aspartate family of amino acids. In: Sonenshein AL, Hock JA & Losick R (Eds) *Bacillus subtilis* and other gram-positive bacteria: biochemistry, physiology, and molecular genetics, (pp 237–267). American Society for Microbiology, Washington, DC

Perego M, Higgins CF, Pearce SR, Gallagher MP & Hoch JA (1991) The oligopeptide transport system of *Bacillus subtilis* plays a role in the initiation of sporulation. Mol Microbiol 5: 173–185

Preininger M & Grosch W (1994) Evaluation of key odorants of the neutral volatiles of Emmentaler cheese by calculation of odour activity values. Lebensm. Wiss. Technol. 27: 237–244

Preininger M, Warmke R & Grosch W (1996) Identification of the character impact flavour compounds of Swiss cheese by sensory studies of models. Z. Lebensm. Unters. Forsch. 202: 30–34

Pritchard GG, Freebairn AD & Coolbear T (1994) Purification and characterization of an endopeptidase from *Lactococcus lactis* subsp. *cremoris* SK11. Microbiol. 140: 923–930

Rallu F, Gruss A & Maguin E (1996) *Lactococcus lactis* and stress. Antonie van Leeuwenhoek 70: 243–251

Rantanen T & Palva A (1997) Lactobacilli carry cryptic genes encoding peptidase-related proteins: Characterization of a prolidase gene (*pep*Q) and a related cryptic gene (*orfZ*) from *Lactobacillus delbrueckii* subsp. *bulgaricus*. Microbiol. 143: 3899–3905

Rawlings ND, Polgar L & Barrett AJ (1991) A new family of serine-type peptidases related to prolyl oligopeptidase [letter]. Biochem J 279: 907-908

Recsei PA, Huynh QK & Snell EE (1983) Conversion of prohistidine decarboxylase to histidine decarboxylase: peptide chain cleavage by nonhydrolytic serinolysis. Proc. Natl. Acad. Sci. USA 80: 973–977

Rescei PA & Snell EE (1981) In: Metabolic interconversion of enzymes 1980 ed. H.Holzen, pp. 335–344. Springer-Verlag: New York

Recsei PA & Snell EE (1984) Pyruvoyl enzymes. Annu. Rev. Biochem. 53: 357–387

Reiter B & Oram JD (1962) Nutritional studies on cheese starters. I. Vitamin and amino acid requirements of single strain starters. J Dairy Res. 29: 63–77

Roig-Sagues AX, Hernandez-Herrero MN, Lopez-Sabater EI, Rodriguez-Jerez JJ & Mora-Ventura MT (1997) Evaluation of three decarboxylating agar media to detect histamine and tyramine-producing bacteria in ripened sausages. Lett. Appl. Microbiol. 25: 309–312

Rollan G, de Nadra MCM, Holgado PR & Oliver G (1985) Aspartate metabolism in *Lactobacillus murinus* CNRS 313. I. Aspartase. J. Gen. Appl. Microbiol. 31: 403–409

Rollan G, de Nadra MCM, Holgado PR & Oliver G (1988) Aspartate aminotransferase of *Lactobacillus murinus*. Folia Microbiol. 33: 344–348

Roudot-Algaron F & Yvon M (1998) Le catabolisme des acides aminés aromatiques et des acides aminés à chaîne ramifiée chez *Lactococcus lactis*. Lait 78: 23–30

Rul F, Gripon J-C & Monnet V (1995) St-PepA, a *Streptococcus thermophilus* aminopeptidase with high specificity for acidic residues. Microbiol. 141: 2281–2287

Rul F, Monnet V & Gripon JC (1994) Purification and characterization of a general aminopeptidase (St-PepN) from *Streptococcus salivarius* ssp. *thermophilus* CNRZ 302. J. Dairy Sci. 77: 2880–2889

Sahlstrøm S, Chrzanowska J & Sorhaug T (1993) Purification and characterization of a cell wall peptidase from *Lactococcus lactis* ssp. *cremoris* IMN-C12. Appl. Environ. Microbiol. 59: 3076–3082

Sanders JW, Leenhouts K, Burghoorn J, Brands JR, Venema G & Kok J (1998) A chloride-inducible acid resistance mechanism in *Lactococcus lactis* and its regulation. Mol. Microbiol. 27: 299–310

Sandine WE & Elliker PR (1970) Microbially induced flavors in fermented dairy products. J. Agric. Food Chem. 18: 557

Santos MH (1996) Biogenic amines: their importance in foods. Int. J. Food Microbiol. 29: 213–231

Sanz Y, Mulholland F & Toldrá F (1998) Purification and characterization of a tripeptidase from *Lactobacillus sake*. J. Agric. Food Chem. 46: 349–353

Sasaki M, Bosman BW & Tan PST (1996) A new, broad-substrate-specificity aminopeptidase from the dairy organism *Lactobacillus helveticus* SBT 2171. Microbiol. 142: 799–808

Schütte H, Hummel W & Kula M-R (1984) L-2-hydroxyisocaproate dehydrogenase-a new enzyme from *Lactobacillus confusus* for the stereospecific reduction of 2-ketocarboxylic acids. Appl. Microbiol. Biotechnol. 19: 167–176

Shankar PA (1977) Interrelationships of *Streptococcus thermophilus* and *Lactobacillus bulgaricus* in yoghurt culture. Thesis, University of Reading, Reading, UK

Shao W, Yüksel GÜ, Dudley EG, Parkin KL & Steele JL (1997) Biochemical and molecular characterization of PepR, a dipeptidase, from *Lactobacillus helveticus* CNRZ32. Appl. Environ. Microbiol. 63: 3438–3443

Sheldon RM, Lindsay RC, Libbey LM & Morgan ME (1971) Chemical nature of malty flavor and aroma produced by *Streptococcus lactis* var. *maltigenes*. Appl. Microbiol. 22: 263–266

Simitsopoulou M, Vafopoulou A, Choli Papadopoulou T & Alichanidis E (1997) Purification and partial characterization of a tripeptidase from *Pediococcus pentosaceus* K9.2. Appl. Environ. Microbiol. 63: 4872–4876

Stepaniak L & Fox PF (1995) Characterization of the principal intracellular endopeptidase from *Lactococcus lactis* subsp. *lactis* MG1363. Int. Dairy. J. 5: 699–713

Stepaniak L, Gobbetti M, Pripp AH & Sorhaug T (1998a) Isolation and characterization of a 67 kDa oligopeptidase from *Propionibacterium freudenreichii* ATCC 9614. Ital. J Food Sci. 10: 117–125

Stepaniak L, Gobbetti M & Sorhaug T (1998b) Isolation and characterization of high molecular mass endopeptidase complex from *Lactococcus lactis*. Milchwissenschaft 53: 255–259

Stepaniak L, Tobiassen RO, Chukwu I, Pripp AH & Sorhaug T (1998c) Purification and characterization of a 33 kDa subunit oligopeptidase from *Propionibacterium freudenreichii* ATCC 9614. Int. Dairy. J. 8: 33–37

Stratton JE, Hutkins RW, Sumner SS &Taylor SL (1992) Histamine and histamine-producing bacteria in retail Swiss and low-salt cheeses. J. Food Prot. 55: 435–439

Straub BW, Kicherer M, Schilcher SM & Hammes WP (1995) The formation of biogenic amines by fermentation microorganisms. Z. Lebensm. Unters. For. 201: 79–82

Strøman P (1992) Sequence of a gene (*lap*) encoding a 95.3-kDa aminopeptidase from *Lactococcus lactis* ssp. *cremoris* Wg2. Gene 113: 107–112

Stucky K, Klein JR, Schueller A, Matern H, Henrich B & Plapp R (1995) Cloning and DNA sequence analysis of *pep*Q, a prolidase gene from *Lactobacillus delbrueckii* subsp. *lactis* DSM7290 and partial characterization of its product. Mol. Gen. Genet. 247: 494–500

Stucky K, Schick J, Klein JR, Henrich B & Plapp R (1996) Characterization of *pep*R1, a gene coding for a potential transcriptional regulator of *Lactobacillus delbrueckii* subsp. *lactis* DSM7290. FEMS Microbiol. Lett. 136: 63–69

Tamime AY & Deeth HC (1980) Yogurt: technology and biochemistry. J. Food Prot. 43: 939

Tan PST & Konings WN (1990) Purification and characterization of an aminopeptidase from *Lactococcus lactis* ssp. *cremoris* Wg2. Appl. Environ. Microbiol. 56: 526–532

Tan PST, Pos KM & Konings WN (1991) Purification and characterization of an endopeptidase from *Lactococcus lactis* ssp. *cremoris* Wg2. Appl. Environ. Microbiol. 57: 3593–3599

Tan PST, Sasaki M, Bosman BW & Iwasaki T (1995) Purification and characterization of a dipeptidase from *Lactobacillus helveticus* SBT 2171. Appl. Environ. Microbiol. 61: 3430–3435

Tan PST, Van Alen Boerrigter IJ, Poolman B, Siezen RJ, De Vos WM & Konings WN (1992) Characterization of the *Lactococcus lactis pep*N gene encoding an aminopeptidase homologous to mammalian aminopeptidase N. FEBS Lett. 306: 9–16

Tan PST, Van Kessel TAJM, Van De Veerdonk FLM, Zuurendonk PF, Bruins AP & Konings WN (1993) Degradation and debittering of a tryptic digest from beta-casein by aminopeptidase N from *Lactococcus lactis* ssp. *cremoris* WG2. Appl. Environ. Microbiol. 59: 1430–1436

Taylor SL, Keefe TJ, Windham ES & Howell JF (1982) Outbreak of histamine poisoning associated with consumption of Swiss cheese. J. Food Prot. 45: 455–457

ten Brink B, Damink C, Joosten HM & Huis in't Veld JH (1990) Occurrence and formation of biologically active amines in foods. Int. J. Food Microbiol. 11: 73–84

Timpone D & Steele JL (1999) Unpublished observations

Tobiassen RO, Pripp AH, Stepaniak L & Sørhaug T (1996) Purification and characterization of an endopeptidase from *Propionibacterium freudenreichii*. J. Dairy Sci. 79: 2129–2136

Tobiassen RO, Sørhaug T & Stephaniak L (1997) Characterization of an intracellular oligopeptidase from *Lactobacillus paracasei*. Appl. Environ. Microbiol. 63: 1284–1287

Tsakalidou E, Anastasiou R, Papadimitriou K, Manolopoulou E & Kalantzopoulos G (1998) Purification and characterisation of an intracellular X-prolyl-dipeptidyl aminopeptidase from *Streptococcus thermophilus* ACA-DC 4. J. Biotech. 59: 203–211

Tsakalidou E, Dalezios I, Georgalaki M & Kalantzopoulos G (1993) A comparative study: Aminopeptidase activities from *Lactobacillus delbrueckii* ssp. *bulgaricus* and *Streptococcus thermophilus*. J. Dairy Sci. 76: 2145–2151

Tsakalidou E & Kalantzopoulos G (1992) Purification and partial characterization of an intracellular aminopeptidase from *Streptococcus salivarius* ssp. *thermophilus* strain ACA-DC 114. J. Appl. Bacteriol. 72: 227–232

Tucker JS & Morgan ME (1967) Decarboxylation of α-keto acids by *Streptococcus lactis* var. *maltigenes*. Appl. Microbiol. 15: 694–700

Tynkkynen S, Buist G, Kunji E, Kok J, Poolman B, Venema G & Haandrikman A (1993) Genetic and biochemical characterization of the oligopeptide transport system of *Lactococcus lactis*. J. Bacteriol. 175: 7523–7532

Ueno Y, Hayakawa K, Takahashi S & Oda K (1997) Purification and characterization of glutamate decarboxylase from *Lactobacillus brevis* IFO 12005. Biosci. Biotech. Biochem. 61: 1168–1171

Umbarger HE (1996) Biosynthesis of the branched-chain amino acids. In: Neidhardt FC (Ed) *Escherichia coli* and *Salmonella*: cellular and molecular biology, 2nd Ed, Vol 1 (pp 442–457). ASM Press, Washington DC

Urbach G (1995) Contribution of lactic acid bacteria to flavor compound formation in dairy products. Int. Dairy J. 5: 877–903

Van Alen-Boerrigter IJ, Baankreis R & De Vos WM (1991) Characterization and overexpression of the *Lactococcus lactis pep*N gene and localization of its product, aminopeptidase N. Appl. Environ. Microbiol. 57: 2555–2561

Van Boven A, Tan PST & Konings WN (1988) Purification and characterization of a dipeptidase from *Streptococcus cremoris* Wg2. Appl. Environ. Microbiol. 54: 43–49

Vanderslice P, Copeland WC & Robertus JD (1986) Cloning and nucleotide sequence of wild type and a mutant histidine decarboxylase from *Lactobacillus* 30a. J. Biol. Chem. 261: 15186–15191

Vangtal A & Hammond EG (1986) Correlation of the flavor characteristics of Swiss-type cheeses with chemical parameters. J. Dairy Sci. 69: 2982–2993

Varmanen P, Rantanen T & Palva A (1996a) An operon from *Lactobacillus helveticus* composed of a proline iminopeptidase gene (*pep*I) and two genes coding for putative members of the ABC transporter family of proteins. Microbiol. 142: 3459–3468

Varmanen P, Rantanen T, Palva A & Tynkkynen S (1998) Cloning and characterization of a prolinase gene (*pep*R) from *Lactobacillus rhamnosus*. Appl. Environ. Microbiol. 64: 1831–1836

Varmanen P, Steele J & Palva A (1996b) Characterization of a prolinase gene and its product and an adjacent ABC transporter gene from *Lactobacillus helveticus*. Microbiol. 142: 809–816

Varmanen P, Vesanto E, Steele JL & Palva A (1994) Characterization and expression of the *pep*N gene encoding a general aminopeptidase from *Lactobacillus helveticus*. FEMS Microbiol. Lett. 124: 315–320

Vesanto E, Peltoniemi K, Purtsi T, Steele JL & Palva A (1996) Molecular characterization, over-expression and purification of a novel dipeptidase from *Lactobacillus helveticus*. Appl. Microbiol. Biotech. 45: 638–645

Vesanto E, Savijoki K, Rantanen T, Steele JL & Palva A (1995) An X-prolyl dipeptidyl aminopeptidase (*pep*X) gene from *Lactobacillus helveticus*. Microbiol. 141: 3067–3075

Vesanto E, Varmanen P, Steele JL & Palva A (1994) Characterization and expression of the *Lactobacillus helveticus pep*C gene encoding a general aminopeptidase. Eur. J. Biochem. 224: 991–997

Voight MN & Eitenmiller RR (1978) Role of histidine and tyrosine decarboxylases and mono- and diamine oxidases in amine build-up in cheese. J. Food Prot. 41: 182–186

Vongerichten KF, Klein JR, Matern H & Plapp R (1994) Cloning and nucleotide sequence analysis of *pep*V, a carnosinase gene from *Lactobacillus delbrueckii* subsp. *lactis* DSM 7290, and partial characterization of the enzyme. Microbiol. 140: 2591–2600

Wijesundera KM & Urbach G (1993) Flavor of Cheddar cheese. Final report to the Dairy Research and Development Corporation, Project CSt66

Wilkins DW, Schmidt RH & Kennedy LB (1986) Threonine aldolase activity in yogurt bacteria as determined by headspace gas chromatography. J. Agric. Food Chem. 34: 150–152

Wohlrab Y & Bockelmann W (1992) Purification and characterization of a dipeptidase from *Lactobacillus delbrueckii* subsp. *bulgaricus*. Int. Dairy. J. 2: 345–361

Wohlrab Y & Bockelmann W (1993) Purification and characterization of a second aminopeptidase (PepC-like) from *Lactobacillus delbrueckii* subsp. *bulgaricus* B14. Int. Dairy. J. 3: 685–701

Wohlrab Y & Bockelmann W (1994) Purification and characterization of a new aminopeptidase from *Lactobacillus delbrueckii* subsp. *bulgaricus* B14. Int. Dairy. J. 4: 409–427

Yan TR, Azuma N, Kaminogawa S & Yamauchi K (1987a) Purification and characterization of a novel metalloendopeptidase from *Streptococcus cremoris* H61: A metalloendopeptidase that recognizes the size of its substrate. Eur. J. Biochem. 163: 259–266

Yan TR, Azuma N, Kaminogawa S & Yamauchi K (1987b) Purification and characterization of a substrate-size-recognizing metalloendopeptidase from *Streptococcus cremoris* H61. Appl. Environ. Microbiol. 53: 2296–2302

Yan TR, Lin MZ, Lin MJ & Sun BJ (1991) Purification and characterization of an X-prolyl-dipeptidyl aminopeptidase from *Streptococcus cremoris* nTR. J. Chin. Biochem. Soc. 20: 21–32

Yang TS & Min DB (1993) Dynamic headspace analysis of volatile compounds of Cheddar and Swiss cheeses during ripening. In: Charalambous (ED) Food Flavors, Ingredients and Composition, (pp 157-174), Elsevier Science Publishers BV, Amsterdam

Yen C, Green L & Miller CG (1980a) Degradation of intracellular protein in *Salmonella typhimurium* peptidase mutants. J. Mol. Biol. 143: 21–33

Yen C, Green L & Miller CG (1980b) Peptide accumulation during growth of peptidase deficient mutants. J. Mol. Biol. 143: 35–48

Yokoyama MT & Carlson JR (1981) Production of skatole and *para*-cresol by a rumen *Lactobacillus* sp. Appl. Environ. Microbiol. 41: 71–76

Yoshpe-Besançon I, Gripon J-C & Ribadeau-Dumas B (1994) Xaa-Pro-dipeptidyl-aminopeptidase from *Lactococcus lactis* catalyses kinetically controlled synthesis of peptide bonds involving proline. Biotech. Appl. Biochem. 20: 131–140

Yüksel GÜ & Steele JL (1996) DNA sequence analysis, expression, distribution, and physiological role of the Xaa-prolyldipeptidyl aminopeptidase gene from *Lactobacillus helveticus* CNRZ32. Appl. Microbiol. Biotech. 44: 766–773

Yüksel GÜ & Steele JL (1997a) Direct sequence submission to GenBank for a prolidase (*pep*Q) from *Lactobacillus helveticus* CNRZ32 (Accession AF012084). Unpublished

Yüksel GÜ & Steele JL (1997b) Direct sequence submission to GenBank for *pep*V from *Lactobacillus helveticus* CNRZ32 (Accession AF012085). Unpublished

Yüksel GÜ & Steele JL (1999) Proline-specific peptidases of *Lactobacillus helveticus* CNRZ32: Cloning and DNA sequence analysis of *pep*Q, a prolidase gene, and characterization of peptidase-deficient mutants. Unpublished results

Yvon M, Thirouin S, Rijnen L, Fromentier D & Gripon JC (1997) An aminotransferase from *Lactococcus lactis* initiates conversion of amino acids to cheese flavor compounds. Appl. Environ. Microbiol. 63: 414–419

Zevaco C, Monnet V & Gripon JC (1990) Intracellular X-prolyl dipeptidyl peptidase from *Lactococcus lactis* ssp. *lactis*: Purification and properties. J. Appl. Bacteriol. 68: 357–366

Zoon P & Allersma D (1996) Eye and crack formation in cheese by carbon dioxide from decarboxylation of glutamic acid. Neth. Milk Dairy J. 50: 309–318

Zúñiga M, Champomier-Verges M, Zagorec M & Pérez-Martínez G (1998) Structural and functional analysis of the gene cluster encoding the enzymes of the arginine deiminase pathway of *Lactobacillus sake*. J. Bacteriol. 180: 4154–4159

Antonie van Leeuwenhoek **76:** 247–261, 1999.

Sulfur metabolism in bacteria associated with cheese

Bart Weimer*, Kimberly Seefeldt & Benjamin Dias
*Western Dairy Center, Department of Nutrition and Food Sciences, Center for Microbe Detection and Physiology, Utah State University, Logan, 84322–8700, USA. (*Author for correspondence; E-mail: Milkbugs@cc.usu.edu)*

Abstract

Metabolism of sulfur in bacteria associated with cheese has long been a topic of interest. Volatile sulfur compounds, specifically methanethiol, are correlated to desirable flavor in Cheddar cheese, but their definitive role remains elusive. Only recently have enzymes been found that produce this compound in bacteria associated with cheese making. Cystathionine β- and γ-lyase are found in lactic acid bacteria and are capable of producing methanethiol from methionine. Their primary function is in the metabolism of cysteine. Methionine γ-lyase produces methanethiol from methionine at a higher efficiency than the cystathionine enzymes. This enzyme is found in brevibacteria, bacilli, and pseudomonads. Addition of brevibacteria containing this enzyme improves Cheddar cheese flavor. Despite recent progress in sulfur metabolism more information is needed before cheese flavor associated with sulfur can be predicted or controlled.

Sulfur-containing compounds & cheese flavor

The chemical basis for Cheddar cheese flavor is not defined. No synthetic mixture has been produced which recreates Cheddar flavor, nor has a set of compounds with the flavor impact of Cheddar been found, despite the existence of many lists of compounds associated with Cheddar cheese flavor (Urbach 1993). Based on these lists, two leading theories for the production of Cheddar flavor were developed. The component balance theory states that many compounds in the appropriate balance lead to typical Cheddar cheese flavor (Müdler 1952), while the single component theory states that a single compound is responsible for Cheddar flavor (Manning 1979). The latter theory is much less accepted, but interestingly, the search continues for the single compound. Some investigators maintain that the extremely elusive nature of the chemical basis of Cheddar flavor supports the component balance theory, and hypothesize that the flavor impact compound is an unstable sulfur-containing compound.

Diverse sets of chemicals are responsible for cheese flavor. Components from the basic, acidic, neutral, polar, nonpolar, and volatile fractions contribute to typical cheese flavor and aroma (Urbach 1995). Interactions between components of these chemical classes are also important, but the mechanisms by which the resulting compounds arise are largely undiscovered. An example of this is the reaction between methanethiol and unidentified components of the acidic and polar fraction of cheese to form compounds associated with the typical aroma of Cheddar cheese, such as thioesters (Wijesundera & Urbach 1993). These result from the reaction of methanethiol and fatty acids (Dumont et al. 1974; Dumont et al. 1976).

The chemical basis of Cheddar flavor has not been determined; however, many long lists of compounds associated with cheese flavor are available. To utilize these lists for predictive value, objective methods to assess cheese flavor are needed. O'Shea et al. (1996) used proteolysis and free amino acid content to devise a scheme to predict age and flavor. Alternatively, the aroma extract dilution analysis (AEDA) technique is designed to suggest compounds that are responsible for cheese aroma based on the dilution to extinction principle (Grosch 1994). In AEDA, an extract of a food is concentrated to a small volume, a series of dilutions are made, and analyzed for specific compounds by high resolution gas chromatography and olfactometry (HRGCO), which involves sniffing of gas chromatograph effluents. The result is expressed as a flavor dilution factor (FD), which is the ratio of the concentration of the odorant in the initial extract to its concentration in the most diluted extract in which odor can still be detected by HRGCO. The FD factor is therefore a relative measure of the odor potency of a compound in a food extract (Grosch 1993). This provides information on what chemicals impart a flavor or aroma at the lowest concentrations, which suggest that these compounds are important in flavor.

The first use of the flavor dilution technique with Cheddar cheese was reported by Christensen and Reineccius (1995). They identified methional, homofuraneol, diacetyl, acetic acid, and butyric acid as high impact compounds in the aroma of Cheddar cheese. This method was subsequently used on Cheddar cheese to demonstrate that the volatile component contains methionethiol and dimethyl sulfide (Milo & Reineccius 1997). These experiments in conjunction with an understanding of microbial physiology will lead to an understanding of how to modify and control cheese flavor development. Initially, a fundamental understanding of the sulfur chemistry and the role of microbes in this cycle must be achieved.

Sulfur compounds comprise a distinct fraction of Cheddar cheese flavor and aroma (Manning 1974). Grill et al. (1966), using reagents to trap specific compounds in Cheddar cheese slurries, found that elimination of sulfur compounds adversely affected the aroma of Cheddar cheese. Methanethiol, in particular, correlates to flavorful Cheddar cheese (Aston & Dulley 1982). Gas chromatographic analysis in conjunction with sensory evaluation of cheese distillates maintained at low temperature also showed that sulfur compounds contribute significantly to the aroma of Cheddar cheese as well (Manning & Robinson 1973). However, alone or in excess, methanethiol does not produce typical Cheddar cheese flavor.

The reducing potential of cheese modulates flavor characteristics. Early experiments demonstrated that reduced glutathione, added to regulate the oxidation/reduction potential, improved Cheddar flavor in cheese and slurries (Harper & Kristoffersen 1970; Kristoffersen 1973). Added glutathione changes the sulfhydryl (–SH) interactions in the cheese matrix and was noted to influence proteolysis of cheese slurries (Harper et al. 1971). Lactic acid bacteria accumulate glutathione, but are not capable of producing this peptide (Wiederholt & Steele 1994). The role of this ability for the improvement of flavor is unknown, but some propose that the starter culture provides a reducing environment for sulfur compound production rather than a direct role in the production of volatile sulfur compounds (VSC). However, this limited role in VSC production seems unlikely, since addition of cysteine and methionine tends to improve flavor. Further work is needed to define the direct role of the oxidation/reduction reactions in cheese for flavor improvement.

Classically, flavor improvement is achieved by supplementing the starter culture with bacteria that are found in high quality aged cheese. This usually includes lactobacilli and some proteinase negative lactococci (Drake et al. 1996; Drake et al. 1997; Lane & Fox 1997). Weimer et al. (1997) observed that addition of *Lactobacillus helveticus* and *Brevibacterium linens* BL2 to cheese milk increased the Cheddar flavor score of 60% reduced-fat Cheddar cheese, while *Lactobacillus casei* diminished the flavor. However, a significant ($p<0.05$) interaction between the starter culture and the flavor adjunct was found, indicating that the addition of flavor adjunct bacteria requires careful pairing with the starter culture to achieve the desired results. While brevibacteria are not typically used as adjuncts in Cheddar cheese production, they produce the most desirable cheese as judged by consumer taste panelists, and they are used commercially in some factories. Of the organisms associated with cheese, *B. linens* and pseudomonads catabolize methionine to form methanethiol most efficiently, while micrococci, lactococci, and lactobacilli also catalyze this reaction (Dias & Weimer 1998a).

At least five low molecular weight compounds occur regularly in the headspace profiles of distinctly Cheddar-flavored aged cheese; hydrogen sulfide, carbonyl sulfide, methanethiol, dimethyl sulfide, carbon disulfide, and dimethyl disulfide (Lindsay & Rippe 1986; Urbach 1995). Methanethiol, dimethyl sulfide, dimethyl trisulfide, dimethyl tetrasulfide, and methional were recently identified as potent odorants in mild regular- and low-fat Cheddar cheese (Milo & Reineccius 1997). The sulfur compounds found in cheese have diverse aromas, but all have low detection thresholds and flavor dilution factors (Table 1).

Methanethiol (CH_3-SH)

Methanethiol is influential in the aroma and flavor compound in many foods, including some surface-ripened cheeses like Limburger, Muenster, and other Trappist-type cheeses. It is accepted that it is a contributor to the characteristic aroma of cheese (Lindsay & Rippe 1986) and is undoubtedly a necessary component of Cheddar aroma (Urbach 1993), but is not solely responsible for Cheddar aroma. Methanethiol is a volatile compound with a boiling point of 6.2 °C possessing a putrid, fecal-like aroma. It is found at low

Table 1. Dectection thresholds and properties of some sulfur compounds found in cheese

Compound	Detection threshold (ppb)	Medium	Flavor/aroma description	Probable precursor	Reference
Hydrogen sulfide	0.18[a]	air	rotten eggs	cysteine	[Amerine, 1965]
Carbonyl sulfide				cysteine	[Aston, 1983]
Methanethiol	2[a,b]	water	cooked cabbage	methionine	[Patton, 1957]
Methional	21[b]	skim milk	boiled potato	methionine	[Patton, 1957]
Dimethyl sulfide	20[a]	milk	cabbage		[Manning, 1973]
Dimethyl disulfide	20[b]	skim milk	cauliflower,	methanethiol	[Patton, 1957]
	1.2[a]	water	garlic, very ripe cheese		[Powers, 1968]
Dimethyl trisulfide	0.01	water		methanethiol	[Chin, 1994]
				methanethiol and formaldehyde	[Urbach, 1995]
S-methyl thioacetate	5[a]	liquid cheese	cooking cauliflower	methanethiol and acetic acid	[Cuer, 1979]
S-methyl thiopropionate	100[a]	liquid cheese	cheesy	methanethiol and propionic acid	[Cuer, 1979]
S-methyl thiobutyrate	100[a]	liquid cheese	chives	methanethiol and butyric acid	[Cuer, 1979]

Dectection by [a]odor or [b]taste.

concentrations in the headspace of Cheddar cheese and is correlated to desirable flavor in Cheddar cheese (Libbey et al. 1963; Libbey & Day 1964; Manning & Robinson 1973; Yang & Min 1994). While known to correlate to flavor, methanethiol has only recently been identified as an essential odorant in Cheddar cheese (Milo & Reineccius 1997). Others have failed to find methanethiol in cheese extracts (Vandeweghe & Reineccius 1990; Arora et al. 1995; Christensen & Reineccius 1995), which is most likely due to the extraction procedure. This points to the difficulties that are encountered when analyzing for VSC, due to their instability, sensitivity to oxidation/reduction reactions, and low concentration in the matrix (Parliment et al. 1982; Chin & Lindsay 1994).

Methanethiol arises in the cheese matrix due to enzymatic or chemical means. Manning (1979) demonstrated the formation of methanethiol and hydrogen sulfide in chemically reduced cheese slurries made without starter cultures. Interestingly, addition of a starter culture, a reducing agent, and methionine or sodium caseinate increases the VSC production. While these experiments suggest a non-microbial mechanism for production of these compounds, contaminating raw milk microflora cannot be ruled out for the production of VSC in the slurries.

Starter cultures, flavor adjunct bacteria, and non-starter bacteria may all contribute to methanethiol formation in cheese and may well be the main reason for methanethiol formation in typical Cheddar cheese. Cheese made with glucono-δ-lactone (GDL) lacks flavor, whereas cheese made with *L. lactis* ssp. *cremoris* develops Cheddar flavor (Manning 1974), and distillates of that cheese contain methanethiol. Dias and Weimer (1999) found the headspace of Cheddar cheese slurries made with lactococci contain methanethiol, while slurries made with GDL do not. Interestingly, slurries made with GDL allowed growth of a spore forming organism after 4 d, corresponding to an increase in VSC, while those made with lactococci did not contain any contaminants.

Dias and Weimer (1998a) demonstrated the ability of certain strains of lactococci and lactobacilli to produce thiols from methionine, indicating metabolism of methionine may be a source of VSC in cheese. In addition, *Brevibacterium linens*, an organism important for flavor development in certain surface-ripened cheeses, produces copious amounts of VSC from methionine (Sharpe et al. 1977; Law & Sharpe 1978; Ferchichi et al. 1985; Dias & Weimer 1998a). The VSC maybe comprised of various compounds with potential for flavor impact.

Dimethyl sulfide (CH_3-S-CH_3)

Dimethyl sulfide is a natural component of skim milk, influenced by the diet of the cow (Forss 1979). It can be synthesized by several species of microorganisms in milk, including *Aerobacter aerogenes* (Kadota & Ishida 1972), is detected in the headspace of Cheddar cheese (Manning & Robinson 1973; Manning 1974; Manning 1976), and is a potent odorant in mild regular-fat and low-fat Cheddar cheese (Milo &

Reineccius 1997). However, Manning et al. (1976) found that an increasing intensity of Cheddar flavor was associated with a decreasing concentration of dimethyl sulfide. This supports Dimos (1992), who found the flavor development of the cheeses did not vary during the year despite wide variations in dimethyl sulfide levels. These observations imply that dimethyl sulfide plays no part or is inversely correlated to Cheddar flavor development. The conflicting results surrounding this compound require further work to define its role in flavor.

Dimethyl disulfide (CH_3-S-S-CH_3) & dimethyl trisulfide (CH_3-S-S-S-CH_3)

Dimethyl trisulfide is a potent odorant in Cheddar cheese, but dimethyl disulfide is not (Milo & Reineccius 1997), and both are found in the headspace of Cheddar cheese (McGugan 1975; Yang & Min 1994; Arora et al. 1995). Methanethiol is readily oxidized to dimethyl disulfide and dimethyl trisulfide (Parliment et al. 1982; Chin & Lindsay 1994). Occurrence of these compounds is a direct result of methanethiol content and is modulated by the low redox potential present in cheese. Dias and Weimer (1999) found methanethiol to decrease as these compounds increased in the headspace of Cheddar cheese slurries. It is also possible that during analysis, methanethiol is oxidized to form these compounds, and this may account for some reports of dimethyl disulfide and dimethyl trisulfide in cheese. Dimethyl disulfide or dimethyl trisulfide levels have not been correlated to Cheddar flavor development, while dimethyl trisulfide was recently noted as an important odorant (Milo & Reineccius 1997). Further work is needed to define the mechanism and cheese conditions needed for production.

Methional (CH_3-S-CH_2-CH_2-CHO)

Methional (β-methyl mercaptopropionaldehyde) can be degraded into methanethiol by either decarboxylation reactions or spontaneous degradation (Ballance 1961). Methional may be formed by the Strecker degradation of methionine (Figure 1), but the exact mechanism is unknown. Methional is reported to be a part of good Cheddar aroma in combination with diacetyl and butyric acid (Dacremont & Vickers 1994). Methional is also a potent odorant of Cheddar in studies using AEDA (Christensen & Reineccius 1995; Milo & Reineccius 1997). Methional formation can also occur via a light catalyzed reaction between methionine and riboflavin in milk, resulting in the 'sunlight' flavor of milk (Allen & Parks 1975). Using NMR, Gao et al. (1998) did not observe methional from methionine nor 4-methylthio 2-oxobutyric acid (KMBA) by lactococci.

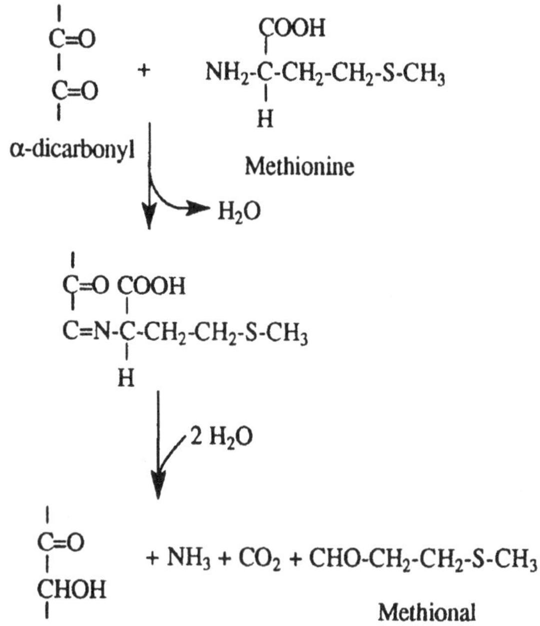

Figure 1. Strecker degradation of methionine.

S-Methylthioesters (CH_3-S-CO-R)

Methylthioesters are important to the characteristic odor of smear-ripened cheese (Dumont et al. 1974; Dumont et al. 1976). The specific flavor of each thioester depends on their chain length and configuration (Cuer et al. 1979). *S*-Methylthioacetate (MTA), which has the flavor of cooked cauliflower, is found in the volatile fraction of Limburger (Parliment et al. 1982) and Swiss cheese (Yang & Min 1994). *S*-Methylthiopropionate has a 'cheesy' odor, and thioesters are patented for use in enhancing dairy-type flavors (van den Bosch et al. 1982).

Precursors for thioester synthesis are methanethiol and free fatty acids, both of which are produced by starter, non-starter, and flavor adjunct organisms during cheese ripening. Species typically found on the surface of Trappist-type cheeses, particularly *B. linens* and micrococci species, are especially good at producing thioesters. MTA is produced by several strains of *B. linens* (Cuer et al. 1979a; Jollivet et al. 1992; Lamberet et al. 1997a; Lamberet et al. 1997b). Some micrococci produce MTA when supplied with methanethiol or grown in conjunction with methanethiol-producing strains of *B. linens* (Cuer et al. 1979b; Lamberet et al. 1997b). Commercial lactococcal starter cultures produce lower levels of MTA than either brevibacteria or micrococci (Lamberet et al. 1997b). Of the starters tested, *L. lactis* ssp. *lactis* biovar. *diacetylactis* produced more MTA then either *Lactococcus lactis* ssp. *lactis* or *Lactococcus lactis* ssp. *cremoris*. Many sulfur compounds are important in cheese flavor and aroma. However, our knowledge of the mechanisms of production is lacking. Therefore, an understanding of the production of methanethiol is important to control and predict cheese flavor.

Methionine in cheese

Methionine catabolism is dependent on its availability from casein. Proteolysis has long been associated with flavor development (Fox et al. 1996). Hence, a great deal of effort was spent developing methods to measure proteolysis in cheese (McSweeney & Fox 1996) and characterizing the lactococcal proteolytic system (Law & Haadrikman 1997) with the aim of accelerating cheese ripening (Fox et al. 1996). To date little success has been achieved in this objective.

Primary and secondary proteolysis, and the intracellular peptidase system of bacteria associated with cheese as a means to provide methionine for VSC is unknown. Casein has only a small amount of cysteine (4 residues in all the caseins) and methionine (16 remain in cheese) for potential metabolism in milk and ripening cheese (Swaisgood 1992). Free methionine is limited during aging and the sulfur-containing amino acids are usually in peptides. The influence of the peptide transport system in lactococci on VSC production is unknown, despite a large amount of information regarding its influence on growth of lactic acid bacteria (Kunji et al. 1996).

Some peptidases have broad activity and release sulfur-containing amino acids (Kunji et al. 1996), but the role of peptidases in VSC production is unknown. Addition of methionine and methionine-containing peptides to a chemically defined growth medium of *L. lactis* ssp. *cremoris* S2 repressed enzymes in the VSC pathways. The same treatment induced VSC production in brevibacteria (Weimer et al. 1997; Dias & Weimer 1998a). This circumstantial evidence suggests that transport and peptidase systems play a role in regulating VSC production, but further investigations are needed to define the function and importance of these reactions.

Methionine transport

Methionine transport has been extensively studied in Gram-negative bacteria and yeast. A high affinity transport system (K_m ranges from 0.1 to 12 μM) for L-methionine is described in *E. coli* (Kadner 1974; Kadner 1975), *Salmonella typhimurium* (Ayling & Bridgeland 1972), and *Yersinia pestis* (Montie & Montie 1975). *E. coli* also contains a second transport system with lower affinity (K_m=40 μM) (Kadner & Watson 1974; Kadner 1975). *S. typhimurium* contains three transport systems for methionine with affinities ranging from 0.1 μM to 18 mM (Ayling et al. 1979).

L-Methionine transport is not as well characterized in Gram-positive organisms compared to *E. coli*. *B. linens* CNRZ 918, of particular interest due to its role in cheese ripening, transports methionine (K_m=55 μM) via a transporter which relies on a Na^+ stimulated proton motive force (Ferchichi et al. 1987). The role of Na^+ in transport is not surprising, as *B. linens* is a species adapted to high sodium levels in its environment. In lactococci, methionine is transported at a low affinity (K_m=100 μM) by the leucine carrier (Driessen et al. 1987) which is primarily driven by the proton-motive force. The concentration of methionine, which is essential for growth of lactococci, is very low in milk. Use of free amino acids in milk as the sole source of nitrogen results in a small amount of lactococcal growth (Juillard et al. 1995).

Lactococci rely on milk peptides for growth, meaning the concerted action of peptide transport systems and peptidases are needed. Di-, tri- and oligopeptide transport systems have been identified in lactococci (Smid et al. 1989; Tynkkynen et al. 1993; Foucaud et al. 1995). However, only the oligopeptide system plays a crucial role in the utilization of milk peptides in growth (Juillard et al. 1995). Transportable milk oligopeptides represent a poor source of methionine, as determined by examining the specificity of peptide utilization by *L. lactis*, and cessation of growth in milk is due to methionine and leucine depletion (Juillard et al. 1998). Further work is needed to define how amino acid transport during cheese ripening influences VSC production.

Methionine metabolism

B. linens was the first cheese-ripening bacterium identified to produce methanethiol from methionine (Sharpe et al. 1977). Gram-negative isolates from raw-milk, including *Pseudomonas fluorescens*, *Pseudomonas fragi*, *Pseudomonas aureofaciens*, *Aeromonas* and *Serratia marcescens* also produce methanethiol (Law et al. 1978). Lactobacilli, *L. lactis* ssp. *lactis*, *L. lactis* ssp. *cremoris*, and micrococci produce lesser amounts of methanethiol (Law et al. 1978; Dias & Weimer 1998a). Methionine plays a central role in the interconversion of sulfur-containing amino acids. Microorganisms metabolize methionine by a variety of different pathways (Soda 1987) (Figure 2). Methionine biosynthesis from homoserine proceeds via a sulfhydrylase pathway (homoserine→O-acetylhomoserine→homocysteine→methionine) and a pathway involving cystathionine (homoserine→O-succinylhomoserine→cystathionine→homocysteine →methionine). The cystathionine pathway is the primary biosynthetic pathway in many bacteria (Soda 1987), however methionine biosynthesis in *Brevibacterium flavum* proceeds via the sulfhydrylase pathway (Ozaki & Shiio 1982).

Methionine and cysteine are linked through homocysteine, which is synthesized from O-succinylhomoserine and cysteine through cystathionine. Methionine is metabolized to *S*-adenosylmethionine (SAM), which is vital for one-carbon transfer reactions within the cell. SAM can be converted to homocysteine and then cystathionine, linking methionine pools to cysteine pools. Methionine can be transaminated to form KMBA, which can be degraded to methanethiol. Additionally, methionine can be catabolized to form methanethiol by a simultaneous deamination and decarboxylation. The amount of methanethiol formed is dependent on the overall cellular methionine concentration available.

In lactococci, transamination of methionine leads to KMBA (Gao et al. 1998), which degrades to form methanethiol. Additionally, cystathionine β-lyase (Alting et al. 1995) and cystathionine γ-lyase (Bruinenberg et al. 1997; Smacchi & Gobbetti 1998) non-specifically degrade methionine, producing methanethiol. Methionine γ-lyase (MGL) directly deaminates and decarboxylates methionine, producing methanethiol, α-ketobutyrate and ammonia. This activity was first found using dried cells of *E. coli* and *Proteus vulgaris* (Ornitake 1938), but has been reported in many organisms including brevibacteria and pseudomonads (Law et al. 1978; Collin et al. 1989; Dias & Weimer 1998a).

Aminotransferases

The first step in the catabolism of many amino acids is transamination. Aminotransferases have broad substrate specificities and can catalyze reversible transamination reactions. Aminotransferases capable of transamination of aromatic amino acids are described in lactococci (Gao et al. 1997; Yvon et al. 1997) and have activity on methionine (Yvon et al. 1997; Gao et al. 1998). Dias and Weimer (1998a) found high levels of methionine aminotransferase activity in lactococci and lactobacilli, but saw no KMBA degradation using cell-extracts. Aminotransferase activity decreased as the levels of methionine in the growth medium increased, with approximately 40% of the activity present with 0.25% methionine in the growth medium, as compared to no methionine in the growth medium. Recently, Gao et al. (1998), using ^{13}C nuclear magnetic resonance, found lactococci catabolize methionine via an aminotransferase, which results in the intermediate KMBA and end product 2-hydroxyl-4-methylthiobutyric acid (HMBA). Significantly greater quantities of methanethiol, as detected by gas chromatography, were produced when whole cells (but not cell-free extracts) were incubated with KMBA rather than with HMBA in cheese-like conditions (pH 5.2, 4% NaCl, 13 °C). Non-enzymatic degradation of KMBA to methanethiol was also observed, as has been noted in previous work (Manning 1979; Soda 1987).

KMBA is degraded to 3-methylthiolpropionic acid (MTPA), which is subsequently degraded to methanethiol in a pathway found in a variety of mammals (Benevenga & Steele 1984; Scislowski et al. 1989; Scislowski & Pickard 1993; Franken et al. 1996). MTPA production from methionine via the KMBA intermediate is characterized in *Xanthomonas campestris* (Ewbank & Maraite 1990), a plant pathogen, and MTPA production from methionine has been noted in many species of fungi, yeast and bacteria (Arima et al. 1970). Aerobically, KMBA is converted to ethylene, CO_2 and methanethiol by many microbes (Ince & Knowles 1986). This conversion is linked to nitrogen limitation (Shipston & Bunch 1989). While these data are encouraging in defining the mechanism of VSC production in lactococci, the role and activity of these pathways in cheese is still unknown.

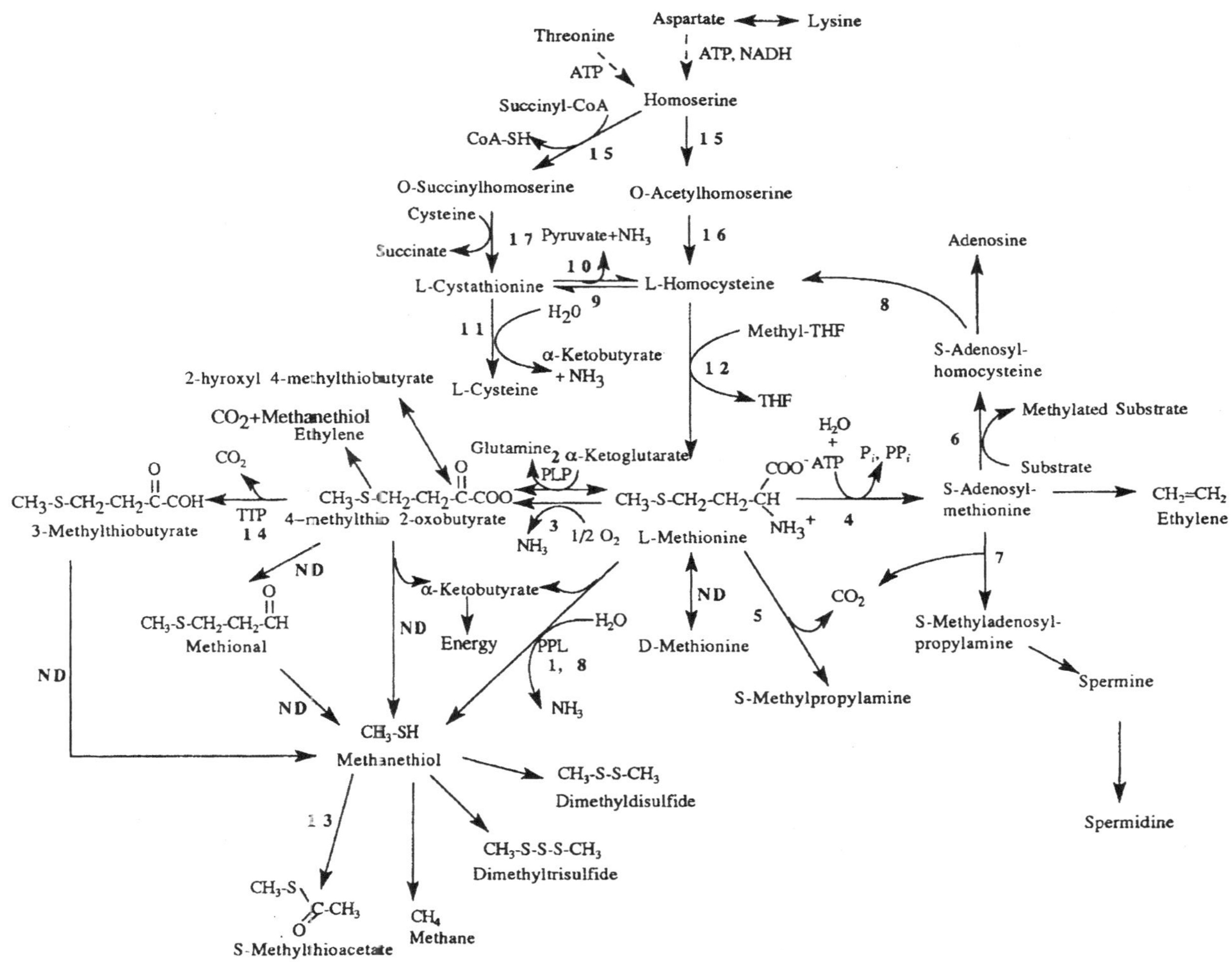

Figure 2. Methionine metabolism. (1) Methionine γ-lyase, (2) L-Amino acid oxidase, (3) Aminotransferases, (4) Methionine adenosyl transferase, (5) Methionine decarboxylase, (6) Adenosylhomocysteinase, (7) S-adenosylmethionine decarboxylase, (8) Adenosyl-homocysteinase, (9) Cystathionine β-synthase, (10) Cystathionine β-lyase, (11) Cystathionine γ-lyase, (12) Homocysteine methyltransferase, (13) Acylenzyme, (14) Decarboxylase, (15) Homoserine O-acetyltransferase, (16) Homocysteine methyl transferase, (17) Cystathionine γ-synthase. (ND) Not demonstrated in microorganisms.

Cystathionine β- and γ-lyase

Cystathionine β-lyase (EC 4.4.1.8) is a pyridoxal phosphate (PLP) dependent enzyme that catalyzes the conversion of cystathionine to homocysteine with the formation of pyruvate and ammonia (Uren 1987). This enzyme is purified and sequenced from *Salmonella typhimurium* and *E. coli* (Guggenheim 1971; Park & Stauffer 1987; Uren 1987; Laber et al. 1996). Cystathionine β-lyase from *L. lactis* ssp. *cremoris* is also purified (Alting et al. 1995). This enzyme shows the ability to catalyze α-, γ-elimination reactions as well, as it produces α-ketobutyrate, methanethiol and ammonia from methionine. It is proposed to play a role in the production of methanethiol in Gouda cheese (Alting et al. 1995). However, Dias and Weimer (1998a), propose a limited role for these enzymes due to repression by methionine in the medium.

Cystathionine γ-lyase (EC 4.4.1.1) also requires PLP and catalyzes the α-, γ-elimination reaction of L-cystathionine producing L-cysteine, α-ketobutyrate, and ammonia. It is a cysteine biosynthetic enzyme involved in the conversion of homocysteine to cysteine which has been cloned and sequenced from *Saccharomyces cerevisiae* (Yamagata et al. 1993). Cystathionine γ-lyase is also purified from *L. lactis* ssp. *cremoris* SK11 (Bruinenberg et al. 1997) and *Lactobacillus fermentum* (Smacchi & Gobbetti 1998). It produces α-ketobutyrate, methanethiol, and ammonia from methionine with slight activity in cheese-like conditions. This enzyme also catalyzes the α-, β-elimination reaction of cysteine, resulting in hydrogen sulfide formation.

The importance of cystathionine β-lyase and cystathionine γ-lyase in VSC production is unknown in cheese. No cystathionine degrading ability was detected in lactococcal cultures tested by Gao et al. (1998), whereas Dias and Weimer (1998a) demonstrated this ability in several lactococcal and lactobacilli strains. The discrepancy between these two groups may be accounted for by the use a more sensitive assay by Dias and Weimer (1998a). Since these are enzymes used in the biosynthetic pathways, it is reasonable to suspect that they may be regulated by extracellular cysteine or methionine concentration and be expressed at low levels in the cell. If this is the case it may have practical implications for cheese ripening for the addition of bacteria, in addition to the starter culture, to boost the amount of these enzymes in the cheese matrix during aging.

The cystathionine degrading activity is negatively influenced by the presence of methionine and cysteine in the growth medium (Dias & Weimer 1998a). Using data obtained for cystathionase activity under cheese-like conditions, they hypothesized that the probable free thiol production in the cheese matrix due to action of these enzymes is insignificant. Very high concentrations of these enzymes in the cell would be needed for significant contribution to free thiols in the cheese matrix based on its stability in cheese-like conditions and the amount of substrate available in the matrix. This suggests enzymes that behave differently from cystathionine β-lyase and cystathionine γ-lyase, such as methionine γ-lyase, may be beneficial in improving Cheddar cheese flavor.

Methionine γ-lyase

L-Methionine γ-lyase (methionase, L-methionine methanethiol-lyase (deaminating), or MGL) (EC 4.4.1.11), is a PLP-dependent enzyme that converts L-methionine into α-ketobutyrate, methanethiol, and ammonia by an α-, γ-elimination reaction. This enzyme plays an important role in the bacterial metabolism of methionine, and is essential for generation of metabolic energy from methionine, α-ketobutyrate via propionyl-CoA, and subsequently succinyl-CoA (Soda et al. 1983). MGL is a multifunctional enzyme, catalyzing the α-, β-elimination and the α-, β-elimination reactions of methionine and its derivatives. Additionally, MGL catalyzes the β-replacement reactions of sulfur amino acids, making this enzyme useful for synthesis of sulfur amino acids, such as ethionine from L-methionine and ethanethiol (Tanaka et al. 1977), and L-selenium amino acids (Esaki et al. 1979). MGL also catalyzes the rapid exchange of the α- and β-hydrogens of methionine and other amino acids with deuterium or tritium. This characteristic provides an enzymatic method for the specific preparation of amino acids deuterated or tritiated at the α- and β-positions (Esaki et al. 1985).

MGL is widely distributed in bacteria, being especially prevalent in pseudomonads, and is induced in the presence of methionine (Tanaka et al. 1985; Dias & Weimer 1998a) and methionine containing peptides from β-casein (Weimer et al. 1997). MGL is purified from *Aeromonas* sp. (Nakayama et al. 1984a), *Pseudomonas putida* (Nakayama et al. 1984b), *Clostridium sporogenes* (Kreis & Hession 1973), *Trichomonas vaginalis* (Lockwood & Coombs 1991), and *B. linens*

BL2 (Dias & Weimer 1998b). Purification and characterization of MGL from a food-grade organism has given us insight as to the important role MGL can play in cheese flavor development. Weimer et al. (1997) and Dias and Weimer (1998a, 1999) demonstrate that the enzyme from *B. linens* BL2 is active under the pH, temperature, and salt concentration conditions which exist in aging Cheddar cheese.

In studies using cheese slurries, addition of MGL or *B. linens* BL2 to the slurries enhanced methanethiol production (Dias & Weimer 1999). MGL is susceptible to proteolysis (Dias & Weimer 1998b), indicating that activity may be lost with cell lysis. This was not demonstrated in Cheddar cheese slurries, however (Dias & Weimer 1999). Weimer et al. (1997) observed *B. linens* BL2 to persist in 60% reduced fat Cheddar cheese for at least 6 months at the same level, indicating lysis of this organism may not readily occur. The ability of this enzyme to produce methanethiol may be linked to the ability of the cell to remain intact and metabolically active during cheese ripening.

The gene for MGL from *Pseudomonas putida* was sequenced by two separate laboratories (Inoue et al. 1995; Hori et al. 1996), and two different MGL genes have been sequenced from *T. vaginalis* (McKie 1998). The two *P. putida* and one of the *T. vaginalis* sequences encode a protein of 398 amino acids, while the other *T. vaginalis* sequence encodes a protein of 396 amino acids. The amino acid sequence of the PLP-binding region is strictly conserved in all cases, and overall they have a high degree of homology (Figure 3). In *P. putida*, MGL (*mdeA*) is found in an operon along with a gene encoding a novel α-ketobutyrate dehydrogenase E1 component (*mdeB*) of an unknown α-keto acid dehydrogenase complex (Inoue et al. 1997). The gene location, structure, and sequence are unknown in bacteria used in cheese making.

Regulation of methionine metabolism

Regulation of methionine metabolism is well characterized in *E. coli*, *Salmonella typhimurium*, and *Bacillus subtilis* (Paulus 1993; Greene 1996). Threonine and the branched chain amino acids, which use homoserine as an intermediate, share the initial steps in production of methionine. Succinylated homoserine is converted to L-cystathionine by cystathionine synthase. Alternatively, via a nucleophilic attack by H_2S or CH_3SH, it is converted to homocysteine or methionine, respectively. β-Cystathionine lyase is used to produce homocysteine as well from L-cystathionine. In lactococci, this enzyme demonstrates demethiolation activity on methionine about 100 fold less efficiently than its intended reaction (Alting et al. 1995). The final step in the synthesis of methionine in these organisms is the methylation of homocysteine (Saint-Girons et al. 1988), which can be catalyzed by two alternate transmethylases (Figure 4). One is a vitamin B_{12}-independent enzyme, the *metE* gene product, and the other is a vitamin B_{12}-dependent enzyme, the *metH* gene product.

Repression of the biosynthetic pathway (except for *metH*) is mediated by a repressor protein (MetJ) and SAM, which functions as a corepressor for MetJ (Saint-Girons et al. 1988). The *metE* gene is repressed by addition of vitamin B_{12} to the growth medium (Greene et al. 1973). Vitamin B_{12} mediated repression requires the *metJ* and *metF* gene products (which provide the methyl donor for homocysteine methylation) as well as the MetH-B_{12} holoenzyme (Dawes & Foster 1971; Mulligan et al. 1982). The mechanism of repression by the MetH-B_{12} holoenzyme requires the MetJ protein, but this repression system is distinct from the repression system mediated by the MetJ protein and SAM (Milner et al. 1982).

Transcription of the *metE* and the *metH* gene is positively regulated by the *metR* gene product (Maxon et al. 1989; Urbanowski & Stauffer 1989), a DNA binding protein (Urbanowski & Stauffer 1989). The *metR* gene is negatively regulated by its own gene product and this autoregulation involves homocysteine as a corepressor (Urbanowski & Stauffer 1987). Homocysteine is required for activation of the *metE* gene (Urbanowski et al. 1987). The vitamin B_{12}-mediated repression of *metE* is due to reduced MetR-homocysteine activation (Wu et al. 1992), which is a result of reduction of MetR activator protein and its coactivator homocysteine by the MetH-B_{12} holoenzyme and the *metF* and *metJ* gene products.

In *B. subtilis*, the acylation of homoserine specifically requires acetyl-CoA, but the homoserine O-acetyltransferase resembles the corresponding O-succinyltransferase of *E. coli* in that it is synergistically inhibited by methionine and SAM (Brush & Paulus 1971). L-cystathionine is formed from O-acetylhomoserine, and can then be converted to homocysteine and finally methionine. In *B. subtilis*, mutants with defects in methionine biosynthesis are known for *metA*, *metB*, *metC*, and *metD*, which map to different sites in the chromosome (Piggot & Hoch 1985). How-

```
1   M H G S N K L P G F A T R A I H H G Y D P Q D H G G A L V P P V Y Q T A T F T F   pseudo-1
1   M R D S H N N T G F S T R A I H H G Y D P L S H G G A L V P P V Y Q T A T Y A F   Pseudo-3
1   M S G H A I D P T H T D T L S I H A N P Q K D Q F G A I V A P I Y Q T S T F L F   Tricho-1
1   M S H E R M T P A - - - T A C I H A N P Q K D Q F G A A I P P I Y Q T S T F V F   Tricho-2

41  P T V E Y G A A C F A G E Q A G H F Y S R I S N P T L N L L E A R M A S L E G G   pseudo-1
41  P T V E Y G A A C F A G E E A G H F Y S R I S N P T L A L L E Q R M A S L E G G   Pseudo-3
41  D N C D Q G G A R F G G K E A G Y M Y T R I G N P T N S A L E G K I A K L E H A   Tricho-1
38  D N C Q Q G G N R F A G Q E S G Y I Y T R L G N P T V S N L E G K I A F L E K T   Tricho-2

81  E A G L A L A S G M G A I T S T L W T L L R P G D E V L L G N T L Y G C T F A F   pseudo-1
81  E A G L A L A S G M G A I T S T L W T L L R P G D E L I V G R T L Y G C T F A F   Pseudo-3
81  E A C A A T A S G M G A I A A S V W T F L K A G D H L I S D D C L Y G C T H A L   Tricho-1
78  E A C V A T S S G M G A I A A T V L T I L K A G D H L I S D E C L Y G C T H A L   Tricho-2

121 L H H G I G E F G V K L R H V D M A D L Q A L E A A M T P A T R V I Y F E S P A   pseudo-1
121 L H H G I G E F G V K I H H V D L N D A K A L K A A I N S K T R M I Y F E T P A   Pseudo-3
121 F E H Q L R K F G V E V D F I D M A V P G N I E K H L K P N T R I V Y F E T P A   Tricho-1
118 F E H A L T K F G I Q V D F I N T A I P G E V K K H M K P N T K I V Y F E T P A   Tricho-2

161 N P N M H M A D I A G V A K I A R K H - G A T V V V D N T Y C T P Y L Q R P L E   pseudo-1
161 N P N M Q L V D I A A V V E A V R G S - D V L V V V D N T Y C T P Y L Q R P L E   Pseudo-3
161 N P T L K V I D I E D A V K Q A R K Q K D I L V I V D N T F A S P I L T N P L D   Tricho-1
158 N P T L K I I D M E R V C K D A H S Q E G V L V I A D N T F C S P M I T N P V D   Tricho-2

200 L G A D L V V H S A T K Y L S G H G D I T A G I V V G S Q A L V D R I R L Q G L   pseudo-1
200 L G A D L V V H S A T K Y L S G H G D I T A G L V V G R K A L V D R I R L E G L   Pseudo-3
201 L G V D I V V H S A T K Y I N G H T D V V A G L V C S R A D I I A K V K S Q G I   Tricho-1
198 F G V D V V V H S A T K Y I N G H T D V V A G L I C G K A D L L Q Q I R M V G I   Tricho-2

240 K D M T G A V L S P H D A A L L M R G I K T L N L R M D R H C A N A Q V L A E F   pseudo-1
240 K D M T G A A L S P H D A A L L M R G I K T L A L R M D R H C A N A L E V A Q F   Pseudo-3
241 K D I T G A I I S P H D A W L I T R G T L T L D M R V K R A A E N A Q K V A E F   Tricho-1
238 K D I T G S V I S P H D A W L I T R G L S T L N I R M K A E S E N A M K V A E Y   Tricho-2

280 L A R Q P Q V E L I H Y P G L A S F P Q Y T L A R Q Q M S Q P G G M I A F E L K   pseudo-1
280 L A G Q P Q V E L I H Y P G L P S F A Q Y E L A Q R Q M R L P G G M I A F E L K   Pseudo-3
281 L H E H K A V K K V Y Y P G L P D H P G H E I A K K Q M K M F G S M I A F D V D   Tricho-1
278 L K S H P A V E K V Y Y P G F E D H E G H D I A K K Q M R M S G S M I T F I L K   Tricho-2

320 G G I G A G R R F M N A L Q L F S R A V S L G D A E S L A Q H P A S M T H S S Y   pseudo-1
320 G G I E A G R G F M N A L Q L F A R A V S L G D A E S L A Q H P A S M T H S S Y   Pseudo-3
321 - G L E K A K K V L D N C H V V S L A V S L G G P E S L I Q H P A S M T H A G V   Tricho-1
318 S G F E G A K K L L D N L K L I T L A V S L G G C E S L I Q H P A S M T H A V V   Tricho-2

360 T P E E R A H Y G I S E G L V R L S V G L E D I D D L L A D V Q Q A L K A S A     pseudo-1
360 T P Q E R A H H G I S E G L V R L S V G L E D V E D L L A D I E L A L E A C A     Pseudo-3
360 P K E E R E A A G L T D N L I R L S V G C E N V Q D I I D D L K Q A L D L V L     Tricho-1
358 P K E E R E A A G I T D G M I R L S V G I E D A D E L I A D F K Q G L D A L L     Tricho-2
```

Figure 3. Sequence alignment of the four known amino acid sequences for L-methionine γ-lyase. Pseudo-1 is from *P. putida* (Inoue et al. 1995), Pseudo-3 is from *P. putida*, cloned and sequenced (Hori et al. 1996), Tricho-1 and 2 are from *T. vaginalis* (McKie et al. 1998).

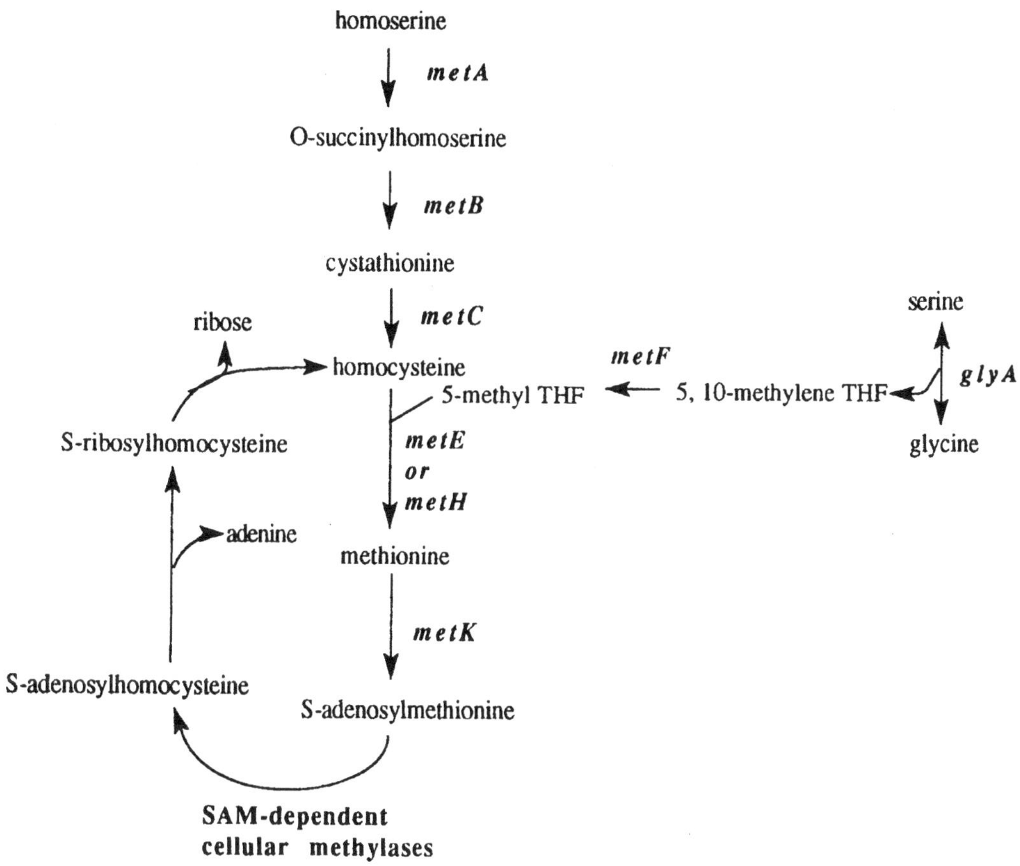

Figure 4. Methionine biosynthetic pathway in *E. coli* and *S. typhimurium*. The *metA*, *metB*, and *metC* genes encode enzymes in the nonfolate branch of the methionine pathway, and the *glyA* and *met*F genes encode enzymes in the folate branch of the methionine pathway. The S-adenosylhomocysteine formed as a by-product of SAM-dependent methylation reactions is salvaged as the adenine base and homocysteine by a regenerative branch of the pathway, which is not genetically characterized. THF, tetrahydrofolate.

ever, no expression or regulation studies with these mutations are available with these mutants. Little is known about the regulation of the methionine biosynthetic genes in *B. subtilus* (Paulus 1993), and only one methionine biosynthetic gene, encoding SAM synthetase (*metE*), is cloned (Yocum et al. 1996). This gene corresponds to *metK* in *E. coli*, and its overexpression leads to methionine auxotrophy. In *E. coli*, SAM acts as a corepressor of genes involved in methionine biosynthesis, and overproduction of SAM in *B. subtilus* likely has the same effect.

In corynebacteria, cystathionine is not an intermediate in methionine biosynthesis, but instead O-acetylhomoserine is directly converted to homocysteine by acetyl homoserine sulfhydrylase (Ozaki & Shiio 1982). Methionine auxotrophs of *B. flavum* are classified into three groups based on their ability to grow on O-acetylhomoserine and/or homocysteine, but no studies on control of expression of the genes responsible for these mutations are published. The first two enzymes of methionine biosynthesis in *B. flavum*, homoserine transacetylase and acetylhomoserine sulfhydrylase, are repressed by methionine (Ozaki & Shiio 1982).

A positive regulator of the MGL (*mde*) operon, which consists of the genes encoding mgL (*mdeA*) and α-ketobutyrate E1 component (*mdeB*) of a novel α-keto acid dehydrogenase complex, is identified in *P. putida* (Inoue et al. 1997). The regulator, encoded by the gene *mdeR*, is on the opposite strand, 127 bp from the translational start of *mdeB*. MdeR is a member of the leucine responsive regulatory protein family (a DNA-binding protein). It is similar to the MetR in the Enterobacteriacea, and may influence this operon by binding to the 127-bp intergenic region between *mdeA* and *mdeR*, which has an unusually low G+C content. This may play a role in the transcriptional regulation by influencing DNA–DNA interactions.

Methanethiol production in *B. linens* CNRZ 918 is induced by methionine, presumably reflecting an induction of MGL (Ferchichi et al. 1985; Collin & Law 1989; Dias & Weimer 1998a). Highest methanethiol production occurs during exponential growth phase (Ferchichi et al. 1985). Glucose favors growth but inhibits production of methanethiol, while lactate favors growth and methanethiol production (Ferchichi et al. 1985). In contrast to what has been reported for other organisms (Ohigashi et al. 1951; Miwatani et al. 1954), the cellular ATP-decreasing compounds azide or malonate favors methanethiol production by resting cells (Ferchichi et al. 1985). A decrease in the ATP concentration in the cell should result in stimulation of MGL activity, as it is involved in producing energy for the cell. Additionally, inhibition of methanethiol production by glucose may represent a catabolite repression response, or modulate the energy balance, making the use of MGL for energy unnecessary.

Clearly, there is much to be learned about the role and metabolism of sulfur-containing amino acids in cheese flavor and cheese making bacteria. Determination of which VSC is critical for flavor is needed. Knowledge of the mechanisms to produce those compounds in bacteria and the genes encoding the enzymes involved is critical to enhance our understanding of how the bacteria associated with cheese impart their beneficial impact during cheese ripening.

Acknowledgements

The research described was supported by the Utah Agricultural Experiment Station as contribution 7140, Dairy Management, Inc., and the Western Dairy Center.

References

Allen C & Parks OW (1975) Evidence for methional in skim milk exposed to sunlight. J. Dairy Sci. 58: 1609–1611

Alting AC, Engels WJM, van Schalkwijk S & Exterkate FA (1995) Purification and characterization of cystathionine β-lyase from *Lactococcus lactis* subsp. *cremoris* B78 and its possible role in flavor development in cheese. Appl. Environ. Microbiol. 61: 4037–4042

Amerine MA, Pangborn RM & Roessler EB (1965) Principles of sensory evaluation of food. Academic Press, New York

Arima K, Yamashita T, Hosoda J & Tamura G (1970) Studies on substances active on the behavior of planarian Part I. Production of trans-3-methylthioacrylic acid and 3-methylthiopropionic acid by microorganisms. Agric. Biol. Chem. 34: 1178–1183

Arora G, Cormier F & Lee B (1995) Analysis of odor-active volatiles in Cheddar cheese headspace by multidimensional GC/MS/sniffing. J. Agric. Food Chem. 43: 748–752

Aston JW & Douglas K (1983) The production of volatile sulphur compounds in Cheddar cheeses during accelerated ripening. Australian J. Dairy Tech. 66–70

Aston JW & Dulley JR (1982) Cheddar cheese flavour. Australian J. Dairy Tech. 37: 59–64

Ayling PD & Bridgeland ES (1972) Methionine transport in wild-type and transport defective mutants of *Salmonella typhimurium*. J. Gen. Microbiol. 73: 127–141

Ayling PD, Mojica AT & Sklopotowski T (1979) Methionine transport in *Salmonella typhimurium*: evidence for at least one low affinity transport system. J. Gen. Microbiol. 114: 227–246

Ballance PE (1961) Production of volatile compounds related to the flavor of foods from the Strecker degradation of DL-methionine. J. Sci. Food and Agric. 12: 532–

Benevenga NJ & Steele RD (1984) Adverse effects of excessive consumption of amino acids. Annual Reviews in Nutrition 4: 157–181

Bruinenberg PG, deRoo G & Limsowtin GK (1997) Purification and characterization of cystathionine γ-lyase from *Lactococcus lactis* subsp. *cremoris* SK11: Possible role in flavor compound formation during cheese maturation. Appl. Environ. Microbiol. 63: 561–566

Brush A & Paulus H (1971) The enzymatic formation of O-acetylhomoserine in *Bacillus subtilis* and its regulation by methionine and S-adenosylmethionine. Biochem. Biophys. Res. Comm. 45: 735–741

Chin H-W & Lindsay RC (1994) Ascorbate and transition-metal mediation of methanethiol oxidation to dimethyl disulfide and dimethyl trisulfide. Food Chem. 49: 387–392

Christensen KR & Reineccius GA (1995) Aroma extract dilution analysis of aged Cheddar cheese. J. Food Sci. 60: 218–220

Collin JC & Law BA (1989) Isolation and characterization of the L-methionine γ-demethiolase from *Brevibacterium linens* NCDO 739. Sciences des Aliments 9: 805–812

Cuer A, Dauphin G, Kergomard A, Dumont JP & Adda J (1979) Production of S-methylthioacetate by *Brevibacterium linens*. Appl. Environ. Microbiol. 38: 332–334

Cuer A, Dauphin G, Kergomard A, Dumont JP & Adda J (1979) Production of S-methylthioacetate by *Micrococcus* cheese strains. Agric. Biol. Chem. 43: 1783–1784

Cuer A, Dauphin G, Kergomard A, Roger S, Dumont JP & Adda J (1979) Flavour properties of some sulphur compounds isolated from cheeses. Lebensmittel-Wissenschaft und Technologie 12: 258–261

Dacremont C & Vickers A (1994) Concept matching technique for assessing importance of volatile compounds for Cheddar cheese aroma. J. Food Sci. 59: 981–985

Dawes J & Foster MA (1971) Vitamin B_{12} and methionine synthesis in *Escherichia coli*. Biochimica et Biophysica Acta 237: 455–464

Dias B & Weimer B (1999) Production of volatile sulfur compounds in Cheddar cheese slurries. Int. Dairy J. (in press)

Dias B & Weimer B (1998a) Conversion of methionine to thiols by lactococci, lactobacilli and brevibacteria. Appl. Environ. Microbiol. 64: 3320–3326

Dias B & Weimer B (1998b) Purification and characterization of L-methionine γ-lyase from *Brevibacterium linens* BL2. Appl. Environ. Microbiol. 64: 3327–3331

Dimos A (1992) A comparative study by GC/MS of the flavor volatiles produced during the maturation of full-fat and low-fat Cheddar cheese. M.S. thesis. Latrobe University, Bundoora, Victoria, Australia

Drake MA, Boylston TD, Spence KD & Swanson BG. (1997) Improvement of sensory quality of reduced fat Cheddar cheese by a *Lactobacillus* adjunct. Food Res. Int. 30: 35–40

Drake MA, Boylston TD, Spence KD & Swanson BG. (1996) Chemical and sensory effects of a *Lactobacillus* adjunct in Cheddar cheese. Food Res. Int. 29: 381–387

Driessen AJM, de Jong S & Konings WN (1987) Transport of branched chain amino acids in membrane vesicles of *Streptococcus cremoris*. J. Bacteriol. 169: 5193–5200

Dumont JP, Degas C & Adda J (1976) L'arome du Pont-l'Eveque: Mise en evidence de constituants votatils quantitativement mineurs. Lait 56: 177–182

Dumont JP, Roger S & Adda J (1974) Etude des composes volatils neutres presents dans les fromages a pate molle et a croute lavee. Lait 54: 31–43

Esaki N, Nakayama T, Sawada S, Tanaka H & Soda K (1985) ^1H NMR studies of substrate hydrogen exchange reactions catalyzed by L-methionine γ-lyase. Biochemistry 24: 3857–3862

Esaki N, Tanaka H, Uemura S, Suzuki T & Soda K (1979) Catalytic action of L-methionine γ-lyase on selenomethionine and selenols. Biochemistry 18: 407–410

Ewbank E & Maraite H (1990) Conversion of methionine to phytotoxic 3-methlthiopropionic acid by *Xanthomonas campestris* pv. *manihotis*. J. Gen. Microbiol. 136: 1185–1189

Ferchichi M, Hemme D & Nardi M (1987) Na^+-stimulated transport of L-methionine in *Brevibacterium linens* CNRZ 918. Appl. Environ. Microbiol. 53: 2159–2164

Ferchichi M, Hemme D, Nardi M & Pamboukdjian N (1985) Production of methanethiol from methionine by *Brevibacterium linens* CNRZ 918. J. Gen. Microbiol. 131: 715–723

Forss DA (1979) Review of the progress of dairy science: Mechanisms of formation of aroma compounds in milk and milk products. J. Dairy Res. 46: 691–706

Foucaud C, Kunji ERS, Hagting A, Richard J, Konings WN, Desmazeaud M & Poolman B (1995) Specificity of peptide transport systems in *Lactococcus lactis*: evidence for a third system which transports hydrophobic di- and tripeptides. J. Bacteriology 177: 4652–4657

Fox PF, Law J, McSweeney PLH & Wallace J (1996) Biochemistry of cheese ripening. In: Fox PF (Ed) Cheese: Chemistry, Physics, and Microbiology, Vol. 1 (pp. 389–438). Chapman & Hall, New York

Fox PF, Wallace JM, Morgan S, Lynch CM, Niland EJ & Tobin J. (1996) Acceleration of cheese ripening. Antonie van Leeuwenhoek 70: 271–297

Franken DG, Blom HJ, Boers GHJ, Tangerman AT, Thomas CMG & Trijbels FJM (1996) Thiamine (B1) supplementation does not reduce fasting blood homocysteine concentration in most homozygotes for homocystinuria. Biochimica et Biophysica Acta 1317: 101–104

Gao S, Mooberry ES & Steele JL (1998) Use of ^{13}C nuclear magnetic resonance and gas chromatography to examine methionine catabolism by lactococci. Appl. Environ. Microbiol. 64: 4670–4675

Gao S, Oh DH, Broadbent JR, Johnson ME, Weimer BC & Steele JL (1997) Aromatic amino acid catabolism by lactococci. Lait 77: 371–381

Greene RC (1996) Biosynthesis of methionine In: Neidhardt FC et al. (Eds) *Escherichia coli* and *Salmonella* Cellular and Molecular Biology Vol 1 (pp 542–560). ASM Press, Washington, DC

Greene RC, Williams RD, Kung HF, Spears C & Weissbach H (1973) Effects of methionine and vitamin B_{12} on the activities of methionine biosynthetic enzymes in *metJ* mutants of *Escherichia coli* K12. Arch. Biochem. Biophys. 158: 249–256

Grill H, Patton S & F. CJ (1966) Aroma significance of sulfur compounds in surface-ripened cheese. J. Dairy Sci. 49: 409–413

Grosch W (1993) Detection of potent odourants in foods by aroma extract dilution analysis. Trends in Food Sci. Technol. 4: 68–73

Grosch W (1994) Determination of potent odourants in foods by aroma extract dilution analysis (AEDA) and calculation of odour activity values (OAVs). Flavour and Fragrance J. 9: 147–158

Guggenheim S (1971) β-Cystathionase (*Salmonella*). Methods in Enzymology 17B: 439–442

Guth H & Grosch W (1993) Identification of potent odourants in static headspace samples of green and black tea powders on the basis of aroma extract dilution analysis (AEDA). Flavour and Fragrance J. 8: 173–178

Harper WJ, Carmona A & Kristoffersen T (1971) Protein degradation in Cheddar cheese slurries. J. Food Sci. 36: 503–506

Harper WJ & Kristoffersen T (1970) Biochemical aspects of flavor development in Cheddar cheese slurries. J. Agric. Food Chem. 18: 563–566

Hori H, Takabayashi K, Orvis L, Carson DA & Nobori T (1996) Gene cloning and characterization of *Pseudomonas putida* L-methionine-α-deamino-γ-mercaptometha ne-lyase. Cancer Res. 56: 2116–2122

Ince JE & Knowles CJ (1986) Ethylene formation by cultures of *Escherichia coli*. Arch. Microbiol. 146: 151–158

Inoue H, Inagaki K, Eriguchi S, Tamura T, Esaki N, Soda K & Tanaka H (1997) Molecular characterization of the *mde* operon involved in L-methionine catabolism of *Pseudomonas putida*. J. Bacteriol. 179: 3956–3962

Inoue H, Inagaki K, Sugimoto M, Esaki N, Soda K & Tanaka H (1995) Structural analysis of the L-methionine γ-lyase gene from *Pseudomonas putida*. J. Biochem. 117: 1120–1125

Jollivet N, Bézenger M-C, Vayssier Y & Belin J-M (1992) Production of volatile compounds in liquid cultures by six strains of coryneform bacteria. Appl. Microbiol. Biotechnol. 36: 790–794

Juillard V, Guillot A, Le Bars D & Gripon J-C (1998) Specificity of milk peptide utilization by *Lactococcus lactis*. Appl. Environ. Microbiol. 64: 1230–1236

Juillard V, Le Bars D, Kunji ERS, Konings WN, Gripon J-C & Richard J (1995) Oligopeptides are the main source of nitrogen for *Lactococcus lactis* during growth in milk. Appl. Environ. Microbiol. 61: 3024–3030

Kadner R (1974) Transport systems for L-methionine in *Escherichia coli*. J. Bacteriol. 117: 232–241

Kadner R (1975) Regulation of methionine transport activity in *Escherichia coli*. J. Bacteriol. 122: 110–119

Kadner RJ & Watson WJ (1974) Methionine transport in *Escherichia coli*: Physiological and genetic evidence for two uptake systems. J. Bacteriol. 119: 401–409

Kadota H & Ishida Y (1972) Production of volatile sulfur compounds by microorganisms. Annual Rev. Microbiol. 26: 127–138

Keith ES & Powers JJ (1968) Determination of flavor threshold levels and subthreshold, additive, and concentration effects. J. Food Sci. 33: 213–218

Kreis W & Hession C (1973) Isolation and purification of L-methionine-α-deamino-γ-mercaptomethane-lyase (L-methionase) from *Clostridium sporogenes*. Cancer Res. 33: 1862–1865

Kunji ERS, Mierau I, Hagting A, Poolman B & Konings WN (1996) The proteolytic system of lactic acid bacteria. Antonie van Leeuwenhoek 70: 187–221

Laber B, Clausen T, Humber R, Messerschmidt A, Egner U, Müller-Fahrnow A & Pohlenz H-D (1996) Cloning, purification, and crystallization of *Escherichia coli* cystathionine β-lyase. FEBS Lett. 379: 94–96

Lamberet G, Auberger B & Bergère JL (1997a) Aptitude of cheese bacteria for volatile S-methyl thioester synthesis I. Effect of substrates and pH on their formation by *Brevibacterium linens* GC171. Appl. Microbiol. Biotechnol. 47: 279–283

Lamberet G, Auberger B & Bergère JL (1997b) Aptitude of cheese bacteria for volatile *S*-methyl thioester synthesis. II. Comparison of coryneform bacteria, Micrococcaceae and some lactic acid bacteria starters. Appl. Microbiol. Biotechnol. 48: 393–397

Lane CN & Fox PF (1997) Role of starter enzymes during ripening of Cheddar cheese made from pasteurized milk under controlled microbiological conditions. Int. Dairy J. 7: 55–63

Law BA & Sharpe ME (1978) Formation of methanethiol by bacteria isolated from raw milk and Cheddar cheese. J. Dairy Res. 45: 267–275

Law J & Haandrikman A (1997) Proteolytic enzymes of lactic acid bacteria. Int. Dairy J. 7: 1–11

Lee CW & Richard J (1984) Catabolism of L-phenylalanine by some microorganisms of cheese origin. J. Dairy Res. 51: 461–469

Libbey LM, Bills DD & Day EA (1963) A technique for the study of lipid-soluble food flavor volatiles. J. Dairy Sci. 28: 329–333

Libbey LM & Day EA (1964) Cheddar cheese flavor: gas chromatography and mass spectral analysis of the neutral components of the aroma fraction. J. Food Sci. 29: 583–589

Lindsay RC & Rippe JK (1986) Enzymic generation of methanethiol to assist in the flavor development of Cheddar cheese and other foods. In: Parliment H & Croteau R (Eds) Biogeneration of Aromas (pp. 286–308). American Chemical Society, Washington, DC

Lockwood BC & Coombs GH (1991) Purification and characterization of methionine γ-lyase from *Trichomonas vaginalis*. Biochem. J. 279: 675–692

Manning DJ & Robinson HM (1973) The analysis of volatile substances associated with Cheddar-cheese aroma. J. Dairy Res. 40: 63–75

Manning DJ (1974) Sulphur compounds in relation to Cheddar cheese flavour. J. Dairy Res. 41: 81–87

Manning DJ, Chapman HR & Hosking ZD (1976) The production of sulfur compounds in Cheddar cheese and their significance in flavor development. J. Dairy Res. 43: 313–320

Manning DJ (1979) Chemical production of essential Cheddar flavour compounds. J. Dairy Res. 46: 531–537

Maxon ME, Redfield B, Cai XY, Shoeman R, Fujita K, Fisher W, Stauffer G, Weissbach H & Brot N (1989) Regulation of methionine synthesis in *Escherichia coli*: effect of the MetR protein on the expression of the *metE* and *metR* genes. Proceedings of the National Academy of Science USA 86: 85–89

McGugan WA (1975) Cheddar cheese flavor. A review of current progress. J. Agric. Food Chem. 23: 1047–1050

McKie AE, Edlind T, Walker J, Mottram JC & Coombs GH (1998) The primitive protozoan *Trichomonas vaginalis* contains two methionine γ-lyase genes that encode members of the γ-family of pyridoxal 5′-phosphate-dependent enzymes. J. Biol. Chem. 273: 5549–5556

McSweeney PLH & Fox PF (1996) Cheese: Methods of chemical analsyis. In: Fox PF (Ed) Cheese: Chemistry, Physics, and Microbiology, Vol. 1 (pp. 389–438). Chapman & Hall, New York

Milner L, Whitfield C & Weissbach H (1982) Effect of L-methionine and vitamin B_{12} on methionine biosynthesis in *E. coli*. Archives of Biochem. Biophys. 133: 413–419

Milo C & Reineccius GA (1997) Identification and quantification of potent odorants in regular-fat and low-fat milk Cheddar cheese. J. Agric. Food Chem. 45: 3590–3594

Miwatani T, Omukai Y & Nakada D (1954) Enzymatic cleavage of methionine and homocysteine by bacteria. Medical Journal of Osaka University 5: 347–352

Montie DB & Montie TC (1975) Methionine transport in *Yersinia pestis*. J. Bacteriol. 124: 296–306

Müdler H (1952) Taste of flavor forming substances in cheese. Netherlands Milk and Dairy Journal 6: 157

Mulligan JT, Margolin W, Krueger JH & Walker GC (1982) Mutations affecting regulation of methionine biosynthetic genes isolated by use of *met-lac* fusions. J. Bacteriol. 151: 609–619

Nakayama T, Esaki N & Lee WJ (1984a) Purification and properties of L-methionine γ-lyase from *Aeromonas* sp. Agric. Biol. Chem. 48: 2367–2369

Nakayama T, Esaki N, Sugie K, Beresov TT, Tanaka H & Soda K (1984b) Purification of bacterial L-methionine γ-lyase. Anal. Biochem. 138: 421–424

Ohigashi K, Tsunetoshi A & Ichihara K (1951) The role of pyridoxal in methyl mercaptan formation, partial purification and resolution of methioninase. Medical Journal of Osaka University 2: 111–117

Ornitake J (1938) On the formation of methylmercaptan from L-cystine and L-methionine by bacteria. J. Osaka Med. Assoc. 37: 263–270

O'Shea BA, Uniacke-Lowe T & Fox PF (1996) Objective assessment of Cheddar cheese quality. Int. Dairy J. 6: 1135–1147

Ozaki H & Shiio I (1982) Methionine biosynthesis in *Brevibacterium flavum*: Properties and essential role of O-acetylhomoserine sulfhydrylase. J. Biochem. 91: 1163–1171

Park YM & Stauffer GV (1987) Cloning and characterization of the *metC* gene from *Salmonella typhimurium* LT2. Gene 60: 291–297

Parliment TH, Kolor MG & Rizzo DJ (1982) Volatile components of Limburger cheese. J. Agric. Food Chem. 30: 1006–1008

Patton S & Josephson DV (1957) A method for determining significance of volatile flavor compounds in foods. Food Res. 22: 316–318

Paulus H (1993) Biosynthesis of the aspartate family of amino acids. In: Sonenshein AL, Hoch JA & Losick R (Eds) *Bacillus* and other Gram-positive bacteria: biochemistry, physiology, and molecular genetics (pp. 237–267). American Society for Microbiology, Washington, DC

Piggot PJ & Hoch JA (1985) Revised genetic linkage map of *Bacillus subtilis*. Microbiol. Rev. 49: 158–179

Saint-Girons I, Parsot C, Zakin MM, Barzu O & Cohen GN (1988) Methionine biosynthesis in *Enterobacteriacea*: Biochemical, regulatory, and evolutionary aspects. Critical Rev. Biochem. 23: S1–S42

Scislowski PWD, Bremer J, Thienen WIAD & Davis EJ (1989) Heart mitochondria metabolize 3-methylthiopropionate to CO_2 and methanethiol. Arch. Biochem. Biophy. 273: 602–605

Scislowski PWD & Pickard K (1993) Methionine transamination-metabolic function and subcellular comparmentation. Molec. Cell. Biochem. 129: 39–45

Sharpe ME, Law BA & Phillips BA (1977) Methanethiol production by coryneform bacteria: Strains from dairy and human skin sources and *Brevibacterium linens*. J. Gen. Microbiol. 101: 345–349

Shipston N & Bunch AW (1989) The physiology of L-methionine catabolism to the secondary metabolite ethylene by *Escherichia coli*. J. Gen. Microbiol. 135: 1489–1497

Smacchi E & Gobbetti M (1998) Purification and characterization of cystathionine γ-lyase from *Lactobacillus fermentum* DT41. FEMS Microbiol. Lett. 166: 197–202

Smid EJ, Driessen JM & Konings WN (1989) Mechanism and energetics of dipeptide transport in membrane vesicles of *Lactococcus lactis*. J. Bacteriol. 171: 292–298

Soda K (1987) Microbial sulfur amino acids: an overview. Methods Enzymol. 143: 453–459

Soda K, Tanaka H & Esaki N (1983) Mulitfunctional biocatalysis: methionine γ-lyase. TIBS 8: 214–217

Swaisgood H (1992) Chemistry of the casiens. In: Fox PF (Ed) Advanced Dairy Chemistry: Proteins, Vol. 1 (pp 63–110). Elsevier Applied Science, New York

Tanaka H, Esaki N & Soda K (1985) A versatile bacterial enzyme: L-methionine γ-lyase. Enzyme Microb. Technol. 7: 530–537

Tanaka H, Esaki N & Soda K (1977) Properties of L-methionine γ-lyase from *Pseudomonas ovalis*. Biochemistry 16: 100–106

Tynkkynen S, Buist G, Kunji W, Kok J, Poolman B, Venema G & Haandrikman A (1993) Genetic and biochemical characterization of the oligopeptide transport system of *Lactococcus lactis*. J. Bacteriol. 175: 7523–7532

Urbach G (1995) Contribution of lactic acid bacteria to flavour compound formation in dairy products. Int. Dairy J. 5: 877–903

Urbach G (1993) Relations between cheese flavour and chemical composition. Int. Dairy J. 3: 389–422

Urbanowski ML & Stauffer GV (1987) Regulation of the *metR* gene of *Salmonella typhimurium*. J. Bacteriol. 169: 5841–5844

Urbanowski ML, Stauffer LT, Plamann LS & Stauffer GV (1987) A new methionine locus, *metR*, that encodes a *trans*-acting protein required for activation of *metE* and *metH* in *Escherichia coli* and *Salmonella typhimurium*. J. Bacteriol. 169: 1391–1397

Urbanowski ML & Stauffer GV (1989) Genetic and biochemical analysis of the MetR activator-binding site in the *metE metR* control region of *Salmonella typhimurium*. J. Bacteriol. 171: 5620–5629

Uren JR (1987) Cystathionine β-lyase from *Escherichia coli*. Methods Enzymol. 143: 483–485

van den Bosch S, van't Land E & Stoffelsma J (1982) US patent 4332829

Vandeweghe P & Reineccius GA (1990) Comparison of flavor isolation techniques applied to Cheddar cheese. J. Agric. Food Chem. 38: 1549–1552

Weimer B, Dias B, Ummadi M, Broadbent J, Brennand C, Jaegi J, Johnson M, Milani F, Steele J & Sisson DV (1997) Influence of NaCl and pH on intracellular enzymes that influence Cheddar cheese ripening. Lait 77: 383–398

Wijesundera C & Urbach G (1993) Flavor of Cheddar cheese. Final report to the Dairy Research and Development Corporation, Project CSt66

Weiderholt KM & Steele JL (1994) Glutathione accumulation in lactococci. J. Dairy Sci. 77: 1183–1188

Wu W-F, Urbanowski ML & Stauffer GV (1992) Role of the MetR regulatory system in vitamin B_{12}-mediated repression of the *Salmonella typhimurium metE* gene. J. Bacteriol. 174: 4833–4837

Yamamagata S, D'Andrea RJ, Fujisaki S, Isaji M & Nakamura K (1993) Cloning and bacterial expression of the CYS3 gene encoding cystathionine γ-lyase of *Saccharomyces cerevisiae* and the physicochemical and enzymatic properties of the protein. J. Bacteriol. 175: 4800–4808

Yang WT & Min DB (1994) Dynamic headspace analyses of volatile compounds of Cheddar and Swiss cheese during ripening. J. Food Sci. 59: 1309–1312

Yocum RR, Perkins JB, Howitt CL & Pero J (1996) Cloning and characterization of the *metE* gene encoding S-adenosylmethionine synthetase from *Bacillus subtilis*. J. Bacteriol. 178: 4604–4610

Yvon M, Thirouin S, Rijnen L, Fromentier D & Gripon JC (1997) An aminotransferase from *Lactococcus lactis* initiates conversion of amino acids to cheese flavor compounds. Appl. Environ. Microbiol. 63: 414–419

Zdych E, Peist R, Reidl J & Boos W (1995) MalY of *Escherichia coli* is an enzyme with the activity of a βC-S lyase (cystathionase). J. Bacteriol. 177: 5035-5039

APPLICATIONS

Antonie van Leeuwenhoek **76:** 265–278, 1999.

Analysis of the intestinal microflora: a renaissance

Gerald W. Tannock
Department of Microbiology, University of Otago, P.O. Box 56, Dunedin, New Zealand

Key words: bacteria, intestine, methodology, microbial ecology

Abstract

The ability of microbial ecologists to analyse the composition of complex bacterial communities has been greatly enhanced by the application of molecular methodologies. The use of these techniques should enable an accurate record of the identity and population dynamics of the inhabitants of the intestinal tract to be obtained, and should promote an improved comprehension of the relationship between the microflora and the human host. This, in turn, will lead to a new concept of the intestinal microflora of humans.

Introduction

Research concerning the intestinal microflora, especially of humans, is in the early stages of a 'renaissance'. The historical event referred to as the 'Renaissance' describes the period, beginning in Europe in the fourteenth century, during which there was a revival of culture and learning, including scientific scholarship (Ferguson 1967). It spans the period between the waning of the Middle Ages (feudal and ecclesiastical) and what might be considered the Modern Age (urban and secular). Long languishing in the equivalent of the Middle Ages, intestinal microflora research seems set to develop new concepts of the large and diverse collection of microbes harboured in the distal intestinal tract (ileum, colon) by healthy humans (Tannock 1995a). At last, we may be entering the Modern Age. This has considerable significance for all those who work with, or are interested in, the Lactic Acid Bacteria (Tannock 1999).

Our picture of the intestinal microflora

A middle aged view

Until about the 1960s, descriptions of the intestinal microflora were relatively simple. Clostridia, lactobacilli, enterococci and *Escherichia coli* were commonly considered the predominant microflora detected in faecal samples, reflecting the relative ease of culture of these bacteria (Rettger et al. 1935). Rumen Microbiologists, thanks to the innovative techniques derived by Hungate, demonstrated the importance of obligately anaerobic bacteria and other microbes in the rumen fermentation, and hence the reliance of the ruminant host on microbial metabolic products for nutritional well-being (Hungate 1966). Holdeman and Moore modified the roll tube techniques of Hungate for use in investigations of anaerobic infections of humans (Holdeman & Moore 1972). It quickly became obvious that the pioneering work of French researchers, such as Veillon and Prevot, on anaerobic infections was accurate and that certain obligate anaerobes were essential components in the pathogenesis of these diseases (Finegold 1977). The derivation of selective media for anaerobes commonly implicated in anaerobic infections, together with the use of anaerobic gloveboxes, accelerated the acquisition of knowledge concerning the reservoir of these anaerobes: the normal microflora (Aranki & Freter 1972; Summanen et al. 1993).

Two major studies were carried out in the United States of America during the 1970s in relation to the influence of environmental factors, particularly diet, on the intestinal microflora and the occurrence of colon cancer (Finegold & Sutter 1978; Moore & Holdeman 1978). While shedding little light on host-microflora relationships, these studies gathered a wealth of information concerning the composition of the intestinal microflora of humans. Using appropriate anaerobic techniques, it was possible to cultivate up to 88% of the total microscopic count in faeces

(Moore & Holdeman 1974). Some bacterial populations were detected at levels of 10^{10} per gram (wet weight) and Moore and Holdeman (1974) estimated that as many as 400 species might be detectable in human faeces. For any given individual, however, 30 to 40 bacterial species constituted 99% of the faecal microflora (Drasar & Barrow 1985).

Space and the intestinal microflora

Where were these microbes living? Elegant studies by Savage showed that there was extensive association of microbial cells with the epithelial surfaces lining the gastrointestinal tract of rats and mice (Savage 1983). The distribution of microbes in the gastrointestinal tract of these animals is unlike that of humans. In mice and rats, a large population of lactobacilli inhabits the gastric region. *Lactobacillus* cells adhere directly (a mucus layer is absent) to the stratified, squamous epithelium of the forestomach where they multiply to form layers of *Lactobacillus* cells. Lactobacilli shed from these layers inoculate the digesta such that these bacteria are detectable throughout the length of the small bowel. Large populations of lactobacilli are also present in the large intestine, although they are outnumbered by obligate anaerobes in this location (Savage 1977). The adherence of lactobacilli to forestomach epithelial cells is specific: they do not adhere to enterocytes or red blood cells obtained from the same animal in which adherence of lactobacilli to forestomach cells has been observed (Tannock 1997).

Segmented, filamentous bacteria (related to the clostridia in terms of 16S rDNA sequence) attach to enterocytes in the distal ileum, particularly in the vicinity of Peyer's Patches (Klaasen et al. 1992; Snel 1997). The microbes have not been cultivated, but ex-germfree mice monoassociated with them have been obtained (Klaasen et al. 1991). The method of attachment involves the insertion of the base of the filament into the murine enterocyte. Actin rearrangements within the enterocyte form a socket by which the filament becomes permanently attached to the mucosal surface (Jepson et al. 1993). Segmented, filamentous microbes have also been detected in the intestinal tracts of chickens where infestation of enterocytes is so heavy that the avian cells are vacuolated, and the underlying tissue oedematous and inflamed, raising the possibility that these bacteria have pathogenic potential (Goodwin et al. 1991; Hill et al. 1992).

Histological sections and scanning electron micrographs of the surface of the colonic mucosa of rats and mice demonstrate large numbers of bacterial cells within the mucus layer overlaying the epithelial surface. Fusiform- and spiral-shaped cells predominate in this layer (Savage et al. 1968). A distribution of bacterial cells of different morphology can be seen in the contents of the intestinal lumen compared to the mucus layer (Tannock 1987).

Epithelium-associated bacteria are also seen in the porcine gastrointestinal tract (lactobacilli colonise the surface of the pars oesophagea) and in the chicken crop (lactobacilli) (Fuller & Brooker 1974; Barrow et al. 1980). It is a consequence of the particular type of epithelium present in these sites. The human gastrointestinal tract is lined by a columnar epithelium and therefore does not provide appropriate sites for the adherence of lactobacilli. Spiral-shaped microbial cells have been demonstrated convincingly in association with human colonic mucosa (photographic evidence has been published), but the evidence for significant association by other bacteria is inconclusive (Croucher et al. 1983; Tannock 1997). There is an absence of evidence that the association of intestinal microflora with mucosal surfaces is beneficial to the host; indeed the converse is probably true. Microbial cells in close proximity to epithelial cells compete with the host for nutrients, should adherence occur in the proximal digestive tract, and some bacterial metabolites may be detrimental to enterocyte structure and function. The persistence of a 'colonic type' microflora in the small bowel of patients suffering from contaminated small bowel syndrome results in malabsorption of lipids, amino acids, and carbohydrates, produces vitamin B12 deficiency, and upsets water and electrolyte transport in enterocytes (Gracey 1983).

Mucus is secreted by epithelium-located (goblet) cells and mucosal glands in order to protect the epithelium underlying the mucus blanket from damage. The mucus flows from the cells of origin, traps particulate material (including bacterial cells), and carries it away from the epithelial surface. It is notable that germfree mice have a mucus layer that is thinner, less dense, and has a lower mucin content than that of conventional animals, probably because of the absence of potentially harmful microbial cells (Szentkuti et al. 1990).

Much overlooked in the case of monogastric animals, the surfaces of plant fragments passing undigested to the large bowel provide an important habitat for intestinal microbes. Faecal preparations examined microscopically show that, as in the rumen, microbial cells adhere to the plant material which is presumably degraded by processes catalysed by bacterial enzymes (Macfarlane & Macfarlane 1995). The extent to which co-aggregation of microbial cells (attachment of cells, belonging to different bacterial species, to one another) occurs in the intestinal tract has not been investigated, although it is crucial to the persistence of some bacteria inhabiting the oral cavity (Kolenbrander 1991).

Our picture of the intestinal microflora is therefore of a microbial soup contained within the distal ileum and colon of healthy humans. Bacterial cells associate extensively with food particles, probably also with each other, and may be trapped in mucus associated with mucosal surfaces. Microbial cells are present in minimal numbers in the stomach and upper to mid-small intestinal contents. They are excluded from colonising these regions because of the acidic pH (stomach) and relatively rapid flow of the digesta (duodenum and jejunum) (Drasar & Hill 1974).

Uncertainty and the intestinal microflora

There are some things that we have never been very clear about. Most investigations of the intestinal microflora of humans have been studies of the microbial content of faeces. Does the study of faecal samples really provide accurate information concerning the intestinal microflora? Based on Savage's studies of the murine gastrointestinal tract, the answer is 'no'. One could not deduce that significant epithelial associations occur in the proximal digestive tract of mice from the microbiological assay of murine faeces (Savage 1977). With respect to humans, the answer is a qualified 'yes' since the microflora is essentially confined to the distal intestinal tract and epithelium-associated populations are apparently absent. Moore and Holdeman (1978) have demonstrated that the composition of the faecal microflora reflects that of the colon. There is a qualification, however, in that the microflora is more metabolically active in the proximal colon compared to the distal colon (Cummings & Macfarlane 1991). This is a reflection of the relative availability of fermentable substrates in the two sites and is of significance with respect to the analysis of the effects of non-digestible food additives (oligosaccharides, resistant starch) on the intestinal ecosystem. Biochemical changes occurring in the contents of the proximal colon may not be detectable in the distal colon or, therefore, in faeces.

Experimental animal studies have been indispensable in formulating basic concepts of the microbial ecology of the intestinal tract of animals. Farm animals have played a similar role, and have emphasised the importance of acquiring fundamental knowledge of the gastrointestinal microflora in order to improve animal husbandry and to promote public health. The development of complex cultures of bacteria from the avian intestine that can increase resistance to colonisation of one-day-old chicks by *Salmonella* species, provides an excellent example (Fuller 1999). The marked differences in the distribution and composition of the microflora in different species of animal must, however, always be taken into account and, ultimately, research aimed at human welfare must always be conducted using human subjects. The usefulness of experimental animal studies depends on the use of an appropriate animal model system which will answer specific questions posed by the researcher. For example, the question 'Do intestinal lactobacilli influence the composition of the lymphocyte population in mesenteric lymph nodes?' can be answered by recourse to the use of an animal model. The question 'Do intestinal lactobacilli promote resistance to infection by *Campylobacter jejuni*?' would have to be tested in humans because a realistic animal model of *Campylobacter* enteritis and of appropriate *Lactobacillus* status, to the author's knowledge, is not available.

Attachment of bacterial cells to cultured mammalian cells has been useful in studying the pathogenesis of intestinal pathogens (Salyers & Whitt 1994). The pathogens concerned are known to adhere to mucosal surfaces *in vivo*, so the mechanisms by which they achieve this can be characterised *in vitro*. Extended to members of the intestinal microflora, *In vitro* adherence assays to columnar cells are inconsequential because they are rarely supported by relevant in vivo observations. What happens in the test tube (agar plate) may be an artefact. For example, it has been demonstrated that susceptibility of lactobacilli or enterococci to the bacteriocin known as lactacin F, assayed on agar plates, does not necessarily predict an antibacterial effect in vivo (Klaenhammer, Allison, Loach and Tannock, unpublished observations).

Microflora-associated characteristics

The presence of enormous numbers of microbial cells, presumably most of which are metabolically active, in the intestinal tract has a major impact on the characteristics of the host. Many features of the biochemistry, physiology and immunology of the animal host are, in fact, microflora-associated characteristics (Luckey 1963; Gordon & Pesti 1971). Even the genetic determinants for L-fucosylation of glycoconjugates associated with the epithelium of the small intestine are up-regulated by the presence of the intestinal microflora (Bry et al. 1996). The impact of the microflora on the host was mostly demonstrated during the 1960s and 1970s when, following the theoretical approach outlined by Pasteur (1885), comparisons of germfree and conventional animals (mostly rats, chickens and mice) were made. The only difference between the groups of animals used in the experiments was the presence (conventional) or absence (germfree) of microbial associates. All other factors being equal between the two groups of animals, any differences observed in the characteristics of the experimental subjects must have been due to microbial activities or lack thereof.

The microflora-associated characteristics that have been described in relation to comparisons of germfree and conventional animals result from the presence and activities of the microflora as a whole. Intestinal microbes, as is found in other ecosystems, doubtless form consortia by which substrates are degraded in sequential processes by different microbial species. Unravelling these sequences of events using microbial isolates from conventional animals has not been attempted to any extent in monogastric ecosystems, except for the degradation of mucins where novel bacterial enzymes have been described (Roberton & Corfield 1999). Other microflora-associated characteristics are likely to result from equally complex interactions.

Monoassociated ex-germfree animals have figured prominently in attempts to define the influences of specific microbes on the animal host. Ex-germfree mice monoassociated with the segmented, filamentous ileal bacterium were, for example, tested for the impact of these microbes on the immune system of the animals. The monoassociated mice, in comparison to germfree animals, had more lymphoid cells in the lamina propria of the ileum and caecum, had more IgA-secreting cells in the intestinal mucosa, higher titres of IgA in intestinal secretions and serum, and an enhanced lymphocyte (mesenteric lymph nodes) proliferative response when induced by concanavalin A (Klaasen et al. 1993).

While ex-germfree animals harbouring a single microbial type have the potential to provide interesting information about host-microflora interactions, the biological significance of the results may be open to question. The microbial population in such animals is generally much higher than that detected for the same organism in a conventional animal. The distribution of the microbes in the gastrointestinal tract of monoassociated animals is also commonly different to that seen in conventional subjects. These differences are due to the absence of the regulatory mechanisms, engendered by the complete microflora, that define population size and distribution in the conventional ecosystem. It would be preferable to determine the influences of specific microbes on host parameters using animals that harboured a complex microflora from which a single bacterial group of interest was absent. The modulatory effects of the remainder of the microflora on the specific microbial population and its activities would then provide information applicable to the conventional animal.

Lactobacillus-free mice have been derived from conventional animals to define the influences of these bacteria on host characteristics (Tannock et al. 1988). Lactobacilli are absent from the gastrointestinal microflora of these animals which is otherwise functionally equivalent, on the basis of twenty-six microflora-associated characteristics, to that of conventional mice. A large population of lactobacilli is present throughout the gastrointestinal tract of conventional mice. If these bacteria have an impact on host characteristics, it should be observable by comparison of mice harbouring a gastrointestinal microflora from which lactobacilli are absent, and otherwise identical animals that have been intentionally colonised with lactobacilli. If all other factors were maintained constant between the groups, *Lactobacillus*-associated characteristics of the animals could be detected. The *Lactobacillus*-free mice that have been derived are maintained in isolators by gnotobiotic technology in order to maintain their *Lactobacillus*-free status, which also permits all of the experiments to be conducted under microbiologically-constant conditions. Biochemical differences have been observed between *Lactobacillus*-free and *Lactobacillus*-colonised mice, but of course some of the microflora-associated characteristics detected by germfree/conventional comparisons are not due to the presence or absence of lactobacilli (Tannock 1995b).

Methodology: this is where the renaissance begins

The analysis of the intestinal microflora has relied, almost completely, on the use of traditional bacteriological methods of culture, microscopy and identification (O'Sullivan 1999). Selective bacteriological culture media are essential for accurate analysis of the microflora because they enable enumeration of specific populations to be made (Summanen et al. 1993). Unfortunately, few culture media used in the analysis of the microflora are absolutely selective and misinterpretations by the novice are easily made. Not all of the species comprising a population may be able to proliferate with equal ease on the selective medium. This introduces bias to the results. Even more significant in preventing the acquisition of an unbiased view of the composition of the intestinal microflora is the inability to cultivate at least 12% of the cells seen in stained, faecal smears viewed microscopically. But the renaissance has begun: analytical methods based on 16S ribosomal RNA gene (16S rDNA) sequences will soon shed new light on the species composition of the intestinal microflora.

16S rDNA contains regions conserved across all bacterial species interspersed with regions (V1 to V9) in which the sequences are variable among bacterial types (Stackebrandt & Rainey 1995). Sometimes, the variable region sequences are highly species-specific. Comparison of 16S rDNA similarities can therefore be used in the identification of bacterial species and consequently, in the analysis of bacterial communities (Raskin et al. 1997). Universal primers can be used in polymerase chain reactions (PCR) to amplify 16S rDNA from bacterial cells in natural samples. The amplified 16S rDNA sequences, hopefully containing copies of the gene from all of the species represented in the sample, are cloned. The 16S rDNA clones are screened (some sequences will have been cloned more than once), and representative clones are sequenced. Since 16S rDNA sequences are one of the cornerstones of bacterial taxonomy, alignment of the sequences with those stored in databanks permits the recognition of which species are represented in the habitat, and detects those that cannot be cultivated by conventional techniques (Amann et al. 1995).

Wilson and Blitchington (1996) analysed the composition of the microflora of human faecal samples using this approach. They concluded, somewhat reassuringly, that the bacterial species detected by non-selective culture, when anaerobic bacteriological methods were of a high quality, gave a good representation of the bacterial types present relative to that reflected by 16S rDNA sequence analyses. The main discrepancies between the two methods was in the detection of members of Gram-positive groups, probably clostridia and eubacteria.

Not so reassuring were the results of studies carried out by Franks and colleagues (1998). They have derived oligonucleotide probes, labelled with fluorescent dyes, that target 16S ribosomal RNA sequences. Although in theory an array of hundreds of probes could be used to enumerate bacterial species in a faecal sample by fluorescent *in situ* hybridisation (FISH), practically, a small collection of oligonucleotide probes that recognise large phylogenetic groups of bacteria has been used so far in studies aimed at monitoring the composition of the microflora. The system is automated: digital images obtained by epifluorescence microscopy are analysed by computer. When used to analyse human faecal microflora, the probes were estimated to detect about two-thirds of the microbial collection. Currently, it appears that the genus *Bacteroides* and members of the *Clostridium coccoides-Eubacterium rectale* group constitute about half of the faecal microflora. In addition to obtaining important observations of this type, the use of oligonucleotide probes might be particularly useful in investigating the spatial relationships of microbes in the intestinal ecosystem. Technical difficulties that could influence accuracy of the results exist. Oligonucleotide probes must reach their target sequence, which is inside the bacterial cell, by passing through the cell wall. This is more easily achieved with some species compared to others. Gram-positive bacterial cells such as those of lactobacilli, for example, are more difficult to permeablise than are others (Welling et al. 1997). Currently, the lowest level of detection (microscopy) using in situ hybridization is 10^6 bacterial cells per gram (Welling et al. 1997).

16S rDNA sequences can also be used as the basis of analysis of bacterial populations using electrophoretic gels. Sequences of no more than 400 bp are suitable for these methods, so fragments of the 16S gene are amplified by PCR in which one of the primers has a GC-rich 5′ end (GC clamp). 16S Fragments are amplified from the microbial community in the sample and the 16S molecular species within the resulting mixture are separated by temperature gradient gel electrophoresis (TGGE) or denaturing gradient gel electrophoresis (DGGE). The double-stranded 16S fragments migrate through a polyacrylamide gel until each kind of fragment is partially denatured by the prevailing temperature or chemical conditions. The fragments do not completely denature because of the GC clamp, and migration is radically slowed when partial denaturation occurs. Because of the variation in the 16S sequences of different bacterial species, thermal and chemical stability is also different, therefore different 16S 'species' can be separated by these electrophoretic methods. A profile of 16S sequences amplified from the sample is thus obtained. TGGE and DGGE can be used to monitor microbial communities over time. DNA fragments can be cut from the gel and re-amplified for sequencing and hence identification of the microbe that the fragment represents can be made. PCR products can be randomly cloned and used as standards in gels, so that at least commonly occurring species can be recognised, and so that different gels can be compared (Muyzer & Smalla 1998; Zoentendal et al. 1998). Additionally, 16S rDNA can be amplified from RNA extracts (reverse transcription and PCR) prepared from samples (Felske et al. 1997). The extracted RNA is essentially ribosomal RNA so subsequent electrophoretic analysis reveals, in relative terms, the 16S sequences of bacteria that were most metabolically active in the ecosystem (more ribosomes).

Quantification of bacterial populations using PCR should enable a further move from reliance on culture-based analysis of the intestinal microflora to be made. Competitive PCR (cPCR) involves the use of an internal DNA standard in the PCR mixture. The internal standard contains the same primer binding sites as the target DNA. The standard and target DNAs compete for reaction reagents to produce PCR products of different sizes which can be separated in an agarose gel. Amounts of target DNA can be calculated by amplification with known amounts of internal standard. The resulting log ratio of intensities of PCR products is compared to a standard curve derived from serial dilutions of target DNA amplified with known amounts of internal standard. As pointed out by Reilly and Attwood (1998), testing primers for specificity and constructing the internal standard are the most time-consuming steps in the development of the method. The results obtained with the technique are considered good, however, showing acceptable levels of variation between replicate PCR reactions carried out on the same sample.

PCR amplifies target DNA exponentially. Instruments, such as the PE-Applied Biosystems detection systems, make fully-automated, real-time detection of PCR products possible. A reporter probe labelled with two fluorescent dyes is used to measure the accumulation of specific PCR product. One of the dyes acts as a 'reporter', the other as a 'quencher'. When both dyes are within a defined distance of each other (at either end of the probe), energy is transferred from the reporter to the quencher. The probe hybridises to the PCR target, but is cleaved by the 5′ nucleolytic activity of *Taq* DNA polymerase so that the reporter and quencher are separated. Cleavage of the probe into short fragments destabilises the portion of the probe still hybridised to the target sequence so that it dissociates back into solution. The resulting separation of the two chromophores due to cleavage of the probe measurably increases the fluorescent signal emitted by the reporter dye. The fluorescent signal is measured as the PCR reaction proceeds. The parameter C_T (threshold cycle) is defined as the fractional cycle number at which the reporter fluorescence generated by cleavage of the probe passes a fixed threshold level above baseline. The higher the starting copy number of nucleic acid target, the fewer the amplification cycles needed before a significant increase in fluorescence is observed and the lower the C_T value. Measuring the starting copy number in an unknown sample is simply done by determining the C_T and reading the corresponding starting copy number of the target from a standard curve of C_T versus copy number (Bassam et al. 1996). This system should lend itself well as a tool for the quantification of intestinal populations since it combines the specificity of fluorescent oligonucleotide probes with the sensitivity of PCR.

PCR-based methods probably hold the key to accurate analysis of the intestinal microflora in the near future. They have the advantage of being relatively easily accessible to many researchers and do not require high levels of technological input to obtain significant data. They will not, however, solve all of the problems. Amplification of 16S rDNA requires that the microbial cells in the sample first be lysed for the extraction of DNA. There is a vast difference in the susceptibility of the cells of different microbial species to lytic procedures. As only one lytic method is usually applied to the sample, it is unlikely that template DNA for the polymerase chain reaction is extracted with equal success from all of the species represented in the habitat. Additionally, polymerase chain reactions are known to amplify rDNA molecules from mixed populations with differing efficiency (Reysenbach et al. 1992). These factors will bias the community profile obtained.

Analysis of the intestinal microflora will be greatly aided by the use of oligonucleotide probes and PCR-based methods of detection and quantification. These methods discriminate bacterial groups and species but, with rare exceptions, will probably not be useful in the analysis of bacterial populations in terms of strains. Molecular typing (genetic fingerprinting) of bacterial isolates to distinguish between strains has been long established as a valuable epidemiological tool in the investigation of infectious disease outbreaks (Swaminathan & Matar 1993). Similarly, molecular typing of *Lactobacillus*, *Bifidobacterium*, and enterobacterial isolates of faecal origin has proven useful in studying the microbial ecology of the intestine of humans and other animals (McCartney et al. 1996; Kimura et al. 1997). Ribotyping and pulsed field gel electrophoresis of digests of genomic DNA are the best tested methods in terms of molecular typing of intestinal bacteria. Perhaps the outstanding observation from this work is that each human subject harbours at least one strain of *Lactobacillus* or *Bifidobacterium* that is unique to that individual (Kimura et al. 1997). Molecular typing of enterobacterial and *Lactobacillus* isolates also detects perturbations and transitions occurring in the intestinal microflora (Bateup et al. 1998; McBurney & Tannock, unpublished observations). A limitation of this method of analysis is that only the numerically predominant strains in the bacterial population can be typed because isolates for typing are obtained by picking discrete colonies from dilution plates.

Despite some inherent technical difficulties, an impressive line-up of analytical methods (microflora-associated characteristics, DNA probes, PCR methods, culture, molecular typing) is now at the fingertips of researchers of the intestinal microflora. A combination of these analytical approaches in studies of the intestinal microflora should rapidly advance knowledge of this complex microbial community. We eagerly await the Modern Age of intestinal microflora research.

The origin and fate of the intestinal microflora: a question of uniqueness

The real intestinal microflora

Analysis of bacterial populations by differentiation of strains using molecular typing methods has revealed a hitherto unrecognised uniqueness in the composition of the intestinal microflora (Kimura et al. 1997). At least among the lactobacilli and bifidobacteria, strains that persist for long periods as members of the intestinal microflora of an individual can be recognised. These strains are not detected in other human subjects and must therefore be highly adapted to the environment of their unique host. Studies carried out at Rockefeller University in the 1960s demonstrated that certain bacterial populations (e.g. lactobacilli) inhabiting the gastrointestinal tract of rodents were acquired soon after birth and persisted throughout the life of the animal (Dubos et al. 1965; Savage 1977). All members of the rodent colony harboured these bacteria in similar numbers. Other bacterial populations were present in only some colonies of animals of the same species (e.g. yeasts, *E. coli*). These observations led Rene Dubos and colleagues (Dubos et al. 1965) to postulate that microbial populations detected in the digestive tract could be divided into three categories:

- the 'autochthonous microbiota' – populations of microbes that were present at high population level, usually throughout the life of the animal. These particular microbes were found in all members of the colony (e.g. lactobacilli).
- the 'normal microbiota' – populations of microbes frequently present in the digestive tract, but of variable population size and absent from some colonies (e.g. *Escherichia coli*).
- 'true pathogens' – accidentally acquired and capable of persisting in the tissues.

The proposed terminology did not become current, mainly because it was difficult to assign more than a few bacterial groups to one category or another, but also because the term 'normal' microflora is commonly used in medical microbiology to denote the complete collection of microbes inhabiting the healthy human body. It may be time to revisit this concept of categories of microbes within the intestinal microflora, however, particularly in the case of humans. It may not be so relevant to discussions of the microflora of experimental or farm animals because they are genetically similar through 'herd improvement' (breeding) practices and because of the maintenance of all members of a given group under similar environmental conditions.

Recent studies, using genetic fingerprinting methods such as ribotyping and pulsed field gel electrophoresis (PFGE) of DNA digests prepared from bacterial isolates, have permitted the analysis of the human faecal microflora at the level of strains. Examination of bifidobacterial and *Lactobacillus* populations in monthly faecal samples collected over a 12-month period has shown that there can be marked variation in the complexity and stability of these bacterial populations between human subjects (McCartney et al. 1996). It is clear from the results of these studies that individual humans harbour unique microfloras when considered at the level of bacterial strains. Bifidobacterial populations were consistently large in all samples that were examined and strains that persisted in the faeces for at least 12 months could be recognised. *Lactobacillus* populations fluctuated in size in the case of most subjects but unique, persisting strains could often be detected. Applying Dubos's terminology in a slightly different way, these persisting strains could be described as 'autochthonous' ('inhabiting a place or region from earliest times') to the intestinal tract of the particular individual.

Analysis of lactobacillus and bifidobacterial populations has also identified the presence of strains that are detected consistently, but unpredictably, in human faecal samples (McCartney et al. 1996). It is probable that these subpopulations vary in size over time so that their detection is intermittent among the numerically predominant strains. They could be termed 'opportunist' strains because their population density may be determined by changes in the intestinal milieu produced by endogenous or exogenous factors. It would be interesting to test this hypothesis by supplementing the diet of human subjects with non-digestible carbohydrates known to be substrates for energy production by specific strains of bacteria residing in the subjects' intestinal tract.

Strains of bacteria incorporated in 'probiotic' products can usually only be detected in the faeces of human subjects while they continue to consume the product (Goldin et al. 1992). Therefore their presence in the intestinal tract is dependent on environmental circumstances (purchase and consumption of the probiotic). They might best be considered 'allochthonous' ('found in a place other than where they were formed') strains in the intestinal tract. They may never become a true component of the intestinal ecosystem of the majority of consumers. If the origin of the probiotic strain was a human intestine, then it must once have been autochthonous or opportunist to an individual, but is allochthonous to all other humans. Instances of probiotic strains persisting in the faeces for more than a week after consumption has ceased have been reported (Goldin et al. 1992). These cases may be evidence that an allochthonous strain can become 'naturalised' and is presumably now 'opportunist'. The reason for introducing this terminology is an attempt to rationalise the effects of probiotic administration on the consumer, relative to that of the strains already resident in the intestinal tract. 'Autochthonous' strains of bacteria must be highly compatible with the intestinal environment of their host since they persist as members of the particular bacterial population. They are tolerated by the immune system of the host to the extent that they are not eliminated from the intestinal ecosystem. Their interactions with the host, and hence their influence on host characteristics, may be entirely different to those exerted by 'allochthonous' strains. Probiotic strains, it can be argued, do not belong to the intestinal ecosystem of the consumer at all. At best they will be 'opportunist' in their relationship with the consumer, but usually 'allochthonous' and therefore will affect the consumer differently to the 'autochthonous' strains. If it is accepted that autochthonous and opportunist categories of intestinal bacteria occur, the search for a 'universal coloniser' (a strain that will persist in the intestine of all consumers), or indeed 'a colonising strain' amongst the lactobacilli and bifidobacteria could cease.

In support of the hypothesis that animals react differently to exposure to autochthonous relative to allochthonous microbes, it can be noted that Berg and Savage (Berg 1983) observed that *Lactobacillus* and *Bacteroides* species isolated from the murine intestine (autochthonous) elicited a relatively weak immune response when administered parenterally to mice. Strains from humans (allochthonous) produced a stronger immune response in the mice (Berg 1983). Similarly, germfree mice monoassociated with autochthonous lactobacilli and bacteroides for 1 to 5 weeks did not show evidence of an immune response to these bacteria. Animals associated with allochthonous lactobacilli and bacteroides, however, exhibited a substantial response (Berg 1983). The enhanced phagocytosis and adjuvant effects produced by the administration of some probiotic strains to rodents and humans may reflect the allochthonous nature of the bacteria (Kato et al. 1981; Perdigon & Alvarez 1992; Schiffrin et al. 1995). It would be interesting to study the effect of long term administration of a probiotic that affects the innate or inducible arms of the immune system. Would the stimulatory effect continue to be exerted in individuals exposed to the probiotic for a long period?

Lactobacilli can be detected in the faeces of the majority of human subjects (Finegold et al. 1983). Yet, the consumption of lactobacillus-containing products has been reported to influence the activity of bacterial enzymes in faecal samples (Goldin et al. 1980). This must mean that the probiotic strain affects the intestinal environment in a different way to that of the autochthonous strains. These results argue that probiotics have a different relationship with the host compared to that of autochthonous strains.

Mutability and stability

In a recent study conducted in collaboration with the New Zealand Dairy Research Institute (Tannock and Munro, unpublished observations), the faecal microflora of ten healthy human subjects was monitored over a six month period to establish the degree of stability or variation in bacterial populations, bacterial enzyme activities, and other microflora-associated characteristics (control period). The composition of the *Lactobacillus* population, where detectable, was analysed by differentiation of strains using pulsed field gel electrophoresis of DNA digests. The subjects then consumed a milk-based product containing viable cells of *Lactobacillus rhamnosus* strain DR20 for six months (test period). A post-test period of three months, during which the probiotic was not administered, followed. The faecal microflora was monitored at monthly intervals during each period. Strain DR20 was detected consistently in faecal samples from six subjects throughout the test period, occasionally in the case of three subjects, and was not detected as a numerically predominant strain in one subject. All of the subjects had detectable numbers of lactobacilli in at least one of their samples during the control period. In all but one of the subjects, strain DR20 could not be detected one month after consumption of the product had ceased. The strain was detected for two months during the post-test period in this exceptional subject. The subjects in whose faeces strain DR20 was detected only occasionally, or not at all, always had a recognisable, autochthonous strain during the control period whose presence may have influenced the fate of DR20 in their intestinal tracts. The intestinal ecosystem therefore has a considerable homeostatic capability. The introduction of a relatively large number of allochthonous bacterial cells will result in the detection of these bacteria in the faeces but, in general, they do not persist in the ecosystem once ingestion has ceased. The composition of the ecosystem, at least in the case of the *Lactobacillus* population, quickly reverts to the 'normal' situation in which autochthonous strains predominate.

Open and closed windows

When does the human host become colonised by these autochthonous strains? Enterobacteria and bifidobacteria detected in infant faeces within a few days of birth can sometimes be demonstrated to be of maternal origin (Tannock et al. 1990). But the acquisition of the intestinal microflora, in terms of strains, has not been investigated thoroughly and a long term, prospective study is indicated. There may be discrete periods during which autochthonous strains become established. These periods may be dictated by the physiological factors in the intestinal tract as much as by exposure of the human to the appropriate strain. Distinct time periods (windows) during which oral streptococci can colonise the human have been reported. *Streptococcus salivarius* colonises during the first week of life, *Streptococcus sanguis* colonises when infants are about 9 months of age, whereas mutans streptococci (particularly *Streptococcus mutans* and *Streptococcus sobrinus*) are acquired around 26 months of age (Tagg et al. 1983; Caufield 1997). This latter 'window' probably relates to the emergence of the first and second molars which provide a large surface area, and particularly fissures, for colonisation. Virgin molars, it is speculated, provide a finite period during which mutans streptococci can colonise the oral cavity. After this period, other members of the dental plaque which has accumulated on the teeth compete with the mutans streptococci and prevent their establishment. A second window for mutans streptococci may open when children are six years of age when the first permanent molars emerge (Caufield 1997). The recognition of 'colonisation windows' in relation to the intestinal tract would provide valuable data upon which to further develop concepts of the microbial ecology of this ecosystem.

Environmental impact reports: understanding your ecosystem

The night before, the morning after

The prevalence of *Lactobacillus* and bifidobacterial species in the intestinal tract of humans is not known accurately. Comprehensive studies of the species composing the human intestinal microflora have seldom been attempted, and those that have were conducted prior to the introduction of genotypic methods for the classification of bacteria (Ballongue 1993). Probably the best investigations available of *Lactobacillus* and *Bifidobacterium* species present in human faeces are those reported by Mitsuoka (1992) and by Biavati and colleagues (1984, 1986). Mitsuoka listed *Lactobacillus crispatus*, *Lactobacillus gasseri*, *Lactobacillus salivarius*, and *Lactobacillus reuteri* as the major species of the *Lactobacillus* microflora. *Lactobacillus johnsonii*, *Lactobacillus ruminis*, *Lactobacillus casei*, and *Lactobacillus brevis* were detected occasionally. Among the bifidobacteria, *Bifidobacterium longum* predominated in adult humans, while *Bifidobacterium bifidum* was detected occasionally. In contrast, *Bifidobacterium infantis* and *Bifidobacterium breve* were the most prevalent species in infant faeces, with *B.*

longum and *B. bifidum* detected occasionally. The identity of the isolates was determined by phenotypic characterisation, and determination of the mole% G+C of genomic DNA in the case of the lactobacilli. Biavati et al. (1984), using DNA-DNA reassociation as the primary method of identification, detected *Bifidobacterium infantis*, *B. breve*, and *B. longum* commonly in the faeces of human infants. *Bifidobacterium bifidum*, *Bifidobacterium pseudocatenulatum*, *Bifidobacterium dentium*, and *Bifidobacterium catenulatum* were isolated less frequently. In the case of samples from adults, *Bifidobacterium adolescentis*, *B. bifidum*, *B. catenulatum*, *B. longum*, and *B. pseudocatenulatum* were detected (Biavati et al. 1986). Differentiation between species, in this latter study, was limited to the use of polyacrylamide gel electrophoresis of soluble cellular proteins.

Probiotics contain species of bifidobacteria and/or lactobacilli that are intended to generate an impact on the intestinal microflora (Fuller 1989). It seems logical that the prevalence of the different *Lactobacillus* and bifidobacterial species in various countries should be known, preferably before the consumption of probiotics becomes even more widespread. As part of the 'environmental impact' report that should be completed for each probiotic that is consumed long term, an accurate assessment of the alterations, in terms of species, to the *Lactobacillus* or bifidobacterial microfloras should be included. Some probiotics have been retailed in particular countries for many years. Has the *Lactobacillus* or bifidobacterial microflora of the inhabitants of these countries altered dramatically since marketing of these products began? What changes have occurred in the proportions of *Lactobacillus* species detected in faeces since the introduction of probiotic products in general? What is the long term effect of any such change on the human population? Comprehensive studies of the species composition of the intestinal microflora will require the application of genotypic tests, rather than the subjective and unreliable results of phenotypic tests, to be applied. PCR-primers based on unique DNA sequences, not just that of 16S rRNA genes, will doubtless be of value in this work. Automated sequencers enable a high throughput of DNA samples to be achieved, so it may be feasible to rely on short, accurately sequenced PCR products (such as 16S-23S intergenic spacer regions; Gurtler & Stanisich 1996) to achieve rapid identification of species by comparison of percent similarities with type culture information held in databanks.

16S sequences: knowing everything but understanding nothing

The plethora of 16S rDNA and other gene sequences likely to be obtained from the application of 16S rDNA cloning and TGGE/DGGE techniques to the investigation of the intestinal ecosystem will threaten to bias studies towards microbial taxonomy, rather than microbial ecology. While knowledge of biodiversity is important, microbial ecology is concerned with investigating the activities of strains in an ecosystem. How strains differ in characteristics that are ecologically important, how these strains interact with other members of the microbial community, and how they interact with their host is of prime importance in microbial ecology. It is the ecology of the members of the intestinal microflora that makes studies of them relevant to human health. What was the precise mechanism, for example, by which the administration of 'Lactobacillus acidophilus' affected the intestinal microflora so that a decrease in β-glucuronidase activity was measured in the faeces of human subjects in the experiment conducted by Goldin and colleagues (Goldin et al. 1980)? How did *Bacteroides thetaiotaomicron* influence the expression of genetic determinants of glycoconjugate fucosylation in enterocytes as reported by Bry and colleagues (1996)? To understand the activities (the workings) of the intestinal microflora *in situ* is a major task confronting microbial ecologists: there must be enormous scope for the application of the talents of microbial physiologists in this work.

Enemies within

The intestinal microflora contains bacterial and fungal species that, under appropriate circumstances, cause disease. These opportunistic pathogens, from an anthropomorphic viewpoint, lurk within our bodies awaiting predisposing conditions to occur that will enable them to escape from the colon and to establish in other body sites. Generally, these opportunistic pathogens are minor members, in terms of population size, of the microflora. Suppressed by the regulatory mechanisms of the intestinal ecosystem, they await liberation by exogenous or endogenous factors that mitigate the factors that normally minimise their multiplication. *Escherichia coli* and *Enterococcus faecalis* are major pathogens of the urinary tract and in post-surgical sepsis; toxins produced by *Clostridium difficile* are responsible for pseudomembranous colitis; *Candida albicans* is an agent of cutaneous and systemic infections; *Bacteroides fragilis* is a major microbial partner with other members of the microflora in the production of mixed infections that result in intra-abdominal abscesses (Finegold 1977; Tannock 1999). The list of opportunistic pathogens is extensive, and even lactobacilli, despite their inclusion in the 'generally recognised as safe' (GRAS) category, have been reported as the cause of bacteraemia, albeit uncommonly (Husni et al. 1997).

Inflammatory bowel diseases (IBD) are chronic, relapsing inflammatory disorders of the intestine. Ulcerative colitis affects the colon, while Crohn's Disease most commonly affects the distal small intestine and proximal large intestine. Clinical features of IBD include diarrhoea and abdominal pain. While some IBD patients enjoy good health and long periods of remission, others suffer chronic ill health and major disruption of lifestyle. Current evidence suggests that IBD results from an interaction between genetic factors and triggering of intestinal inflammation early in life. A resulting dysregulation of the immune system produces the chronic, relapsing conditions. Accumulated observations from human cases and experimental animal studies suggest that the intestinal microflora, or their metabolic products, play an integral role either as targets for immune cells, or facilitators or amplifiers of inflammation, though the specific aetiological agents are, as yet, unknown (Chadwick & Chen 1999).

The intestinal microflora also has a role in the aetiology of inflammatory conditions of the joints. Ankylosing spondylitis is a chronic inflammatory disease involving principally the sacro-iliac joints and spine. The disease has a high association with humans having human leucocyte antigen B27 (HLA-B27). Inflammation occurs at the insertion of ligaments into bone. This results in the erosion of tendon insertions and new bone formation. Progressive ossification of ligaments produces bony bridging of joints. Back pain and stiffness are common clinical presentations. The disease is of unknown aetiology, but persistent inflammation of the intestine has been demonstrated in ankylosing spondylitis patients, and rats transgenic for human HLA-B27 develop a spondyloarthritis, but only if they harbour an intestinal microflora. Germfree, transgenic rats do not suffer from the disease (Gastone 1997). A microbial aetiology, involving the intestinal tract, in the initiation or progression of ankylosing spondylitis is therefore likely and worthy of investigation.

Cell wall material from *Bifidobacterium breve*, *Bifidobacterium adolescentis*, *Eubacterium aerofaciens*, and *Lactobacillus casei* have been reported to produce moderate to severe arthritis in rats inoculated intraperitoneally with the bacterial substances though the relevance of this information, produced by systemic inoculation of rats, to human subjects is not clear (Lehman et al. 1987; Severijnen et al. 1989).

Altering the composition of the intestinal microflora is obviously not something to be tinkered with in a light moment. Extensive knowledge of the long term consequences of such modification is required, and the techniques by which the necessary analyses of the microflora can be accomplished must be available.

The modern world: microbial ecology and anthropology

What determines the composition of the intestinal microflora? What factor(s) regulates the colonisation of a human by their authochthonous bacteria? There are many interesting facets of microbial ecology to study in regard to these questions. The importance of the genetic constitution of the host in the proclivity of specific strains to colonise its intestinal tract should be investigated, as should the impact of environmental factors. Both of these aspects could be studied using a large number of identical and non-identical twins, raised together or raised apart, as subjects. They would need to be monitored from infancy and far into adulthood if worthwhile results were to be obtained. Analytical techniques to carry out such studies are now available, but the logistics of the study would be complex. On the same note, it would indeed be wonderful if the faecal samples collected during the 1970s in order to compare the microfloras of rural Japanese and Japanese-Hawaiians who differed in colon cancer risk, were still extant in laboratory freezers (Moore et al. 1978). They could now be examined more sensitively using PCR and oligonucleotide probes.

While much of the world's population now embraces satellite television culture, and the traveller senses an increasing sameness in almost every country, unique lifestyles are still followed by distinct ethnic groups. It is time for a large team of microbial ecologists and anthropologists to investigate the intestinal microfloras of human societies in various geographical areas. This must be done before the modern age affects all communities with convenience foods and canned drinks, and above all, probiotics. Before long, we will never again have the opportunity to know what the pristine intestinal microflora of a human really was like.

Acknowledgements

The author's recent research described in this article has been supported by the New Zealand Dairy Research Institute, Meiji Milk Products, Japan, and the Otago University Research Committee. The support of technical and student associates throughout these studies has been greatly appreciated.

References

Amann RI, Ludwig W & Schleifer K-H (1995) Phylogenetic identification and *in situ* detection of individual microbial cells without cultivation. Microbiol. Rev. 59: 143–169

Aranki A & Freter R (1972) Use of anaerobic glove boxes for the cultivation of strictly anaerobic bacteria. Am. J. Clin. Nutr. 25: 1329-1334

Ballongue J (1993) Bifidobacteria and probiotic action. In: Salminen S & von Wright A (Eds) Lactic Acid Bacteria (pp 357–428). Marcel Dekker, New York

Barrow PA, Brooker BE, Fuller R & Newport MJ (1980) The attachment of bacteria to the gastric epithelium of the pig and its importance in the microecology of the intestine. J. Appl. Bacteriol. 48: 147–154

Bassam BJ, Allen T, Flood S, Stevens J, Wyatt P & Livak KJ (1996) Nucleic acid sequence detection systems: revolutionary automation for monitoring and reporting PCR products. Australasian Biotechnol. 6: 285–294

Bateup JM, Dobbinson S, McConnell MA, Munro K & Tannock GW (1998) Molecular analysis of the composition of *Lactobacillus* populations inhabiting the stomach and caecum of pigs. Microbial Ecology in Health and Disease 10: 95–102

Berg RD (1983) Host immune response to antigens of the indigenous intestinal flora. In: Hentges DJ (Ed) Human Intestinal Microflora in Health and Disease (pp 101–126). Academic Press, New York

Biavati B, Castagnoli P, Crociani F & Trovatelli LD (1984) Species of the genus *Bifidobacterium* in the feces of infants. Microbiologica 7: 341–345

Biavati B, Castagnoli P & Trovatelli LD (1986) Species of the genus Bifidobacterium in the feces of adults. Microbiologica 9: 39–45

Bry L, Falk PG, Midtvedt T & Gordon JI (1996) A model of host-microbial interactions in an open mammalian ecosystem. Science 273: 1380–1383

Caufield PW (1997) Dental caries – a transmissible and infectious disease revisited: a position paper. Pediatric Dentitstry 19: 491–498

Chadwick VS & Chen W (1999) The intestinal microflora and inflammatory bowel disease. In: Tannock GW (Ed) Medical Importance of the Normal Microflora (pp 177–221). Kluwer Academic Publishers, Dordrecht, The Netherlands

Croucher SC, Houston AP, Bayliss CE & Turner RJ (1983) Bacterial populations associated with different regions of the human colon wall. Appl. Environ. Microbiol. 45: 1025–1033

Cummings JH & Macfarlane GT (1991) The control and consequences of bacterial fermentation in the human colon. J. Appl. Bacteriol. 70: 443–459

Drasar BS & Barrow PA (1985) Intestinal Microbiology. American Society for Microbiology, Washington DC

Drasar BS & Hill MJ (1974) Human Intestinal Flora. Academic Press, London

Dubos R, Schaedler RW, Costello R & Hoet P (1965) Indigenous, normal, and autochthonous flora of the gastrointestinal tract. J. Exper. Med. 122: 67–76

Felske A, Rheims H, Wolterink A, Stackebrandt E & Akkermans ADL (1997) Ribosome analysis reveals prominent activity of an uncultured member of the class of Actinobacteria in grassland soils. Microbiology 143: 2983–2989

Ferguson WK (1967) The Renaissance. Holt, Rinehart and Winston, New York

Finegold SM (1977) Anaerobic Bacteria in Human Disease. Academic Press, New York

Finegold SM, Attebury R & Sutter VL (1974) Effect of diet on human fecal flora: comparison of Japanese and American diets. Am. J. Clin. Nutr. 27: 1456–1469

Finegold SM, Sutter VL & Mathisen GE (1983) Normal indigenous intestinal flora. In: Hentges DJ (Ed) Human Intestinal Microflora in Health and Disease (pp 3–31). Academic Press, New York

Finegold SM & Sutter VL (1978) Fecal flora in different populations, with special reference to diet. Am. J. Clin. Nutr. 31: S116–S122

Franks AH, Harmsen HJM, Raangs GC, Jansen GJ, Schut F & Welling GW (1998) Variations of bacterial populations in human feces measured by fluorescent *in situ* hybridization with group-specific 16S rRNA-targeted oligonucleotide probes. Appl. Environ. Microbiol. 64: 3336–3345

Fuller R (1989) Probiotics in man and animals. J. Appl. Bacteriol. 66: 365–378

Fuller R (1999) Probiotics for farm animals. In: Tannock GW (Ed) Probiotics: a Critical Review (pp 15–22). Horizon Scientific Press, Wymondham, UK

Fuller R & Brooker BE (1974) Lactobacilli which attach to the crop epithelium of the fowl. Am. J. Clin. Nutr. 27: 1305–1312

Gastone JSH (1997) Pathogenic role of gut inflammation in the spondyloarthropathies. Current Opinion in Rheumatology 9: 302–307

Goldin BR, Gorbach SL, Saxelin M, Barakat S, Gualtieri L & Salminen S (1992) Survival of *Lactobacillus* species (strain GG) in human gastrointestinal tract. Digestive Diseases and Sciences 37: 121–128

Goldin BR, Swenson L, Dwyer J, Sexton M & Gorabch SL1 (980) Effect of diet and *Lactobacillus acidophilus* supplements on human fecal bacterial enzymes. J. National Cancer Institute 64: 255–261

Goodwin MA, Cooper GL, Brown J, Bickford AA, Waltman WD & Dickson TG (1991) Clinical, pathological, and epizootiological features of the long-segmented filamentous organisms (Bacteria, LSFOs) in the small intestines of chickens, turkeys, and quails. Avian Diseases 35: 872–876

Gordon HA & Pesti L (1971) The gnotobiotic animal as a tool in the study of host-microbial relationships. Bacteriol. Rev. 35: 390–429

Gracey M (1983) The contaminated small bowel syndrome. In: Hentges DJ (Ed) Human Intestinal Microflora in Health and Disease (pp 495–513). Academic Press, New York

Gurtler V & Stanisich VA (1996) New approaches to typing and identification of bacteria using the 16S-23S rDNA spacer region. Microbiology 142: 3–16

Hill JE, Kelley LC & Langheinrich KA (1992) Visceral granulomas in chickens infected with a filamentous bacteria. Avian Diseases 36: 172–176

Holdeman LV & Moore WEC (1972) Roll-tube technique for anaerobic bacteria. Am. J. Clin. Nutr. 25: 1314–1317

Hungate RE (1966) The Rumen and its Microbes. Academic Press, New York

Husni RN, Gordon SM, Washington JA & Longworth DL (1997) Lactobacillus bacteremia and endocarditis: review of 45 cases. Clinical Infectious Diseases 25: 1048–1055

Jepson MA, Clark MA, Simmons NL & Hirst BH (1993) Actin accumulations at sites of attachment of indigenous apathogenic segmented filamentous bacteria to mouse ileal epithelial cells. Infection and Immunity 61: 4001–4004

Kato I, Kobayashi S, Yokokura T & Mutai M (1981) Antitumor activity of *Lactobacillus casei* in mice. Gann 72: 517–523

Kimura K, McCartney AL, McConnell MA & Tannock GW (1997) Analysis of fecal populations of bifidobacteria and lactobacilli and investigations of the immunological responses of their human hosts to the predominant strains. Appl. Environ. Microbiol. 63: 3394–3398

Klaasen HLBM, Koopman JP, van den Brink ME, van Wezel HPN & Beynen AC (1991) Mono-association of mice with non-cultivable, intestinal, segemented, filamentous bacteria. Archives of Microbiology 156: 148–151

Klaasen HLBM, Koopman JP, Poelma FGJ & Beynen AC (1992) Intestinal, segmented, filamentous bacteria. FEMS Microbiol. Rev. 88: 165–180

Klaasen HLBM, van der Heijden PJ, Stok W, Poelma FGJ, Koopman JP, van den Brink ME, Bakker MH, Eling WMC & Beynen AC (1993) Apathogenic, intestinal, segmented, filamentous bacteria stimulate the mucosal immune system of mice. Infection and Immunity 61: 303–306

Kolenbrander PE (1991) Coaggregation: adherence in the human oral microbial ecosystem. In: Dworkin M (Ed) Microbial Cell-Cell Interactions (pp 303–329). American Society for Microbiology, Washington DC

Lehman TJ, Cremer MA, Walker SM & Dillon AM (1987) The role of humoral immunity in *Lactobacillus casei* cell wall induced arthritis. J. Rheumatology 14: 415–419

Luckey TD (1963) Germfree Life and Gnotobiology. Academic Press, New York

Macfarlane GT & Macfarlane S (1995) Human intestinal 'biofilm' communities. In: Wimpenny J, Handley P, Gilbert P & Lappin-Scott H (Eds) The Life and Death of Biofilm (pp 83–87). BioLine, Cardiff

McCartney AL, Wang W & Tannock GW (1996) Molecular analysis of the composition of the bifidobacterial and lactobacillus microflora of humans. Appl. Environ. Microbiol. 62: 4608–4613

Mitsuoka T (1992) The human gastrointestinal tract. In: Wood BJB (Ed) The Lactic Acid Bacteria, Volume 1, The Lactic Acid Bacteria in Health and Disease (pp 69–114). Elsevier Applied Science, London

Moore WEC, Cato EP & Holdeman LV (1978) Some current concepts in intestinal bacteriology. Am. J. Clin. Nutr. 31: S33–S42

Moore WEC & Holdeman LV (1974) Special problems associated with the isolation and identification of intestinal bacteria in fecal flora studies. Am. J. Clin. Nutr. 27: 1450–1455

Muyzer G & Smalla K (1998) Application of denaturing gradient gel electrophoresis (DGGE) and temperature gradient gel electrophoresis (TGGE) in microbial ecology. Antonie van Leewenhoek 73: 127–141

O'Sullivan DJ (1999) Methods of analysis of the intestinal microflora. In: Tannock GW (Ed) Probiotics: a Critical Review (pp 23–44). Horizon Scientific Press, Wymondham, UK

Pasteur L (1885) Observations relatives a la note precedente de M. Duclaux. Academies des Sciences, Comptes Rendues (Paris) 100: 68

Perdigon G & Alvarez S (1992) Probiotics and the immune state. In: Fuller R (Ed) Probiotics. The Scientific Basis (pp 146–180). Chapman and Hall, London

Raskin L, Capman WC, Sharp R, Poulsen LK & Stahl DA (1997) Molecular ecology of gastrointestinal ecosystems. In: Mackie RI, White BA & Isaacson RE (Eds) Gastrointestinal Microbiology, Volume 2, Gastrointestinal Microbes and Host Interactions (pp 243–298). Chapman and Hall, New York

Reilly K & Attwood GT (1998) Detection of *Clostridium proteoclasticum* and closely related strains in the rumen by competitive PCR. Appl. Environ. Microbiol. 64: 907–913

Rettger LF, Levy MN, Weinstein L & Weiss JE (1935) *Lactobacillus acidophilus* and its Therapeutic Application. Yale University Press, New Haven, USA

Reysenbach A-L, Giver LJ, Wickham GS & Pace NR (1992) Differential amplification of rRNA genes by polymerase chain reaction. Appl. Environ. Microbiol. 58: 3417–3418

Roberton AM & Corfield AP (1999) Mucin degradation and its significance in inflammatory conditions of the gastrointestinal tract. In: Tannock GW (Ed) Medical Importance of the Normal Microflora (pp 222–261). Kluwer Academic Publishers, Dordrecht, The Netherlands

Salyers AA & Whitt DD (1994) Bacterial Pathogenesis. American Society for Microbiology, Washington DC

Savage DC (1977) Microbial ecology of the gastrointestinal tract. Annual Reviews of Microbiology 31: 107–133

Savage DC (1983) Morphological diversity among members of the gastrointestinal microflora. International Review of Cytology 82: 305–334

Savage DC, Dubos R & Schaedler RW (1968) The gastrointestinal epithelium and its autochthonous bacterial flora. J. Exper. Med. 127: 67–76

Schiffrin EJ, Rochat F, Link-Amster H, Aeschlimann JM & Donnet-Hughes A (1995) Immunomodulation of human blood cells following the ingestion of lactic acid bacteria. J. Dairy Sci. 78: 491–497

Severijnen AJ, van Kleef R, Hazenberg MP & van de Merwe J (1989) Cell wall fragments from major residents of the human intestinal flora induce chronic arthritis in rats. J. Rheumatology 16: 1061–1068

Snel J (1997) Symbiosis between the Mouse and Segmented Filamentous Bacteria: a Gnotobiotic Study. PhD thesis, University of Nijmegen

Stackebrandt E & and Rainey FA (1995) Partial and complete 16S rDNA sequences, their use in generation of 16S phylogenetic trees and their implications in molecular ecological studies. In: Akkermans DL, van Elsas JD & de Bruijn FJ (Eds) Molecular Microbial Ecology Manual (pp 1–17). Kluwer Academic Publishers, Dordrecht, The Netherlands

Summanen P, Baron EJ, Citron DM, Strong C, Wexler HM & Finegold SM (1993) Wadsworth Anaerobic Bacteriology Manual. Star Publishing Company, Belmont, USA

Swaminathan B & Matar GM (1993) Molecular typing methods. In: Persing DH, Smith TF, Tenover FC & White TJ (Eds) Diagnostic Molecular Microbiology. Principles and Applications (pp 26–50). American Society for Microbiology, Washington DC

Szentkuti L, Riedesel H, Enss M-L, Gaertner K & von Engelhardt W (1990) Pre-epithelial mucus layer in the colon of conventional and germfree rats. Histochemistry J. 22: 491–497

Tagg JR, Pybus V, Phillips LV & Fiddes TM (1983) Application of inhibitor typing in a study of the transmission and retention in the human mouth of the bacterium *Streptococcus salivarius*. Archives of Oral Biology 28: 911–915

Tannock GW (1987) Demonstration of mucosa-associated microbial populations in the colons of mice. Appl. Environ. Microbiol. 53: 1964–1968

Tannock GW (1995a) Normal microflora. An introduction to microbes inhabiting the human body. Chapman and Hall, London

Tannock GW (1995b) Microecology of the gastrointestinal tract in relation to lactic acid bacteria. International Dairy J. 5: 1059–1070

Tannock GW (1997) Influences of the normal microbiota on the animal host. In: Mackie RI, White BA & Isaacson RE (Eds) Gastrointestinal Microbiology, Volume 2, Gastrointestinal Microbes and Host Interactions (pp 466–497). Chapman and Hall, New York

Tannock GW (Ed) (1999) Medical Importance of the Normal Microflora. Kluwer Academic Publishers, Dordrecht, The Netherlands

Tannock GW (Ed) (1999) Probiotics: a Critical Review. Horizon Scientific Press, Wymondham, UK

Tannock GW, Crichton C, Welling GW, Koopman JP & Midtvedt T (1988) Reconstitution of the gastrointestinal microflora of lactobacillus-free mice. Appl. Environ. Microbiol. 54: 2971–2975

Tannock GW, Fuller R, Smith SL & Hall MA (1990) Plasmid profiling of members of the family *Enterobacteriaceae*, lactobacilli, and bifidobacteria to study the transmission of bacteria from mother to infant. J. Clin. Microbiol. 28: 1225–1228

Welling GW, Elfferich P, Raangs GC, Wildeboer-Veloo ACM, Jansen GJ & Degener JE (1997) 16S ribosomal RNA-targeted oligonucleotide probes for monitoring of intestinal tract bacteria. Scandanavian J. Gastroenterology 32 Supplement 222: 17–19

Wilson KH & Blitchington RB (1996) Human colonic biota studied by ribosomal DNA sequence analysis. Appl. Environ. Microbiol. 62: 2273–2278

Zoetendal E, Akkermans AD & De Vos WM (1998) Temperature gradient gel electrophoresis analysis of 16S rRNA from human fecal samples reveals stable and host-specific communities of active bacteria. Appl. Environ. Microbiol. 64: 3854–3859

Antonie van Leeuwenhoek **76**: 279–292, 1999.

Probiotics: from myth to reality. Demonstration of functionality in animal models of disease and in human clinical trials

Colum Dunne[1,2], Lisa Murphy[1], Sarah Flynn[1], Liam O'Mahony[1,2], Sile O'Halloran[1], Maria Feeney[1], Darrin Morrissey[1,2], Gerardine Thornton[1], Gerald Fitzgerald[1,2], Charles Daly[1,2], Barry Kiely[2], Eamonn M. M. Quigley[4], Gerald C. O'Sullivan[3], Fergus Shanahan[4] & J. Kevin Collins[1,4,*]
[1]*Department of Microbiology, University College, Cork, Ireland;*
[2]*National Food Biotechnology Centre, University College, Cork, Ireland;*
[3]*Department of Surgery, Mercy Hospital, Cork, Ireland;*
[4]*Department of Medicine, University College, Cork, Ireland.*
*(*Author for correspondence; E-mail: nfbc@ucc.ie)*

Abstract

The enteric flora comprise approximately 95% of the total number of cells in the human body and are capable of eliciting immune responses while also protecting against microbial pathogens. However, the resident bacterial flora of the gastrointestinal tract (GIT) may also be implicated in the pathogenesis of several chronic conditions such as inflammatory bowel disease (IBD). The University College Cork-based Probiotic Research Group has successfully isolated and identified lactic acid bacteria (LAB) which exhibit beneficial probiotic traits. These characteristics include the demonstration of bile tolerance; acid resistance; adherence to host epithelial tissue; and *in vitro* antagonism of potentially-pathogenic micro-organisms or those which have been implicated in promoting inflammation. The primary objective of this report is to describe the strategy adopted for the selection of potentially effective probiotic bacteria. The study further describes the evaluation of two members of the resulting panel of micro-organisms (*Lactobacillus salivarius* subsp. *salivarius* UCC118 and *Bifidobacterium longum infantis* 35624) under *in vitro* conditions and throughout *in vivo* murine and human feeding trials. Specifically, an initial feeding study completed in Balb/c mice focused upon (i) effective delivery of the probiotic micro-organisms to the GIT and evaluation of the ability of the introduced strains to survive transit through, and possibly colonise, the murine GIT; (ii) accepting the complexity of the hostile GIT and faecal environments, development of a method of enumerating the introduced bacterial strains using conventional microbiological techniques; and (iii) assessment of the effects of administered bacterial strains on the numbers of specific recoverable indigenous bacteria in the murine GIT and faeces. Additional research, exploiting the availability of murine models of inflammatory bowel disease, demonstrated the beneficial effects of administering probiotic combinations of *Lactobacillus salivarius* UCC118 and *Bifidobacterium longum infantis* 35624 in prevention of illness-related weight loss. A further ethically-approved feeding trial, successfully conducted in 80 healthy volunteers, demonstrated that yoghurt can be used as a vehicle for delivery of *Lactobacillus salivarius* strain UCC118 to the human GIT with considerable efficacy in influencing gut flora and colonisation.

Introduction

Criteria for the selection of probiotic micro-organisms

Diet, stress, and modern medical practices have been implicated as factors capable of exerting an influence on human health and nutrition. As a result of these and other factors, the incidence of illnesses which may be induced by deficient or compromised microflora (e.g., gastrointestinal tract infections; constipation; irritable bowel syndrome (IBS); inflammatory bowel disease (IBD) – Crohn's disease and ulcerative colitis; food allergies; antibiotic-induced diarrhoea;

cardiovascular disease; and certain cancers) is increasing (Salminen et al. 1995, 1998a; Schaafsma 1995). Probiotics have been defined as live microbial food supplements which beneficially affect the host by improving the intestinal microbial balance (Fuller 1989), or more broadly, as 'living micro-organisms, which upon ingestion in certain numbers, exert health affects beyond inherent basic nutrition' (Guarner & Schaafsma 1998). Combinations of various micro-organisms, particularly species of *Lactobacillus* and *Streptococcus*, have traditionally been used in fermented dairy products to promote human health through the influences they may exert on the microbial ecology of the host, lactose intolerance, incidence of diarrhoea, mucosal immune response, levels of blood cholesterol, and cancer (Salminen et al. 1998a,b). Indeed, a number of probiotic bacteria are now being successfully exploited commercially (e.g., *Lactobacillus rhamnosus* GG (Saxelin 1997), *L. casei* Shirota (Aso & Akazan 1992), and *L. acidophilus* LA-1 (Bernet et al. 1994)). However, many consumers, consumer organisations, and members of the scientific community remain sceptical of such products and their associated probiotic claims. As a result, the food industry (in collaboration with academic scientists, clinicians, and workshops such as The Lactic Acid Bacteria Industrial Platform (LABIP) and PROBDEMO (FAIR CT-96 1028) (Mattila-Sandholm 1997) supported by European Union-funded programmes) aims to promote the generation and dissemination of consensus opinions obtained through the direct interaction of research institutions and universities with industry.

Such a consensus was reached when the participants involved in the LABIP workshop on probiotics concluded that 'probiotics may be consumed either as a food component or as a non-food preparation'. The LABIP participants further supported criteria outlined by others for the selection and assessment of probiotic LAB (Guarner & Schaafsma 1998). In summary, to fulfil these criteria (Marteau & Rambaud 1993; Huis in't Veld & Shortt 1996; Salminen et al. 1996; Tannock 1997; Brassart et al. 1998; Guarner & Schaafsma 1998) probiotic micro-organisms should:

- Be of human origin.
- Demonstrate non-pathogenic behaviour.
- Exhibit resistance to technological processes (i.e., viability and activity in delivery vehicles).
- Prove resistant to gastric acid and bile.
- Adhere to gut epithelial tissue.
- Be able to persist, albeit for short periods, in the gastrointestinal tract.
- Produce antimicrobial substances.
- Modulate immune responses.
- Have the ability to influence metabolic activities (e.g., cholesterol assimilation, lactase activity, vitamin production).

In addition, these requirements were further expanded by Berg (Berg 1998) and Salminen et al. (Salminen et al. 1996, 1998a) who stated that:

a) each potential probiotic strain should be documented and assessed independently.
b) extrapolation of data from closely related strains is not acceptable.
c) only well defined strains, products and study populations should be used in trials.
d) where possible all human studies should be randomised, double-blind, and placebo-controlled.
e) results should be confirmed by independent research groups.
f) preferentially, the study should be published in a peer-reviewed journal.

Isolation of potential probiotic bacteria

In the development of probiotic foods intended for human consumption, strains of lactic acid bacteria such as *Lactobacillus*, *Bifidobacterium*, *Enterococcus* have been most commonly used. This is primarily due to the perception that they are desirable members of the intestinal microflora (Table 1) (Goldin & Gorbach 1992; Berg 1998). In addition, these bacteria have traditionally been used in the production of fermented dairy products and have 'GRAS: generally regarded as safe' status (O'Sullivan et al. 1992). However, a subset of the probiotic micro-organisms currently employed in the dairy-food industry are not of human origin and, therefore, do not meet the criteria for the selection of probiotic microbes acceptable for human consumption, as outlined above (Daly & Davis 1998).

The enteric flora, including greater than 500 bacterial species, comprise approximately 95% of the total number of cells in the human body and contribute significantly to the host's resistance to infectious disease. Furthermore, changes in the composition of the intestinal flora are often associated with disease and may, in some cases, be a causative factor (Morotomi et al. 1975). At University College Cork, it was recognised that while lactobacilli isolated from human faeces have been relatively well characterised (Kandler & Weiss 1986)few attempts had been made to isolate potential adherent probiotic bacteria directly

Table 1. Probiotic bacteria and their reported properties

Strain	Reported effects in clinical studies	Selected references
Lactobacillus acidophilus LC1	Immune enhancing; vaccine adjuvant; adherence to human intestinal cells; balancing of intestinal microflora	(Bernet et al. 1994)
Lactobacillus acidophilus NCFO1748	Lowering of faecal enzymes; prevention of radiotherapy -related diarrhoea; treatment of constipation	(Lidbeck et al. 1992)
Lactobacillus rhamnosus GG	Prevention of antibiotic associated diarrhoea; treatment and prevention of rota-virus diarrhoea; treatment of relapsing *Clostridium difficile* diarrhoea; prevention of acute diarrhoea; alleviation of Crohn's disease; antagonistic against anticarcinogenic bacteria	(Salminen et al. 1993)
Lactobacillus casei Shirota	Prevention of intestinal disturbances; balancing of intestinal bacteria; lowering of faecal enzymes; inhibition of superficial bladder cancer	(Aso & Akazan 1992)
Lactobacillus gasseri	Faecal enzyme reduction; survival in the intestinal tract.	(Pedrosa et al. 1995)
Bifidobacterium bifidum	Treatment of rota-virus diarrhoea; balancing of intestinal microflora; treatment of viral diarrhoea	(Marteau et al. 1990)

(Adapted from Lee & Salminen (1995)).

from the human intestinal mucosa (the environment into which they may subsequently be re-introduced and required to function). Therefore, bacteria were isolated from resected human terminal ileum (the small intestine consists of the duodenum, jejunum, and ileum) obtained from patients undergoing urinary tract reconstructive surgery. Approximately 1500 catalase negative bacterial isolates were chosen and further characterised. Of these, greater than 60% were Gram positive, homofermentative cocci; approximately 18% proved to be Gram negative rods and heterofermentative coccobacilli; while 22% were predominantly homofermentative coccobacilli. Of the latter group, thirty-eight Gram-positive isolates were further characterised by physiological and biochemical means. Of these micro-organisms, 28 were subsequently found to belong to the *Lactobacillus* genus (mainly *L. paracasei* and *L. salivarius*) (Thornton 1996). *Lactobacillus acidophilus* has previously been recovered in relatively high numbers from the GIT. However, additional reports have demonstrated that other *Lactobacillus* groups (*L. crispatus*, *L. gasseri*, *L. salivarius*, and *L. reuteri*) can also be isolated at similar levels (Molin et al. 1993), indicating the lack of a single dominant species.

In addition to lactobacilli, presumptive strains of *Bifidobacterium* were isolated from ileum tissue samples which had been homogenised. Using molecular analysis, these bifidobacteria were classified as *B. infantis* (O'Riordan & Fitzgerald 1997). Interestingly, these results differ significantly from previous observations that adults do not possess detectable levels of *B. infantis* but rather *B. longum* and *B. bifidum* (Mitsouka 1984).

Assessment of potential probiotic bacteria *in vitro*

A selection of the bacterial strains isolated from the surgically-removed segments of human ileum were chosen for *in vitro* evaluation, within the parameters outlined above (Marteau & Rambaud 1993; Huis in't Veld & Shortt 1996; Salminen et al. 1996; Tannock 1997; Brassart et al. 1998; Collins et al. 1998; Guarner & Schaafsma 1998).

Resistance to gastric acid

Prior to reaching the intestinal tract, probiotic bacteria must first survive transit through the stomach where the secretion of gastric acid constitutes a primary defence mechanism against the majority of ingested micro-organisms. Therefore, preliminary experiments were performed to determine the level of acid resistance exhibited by a number of both *Lactobacillus* and *Bifidobacterium* strains isolated from the human ileum. Survival of the bacterial strains was initially assessed by adding approximately 10^8 cfu/ml and 10^6 cfu/ml lactobacilli and bifidobacteria, respectively, to MRS medium (De Man et al. 1960) amended with HCL to between pH2.0 and pH3.4. However, survival of bacterial strains in human gastric juice is a more accurate indication of the ability to survive passage through the stomach. For this reason, human gastric juice was obtained from healthy subjects by aspiration through a nasogastric tube. As the pH of the stomach is known to fluctuate (e.g., when an individual is fasting the stomach pH may be as low as 1.5 (Draser et al. 1969)) the pH of the obtained samples were quantified prior to use. The gastric juice was then added to MRS medium as described above. The observed results indicated that the lactobacilli isolates (including *Lactobacillus salivarius* UCC118) have the potential to successfully transit the human stomach and may possess the ability to reach the intestinal environment in which they may function effectively. However, when exposed to human gastric juice the bifidobacteria strains proved significantly less acid resistant than the lactobacilli (Thornton 1996).

Bile acid resistance

Whilst evaluating the potential of utilising lactic acid bacteria as effective probiotics it is generally considered necessary to evaluate their ability to resist the effects of bile acids (Lee & Salminen 1995). Bile acids are synthesised in the liver from cholesterol and are secreted from the gall-bladder into the duodenum in the conjugated form (500-700 ml/day) (Hoffman et al. 1983). These acids then undergo extensive chemical modifications (deconjugation, dehydroxylation, dehydrogenation, and deglucuronidation) in the colon due almost solely to microbial activity (Hill & Draser 1968; Shimada et al. 1969; Hylemon & Glass 1983). Both conjugated and deconjugated bile acids exhibit anti-bacterial activity inhibiting the growth of *E. coli* strains, *Klebsiella* sp., and *Enterococcus* sp. *in vitro* (Lewis & Gorbach 1972; Stewart et al. 1986). However, the deconjugated forms are more inhibitory (Floch et al. 1972; Percy-Robb & Collee 1972; Stewart et al. 1986), while Gram positive bacteria are found to be more sensitive than Gram negative (Floch et al. 1972).

In these preliminary studies, the objective was to determine the level of bile acid resistance exhibited by a number of both *Lactobacillus* and *Bifidobacterium* strains isolated from the human ileum. In short, solid media were supplemented with bovine bile (Sigma Chemicals Co. Ltd., Poole, UK), porcine bile (Sigma), and human bile (obtained at laproscopic cholecystectomy from human gall-bladders) to final concentrations of between 0.3 and 7.5%. These plates were then incubated at 37 °C under anaerobic conditions and growth was recorded after 24–48 hours (Thornton 1996). Following analysis, it was apparent that while the tested lactobacilli and *Bifidobacterium* strains exhibited resistance to the bovine bile used, the porcine bile employed in these assays proved significantly more inhibitory to both of the bacterial groups (Thornton 1996). However, in relation to the assessment of probiotic strains intended for human consumption, the most relevant determination is that of their ability to grow in the presence of human bile. Interestingly, regardless of the resistance patterns observed in the presence of either bovine or porcine bile, all of the assayed bacteria were capable of growth in physiologically relevant concentrations of human bile (approximately 0.3%) (Thornton 1996).

Adherence of bacterial isolates to HT-29 and Caco-2 cell-lines

As outlined above, the consensus opinion of the LABIP participants in relation to the selection of probiotic bacterial strains is that they be capable of adhesion to gut epithelial tissue and possess the ability to colonise the gastrointestinal tract.

HT-29 and Caco-2 cells are human intestinal cell-lines expressing morphological and physiological characteristics of normal human enterocytes (Brassart et al. 1998). These cell-lines have been previously exploited to elucidate the mechanisms mediating enteropathogen adhesion (Neeser et al. 1989; Kerneis et al. 1991; Bernet et al. 1994). In more recent studies, however, they have been employed in order to select for, and subsequently assess, lactic acid bacteria on the basis of their adherence properties (Coconnier et al. 1992; Bernet et al. 1993; Greene & Klaenhammer 1994; Crociani et al. 1995; Sarem et al. 1996; Tuomola & Salminen 1998). In our studies, completed using both HT-29 and Caco-2 cell-lines, the observed adherence of the *Lactobacillus* strains (including *Lactobacillus salivarius* UCC118) compared well with that of the well-characterised adherent strain *Lactobacillus rhamnosus* GG (Dunne et al. 1999). However, significantly lower levels of adherence were observed when the *Bifidobacterium* strains were assessed, regardless of the cell-line used (Dunne et al. 1999).

Antimicrobial activity

Several metabolic compounds produced by lactic acid bacteria demonstrate antimicrobial effects, including organic acids, fatty acids, and hydrogen peroxide (Ouwehand 1998). However, bacteriocins or proteinaceous substances with specific inhibitory activity against closely related species are perhaps the most extensively studied (Abee et al. 1994; McAuliffe et al. 1998; Ouwehand 1998). Of these, nisin which is produced by some *L. lactis* subsp. *lactis* strains is the only purified bacteriocin approved for use in products intended for human consumption (Dodd & Gasson 1994; Jack et al. 1995).

In vitro *antagonism of selected micro-organisms by* Lb. salivarius *UCC118 or its antimicrobial product, ABP118*

In our studies, the lactobacilli and bifidobacteria isolated from the human ileum were assayed for antimicrobial activity against a range of indicator bacteria including strains of *Listeria*, *Bacillus*, *Enterococcus*, *Staphylococcus*, *Clostridium*, *Pseudomonas*, *E. coli*, *Lactobacillus*, *Streptococcus*, *Bifidobacterium*, and *Lactococcus*. Antimicrobial activity of *Lb. salivarius* UCC118 was detected using the method described by Tagg et al. (1971). In summary, strain UCC118 was incubated anaerobically in MRS broth for 12–16 h at 37 °C. Following centrifugation at 3000 *g* and washing in quarter-strength Ringer's solution, strain UCC118 was resuspended in an equal volume of quarter-strength Ringer's solution. *Lb. salivarius* UCC118 was then inoculated (100 μl) onto solid MRS medium (supplemented with 2% ß-glycerophosphate (Sigma Chemicals, Poole)) and incubated anaerobically overnight at 37° C. The grown cultures were overlaid with the test micro-organisms in sloppy agar, incubated overnight and zones of inhibition (greater than 1 mm in radius) were recorded. To determine whether the observed antimicrobial activity of *Lb. salivarius* UCC118 was secretory in nature, aliquots (500 μl) of the 12–16 h UCC118 culture were filter-sterilised and the cell-free supernatant subsequently assayed for the ability to inhibit the growth of selected potential pathogens, as described above.

Both *Lb. salivarius* UCC118 and its cell-free supernatant were found capable of significantly inhibiting the *in vitro* growth of a range of both Gram positive and Gram negative bacterial cultures (Table 2). Significantly, neither *Lb. salivarius* UCC118 nor its supernatant were found capable of impairing the *in vitro* growth of closely related lactic acid bacteria (such as *Lb. acidophilus* or *Lb. reuteri*) with the exception of *Lb. fermentum* KLD (Table 2). The growth of strains

Table 2. Inhibition of the growth of selected micro-organisms by *Lactobacillus salivarius* subsp. *salivarius* strain UCC118 or its derived antimicrobial factor, ABP118, under *in vitro* conditions

Assayed micro-organisms	*In vitro* inhibition of growth by *Lb. salivarius* UCC118	*In vitro* inhibition of growth by *Lb. salivarius* UCC118-derived ABP118
Lactobacillus acidophilus	–	–
Lactobacillus bulgaricus	–	–
Lactobacillus rhamnosus GG	–	–
Lactobacillus fermentum KLD	+	+
Lactobacillus plantarum	–	–
Streptococcus thermophilus	–	–
Streptococcus mutans	–	–
Bifidobacterium longum	+/–	–
Bifidobacterium bifidum	+/–	–
Bifidobacterium breve	+/–	–
Bifidobacterium infantis	+/–	–
Bacillus subtilus	+	+
Bacillus cereus	+	+
Bacillus thuringiensis	+	+
Bacillus coagulans	+	–
Enterococcus faecalis	+	+
Enterococcus faecium	+	+
Escherichia coli	+	–
Listeria monocytogenes	+	+
Pseudomonas putida	+	–
Pseudomonas aeruginosa	+	–
Pseudomonas fluorescens	+	+
Salmonella typhimurium	+	–
Salmonella enteriditis	+	–
Staphylococcus aureus (non-methicillin resistant)	+	–
Staphylococcus aureus (methicillin resistant)	+	+
Candida albicans	–	–

of *Leuconostoc*, *Lactococcus*, and *Pediococccus* was also uninhibited (data not shown).

Purification and partial sequencing of the Lb. salivarius *UCC118-derived proteinaceous antimicrobial factor ABP118*

Lb. salivarius UCC118 was incubated anaerobically in 2 L of liquid MRS medium at 37 °C for approximately 6–8 h. Following centrifugation at 3000 *g*, the *Lb. salivarius* UCC118-produced antimicrobial agent, ABP118, was purified to homogeneity in a sequential process of ammonium sulphate precipitation, hydrophobic interaction chromatography, cation exchange chromatography, and C_2/C_{18} reverse-phase chromatography. Antimicrobial activity of the purified factor was confirmed using the method described above (Table 2).

The purified antimicrobial factor was hydrolysed and analysed on an amino acid analyser, as described previously (Fykse et al. 1988). The amino acid sequence was also analysed by Edman degradation (Cornwell et al. 1988). The C-terminal sequence of ABP118 was determined following cleavage of the antimicrobial factor with cyanogen bromide (CnBr) as described previously (Sletten & Husby 1974).

Strain UCC118-derived ABP118 was found to be heat stable, resistant over a wide pH range, and resistant to treatment with a number of detergents and organic solvents (data not shown). On the basis of the amino acid composition the ABP118 peptide has a molecular weight of between 4.8 and 5.2 kDa. Although, the N-terminus was blocked for sequencing by Edman degradation, following cyanogen bromide cleavage residues (K-R-G-P-N) were sequenced at the C-terminus. Similarities in the non-redundant protein databases at the National Centre for Biological Information (NCBI) were searched for using the BLAST algorithm (Altschul et al. 1990). However, the sequenced residues did not demonstrate any significant homology with other antimicrobial factors in the databases.

Probiotic modulation of host immune responses

Both immunological and non-immunological defence mechanisms protect the human gastrointestinal tract, considered to be the largest immune organ in the body due to the extensive surface area of the small bowel, from colonisation by intestinal bacteria (McCracken & Gaskins 1999). Innate defence mechanisms include the low pH of the stomach, bile salts, peristalsis, mucin layers and anti-microbial compounds such as lysozyme (Savage 1997). Immunological mechanisms include specialised lymphoid aggregates termed Peyers patches which are distributed throughout the small intestine and colon. Luminal antigens presented at these sites can result in the stimulation of appropriate T and B cells, the establishment of cytokine networks, and the secretion of antibodies into the gastrointestinal tract (Neutra & Kraehenbuhl 1996). Notably however, as the gastrointestinal mucosa is the largest surface at which the host interacts with the external environment, specific control mechanisms must be in place to regulate immune responsiveness to the 100 tons of food handled by the gastrointestinal tract over an average lifetime (Shanahan 1994). Furthermore, as the gastrointestinal tract is colonised by over 500 species of bacteria (numbering 10^{11}–10^{12}/g in the colon) these control mechanisms must be capable of distinguishing non-pathogenic adherent bacteria from invasive pathogens potentially capable of causing significant damage to the host.

Immune education

The enteric flora are an important aspect of the development and appropriate function of the intestinal immune system. In the absence of an enteric flora, the intestinal immune system is underdeveloped, as demonstrated in germ free animal models, and certain functional parameters are diminished (e.g., macrophage phagocytic ability and immunoglobulin production) (Wostman 1996). Furthermore, the importance of the gut flora in stimulation of non-damaging immune responses is becoming more evident. In short, the increases in observed incidence and severity of allergies (and conditions such as inflammatory bowel disease – Crohn's disease and ulcerative colitis) in the Western world has been linked with increases in standards of hygiene and sanitation, concomitant with a decrease in the number and range of infectious challenges encountered by the growing and developing host. This lack of immune education may allow the host to over-react to non-pathogenic antigen-containing commensal flora, resulting in inflammatory damage and/or allergy and autoimmunity.

The deliberate consumption of non-pathogenic bacteria, such as probiotic bacteria, may provide a health promoting immune-educating challenge to the host. The promotion of broadspectrum immunoglobulin secretion into the lumen, following consumption of probiotic bacteria, may result in binding of allergens and prevention of their transmucosal uptake by the host. In addition, interaction with certain intraepithelial lymphocyte subsets may suppress immune responses to allergens within the gastrointestinal tract possibly resulting in tolerance to perceived antigens. Thus, the deliberate consumption of probiotic bacteria in order to replace immune stimuli artificially is being researched extensively.

Oral vaccination using recombinant probiotic bacteria

The majority of pathogenic organisms gain entry via mucosal surfaces. Efficient vaccination of these sites may protect against invasion by particular infectious agents. Oral vaccination strategies have concentrated, to date, on the use of attenuated live pathogenic organisms or purified encapsulated antigens. However, lactic acid bacteria, modified to deliver antigens derived from infectious agents, may provide an attractive alternative as these bacteria are considered to be safe for human consumption (GRAS status). The European Commission-sponsored 'LAB-VAC' programme (BIO4-CT96-0542: representing the collaboration of 10 research groups from France, Belgium, the Netherlands, Ireland, Italy, England, and Australia) is specifically focused upon the development of lactic acid bacteria as non-pathogenic viable antigen delivery systems. Notably, within this programme, particular emphasis has been placed on the immunological evaluation of such bacteria as potential vectors for the production and local delivery of protective antigens in the gastrointestinal tract.

In vivo assessment of potential probiotic bacterial strains

To date, the benefits of probiotics have predominantly been demonstrated under defined and well-controlled laboratory conditions. In response, the LABIP workshop participants, amongst others, have stated that while 'several *in vitro* assays or animal studies ... are very useful in the preselection of (probiotic) bacterial strains ... the proof of efficacy in humans should be granted by at least one well-designed human study (Berg 1998; Collins et al. 1998; Guarner & Schaafsma 1998). Recently, several prospective studies have demonstrated the efficacy of lactic bacteria administration for both prophylactic and therapeutic use against diarrhoea in premature infants (Millar et al. 1993), new-borns (Sepp et al. 1993), children (Isolauri et al. 1995) and in the therapy of antibiotic-related diarrhoea (Siitonen et al. 1990), and traveller's diarrhoea (Oksanen et al. 1990).

Murine feeding studies

As described above, the criteria used in our studies for the selection and assessment of potential probiotic bacterial strains resulted in the identification of a small number of *Lactobacillus* and *Bifidobacterium* strains capable of resisting the effects of bile and low pH, of adhering to human epithelial cell-lines, and of exhibiting antimicrobial activity *in vitro* (Thornton 1996). Of these bacterial isolates, *Lb. salivarius* UCC118 and *Bifidobacterium longum infantis* UCC35624 were selected for assessment of their ability to survive transit through the murine gastrointestinal tract.

Feeding trial in Balb/c mice

Due to the fact that the gastrointestinal tract is a complex and hostile environment, it appears unlikely that a single probiotic bacterial strain will be capable of significantly influencing the microbial ecology of the host. However, for any micro-organism to effectively fulfill a prophylactic role, it must be capable of surviving and colonising this environment at least transiently. For instance, while *Lb. rhamnosus* GG is easily enumerated, it shows limited persistence in faeces after termination of feeding and was found to cause no significant changes in the population of lactobacilli over a five week study (Goldin et al. 1992). Recently, however, Alander et al. (1997) have shown that *Lb. rhamnosus* GG is recoverable from colonic biopsies obtained from humans administered the probiotic strain, indicating that faecal enumeration is not truly reflective of what occurs in the human gut.

The LABIP workshop participants, amongst others, have further stated that *in vitro* assays or animal studies are useful in the preselection of (probiotic) bacterial strains (Berg 1998; Collins et al. 1998; Guarner & Schaafsma 1998). Therefore, the objectives of this study completed in pathogen-free mice focused upon (i) effective delivery of the probiotic micro-organisms to the GIT and evaluation of the ability of the strains to survive transit through, and possibly colonise, the murine GIT; (ii) accepting the complexity of the hostile GIT and faecal environments, development of a method of enumerating the

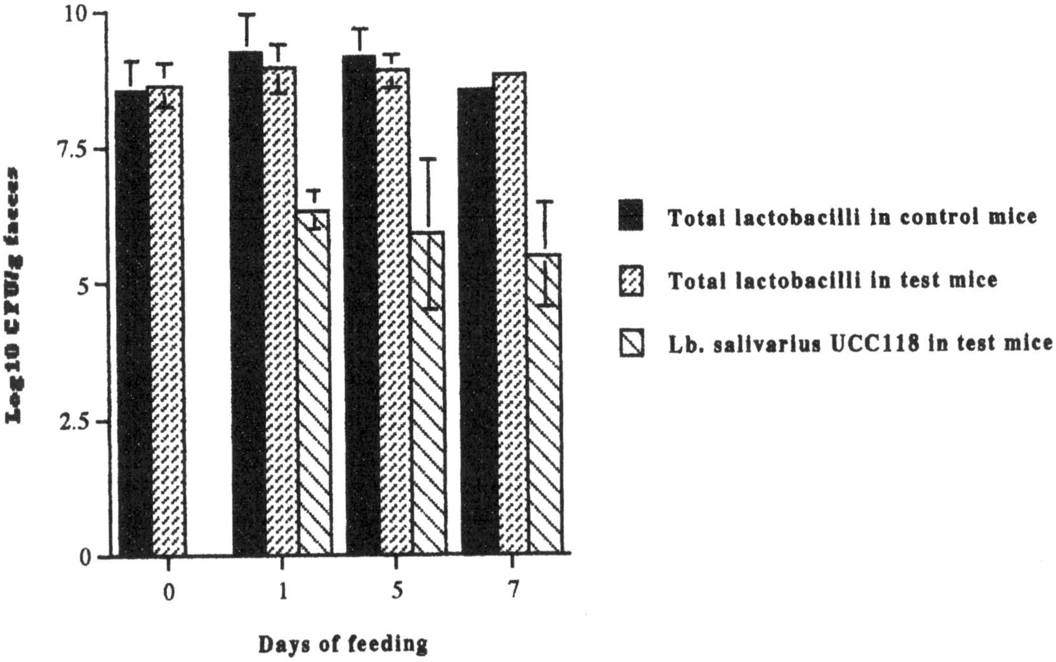

Figure 1. Enumeration of total lactobacilli and *Lb. salivarius* subsp. *salivarius* strain UCC118 in mice during feeding trials. Results are expressed as $\log_{10}$ CFU/g faeces.

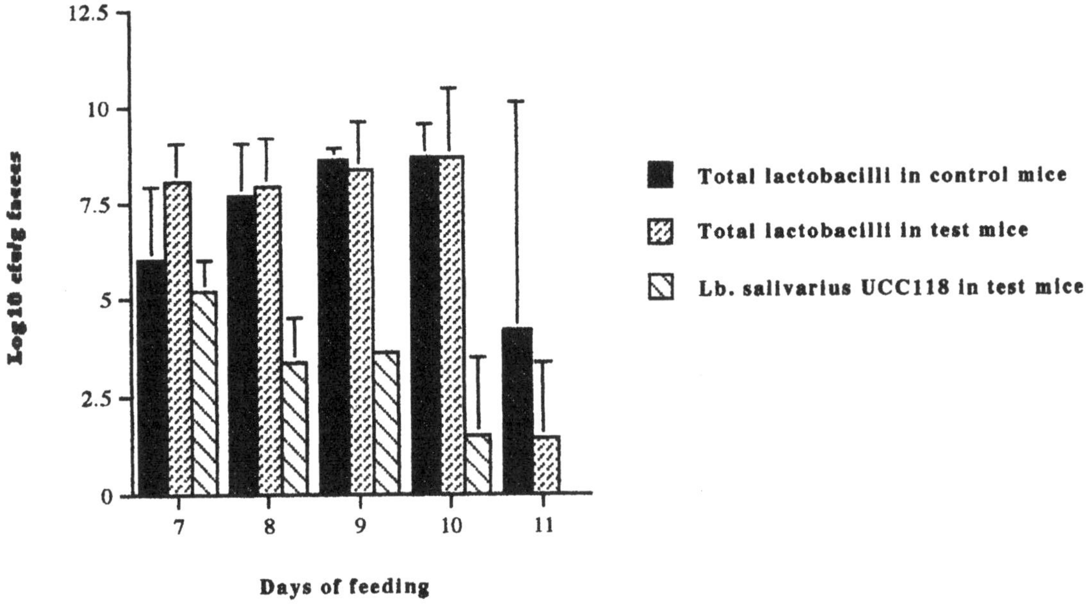

Figure 2. Enumeration of total lactobacilli and *Lb. salivarius* subsp. *salivarius* strain UCC118 following cessation of feeding at day 7 of the experiment. Results are expressed as $\log_{10}$ CFU/g faeces.

introduced bacterial strains using conventional microbiological techniques; and (iii) assessment of their effect on the numbers of indigenous bacteria in the murine GIT and faeces. *Lb. salivarius* strain UCC118 was administered (4×10^9 CFU/day) to Balb/c mice in skim milk diluted with sterile drinking water and enumerated from collected faeces (Figure 1) due to their previously-generated resistance to rifampicin. Importantly, examination of the survival of strain UCC118 indicated that rifampicin resistance is a suitable non-genetic method of enumerating introduced bacterial strains and that skim milk proved an effective delivery vehicle. A further objective of this study was to investigate the length of time administered *Lb. salivarius* UCC118 would persist in faeces following termination of feeding. In this study, it was found that the probiotic strain could be recovered (with no oral supplementation) from faeces for up to three days after cessation of feeding (Figure 2). Strain UCC118(LM2) was no longer recoverable from faeces after 4 days (Figure 2). These results suggest that *Lb. salivarius* strain UCC118(LM2) is capable of persistence in the mouse gut, at least for a limited period. This observation is supported by the results of a study completed by Goldin et al. (Goldin et al. 1992) where it was demonstrated that 60–80% of individuals consuming *Lb. rhamnosus* GG excreted this strain for 3–4 days, but only 33% of the population after 7 days. Therefore, it appears likely that daily administration of the preferred strain is necessary for maintenance of high probiotic levels.

It is not difficult to understand how a single bacterial strain, introduced into an environment such as the human GIT containing greater than 500 bacterial species, may be unable to effectively compete and become established. Fonty et al. (Fonty et al. 1993), suggested that the establishment of an introduced strain is governed, not only by the mode of administration of the strain, but also by the interaction of micro-organisms within the gut environment. Throughout this study, the levels of indigenous enterococci, bifidobacteria, coliforms, and *Bacteroides* culturable from mouse faeces were not significantly modified by administration of *Lb. salivarius* UCC118.

Feeding trial in a murine model of inflammatory bowel disease

Inflammatory Bowel Disease (IBD) – ulcerative colitis and Crohn's disease are immune mediated diseases of the intestine where inappropriate T cell reactivity to constituents of the normal intestinal flora may be of etiological significance. Although no specific bacterial species have been directly associated with the pathogenesis of human IBD, analysis of the luminal enteric flora has revealed alterations in the composition of the bacterial species compared to healthy individuals. In Crohn's disease, concentrations of bacteroides, Eubacteria and *Peptostreptococcus* are increased, whereas bifidobacteria numbers are significantly reduced. Furthermore, in ulcerative colitis concentrations of facultative anaerobic bacteria are increased. These results indicate that in the setting of intestinal inflammation, bacterial populations can be significantly altered and that there may be reductions in the numbers of bifidobacteria or lactobacilli which are considered healthy or potentially beneficial components of the normal flora.

In this study, the potential benefits of administering a combination of the probiotic strains *Lb. salivarius* UCC118 and *Bifidobacterium longum infantis* UCC35624 in modifying the effects of IBD were evaluated in a laboratory model. This model was developed through the transfer of spleen- or lymph node-derived $CD4^+$ T lymphocytes, expressing high levels of $CD45RB^{high}$, from normal mice into immunodeficient SCID (Severe Combined ImmunoDeficient) mice. The recipient mice subsequently suffer from a wasting disease of chronic intestinal inflammation, most severe in the colon (Morrissey et al. 1993; Aranda et al. 1997).

The results of this study demonstrated that oral administration of *Bifidobacterium* strain UCC35624 combined with *Lb. salivarius* UCC118 significantly reduced disease severity as manifested by reduced weight loss (Figure 3), improvement of colon pathology and markedly improved appearance of the mice over a six week period. All the control mice developed a chronic wasting disease which was also observed in the mice administered a non-probiotic dairy product (Figure 3). The significance of these results lie in the fact that while bifidobacteria are one of several predominant culturable bacteria present in the colonic microflora, their functions in the colon have not been completely elucidated. However, it has recently been found that patients suffering active Crohn's disease have significantly less recoverable bifidobacteria in their faeces compared to healthy individuals and that this reduction in bifidobacteria numbers was observed to be directly correlated with decreased levels of β-D-galactosidase (produced by bifidobacteria) activity (Favier et al. 1997). Therefore, the data presented

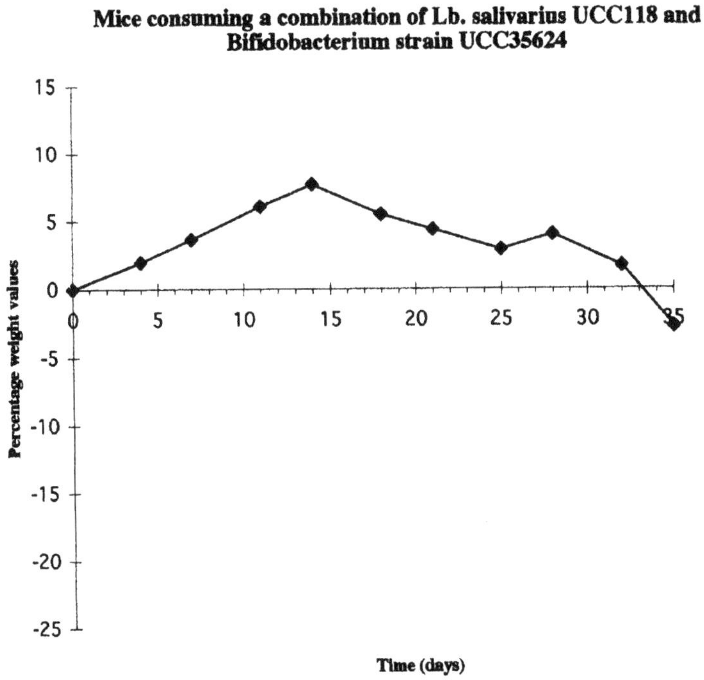

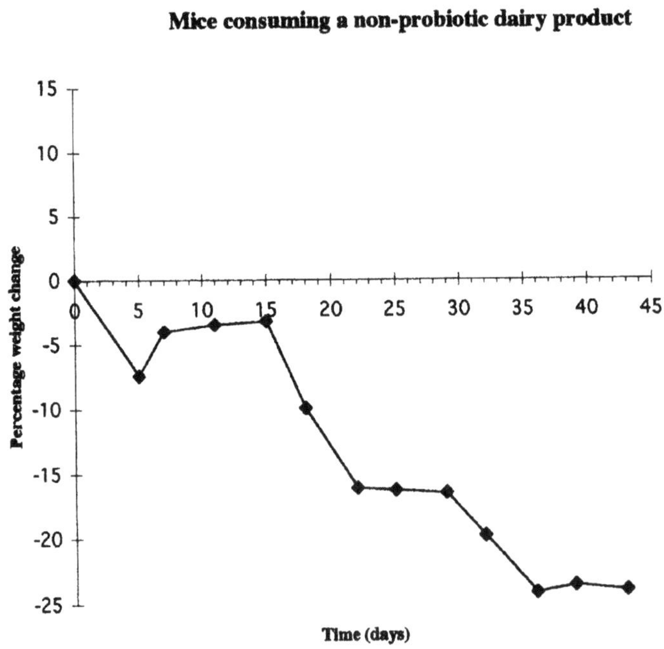

Figure 3. Assessment of weight loss in CD45RB^{high} SCID (Severe Combined ImmunoDeficient) mice consuming a probiotic combination of *Lb. salivarius* subsp. *salivarius* UCC118 and *Bifidobacterium longum infantis* UCC35624 compared with those fed a non-probiotic dairy product.

here support suggestions proposed in other studies (Bartram et al. 1994) that strains of bifidobacteria may play important roles in maintaining a balanced healthy intestinal microflora.

Feeding studies involving healthy human adults

Following the successful completion of the mouse feeding trial (Murphy et al. submitted), an eighty-volunteer, double-blinded, placebo-controlled, ethically approved feeding trial was performed which compared the efficacy of two standard food systems (fermented milk and fresh milk) in delivering *Lactobacillus salivarius* strain UCC118 to the human gastrointestinal tract. The results of this trial have indicated that both fresh milk and yoghurt are effective vehicles for the bacterial strain (Morrissey et al. unpublished observations). In addition, in a proportion of volunteers (<10%) significant numbers of strain UCC118 were still detectable in faeces 3 weeks after cessation of its administration, indicating that the strain was capable of persisting within the human GIT (Morrissey et al. unpublished observations). Interestingly, strain UCC118 did not obviously influence the numbers of other lactobacilli in the gut, but did cause significant alterations in the numbers of excreted clostridia and enterococci. Notably, these results correlate well with those observed by Spanhaak et al. (1998) upon completion of a study assessing the survival of the commercially-available probiotic strain *Lactobacillus casei* Shirota during passage through the gastrointestinal tract of healthy human males. However, while the *Lb. casei* Shirota strain was also found capable of surviving transit through the human intestinal environment, in this feeding study a significant increase was observed in the numbers of excreted faecal-borne bifidobacteria following four weeks of probiotic intake. In addition, it was found that the numbers of *Clostridium* and *Enterococcus* were not significantly influenced by consumption of the *Lb. casei* Shirota probiotic product (Spanhaak et al. 1998).

Several immunological parameters were assessed during the completion of both of the described human feeding studies. During the administration of *Lb. casei* Shirota there were no distinct effects on assessed immune responses (which included natural killer cell activity; phagocyte functions; delayed-type hypersensitivity; and cytokine and humoral parameters) (Spanhaak et al. 1998). Analysis of immune parameters in the *Lb. salivarius* UCC118 feeding study demonstrated that macrophage phagocytic activity also remained unchanged over the period of probiotic consumption, while granulocyte phagocytic activity significantly increased in those volunteers consuming yoghurt-borne strain UCC118. Systemic antibody titres (both total antibody levels and antibodies specific to strain UCC118) remained constant. In addition, levels of systemic IL-1α, IL-1β, IL-2 soluble receptor, IL-6, TNFα and IFNγ remained unaltered throughout the trial period, indicating that consumption of *Lb. salivarius* strain UCC118 did not induce significant inflammatory responses in healthy subjects. Mucosal IgA antibodies titres, specific to the probiotic strain, increased significantly in volunteers consuming the probiotic yoghurt product, while total IgA levels remained unchanged. Thus, consumption of strain UCC118 in a fermented dairy product resulted in stimulation of the mucosal immune system, but did not appear to induce systemic responses.

Conclusions

In this report, we have attempted to describe the logical criteria adopted for *in vitro* selection of probiotic bacteria. However, given that the human gastrointestinal tract is a complex and hostile environment, we acknowledge that it is unlikely that a single probiotic bacterial strain will have the capacity to influence the microbial ecology of the host and, by doing so, beneficially affect lactose intolerance, the incidence of diarrhoea, mucosal immune responses, levels of blood cholesterol, and the induction of cancer. More likely, we feel that these effects will require the introduction of combinations of strains, such as those described above in relation to the prevention of weight-loss in diseased mice. Furthermore, it is essential that these probiotic strains are not developed as individual entities but rather as the active ingredients of the food products which are ultimately intended for human consumption.

Acknowledgements

This work was supported in part by grants awarded from Bio Research Ireland; the Irish Department of Agriculture and Food non-commissioned food research programme; and DGXII of the European Commission (PROBDEMO: FAIR-CT96-1028 and LAB-VAC: BIO4-CT96-0542).

References

Abee T, Klaenhammer TR & Letellier L (1994) Kinetic studies of the action of Lactacin F, a bacteriocin produced by *Lactobacillus johnsonii* that forms poration complexes in the cytoplasmic membrane. Appl. Environ. Microbiol. 60: 1006–1013

Alander M, Korpela R, Saxelin M, Vilpponen-Salmela T, Mattila-Sandholm T & von Wright A (1997) Recovery of *Lactobacillius rhamnosus* GG from human colonic biopsies. Lett. Appl. Microbiol. 24: 361–364

Altschul SF, Gish W, Miller W, Myers EW & Lipman DJ (1990)Basic local alignment search tool. J. Mol. Biol. 215: 403–410

Aranda R, Sydora B, McAllister PL, Binder SW, Ying Yang H, Targan SR & Kronenberg M (1997) Analysis of intestinal lymphocytes in mouse colitis mediated by transfer of $CD4^+$, CD45 RB ^{high}T cells to SCID recipients. J. Immunol. 3464–3473

Aso Y & Akazan H (1992) Prophylactic effect of a *Lactobacillus casei* preparation on the recurrence of superficial bladder cancer. Urol. Int. 49: 125–129

Bartram HP, Scheppach W, Gerlach S, Ruckdeschel G, Kelber E & Kasper H (1994) Does yogurt enriched with *Bifidobacterium longum* affect colonic microbiology and fecal metabolites in healthy subjects? Am. J. Clin. Nutr. 59: 428–432

Berg RD (1998) Probiotics, prebiotics or 'conbiotics'. Trends Microbiol. 6: 89–92

Bernet M-F, Brassart D, Neeser J-R & Servin AL (1993) Adhesion of human bifidobacterial strains to cultured human intestinal epithelial cells and inhibition of enteropathogen-cell interactions. Appl. Environ. Microbiol. 59: 4121–4128

Bernet MF, Brassart D, Neeser JR & Servin AL (1994) *Lactobacillus acidophilus* LA-1 binds to cultured human intestinal cell-lines and inhibits cell attachment and cell invasion by enterovirulent bacteria. Gut 35: 483–489

Brassart D, Schiffrin E, Rochat F, Offord EA, Macé C & Neeser J-R (1998) The future of functional foods: Scientific basis and future requirements. Lebensmittel Technol. 7–8: 258–266

Coconnier M-H, Klaenhammer TR, Kerneis S, Bernet M-F & Servin AL (1992) Protein-mediated adhesion of *Lactobacillus acidophilus* BG2FO4 on human enterocyte and mucus-secreting cell lines in culture. Appl. Environ. Microbiol. 58: 2034–2039

Collins JK, Thornton G & Sullivan GD (1998) Selection of probiotic strains for human applications. Int. Dairy J. 8: 487–490

Cornwell GG, Sletten K, Johansson B & Westermark P (1988) Evidence that the amyloid fibril protein in senile systemic amyloidosis is derived from normal prealbumin. Biochem. Biophys. Res. Commun. 154: 648–653

Crociani J, Grill J-P, Huppert M & Ballongue J (1995) Adhesion of different bifidobacteria strains to human enterocyte-like Caco-2 cells and comparison with *in vivo* study. Lett. Appl. Microbiol. 21: 146–148

Daly C & Davis R (1998) The biotechnology of lactic acid bacteria with emphasis on applications in food safety and human health. Agricultural Food Sci. in Finland 7: 251–265

De Man JC, Rogosa M & Sharpe MT (1960) A medium for the cultivation of lactobacilli. J. Appl. Bacteriol. 23: 130–135

Dodd HM & Gasson MJ (1994) Bacteriocins of lactic acid bacteria. In: Gasson MJ & de Vos WM (Eds) Genetics and biotechnology of lactic acid bacteria (pp 211–251). Blackie Academic and Professional, Glasgow

Draser BS, Shiner M & McLeod GM (1969) Studies on the intestinal flora 1. The bacterial flora of the gastrointestinal tract in healthy and achlorhydric persons. Gastroenterol. 56: 71–79

Dunne C, O'Halloran S, O'Mahony L, Feeney M, Kiely B, von Wright A, Vilpoonen-Salmela T, Shanahan F, O'Sullivan G & Collins JK (1999) Epithelial adhesion of probiotic microorganisms *in vitro* and *in vivo*. Gastroenterol. 116: 4 (2): G3058

Favier C, Neut C, Mizon C, Cortot A, Colombel JF & Mizon J (1997) Fecal B-D-Galactosidase production and bifidobacteria are decreased in Crohn's disease. Dig. Dis. Sci. 42: 817–822

Floch MH, Binder HJ, Filburn B & Gershengoren W (1972) The effect of bile acids on intestinal microflora. Am. J. Clin. Nutr. 25: 1418–1426

Fuller R (1989) A review: probiotics in man and animals. J. Appl. Bacteriol. 66: 365–378

Fonty G, Raibaud P & Gouet P (1993) Manipulation of the gut microflora: experimental approach in animals. Proc. Nutr. Soc. 52: 345–356

Fykse EM, Sletten K, Husby G & Cornwell GG (1988)The primary structure of the variable region of an immunoglobulin IV light-chain amyloid-fibril protein (AL GIL). Biochem. J. 256: 973–980

Goldin BR & Gorbach SL (1992) Probiotics for humans. In: Fuller R (Ed) Probiotics, the scientific basis (pp 355–376). Chapman and Hall, London

Goldin BR, Gorbach SL, Saxelin M, Barakat S, Gualtieri L & Salminen S (1992) Survival of *Lactobacillus* species (Strain GG) in human gastrointestinal tract. Dig. Dis. Sci. 37: 121–128

Greene JD & Klaenhammer TR (1994) Factors involved in adherence of lactobacilli to human Caco-2 cells. Appl. Environ. Microbiol. 60: 4487–4494

Guarner F & Schaafsma GJ (1998) Probiotics. Int. J. Food Microbiol. 39: 237–238

Hill MJ & Draser BS (1968) Degradation of bile salts by human intestinal bacteria. Gut 9: 22–27

Hoffman AF, Molino G, Milanese M & Belforte G (1983) Description and stimulation of a physiological pharmokinetic model for the metabolism and enterohepatic circulation of bile acids in man. J. Clin. Invest. 71: 1003–1022

Huis in't Veld J & Shortt C (1996) Selection criteria for probiotic micro-organisms. In: Leeds AR & Rowland IR (Eds) Gut Flora and Health – Past, Present and Future (pp 19–26). The Royal Society of Medicine Press Ltd, London

Hylemon PB & Glass TL (1983) Biotransformation of bile acids and cholesterol by the intestinal microflora. In: Hentges DJ (Ed) Human intestinal microflora in health and disease (pp 189–213). Academic Press, New York

Isolauri E, Juntunen M, Saxelin M & Vesikari T (1995) Lactic acid bacteria in the treatment of acute rotavirus gastroenteritis. J. Pediatr. Gastroenterol. 20: 333–336

Jack RW, Tagg JR & Ray B (1995) Bacteriocins of Gram positive bacteria. Microbiol. Rev. 59: 171–200

Kandler O & Weiss N (1986) Regular, non-sporing Gram positive rods. In: Sneath PHA, Mair N, Sharpe ME & Holt JG (Eds) Bergey's Manual of Systematic Bacteriology (pp 2: 1208–1234) Williams and Wilkins, Baltimore

Kerneis S, Bilge SS, Fourel V, Chauviere G, Coconnier M-H & Servin AL (1991) Use of purified F1845 fimbrial adhesin to study localization and expression of receptors for diffusely adhering *Escherichia coli* during enterocytic differentiation of human colon carcinoma cell lines HT-29 and Caco-2 in culture. Infect. Immun. 59: 4013–4018

Lee Y-K & Salminen S (1995) The coming age of probiotics. Trends Food Sci. Technol. 6: 241–245

Lewis R & Gorbach S (1972) Modification of bile acids by intestinal bacteria. Arch. Intern. Med. 130: 545–549

Lidbeck A, Overnick E, Rafter J, Nord CE & Gustaffsson JA (1992) Effect of *Lactobacillus acidophilus* supplements on mutagen excretion in faeces and urine in humans. Micro. Ecol. Health. Dis. 5: 59–67

Marteau P, Flourie B, & Pochart P et al. (1990) Effect of chronic ingestion of a fermented dairy product containing *Lactobacillus acidophilus* and *Bifidobacterium bifidum* on metabolic activities of the colonic flora in humans. Am. J. Clin. Nutr. 52: 685–688

Marteau P & Rambaud J-C (1993) Potential of using lactic acid bacteria for therapy and immunomodulation in man. FEMS Microbiol. Rev. 12: 207–220

Mattila-Sandholm T (1997) Demonstration Project FAIR CT96-1028. In: Alander M, Kauppila T & Mattila-Sandholm T (Eds) Novel methods for probiotic research: 2nd workshop demonstration of the nutritional functionality of probiotic foods FAIR CT96-1028 (pp 11–17). Technical Research Centre of Finland (VTT)

McAuliffe O, Ryan MP, Ross RP, Hill C, Breeuwer P & Abee T (1998) Lanticin 3147, a broad-spectrum bacteriocin which selectively dissipates the membrane potential. Appl. Environ. Microbiol. 64: 439–445

McCracken VJ & Gaskins HR (1999) Probiotics a critical review. Horizon Scientific Press, UK

Millar MR, Bacon C, Smith SL, Walker V & Hall MA (1993) Enteral feeding of premature infants with *Lactobacillus* GG. Arch. Dis. Child. 69: 483–487

Mitsuoka T (1984) Taxonomy and ecology of bifidobacteria. Bifidobacteria Microflora 3: 11–28

Molin G, Jeppsson B & Johansson ML et al. (1993) Numerical taxonomy of *Lactobacillus* sp. associated with healthy and diseased mucosa of the human intestines. J. Appl. Bacteriol. 1993: 74: 314–323

Morotomi M, Watanabe T, Suegara N, Kawai Y & Mutai M (1975) Distribution of indigenous bacteria in the digestive tract of conventional and gnotobiotic rats. Infect. Immun. 11: 962–968

Morrisey PJ, Charrier K, Braddy S, Liggitt D & Watson JD (1993) $CD4^+$ T cells that express high levels of CD45RB induce wasting disease when transferred into congenic severe combined immunodeficient mice. Disease development is prevented by cotransfer of purified $CD4^+$ T cells. J. Exp. Med. 178: 237–244

Murphy L, Dunne C, Kiely B, Shanahan F, O'Sullivan GC & Collins JK (Submitted)*In vivo* assessment of potential probiotic *Lactobacillus salivarius* strains: evaluation of their establishment, persistence, and localisation in the murine gastrointestinal tract

Neeser J-R, Chambaz A, Golliard M, Link-Amster H, Fryder V & Kolodziejczyk E (1989) Adhesion of colonisation factor antigen II-positive enterotoxigenic *Escherichia coli* strains to human enterocytelike differentiated HT-29 cells: a basis for host-pathogen interactions in the gut. Infect. Immun. 57: 3727–3734

Neutra MR & Kraehenbuhl J-P (1996) Essentials of mucosal immunology. Academic Press, San Diego

Oksanen PI, Salminen S & Saxelin M et al. (1990) Prevention of traveller's diarrhea by *Lactobacillus* GG. Ann. Med. 22: 53–56

O'Riordan K & Fitzgerald GF (1997) Determination of genetic diversity within the genus *Bifidobacterium* and estimation of chromosomal size. FEMS Microbiol. Lett. 156: 259–264

O'Sullivan MG, Thornton GM, O'Sullivan GC & Collins JK (1992) Probiotic bacteria: myth or reality? Trends Food Sci. Technol. 3: 309–314

Ouwehand AC (1998) Antimicrobial components from lactic acid bacteria. In: Salminen S & von Wright A (Eds) Lactic acid bacteria: Microbiology and functional aspects, 2nd edition (pp 139–160). Marcel Dekker Inc., New York

Pedrosa MC, Golner BB, Goldin BR, Barakat S, Dallal GE & Russell RM (1995) Survival of yogurt-containing organisms and *Lactobacillus gasseri* (ADH) and their effect on bacterial enzyme activity in the gastrointestinal tract of healthy and hypochlorhydric elderly subjects. Am. J. Clin. Nutr. 61: 353–359

Percy-Robb IW & Collee JG (1972) Bile acids: a pH dependent antibacterial system in the gut? Brit. Med. J. 3: 813–815

Salminen S, Deighton MA & Gorbach SL (1993) Lactic acid bacteria in health and disease. In: Salminen S. & von Wright A (Eds) Lactic acid bacteria (pp 199–225). Marcel Dekker Inc., New York

Salminen S, Isolauri E & Onnela T (1995) Gut microflora in health and disease. Chemotherapy 41(suppl 1): 5–15

Salminen S, Isolauri E & Salminen E (1996) Clinical uses of probiotics for stabilising the gut mucosal barrier: successful strains and future challenges. Antonie van Leeuwenhoek 70: 251–262

Salminen S, Deighton MA, Benno Y & Gorbach SL (1998a) Lactic acid bacteria in health and disease. In: Salminen S. & von Wright A (Eds) Lactic acid bacteria: Microbiology and functional aspects, 2nd edition (pp 211–254). Marcel Dekker Inc., New York

Salminen S, Bouley C, Boutron-Ruault M-C, Cummings JH, Franck A, Gibson GR, Isolauri E, Moreau M-C, Roberfroid M & Rowland I (1998b) Functional food science and gastrointestinal physiology and function. Brit. J. Nutr. 80 (suppl 1): s147–s171

Sarem F, Sarem-Damerdji LO & Nicolas JP (1996) Comparison of the adherence of three *Lactobacillus* strains to Caco-2 and Int-407 human intestinal cell lines. Lett. Appl. Microbiol. 22: 439–442

Savage DC (1997) Microbial Ecology of the Gut. Academic Press, London

Saxelin M (1997) *Lactobacillus* GG – a human probiotic strain with thorough clinical documentation. Food Reviews Int. 13: 293–313

Schaafsma G (1995) Application of lactic acid bacteria in novel foods from a nutritional perspective. In: Novel G & Le Querler J-F (Eds) Lactic acid bacteria: Actes du Colloque LACTIC 94 (pp 85–93). Presses Universitaires de Caen

Sepp E, Mikelsaar M & Salminen S (1993) Effect of administration of *Lactobacillus casei* strain GG on the gastrointestinal microbiota of newborns. Microbiol. Ecol. Health Dis. 6: 309–314

Shanahan F (1994) Physiology of the gastrointestinal tract. Raven Press

Shimada K, Bricknell KS & Finegold SM (1969) Deconjugation of bile acids by intestinal bacteria: a review of literature and additional studies. J. Infect. Dis. 119: 73–81

Siitonen S, Vapautalo H & Salminen S et al. (1990) Effect of Lactobacillus GG yogurt in prevention of antibiotic-associated diarrhea. Ann. Med. 22: 57–59

Sletten K & Husby G (1974) The complete amino acid sequence of non-immunoglobulin amyloid proteins AS in rheumatoid arthritis. Eur. J. Biochem. 41: 117–125

Spanhaak S, Havenaar R & Schaafsma G (1998) The effect of consumption of milk fermented by *Lactobacillus casei* strain Shirota on the intestinal microflora and immune parameters in humans. Eur. J. Clin. Nutr. 52: 899–907.

Stewart L, Pellegrini CA & Way LW (1986) Antibacterial activity of bile acids against common biliary tract organisms. Surg. Forum 37: 157–159

Tagg JR & McGiven AR (1971)Assay systems for bacteriocins. App. Microbiol.21:943

Tannock GW (1997) Probiotic properties of lactic-acid bacteria: plenty of scope for fundamental R & D. TIBTECH 15: 270–274

Thornton GM (1996) Probiotic bacteria. Selection of *Lactobacillus* and *Bifidobacterium* strains from the healthy human gastrointestinal tract; characterisation of a novel *Lactobacillus*-derived antibacterial protein. (Thesis). National University of Ireland

Tuomola (née Lehto) EM & Salminen SJ (1998) Adhesion of some probiotic and dairy *Lactobacillus* strains to Caco-2 cell cultures. Int. J. Food Microbiol. 41: 45–51

Wostmann BS (1996) Germfree and gnotobiotic animal models. CRC Press, Boca Raton

Antonie van Leeuwenhoek **76:** 293–315, 1999.

Bringing a probiotic-containing functional food to the market: microbiological, product, regulatory and labeling issues

Mary Ellen Sanders[1,2,*] & Jos Huis in't Veld[3,4]
[1]*Dairy and Food Culture Technologies, 7119 S. Glencoe Ct., Littleton, Colorado 80122, USA;*
[2]*Dairy Products Technology Center, California Polytechnic State University, San Luis Obispo, California 93407, USA;* [3]*School of Veterinary Medicine, Utrecht University, Utrecht, The Netherlands;*
[4]*Yakult Europe B.V., Prof. J.H. Bavincklaan 5, 1183 AT Amstelveen, The Netherlands*
(**Author for correspondence: E-mail: mesanders@msn.com*)

Abstract

Properly formulated probiotic-containing foods offer consumers a low risk, low cost dietary component that has the potential to promote health in a variety of ways. Several such products are available commercially, although markets in Japan and Europe are more developed than in the USA. Once healthful attributes of a probiotic product have been identified, there remain microbiological, product, regulatory and labeling issues to be addressed prior to marketing. Microbiological and product issues include safety, effective scale-up for manufacturing, definition of probiotic activity, probiotic stability in the product over the course of product manufacture, shelf-life and consumption, definition of effective dose and target population(s), and development of quality assurance approaches. Examples of probiotic-containing foods are given. Regulatory and labeling issues are complicated because they differ for each country, but are likewise critical because they provide the means for communication of the product benefits to the consumer. The regulatory climate worldwide appears to be one of caution about overstating the benefits of such products but at the same time not preventing corporate commitment to marketing.

Introduction

The increasing commercial interest in probiotic products is the rationale for this paper which will review the microbiological, product and regulatory issues around marketing a probiotic product, and will identify some critical questions for future investigation into probiotics.

Current science-based investigations and research in Asian, North American, Australian and European countries are helping to build a sound body of evidence regarding diet and health. The primary role of our diet is to provide enough nutrients to fulfill requirements while giving us a feeling of satisfaction and well-being. However, recent discoveries in several aspects of bioscience support the hypothesis that, beyond nutrition, diet may modulate various functions in the body. These new concepts have lead to the introduction of functional foods encompassing a wide range of ingredients and functional aspects (Roberfroid 1998). Specific dairy products fermented by selected lactic acid ('probiotic') bacteria which function in intestinal ecology and physiology belong to these categories and form the basis for this paper.

Probiotics for gastrointestinal implication are defined as living microorganisms which upon ingestion in certain numbers exert health effects beyond basic nutrition (Guarner & Schaafsma 1998). Several scientific works (Tomera 1996; Fonden et al. 1999; Tannock 1999) review of the area of physiological activity, mechanisms and clinical studies. The key points of accepted and potential benefits are briefly summarized in Table 1.

How and to what extent different strains and doses of exogenously supplied microbes influence human health is still an area under investigation. However, there is no question that the presence and activity of the 10^{14} microbes colonizing the human body lead to profound influences on human physiology (Holzapfel et al. 1998; Tannock 1998). The ability of probiotic bacteria entering this ecosystem to influence the activities of colonizing flora, to exert direct effects on the intestinal mucosa or to deliver functional enzymes or other cellular products or byproducts is the scientific

Table 1. Researched effects of probiotic bacteria

Target Health Benefit	Suspected mechanism
Aid in lactose digestion	Bacterial lactase acts on lactose in the small intestine
Resistance to enteric pathogens	– Adjuvant effect increasing antibody production – Systemic immune effect – Colonization resistance – Alteration of intestinal conditions to be less favorable for pathogenicity (pH, SCFA, bacteriocins)
Anti-colon cancer effect	– Antimutagenic activity – Alteration of pro-cancerous enzymatic activity of colonic microbes – Stimulation of Immune function – Influence on 2° bile salt concentration
Small bowel bacterial overgrowth	– Lactobacilli influence activity of overgrowth flora, decreasing toxic metabolite production – Antibacterial characteristics
Immune system modulation	– Strengthening of non-specific and antigen-specific defense against infection and tumors – Adjuvant effect in antigen-specific immune responses – Regulating/influencing Th1/Th2 cells, production cytokines
Allergy	Prevention of antigen translocation into blood stream
Blood lipids, heart disease	– Assimilation of cholesterol by bacterial cell – Alteration of activity of bile salt hydrolase enzyme – Antioxidative effect
Antihypertensive effect	– Bacterial peptidase action on milk protein results in antihypertensive tripeptides – Cell wall components act as ACE inhibitors
Urogenital Infections	– Adhesion to urinary and vaginal tract cells – Competitive exclusion – Inhibitor production (H_2O_2, biosurfactants)
Infection caused by *Helicobacter pylori*	Competitive colonization
Hepatic encephalopathy	Competitive exclusion or inhibition of urease-producing gut flora

basis of the probiotic theory and the rationale for the development of probiotic-containing products.

Frequently, probiotic microbes are members of the genus *Lactobacillus* or *Bifidobacterium* (Table 2), but other non-pathogenic bacteria (including members of the genera *Enterococcus* and yeast) are also used as probiotics. Daily ingestion in minimum doses of 10^{9-10} viable microbes has often been the minimum required to show a health effect, although dose response studies are often not included in clinical evaluations of probiotics.

To study health effects of probiotics , the incidence and/or duration of an acute, short term disease, such as diarrhea, are frequently measured. The effect of probiotic bacteria on the incidence of diseases with a protracted etiology, such as cancer or heart disease,

Table 2. Commercially used probiotic lactobacilli and bifidobacteria with some measure of health characteristics documentation

Company	Strain
Valio Dairy, Helsinki, Finland	*Lactobacillus rhamnosus* GG
Nestle, Lausanne, Switzerland	*Lactobacillus johnsonii* La1 (Lj1)
Yakult, Tokyo, Japan	*Lactobacillus casei* strain Shirota *Bifidobacterium breve* strain Yakult
Morinaga Milk Industry Co. Ltd., Zama City, Japan	*Bifidobacterium longum* BB536
Rhodia, Madison, Wisconsin, USA	*Lactobacillus acidophilus* NCFM®
Chr. Hansens, Milwaukee, Wisconsin, USA	*Lactobacillus casei* CRL 431 Gilliland (La-Mo)
BioGaia, Raleigh, North Carolina, USA	*Lactobacillus reuteri* SD2112
Probi, Lund, Sweden	*Lactobacillus plantarum* 299V *Lactobacillus rhamnosus* 271
Urex Biotech Inc., London, Ontario, Canada	*Lactobacillus fermentum* *Lactobacillus rhamnosus*
Danone, Le Plessis-Robinson, France	*Lactobacillus casei* DN014 001 (*Immunitas*)
Biocodex, Inc., Seattle, Washington, USA	*Saccharomyces boulardii*
Meiji Milk Products, Tokyo, Japan	*Lactobacillus delbrueckii* subsp. *bulgaricus* 2038 *Streptococcus thermophilus* 1131
Snow Brand Milk Products, Tokyo, Japan	*Lactobacillus acidophilus* SBT-2062 *Bifidobacterium longum* SBT-2928

have generally not been measured. [One exception is the study by Aso & Akazan (1992) on the influence of probiotic bacteria on recurrence of bladder cancer.] Instead, studies have focused on biomarkers of these diseases, or animal studies. Epidemiological evidence comparing the health of sizable populations that consume foods containing defined doses of specific probiotic strains with those that don't is lacking. A few case-controlled epidemiology studies have shown an inverse relationship between yoghurt or fermented milk consumption and breast or colorectal cancers (Monique et al. 1986; van't Veer et al. 1989; Peters et al. 1992; Kampman et al. 1994), but conclusions on probiotic effects on these diseases cannot be made from these studies, partially due to lack of definition of probiotic content of products, and partly due to the multitude of other dietary factors in fermented milks or yoghurts that may play a role in the etiology of these diseases. A long-term prospective study following probiotic consuming and non-consuming populations would be more telling. However, epidemiology studies are not useful for detecting small effects (Langseth 1996), and the results of such a study might be inconclusive.

Probiotics may have health benefits and be particular candidates for consideration or authorization of health claims. Health claims should be based only on sound science to ensure the flow of truthful, non-misleading, clearly understandable information to the consumer in order to aid choice of healthier diets and, thereby, maintain or improve their health.

This article will discuss the microbiological, product, regulatory and labeling issues involved in the introduction of a probiotic product in different parts of the world.

Strain selection

Although progress in probiotic research has been achieved over the last few years, not all of the available probiotic bacteria which are on the market have adequate scientific documentation. If nutritional and health benefits are to be derived from products containing probiotic bacteria, it is desirable that we understand the mechanisms by which these benefits are derived and use those strains demonstrating the most promise in this regard. The probiotic concept will only gain acceptance if these underlying mechanisms are elucidated. Consequently, it is necessary to establish rational criteria for the screening and selection of candidate microorganisms (Havenaar et al. 1992) and also to evaluate the efficacy of the selected strains or the food products in well-controlled human clinical trials.

In general, it cannot be assumed that activities exhibited by one probiotic strain are going to be the same for another strain, even of the same genus and species. This is not unexpected considering the wide ranges of results observed from *in vitro* tests and can be explained by strain-specific differences in biochemistry or physiology. These strain-specific activities justify efforts to identify the best strain or combination of strains for specific health targets. One decision making scheme for selecting strains for probiotic use is shown in Figure 1.

Several efforts to compare laboratory or commercial strains across a range of *in vitro* activities have been reported and recently reviewed (Collins et al. 1998; Collins, this journal). Although numerous *in vitro* strain comparison efforts have been published, there are few human clinical studies which compare strains. Clinical evaluations are usually designed to compare a test strain to a placebo control, not to compare efficacy of more than one strain. This is motivated by cost, and by the fact that many clinical studies are funded by commercial enterprises with specific interest in only one strain. *In vitro* screening of candidate probiotic strains is thought to provide some insight into wise choices for *in vivo* functionality (Huis in't Veld & Havenaar 1997; Collins et al. 1998). At best, expectations from these *in vitro* efforts eliminate strains which are unlikely to perform well *in vivo*. At worst, *in vitro* analysis leads to conclusions based on flawed logic stemming from limited knowledge of the complex human system.

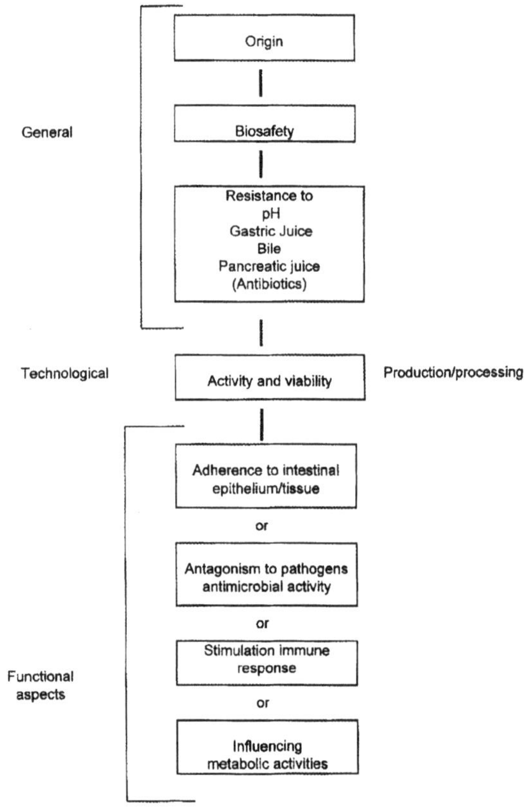

Figure 1. Strategy for selection of probiotic strains.

The ability of in vitro screening to provide physiologically-relevant results may be improved by use of a dynamic rather than static model approach. Such a model has been developed by researchers at TNO Nutrition and Food Research at Zeist in The Netherlands and described in detail by Minekus et al. (1995), validated and applied in relation to the survival of ingested lactic acid bacteria (Marteau et al. 1997). The model is composed of all the major constituents of the human gastrointestinal system (stomach, small intestinal and colon) and has a number of extra features in comparison to other existing models, such as simulation of peristaltic movement, physiological pH curves, gastrointestinal secretion, intestinal absorption of nutrients and water, physiological gastric emptying, intestinal transport patterns, and high efficiency water absorption in colon. Recently a new type of large intestinal model also has been developed (Minekus et al., in press), combining removal of metabolites and water, which is necessary to obtain physiological concentrations of micro-organisms, dry matter and microbial metabolites. High densities of micro-organisms, comparable to that found in the colon *in vivo*, were achieved by absorption of water and dialysis of metabolites through hollow fibres inside the compartments. Incubation with faecal flora resulted in total anaerobic bacterial counts of above 10^{10}CFU/ml with stable physiological levels of *Bifidobacterium, Lactobacillus, Enterobacteriaceae* and *Clostridium*. The dry matter content was approximately 10%, while the total short chain fatty acids concentration was maintained at physiological levels with similar molar ratios of acetic acid, propionic acid and butyric acid as measured *in vivo*. Studies, so far, have shown that this dynamic TNO model accurately reproduces the human conditions supporting its validity.

Kinetic parameters are very important for studying the survival of micro-organisms in the gastrointestinal tract. The influence of separate gastrointestinal factors on gastric and intestinal delivery of viable lactic acid bacteria were investigated using this dynamic model (Marteau et al. 1997). Survival data obtained for *L. acidophilus* and *Bifidobacterium* species were similar to those obtained for the same strains in human intubation studies, showing the value of this dynamic model.

As in other *in vitro* models, this dynamic model also has certain disadvantages. For example there are no epithelium, active transport of metabolites or direct immune response. However, compared to static *in vitro* systems this model seems to be more realistic. Finally the model can be used for studying the ecology of the intestinal flora, including the role of non-digestible oligosaccharides, (pro)carcinogen bioconversions, and the role of antimicrobial components.

Definition of probiotic activity

In general, the presumption is that probiotic viability is a reasonable measure of activity. This is certainly a defensible assumption, considering the definition of probiotics offered earlier in this paper (see Introduction). In most cases, even if viability is not required, it is likely correlated with most effects since it is a useful indicator of the number of cells present, regardless of what cell component may be active. Situations where viability is not required for probiotic activity include improved digestion of lactose (Vesa et al. 1996), some immune system modulation (Perdigon et al. 1986; Tomioka & Saito 1992; Solis Pereyra & Lemonnier 1993; Hosono et al. 1997; Marin et al. 1997) activities, and anti-hypertensive (Maeno et al. 1996) effects. Effects have been linked to non-viable cells (cell components, enzyme activities or fermentation products). However, the method of killing probably influences the result. In the case of β-galactosidase activity, one study showed that cells killed using gamma-irradiation maintained the ability to promote lactose digestion in rats and piglets (Titze et al. 1996). This result is not contradictory to reports which demonstrate decreased effectiveness of pasteurized yoghurt on reducing breath hydrogen levels (Savaiano et al. 1984). Cells killed by treatment which decreases lactase activity would be expected to yield different results than a killing method which maintained lactase activity. In the case of immune system effects, bacterial cell wall components or heat-killed whole cells have been identified as active biological response modifiers (Perdigon et al. 1986; Hosono et al. 1997; Marin et al. 1997). An interesting case of a functional effect from a pasteurized probiotic product is the FOSHU-approved Japanese drink, Ameal-S, produced by the Calpis Co. Ltd., Kanagawa, Japan (Hata et al. 1996). This product is made by the fermentation of milk by *Lactobacillus helveticus* CP790. A cell-wall-associated proteinase yields at least two anti-hypertensive tripeptides from digestion of casein (Maeno et al. 1996). Viable cells are not required for functionality in this case. Definition of the active property of a probiotic product is important in developing quality control parameters of production and storage. Efforts to maximize shelf-life must be focused on maintaining optimal levels of this ingredient, be it the intact, viable cell or some cell component(s).

Consumption level

A related effort is the definition of an efficacious consumption level of the probiotic. Dose studies conducted without a clear understanding of the active principle could be futile. Having said this, the fact remains that for the most part, probiotic products are standardized based on viable count, with the assumption that this is the important factor to consider in product functionality.

As yet not very much is known about the minimal dose and/or frequency of probiotic consumption required for a physiological effect. Several clinical trials have shown that consumption of probiotics results in recovery of 10^6–10^8 per gram of the specific probiotic lactobacilli in human faeces (Goldin 1992; Spanhaak et al. 1998; Dunne et al., these proceedings). They have also shown that these strains were the most dominant faecal lactobacilli. The TNO *in vitro* gastrointestinal model showed survival of some probiotic bacteria between 10–40% of the total numbers. For functionality of probiotics it is thought that at least 10^8–10^9 live bacteria should reach the small intestine daily. When this is true the actual total daily dose must be about 10^9–10^{10} probiotic bacteria. However, comparison of the total dose and survival rate (10–40%) to the measured number of probiotic strains in faecal samples suggests that growth of probiotic bacteria in the intestinal tract will also occur. It is essential that future research further clarify this important issue.

Factors contributing to probiotic biological activity

Since probiotics are biological entities, there are many factors that might contribute to their biological activity.

1. Conditions of the consuming host. The goal of high potency probiotic products is to maximize the likelihood of delivering high populations of viable cells to the target site, usually the small or large intestine, but also including sites such as the mouth, stomach, vagina or urinary tract. Therefore, the relevant characteristic is not necessarily dose consumed, but dose reaching the target site in the body. In consideration of this fact, it is conceivable that strains particularly well-suited to survival of passage through the gastrointestinal tract could be provided in lower numbers in a product than another, more poorly adapted strain. An additional complication is that survival and growth of probiotics once consumed are certainly dependent on what conditions are encountered inside the human body, and these conditions cannot be expected to be the same for all individuals. *In vivo* probiotic survival would not be the same, for example, in elderly consumers with atrophic gastritis compared to young adult consumers in prime health. Products targeted for specific populations perhaps suffering from compromised health may be designed to take these factors into consideration. However, this type of product targeting is less feasible for probiotic food products.
2. Genetic makeup of the strain.
3. Conditions of growth. The fermentation technology used during strain or product production is of major importance in microbiological stability. The effects of processing (mixing, packaging, homogenization), subsequent handling, the composition of the bacterial growth medium, temperature of growth, use of techniques to induce a stress response, length of time of fermentation, and other conditions all contribute to the ability of probiotic cells to maintain viability and activity during storage.
4. Composition and processing history of the growth medium for strain or food product production.
5. Conditions of preservation and storage. The use of protective chemicals, encapsulation, or microencapsulation during the freezing, drying or storage of probiotic strains can promote stability greatly. Temperature, water activity, levels of oxygen and the presence of stabilizing chemicals can greatly influence probiotic stability during storage. Figure 2 establishes the effect of temperature on short term stability of a probiotic strain, and demonstrates the importance of refrigerated temperature-stable cultures. This product has a 6-day fermentation time in order to promote survival by adaptation of probiotic microorganism to acid conditions. The shelf life of this product is 30 days.
6. Methods of measuring viability. Different microbiological enumeration conditions can yield very different results on levels of viable cells. The presence of selective agents, which may be required for differential enumeration of component strains in a probiotic strain blend, can reduce the levels of cells recovered (Sanders et al. 1996), especially when sublethally injured cells are present. Since the ability of injured cells to resist acid and bile will be compromised, it can certainly be postulated that use of bile or low pH in a medium used for enumerating probiotic microbes would be more likely to reflect a physiologically relevant, albeit lower, count of probiotic. Even better is the use of a dynamic model to measure viability under physiological conditions.

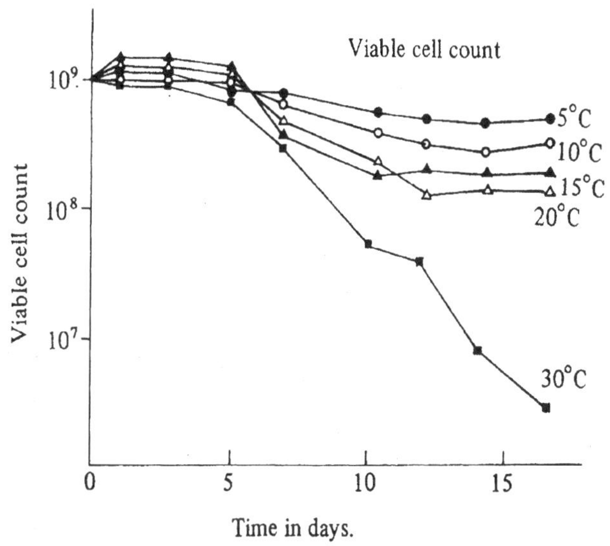

Figure 2. Changes in viable cell count of Yakult beverage containing *Lactobacillus casei* Shirota during storage at various temperatures at pH 3,6.

Shelf-life

In general, a comprehensive approach to shelf-life of probiotic bacteria is tied to maintenance of efficacy. This implies knowledge of factors responsible for efficacy and how they are affected by shelf-life. Assuming efficacy is tied to viability, the literature suggests that the performance of probiotic-containing products with regard to shelf stability is mixed. These studies provide 'snapshot' images of commercial products, some with little knowledge of storage history. Reuter (1997) and Holzapfel et al. (1998) have measured the composition and viability of various commercial probiotics (Table 3). Hamilton-Miller et al. (1999) tested probiotic-containing European yoghurts and dried supplements for quantity and species of probiotic bacteria, and compared laboratory results with label claims. Five of eleven yoghurts tested displayed labels which specified genus and species or strain of the live cultures contained within, but no statement of levels of microbes was made on any product. In general, the species identified in the yoghurts were consistent with the labeling. All 29 of the supplement products indicated genus and species, and 25 of these made some statement on product label or in the catalogue of expected quantities. The authors considered the labeling of 9 of 29 of these products to be 'acceptable' or 'satisfactory' based on agreement with stated species and levels. In an earlier study, Hamilton-Miller et al. (1996) analyzed 13 dietary supplement products in Britain. In addition to discrepancies on species reported, levels of lactobacilli in six of the products were greater than 2 log cycles lower than reported on the label. Canganella et al. (1997) evaluated 15 commercial probiotic pharmaceutical products from Europe containing one or more strains of *Enterococcus*, *Lactobacillus*, *Bifidobacterium*, *Escherichia coli*, *Bacillus subtilis*, and yeast. Data on comparison of expected counts and recovered counts were reported on only three products containing lactobacilli. In these three cases, one product contained only ~20% of expected count, another showed 2 log cycles lower count, and a third showed a 300-fold higher count. In only the first case was the *Lactobacillus* recovered of the same species listed on the label (*L. acidophilus*) as indicated by API testing. Note that the failure to use DNA-based methods for speciation of the lactobacilli as reviewed by Tannock (1998) in this study bring authors' conclusions about species diversion from label claim into question. The Tufts University Diet

Table 3. Characteristics of probiotic products on the European Market, analyzed in Germany from grocery store shelfs by classical methods. Adapted from Reuter (1997) and Holzapfel et al. (1998)

Producer (country)	Product name	Probiotic organism as stated on the label	Probiotic organism identified in product	Number of bacteria/ml	pH
Drink type products					
Yakult (NL)	Yakult	*L. casei* Shirota	*L. casei*	9.2×10^8	3.77
Südmilch (GER)	Vifit	*L. casei* LGG	*L. rhamnosus*	3.1×10^7	3.91
			S. thermophilus	3.3×10^7	
			L. bulgaricus	5.1×10^7	
Chambourcy (F)	LC1 drink	*L. acidophilus*1	*L. johnsonii*	2.4×10^4	4.23
			S. thermophilus	8.5×10^8	
Mona, Campina Melkunie (NL)	Fysiq drink	*L. acidophilus* Gilliland	*S. thermophilus*	5.6×10^8	4.34
Molkerei Schöppingen (GER)	Feine Schwedenmilch	BB 536	*B. longum*	1.9×10^4	4.48
			S. thermophilus	2.2×10^4	
Danone (F)	Actimel	*L. casei* Imunitas	*L. casei*	1.5×10^8	4.05
			L. bulgaricus	2.8×10^4	
Emmi (CH)	Aktifit plus	*L.casei* LGG, *L. acidophilus*, *S. thermophilus* *B. bifidum*	*L. rhamnosus*	1.8×10^8	4.03
Nestlé (CH)	Lc1 Go	*L. helveticus*	ND	ND	ND
Yoghurt type products					
Südmilch (GER)	Vifit	*L. casei* LGG	*L. rhamnosus*	1.4×10^7	4.08
			L. acidophilus	8.2×10^6	
			S. thermophilus	4.2×10^7	
			L. bulgaricus	1.5×10^7	
Toni Lait (CH)	SymBalance	*L. casei* 01	*L. casei*	9.5×10^8	4.45
		L. acidophilus La5	*L. acidophilus*	3.6×10^7	
		L. reuteri	*L. reuteri*	1.7×10^5	
		Bifidobacterium	*S. thermophilus*	5.2×10^8	
			L. bulgaricus	1.7×10^7	
Nestlé (CH) Chambourcy (F)	LC1 (vanille, nature, pur 1)	*L. acidophilus* La1	*L. johnsonii*	2.6×10^7 est	4.14–4.25
			S. thermophilus	2.5×10^8	
Mona, Campina Melkunie (NL)	Fysiq yoghurt Gilliland	*L. acidophilus* *L. acidophilus*	*L. casei*	2.6×10^6	4.21
			6.0×10^6		
			S. thermophilus	9.6×10^8	
			L. bulgaricus	7.2×10^6	
Danone (F)	Bio au casei actif	*L. casei*	*L. casei*	2.0×10^8	4.33
			L. bulgaricus	6.0×10^5	
Molkerei Müller (GER)	Bi'Ac	*Bifidus*[a] *longum* 536	*L. acidophilus*	1.9×10^7	4.20
			S. thermophilus	2.3×10^8	
Danone (F)	Actimel joghurt	*L. casei* Imunitas	*L. casei*	1.2×10^8	4.16
			L. bulgaricus	5.0×10^5	

Abbreviations: *L.=Lactobacillus*, *S.=Streptococcus*, *B.=Bifidobacterium* ([a]*Bifidus* is not a correct name but should be *Bifidobacterium*, ND=not determined).
Underlined species show the difference between name on label and identified.

& Nutrition Letter (Anon. 1996) reported analysis of enumeration of yoghurt starter bacteria and *L. acido-*

philus in USA yoghurt products three weeks before and at the end of shelf-life. Although methods weren't reported, results indicated that 7 of 8 yoghurts exceeded 10^8/gm of at least *S. thermophilus* at the end of shelf-life (*L. bulgaricus* numbers were 1–2 log cycles lower than *S. thermophilus* in all yoghurts). Three weeks prior to expiration date, all yoghurts contained $<10^5$ cfu/gm of *L. acidophilus*. From these data, it is not clear if this is a stability problem or an inoculation problem in the yoghurt, but the acidic nature of yoghurt is often not favorable for long term stability. Micanel et al. (1997) reported enumerations on commercial Australian yoghurts. Less than a one log cycle loss in counts of *L. acidophilus* was seen in 4 brands of full-fat and reduced-fat yoghurts tested and stored at 4 °C, although one brand contained <1000 cfu/gm of bifidobacteria or *L. acidophilus* even at time 0, suggesting that this yoghurt was not supplemented with these probiotic bacteria. Neither storage at 10 °C nor fat level appeared to significantly affect survival, but fruited yoghurts showed poorer stability, especially for bifidobacteria. The maximum level of *L. acidophilus* reported was 1.2×10^8 cfu/gm. At this level, a 200 g serving of yoghurt would provide over 10^{10} cfu, a level suggested by many studies as being functional.

Activities of bacteria as they are assessed in the stored product may be less important as residence time, accompanied by probiotic growth, in the body increases. Emergence from a resting state, or repair of injury suffered during production or storage (e.g., acid damage), to an actively growing state *in vivo*, will no doubt result in cellular functions based more on the *in vivo* environment encountered by the probiotic than the environment during storage. Predicting the nature and extent of these *in vivo* influences is one of the challenges of *in vitro* probiotic strain characterization efforts.

Probiotic functional foods: concepts and examples

Probiotic-containing products come in a variety of formats: conventional foods (probiotic-containing yoghurts, fluid milk and cottage cheese, consumed primarily for nutritional purpose, but also for probiotic benefits), food supplements or fermented milks (food formulations whose primary purpose is as a delivery vehicle for probiotic bacteria and their fermentation endproducts, consumed for health effects, in monoculture [Yakult, Japan] or mixed cultures [LC1, Nestlé, CH; Actimel, Danone, F, and others]), and dietary supplements (capsules and other formats designed to be taken by healthy individuals looking to enhance health).

In the development of a probiotic-containing food product, the product concept must take into account the desire of consumers for health-enhancing foods, but not at the expense of taste, convenience and pleasurable nature of product. As important is the ability to communicate messages on health to the consumer. These communications, of course, must not only address health messages that are meaningful and appealing to the consumer (e.g., promotes gastrointestinal health, supports the body's natural immune function), but must be scientifically accurate, and in accordance with applicable legislation, which varies from country to country. In addition to communications on health benefits of probiotic products, another important area of communication is on viable count or active ingredients in probiotic products. As previously indicated, statements as to content and counts of bacteria in commercial products are frequently not accurate or are inadequate. Food products, although perhaps claiming the presence of certain genera, species or strain of probiotic, rarely harbor any messages on probiotic potency. In countries where there is no legislation requiring this type of labeling, this leaves the consumer unable to make an informed choice based on a guaranteed probiotic content of a food product. In the long run, the failure of the industry to self-regulate in this regard may be its undoing. Consumers, unable to identify a potent product, may find a specific probiotic product ineffective, and may turn away from the entire product category.

Probiotic products in the USA are limited in diversity and market penetration. The situation is different in Japan and Europe, where probiotic containing yoghurts, fermented milks and dried supplements are more commonplace and accepted (Tables 2 and 3). In the USA, most national brands of yoghurt are supplemented with probiotic bacteria, usually *L. acidophilus* or related *Lactobacillus*, and sometimes with a *Bifidobacterium* strain. Rarely are strain designations or claims of viable counts used on the labels. National brands of yoghurt generally do not make health claims, although some regional brands do (Sanders 1998). Stonyfield Farm®, based in Londonderry, New Hampshire, includes six lactic acid bacterial species in all its yoghurts, including, in addition to the yoghurt starter strains, *L. acidophilus, L. casei, L. reuteri* and *Bifidobacterium* ssp. Danone, Paris, F has introduced Actimel in the USA containing 10^{10} *L. casei* per serving. Consumption of fermented milks in the USA is quite low (fermented milk as a category of fluid milk items is so low it is not even tracked by the United States Department of Agriculture Economic Research Service, and this source identifies the per capita sales of yoghurt as a mere 5.1 pounds for 1997), and not a significant outlet for probiotic delivery. Two new fermented milk products, one marketed as a dietary supplement, the other as a drinkable yoghurt are also available in the USA. Bona, Inc. (Whitehouse, New Jersey) sells a fruit-flavored drinkable yoghurt containing *L. acidophilus, Bifidobacterium* ssp., and *L. casei* in addition to yoghurt starters. Alta Dena (City of Industry, California), along with other smaller dairies, sells a similar product. Lifeway Foods (Skokie, Illinois) and GalaGen Inc. (Arden Hills, Minnesota) cooperated to sell a kefir-style beverage (albeit without the yeast) containing antibodies isolated from bovine colostrum and an undefined blend of probiotic lactobacilli. Dietary supplement capsules and other non-food preparations are a small, but growing, product area in the USA.

Product efficacy

Probiotic consumption can be justified only if a health effect or reduction in risk of disease can be realized by the consumer. Although the studies targeting many of these effects are disease-focused and claims directly related to disease may never be allowed on probiotic foods, taken together these studies build a case for the potential impact and mechanisms of probiotic bacteria on intestinal health.

Substantiation of health effects is a challenge facing the probiotics industry, in large part because of the high cost associated with conducting controlled clinical or epidemiological evaluations. These health effects are believed to be genera, species and strain specific. Scientifically speaking, the absence of proof of efficacy of commercial products does not disprove efficacy. In many cases, the controlled studies have just not been done. However, a lack of proof has a great influence on the use of health claims throughout the world. There is no motivation for companies or consumers to invest in technology supported mainly by testimonial, inference or supposition. Therefore, the clinical evaluations of commercial products must be conducted and are essential for labeling health claims. The few probiotic strains whose efficacy in human health is supported by clinical evaluations have a legitimate marketing advantage, even though the strains may be no more effective than other, untested bacteria.

The number of variables needing to be controlled for clinical evaluation of probiotics is quite large and includes:

1. Strain or combination of strains
2. Growth conditions of strain
3. Format for delivery (dried, supplemented in a food product, grown in a food product)
4. Consumption level of active ingredient
5. Test population
6. Validated biomarkers in combination with clinical endpoint
7. Clinical protocol, with preference to blinded, placebo-controlled formats

A thorough review of clinical outcome studies conducted with probiotics is beyond the scope of this review, and the reader is referred to other reviews on this topic (Salminen et al. 1996; Fonden 1999).

Safety

Although the safety of traditional lactic starter bacteria has never been in question, the more recent use of intestinal isolates of bacteria (bifidobacteria, intestinal lactobacilli and enterococci) to be delivered as probiotics in high numbers to consumers with potentially compromised health, has raised the question of safety. These intestinal isolates do not share the centuries-old tradition of being consumed as components of fermented dairy products. However, their presence in commercial products over the past few decades has not given any indication of safety concern.

In the USA, Title 21 of the Code of Federal Regulations (21 CFR) and the FDA Office of Premarket Approval lists microorganisms which are approved food additives or which enjoy Generally Recognized as Safe (GRAS) status. This list specifies that harmless lactic acid producing bacteria, including *Lactobacillus acidophilus*, used as optional ingredients in specified standardized foods (such as yoghurt and buttermilk) are sanctioned. In addition, this list specifically includes *Streptococcus* (*thermophilus* and the species *cremoris, lactis*, and *lactis* subsp. *diacetylactis*, now recognized as *Lactococcus* species) *Lactobacillus* (*bulgaricus, fermentum, lactis*), and three *Leuconostoc* species. Note that several probiotic lactobacilli species (e.g., *reuteri, casei, paracasei, gasseri, plantarum*) and bifidobacteria are not specifically on this GRAS list. However, the acknowledged nature of this list is that it is partial and absence of a microbe on this list does not imply it is not safe for use. It simply means application for inclusion in this list has not been made. These species are accepted as safe in Japan and Europe.

The safety of lactobacilli and bifidobacteria has been reviewed (Aguirre & Collins 1993; Donohue et al. 1993; Molin et al. 1993; Gasser 1994; Adams & Marteau 1995; Saxelin et al. 1996a,b; Salminen et al. 1998). The general conclusion is that the pathogenic potential of lactobacilli and bifidobacteria is quite low. This is based on the prevalence of these microbes in fermented food, as commensal colonizers of the human body, and the concomitant low level of infection attributed to them. However, isolated reports of association of lactobacilli (Aguirre & Collins 1993; Saxelin et al. 1996b) and bifidobacteria (Gasser 1994) with human infection (commonly endocarditis) in patients with compromised health suggest that these microbes have opportunistic capability. One report suggests that lactobacilli from clinical blood samples (twelve isolates) from Finland were distinct (based on carbohydrate fermentation patterns) from the prevalent lactobacilli available commercially in dairy products (nine strains tested) and one pharmaceutical product (Saxelin et al. 1996a), suggesting that lactobacilli involved in clinical infection did not originate from the food supply. However, this negative result cannot be construed as a definitive proof. A report from the Lactic Acid Bacteria Industrial Platform (Guarner & Schaafsma 1998) indicated that the overall risk of infection from lactic acid bacteria (excluding enterococci) is very low, but that *Lactobacillus rhamnosus* deserved particular surveillance due to the greater proportion of infections attributed to this species compared to other lactobacilli. One commercial strain of *Lactobacillus rhamnosus*, designated GG, has been used repeatedly in clinical trials and human studies, including one involving enteral feeding of premature infants (Millar et al. 1993) with no indication of pathogenic potential (Saxelin 1997).

Regarding the safety of enterococci, the picture is less clear-cut. Foods containing enterococci are consumed on a regular basis. However, safety reports seem to agree that the enterococci pose a greater threat than other lactic acid bacteria. Giraffa et al. (1997) conclude that the enterococci, although a group of phenotypically heterogeneous organisms, display potential pathogenic properties. Adams & Marteau (1995) exclude the enterococci from the lactic acid bacteria which they regard as having low pathogenic potential. Enterococci have been isolated from clinical infections as well, and not only bloodstream and endocarditis infections. *Enterococcus*-mediated infections of the biliairy tract, the abdomen, burn wounds, surgical wounds, and many others have been reported (Aguirre & Collins 1993; Jett et al. 1994). Often, enterococci are isolated as pure cultures from these infections, showing their primary pathogenic nature (Aguirre & Collins 1993). In the USA, vancomycin resistant enterococci comprise a worrisome source of nosocomial infections since 1986 (Arthur & Courvalin 1993). This forms a serious threat to mankind as it is known that gene transfer between enterococci and related organisms can occur rather easily. In this respect, transfer of the *VanA* gene from *Enterococcus faecium* to *Staphylococcus aureus* has already been observed *in vitro* (Noble et al. 1992). These observations and contradictory characteristics of enterococci make the use of these microorganisms more and more questionable as food fermentation agents (Arthur et al. 1996).

Regulatory and product labeling issues

Introduction. Regulatory and product labeling issues in the functional foods area primarily revolve around two issues, safety and assuring that product labeling and promotion, while communicating healthful product effects, is not misleading. This second item involves regulation of health claims. In general, a health claim can be defined as a direct, indirect or implied claim in food labeling, advertising and promotion that consumption of a food carries a specific health benefit or avoids a specific health detriment. This can be further divided into two subclasses, a *generic health claim*, which is based on well-established, generally accepted knowledge from evidence in the scientific literature and/or from recommendations from national or international public health bodies, and an *innovative health claim*, which is based on scientific evidence on specific existing or new foods. At this moment there is in many parts of the world active discussion about what is needed to establish a health claim. Generally, it can be concluded that the data should be based on:

1. The totality of the evidence, not just data which support the health claim;
2. Human studies or evidence, not just biochemical, cellular or animal studies;
3. Studies in humans which are the most methodologically sound available.

However, there is a need to clarify and strengthen the requirements for evidence to substantiate health claims. Furthermore, once scientific validity has been established, claims should be able to express clearly and more directly the variety of relationships between function-driven foods and human health. Its purpose is to prevent the use of misleading, unsubstantiated and false health claims and to clarify the distinction between health claims and medicinal claims, the latter generally prohibited on foods worldwide.

The approaches used for functional food science, especially in Japan and to a lesser extent in the USA, to support these new concepts have mostly been product or food component-driven and they are likely to be very much influenced by local, traditional or cultural characteristics. To be more universal, especially in acceptance of probiotic lactic acid bacteria-containing products, a science based, function-driven approach is preferable and forms a basis for the regulatory and labeling issues.

It is beneficial to make the consumer aware, in clear language, of the role of a food in maintaining good health and reducing the risk of disease. However, there is a fundamental difference between a food product which has a direct preventative effect on a disease and a food product which maintains good health as part of a life style and may reduce the risk of disease occurring. The former is a medicine whereas the latter is likely to be a food. This is one of the major discussion points which should be reflected in the claims which can be lawfully and truthfully made for a food. The distinction between the two is, however, far from clear in the mind of the consumer and may often be overlooked. Some health claim committees want industry to be able to clearly communicate the fundamental message that certain foods as part of a healthy diet will reduce the risk of certain diseases but without giving the impression that any food is capable of preventing, treating or curing a disease.

Japan. The concept of functional foods was first used in Japan in the late 1980s. The term indicated foods fortified with specific ingredients having certain health benefits. Due to the fact that the market of functional foods was growing rapidly, the demand emerged to have legal support in order to control the unjustified health claims that were suggested by some products. In Japan, as in Europe today, it was the industry that pushed the regulator to have legal supervision of the claims. This was to assure consumer confidence in claims made by legitimate functional foods and to differentiate these from 'miracle foods'. Since the 1980s the Japanese government under the Nutritional Improvement Law has stimulated the development of knowledge on function of food by providing financial grants. This effort has resulted in the differentiation of foodstuffs in three categories, namely nutritional function, palatability, and the ability to prevent disease. The term functional food describes food with attributes from the third category. During the following years, the government subsidized several academic research projects to further investigate the physiological function of foods and molecular basis in order to provide a regulatory framework for functional foods and to rationalize the means of evaluation of effectiveness of foods with potential health benefits.

It took the Functional Foods Discussion Group approximately four years before the Ministry of Health and Welfare issued in 1991 the 'Labeling Regulation for Foods for Specified Health Use' (FOSHU). With the implementation of this regulation, Japan was the first nation formally accepting the functional food concept. The term 'functional food' was not used in the Japanese regulation due to the fact that the word 'function' was previously used in the description and definition of pharmaceuticals. Using the same term for depicting properties of food and pharmaceuticals may have caused confusion in Japan, where the awareness of 'Ishokudogen' (medicine and food from the same origin) risks blurring the borders between food and medicine.

The FOSHU-regulatory aspects in Japan. In September 1991 the FOSHU decree became officially in force, providing a voluntary system for approval to which industry could apply when introducing products that could bear a health claim. These foods are defined in the Nutrition Improvement Law Article 12 as member of the group of Special Nutritive Foods (Figure 3), being foods consumed in the normal way as part of the daily diet. It therefore excludes food or food supplements in pills, tablet or capsule forms. By definition, by consuming these foods 'certain effects to maintain health can be expected'. Although this definition is quite loose, FOSHU approval requires a collection of information, including safety reports, scientific evidence of the effect in humans, and nutritional analysis. A FOSHU product is therefore food that is approved by the Minister of Health and Welfare as having established proof of being effective in the maintenance and the improvement of health. They bear a permit mark (a distinctive logo) along with a general de-

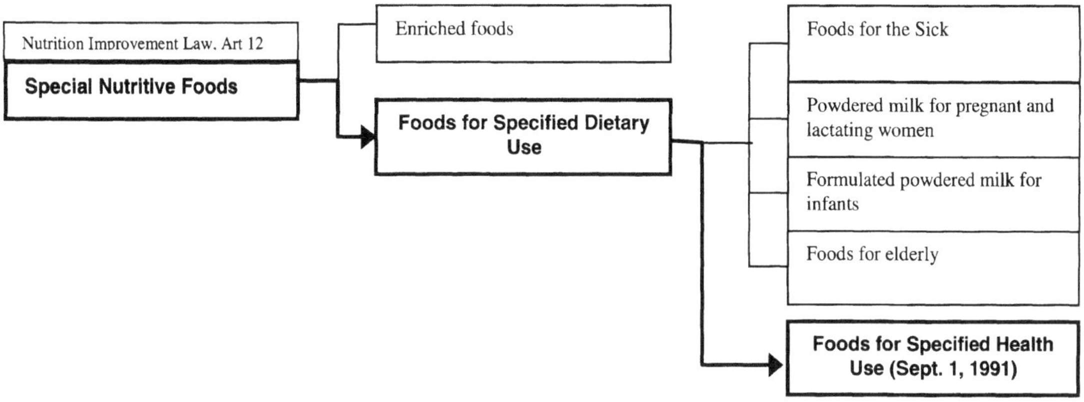

Figure 3. System of food types in the Japanese legislation.

Table 4. Probiotic bacteria with FOSHU clearance and products based on these bacteria

Probiotic strain with FOSHU clearance	Example of product based on this strain	Statements allowed (see Table 5)	Manufacturer
L. rhamnosus GG	ONAKA-HE-GG (drink yoghurt)	1–4	Takanashi milk products, Tokyo
B. longum BB536	Bifidus plain yoghurt	2	Morinaga milk industries, Tokyo
L. bulgaricus+S. thermophilus	1. Meiji Bulgaria yoghurt LB81 2. Meiji Bulgaria nomu yoghurt LB81 plain	4, 6	Meiji milk product, Tokyo
L. acidophilus SBT-2062 *B. longum* SBT-2928	Yukijirushi Nachure (yoghurt)	3, 4	Snowbrand milk products, Tokyo
L. casei Shirota	1. Yakult (fermented milk drink), 3 variants 2. Joie (fermented milk drink), 8 flavours	1–5, 7	Yakult Honsha, Tokyo
B. breve strain Yakult	1. Mil Mil 2. Mil Mil E	1–5, 7	Yakult Honsha, Tokyo

scription by which the health maintenance property is described together with the suggested dose and other suggestions to users. Probiotic strains with FOSHU clearance are listed in Table 4. The approval system gives permission to make certain claims in labeling a FOSHU food and is the first regulatory system for functional foods. Statements currently allowed on

Table 5. Statements currently allowed on FOSHU cleared products with probiotic bacteria

1.	Reaches the intestines alive
2.	Increases the intestinal lactobacilli/bifidobacteria/beneficial bacteria
3.	Promotes the maintenance of good intestinal environment
4.	Regulates/helps maintain good gastrointestinal condition
5.	Maintains the intestine in good health
6.	Helps balance the intestinal flora
7.	Reduces harmful bacteria

FOSHU cleared products with probiotic bacteria are shown in Table 5.

The FOSHU-approval system in Japan. Each food product that is submitted for FOSHU approval is evaluated on an individual basis. Since the initial procedure was felt to be a burden by the industry due to its lengthy nature, the FOSHU approval system was revised in November 1997. However still there are no clear evaluation standards (e.g., for scientific level of documentation). This may also be the reason that the number of FOSHU approved foodstuffs initially remained quite low. In 1993 and 1994 less than 15 products per year were approved, but after 1997 this level increased.

Essentially there are three parties involved in the procedure for FOSHU approval. The manufacturer of the product, the Ministry of Health and Welfare which finally decides about the approval and the FOSHU Department of Japan Health and Nutritional Food Association (JHNFA) which is entitled to perform a pre-registration and to assist the manufacturer in the registration process. The JHNFA is an industrial association, approved and supported by the Ministry of Health and Welfare. Originally working to set standard specifications for health food and promotion of healthy life standards, its activities were extended in 1992 and now include assistance to the applicant for FOSHU approval. The internal evaluation of this organization provides an a priori scientific assessment, which can be used to evaluates the dossier before submission. The applicant for FOSHU approval should be a member of this association.

The approval procedure consists of two consecutive steps (Figure 4). Before a product can be submitted for FOSHU clearance, the **ingredient** to which the beneficial effect is attributed should be first evaluated by the JHNFA-FOSHU subcommittee. As a part of the evaluation process by the JHNFA-FOSHU department, a scientific and a specialists committee evaluate the submitted dossier. The applicant, with the assistance of the JHNFA-FOSHU department, does the preparation of the final application form, including the evaluation report, which is to be used in the second step of the application. Subsequently, the approved application of the ingredient is filed at the Public Health Centre.

Table 6. Documents to be submitted for FOSHU application procedure (From ILSI Japan, Study Committee on Functional Foods, June 1998)

1.	Sample of the label
2.	Clinical and nutritional documentation demonstrating the health effect of the food or its constituents (including test results on 20–30 Japanese subjects)
3.	Clinical and nutritional documentation showing determination of the daily intake of the food or its constituents
4.	Documentation demonstrating the safety of the food or its constituents
5.	Documentation on the stability of the food or its constituents
6.	Results of the physicochemical properties of the food or its constituents
7.	Results of the qualitative and quantitative determination of the foods constituents, and description of the analytic methods used
8.	Results of tests determining the constituents and their energy value

The second step for FOSHU approval is to obtain the permit for making health claims on a food product by the Ministry of Health and Welfare. Through the public health centre, the applicant requests to the Ministry of Health and Welfare approval to make specific claims on a **product** that contains a previously approved ingredient. The dossier should contain clinical proof, with studies done on Japanese populations, of the effect and substantial information on the safety of the ingredient and the product. Furthermore nutritional, compositional and physicochemical documentation should be submitted (Table 6). Data on safety of the product can be derived from history of safe use, although debate is still ongoing whether history alone is enough to prove safety of specific ingredients, in particular if the concentration of these ingredients is increased in the final product. The approval system will also state the specific health claims that can be

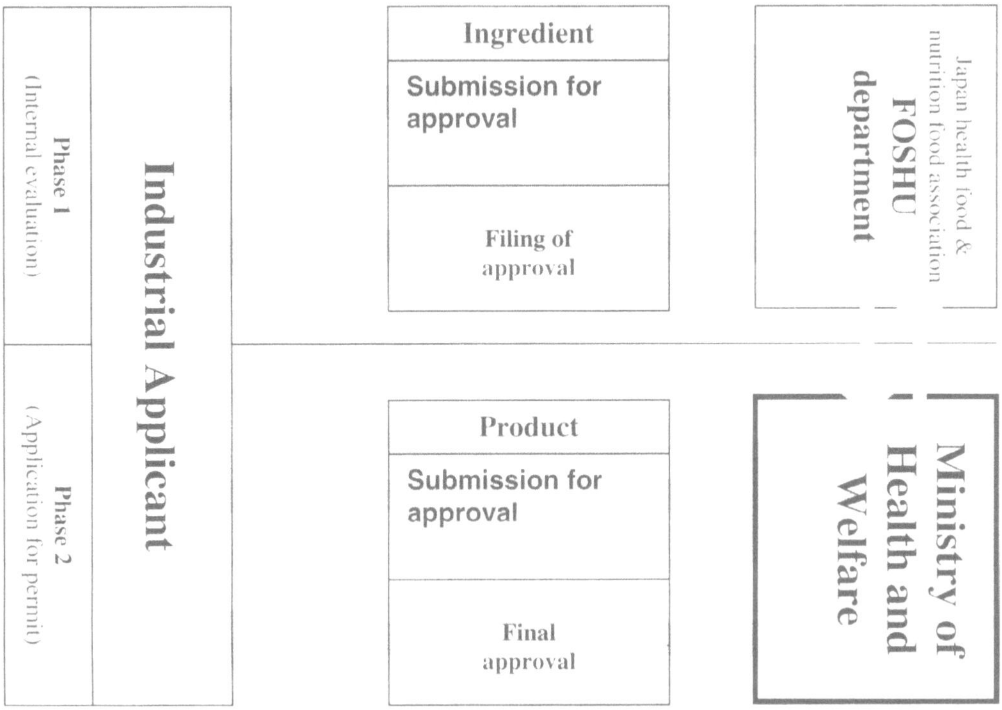

Figure 4. Foods for specified health use (FOSHU) application system in Japan.

made on the submitted product. In 1996 only 3 probiotic products obtained the FOSHU approval. This increased to 8 in 1997 and about 21 in 1998. In total, about 126 functional foods had FOSHU approval by 1998.

United States. Functional foods have no legal definition in USA food law (Berner & O'Donnell 1998). These foods are legally lumped together with conventional foods, or if meeting specific labeling requirements, may be presented as one subclass of conventional foods known as dietary supplements. For the most part in the USA, conventional food products do not carry labels that suggest a relationship between a food and the diagnosis, cure, mitigation, treatment, or prevention of any disease. In general, doing so represents the product as either a misbranded food or an illegal drug, and the food can be subject to recall and the manufacturer to penalties. However, there are legal exceptions to this general rule which manufacturers of functional foods can explore for options in communicating health benefits of their products. McNamara (1998) nicely summarizes the situation in the USA on allowable communication of health information on food products.

Prior to 1990, no health statements were allowed on conventional foods, but the passage of the 1990 Nutrition Labeling and Education Act (NLEA) allowed foods to bear a health claim under the condition that the claim was approved by the FDA. To date, eleven such claims have been evaluated and approved by the FDA (Table 7). The FDA Modernization Act (FDAMA) of November 21, 1997 further amended the FDCA providing limited authorization of food labeling to include certain health claims without FDA pre-approval. This provision required that the claims be based on an 'authoritative statement' by a 'scientific body of the U.S. Government with official responsibility for public health protection or research', such as the National Institutes of Health, the National Academy of Health or the Centers for Disease Control and Prevention. Although premarket approval is not required for use of such a health claim, premarket notification (120 days prior to use) of the FDA is required. The FDA has authority to reject use of the claim after review of the submission.

Table 7. Health claims approved by the FDA for use on foods sold in the United States. Claims 1–5 are also approved for use on dietary supplements (Storlie et al. 1998)

1.	Calcium and osteoporosis
2.	Folate and neural tube defects
3.	Soluble fiber from whole oats and coronary heart disease
4.	Soluble fiber from psyllium husk and coronary heart disease
5.	Sugar alcohols and dental caries
6.	Dietary lipids and cancer
7.	Dietary saturated fat and cholesterol and coronary heart disease
8.	Fiber-containing grain products, fruits and vegetables and cancer
9.	Fruits and vegetables and cancer, for foods that are naturally a 'good source' of vitamin A, vitamin C, or dietary fiber
10.	Fruits, vegetables, and grain products that contain fiber, particularly soluble fiber, and coronary heart disease
11.	Sodium and hypertension

In addition to this class of health claims, another type of health statement is allowable on a class of food known as dietary supplements (Table 8) as stipulated in the Dietary Supplement Health and Education Act of 1994. This statement, termed a 'structure/function claim' (not a health claim), relates a substance to normal, healthy functioning of the human body. For example, 'probiotic lactobacilli help maintain a healthy intestinal tract' would be allowable as long as it is truthful and not misleading. No prior approval of these statements is required by the FDA (notification of the FDA within 30 days of marketing such a product is required), although companies may be required to provide scientific substantiation upon request of the FDA (Public Law 103-417).

Structure/function statements are not commonly used on conventional foods. However, McNamara offers the opinion that structure/function claims do not appear to meet the legal definition of health claims and therefore should not require FDA approval for use on a food. This sentiment was echoed by Storlie et al. (1998) who stated that the 'FFDCA (Federal Food Drug and Cosmetic Act) permits both foods and dietary supplements to carry 'structure/function' claims, but not claims purporting drug-like effects'. In addition, Kurtzweil (1998), in a publication from the FDA, states that manufacturers may make statements about a food substance's effect on the structure or function of the body. However, most labeling of conventional foods (including 'functional foods') in the USA avoids such statements, perhaps reflecting a common, cautious attitude about FDA scrutiny of such statements, and reflecting uncertainty about how to word structure/function claims so as not to invoke FDA action. Berner & O'Donnell (1998) indicate that the regulatory position on the use of the structure/function claim on conventional foods is 'unclear', and McNamara admits uncertainty 'about how far this type of 'structure/function' claim can be 'pushed' before FDA will assert either drug status or health claim status'. The content and wording of the statement will be a subtle but important factor in satisfying FDA requirements. A proposal announced April 1998 from the Commission on Dietary Supplement Labels aims to define criteria for structure/function claims. This committee's recommendations will surely apply to structure/function claims used on conventional foods as well.

Since dietary supplements are allowed to take the form of a conventional food, in the end, the differentiation of a product as a food or as a dietary supplement is based on how the product is represented to the consumer. Storlie et al. (1998) discuss this issue in detail, and provide useful examples of products labeled both as a food and as a dietary supplement. Some issues pertaining to this distinction are shown in Table 9.

Currently, probiotic lactobacilli and bifidobacteria are primarily sold in the USA as dietary supplements or as components of foods, fluid milk, cottage cheese and yoghurt. As such, they fall under two separate regulatory categories: food ingredients and dietary supplements (Table 9). The approach to product labeling used by most major food manufacturers using probiotic bacteria is to list the genus and species (species is sometimes omitted on bifidobacteria) on the label, but make no inference as to its function or purpose. The approach to labeling by the dietary supplement manufacturer of probiotics has been more aggressive. Statements which link probiotic bacteria to the maintenance of intestinal health and improved digestion are common. The packaging of Culturelle™, a dietary supplement capsule product containing *Lactobacillus* GG states 'Helps You Feel Good!' and '...Maintains a Healthy Intestinal Tract'. Some probiotic products contain statements about support of a strong immune system. However, FDA scrutiny recently befell Weider Nutrition International, Inc. (Salt Lake City, Utah) with the challenge by the FDA's Center for Food Safety and Applied Nutrition that the labeling of their

Table 8. Some definitions important in USA food regulations

Conventional food	Generally consumed for taste, aroma or general nutrition.
Dietary supplement	A dietary supplement is a product (other than tobacco) intended to supplement the diet that contains one or more of the following dietary ingredients: a vitamin, a mineral, an herb or other botanical, an amino acid, a dietary substance for use by man to supplement the diet by increasing total dietary intake, a concentrate, metabolite, constituent, extract, or combination of any of these ingredients. The product may be in tablet, capsule, powder, softgel, gelcap, liquid, or conventional food form but may not be represented for use as a conventional food or as a sole item of a meal or of the diet. 'Dietary ingredient' notably refers to both nutrients and other dietary ingredients.
Drug	A product recognized officially as a drug; a product intended for use in the diagnosis, cure, mitigation, treatment, or prevention of disease; or a product (other than food) intended to affect the structure or any function of the body.
Nutraceutical	No legal definition for this term is found in USA food law. It is commonly recognized as a food, not a conventional food, providing one or more supplemental dietary ingredients for the purpose of improving health or reducing the risk of disease.
Disease claim	A claim that a substance can be used to diagnose, mitigate, treat, cure, or prevent a specific disease or class of diseases.
Health claim	A statement which characterizes the relationship of any substance to a disease or health-related condition. Examples include 'protects against cancer' or 'reduces nausea associated with chemotherapy'. These claims differ from nutritional content claims (i.e., 'low fat', 'high fiber') and structure claims (see below).
Structure/function claim	A statement of nutritional support that describes the role of a nutrient or dietary ingredient intended to affect the structure or functioning of the human body, or characterizes the documented mechanism by which a nutrient or dietary ingredient acts to maintain such structure or function. Examples include 'supports the immune system', 'promotes regularity', or 'helps maintain cardiovascular health'. Claims that substances can treat, diagnose, cure or prevent a disease are not structure/function claims.
Food label	A display of written, printed, or graphic matter upon the immediate container of a product.
Labeling	All labels and other written, printed or graphic matter upon any product or any of its containers or wrappers or accompanying such product.
DSHEA	Dietary Supplement Health and Education Act, October 1994.
FDAMA	Food and Drug Administration Modernization Act, November 1997.
NLEA	Nutrition Labeling and Education Act of 1990.

L. acidophilus-containing products constituted illegal drug claims. The labels read that these products were 'necessary for successful implantation and competitive exclusion of pathogenic bacteria' (Anon. 1998). The claims were perceived by this agency as implying prevention or treatment of 'various digestive diseases caused by pathogenic microorganisms'.

In summary, federal food regulations control the labeling of foods with health claims in all 50 states (states are allowed to have more, but not less, restrictive regulations, and regarding health claims on foods, the federal regulations prevail). Health claims are not allowed on foods unless previously approved by the FDA or unless derived from an authoritative

Table 9. Comparison of conventional food and dietary supplements as regulated by USA food law (compiled from Storlie et al. 1998)

	Food	Dietary Supplement
Product label	Identified by common name	Must be identified as 'supplement', and cannot be represented as a meal or meal replacement (although may be in conventional food form)
Nutrition label	'Nutrition Facts'; May not contain ancillary, functional ingredients beyond core nutrients	'Supplement Facts'; May include ancillary ingredients
Safety	All ingredients and final product must be safe, but burden of proof of safety rests with manufacturer	All ingredients and final product must be safe, but burden of proof of safety risk rests with FDA
Health Statements	FDA approved health claims only	FDA approved health claims for dietary supplements (5 of 11 approved for foods) and structure/function claim allowed
New dietary ingredient	Must use food additive petition process to secure agency preapproval for new dietary ingredient[a]	May employ a 75-day premarket notification process to inform FDA about use of new dietary ingredients

[a] A New Dietary Ingredient refers to any ingredient that was not marketed in the United States prior to October 15, 1994.

USA government body involved in health protection or research. Approval of a new health claim through the FDA petitioning process is time consuming, requires significant scientific documentation and agreement among experts, and has only been allowed on 11 claims to date (Table 9). While products presented to consumers as dietary supplements are allowed to bear structure/function statements, it is a matter of legal opinion what is allowable on conventional foods. The climate in the USA has been conservative with regard to labeling of foods with health statements.

Current legal situation at European-level. The development of probiotics in Europe for the European regulator is a new concept. Past legislation did not anticipate any specific legal requirement of functional foods. Therefore, no specific category exists for functional foods or foods on the border between food and medicines. To complicate matters, interpretation of the EU-Medicine Directive 65/65 is different according to member states. In a jurisprudence it is stated that a member state can consider a product as a medicine according to its composition or presentation. The regulator point of view on the approval of market introduction of a new functional food will therefore be regarded mainly through two focus points.

The first focus point is the safety of these foods. Safety addresses any chemical, physical or microbiological hazard and is regulated through the general EU-Directive 92/59 on Product Safety, Directive 85/374 on Product Liability, and by the product- or hygiene-specific directives. Safety also encompasses the demonstration of the safety of daily product consumption. Foods with a long history of consumption are presumed to be safe. Therefore the safety of probiotic products using well known lactobacilli or bifidobacteria strains already used in large quantities will likely never be questioned. However for foodstuffs or ingredients which where not marketed in substantial amounts in the EU market before April 1997 and which meet some criteria of novelty in production, genetic structure, raw material use or consumption use the EU-Novel Food regulation 258/97 of January 27, 1997 applies. In the Scientific Committee of Food guidelines, strict safety and monitoring requirements are set. Introduction of a new strain first needs to meet the novel foods requirements. The Novel Foods Regulation 258/97 defines novel foods in article 1.2 as 'foods which have not hitherto been used for human consumption to a significant degree within the Community and which fall under the following categories:

1. Foods and food ingredients containing or consisting of genetically modified organisms within the meaning of directive 90/220/EEG,
2. Foods and food ingredients produced from, but not containing, genetically modified food organisms,
3. Foods and food ingredients with a new or intentionally modified primary molecular structure,
4. Food and food ingredients consisting of or isolated from microorganisms, fungi or algae,
5. Foods and food ingredients consisting of or isolated from plants and food ingredients isolated from animals, except for food and food ingredients obtained by traditional propagating or breeding practices and having a history of safe food use,
6. Foods and food ingredients to which has been applied a production process nor currently used, where that process gives rise to significant changes in the composition or structure of the foods or food ingredients which affect their nutritional value, metabolism or level of undesirable substances'.

The second focus point of the regulator will be the issue of labeling and health claims. In Europe no specific legislation regarding health claims exists, but the general labeling directive 79/112/EU of 18.12.1978 applies. Article 2 of the directive states that a label cannot attribute properties to the foodstuff regarding prevention, treatment or curing of a disease or make allusion of having those properties.

As the general knowledge in Europe on lactic acid bacteria and their role in the maintenance of health is high, industry is pushing to be able to use this knowledge in communications to consumers. The regulator can be expected to reasonably question, however, if the properties claimed for the ingredient are still valid if no guarantee is made as to the survival of the probiotic in the product, the amount of the ingredient in the product which could have a health effect, survival in the gastrointestinal tract, and other such 'active ingredient' concerns.

The result of these issues is that in Europe, national authorities do not always know how to deal with these foods regarding authorization, certification or approval of labeling requirements. This makes it also very difficult to present a uniform strategy to producers wanting to introduce a new functional food in the EU-market. The application of marketing of the same strain of bacteria can be different according to the presentation form of the product. The use of a probiotic bacteria will undergo less scrutiny when used as a food ingredient than if it was used in its freeze-dried form in a food supplement. The reasoning is obvious as the concentration of the probiotic used in a daily consumable foodstuff will be far less than the concentration of a probiotic in a food supplement product.

Possible developments in Europe. An imminent change in the labeling legislation of the European Commission is not foreseen. This supports the idea of voluntary codes at the national or European levels. A change in the labeling directive accompanied by guidelines on interpretation at the national level might be another approach to clarifying labeling issues around functional foods. However, considering the normal time spent for the set up of legislation and in view of the forthcoming elections and changes at EU-Parliament and EU-Commission in 1999, it is not expected that the problem will be resolved quickly. In the green book, a discussion document of the European Commission in which future policies regarding food legislation are taken up priority is to consumer protection and communication. It was therefore believed that health claims would be also a priority. However the Commission has repeatedly stated that it would not initiate discussion on this field. It is likely that the European Commission will also await the outcome of the discussions at Codex level in order to have an internationally recognized guideline. France plays a leading role in the discussion and is expected to submit draft guidelines for the use of health claims to the Codex Committee meeting on food labeling in 1999.

ILSI (International Life Science Institute, Washington, D.C.) is a non-profit, world-wide foundation established in 1978 to advance the understanding of scientific issues related to nutrition, food safety, toxicology and the environment. It has different branches all over the world and is affiliated with the WHO. In the consensus document resulting in Functional Food Science in Europe (FUFOSE) funded by the European Commission and coordinated by ILSI Europe branch, a Functional Food is defined as being functional 'if it contains a food component (whether a nutrient or not) which affects one or a limited number of function(s) in the body in a targeted way so as to have positive effects'. Furthermore 'a functional food must remain food and they must demonstrate their effects in amounts which can normally be expected to be consumed in the diet. They are not pills or capsules but part of a normal food pattern'. According to the conclusion of this project the scientific evidence should be focused on a function-driven approach rather than a product claim-driven approach. Research is needed to identify the validated markers (whether they are markers of exposure to the functional food component, markers which relate to target function or biological response or intermediate markers of the actual endpoint or health outcome).

The CIAA (Confederation of the Food and Drink Industry) is representing 15 National Food Federations and 35 branch organizations from various sectors of the food and drink industry. The CIAA opinions and recommendations are forwarded to the European administration and institutions. It has issued an internal position paper regarding health claims. This consensus document separates 5-claims which the industry feels are distinct and reasonable for foods: Nutrition Function Claim, which describes the role of a nutrient or nutrients in the organism's normal functioning; Enhanced Function Claim, which describes a specific effect of food or food constituents on physiological functions or biological activities that can be distinguished for normal growth, development and functions of the body; Health Effect Claim, which refers to a specific positive health effect on the human organism (i.e., a physiological function or biological parameter); Healthy Eating Pattern Claim, which refers to official recommendations relating to healthy eating habits; and a Disease Risk Reduction Claim, which refers to reduction in risk of contracting a particular illness with product consumption. This last claim delineates an indirect relationship, and does not directly relate to prevention of an illness. Furthermore, it is the last claim which is mainly considered inadmissible by EU regulators.

The council of Europe is an independent body which has no legislative power, but initiates harmonization through position papers. (The Council of Europe is not the same as the Council of Ministries.) The Council of Europe has now conceived a specific working group to draft a position paper on the issue of functional foods. The aim is also to initiate harmonized legislation all over Europe.

The basis of labeling legislation in all members states derives from the European labeling directive 79/112. A directive (contrary to a regulation) needs to be implemented in national legislation. Therefore some stricter requirements can be found at the national level but all enact the basic article 2 stating that it is prohibited to claim that a food is capable of preventing, treating or curing a disease. As an example of deviation, in Belgium and Germany mentioning the name of human organs, or the representation of medicinal devices, materials, doctors or nurses on labels is prohibited. In the UK food labeling regulations forbid implied claims.

Comparison of EU member state regulatory requirements. The seven EU-member states already initiating proposals, or having installed codes or legislation regarding health claims are Sweden, UK, Belgium, The Netherlands, Spain, France and Finland. In Spain and Finland, codes are taken up in a legal reference text. In all other countries the industry initiated or was involved in developing the code. Also in Spain and Finland, the disease risk reduction claim is not allowed due to concern from authorities that food not be represented closely as a pharmaceutical. All codes define allowed health claims differently, making it very difficult for a producer to develop a uniform admissible claim which can be used in these member states with one application dossier. Every code also has different requirements for the level of scientific substantiation included in the dossier. The UK, Belgium and the Netherlands have included principles for (clinical) trials in the codes.

All codes with the exception of the Swedish code, have a clause to have the scientific validation be carried out by an independent scientific advisory body. The co-ordination of the introduction of the dossier is however done by an official recognized party, like the Food Center in the Netherlands and the high Food Council in Belgium.

The idea of having a positive list of allowed generic claims is not always taken up in the codes. The industry's opinion is also to avoid lists of authorized claims as it will discourage food manufactures from researching and developing innovative products with new health benefits.

France is the only member state in which premarket advice is required by obtaining the 'Visa PP' (Visa for Publicity and Promotion). This makes it difficult for producers marketing in other member states to be assured that the claims they want to use will not be problematic. The actions taken against producers using claims which are not in compliance with the requirements are different in the members states, but do sometimes include the publication of the negative advice. Hence, producers will have to reconsider seriously in advance which claim they want to use in consumer communications.

Australia and New Zealand. As in many countries, health claims have been prohibited from being used on food labels or in advertising and publicity in Australia and New Zealand. However as a first step in lifting the ban on the use, a pilot health claim trial started last year under the responsibility of the government, the Australia New Zealand Food Authority (ANZFA), and the Australian Food Council which is representing the Australian food and drink companies.

The controlling authority is responsible for assuring that no product is displaying a health claim until it has fulfilled several requirements. Not all food products can be included in this pilot study and moreover food products which may be authorised by the Australia New Zealand Food Standard Council to bear a claim will be those containing an appropriate level of a least 40 mg of folate per serving. Approval can also only be given if the ANZFA is satisfied on the safety aspect of the product.

It is believed that the pilot study represents an opportunity to prove that industry can take responsibility in providing credible and scientifically substantiated information without misleading the consumer. If the study is successful it could lead to adaptations of the clause of the Australian Food Standards Code, permitting then general health claims on other products.

Critical questions related to probiotics

Although holding much promise, there are many key issues in the development and marketing of probiotic products that need to be addressed, as indicated in the following list.

1. Identification of physiologically-relevant biomarkers which can be used to assess parameters of probiotic effectiveness (strain, dose, strain growth and delivery conditions, etc.) in humans.
2. Definition of active principle(s).
3. Stability parameters for active principle(s).
4. Resolution of taxonomy of probiotic species and development of user-friendly methods for conducting speciation assessments.
5. Science-driven implementation of findings in commercial products.
6. Epidemiology and properly controlled human intervention studies to confirm probiotic efficacy.

Conclusion

The microbiological and regulatory issues involved in bringing a probiotic product to market are vast and diverse. Future research will help clarify how and to what extent specific probiotic strains influence human health. Future attention by government regulatory bodies, in conjunction with industry and academia, will simplify and improve, regulatory and labeling issues associated with marketing probiotic products worldwide. The consumer will benefit from these developments with more efficacious and more accurately labeled products.

References

Adams MR & Marteau P (1995) On the safety of lactic acid bacteria from food. Int. J. Food Microbiol. 27: 263–264

Aguirre M & Collins MD (1993) Lactic acid bacteria and human clinical infection. J. Appl. Bacteriol. 75: 95–107

Anon. (1996) Is yogurt the health food it's cracked up to be? Tufts Univ. Diet Nutr. Let. November, pp. 3–5

Anon. (1998) FDA denies numerous structure/function claims. Food Regulation Weekly, Nov. 23, 1998

Arthur M & Courvalin P (1993) Genetics and mechanisms of glycopeptide resistance in enterococci. A minireview. Antimicrob. Agents Chemother. 37: 1563–1571

Arthur M, Reynolds P & Courvalin P (1996) Glycopeptide resistance in enterococci. Trends in Microbiol. 4: 401–407

Aso Y & Akazan H (1992) Prophylactic effect of *Lactobacillus casei* preparation on the recurrence of superficial bladder cancer. BLP study group. Urol. Int. 49: 125–129

Berner LA & O'Donnell JA (1998) Functional foods and health claims legislations: applications to dairy foods. Int. Dairy J. 8: 355–362

Canganella F, Paganini S, Ovidi M, Vettraino AM, Bevilacqua L, Massa S & Trovatelli LD (1997) A microbiological investigation on probiotic pharmaceutical products used for human health. Microbiol. Res. 152: 171–179

Collins JK, Thornton G & Sullivan GO (1998) Selection of probiotic strains for human applications. Int. Dairy J. 8: 487–490

Donohue DC, Deighton M, Ahokas JT & Salminen S (1993) Toxicity of lactic acid bacteria. In: Salminen and von Wright (Eds.) Lactic Acid Bacteria. (pp 307–313). Marcel Dekker, Inc., New York

Dunne C, Murphy L, Flynn S, O'Mahony L, O'Halloran S, Feeney M, Morrisey D, Thornton G, Fitzgerald G, Daly C, Kiely B, Quigley EMM, O'Sullivan GC, Shanahan F & Collins JK (1999) Probiotics: from myth to reality. Demonstration of functionality in animal models of disease and in human clinical trials. These proceedings.

Fonden R, Mogensen G, Tanaka R & Salminen S (1999) Effect of fermented dairy products on intestinal microflora, human nutrition and health: current knowledge and future perspectives. IDF Document. (In press)

Gasser F (1994) Safety of lactic acid bacteria and their occurrence in human clinical infections. Bull. Inst. Pasteur 92: 45–67

Giraffa G, Carminati D & Neviani E (1997) Enterococci isolated from dairy products: a review of risks and potential technological use. J. Food Protect. 60: 732–738

Goldin BR, Gorbach SL, Saxelin M, Barakat S, Gualtieri L & Salminen S (1992) Survival of *Lactobacillus* species (Strain GG) in human gastrointestinal tract. Dig. Dis. Sci. 37: 121–128

Guarner F & Schaafsma GJ (1998) Probiotics. Int. J. Food Microbiol. 39: 237–238

Hamilton-Miller JMT, Shah S & Smith CT (1996) 'Probiotic' remedies are not what they seem. Brit. Med. J. 312: 55–56

Hamilton-Miller JMT, Shah S & Winkler JT (1999) Public health issues arising from microbiological and labelling quality of foods and supplements containing probiotic microorganisms. Public Health Nutr. (in press)

Hata Y, Yamamoto M, Ohni M, Nakajima K, Nakamura Y & Takano T (1996) A placebo-controlled study of the effect of sour milk on blood pressure in hypertensive subjects. Amer. J. Clin. Nutr. 64: 767–771

Havenaar R & Huis in't Veld JHJ (1992) Probiotics: a general view. In: Wood BJB (Ed.) The lactic acid bacteria: Vol. I The lactic acid bacteria in health and disease. (pp 151–170). Elsevier Applied Science, London

Holzapfel WH, Haberer P, Snel J, Schillinger U & Huis in't Veld JHJ (1998) Overview of gut flora and probiotics. Int. J. Food Microbiol. 41: 85–101

Hosono A, Lee J, Ametani A, Natsume M, Hirayama M, Adachi T & Kaminogawa S (1997). Characterization of a water-soluble polysaccharide fraction with immunopotentiating activity from *Bifidobacterium adolescentis* M101-4. Biosci. Biotech. Biochem. 61: 312–316

Huis in't Veld JHJ & Havenaar R (1997) Selection criteria and application of probiotic microorganisms in man and animal. Microecol. Ther. 26: 43–57

Jett BD, Huycke MM & Gilmore MS (1994) Virulence of enterococci. Clin Microbiol. Rev. 7: 462–478

Kampman E, Goldbohm RA, van den Brandt PA & van't Veer P (1994) Fermented dairy products, calcium and colorectal cancer in the Netherlands cohort study. Cancer Res. 54: 3186–3190

Kurtzweil P (1998) Staking a claim to good health. FDA and science stand behind health claims on foods. FDA Consumer, Nov-Dec. http://vm.cfsan.fda.gov/~dms/

Langseth L (1996) Nutritional epidemiology: possibilities and limitations. ILSI Press, Washington, DC

Maeno M, Yamamoto N & Takano T (1996) Identification of antihypertensive peptides from casein hydrolysate produced by a proteinase from *Lactobacillus helveticus* CP790. J. Dairy Sci. 73: 1316–1321

Marin ML, Lee JH, Murtha J, Ustunol Z & Pestka JJ (1997) Differential cytokine production in clonal macrophage and T-cell lines cultured with bifidobacteria. J. Dairy Sci. 80: 2713–2720

Marteau P, Minekus M, Havenaar R & Huis in't Veld JHJ (1997) Survival of lactic acid bacteria in a dynamic model of the stomach and small intestine: validation and the effect of bile. J. Dairy Sci. 80: 1031–1037

McNamara SH (1998) So you want to market a food and to make health-related claims – How far can you go? What rules of law will govern the claims you want to make? Food Drug Law J. 53: 421–436

Micanel N, Haynes IN & Playne MJ (1997) Viability of probiotic cultures in commercial Australian yogurts. Aust. J. Dairy Technol. 52: 24–27

Millar MR, Bacon C, Smith SL, Walker V & Hall MA (1993) Enteral feeding of premature infants with *Lactobacillus* GG. Arch. Dis. Childhood 69: 483–487

Minekus M, Smeets-Peeters M, Bernalier A, Marol S, Havenaar R, Marteau P, Alric M, Fonty G & Huis in't Veld JHJ (1999) A computer-controlled model of the large intestine with peristaltic mixing, absorption of fermentation products and a high-density microflora. (in press)

Minekus M, Marteau P, Havenaar R & Huis in't Veld JHJ (1995) A multicompartmental dynamic computer-controlled model simulating the stomach and small intestine. ATLA (Alternative To Laboratory Animals) 23: 197–209

Molin G, Jeppsson B, Johansson M-L, Ahrne S, Nobaek S, Stahl M & Bengmark S (1993) Numerical taxonomy of *Lactobacillus* spp. associated with healthy and diseased mucosa of the human intestines. J. Appl. Bacteriol. 74: 314–323

Monique GLe, Moulton LH, Hill C & Kramar A (1986) Consumption of dairy produce and alcohol in a case-control study of breast cancer. J. Nat. Cancer Instit. 77: 633–636.

Noble WC, Virani Z & Cree RGA (1992) Co-transfer of vancomycin and other resistance genes from *Enterococcus faecalis* NCTC 12201 to *Staphylococcus aureus*. FEMS Microbiol. Lett. 93: 195–198

Perdigon G, Nader de Macias ME, Alvarez S, Oliver G & Pesce de Ruiz Holgado AA (1986) Effect of perorally administered lactobacilli on macrophage activation in mice. Infect. Immun. 53: 404–410

Peters RK, Pike MC, Garabrant D & Mack TM (1992) Diet and colon cancer in Los Angeles County, California. Cancer Causes and Control 3: 457–473

Reuter G (1997) Present and future of probiotics in Germany and in Central Europe. Biosci. Microflora 16: 43–51

Roberfroid MB (1998) Prebiotics and synbiotics: concepts and nutritional properties. Brit. J. Nutr. 80: S197–S202

Salminen S, von Wright A, Morelli L, Marteau P, Brassart D, de Vos WM, Fonden R, Saxelin M, Collins K, Mogensen G, Birkeland S-E & Mattila-Sandholm T (1998). Demonstration of safety of probiotics – a review. Int. J. Food. Microbiol. 44: 93–106

Salminen S, Isolauri E & Salminen E (1996) Clinical uses of probiotics for stabilizing the gut mucosal barrier: successful strains for future challenges. Antonie Van Leeuwenhoek 70: 347–358

Sanders ME (1998) Development of consumer probiotics for the US market. Brit. J. Nutr. 80: S213–S218

Sanders ME, Walker DC, Walker KM, Aoyama K & Klaenhammer TR (1996) Performance of commercial cultures in fluid milk applications. J. Dairy Sci. 79: 943–955

Savaiano DA, AbouElAnouar A, Smith DE & Levitt MD (1984) Lactose malabsorption from yogurt, pasteurized yogurt, sweet acidiohilus milk, and cultured milk in lactase-deficient individuals. Amer. J. Clin. Nutr. 40: 1219–1223

Saxelin M, (1997) *Lactobacillus* GG – a human probiotic strain with thorough clinical documentation. Food Rev. Int. 13: 293–313

Saxelin M, Rautelin H, Salminen S & Makela PH (1996a) Safety of commercial products with viable lactobacillus strains. Infect. Dis. Clin. Pract. 5: 331–335

Saxelin M, Chuang MH, Chassy B, Rautelin H, Makela PH, Salminen S & Gorbach SL (1996b) Lactobacilli and bacteremia in southern Finland, 1989-1992. Clin. Infect. Dis. 22: 564–566

Solis Pereyra B & Lemonnier D (1993) Induction of human cytokines by bacteria used in dairy foods. Nutr. Res. 13: 1127–1140

Spanhaak S, Havenaar R & Schaafsma G (1998) The effect of consumption of milk fermented by *Lactobacillus casei* strain Shirota on the intestinal microflora and immune parameters in humans. Europ. J. Clin. Nutr. 52: 899–907

Storlie J, O'Flaherty MJ & Hare K (1998) Food or Supplement? Choosing the appropriate regulatory path. Food Technol. 52: 62–69

Tannock GW (1999) Probiotics: A critical review. Horizon Press, Wymondham (Norfolk), UK

Tannock GW (1998) Studies of the intestinal microflora: a prerequisite for the development of probiotics. Int. Dairy J. 8: 527–533

Titze A, Kuhn C, Lorenz A, de Vrese M & Barth C (1996) The influence of viable lactobacilli on lactose degradation in the gut of gnotobiotic animals. XIIth International Symposium on Gnotobiology, Honolulu, pp. 43

Tomera JF (1996) Yogurt: probiotic pharmacology. Drugs of Today 32: 539–551

Tomioka H & Saito H (1992) Lactic acid bacteria in the support of immuno-compromised hosts. In: Wood BJB (Ed.) The lactic acid bacteria: Vol. I The lactic acid bacteria in health and disease. (pp 263–296) Elsevier Applied Science, London

van't Veer P, Dekker JM, Lamers JWJ, Kok FJ, Schouten EG, Brants HAM, Sturmans F & Hermus RJJ (1989) Cancer Res. 49: 4020–4023

Vesa TH, Marteau Ph, Zidi S, Briet F, Pochart Ph & Rambaud JC (1996) Digestion and tolerance of lactose from yoghurt and different semi-solid fermented dairy products containing *Lactobacillus acidophilus* and bifidobacteria in lactose maldigesters – Is bacterial lactase important? Euro. J. Clin. Nutr. 50: 730–733

Antonie van Leeuwenhoek **76:** 317–331, 1999.

Lactic acid bacteria in the quality improvement and depreciation of wine

Aline Lonvaud-Funel
Biotechnology and Applied Microbiology Department, Faculté d'œnologie, Unité associée INRA/Université Victor Segalen Bordeaux 2, 351, Cours de la Libération, 33405 Talence, Cedex, France

Key words: lactic acid bacteria, wine

Abstract

The winemaking process includes two main steps: lactic acid bacteria are responsible for the malolactic fermentation which follows the alcoholic fermentation by yeasts. Both types of microorganisms are present on grapes and on cellar equipment. Yeasts are better adapted to growth in grape must than lactic acid bacteria, so the alcoholic fermentation starts quickly. In must, up to ten lactic acid bacteria species can be identified. They belong to the *Lactobacillus, Pediococcus, Leuconostoc* and *Oenococcus* genera. Throughout alcoholic fermentation, a natural selection occurs and finally the dominant species is *O. oeni*, due to interactions between yeasts and bacteria and between bacteria themselves. After bacterial growth, when the population is over 10^6CFU/ml, malolactic transformation is the obvious change in wine composition. However, many other substrates can be metabolized. Some like remaining sugars and citric acid are always assimilated by lactic acid bacteria, thus providing them with energy and carbon. Other substrates such as some amino acids may be used following pathways restricted to strains carrying the adequate enzymes. Some strains can also produce exopolysaccharides. All these transformations greatly influence the sensory and hygienic quality of wine. Malic acid transformation is encouraged because it induces deacidification. Diacetyl produced from citric acid is also helpful to some extent. Sensory analyses show that many other reactions change the aromas and make malolactic fermentation beneficial, but they are as yet unknown. On the contrary, an excess of acetic acid, the synthesis of glucane, biogenic amines and precursors of ethylcarbamate are undesirable. Fortunately, lactic acid bacteria normally multiply in dry wines; moreover some of these activities are not widespread. Moreover, the most striking trait of wine lactic acid bacteria is their capacity to adapt to a hostile environment. The mechanisms for this are not yet completely elucidated . Molecular biology has provided some explanations for the behaviour and the metabolism of bacteria in wine. New tools are now available to detect the presence of desirable and undesirable strains. Even it much remains unknown, winemakers and oenologists can nowadays better control the process. By acting upon the diverse microflora and grape musts, they are more able to produce healthy and pleasant wines.

Introduction

At harvest, grape berries carry lactic acid bacteria, besides yeasts, acetic acid bacteria and molds. Only yeasts and lactic acid bacteria are involved in wine-making. Acetic acid bacteria survive in low concentrations and molds are discarded, both because of the low redox potential of the medium as soon as grapes are poured into fermentation tanks. Yeasts are better adapted than lactic acid bacteria to growth in grape musts, which are very high in sugar concentrations (>210 g l^{-1}) and have a low pH 3.0–3.3. Therefore, alcoholic fermentation starts very quickly. When all reducing sugars are fermented to ethanol, yeast levels decline and lactic acid bacteria growth occurs. Malolactic fermentation then follows the alcoholic fermentation. Once all malic acid is degraded, lactic acid bacteria are eliminated by sulfiting. Sulfur dioxide is at this stage of winemaking the only authorized and efficient agent for the microbial stabilization of wine. Nearly all red wines and many white wines are obtained by these two fermentation steps. The term malolactic fermentation however is not correct since the transformation of L-malic acid to L-lactic acid is not a fermentative pathway, but a decarboxylation. Yet winemakers gave this name to the event by analogy with alcoholic fermentation; indeed during both steps, CO_2 escapes from wine by bubbling. In ideal conditions, malolactic fermentation follows alcoholic fermentation within a few days (Lafon-Lafourcade 1983).

During their growth, lactic acid bacteria ferment residual sugars, hexoses and pentoses left by yeasts and transform numerous wine components (Ribéreau-Gayon et al. 1998). Even if malic acid is the most important in terms of quantity and influence on wine composition, other known and unknown substrates are metabolized and probably have a decisive effect on wine quality. Depending on the species and even on the strains, and on the moment they multiply, lactic acid bacteria may be beneficial or detrimental to quality. The control of yeast and bacterial growth is therefore necessary in winemaking. During maturation and aging they must be eliminated as much as possible. During these phases, it is mainly chemical reactions which modify aromas and wine color by oxidation, esterification and polymerization of the chemical active components.

Due to the variety of lactic acid bacteria species and strains in grape must and wine, and to the wine composition, various factors may disturb the harmonious equilibrium of wine microflora and induce slight or pronounced alterations. Basic knowledge of this has increased considerably, from the moment when enologists identified lactic acid bacteria in wine 40 years ago (Peynaud 1956). However many gaps still exist and one of the most evident concerns the natural adaptation of these bacteria to such a harsh medium. Moreover, only a few of the metabolic pathways of wine components are known; those participating in energy provision and sensorial profile changes are still poorly understood.

Populations of lactic acid bacteria in grape musts and wines

On grape berries the bacterial population is low, or at least only a few bacteria are capable of growth on nutritional medium such as MRS in laboratory conditions. From crushed grapes poured into fermentation tanks, many more bacteria can be enumerated and isolated. The population varies from about 10^2 CFU/ml to 10^4 CFU/ml, depending on climatic conditions during the final days of grape maturation. It is mainly correlated with pH; the higher the pH, the higher the total lactic acid bacteria population. At this stage up to 8 or 9 species can be identified. Four genera are represented: *Lactobacillus*, *Pediococcus*, *Leuconostoc* and *Oenococcus*. Lactobacilli belongs to facultative (*Lactobacillus plantarum*, *L. casei*) and obligatory (*L. hilgardii*, *L. brevis*, *L. fructivorans*) heterofermentative species. The homofermentative cocci are mainly *Pediococcus damnosus* and *P. pentosaceus*. *P. parvulus* has also been isolated from Washington state wines (Edwards et al. 1994) and Australian wines (Davis et al. 1988). Heterofermentative cocci of wines were classified in the *Leuconostoc* genus until 1995 in the species *Leuconostoc mesenteroides* and *Leuconostoc oenos*. The latter was the only representative of the acidophilic branch of the genus with low relationships in term of RNA/DNA hybridization with the other leuconostocs (Garvie 1981). The molecular approach to taxonomy, based on 16S r-DNA and 23S r-DNA sequencing, led to the creation of a new genus *Oenococcus*, with the sole species *O. oeni* (Dicks et al. 1995). This new classification finally accounts for the originality of such bacteria.

During the first days of alcoholic fermentation, the lactic acid bacteria population generally increases to a maximum of 10^4 CFU/ml, and then decreases to around 10^2 CFU/ml at the end of alcoholic fermentation. Ethanol can be around 5–6% at the peak of the lactic acid bacteria population. Most importantly, not only total bacterial counts diminish but the original diversity of the species does also. In most cases *O. oeni* predominates at the end and after alcoholic fermentation. Identification by colony hybridization using total genomic DNA probes clearly shows that *Lactobacillus* species, *Pediococcus* and *L. mesenteroides* progressively disappear, or at least are at too low a concentration to be detected, while *O. oeni* is the only species identified when fermentation is finished (Table 1) (Lonvaud-Funel et al. 1991). However, obviously some of the other species are not completely eliminated. Malolactic fermentation starts when the population reaches 10^6 CFU/ml after a fast growth rate which may begin sooner or later after the end of alcoholic fermentation. The lag phase between the two fermentations mainly depends on temperature, pH and ethanol contents, the main factors determining bacterial growth in wine. The natural selection of *O. oeni* during fermentation is mainly due to the progressive increment in ethanol and other products of yeast metabolism. Fatty acids such as decanoic and dodecanoic acid are powerful inhibitors of lactic acid bacteria growth (Edwards et al. 1990). Like ethanol they alter

Table 1. Specific enumeration of lactic acid bacteria in cabernet sauvignon fermenting must (CFU/ml) (Lonvaud-Funel et al. 1991)

Day	Alcohol	*O. oeni*	*L. mesenteroides*	*P. damnosus*	*L. hilgardii*	*L. brevis*	*L. plantarum*	*L. casei*	Total
0	0	nd	2.9×10^{2}	6.0×10^{2}	1.1×10^{3}	nd	7.5×10^{1}	7.7×10^{1}	2.5×10^{3}
3	7	nd	1.7×10^{4}	3.8×10^{4}	8.0×10^{4}	2.0×10^{4}	2.0×10^{4}	2.0×10^{4}	1.7×10^{5}
6	9	nd	9.6×10^{4}	3.7×10^{4}	4.0×10^{4}	4.5×10^{3}	nd	nd	1.5×10^{5}
10	13	4.2×10^{3}	3.2×10^{3}	4.9×10^{3}	4.4×10^{3}	nd	nd	nd	1.8×10^{4}
18	13	1.4×10^{6}	nd	nd	nd	nd	nd	nd	3.4×10^{6}

nd: not detected.

the bacterial membrane (Lonvaud-Funel et al. 1988). Some yeast strains may also produce relatively large amounts of sulfur dioxide from their sulfur compound metabolism (King & Beelman 1986). In addition to the increasing toxicity of the medium, some nutrient nitrogen lack can occur, at least during the explosive phase of yeast growth. However, this relative starvation is transitory since yeasts release aminoacids, especially at the end of fermentation. Of the variety of lactic acid bacteria, species *O. oeni* is probably the best adapted to overcome these obstacles. However, some *Pediococcus* and *Lactobacillus* strains can survive. They are potent spoilage agents after winemaking. Interactions between lactic acid bacteria species have also been demonstrated. They are involved in the selection during and after winemaking (Lonvaud-Funel & Joyeux 1993).

Other difficulties in malolactic fermentation have been ascribed by Sozzi et al. (1976, 1982) to phage attack. Phages have also been found in Australian, South-African, German and French wines. Lysogeny is widespread in the species *O. oeni* (Poblet et al. 1998). Protein composition, restriction profile, genome size and DNA-DNA homologies have been studied for some phages (Nel et al. 1987; Arendt & Hammes 1992, Santos et al. 1996). Gindreau et al. (1997) studying a temperate bacteriophage showed that the organization of the site-specific integration system is similar to other phage systems. The bacterial attachment site was the same for all the strains compared. Moreover, spontaneous phage induction occurs in wine. However the phages readily disappear, through inactivation by wine components, and their population cannot reach a sufficient level to destroy all the bacterial population. Moreover, since the indigenous microflora of wines comprises a variety of *O. oeni* strains, and given the specific sensitivity to phage infection, only part of the population dies. This mechanism also is involved in natural strain selection. In summary, in spite of the presence of phages in wine

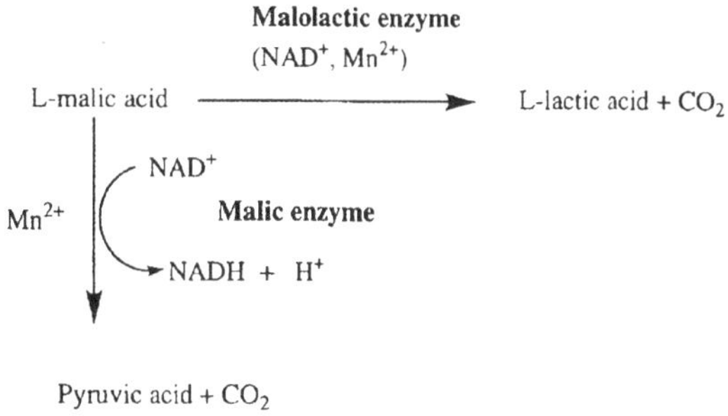

Figure 1. Transformation of L-malic acid by the malolactic and malic enzymes.

and of their possible activity during a short period, it seems that they are not responsible for influencing malolactic fermentation.

When all malic acid is degraded, wines are stabilized by sulfiting. Most of the bacteria and possibly the remaining yeasts are sensitive to sulfur dioxide. However its activity is related to pH. In high pH wines, the ineffectiveness of SO_2 and the lesser stringency of the medium mean that bacteria survive more easily. In such wines, relatively high levels of viable cells may be found reaching sometimes 10^5 or 10^6 CFU/ml, even many months after winemaking. In these conditions *Lactobacillus* and *Pediococcus* are frequently predominant and induce spoilage. Only physical treatments such as heat treatment or filtration can eventually eliminate all the viable bacteria.

The principal reaction of malolactic fermentation: transformation of L-malic acid

The main value of malolactic fermentation is the biological deacidification which results from the transformation of L-malic acid (dicarboxylic acid) to L-lactic acid (monocarboxylic acid). All malic acid is degraded, depending on the production area from 2 to 10 g/l of L-malic acid. This induces an increase

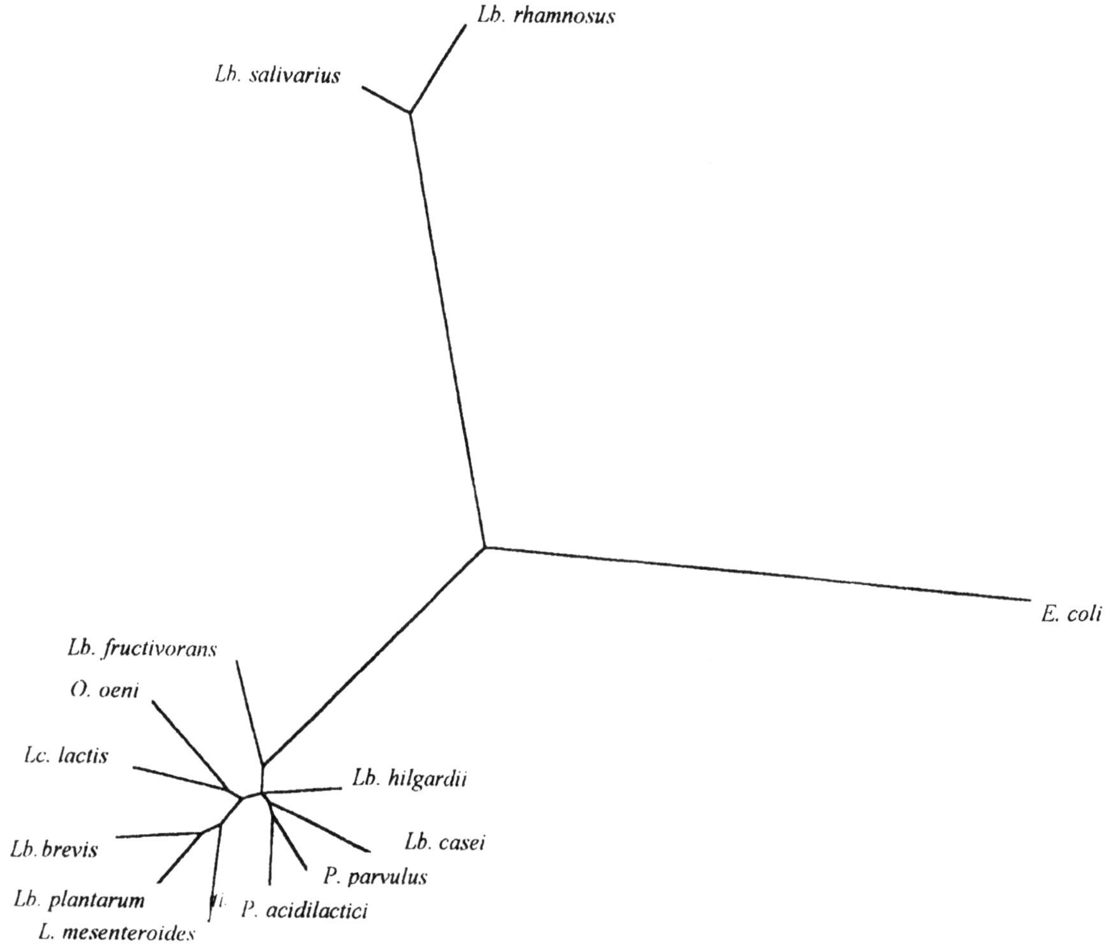

Figure 2. Fitch distance matrix tree of malolactic gene sequences of 12 lactic acid bacteria species. *E. coli* malic enzyme was used as the outgroup.

in pH, and change in wine taste. Decarboxylation of malic acid into L-lactic acid is catalysed by the malolactic enzyme (MLE) which is different from the malic enzyme leading to pyruvate (Figure 1). First purified from *Lactobacillus plantarum* (Schütz & Radler 1973; Lonvaud 1975), this enzyme was further purified in many other wine lactic acid bacteria species of all the genera represented in wine (Lonvaud-Funel 1995). The kinetic parameters where first thoroughly studied for the enzyme of *L. mesenteroides* (Lonvaud-Funel & Strasser de Saad 1982) and the first amino acid sequence was obtained in a non-enological strain of *Lactobacillus lactis* (Lonvaud-Funel 1992, unpublished). The malolactic activity of the entire cells is strictly dependent on the integrity of the bacterial membrane. Because of the optimum pH of the enzyme (around 5.8), the need for cofactors (Mn^{2+}, NAD^{+}), and the inhibitory effects of many wine components (carboxylic acids, polyphenols), the protein must be protected from the medium by the cell membrane. This also explains why in spite of basic research, an enzyme reactor usable for wine treatment has never been developed (Festaz-Furet 1991).

The N-ter amino acid sequence of the MLE protein gave Denayrolles et al. (1994) the starting point for isolation of *mle*S, the gene encoding the MLE protein. Later the *mle*S gene of *O. oeni* was sequenced (Labarre et al. 1996) and proved to be very close to the *L. lactis* gene. The deduced aminoacid sequence of the *mle*S gene revealed that the protein is highly similar to the malic enzyme of many other organisms. The best score is obtained with the NAD^{+}-dependent malic enzyme of *E. coli*. A phylogenetic approach on the central region representing 36% of the gene for twelve lactic acid bacteria species has been done. Figure 2 is the phylogenetic tree showing the proximity between some species and the distance between others (Groisillier & Lonvaud-Funel, unpublished).The alignment of the prokaryotic and eukaryotic malic enzyme aminoacid sequences and of the malolactic enzyme of *L. lactis* revealed highly conserved regions. The nucleotide binding domains and the malate binding site are clearly identified. Other regions of unknown functions are also conserved (Denayrolles et al. 1994). Recent results show that the same is true when comparison concerns the *mle*S sequence of twelve other species Thus the malolactic enzyme is very close to the malic enzyme is spite of a fundamental specificity in the reaction. Malic enzyme produces pyruvate, while malolactic enzyme leads to L-lactate (Figure 1). So far the difference in activity has not been explained by the comparison of the protein sequence.

The *mle*S gene was functionnally expressed in *Saccharomyces cerevisiae* (Ansanay et al. 1993; Denayrolles et al. 1995). The idea was to transform wine yeast strains with the *mle*S gene in order that malolactic transformation might occur during alcoholic fermentation. However, the engineered yeast could not degrade all the malic acid due to a limitation in the transport of the substrate inside the cell. Cloning and expression of the *mae*1 gene that encodes the malate permease of *Schizosaccharomyces pombe*, together with the *mle*S gene, finally led to a recombinant malolactic *S. cerevisiae* strain that fermented malic acid of grape must in a few days (Volschenk et al. 1997). The integration of *mae*1 and *mle*S genes into the genome of wine yeasts is the next step to stabilize the activities which were initially preliminary cloned and transferred onto plasmids. The use of such a recombinant yeast strain would be a technological solution to the problem of malolactic fermentation. However, it would not be applicable to all wines since malolactic fermentation is not only the transformation of malic acid but also a set of numerous biochemical reactions and bacterial metabolisms.

To overcome the difficulties in triggering malolactic fermentation, winemakers and wine microbiologists came in the late 70s to the idea of inoculation with selected malolactic starters. The defect of the lactic acid bacteria population responsible for the absence of malolactic fermentation would be suppressed by the addition of high cell numbers. However, many unsuccessful trials were done until it was found that survival of bacteria freshly inoculated in wine could be improved by a 're-activation' step (Lafon-Lafourcade et al. 1983). Since that time, several selected starters have been commercialized. The lyophilized or frozen preparations are incubated in a suitable medium, mainly composed of grape juice, or wine, and yeast extract, for one or two days before use. This incubation does not result in multiplication of the bacteria but rather in an adaptation step. The efficiency of such starters was also improved by the preliminary addition of yeast cell walls as detoxication material. Indeed, the macromolecules of 'yeast ghosts' fix toxic compounds such as fatty acids and probably add activating factors (Paraskevopoulos 1988). However, the use of such starters was too time-consuming and unreliable.

Table 2. Changes in protein and phospholipid composition of *Oenococcus oeni* cells exposed to heat for 30 min at various temperatures (Garbay & Lonvaud-Funel, 1996)

Temperature °C	Proteins (mg/ml)	Phospholipids (μmolesPi/ml)	(Phospholipid/ Protein)
25	2.2	0.66	0.30±0.018
37	2.6	0.68	0.26±0.020
42	2.8	0.73	0.26±0.020
50	6.0	0.65	0.18±0.015
60	12.0	1.05	0.08±0.010

Basic research was performed to interpret the loss of adaptation to wine conditions during the isolation of strains from fermenting wines. As fatty acids and ethanol were shown to inhibit the growth and viability of bacteria, it was supposed that the first target of their toxic effect was the membrane (Lonvaud-Funel et al 1988). Indeed, several stresses such as heat shock, ethanol and fatty acids and acidic pH conditions result in a clear response of *O. oeni*. On one hand, their survival after stress is better when they are inoculated into wine. On the other, the membrane composition varies. Any kind of stress leads to the significant decrease in the phospholipid to protein ratio. Not only the phospholipid content decreases, but also and above all the protein content increases up to fivefold (Table 2). The electrophoretic analysis of membrane protein reveals 17 bands. Whatever the stress, three proteins are overexpressed. The analysis of the membrane composition over a cell cycle also showed that the same variation occurs when the cells enter the stationary phase (Garbay & Lonvaud-Funel 1996). This might explain the better survival of commercial starters after reactivation; the incubation brings the cells to a similar state. In addition, Guzzo et al. (1997) found that *O. oeni* also responds to stress by synthesis of 6 stress proteins. Among these, the 18 kDa protein named LO18 was purified. It is associated to the membrane by weak binding and acts as a chaperon protein. In summary, like every viable cell, *O. oeni* responds to environmental changes and in particular to hostile conditions by the ubiquitous synthesis of stress proteins. Reliable ready-to-use starters were finally obtained when research on adaptation to harsh conditions was undertaken. The first malolactic starter for direct inoculation was evaluated in many wine producing areas (Nielsen et al. 1996). Its efficacy was successfully demonstrated in spite of some failures in the most difficult wines such as acidic white wines.

Influence of malolactic fermentation on the sensory quality of wine

In terms of quantity, there is no doubt that the transformation of malic acid is the major event during malolactic fermentation. The total acidity is lowered and the L-malic acid molecule is stoichiometrically replaced by L-lactic acid. Both result in loss of acidity not only by loss of the acidic equivalent but also because the strong green taste of malic acid is replaced by the less aggressive taste of lactic acid. However, more subtle changes occur: some aromas increase with a flavor complexity. Some varietal aromas revealed during alcoholic fermentation by yeast disappear or change after malolactic fermentation. This explains why malolactic fermentation is favourable for most red and white wines where the fruitiness attributed to the grape variety is not essential. These wines more complex in taste, with an improved mouthfeel, are often subjected to aging for a long time in barrels, and need bottle-aging to reach their plenitude. On the contrary, light red wines and some white wines, are characterized by the grape aromas and by their vivacity which fades with malolactic fermentation.

Numerous substances produced by bacteria are involved in the aroma changes of wine during malolactic fermentation. However, to date few of them have been identified. One of the most significant descriptors for sensory panelists is buttery. This is directly linked to the increased concentration of diacetyl which is recognized as the major contributor to aroma change during malolactic fermentation (Bertrand et al. 1984; Davis et al. 1985). Acetoin is also produced but its threshold is higher. Acetoinic compounds which comprise diacetyl, acetoin and 2,3-butanediol result from citric acid metabolism (Figure 3). All heterofermentative cocci (*Leuconostoc* and *Oenococcus*) and facultative heterofermentative lactobacilli (*L. plantarum*, *L. casei*) of wines degrade citric acid. During malolactic fermentation, *O. oeni*, which is dominant, metabolizes citric acid more slowly than it degrades malic acid. The initial citric acid concentration in wine is about 250–300 mg/l, and according to the rate of malolactic fermentation, the final concentration is between 0 and 100 mg/l. However, even after sulfiting this metabolism goes on and finally citric acid completely disappears. The major products of this degradation are acetic acid and acetoinic compounds (C4 compounds), two groups of substances which significantly affect wine taste. Acetic acid is produced in the first reaction catalyzed by the citrate lyase. Acetoinic compounds are produced from pyruvate resulting from the activity of citrate lyase and oxaloacetate decarboxylase. Pyruvate is usually reduced to lactate to reoxidize NADH in the fermentative pathway. However, when additional pyruvate is produced from citric acid in the absence of sugar, it is shifted to the production of acetoin and butanediol. The admitted pathway for the production of C4 compounds is that the C5 α-acetolactate is produced from two pyruvate molecules. Then the α-acetolactate decarboxylase catalyzes the reaction to acetoin, which accumulates in the medium or is reduced to butanediol. Diacetyl could be produced by the chemical decarboxylation of α-acetolactate (Hugenholtz 1993). Finally, the diacetyl reductase activity can also lead to acetoin. The key enzymes citrate lyase, α-acetolactate synthase and α-acetolactate decarboxylase, and the related genes have been studied in several lactic acid bacteria and non-lactic acid bacteria species and notably in *O. œni* (Garmyn et al. 1993).

The accumulation of diacetyl, acetoin and acetic acid in wine varies according to the rate of malolactic fermentation. When malolactic fermentation is fast, acetic acid production from a given amount of citric acid is relatively high and diacetyl+acetoin is low. On the contrary, when bacteria multiply more slowly, less acetic acid and more diacetyl+acetoin are excreted. This observation was also verified in laboratory cultures of *L. mesenteroides* (Table 3). Labelled citrate also showed that some pyruvate can be shifted to the production of acetyl CoA and finally to fatty acids, since radioactivity is found in the bacterial membrane

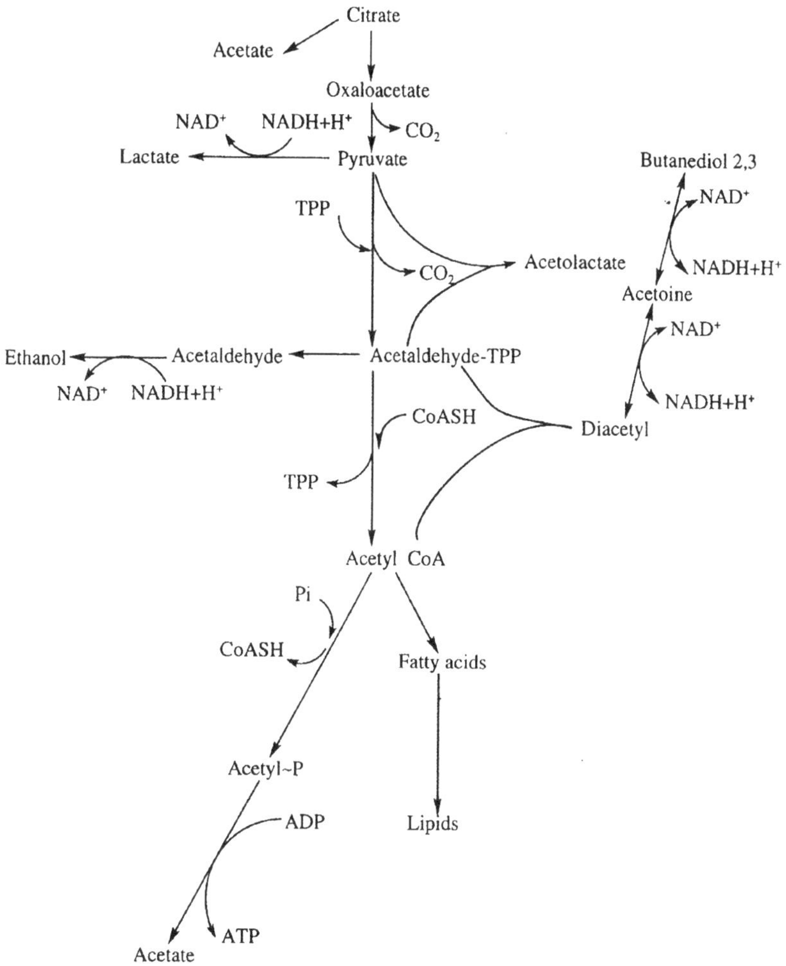

Figure 3. Citric acid metabolism by lactic bacteria of wine.

Table 3. Influence of pH and temperature on the production of acetic acid and acetoine+diacetyl from citric acid by *Leuconostoc mesenteroides* cultured in Carr medium (Bertrand et al. 1984)

(A) role of pH:

pH	Citric acid degraded (g/l)	Acetic acid (g/l)	Acetoine+diacetyl(mg/l)
4.8	0.95	0.68	28
4.1	1.00	0.34	124

(B) role of temperature:

acetoine+diacetyl (mg/l) produced from 0.5 g/l citric acid+0.1 g/l pyruvic acid

pH	Temp.=18 °C	Temp.=30 °C
4.1	68	32
4.8	58	23

(Harvey & Collins 1963). It may be assumed that in fast growing conditions (high pH and temperature), the need for the molecule to synthesize fatty acids and lipids is much higher. In these conditions, pyruvate is mainly oriented towards acetyl CoA and less to acetoinic compounds. The contrary is true when growth is limited (acidic pH and low temperature); an excess of pyruvate leads to C4 compounds. Thus, in addition to the initial amount of citric acid in wine, the diacetyl level varies with the conditions of malolactic fermentation. Moreover, the descriptive analysis of the aroma of wine fermented by different bacteria shows that significant differences exist between wine strains, as demonstrated for *L. lactis* (McDaniel et al 1987; Hugenholtz 1993). Moreover, the final concentration of diacetyl in wine is also affected by the diacetyl reductase activity of both yeast and bacteria lees. In addition it combines reversibly with sulfur dioxide like many other ketonic compounds of wine. Therefore the influence of the citric acid metabolism on the sensory quality of wine, due to the complexity caused by diacetyl, is determined by several technological factors. The prolonged contact of wine with lees reduces its role, while early racking or clarification enhances its participation in wine aroma (Nielsen & Prahl 1996). In white and red wines the average thresholds are respectively 4.5–9.5 mg/l and 12–14 mg/l for diacetyl, 430–600 mg/l and 2000 mg/l for acetoin. Usually diacetyl concentrations in wine are about 5–10 mg/l. Concentrations higher than the thresholds are not appreciated by all people. Two groups of tasters were revealed in an experiment: those who liked the butter flavor of diacetyl, and the others who did not (Bertrand et al. 1984). However, the result is also affected by the wine composition and structure and by the grape variety, especially for aromatic white wines. Besides diacetyl, lactic acid bacteria might be involved in the production of methylglyoxal. The concentration of this C3 compound (analogous to diacetyl) increases during malolactic fermentation (De Revel & Bertrand 1993). The aroma impact is probably much lower than diacetyl since C3 compounds are lighter in aroma intensity than C4.

Not all the aroma changes occurring during malolactic fermentation are attributed only to citrate metabolism. Henick-Kling et al. (1993) categorized aromas which did not change, which increased and which decreased during malolactic fermentation of Chardonnay wine. Both for white wines (Sauvignon and Semillon) and for red wines (Cabernet Sauvignon), fruitiness is enhanced while vegetative aromas are reduced. To date very few substances have been identified to explain such variations.

After malolactic fermentation, red wine color and body are affected as a result of modifications in phenolic compounds. Malolactic fermentation significantly decreases free anthocyans and astringency by increasing the reaction between tannins and anthocyans. During malolactic fermentation, some of the phenolic compounds precipitate or undergo structural change. Malolactic fermentation in barrels confers specific modifications which all combine for a better stabilization of color (Vivas et al. 1995).

Lactic acid bacteria as spoilage agents of wine

Increase of wine acidity

In most cases, when winemaking is well controlled the multiplication and subsequent biochemical reactions supported by lactic acid bacteria improve wine quality and stability. However, they may also depreciate it sometimes enough to make it unmarketable. Lactic acid bacteria can spoil wine during winemaking or during maturation and bottle aging. In the first case, bacteria which are going to perform the malolactic fermentation grow too early. Their multiplication takes place at the end of alcoholic fermentation, not after as it should do. Therefore bacteria ferment carbohydrates, particularly hexoses, which have not been totally fermented by yeasts. At this stage, most of the bacterial population is composed of heterofermentative strains, mainly *O. oeni*. Besides ethanol and CO_2 which are also formed by yeasts, the major products of fermentation are, acetic acid and D-lactic acid. In consequence, the volatile acidity of wine increases. In limited amounts the wine is more or less depreciated. If it exceeds the limit of 1 g/l volatile acidity (expressed as acetic acid) the wine is unmarketable. The quantity of acetic acid produced depends on the amount of hexoses fermented and on the total bacterial population. This accident named 'piqûre lactique' occurs when the end of alcoholic fermentation is too slow or when its stops. Monitoring the evolution of yeasts and alcoholic fermentation clearly shows that most of the latter takes place during the stationary phase of the yeast cycle. The high sugar concentration, the low pH and nitrogen deficiency, together with the excretion of toxic yeast metabolites, make the yeast activity very difficult. In these conditions growth of bacteria might start. They may ferment several g/l of sugars together with malic acid. The malolactic fermentation may be more or less finished, while there may still be some unfermented sugar. Normally the interactions between yeasts and bacteria, together with the effect of sulfur dioxide added to grape must, prevent such premature bacterial growth. The volatile acidity resulting from a 'piqûre lactique' can be ascribed to lactic acid bacteria by determination of D-lactic acid, which is the only stereoisomer formed, besides acetic acid, by the dominant heterofermentative cocci. This major incident of winemaking is not rare. The technological operations and monitoring of alcoholic fermentation must be strict. Difficulties are predictable when grapes are very mature, particularly when the climate is hot and dry before harvest. In these conditions, the high sugar and low acidity of must are both factors which are respectively in favor of stuck in yeast activity and of bacteria growth.

Fortified wines are also affected by 'piqûre lactique'. These products are prepared by the addition of brandies or wine spirits (Cognac, Armagnac) to grape must more or less fermented by yeast. They have a 16 to 20% ethanol level and contain variable but rather high concentrations of hexoses. Port, Sherry, Pinaud des Charentes are some examples. In spite of their high level in ethanol, the heterofermentative lactobacilli can multiply. Since the generation time of bacteria in such an environment is long, the multiplication becomes evident most often during barrel aging or after bottling. Volatile acidity is unacceptable, and the wine is gaseous and cloudy. Most of the time *L. hilgardii* and *L. fructivorans* strains are involved. Some strains are not only tolerant to ethanol, but also dependent on ethanol for growth (Couto & Hogg 1994).

Ropiness of wine

The other spoilages of taste and aromas may be attributed to the development of special lactic acid bacteria strains instead of species. In 1860 Pasteur described three 'wine diseases' named 'amertume', 'tourne' and 'graisse'. The microorganisms responsible for these alterations were lactic acid bacteria. The first two diseases are not widespread in wines, and they result from the glycerol and tartaric acid metabolism by lactobacilli. The third is more usual and a real concern for many wines of the 1997 vintage. This deterioration is characterized by an abnormal increase in viscosity, to such an extent that the wine runs thicker than oil. The ropy wines contain polysaccharides produced from residual sugars (less than 1 g/l) by strains of *Pediococcus damnosus*. This species is normally present in grape must and disappears almost completely during the winemaking process. Sometimes *P. damnosus* also plays a large part in malolactic fermentation. Most of the *P. damnosus* strains are not spoilage agents. Only some of them can synthetize exocellular polysaccharides (EPS). The glucan produced from glucose has a trisaccharides repeating unit structure (Llauberes et al. 1990) (Figure 4).

Concentrations of the glucan around 100 mg/l are high enough to give the wine the abnormal and unacceptable viscosity. *P. damnosus* strains which produce the EPS differ from ordinary strains by the presence of a 4 kb plasmid. During repeated transfers of ropy *P. damnosus* strains in culture media added with both 8 to 11% ethanol, the ropy character is conserved and the plasmid is stable. The ropy strains are much more tolerant to ethanol than the others. The loss of the plasmid by repeated subcultures in laboratory broth is correlated to the loss of the ability to produce the EPS. Thus, although it has not been verified that the transformation of unropy strains by the plasmid could transfer the EPS synthesis ability, several observations support the hypothesis that it is really involved in the phenomenon. A DNA probe obtained by labelling a 1.2 kb part of the plasmid is routinely used to detect the presence of such strains in wine (Lonvaud-Funel et al. 1993). The function encoded by the plasmid is unknown.

The deterioration may occur when wine is still in vats. In this case, it is possible to recover a normal viscosity by mechanical treatments. Generally the wine does not have other defaults except if other microorganisms, yeast or bacteria are also present. However, in most cases, the ropiness develops very slowly and becomes evident several weeks or months after bottling. Knowledge on these particular strains is not yet sufficient to understand their behaviour in wine, their possible relations with other microorganisms and the factors involved in EPS synthesis, in order to predict their development. In consequence, today the detection of such bacteria in wine before bottling requires treatment with very drastic filtration steps or by heating. In addition as they are very tolerant to hostile conditions and even to SO_2 (perhaps because of the EPS layer around the cell), the contamination of a whole cellar is very easy. Rigorous cleaning of all the equipment is the only means to eliminate them.

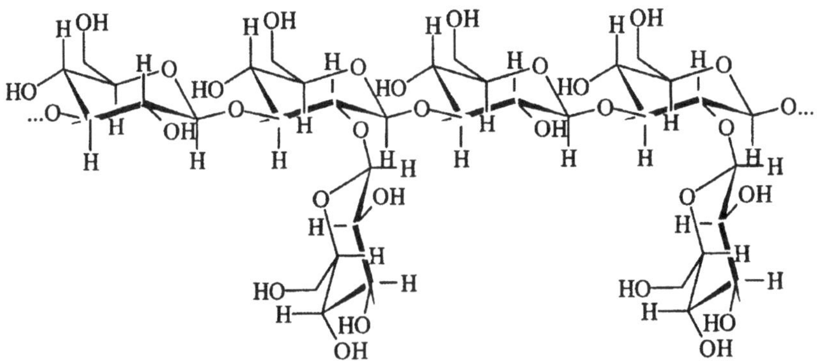

Figure 4. Structure of the polysaccharide produced by *Pediococcus damnosus*.

Production of off-flavors

During and after winemaking, off-flavors sometimes appear. These defects are not exactly characterized, either by the substances concerned or by the reason for their formation. Special attention has been given in recent years to the 'animal' phenolic odors due to excessive amounts of volatile phenols in red wines, 4-ethylphenol and 4-ethyl gaïacol. Both substances are produced by decarboxylation and subsequent reduction of p-coumaric and ferulic acid which are components of grapes and wines. Concentrations of volatile phenols higher than the taste thresholds gives the wine the phenolic character. Most often this degradation occurs during barrel aging and it is attributed to the activity of *Brettanomyces* sp. strains. A strong correlation is shown between their population and the level of volatile phenols. Normally this problem can be avoided since these yeasts cannot survive in wine due to their sensitivity to sulfur dioxide (Chatonnet et al. 1995). However, Cavin el al. (1993) showed that lactic acid bacteria (*L. plantarum, L. brevis* and *Pediococcus*) could metabolize phenol carboxylic acids, ferulic anc p-coumaric acids. The p-coumaric acid decarboxylase activity has been characterized in *L. plantarum* (Cavin et al. 1997). The *Pediococcus* gene encoding the enzyme is transcripted only when p-coumaric acid is added as inductor. However, *L. plantarum* might not be involved in the deterioration due to its low frequency in wine during malolactic fermentation and later. The possible existence of *Pediococcus* strains capable of such activity would be much more worrying.

'Mousy' off-flavors are also produced by lactic acid bacteria in wine and three heterocyclic volatile bases are responsible: 2-acetyltetrahydropyridine, 2-ethyltetrahydropyridine and 2-acetyl1-pyrroline. Costello et al. (1996) studied the ability for wine lactic acid bacteria to produce such compounds in grape juice and wine-based medium. Heterofermentative species, lactobacilli and *O. oeni* gave positive results. The preliminary results showed that the production involves a complex interaction between the carbohydrate and aminoacid metabolism in the presence of ethanol.

Influence of lactic acid bacteria on the hygienic quality of wines

Two groups of substances possibly released in wine by lactic acid bacteria are undesirable with respect to the hygienic quality: biogenic amines and ethylcarbamate. They are produced during and after winemaking. Some are also present in low amounts in grape juice. Biogenic amines are physiologically active in the human metabolism. They are necessary for several functions and, most of the time, their uptake does not produce any disturbances. However, if too high amine concentrations are ingested, or if the detoxication process is inhibited either by drugs or genetically, possible toxicological effects may occur. Biogenic amines are present in fermented foods, vegetables, meat products and fish often in higher concentrations than in wine (Ten Brink et al. 1990; Halasz et al. 1994; Silla-Santos 1996). However, the addition of all these and the possible contribution of ethanol can exceed the limits for sensitive people.

In wine most of the aliphatic, aromatic and heterocyclic amines have been identified: ethylamine, isoamylamine, diaminobutane (putrescine), diaminopentane (cadaverine), tyramine, phenylethyl-

Table 4. Production of biogenic amines in a red wine during malolactic fermentation. Influence of the pH (after Lonvaud-Funel & Joyeux 1994)

pH	Histamine (mg/l)	Tyramine (mg/l)	Putrescine (mg/l)
3.5	1.2	1.2	7.0
3.7	3.2	2.9	17.0

amine and histamine. They are produced by the decarboxylation of the corresponding aminoacid; isoamylamine, methylamine and putrescine are also in low concentrations in grape musts. The first study in oenology concerned histamine. Several papers have found conflicting results. The most common idea was that in wine *Pediococcus* sp. was responsible for histamine production, and that this was related to bad winemaking control. However, the analysis of many wines showed that many of them contained biogenic amines and the discussion on the implication of *Pediococcus* sp. began again. From a wine containing high concentrations of biogenic amines, the bacterial biomass was harvested. It was composed only of *O. oeni* strains, or at least other species were in too low concentrations to be detected. These bacteria were inoculated into wines without amines. After growth the analyses showed that they actually were able to produce amines (Table 4). At the highest pH where bacteria could grow faster, the production was the greatest. If yeast lees were conserved as often in practice, the production was higher since precursors were secreted or released during yeast autolysis (Lonvaud-Funel & Joyeux 1994).

One *O. oeni* strain has been isolated and characterized for its histidine decarboxylase activity. The protein was purified to homogeneity; it is an hexameric system composed of two subunits $[\alpha\beta]_6$. Like some other histidine decarboxylases it is a pyruvoyl enzyme. It is synthesized as an inactive proenzyme, and cleaved by 'serinolysis' into the subunits forming the active protein HDC (Coton et al. 1998a). Experiments proved that *O. oeni* could take advantage of histidine decarboxylation. The energy conservation mechanisms in the electrogenic transport of substrate and product and the decarboxylation demonstrated in other lactic acid bacteria are possibly responsible (Poolman 1993). The proenzyme is encoded by the*hdc* gene which is immediately followed, as in an operon, by another open reading frame possibly coding for a transport system. Based on the comparison of the nucleotide sequences of the *hdc* gene of several bacterial species, three primer sets were defined (Le Jeune et al. 1995) and a DNA probe was designed for detecting the strains which carry the *hdc* gene. It was then found that all the *O. oeni* strains positive to PCR test or to DNA hybridization could produce histamine from histidine. These two molecular tools proved to be usable not only for *O. oeni* but also for other lactic acid bacteria species.

CL1 5' CCA/T GGT/A AAA/T ATA/T GGA AAT GGA/T TA 3'

JV17 5' AGA CCA TAC ACC ATA ACC TTG 3'

Figure 5. Sequences of the primers used to amplify a 500 bp region of the *hdc* gene.

The PCR test was used in a study of red wines produced in the South-West of France (Coton et al. 1998b). The primers CL1 and JV17 as described by Le Jeune et al. (1995) amplified a 500 bp region of the *hdc* gene (Figure 5). The results clearly showed that histamine-producing strains of *O. oeni* were very frequent in wine. Lactic acid bacteria strains isolated from Argentinian wines also produced histamine and were identified as *L. hilgardii* (Farias et al. 1993). The time of production is not restricted to the duration of malolactic fermentation. Histamine levels increase during wine storage (Coton et al. 1998) . So far no study has been completed on the tyrosine and ornithine decarboxylase of wine lactic acid bacteria. Since not all wines contain the corresponding tyramine and diaminobutane, it may be assumed that, like for histamine, only some strains carry the enzyme. As yet, the concentration of biogenic amines in wine has not been officially set. However, in some transactions they already have caused problems. In consequence, more work is necessary to prevent their production.

Another concern in wine is ethylcarbamate known as a possible carcinogen. Precursors in wine are urea, produced by yeasts, citrulline and carbamyl phosphate produced by lactic acid bacteria by the arginine deiminase pathway. Heterofermentative lactobacilli such as *L. hilgardii* frequently isolated in wines are well known for this metabolism. This is one of the phenotypic characters taken into account in the Bergey's manual for classification. In fortified wines which are often spoiled by *L. hilgardii* this metabolism is obvious (Hogg et al. 1996). Although *O. oeni* is described in the classification as unable to degrade arginine, Liu et al. (1994) clearly demonstrated that strains of this species, like a strain of *L. buchneri*, could actually produce ethylcarbamate precursors from arginine

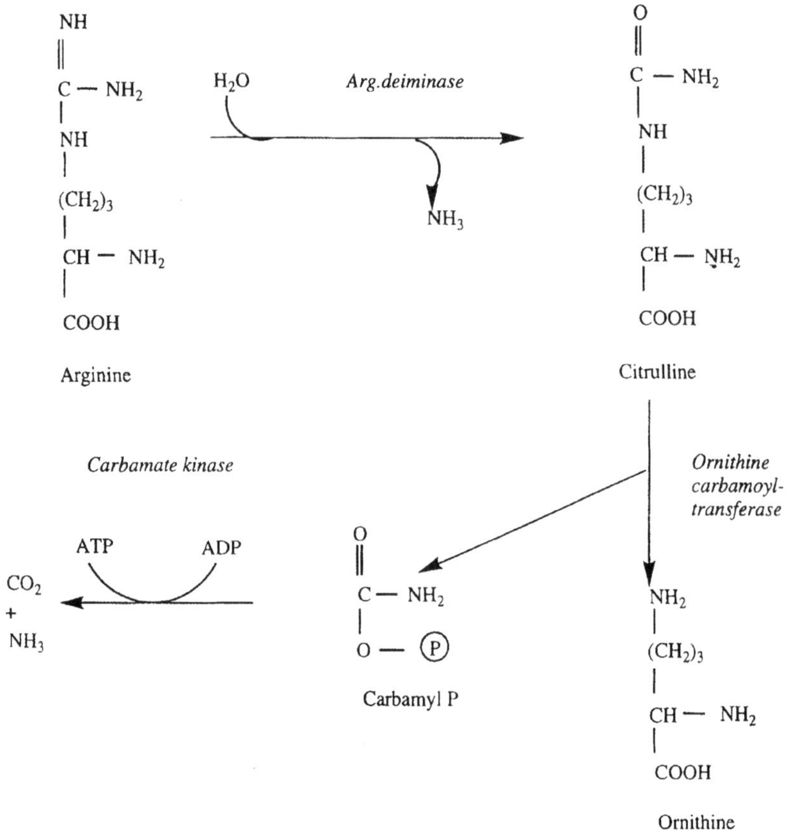

Figure 6. Metabolic pathway of arginine by *Oenococcus œni.*

(Figure 6). The formation of ethylcarbamate was accelerated by heating wine and media after growth of the strains studied. It was correlated with arginine degradation and citrulline excretion. Furthermore, Liu et al. (1995) studied the activities of the three enzymes of the pathway: arginine deiminase (ADI), ornithine transcarbamylase (OTC) and carbamate kinase (CK). No urea was produced from arginine: ornithine and ammonia were the major products, and some citrulline was excreted in the medium (Liu et al. 1996). The existence of this pathway in *O. oeni* has recently been confirmed by cloning and sequencing *arc*A, *arc*B and *arc*C, the respective genes encoding the three enzymes (Tonon & Lonvaud-Funel, unpublished). Not all *O. oeni* strains can degrade arginine. However most of the operon seems to be present, even if there is no activity. Moreover, strains which are able to metabolize arginine might take advantage of this additional energy source for growth.

Biogenic amines and ethylcarbamate are undesirable compounds in wine. Since consumers understandably look for healthy and hygienic quality products, in addition to high sensory quality, wine producers and winegrowers must pay increasing attention to such concerns. When the indigenous lactic microflora of a cellar is known to comprise undesirable strains, their participation in malolactic fermentation can be prevented by the use of malolactic starters. The ability to decarboxylate aminoacids and to degrade arginine must now be included in the selection criteria for new starters.

Conclusion

In spite of their relatively discrete presence, lactic acid bacteria multiply in wines and metabolize many components of the medium. For a long time, the only role that lactic acid bacteria were considered to play in winemaking was malolactic fermentation. Obviously several g/l are transformed while the other reactions only affect a few mg/l or less of sustrates. However, sensory more than chemical analyses have shown that 'secondary' metabolisms occur and positively or negatively influence wine quality. The question is, which species and which strain grows, at what time during the winemaking process, and finally what are the substances produced. The first answer is provided by the environmental conditions in which the very diverse natural microflora evolve. The other questions concern the specific enzymatic activities which depend on species or on strains. Of the great variety of species and strains contained in grapes berries and cellar equipment, only some of them will survive the harsh conditions encountered during alcoholic fermentation.

Ideally during wine aging, no yeasts or bacteria should survive in wine. The only time where lactic acid bacteria growth is admitted is soon after alcoholic fermentation. Many factors influence their growth, mainly the multitude of wine components from grapes and the microbial metabolism, and the temperature. Temperature can be controlled, not the others. Only sulfur dioxide, first on grapes then in wine after malolactic fermentation, gives the winemaker a means to avoid bacteria. Fortunately, on the whole, the tolerance of lactic acid bacteria to the process and to the medium means that their activity may be harmoniously controlled. The wine is less acidic, more complex and the fruitiness and color are enhanced by malolactic fermentation. *O. oeni* strains are the most desirable required for malolactic fermentation and so far no sensorial defect has been attributed to them. Malolactic starters are produced with selected strains of this species. However, since it has just been shown that some strains can produce undesirable products from the hygienic point of view, care must be taken and selection must include the relevant activities.

Much knowledge is still needed on *O. oeni*. Apart from its originality in the lactic acid bacteria group, the most interesting aspect is the identification of its adaptative response to the environment. It has scientific and technological impacts. All the other species are undesirable. Although not all the strains spoil wine, most depreciations and diseases are related to lactobacilli and pediococci, but they normally are destroyed during wine production. However, some strains demonstrate abnormal tolerance to the medium, especially to the ethanol concentration. *L. fructivorans* and *L. hilgardii* can even spoil fortified wines. The ropy *P. damnosus* strains seem to be protected from wine inhibitors when compared to non-ropy strains. All are well adapted to growing during storage even after bottling in spite of sulfiting. All make the wine cloudy; in addition the heterofermentative lactobacilli increase volatile acidity and *Pediococcus* the viscosity. Their economic impact is significant. The importance of the problem may be roughly related to the vintage which determines the wine composition, including notably pH, ethanol, phenolic compounds and the diversity of the initial lactic acid bacteria population. However it may also depend on cellar hygiene and on the technological processes used.

As for other fermented food, lactic acid bacteria are necessary to obtain quality but they must be well-controlled. Molecular biology together with classical microbiology on one hand, and the accurate observation of practical work in the cellar on the other hand, are now providing more and more data on the biology of these bacteria in such hostile conditions . Yet much remains to be discovered.

References

Arendt EK & Hammes WP (1992) Isolation and characterization of *Leuconostoc oenos* phages from German wines. Appl. Microbiol. Biotechnol 37: 643–646

Bertrand A, Smirou-Bonnamour C & Lonvaud-Funel A (1984) Aroma compounds formed in malolactic fermentation. Proceedings of the Alko Symposium on flavour research of alcoholic beverages. Helsinki 1984. Biotech. Indus. Ferm. Res. 3: 39–49

Cavin JF, Andioc V, Etievant PX, & Divies C (1993) Ability of wine lactic acid bacteria to metabolize phenol carboxylic acids. Am. J. Enol. Vitic. 44: 76–80

Cavin JF, Barthelmebs L & Divies C (1997) Molecular characterization of an inducible p-coumaric and decarboxylase from *Lactobacillus plantarum*: gene cloning, transcriptional analysis, over expression in *E. coli*, purification and characterization. Appl. Environ. Microbiol. 63: 1939–1944

Chatonnet P, Dubourdieu D & Boidron JN (1995) The influence of *Brettanomyces/Dekkera* sp. yeasts and lactic acid bacteria on the ethylphenol contents of red wines. Am. J. Enol. Vitic. 46: 463–468

Couto JA & Hogg TA (1994) Diversity of ethanol-tolerant lactobacilli isolated from Douro fortified wines: clustering and identification by numerical analysis of electrophoretic protein profiles. J. Appl. Bacteriol. 76: 487–491

Costello PJ, Lee TH & Henschke P (1996) Recent aspects of mousy taint wine spoilage by lactic acid bacteria. Vth Symposium on lactic acid bacteria. Veldhoven, September 8/12, The Netherlands

Coton E, Rollan GC & Lonvaud-Funel A (1998a) Histidine decarboxylase of *Leuconostoc oenos* 9204: purification, kinetic properties, cloning and nucleotide sequence of the *hdc* gene. J. Appl. Bacteriol. 84: 143–151

Coton E, Rollan GC, Bertrand A & Lonvaud-Funel A (1998b) Histamine-producing lactic acid bacteria in wines; early detection frequency and distribution. Am. J. Enol. Vitic. 49: 199–204

Davis C, Silveira NFA & Fleet GH (1985) Occurrence and properties of bacteriophages of *Leuconostoc oenos* in Australian wines. Appl. Environ. Microbiol. 50: 872–876

Davis CR, Wibowo D, Eschenbruch R, Lee TH & Fleet GH (1985) Practical implications of malolactic fermentation: a review. Am. J. Enol. Vitic. 36: 290–301

Davis CR, Wibowo D, Fleet GH & Lee TH (1988) Properties of wine lactic acid bacteria: their potential enological significance. Am. J. Enol. Vitic. 39: 137–142

Denayrolles M, Aigle M & Lonvaud-Funel A (1994) Cloning and sequence analysis of the gene encoding *Lactococcus lactis* malic enzyme: relationship with malic enzymes. FEMS Microbiol. Lett. 116: 79–86

Denayrolles M, Aigle M & Lonvaud-Funel A (1995) Functional expression in *Saccharomyces cerevisiae* of the *Lactococcus lactis mleS* gene encoding the malolactic enzyme. FEMS Microbiol. Lett. 125: 37–44

De Revel G & Bertrand A (1993) Dicarbonyl compounds and their reduction products in wine: identification of wine aldehydes. 7th Weurman Flavor Research Symposium. TNO. June 1993

Dicks LM, Dellaglio F & Collins MG (1995) Proposal to reclassify *Leuconostoc oenos* as *Oenococcus oeni* [corrig.]. Gen. Nov., comb. Nov., Int. J. Syst. Bacteriol. 45: 395–397

Edwards CG, Beelman RB, Bartley CE & McConnell A (1990) Production of decanoic acid and other volatile compounds and the growth of yeast and malolactic bacteria during vinification. Am. J. Enol. Vitic. 41: 48–56

Edwards CG & Jensen KA (1992) Occurrence and characterization of lactic acid bacteria from Washington State wines: *Pediococcus* spp. Am. J. Enol. Vitic. 43: 233–238

Farias ME, Manca De Nadra MC, Rollan GC & Strasser De Saad AM (1993) Histidine decarboxylase activity in lactic acid bacteria from wine. J. Int. Sci. Vigne Vin 27: 191–199

Festaz-Furet B (1991) Réacteur à enzyme malolactique et NAD. Thesis. University of Saint Etienne

Garbay S & Lonvaud-Funel A (1996) Response of *Leuconostoc oenos* to environmental changes. J. Appl. Bacteriol. 81: 619–625

Garmyn D, Monnet C, Martineau B, Guzzo J, Cavin JF & Divies C (1996) Cloning and sequencing of the gene encoding α-acetolactate decarboxylase from *Leuconostoc oenos*. FEMS Microbiol. Lett. 145: 445–450

Garvie EI (1981) Sub-divisions within the genus *Leuconostoc oenos* as shown by RNA/DNA hybridization. J. Gen. Microbiol. 127: 209–212

Gindreau E, Torlois S & Lonvaud-Funel A (1997) Identification and sequence analysis of the region encoding the site-specific integration system from *Leuconostoc oenos* (*Oenococcus oeni*) temperate bacteriophage ϕ10MC. FEMS Microbiol. Lett. 147: 279–285

Guzzo J, Delmas F, Pierre F, Jobin MP, Samyn B, Van Beewmer I, Cavin JF & Divies C (1997) A small heat-shock protein from *Leuconostoc oenos* induced by multiple stresses and during stationary growth phase. Lett. Appl. Microbiol. 24: 393–396

Halasz A, Barath A, Simon-Sarkadi L & Holzapfel W (1994). Biogenic amines and their production by microorganisms in food. Trends Food Sc. Technol. 5: 42–49

Harvey RJ & Collins EB (1963) Roles of citrate and acetoin in the metabolism of *Streptococcus diacetilactis*. J. Bacteriol. 86: 1301–1307

Henick-Kling T, Acree TE, Gavit BK, Kreiger SA & Laurent MH (1993) Sensory aspects of malolactic fermentation. Stockley CS & Lee TH (Eds.). Proceedings of the eighth Australian wines industry technical conference, October 1992, Melbourne

Hogg T, De Revel G, Couto J, Capela A & Pintado M (1996) *Lactobacillus hilgardii*. Deterioration of fortified wines and related effects. In: Lonvaud-Funel A (Ed.) 'Oenologie 95' (pp. 337–340). Technique et Documentation Lavoisier, Paris

Hugenholtz J (1993) Citrate metabolism in lactic acid bacteria. FEMS Microbiol. Rev. 12: 165–178

King SW & Beelman RB (1986) Metabolic interactions between *Saccharomyces cerevisiae* and *Leuconostoc oenos* in a model grape juice/wine system. Am. J. Enol. Vitic. 37: 53–60

Labarre C, Guzzo J, Cervin JF & Divies C (1996) Cloning and characterization of the genes encoding the malolactic enzyme and the malate permease of *Leuconostoc oenos*. Appl. Environ. Microbiol. 62: 1274–1282

Lafon-Lafourcade S (1983) Wine and brandy in Biotechnology. Vol. 5. HJ Rehm & G Reed (Eds) (pp. 82–163). Verlag Chemie

Lafon-Lafourcade S, Carre A, Lonvaud-Funel A & Ribereau-Gayon P (1983) Induction de la fermentation malolactique des vins par inoculation d'une biomasse congelée de *Leuconostoc oenos* après réactivation. Conn. Vigne Vin 17: 55–71

Le Jeune C, Lonvaud-Funel A, Ten Brink B, Hofstra H & Van Der Vossen JMBM (1995) Development of a detection system for histidine decarboxylating lactic acid bacteria based on DNA probes, PCR and activity test. J. Appl. Bacteriol. 78: 316–326

Liu SO, Pritchard GG, Hardman J & Pilone G (1994) Citrulline production and ethylcarbamate (urethane) precursor formation from arginine degradation by wine lactic acid bacteria, *Leuconostoc oenos* and *Lactobacillus buchneri*. Am. J. Enol. Vitic. 45: 235–242

Liu SO, Pritchard GG, Hardman MJ & Pilone GJ (1995) Occurrence of arginine deiminase pathway enzymes in arginine catabolism by wine lactic acid bacteria. Appl. Environ. Microbiol. 61: 310–316

Liu SO, Pritchard GG, Hardman MJ & Pilone GJ (1996) Arginine catabolism in wine lactic acid bacteria: is it via the arginine deiminase pathway or the arginase-urease pathway. J. Appl. Bacteriol. 81: 486–492

Llauberes RM, Richard B, Lonvaud-Funel A & Dubourdieu D (1990) Structure of an exocellular β-D-glucan from *Pediococcus* sp., a wine lactic acid bacterium. Carbohyd. Res. 203: 103–107

Lonvaud M (1975) Recherches sur l'enzyme des bactéries lactiques assurant la transformation du malate en lactate. Thesis. University of Bordeaux 2

Lonvaud-Funel A (1995) Microbiology of malolactic fermentation. Molecular aspects. FEMS Microbiol. Lett. 125: 37–44

Lonvaud-Funel A & Strasser De Saad AM (1982) Purification and properties of a malolactic enzyme from strain of *Leuconostoc mesenteroides* isolated from grapes. Appl. Environ. Microbiol. 43: 357–361

Lonvaud-Funel A, Joyeux A & Desens C (1988) The inhibition of malolactic fermentation of wines by products of yeast metabolism. J. Food Sci. Agri. 44: 183–191

Lonvaud-Funel A, Joyeux A & Ledoux O (1991) Specific enumeration of lactic acid bacteria in fermenting grape must and wine by colony hybridization with non-isotopic DNA probes. J. Appl. Bacteriol. 71: 501–508

Lonvaud-Funel A & Joyeux A (1993) Antagonisms between lactic acid bacteria of wines: inhibition of *Leuconostoc oenos* by *Lactobacillus plantarum* and *Pediococcus pentosaceus*. Food microbiol. 10: 411–419

Lonvaud-Funel A, Guilloux Y & Joyeux A (1993) Isolation of a DNA probe for identification of glucan producing *Pediococcus damnosus* in wine. J. Appl. Bacteriol. 74: 41–47

Lonvaud-Funel A & Joyeux A (1994) Histamine production by wine lactic acid bacteria: isolation of a histamine producing strain of *Leuconostoc oenos*. J. Appl. Bacteriol. 77: 401–407

McDaniel M, Henderson LA, Watson BT Jr & Heatherbell D (1987) Sensory panel training and screening for descriptive analysis of the aroma of Pinot Noir wine fermented by several strains of malolactic bacteria. J. Sensor Studies 149–167

Nel L, Wingfield BD, Van Der Meer LJ & Van Vuuren HJJ (1987) Isolation and characterization of *Leuconostoc oenos* bacteriophages from wine and sugar cane. FEMS Microbiol. Lett. 44: 63–67

Nielsen JC & Prahl C (1996) Metabolism of citric acid by *Leuconostoc oenos* in direct inoculation, effect on wine flavour. In: Lonvaud-Funel A (Ed.) 'Oenologie 95' (pp. 317–320). Technique et Documentation Lavoisier, Paris.

Nielsen JC, Prahl C & Lonvaud-Funel A (1996) Malolactic fermentation in wine by direct inoculation with freeze-dried *Leuconostoc oenos* cultures. Am. J. Enol. Vitic. 47: 42–48

Paraskevopoulos Y (1988) Utilisation des enveloppes cellulaires de levure pour la stimulation de la fermentation malolactique. Interprétation de leur mode d'action. Thesis University of Bordeaux 2

Peynaud E (1956) New information concerning biological degradation of acids. Am. J. Enol. Vitic. 7: 150–156

Poblet-Icart M, Bordons A & Lonvaud-Funel A (1998) Lysogeny of *Oenococcus oeni* (Syn. *Leuconostoc oenos*) and study of their induced bacteriophages. Curr. Microbiol. 36: 365–369

Poolman B (1993) Energy transduction in lactic acid bacteria. FEMS Microbiol. Rev 12: 125–148

Ribereau-Gayon P, Dubourdieu D, Doneche B & Lonvaud A (1998) Le métabolisme des bactéries lactiques. In Traité d'œnologie. Tome 1. Microbiologie du vin.Vinifications. (pp. 171–196). Dunod Ed., Paris

Santos R, Vieira G, Santos MA & Pavela H (1996) Characterization of temperate bacteriophages of *Leuconostoc oenos* and evidence for two prophage attachment sites in the genome of the starter strain PSU-1. J. Appl. Bacteriol. 81: 383–392

Schutz M & Radler F (1973) Das 'Malatenzym' von *Lactobacillus plantarum* and *Leuconostoc mesenteroides*. Arch. Mikrobiol. 91: 183–202

Silla-Santos MH (1996) Biogenic amines: their importance in foods. Int. J. Food Microbiol. 29: 213–231

Sozzi T, Maret R & Poulain JM (1976) Mise en évidence de bactériophages dans le vin. Experimenta 32: 568–569

Sozzi T, Gnaegi F, D'Amico N & Hose H (1982) Difficultés de fermentation malolactique du vin dues à des bactériophages de Leuconostoc oenos. Rev. Suisse Vit. Arb. Hort. 14: 17–23

Ten Brink B, Damink C, Joosten HML, Huis IN, 't Veld JHJ (1990) Occurrence and formation of biologically active amines in foods. Int. J. Food Microbiol. 11: 73–84

Vivas N, Lonvaud-Funel A, Glories Y & Augustin M (1995) The effect of malolactic fermentation in barrels and in tanks on the composition and the quality of red wines. J. Sci. Techn. Tonnell. 1: 65–80

Vivas N, Lonvaud-Funel A & Glories Y (1997) Effect of phenolic acids and anthocyanins on growth, viability and malolactic activity of a lactic acid bacteria. Food Microbiol. 14: 291–300

Volschenk H, Viljoen M, Grobler J, Bauer F, Lonvaud-Funel A, Denayrolles M, Subden RE & Van Vuuren HJJ (1997) Malolactic fermentation in grape musts by a genetically engineered strain of *Saccharomyces cerevisiae*. Am. J. Enol. Vitic. 48: 193–197

SHORT LECTURES

Antonie van Leeuwenhoek **76:** 335, 1999.

Overview on applications for bacteriocin-producing lactic acid bacteria and their bacteriocins

John B. Luchansky
United States Department of Agriculture, Agricultural Research Service, Eastern Regional Research Center, 600 East Mermaid Lane, Wyndmoor, PA 19038, U.S.A.

Improvements in the production, manufacture, and/or storage of foods by both the food industry and the consumer have significantly diminished, but not eliminated, the likelihood of food-related illness and product spoilage. Several highly publicized and very costly outbreaks and recalls due to contamination of foods with pathogenic bacteria have resulted in greater awareness of food safety for consumers, producers, processors, and regulators. In the United States alone, food borne illness is responsible for about 9,000 deaths each year, with yearly incidences ranging from 24 to 81 million cases and yearly costs ranging from about $5 to $25 billion. In Europe, morbidity from food borne illness is second only to respiratory diseases, with estimates of 50,000 to 300,000 cases of acute gastro-enteritis per million population per year. Coupled with an enhanced awareness of food safety, more interest and research has also been directed toward developing alternatives to antibiotics in animal feeds, particularly throughout Europe. As an alternative to antibiotics, information about the contributions of the LAB to animal health and feed safety are the subject of much debate due, in part, to a paucity of scientifically-controlled studies, particularly studies evaluating their performance in feeds. For these reasons, there remains a critical need to develop, implement, and optimize strategies to better control undesirable microbes in foods both pre- and post-harvest, as well as in animals targeted for human consumption.

The production of antimicrobials and the lowering of pH during the transformation of raw foodstuffs into derivatives that are more stable by the lactic acid group of bacteria (LAB) is an approach to food preservation that has been practiced for centuries. The time, temperature, and pH of the fermentation, as well as the natural antimicrobials produced by the LAB, have proven effective for controlling both pathogenic and spoilage microbes in a variety of foods. The development and optimization of the LAB and attendant 'natural' inhibitors as biopreservatives to control undesirable bacteria remains a primary focus of several laboratories involved in research related to food safety and quality. Several antimicrobials produced by the LAB have been shown to inhibit undesirable bacteria in synthetic media. In recent years, biopreservatives have also been used effectively to extend shelf life and ensure the safety of foods; however, the efficacy of most antimicrobials of LAB origin, notably bacteriocins, has not yet been fully explored in foods. Despite the potential negative impact of various food components on biological activity, biopreservatives have found applications in dairy, meat, vegetable, fruit, bakery, and fishery products, as well as in certain beverages. Strategies for incorporating biopreservatives into foods include the use of LAB with proven antimicrobial activity as starter cultures or starter adjuncts, the use of a biopreservative preparation in the form of a previously fermented product, and/or the use of semipurified, purified, or chemically synthesized bacteriocins. There is also considerable interest in using biopreservatives in concert with food-grade chemicals and/or new-generation non-thermal intervention systems for foods and/or food contact surfaces. The available literature substantiates reductions of undesirable bacteria in the range of 100 to 1000 CFU per ml or gram of food that are directly attributable to biopreservatives.

The purpose of this presentation is to summarize recent advances in the development and applications of the LAB and associated biopreservatives to enhance food quality and safety. Examples will be provided on the development and validation of bacteriocins or bacteriocinogenic cultures to control a variety of food-borne pathogens, with special emphasis on *Listeria monocytogenes*.

Antonie van Leeuwenhoek **76:** 337–346, 1999.

Developing applications for lactococcal bacteriocins

R. Paul Ross[1,*], Mary Galvin[2,3], Olivia McAuliffe[2,3], Sheila M. Morgan[1,2], Maire P. Ryan[1,2], Denis P. Twomey[2,4], William J. Meaney[4] & Colin Hill[2,3]
[1]*Teagasc, Dairy Products Research Centre, Moorepark, Fermoy, Co. Cork, Republic of Ireland*
[2]*National Food Biotechnology Centre*
[3]*Department of Microbiology, University College Cork, Cork, Republic of Ireland*
[4]*Dairy Production Centre, Moorepark, Fermoy, Co. Cork, Republic of Ireland*
*(*Author for correspondence; E-mail: pross@moorepark.teagasc.ie)*

Key words: bacteriocins, lacticin 3147, *Lactococcus*, lactococcins

Abstract

While much of the applied research carried out to date with bacteriocins has concerned nisin, lactococci produce other bacteriocins with economic potential. An example is the two component bacteriocin lacticin 3147, which is active over a wide pH range and has a broad spectrum of activity against Gram-positive bacteria. Since the genetic determinants for lacticin 3147 are encoded on a large self-transmissible plasmid, the bacteriocin genes may be conveniently transferred to different lactococcal starters. The resulting food-grade strains can then be used to make a significant impact on the safety and quality of a variety of fermented foods, through the inhibition of undesirable microflora. The bacteriocin is heat stable so it can also be used as an ingredient in a powdered form such as a spray-dried fermentate. Given the observation that lacticin 3147 is effective at physiological pH, there is also considerable potential for biomedical applications. Field trials have demonstrated its efficacy in the prevention of mastitis infections in dairy cows. In contrast to lacticin 3147, the lactococcin bacteriocins A, B and M have a narrow spectrum of activity limited to lactococci. Strains which produce these inhibitors can be exploited in the acceleration of cheese ripening by assisting the premature lysis of starter cultures.

Introduction

Bacteriocins produced by lactic acid bacteria have been the subject of intensive scientific scrutiny during the past 20 years, leading to the detailed characterization of a wide range of these natural inhibitors (De Vuyst & Vandamme 1994). Despite these investigations, only a small number have demonstrated significant economic potential in food and biomedical fields. The best known bacteriocin is nisin, a lanthionine-containing inhibitor produced by *Lactococcus lactis* which is bactericidal against Gram-positive bacteria, including food pathogens such as *Listeria monocytogenes* (Hurst 1983). Nisin has a long history of food use in at least 48 different countries and was awarded FDA approval in 1988 (U.S. Government Federal Register, 1988). More recently, biomedical applications have been proposed which include its potential to treat mastitis infections in cows (Sears et al. 1995) and gastrointestinal infections in humans (Blackburn et al. 1994). In general, the potential applications of specific bacteriocins can be predicted by their properties; with characteristics such as host range, pH and heat stability being among the most important. For example, nisin is most effective in acidic environments, given its reported poor solubility at neutral and alkaline pH (Hurst 1983).

In an effort to isolate bacteriocins with commercially useful properties, many laboratories have undertaken screening programmes which investigate the antimicrobial potential of strains isolated from a variety of food sources. It was from such an approach that we isolated and characterized two bacteriocin systems from *Lactococcus* and we have identified a number of possible applications. The first of these is lacticin 3147, a two-component bacteriocin produced by an isolate from an Irish kefir-like grain (Ryan et al. 1996). This bacteriocin has a broad spectrum of activity against Gram-positive bacteria and has several potential uses both in food and biomedical applications. The second producer is similar to many strains of lactococci which produce three narrow spectrum bacteriocins designated lactococcins A, B and M. These bacteriocins have activity only against other lactococci and consequently are ineffective as preservatives in food. However, in a very distinct application, we have demonstrated that strains which produce lactococcin A, B and M can exhibit a bacteriolytic effect which induces lysis of starter cultures during cheese manufacture and ripening resulting in improved cheese flavour and quality (Morgan et al. 1995, 1997).

This paper will review the applications of lacticin 3147 in food systems where it is delivered either as a food ingredient or through the action of a bacteriocin-producing culture. Secondly, the potential biomedical applications of lacticin 3147 will be discussed including its exploitation for mastitis control. Lastly, the lactococcin-based system for acceleration of cheese ripening, through starter cell lysis, will be considered in detail.

Applications of lacticin 3147

Lacticin 3147 is a broad-host range, two-component bacteriocin produced by *Lactococcus lactis* subsp. *lactis* DPC3147. The ability to produce lacticin 3147 was first recognized in an isolate obtained from an Irish kefir-like grain which had been used domestically for the production of buttermilk (Rea et al. 1996). It is a broad host range bacteriocin which acts on the cytoplasmic membrane of sensitive cells, forming pores which leak potassium and phosphate ions (McAuliffe et al. 1998). Lacticin 3147 has been shown to inhibit a wide range of Gram-positive bacteria including *Listeria, Clostridium, Staphylococcus* and *Streptococcus* species, with no activity detected against Gram-negative bacteria (Ryan et al. 1996). Moreover, a number of Gram-positive strains which exhibited reduced sensitivity to lacticin 3147 in agar plate assays were efficiently killed in broth, highlighting the broad spectrum of inhibition of this bacteriocin (Galvin et al. 1999). Purification of the bacteriocin from the supernatant of the producing strain revealed its two-component nature, with both components required for full activity (McAuliffe et al. 1998). The genetic determinants for lacticin 3147 production and immunity are encoded on a 60.2 kb conjugative plasmid, pMRC01, the sequence of which has been fully elucidated (Dougherty et al. 1998). Sequence analysis identified the bacteriocin coding region on the basis of homologies to genes known to be involved in the production and export of the class of bacteriocins known as lantibiotics (Klaenhammer 1993).

Lacticin 3147-producing cultures in fermented foods

Although the producing strain of lacticin 3147, *L. lactis* DPC3147, was isolated from a dairy source, it was found to be unsuitable as a starter culture for production of fermented foods, such as cheese, due to a strain-associated off-flavour in the final product. Therefore, before the initiation of food production trials, it was necessary to create a number of specialised bacteriocin-producing starter cultures suitable for cheese-making. At present, conjugal transfer of naturally-occurring plasmids is the only generally accepted approach for the genetic improvement of starter cultures destined for use in foods (Gasson & Fitzgerald 1994). Due to the conjugative nature of pMRC01, it proved possible to transfer this plasmid into a number of commercial lactococcal starters (Coakley et al. 1997; O'Sullivan et al. 1998) and to date, as many as twenty five lacticin 3147-producing transconjugants have been generated. Since no spontaneous resistance to lacticin 3147 has been observed in lactococci, bacteriocin immunity encoded by pMRC01 was used as a food-grade marker for selection of transconjugants. However, a major concern when modifying such valuable strains is that their industrially important traits be retained. In addition to now producing lacticin 3147, the transconjugants constructed were shown to produce acid in pasteurised milk at rates comparable to those of the commercial strains. These transconjugants were also improved with respect to phage resistance, since this trait is also linked to the plasmid pMRC01 (Ryan et al. 1996; Coakley et al. 1997; O'Sullivan et al. 1998).

The ultimate goal of our starter improvement programme is to offer the cheesemaker greater control over the cheese flora using bacteriocins as biological tools. For example, the lacticin 3147-producing starters may be used to control cheese quality through reducing non-starter lactic acid bacteria (NSLAB) populations during ripening. Cheddar cheese is manufactured using one or more mesophilic *Lactococcus* as a starter culture, and has a ripening time of at least 6 months. During ripening, NSLAB populations can reach levels of 10^7–10^8 cfu/g of cheese, sometimes attaining numbers significantly higher than the starter

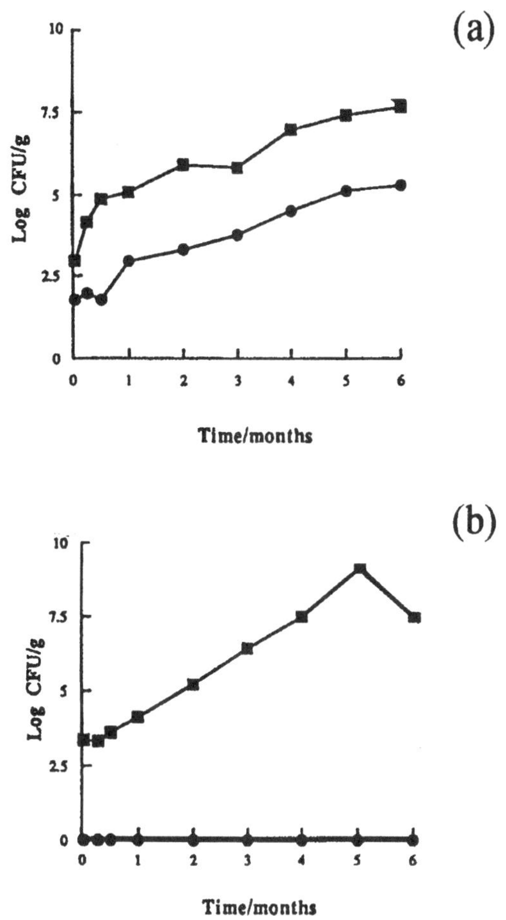

Figure 1. Monitoring the inhibition of NSLAB by lacticin 3147-producing starter cultures over a 6 month ripening period in Cheddar cheese. (a) Growth of NSLAB using either the lacticin 3147-producing transconjugant DPC4275 (●) or DPC4268 (■, Bac⁻) as a starter culture. (b) Growth of NSLAB using either a blend of three natural lacticin 3147-producers (●) or DPC4268 (■, Bac⁻) as starter cultures.

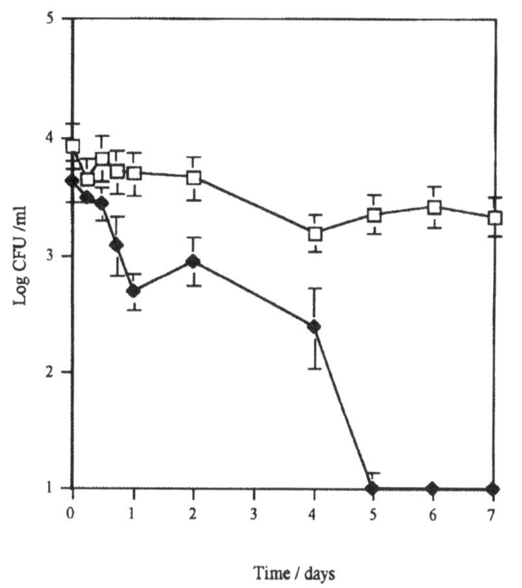

Figure 2. Inhibition of *L. monocytogenes* Scott A in cottage cheese produced with a lacticin 3147 starter culture stored at 4 °C. (□) Cheese manufactured with DPC4268, no bacteriocin; (♦) cheese manufactured with DPC4275 producing lacticin 3147.

(Fox et al. 1993; McSweeney et al. 1995). The contribution of NSLAB to Cheddar cheese quality is as yet unclear but they are thought to play a role in the proteolysis and flavour development which occurs during ripening. A major defect associated with NSLAB populations in Cheddar cheese is the development of calcium lactate crystals caused as a consequence of their ability to racemise L-lactate to D-lactate (Thomas & Crow 1983). These undesirable populations are significantly reduced throughout ripening when lacticin 3147-producing starters, such as the transconjugant DPC4275, are employed during manufacture (Figure 1a). Cheese manufactured with this strain contained 100-fold less NSLAB than control cheese following 6 months of ripening. Moreover, cheese manufactured with three natural lacticin 3147 producers (strains isolated from kefir-like grains which produce higher levels of lacticin 3147 than transconjugants) had no detectable NSLAB over the same time period (Figure 1b). Importantly, the gross composition of the cheese produced with the transconjugant starter was found to be within the parameters set by Giles and Lawrence (1973) for Cheddar cheese, with no significant differences between control cheese and cheese made with bacteriocin-producing starters. Lacticin 3147-producing starters may also be particularly valuable in the supression of NSLAB populations in reduced fat Cheddar cheese ripened at elevated temperatures (Fenelon et al. 1999).

In an obvious extension of this approach, lacticin 3147-producing starters have also been used to improve the safety of fermented dairy foods. *L. monocytogenes* has been implicated in a number of outbreaks of foodborne disease resulting from the consumption of freshly fermented dairy products, such as cottage cheese (Farber & Peterkin 1991). Cottage cheese manufactured with the lacticin 3147-producing transconjugant *L. lactis* DPC4275 was inoculated with *L. monocytogenes* Scott A at a level of 10^4 cfu/g cheese. The numbers of *Listeria* in the control cheese, manufactured with a non-lacticin 3147-producing starter, remained unchanged over the one week storage period. In contrast, there was a 1000-fold reduction in *Listeria* numbers following the 5 day ripening period (Figure 2; McAuliffe et al. 1999). It is important to note that this application differs significantly from the Cheddar cheese example since, as a consequence of the heating step in cottage cheese manufacture, the starter does not survive in the product. Thus, the protective effect is due to residual lacticin 'seeded' into the cheese during manufacture.

A lacticin 3147-producing transconjugant has also been used as a protective culture to inhibit *Listeria* on the surface of a mould-ripened cheese. The pH on the surface of such cheeses can range from 6.5 to 8.0, such conditions being favourable for the growth of *L. monocytogenes*. It was observed that the presence of the lacticin 3147 producer on the cheese surface reduced the numbers of deliberately inoculated *L. monocytogenes* Scott A 1000-fold, while in a parallel experiment a nisin-producer caused no reduction in *Listeria* numbers compared with control data (unpublished). This is most likely due to the fact that while nisin is more active in acidic conditions, lacticin 3147 is effective over a broad pH range.

The incorporation of bacteriocin-producing cultures into foods has certain advantages over using concentrated bacteriocin preparations. The lacticin 3147 producing organism carries a Generally Regarded As Safe (GRAS) status and was isolated from a food source, namely an Irish kefir-like grain. One of the main advantages of the culture approach is that bacteriocin is produced during the manufacture of the product rather than being added as a concentrate where it may be considered as an additive. Indeed, cheese manufactured with lacticin 3147-producing starters retains significant bacteriocin activity throughout ripening (Ryan et al. 1996; Figure 3). Therefore fermented foods manufactured with bacteriocin producing cultures could in turn be used as ingredients with preservative qualities in more complex foods.

Lacticin 3147 as a powdered biopreservative

The development of a bioactive powdered ingredient containing lacticin 3147 has also been investigated for improvement of food safety and quality (Morgan et al. 1999). Natural preservatives such as lacticin 3147 are now in high demand, given consumer concerns in the area of food safety, particularly with regard to chemical additives. Lacticin 3147 is produced very efficiently in dairy-based media, achieving high activity (2560 AU/ml) when produced in Cheddar cheese whey and whole milk, and significant activity (1280 AU/ml) in 10% reconstituted demineralized whey powder and 10% reconstituted skim milk powder. Given its widespread use as an inexpensive food ingredient, demineralized whey powder was chosen as a suitable growth medium in which to develop a food grade additive. *L. lactis* DPC3147 growth and lacticin 3147 production were optimised in this medium, and pilot scale spray dried preparations of lacticin 3147 were manufactured and assessed. Importantly, no loss in bacteriocin activity was observed during the spray drying process. The resulting powder exhibited significant activity against two foodborne pathogens, *L. monocytogenes* and *S. aureus*, at pH 5 and pH 7. Indeed, a 10% solution of the powder proved sufficient to cause greater than a 99% kill of *L. monocytogenes* over a 3 hour period at pH 5 and pH 7 (Figure 4a, b). Surprisingly, however, the lacticin 3147-enriched powder proved to be more effective at pH 7 than at pH 5 against *S. aureus*. A 15% solution of the powder was sufficient to kill 90% of staphylococcal cells in 3 hours, while a 99.98% kill was achieved at neutral pH (Figure 4c, d).

The development of a lacticin 3147-enriched powder from demineralized whey could have immediate applications in foods wherein whey is an existing ingredient. We have evaluated the performance of the lacticin 3147 powder against *L. monocytogenes* in one such food; infant formula. Results show that the powder effectively eliminated 99.9% of the food pathogen in 3 hours, when used at concentrations of 30% (w/w) or greater. In contrast, *L. monocytogenes* grew in the control preparations to which no lacticin 3147 had been added. The inhibitory effect observed in the infant formulation experiments may be indicative of the type of spoilage control and safety enhancement that a lacticin 3147 powder may confer in foods. Nisaplin®, a food ingredient for prevention of food spoilage, containing nisin, is most effective at acidic pH (below pH 6.0). The discovery that the lacticin 3147-enriched powder had a greater inhibitory effect at pH 7 is very significant and manufacturers of neutral pH foods which rely on chemical preservatives should consider the incorporation of this bioactive powder as an alternative.

Biomedical applications: mastitis control

While veterinary medicine has predominantly relied on traditional antibiotics for prophylactic and treat-

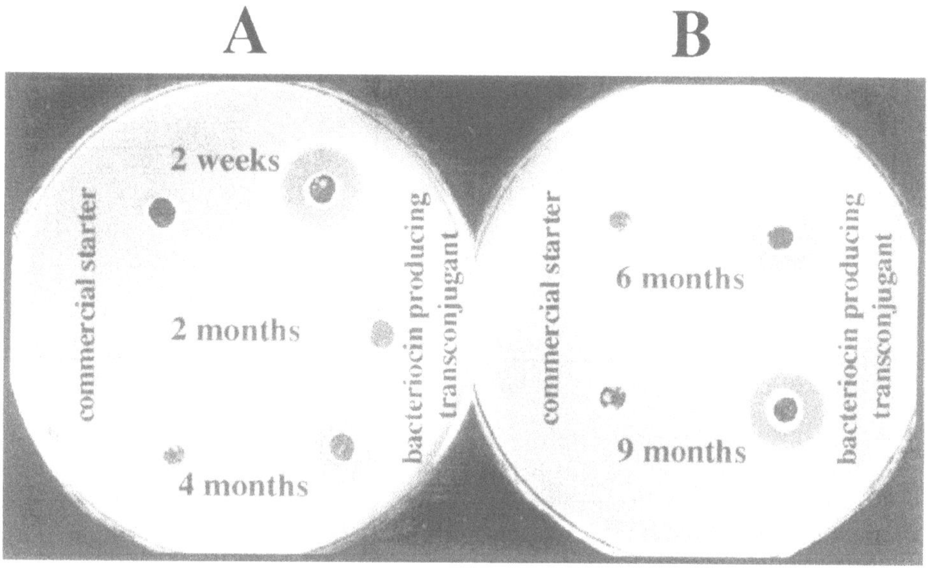

Figure 3. Lacticin 3147 activity detectable in Cheddar cheese manufactured with a commercial starter and its bacteriocin producing transconjugant over a nine-month period.

ment therapies, bacteriocins offer an attractive natural alternative and indeed may provide a number of distinct advantages. The concept of using bacteriocins in veterinary applications is not a novel one and since the 1940's, nisin has been evaluated as an alternative to intramammary antibiotics in the treatment of bovine mastitis. While some of the earliest intramammary work with nisin indicated that it was potentially effective in treating mastitis, some problems with irritancy were encountered, which were possibly due to impurities and particle size of the nisin preparations (Taylor et al. 1949). More recent work has demonstrated that nisin is inhibitory to the principal Gram-positive mastitic pathogens (Broadbent et al. 1989). There remains considerable interest in the exploitation of nisin for the treatment and prevention of bovine mastitis. As a result, two products, a pre- and post-milking dip and cleaning wipes, incorporating nisin as an active ingredient have emerged recently on the market. In addition, nisin has also been used in conjunction with another anti-staphylococcal agent, lysostaphin, in the treatment of bovine mastitis (Sears et al. 1995). These two antimicrobials displayed complementary and synergistic bactericidal activity against prominent mastitic organisms with cure rates of 66%, 95% and 100% against *S. aureus*, *Streptococcus agalactiae* and *Streptococcus uberis*, respectively. The combination of these bacteriocins did not elicit an adverse physiological reaction when infused into the teat sinus using an aqueous base formulation.

Table 1. Lacticin 3147 can prevent bovine mastitis. Results of an animal trial demonstrates the ability of teat seal containing lacticin 3147 to prevent deliberately infected teats from shedding the challenge organism (*Str. dysgalactiae*) or showing signs of clinical mastitis

Treatment	Number of teats	Teats shedding challenge organism+clinical signs of mastitis
Teat seal	33	20 (61.0%)
Teat seal+lacticin 3147	35	2 (6.0%)

Like nisin, lacticin 3147 has been shown to be bactericidal to all the major Gram-positive mastitic pathogens (Ryan et al. 1998). We have evaluated the efficacy of using lacticin 3147 in a commercial dry cow product called 'Teat seal' (Cross Vetpharm Group Ltd, Dublin, Ireland). The teat seal is a non-antibiotic formulation, comprising a heavy inorganic salt in a paraffin/wax base, which forms a plug in the teat sinus and acts as a physical barrier against infection. The addition of surfactants such as Tween 80 was found to be necessary in order to release bacteriocin activity from the lacticin 3147-containing seals, presumably due to the partitioning of the bacteriocin into

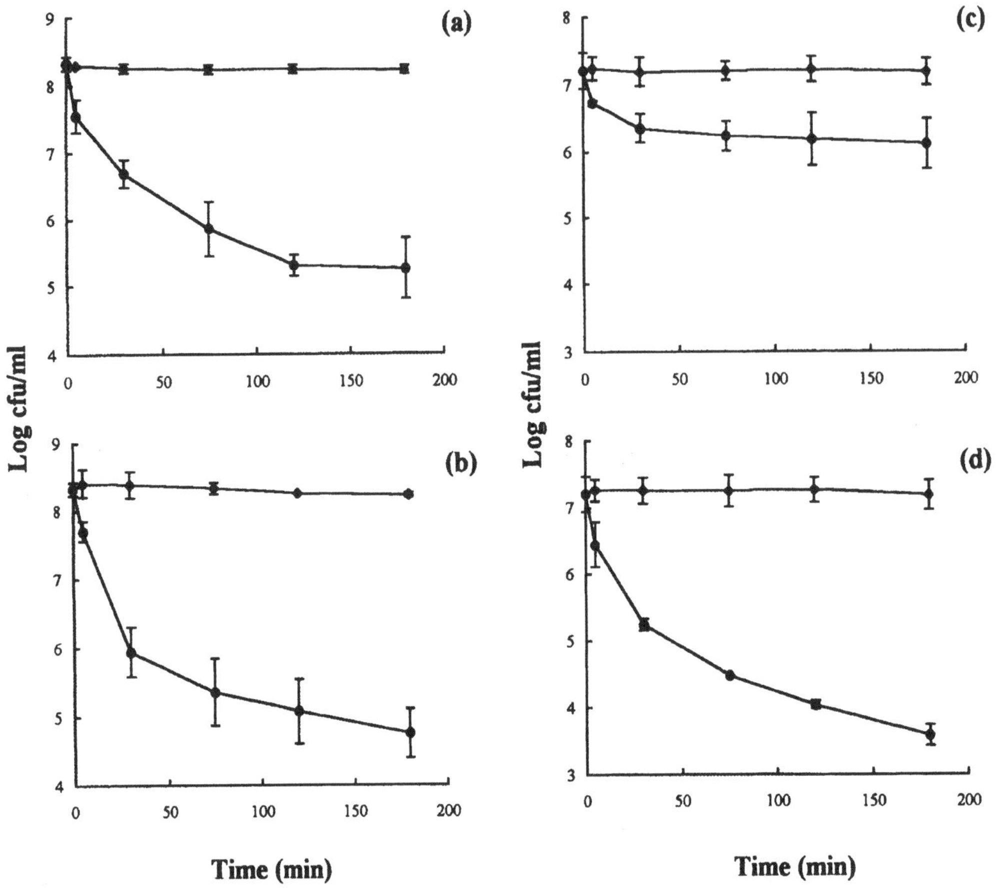

Figure 4. Effect of lacticin 3147 powder on the viability of *Listeria monocytogenes* Scott A in buffer at 30 °C (a) at pH 5 and (b) at pH 7, (♦) no addition, (●) addition of 10% lacticin 3147 powder. The effect of lacticin 3147 on the viability of *Staphylococcus aureus* 10 (c) at pH 5 and (d) at pH 7, (♦) no addition, (●) addition of 15% lacticin 3147 powder. Error bars represent the standard deviation for duplicate experiments.

the hydrophobic environment of the seal in the absence of surfactant. This novel non-antibiotic formulation did not cause any significant irritancy problems when infused into the udder. The effectiveness of the teat seal and lacticin 3147 formulation was evaluated in animal trials at the drying-off stage by infusing the lacticin 3147-containing teat seal, followed by artificially challenging teats with a relatively high inoculum ($\sim 1.5 \times 10^4$ cfu/teat) of *Streptococcus dysgalactiae* M selected to cause the failure of the teat seal control. The teat seal and lacticin 3147 treatment was found to significantly reduce the incidence of teats shedding the infused organism, from 61% in control teats, to just 6.0% in the treated quarters (Table 1; Ryan et al. 1999).

This study clearly demonstrated the potential for a bacteriocin-based non-antibiotic formulation in preventing the onset of mastitis during the non-lactating period. Therefore, while teat seal alone is known to provide an effective physical barrier to infection (Woolford et al. 1998), the addition of lacticin 3147 significantly enhances the product by providing an extra anti-microbial barrier to infection. In addition, the localization of lacticin 3147 in teat seal limits significant amounts of the bacteriocin being found in milk products as it is entrapped in the teat seal and removed once milking begins. Furthermore, the minor residues of the lacticin 3147 preparation which may be encountered in milk should pose minimal risks to the consumer as it is a natural food-grade peptide, which can be manufactured in a food-grade manner and is degraded by digestive enzymes such as trypsin and α-chymotrypsin (Ryan et al. 1996).

As a preventative treatment, intramammary antibiotics are routinely administered to dairy cows at drying-off, even when no symptoms of clinical or sub-clinical infections are evident. Considering the potential health threat antibiotic treatments may pose in promoting the emergence of antibiotic resistant microorganisms, and in light of the negative perception towards overuse of antibiotics in animals, it is envisaged that this practice may be limited to animals displaying signs of infection in the future. Therefore, a product such as the teat seal and lacticin 3147 formulation provides an alternative antibiotic-free therapy which should help maintain the infection-free status of the teats during the dry period.

Lacticin 3147 also has potential in the treatment of clinical cases of mastitis provided that it can be produced in a sufficiently potent form and is non-irritating when introduced into the udder. If this can be accomplished in a food-grade manner, then the withholding period for milk from treated quarters may be reduced or eliminated, thus providing significant economic savings to the producer. Although, it is highly unlikely that residues of lacticin 3147 in the milk would have a negative impact on consumer health, it is possible that the addition of a concentrated form of lacticin 3147 to the milk could adversely effect starter bacteria in downstream applications such as cheese or yogurt manufacture and therefore some withholding of milk may be necessary.

Biomedical applications: human potential

Antibiotic resistance amongst human pathogens has become a serious public health issue, and intensive research efforts have been devoted to developing alternative antimicrobial agents. Such concerns have prompted increased interest in the possible therapeutic value of natural antimicrobials such as bacteriocins. In this respect, lacticin 3147 has been investigated for its ability to inhibit a number of human pathogens including methicillin-resistant *S. aureus*, vancomycin-resistant *Enterococcus faecalis*, penicillin-resistant *Pneumococcus, Propionibacterium acne* and *Streptococcus mutans* (Galvin et al. 1999). In all cases, a population of over 10^4 cells was eliminated within two hours, by using a concentrated preparation of lacticin 3147 (20,000 AU/ml). Interestingly, *P. acne* exhibited particular sensitivity, suggesting that lacticin 3147 may have potential in the treatment of acne when used in topical formulations.

Although further studies are necessary to assess the full potential of lacticin 3147 as an alternative therapeutic agent, our preliminary findings suggest that this bacteriocin may provide an attractive addition to the range of antimicrobials currently available for antibiotic-resistant Gram-positive infections.

Lactococcins in cheese manufacture

As stated above, virtually all industrial applications proposed for bacteriocins involve the exploitation of those exhibiting a broad spectrum of inhibition. An exception is the use of bacteriocin-induced lysis to improve the quality of Cheddar cheese and to accelerate the rate of ripening. Lactococcin producers have been isolated by a number of groups (Geis et al. 1983; Scherwitz et al. 1983; van Belkum et al. 1989; Morgan et al. 1995), indicating that their occurrence is widespread. The genetic determinants in each case are plasmid located and it appears that the genetic organization of the region encoding lactococcin A, B and M is highly conserved among the various producing strains.

Lactococcin producers exhibit a narrow spectrum of inhibition confined to lactococci, and as such their commercial applications are limited. A novel application for lactococcin producers was initially mooted when it was observed that *L. lactis* ssp. *lactis* biovar. *diacetylactis* DPC3286 (encoding lactococcins A, B and M) demonstrated a lytic effect on sensitive strains, including mesophilic lactococci used in the manufacture of Cheddar cheese (Morgan et al. 1995; Figure 5). In addition, we have observed that another strain, *L. lactis* subsp. *cremoris* 9B4 (lactococcin A, B and M producer) was also bacteriolytic to lactococci (unpublished results), suggesting that bacteriolysis is a general property associated with lactococcin production. Such a property may be of great benefit during cheese-manufacture, since various studies have shown that lactococcal starter lysis contributes to cheese quality and flavour (Chapot-Chartier et al. 1994; Wilkinson et al. 1994; Crow et al. 1995). This is thought to be due to the associated release of intracellular proteinases and peptidases which increase the rate of proteolysis, a rate-limiting step in the ripening process. For this reason the strain DPC3286 was evaluated for use as an adjunct culture during Cheddar cheese manufacture to elicit premature starter lysis.

Two large scale Cheddar cheese making trials (each involving three 500 L vats) were performed,

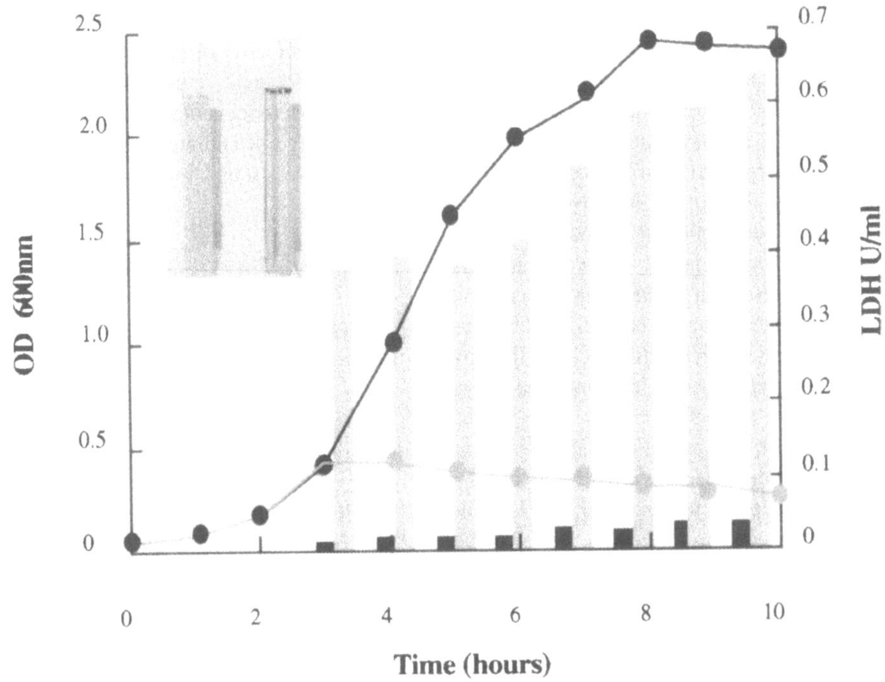

Figure 5. Lactococcin A, B and M induced lysis of a lactococcal strain in LM17 broth. OD of control (●), OD in the presence of lactococcin A, B and M (○), LDH release in control (■), LDH release in the presence of lactococcin A, B and M (□). Insert demonstrates the lytic ability of lactococcin A, B and M, where a culture clears following overnight incubation in the presence of these bacteriocins.

using milk fermented with *L. lactis* subsp. *lactis* HP alone, or HP and a bacteriocin negative strain (missing the plasmid pSM78) as controls (Morgan et al. 1997). Experimental cheeses were manufactured with varying levels of the adjunct culture DPC3286. Results demonstrated that there was increased release of intracellular enzymes. For example, levels of LDH detected in cheese juice (extracted in accordance with the method outlined by Morris et al. 1988) were between 26% and 66% higher than in control cheeses, and these higher enzyme levels persisted throughout the 6 month ripening period. In addition, a greater degree of secondary proteolysis was observed and the overall concentration of free amino acids was greater, with between 22% and 47% higher levels detected in cheeses manufactured using DPC3286. The higher levels of proteolysis were positively correlated with a reduction in bitterness (associated with a decrease in hydrophobic peptides, as measured by reverse phase HPLC). The reduction in bitterness was substantiated by a commercial grader when the cheeses were graded at 3 and 6 months. Importantly, all cheeses manufactured with the bacteriocin-producing adjunct graded best for flavour and aroma, as well as body and texture. In addition, the grader's comments indicated that these cheeses were of better quality than the control cheeses.

These results demonstrate that there is an application for a narrow spectrum bacteriocin producer, producing lactococcins A, B and M and demonstrating a lytic effect on sensitive starter cultures. The use of this culture as an adjunct during Cheddar cheese manufacture resulted in higher levels of starter cell lysis, greater proteolysis and a reduction in bitterness. Much research effort has gone into devising methods to accelerate cheese ripening, particularly of hard and semi-hard varieties, since it is both a lengthy and costly process. Although commercial enzyme preparations have been found to accelerate cheese ripening, these are not in widespread use due to their cost and the difficulty of distribution throughout the curd, in addition to the fear of over-ripening. Perhaps the most appropriate and practical methods for the acceleration of cheese ripening are focused on the starter cultures and their modification. These methods involve no additional cost for the manufacturer and have no legal limitations. Methods focusing on increasing starter cell lysis resulting in early enzyme release may offer the best solution to the problems of lengthy ripening times. The use of DPC3286, a lactococcin producer in such an application has already proved successful. The strain is lactose and proteinase negative and as such does not grow to high cell numbers during Cheddar cheese manufacture. Importantly, DPC3286 does not out-compete the starter flora during the fermentation process. In addition, since DPC3286 exhibits a narrow spectrum of inhibition, it does not prevent the growth of NSLAB, which are thought to contribute to flavour development during the maturation process (Martley and Crow 1993; McSweeney et al. 1995).

There are drawbacks associated with the use of bacteriocins in lysing starter cells, in that the rate of acidification can be retarded, particularly in situations where sensitive starters are employed. In practice, a fine balance is required, such that significant cell lysis occurs, but without slowing cheese production to a commercially unacceptable level. In an effort to overcome such problems, we are currently developing a system which utilizes an additional-acid producing strain (more tolerant to the effect of the adjunct DPC3286) in conjunction with a sacrificial strain designed to release ripening enzymes. This three strain starter system is being designed using commercial starter cultures and should yield increased lysis without compromising the rate of acidification.

Conclusions

The examples presented in this paper have outlined possibilities for future exploitation of two contrasting lactococcal bacteriocin systems. In both systems, the bacteriocin can be considered a useful tool which influences the microbial flora of the product, lacticin 3147 for inhibiting food pathogens and spoilage organisms in a variety of products or the lactococcins for lysing starters in cheese. The two systems are quite different, lacticin 3147 has a broad spectrum of inhibition and may be of use as an alternative to nisin in many applications while the lactococcins only inhibit lactococci. Importantly, lacticin 3147 does have a number of advantages over nisin including its effectiveness over a broad pH range, which suggests additional possibilities in non-acid foods and in biomedical applications. In addition, we have never observed resistance to lacticin 3147 in sensitive Gram-positive target strains, a significant observation given recent concerns regarding the development of resistance to antibiotics among many human pathogenic species In this respect, it is useful to debate whether biomedical applications for bacteriocins should be pursued concurrently with food applications. The argument against would highlight the possibility that continued exposure to the bacteriocin in a food could compromise its clinical effectiveness as an antimicrobial agent due to the emergence of resistant populations. Of course, this assumes that the potential benefits to human health would be better served using bacteriocins as 'drugs' rather than as food preservatives. However, such dilemmas may be of less concern for the exploitation of lacticin 3147 given that sensitive strains apparently do not develop resistance.

Acknowledgements

This research has been part funded by grant aid under the Food Sub-Programme of the Operational Programme for Industrial Development which is administered by the Department of Agriculture, Food and Forestry and is supported by national and EU funds, by Cross Vetpharm Group Ltd., (Dublin, Ireland), by the Teagasc Walsh Fellowship programme and by Dairy Levy funding.

References

Blackburn P, Projan SJ & Goldberg EB (1994) Pharmaceutical bacteriocin compositions and methods for using the same. United States Patent 5: 304,540

Broadbent JR, Chou C, Gillies K & Kondo JK (1989) Nisin inhibits several gram-positive, mastitis-causing pathogens. J. Dairy Sci 72: 3342–3345

Chapot-Chartier M-P, Deniel C, Rousseau M, Vassal L & Gripon J-C (1994) Autolysis of two strains of *Lactococcus lactic* during cheese ripening. Int. Dairy J. 4: 251–269

Coakley M, Fitzgerald GF & Ross RP (1997) Application and evaluation of the phage resistance- and bacteriocin-encoding plasmid pMRC01 for the improvement of dairy starter cultures. Appl. Environ. Microbiol. 63: 1434–1440

Crow VL, Martley FG, Coolbear T & Roundhill S (1995) The influence of phage-assisted lysis of *Lactococcus lactis* subsp. *lactis* ML8 on Cheddar cheese ripening. Int. Dairy J. 5: 451–472

De Vuyst L. & EJ Vandamme (Eds) (1994) Bacteriocins of lactic acid bacteria. Chapman and Hall, London

Dougherty BA, Hill C, Weidman JF, Richardson DR, Venter JC & Ross RP (1998) Sequence analysis of the 60kb conjugative, bacteriocin-producing plasmid pMRC01 from *Lactococcus lactis* DPC3147. Mol. Microbiol. 29: 1029–1038

Farber JM & Peterkin PI (1991) *Listeria monocytogenes*, a food-borne pathogen. Microbiol. Rev. 55: 476–511

Fenelon MA, Ryan MP, Rea MC, Guinee TP, Ross RP, Hill C & Harrington D (1999) Elevated temperature ripening of reduced fat

Cheddar cheese made with or without lacticin 3147-producing starter culture. J. Dairy Sci. 82: 10–22

Fox PF, Law J, McSweeney PLH & Wallace J (1993) Biochemistry of cheese ripening. In: Fox PF (Ed) Cheese – Chemistry, Physics and Microbiology, 2nd ed, Vol 1, (pp 389–438). Chapman and Hall, London

Galvin M, Hill, C & Ross RP (1999) Lacticin 3147 displays activity in buffer against Gram-positive pathogens which appear insensitive in standard plate assays. Lett. Appl. Microbiol. 28: 355–358

Gasson MJ & Fitzgerald GF (1994) Gene transfer systems and transposition. In: Gasson MJ & deVos WM (Eds.) Genetics and biotechnology of lactic acid bacteria (pp 1–51). Chapman and Hall, London, UK

Geis A, Singh J & Teuber M (1983) Potential of lactic streptococci to produce bacteriocin. Appl. Environ. Microbiol. 45: 205–211

Giles J & Lawrence RC (1973) The assessment of cheddar cheese quality by compositional analysis. N.Z.J. Dairy Sci. Technol. 8: 148–151

Hurst A (1983) Nisin and other inhibitory substances from lactic acid bacteria. In: Davidson PM & Branen AL (Eds) Antimicrobials in food (pp 327–351). Marcel Dekker, New York

Klaenhammer TR (1993) Genetics of bacteriocins produced by lactic acid bacteria. FEMS Microbiol. Rev. 2: 39–86

Martley FG & Crow VL (1993) Interactions between non-starter micro-organisms during cheese manufacture and ripening. Int. Dairy J. 3: 461–483

McAuliffe O, Hill C & Ross RP (1999) Inhibition of *Listeria monocytogenes* in cottage cheese manufactured with a lacticin 3147-producing starter culture. J. Appl. Microbiol. 86: 251–256

McAuliffe O, Ryan MP, Ross RP, Hill C, Breeuwer P & Abee T (1998) Lacticin 3147, a broad-spectrum bacteriocin which selectively dissipates the membrane potential. Appl. Environ. Microbiol. 64: 439–445

McSweeney PLH, Lynch CM, Walsh EM, Fox PF, Jordan KN, Cogan TM & Drinan FD (1995) Role of non-starter lactic acid bacteria in Cheddar cheese ripening. In: Cogan T, Fox PF & Ross RP (Eds) 4th Cheese Symposium (pp 32–45). Teagasc, Fermoy, Ireland

Morgan S, Ross RP & Hill C (1995) Bacteriolytic activity caused by the presence of a novel lactococcal plasmid encoding lactococcin A, B and M. Appl. Environ. Microbiol. 61: 2995–3001

Morgan S, Ross RP & Hill C (1997) Increasing starter cell lysis in Cheddar cheese using a bacteriocin producing adjunct. J. Dairy Sci. 80: 1–10

Morgan SM, Galvin M, Kelly J, Ross RP & Hill C (1999) Development of a lacticin 3147 enriched whey powder with inhibitory activity against foodborne pathogens. J. Food Protection (in press)

Morris HA, Holt C, Brooker BE, Banks JM & Manson W (1988) Inorganic constituents of cheese: analysis of juice from a one month old Cheddar cheese and the use of light and electron microscopy to characterize the crystalline phases. J. Dairy Res. 55: 255–268

O'Sullivan D, Coffey A, Fitzgerald GF, Hill C & Ross RP (1998) Design of a phage-insensitive lactococcal dairy starter via sequential transfer of naturally occurring conjugative plasmids. Appl. Environ. Microbiol. 64: 4618–4622

Rea MC, Lennartsson T, Dillon P, Drinan FD, Reville WJ, Heapes M & Cogan TM (1996) Irish kefir-like grains: their structure, microbial composition and fermentation kinetics. J. Appl. Bact. 81: 83–94

Ryan MP, Flynn J, Hill C, Ross RP & Meaney WJ (1999) The natural food grade inhibitor, lacticin 3147, reduced the incidence of mastitis after experimental challenge with *Streptococcus dysgalactiae* in non-lactating dairy cows. J. Dairy Sci. Accepted for publication.

Ryan MP, Meaney WJ, Ross RP & Hill C (1998) Evaluation of lacticin 3147 and teat seal containing this bacteriocin for inhibition of mastitis pathogens. Appl. Environ. Microbiol. 64: 2287–2290

Ryan MP, Rea MC, Hill C & Ross RP (1996) An application in cheddar cheese manufacture for a strain of *Lactococcus lactis* producing a novel broad-spectrum bacteriocin, lacticin 3147. Appl. Environ. Microbiol. 62: 612–619

Scherwitz KM, Baldwin KA & McKay LL (1983) Plasmid linkage of a bacteriocin-like substance in *Streptococcus lactis* subsp. *diacetylactis* strain WM_4: transferability to Streptococcus lactis. Appl. Environ. Microbiol. 45: 1506–1512

Sears PM, Peele J, Lassauzet M & Blackburn P (1995) Use of antimicrobial proteins in the treatment of bovine mastitis. In: Proceedings of the 3rd International Mastitis seminar (p 17–18)

Taylor JI, Hirsch A & Mattick TR (1949) The treatment of bovine streptococcal and staphylococcal mastitis with nisin. The Vet Record 61: 197–198

Thomas TD & Crow VL (1983) Mechanism of D(-)-lactic acid formation in cheddar cheese. N.Z.J. Dairy Sci. Technol. 18: 131–141

United States Government Federal Register (1988) Nisin preparation: affirmation of GRAS status as a direct human food ingredient. 53: 11247–11313

van Belkum MJ, Hayema BJ, Geis A, Kok J & Venema G (1989) Cloning of two bacteriocin genes from a lactococcal bacteriocin plasmid. Appl. Environ. Microbiol. 55: 1187–1191

Wilkinson MG, Guinee TP, O'Callaghan DM & Fox PF (1994) Autolysis and proteolysis in different strains of starter bacteria during Cheddar cheese ripening. J. Dairy Res. 61: 249–262

Woolford MW, Williamson JH, Day AM & Copeman PJA (1998) The prophylactic effect of a teat sealer on mastitis during the dry period and the following lactation. New Zealand Vet. J. 46: 12–19

Antonie van Leeuwenhoek **76:** 347–352, 1999.

Multidrug resistance in lactic acid bacteria: molecular mechanisms and clinical relevance

Hendrik W. van Veen*, Abelardo Margolles, Monique Putman, Kanta Sakamoto & Wil N. Konings

*Department of Microbiology, Groningen Biomolecular Sciences and Biotechnology Institute, University of Groningen, Kerklaan 30, NL-9751 NN Haren, The Netherlands (*Author for correspondence; E-mail: h.w.van.veen@biol.rug.nl)*

Key words: ABC transporter, drug binding sites, drug efflux, membrane protein, secondary transporter

Abstract

The active extrusion of cytotoxic compounds from the cell by multidrug transporters is one of the major causes of failure of chemotherapeutic treatment of tumor cells and of infections by pathogenic microorganisms. The secondary multidrug transporter LmrP and the ATP-binding cassette (ABC) type multidrug transporter LmrA in *Lactococcus lactis* are representatives of the two major classes of multidrug transporters found in pro- and eukaryotic organisms. Therefore, knowledge of the molecular properties of LmrP and LmrA will have a wide significance for multidrug transporters in all living cells, and may enable the development of specific inhibitors and of new drugs which circumvent the action of multidrug transporters. Interestingly, LmrP and LmrA are transport proteins with very different protein structures, which use different mechanisms of energy coupling to transport drugs out of the cell. Surprisingly, both proteins have overlapping specificities for drugs, are inhibited by the same set of modulators, and transport drugs via a similar transport mechanism. The structure-function relationships that dictate drug recognition and transport by LmrP and LmrA will represent an intriguing new area of research.

Introduction

The nutritive and therapeutic treatment of farm animals with antibiotics, amounting to half of the world's antibiotic output, has selected for drug resistant microorganisms that contaminate the food produced (Perreten et al. 1997). Likewise, the selection of drug-resistant pathogenic microorganisms in hospitalized patients with serious infections such as pneumonia, urinary tract infections, skin infections, and bacteremia has generally been ascribed to the widespread use of antimicrobial agents (Hughes & Tenover 1997). Issues of concern regarding Gram-negative bacteria include the extended drug resistance spectrum of ß-lactamase-producing *Escherichia coli*, *Salmonella typhimurium* and *Klebsiella pneumonia*. Current concerns with Gram-positive pathogens are increasing multidrug resistance in *Listeria monocytogenes*, methicilline-resistant *Stalphylococcus aureus* and penicilline-resistant *Streptococcus pneumonia*. Besides bacteria, parasitic protozoa are responsible for some of the most devastating and prevalent diseases of humans and domestic animals, such as malaria [*Plasmodium* spp.] and toxoplasmosis [*Toxoplasma* spp.] (for review, see Borst & Ouelette 1995). Chemotherapeutic treatment of infections by pathogens is being eroded by multidrug resistance. With few new drugs in the pipeline, prevention and circumvention of microbial multidrug resistance are medical and veterinary priorities, which require insight into the molecular mechanisms of microbial multidrug resistance.

Mechanisms of drug resistance

The presence of multidrug resistance among pathogenic and non-pathogenic microorganisms suggests that multidrug resistance mechanisms did not arise recently in pathogens in response to antimicrobial chemotherapy. Toxic compounds have always been part of the natural habitat of microorganisms. As a result, microorganisms have developed various mechanisms to resist cytotoxic compounds. Microorganisms can eliminate the drug target in the cell through the alteration or replacement of molecules that are normally bound by the antibiotic (Spratt 1994).

Alternatively, microorganisms can reduce the intracellular concentration of drugs by: (i) synthesizing enzymes that degrade antibiotics or that chemically modify, and so inactivate, the drugs (Davies 1994), (ii) eliminating entry ports for hydrophilic drugs, such as outer membrane porins in Gram-negative bacteria (Nikaido 1994), and (iii) manufacturing drug efflux systems that export lipophilic drugs before these compounds have the chance to find their cellular targets (Lewis 1994; van Veen & Konings 1997). Some of these drug efflux systems are fairly specific for a given drug or class of drugs, but the so-called multidrug transporters have specificity for compounds with very different chemical structures and cellular targets. Multidrug transporters can be amplified in drug resistant pathogenic microorganisms, and can shift their drug profiles, making them a menace to drug treatment.

Multidrug transporters LmrP and LmrA

Lactococcus lactis has developed fascinating antibiotic resistance mechanisms (Figure 1). A membrane protein, LmrP, mediates drug extrusion by a proton motive force-driven, electrogenic proton/drug exchange reaction (Bolhuis et al. 1996a). LmrP contains 12 membrane-spanning segments which are connected by hydrophilic loops protruding into the cytoplasmic and extracellular space, and is a member of the Major Facilitator Superfamily (Bolhuis et al. 1995). LmrP is homologous to various multidrug transporters in pathogenic bacteria and to the mammalian organic cation transporter OCT1 in kidney epithelial cells (van Veen & Konings 1997). Another drug transport system in *L. lactis*, LmrA, belongs to the ATP-Binding Cassette (ABC) Superfamily and is driven by ATP-hydrolysis. LmrA contains an N-terminal membrane domain with six membrane spanning segments followed by the ABC domain (van Veen et al. 1996). LmrA is homologous to prokaryotic ABC transporters such as the hop-resistance protein HorA in *Lactobacillus brevis* (Sami et al. 1997) and ABC proteins in *Bacillus subtilis*, *Staphylococcus aureus*, *Escherichia coli*, *Helicobacter pylori*, *Haemophilus influenzae*, and *Mycoplasma genitalium* (van Veen and Konings 1998). LmrA is also homologous to eukaryotic ABC transporters, and is a half-molecule version of the human multidrug resistance P-glycoprotein which is a cause of multidrug resistance of tumor cells during chemotherapy (Gottesman et al. 1995). Recent studies of the expression of covalently linked dimers of LmrA in *L. lactis* suggests that, by analogy with P-glycoprotein, LmrA is transport-active as a homodimer (van Veen, unpublished data).

From cells to proteoliposomes

LmrP and LmrA were initially characterized in a number of mutants of *L. lactis* MG1363 which were isolated on the basis of resistance to high concentrations of ethidium, daunomycin, and rhodamine 6G, and which showed an increased rate of energy-dependent ethidium and daunomycin efflux compared to the wildtype (Bolhuis et al. 1994). Subsequently, the genes encoding these transport proteins were identified and expressed in *E. coli* and *L. lactis*. Gene inactivation studies in *L. lactis* have established that ATP-dependent drug extrusion via LmrA still occurs in a *L. lactis*Δ*lmrP* strain (Bolhuis et al. 1995). Methods have been developed to routinely overexpress histidine-tagged LmrP and LmrA in *L. lactis*, for LmrA up to levels of 35% of total membrane protein (Putman et al. 1999; Margolles et al. submitted). For this expression, a novel protein-expression system for cytotoxic proteins is used which is based on the nisin-inducible *nisA* promoter (de Ruyter et al. 1996). LmrP and LmrA were efficiently solubilized with dodecyl maltoside (DDM), purified by Ni-affinity chromatography, and reconstituted in DDM-destabilized preformed liposomes prepared from *E. coli* phospholipids and egg phosphatidylcholine. Interestingly, the inhibition of the activity of LmrP by detergents restricted the range of compounds that could be used for solubilization and reconstitution of the protein, because low concentrations of detergent are retained in proteoliposomes (Putman et al. 1999a). Currently, the drug specificity and other functional properties of LmrP and LmrA are studied in whole cells, plasma membrane vesicles, and proteoliposomes containing the purified and reconstituted proteins.

Drug specificity

The drug specificity is one of the most intriguing aspects of LmA and LmrP. The number of compounds that are recognized by these proteins is vast.

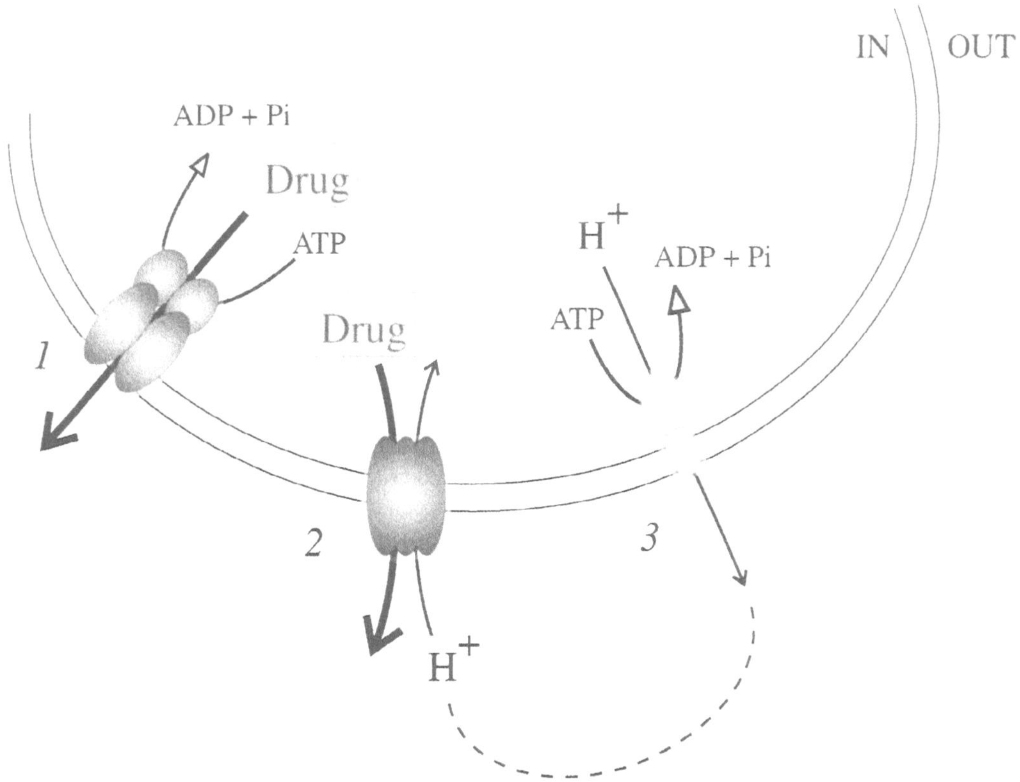

Figure 1. Schematic presentation of the multidrug transporters LmrP and LmrA in *Lactococcus lactis*. (*1*) LmrA is an ABC-type multidrug transporter that mediates the extrusion of drugs from the cell at the expense of ATP. (*2*) LmrP is a secondary multidrug transporter that mediates the extrusion of drugs from the cell in antiport with protons. The driving force for drug extrusion via proton-dependent drug antiport systems is supplied by the proton motive force, which is generated via proton extrusion by (*3*) the F_1F_0 H^+-ATPase. IN and OUT refer to the inside and outside of the cytoplasmic membrane, respectively.

To compare the pharmacological properties of bacterial LmrA with those of its human homolog P-glycoprotein, LmrA was functionally expressed in human lung fibroblast cells and insect cells which previously had been used for the characterization of P-glycoprotein (van Veen et al. 1998). Lung fibroblast cells expressing LmrA protein showed a 10 to 60-fold increased resistance to a variety of natural product drugs and synthetic chemotherapeutic drugs which are typical substrates of P-glycoprotein (Endicott & Ling 1989): (i) anthracyclines such as daunomycin and doxorubicin; (ii) *vinca*-alkaloids such as vinblastine and vincristine; and (iii) cytotoxic agents such as ethidium bromide, rhodamine 6G and 123, and colchicine. P-glycoprotein function can be inhibited by a large group of structurally unrelated modulators (Endicott & Ling 1989). All of these modulators tested also reversed drug resistance generated by expression of LmrA in lung fibroblast cells: (i) calcium channel blockers such as verapamil and its analog CP100-356; (ii) 1,4-dihydropyridines such as nicardipine; (iii) indolizine sulfones such as SR33557; (iv) antimalarials such as quinine and quinidine, (v) immunosuppressants such as cyclosporin A, (vi) the *Rauwolfia* alkaloid reserpine, and (vii) the phenylalkylamine verapamil. Thus, the drug and modulator specificity of LmrA is very similar to that of P-glycoprotein. Interestingly, all compounds mentioned above also interact with the secondary multidrug transporter LmrP. In addition, LmrA and LmrP are able to transport antibiotics. Recently, this phenomenon has been studied in detail for LmrP (Table 1). From toxicity assays in whole cell of *L. lactis* and *E. coli*, and transport studies in membrane vesicles and proteoliposomes it was concluded that LmrP mediates the transport of various members of the lincosamides, macrolides, streptogramins, and

Table 1. Binding and transport of clinically relevant antibiotics by LmrP[1]

Group	Antibiotic	Binding to LmrP	Transport by LmrP
Aminoglycosides	gentamycin	–	–
	kanamycin	–	–
β-Lactams	meropenem	–	–
Cephalosporins	ceftazidime	n.t.[2]	–
Lincosamides	clindamycin	+	+
Macrolides	azithromycin	+	+
	clarithromycin	+	+
	dirithromycin	+	+
	erythromycin	+	+
	roxithromycin	+	+
	spiramycin	–	–
Penicillins	ampicillin	–	–
Quinolones	ciprofloxacin	n.t.	–
	ofloxacin	n.t.	–
Streptogramins	dalfopristin	+	+
	quinupristin	+	–
Tetracyclins	tetracyclin	n.t.	+
Others	chloramphenicol	–	–
	sulfamethoxazol	–	n.t.
	trimethoprim	–	–
	vancomycin	–	–

1. Putman et al., submitted.
2. Not tested.

tetracyclins (Putman et al. submitted). These observations demonstrate the important role that multidrug transporters can play in microbial antibiotic resistance.

Transport models

Several transport models have been postulated for LmrP, LmrA and P-glycoprotein pump function to explain the broad specificity for chemically unrelated compounds. Drug translocation may involve substrate transport from the cytoplasm to the exterior [conventional transport hypothesis (Altenberg et al. 1994)] which would require flexibility of an 'enzyme-like' drug recognition site. However, a property common to most LmrP, LmrA and P-glycoprotein substrates is their ability to intercalate between the phospholipids of biological membranes. This notion has led to the suggestion that these transporters recognize drugs within the membrane [vacuum cleaner and flippase hypotheses (Raviv et al. 1990; Higgins & Gottesman 1992; Bolhuis et al. 1996a, b). Drug recognition within the membrane is supported by a number of observations: (i) photo-affinity analogs of P-glycoprotein substrates label P-glycoprotein predominantly in or near the transmembrane helices 4-6, and 11-12 (Greenberger 1993; Morris et al. 1994; Germann 1996), (ii) point mutations resulting in the alteration in drug specificity of P-glycoprotein are frequently found in transmembrane helices (Gottesman et al. 1995), (iii) acetoxymethyl esters of several fluorescent probes accumulate less in P-glycoprotein or LmrA-expressing cells, despite the fact that the ester moieties are rapidly cleaved by intracellular esterases and the resulting carboxylates are not substrates for P-glycoprotein and LmrA (Homolya et al. 1993; Bolhuis et al. 1996b), (iv) the kinetics of ATP-dependent transport of Hoechst 33342 by P-glycoprotein and LmrP, and of TMA-DPH by LmrA and LmrP in membrane vesicles is consistent with transport of the compounds from the inner, but not from the outer leaflet of the lipid bilayer (Bolhuis et al. 1996a, b; Shapiro & Ling 1997a, b). This transport mechanism is likely to be a more general mechanism for membrane transporters with hydrophobic substrates. The human *MDR2* gene-encoded P-glycoprotein transports phosphatidylcholine from the cytoplasmic leaflet of the bile canicular membrane of hepatocytes into the bile (Smit et al. 1993; Ruetz & Gros 1994). In addition, the *E. coli* α-hemolysine transporter HlyB most likely binds the transport signal sequence of α-hemolysine, when the signal sequence forms an amphiphilic helix that binds to the cytoplasmic leaflet of the plasma membrane (Sheps et al. 1995; Zhang et al. 1995). Thus, a main determinant of specificity for multidrug transporters will be the ability of drugs to be intercalated into the lipid bilayer. The subsequent interactions between drug molecules and drug binding sites on the transport proteins will be the second determinant of drug specificity.

Multiple drug binding sites

There is increasing evidence for the presence of more than one drug binding site on LmrP, LmrA and P-glycoprotein. For LmrP, transport studies in membrane vesicles of *L. lactis* suggest that some drugs inhibit LmrP-mediated Hoechst 33342 transport through competition with Hoechst 33342 for binding to the same drug binding site on LmrP, whereas other drugs inhibit LmrP-mediated Hoechst 33342 transport through binding to a drug binding site different from the Hoechst 33342 binding site (Putman et al. 1999b).

For LmrA expressed in plasma membranes of *Spodoptera frugiperda* insect cells (van Veen et al. 1998) and P-glycoprotein expressed in plasma membranes of Chinese Hamster Ovary B30 cells (Ferry et al. 1992; Ferry & Kerr 1994; Martin et al. 1997), kinetic analysis of drug dissociation revealed the presence of two non-identical, allosterically-linked drug binding sites in both proteins. As yet there is no structural information of sufficient detail to allow structure-function correlations and detailed insight into mechanistic aspects of multidrug transporters. A recent report on the structure of P-glycoprotein at 2.5 nm resolution showed a monomeric molecule with discrete domains (Rosenberg et al. 1997), giving a promising basis for the future.

Concluding remarks

Infectious disease caused by multidrug resistant microorganisms is an increasingly complex clinical and public health problem. Prevention and control strategies will require the application of epidemiological and behavioral approaches, as well as research technologies aimed at the basic mechanisms of drug resistance. This paper summarizes some molecular properties of the multidrug transporters LmrP and LmrA in *L. lactis*, and shows that these systems are very similar in function to those found in mammalian tumor cells. Unfortunately, multidrug transporters have an exceptionally broad drug specificity, and can shift their specificities in the course of drug treatment. The structure-function relationships that dictate drug recognition and transport by multidrug transporters will represent an intriguing new area of inquiry, and may allow the rational development of inhibitors of these proteins, and of new drugs which are not extruded from the cell.

Acknowledgements

We would like to thank Wim van Klompenburg, Atsushi Yokota, Marloes Veenstra, Catherine Martin and Richard Callaghan for discussions, and Chris Higgins for advice and encouragement. The research described in these proceedings was supported by the University of Groningen, the Biotechnology and Structural Biology programmes of the Commission of the European Communities, and the Dutch Cancer Society. H.W.V.V. is a fellow of the Royal Netherlands Academy of Arts and Sciences.

References

Altenberg GA, Vanoye CG, Horton JK & Reuss L (1994) Unidirectional fluxes of rhodamine 123 in multidrug resistant cells: Evidence against direct extrusion from the plasma membrane. Proc. Natl. Acad. Sci. USA 91: 4654–4657

Bolhuis H, Molenaar D & Poelarends G, et al. (1994) Proton motive force-driven and ATP-dependent drug extrusion systems in multidrug resistant *Lactococcus lactis*. J. Bacteriol. 176: 6957–6964

Bolhuis H, Poelarends G & van Veen HW, et al. (1995) The lactococcal *lmrP* gene encodes a proton motive force-dependent drug transporter. J. Biol. Chem. 270: 26092–26098

Bolhuis H, van Veen HW & Brands JR et al. (1996a) Energetics and mechanism of drug transport mediated by the lactococcal MDR transporter LmrP. J. Biol. Chem. 271: 24123–24128

Bolhuis H, van Veen HW & Molenaar D et al. (1996b) Multidrug resistance in *Lactococcus lactis*: evidence for ATP-dependent drug extrusion from the inner leaflet of the cytoplasmic membrane. EMBO J. 15: 4239–4245

Borst P & Ouellette M (1995) New mechanisms of drug resistance in parasitic protozoa. Annu. Rev. Microbiol. 49: 427–460

Davies J (1994) Inactivation of antibiotics and the dissemination of resistance genes. Science 264: 375–382

Endicott JA & Ling V (1989) The biochemistry of P-glycoprotein mediated multidrug resistance. Annu. Rev. Biochem. 58, 137–171

Ferry DR & Kerr DJ (1994) P-glycoprotein, a transporter with allosterically coupled drug-acceptor sites as a target for rational drug design. In: Kerr DJ & Workman P (Eds) New Molecular Targets For Cancer Chemotherapy (pp 177–193). CRC Press, London

Ferry DR, Russell MA & Cullen MH (1992) P-glycoprotein possesses a 1,4-dihydropyridine-selective drug acceptor site which is allosterically coupled to a vinca-alkaloid selective bindign site. Biochem. Biophys. Res. Commun. 188: 440–445

Germann UA (1996) P-glycoprotein – a mediator of multidrug resistance in tumour cells. Eur. J. Cancer 32A: 927–944

Gottesman MM, Hrycyna CA, Schoenlein PV et al. (1995) Genetic analysis of the multidrug transporter. Annu. Rev. Genet. 29: 607–649

Greenberger LM (1993) Major photoaffinity drug labeling sites for iodoaryl azidoprazosin in P-glycoprotein are within, or immediately C-terminal to, transmembrane domain 6 and 12. J. Biol. Chem. 268: 11417–11425

Higgins CF & Gottesman MM (1992) Is the multidrug transporter a flippase? TIBS 17: 18–21

Homolya L, Holló Z & Germann UA et al. (1993) Fluorescent cellular indicators are extruded by the multidrug resistance protein. J. Biol. Chem. 268: 21493–21496

Hughes JM & Tenover FC (1997) Approaches to limiting emergence of antimicrobial resistance in bacteria in human populations. Clin. Infect. Diseases 24: S131–135

Lewis K (1994) Multidrug resistance pumps in bacteria: variations on a theme. Trends in Biochemical Sciences 19: 119–123

Margolles M, Putman M, van Veen HW & Konings WN. Functional reconstitution into proteoliposomes of LmrA, the bacterial homolog of the human multidrug resistance P-glycoprotein. (Submitted.)

Martin C, Berridge G, Higgins CF & Callaghan R (1997) The multidrug resistance reversal agent SR33557 and modulation of vinca alkaloid binding to P-glycoprotein by an allosteric interaction. Br. J. Pharmacol. 122: 765–771

Morris DI, Greenberger LM & Bruggeman EP et al. (1994) Localization of the forskolin labeling sites for both halves of P-

glycoprotein: similarity of the sites labeled by forskolin and prazosin. Mol. Pharmacol. 46: 329–337
Nikaido H (1994) Prevention of drug access to bacterial targets: permeability barriers and active efflux. Science 264: 382–388
Perreten V, Schwarz F & Cresta L et al. (1997) Antibiotic resistance spread in food. Nature 389: 801–802
Putman M, van Veen HW, Degener JE & Konings WN. The lactococcal multidrug transporters LmrA and LmrP confer resistance to a broad range of antibiotics. (Submitted)
Putman M, van Veen HW, Poolman B & Konings WN (1999a) Restrictive use of detergents in the functional reconstitution of the secondary multidrug transporter LmrP. Biochemistry 38: 1002–1008
Putman M, Koole L, van Veen HW & Konings WN. (1999b) The secondary multidrug transporter LmrP contains multiple drug interaction sites. Biochemistry (In press)
Raviv Y, Pollard HB & Bruggeman EP et al. (1990) Photosensitized labeling of a functional multidrug transporter in living drug resistant tumor cells. J. Biol. Chem. 265: 3975–3980
Rosenberg MF, Callaghan R, Ford RC & Higgins CF (1997) Structure of the multidrug resistance P-glycoprotein to 2.5 nm resolution determined by electron microscopy and image analysis. J. Biol. Chem. 272: 10685–10694
Ruetz S & Gros P (1994) Phosphatidylcholine translocase: a physiological role for the *mdr2* gene. Cell 77: 1071–1081
de Ruyter PGGA, Kuipers O & de Vos WM (1996) Controlled gene expression systems in *Lactococcus lactis* with the food-grade inducer nisin. Apl. Environ. Microbiol. 62: 3662–3667
Sami M, Yamashita H & Hirono T et al. (1997) Hop-resistant *Lactobacillus brevis* contains a novel plasmid harboring a multidrug resistance-like gene. J. Ferment. Bioeng. 84: 1–6
Shapiro AB & Ling V (1997a) Extraction of Hoechst 33342 from the cytoplamsic leaflet of the plasma membrane by P-glycoprotein. Eur. J. Biochem. 250: 122–129
Shapiro AB & Ling V (1997b) P-glycoprotein-mediated Hoechst 33342 transport out of the lipid bilayer. Eur. J. Biochem. 250: 115–121
Sheps JA, Cheung I & Ling V (1995) Hemolysin transport in *Escherichia coli*: point mutations in HlyB compensate for a deletion in the predicted amphiphilic helix regions of the HlyA signal. J. Biol. Chem. 270: 14829–14834
Smit JJM, Schinkel AH & Oude Elferink RPJ et al. (1993) Homozygous disruption of the murine mdr2 P-glycoprotein gene leads to a complete absence of phospholipid from bile an to liver disease. Cell 75: 451–462
Spratt BG (1994) Resistance to antibiotics mediated by target alterations. Science 264: 388–393
van Veen HW, Callaghan R & Soceneantu L et al. (1998) A bacterial antibiotic resistance gene that complements the human multidrug resistance P-glycoprotein gene. Nature 391: 291–295
van Veen HW & Konings WN (1997) Multidrug transporters from bacteria to man: similarities in structure and function. Sem. Cancer Biol. 8: 183–191
van Veen HW & Konings WN (1998) The ABC family of multidrug transporters in microorganisms. Biochim. Biophys. Acta 1365: 31–36
van Veen HW, Venema K & Bolhuis H et al. (1996) Multidrug resistance mediated by a bacterial homolog of the human multidrug transporter MDR1. Proc. Natl. Acad. Sci. USA 93: 10668–10672
Zhang F, Yin Y, Arrowsmith CH & Ling V (1995) Secretion and circular dichroism analysis of the C-terminal signal peptides of HlyA and LktA. Biochemistry 34: 4193–4201

Antonie van Leeuwenhoek **76:** 353–355, 1999.

DNA-microarrays and food-biotechnology

Oscar P. Kuipers*, Anne de Jong, Siger Holsappel, Sierd Bron, Jan Kok & Leendert W. Hamoen
*Department of Genetics, Groningen Biomolecular Sciences and Biotechnology Institute (GBB), University of Groningen, PO Box 14, 9750 AA Haren, The Netherlands. (*Author for correspondence; Email: o.p.kuipers@biol.rug.nl)*

Key words: DNA microarrays, food, functional genomics, Gram-positive bacteria

The recent boost in bacterial genome sequences (e.g. Kunst et al. 1997; Bolotin et al. 1999) now enables true 'functional genomics' studies, including transcriptome- proteome- and metabolome analyses as well as structural genomics. For these areas novel high-throughput analytical methods have been developed, such as DNA-microarrays and improved 2D-electrophoresis methods combined with MALDI-TOF analyses. The large number of experimental data generated are gathered into large data-bases, the interpretation of which will greatly depend on novel bioinformatics methodologies, which should enable linking of the databases and facilitate classification and interpretation of the results.

DNA-microarrays provide a powerful tool for analyzing transcription profiles of whole genomes of any organism (Schena et al. 1998, Graves 1999). An effective way of producing DNA-microarrays is to spot amplicons of each ORF annotated in the genome sequence of interest on a defined support material. Another type of DNA-chip is provided by Affymetrix, a company that synthesizes oligonucleotides on solid materials by photolithography (for example De Saizieu et al. 1998). Bacterial genomes of Gram-positive bacteria typically harbour between 2000 and 5000 genes. Amplicons can conveniently be generated by PCR, and subsequently spotted onto membrane filters or glass-slides. On glass-slides over 100.000 spots per cm^2 can be accommodated, providing sufficient combinatorial possibilities for bacterial genome applications. Total RNA is isolated from the wild-type strain and from the isogenic strain under investigation, e.g. a strain containing a specific mutation or grown under different conditions. Subsequently, cDNA is made while incorporating either α-^{33}P-dCTP label (for membrane filters) or Cy3/Cy5 fluorescent label (for glass-slides). The labeled cDNA is then used for hybridization to the DNA-arrays and the signals are detected by phosphor-imaging (filters) or confocal laser scanning (glass-slides). In the latter case the wild-type and mutant-strain cDNA's are differentially labelled and these can be used in one combined sample for hybridization, providing the attractive possibility of multiplexing. The data obtained provide information on differential gene expression, which is extremely useful for studying pleiotropic regulatory effects, e.g. those involved in stress response, in complex regulatory cascades and in resistance mechanisms against antibiotics, antimicrobial peptides, phages and/or stress conditions.

We recently applied the DNA-array screening technology as a new and powerful tool to study gene regulation in *Bacillus subtilis*. This Gram-positive soil bacterium has developed a wide variety of adaptations to cope with the strong nutritional and physico-chemical fluctuations which occur in its environment. A motility system in concert with a chemotactic sensing device offers the bacterium the possibility to search for better growth and/or survival conditions. When nutrient limitations continue, *B. subtilis* cells start to secrete degradative enzymes and antibiotics, and acquire the ability to differentiate into cells competent for genetic transformation. Prolonged nutritional deprivation ultimately results in the production of dormant spores. *B. subtilis* employs elaborate systems for sensing the optimal conditions to initiate and regulate these different adaptations. The regulation of competence development has been studied in our laboratory for many years. Competence development is initiated when exponential growth ceases, and is optimal in minimal medium with glucose as the sole carbon source. A sufficiently high cell density is another prerequisite for optimal competence. The various environmental signals are interpreted by a complex signal transduction pathway and ultimately result in the synthesis of the competence transcription factor, encoded by *comK*.

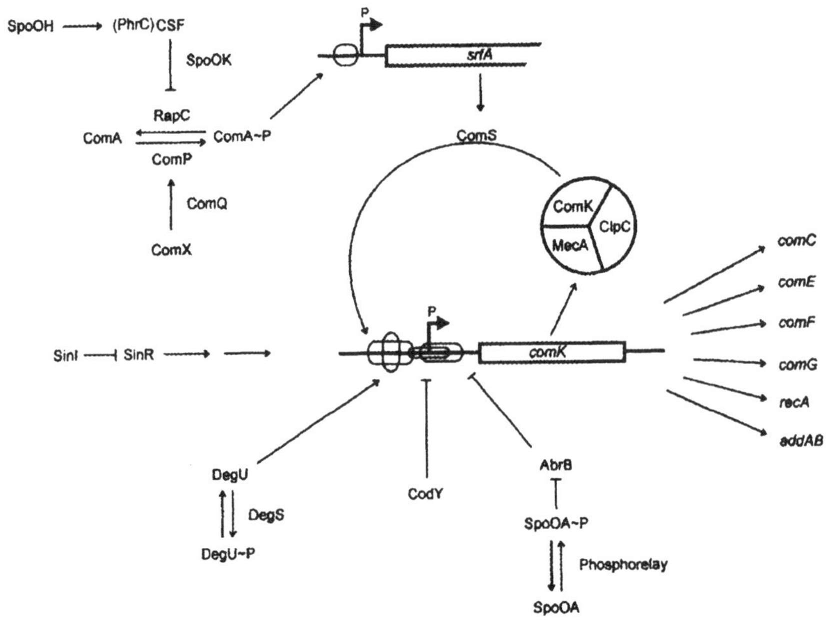

Figure 1. Regulation via the competence signal transduction cascade ultimately results in the synthesis of the competence transcription factor, ComK, which activates the transcription of the late competence genes encoding the DNA binding and uptake machinery (*comC, comE, comF, comG*), as well as of genes necessary for recombination (*recA, addAB*). During exponential growth, ComK is kept inactive by the formation of a ternary complex with MecA and the chaperone-like protein ClpC. Towards the end of the exponential growth phase, two quorum sensing pheromones, ComX and CSF, accumulate in the medium, activating a pathway which results in the phosphorylation of the response regulator ComA. ComA P activates the transcription of *srfA*, which encodes the synthetase of the lipopeptide antibiotic surfactin. Embedded within the *srfA* operon is a small gene,*comS*. The synthesis of ComS in response to the quorum-sensing signal transduction pathway destabilizes the ternary ComK/MecA/ClpC complex in which ComK is held inactive. Since ComK synthesis is positively autoregulated, the release of active ComK from the ternary complex results in the rapid accumulation of ComK. In addition to ComK itself, the DNA binding proteins DegU, AbrB and SinR are also required for the transcriptional activation of *comK*. For full competence development repression at the *comK* promoter due to the binding of AbrB and CodY must be relieved as well.

ComK activates the transcription of the late competence genes, encoding the DNA binding and uptake machinery (*comC*, *comE*, *comF*, *comG*), as well as transcription of genes necessary for DNA recombination, such as *recA* and *addAB*. In addition, ComK is also required for the expression of its own gene. The ComK-binding motif has been identified (Hamoen et al. 1998), and the signal transduction cascade leading to the expression of ComK has been elucidated in great detail (Figure 1).

However, several important questions remain and genome-wide expression profiling experiments are currently being used to obtain answers to these questions. To fully understand the mechanism of competence it is necessary to identify the proteins of the DNA uptake and integration systems which are activated during competence. As such genes are regulated by ComK, we searched for ComK activated genes. Several of these genes have been identified (Figure 1). When the entire genome sequence of *B. subtilis* was screened for possible functional ComK binding sites, more than hundred potential ComK-activated genes were found. DNA-microarrays comprising all open reading frames of *B. subtilis* were used to determine additional ComK-regulated genes. By comparing the genome-wide expression profiles of a wild type *B. subtilis* strain with a strain in which the *comK* gene is deleted, we are gaining insight in the gene products likely to be required for the DNA binding, translocation, and recombination processes that make *B. subtilis* cells competent for genetic transformation.

Of course, specific DNA-arrays can be developed for various other purposes, e.g. rapid identification of pathogens and spoilage bacteria, mutation analyses, and for studying DNA-protein interactions. Morever, also RNA of related or even less-related species can be used for differential gene expression, using less stringent hybridization conditions, providing for virtual expression arrays (VEA, Garner 1999). The combination of DNA-microarrays with proteome research and the elucidation of structure/function relationships of biomolecules will eventually result in a true understanding of whole cell functioning. On the short term it will stimulate various areas of microbiological, genetic and biochemical research, directed towards a better understanding of metabolic pathways, complex regulatory mechanisms and novel bioactivities, forming an indispensable knowledge-base in modern biotechnology. Finally, it is expected that many more genome sequences of lactic acid bacteria will become available in the near future, including those of several *Lactobacillus* species and that of *Streptococcus thermophilus*, which creates attractive possibilities for studying the latter organisms in their industrial environments or in their host. On the side of pathogens and food spoilage bacteria a fast increase in sequencing data is also taking place, with organisms as *Listeria monocytogenes, Bacillus* sp. and *Clostridium* sp. being sequenced at the moment, creating exciting possibilities for the development of novel antimicrobial strategies, and the study of their *in situ* activities.

References

Bolotin A, Mauger S, Malarme K, Sorokin A & Ehrlich SD (1999) *Lactococcus lactis* IL1403 diagnostic genomics, http://spock.jouy.inra.fr

De Saizieu A, Certa U, Warrington J, Gray C, Keck W & Mous J (1998) Nature Biotechnology 16, 45–48

Garner S (1999) Lab Chips and Microarrays for Biotechnological Applications, Conference January 13–15, 1999 Zurich Switzerland: 1. Virtual Expression Arrays (VEAs) – assist design of future arrays., Center for Biomedical Inventions & McDermot Center for Human Growth and Development, 2. Monitoring the expression of ripening associated strawberry genes using cDNA microarrays. Asaph Aharoni et al., Cell Biology Department, CPRO-DLO

Graves DJ (1999) Powerful tools for genetic analysis come of age. TIBTECH 17: 127–134

Hamoen LW, van Werkhoven AF, Bijlsma JJE, Dubnau D & Venema G (1998) The competence transcription factor of *Bacillus subtilis* recognizes short A/T-rich sequences arranged in a unique flexible pattern along the DNA-helix. Genes & Development 12: 1539–1550

Kunst F et al. (1997) The complete genome sequence of the Gram-positive bacterium *Bacillus subtilis*. Nature 390: 249–256

Schena M, Heller RA, Theriault TP, Konrad K, Lachenmeier E & Davis RW (1998) Microarrays: biotechnology's discovery platform for functional genomics. TIBTECH 16: 301–306

Antonie van Leeuwenhoek **76:** 357–365, 1999.

Exopolysaccharides produced by *Lactococcus lactis*: from genetic engineering to improved rheological properties?

Michiel Kleerebezem[1,2,*], Richard van Kranenburg[1,2], Remco Tuinier[3,4], Ingeborg C. Boels[1,2], Pieternella Zoon[3], Ellen Looijesteijn[2], Jeroen Hugenholtz[1,2] & Willem M. de Vos[1,2]
Wageningen Centre for Food Sciences; [1]NIZO food research, [2]Microbial Ingredients Section and [3]Fermented Products Section, P.O. Box 20, 6710 BA Ede, The Netherlands, [4]Present address: University of Utrecht, Van 't Hoff Laboratory for Physical and Colloid Chemistry, Padualaan 8, 3584 CH Utrecht, The Netherlands

Abstract

Over the last years, important advances have been made in the study of the production of exopolysaccharides (EPS) by several lactic acid bacteria, including *Lactococcus lactis*. From different EPS-producing lactococcal strains the specific *eps* gene clusters have been characterised. They contain *eps* genes, which are involved in EPS repeating unit synthesis, export, polymerisation, and chain length determination. The function of the glycosyltransferase genes has been established and the availability of these genes opened the way to EPS engineering. In addition to the *eps* genes, biosynthesis of EPS requires a number of housekeeping genes that are involved in the metabolic pathways leading to the EPS-building blocks, the nucleotide sugars. The identification and characterisation of several of these housekeeping genes (*galE, galU, rfbABCD*) allows the design of metabolic engineering strategies that should lead to increased EPS production levels by *L. lactis*. Finally, model development has been initiated in order to predict the physicochemical consequences of the addition of a EPS to a product.

Microbial exopolysaccharides

Polysaccharides are a highly diverse group of polymers, of which the functional features are determined by their structural characteristics that may differ in molecular weight, saccharide-linkage type, degree of branching and chemical composition. This diversity has led to very broad application possibilities in industry, ranging from paper manufactury to oil recovery and food production. High molecular weight polysaccharides are indispensable ingredients in a large variety of food products, where they are used as thickeners, stabilisers, emulsifiers, fat-replacers or gelling agents. Currently, most of the polysaccharides used in food-industry are derived from plants (e.g. starch, pectin, cellulose) and seaweeds (e.g. alginate, carrageenan). Alternative sources of biothickeners are found among the microbial extracellular polysaccharides (Sutherland 1998), which can either be present as capsular polysaccharide (CPS) associated with the cell surface or secreted as extracellular polysaccharide (EPS) in the environment of the cell. Four classes of microbial polysaccharides can be distinguished on the basis of their composition and mechanism of biosynthesis (Sutherland 1993). In this review we will only discuss the heteropolysaccharides that consist of repetitive, regular repeating units. In terms of application in food products, one of the most important examples of this class of bacterial polysaccharides is xanthan, which is produced by the phytopathogenic bacterium *Xanthomonas campestris*. Despite the fact that *X. campestris* is not an acceptable bacterium in food products (food-grade), the rheological properties of xanthan have led to its wide application as food-additive. Preferentially, bacterial polysaccharides to be used as additives in food products should be produced by non pathogenic, 'safe' bacteria. Lactic acid bacteria (LAB) in general have a 'food-grade' status and EPSs produced by these bacteria can be considered as 'food-grade' additives. Therefore, EPS production by LAB has received considerable attention since they appear to be relevant for dairy-product properties like texture and mouthfeel. Moreover, it has been suggested that they are active as prebiotic (Gibson & Robertfroid 1995), cholesterol lowering neutraceutical (Nakajima et al. 1992a) or immun-

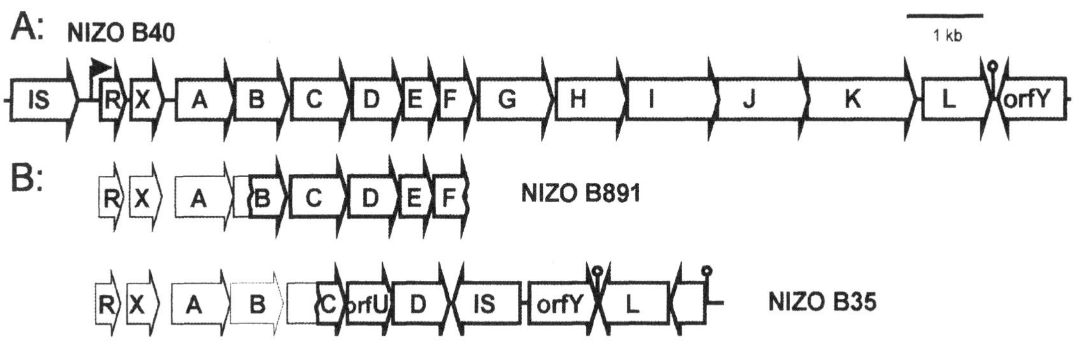

Figure 1. Genetic organisation of the *L. lactis* NIZO B40 (A), B891 (B) and B35 (B) *eps* gene cluster. In Figure 2B, PCR-product based gene annotations are indicated as thin arrows while sequences and annotations derived from cloned regions are indicated as bold arrows (van Kranenburg et al. 1999b).

omodulant (Kitazawa et al. 1993; Hosono et al. 1997). Over the last years, the chemical structure, and the molecular biology and genetics of the biosynthesis, of various repetitive heteropolysaccharide EPSs produced by LAB have been studied (Stingele et al. 1996; Stingele 1998; van Kranenburg et al. 1997, 1999a,b; Lamothe et al. 1998). Here, we review research on EPS production by the mesophilic LAB *Lactococcus lactis*. Both EPS-specific genetics and the general metabolism that is required for EPS precursor biosynthesis will be discussed. Finally, we will give an impression of recent advances of model-forming research considering the physicochemical properties of EPS in watery solution and its physical interactions with milk components.

EPS genetics in Lactococcus lactis

The biosynthesis of EPSs that consist of repetitive, regular repeating units appears to follow a common mechanism. The repeating units are assembled at the membrane by the sequential addition of sugars to the growing repeating unit that is anchored on the lipid carrier. After completion of an EPS repeating unit, it is supposed to be exported and polymerised to form the cell-surface polysaccharides (Sutherland 1993). The production of EPS by *L. lactis* is associated with strains that are isolated from highly viscous Scandinavian fermented milk products. The EPS specific *eps* genes are encoded on large plasmids (>20 kb) that can be transferred from one lactococcal strain to the next by conjugation, which concomitantly results in transfer of EPS production (Vedamuthu and Neville 1986; von Wright and Tynkkynen 1987; van Kranenburg et al. 1997, 1998, 1999a,b). The best characterised EPS producing *L. lactis* strain is NIZO B40 that produces a phosphopolysaccharide that is structurally identical to that produced by *L. lactis* strain SBT 0495, with a repeating unit consisting of $\rightarrow$4)[α-L-Rha*p*-(1$\rightarrow$2)][α-D-Gal*p*-1-PO_4-3]-β-D-Gal*p*-(1$\rightarrow$ 4)-β-D-Glc*p*-(1$\rightarrow$4)-β-D-Glc*p*-(1$\rightarrow$ (Figure 2) (Nakajima et al. 1990, 1992b; van Kranenburg et al. 1997, 1999a). A 12 kb gene cluster that is specific for NIZO B40 EPS production contains 14 coordinately expressed genes, *epsRXABCDEFGHIJKL* (Figure 1A), and is localised on a 40 kb plasmid designated pNZ4000 (van Kranenburg et al. 1997). Based on homology to gene products encoded by other polysaccharide biosynthetic gene clusters, putative functions in polysaccharide export, polymerisation, chain length determination and subunit biosynthesis (glycosyltransferase; GTF) could be assigned to several gene products encoded within the B40 gene cluster (Figure 1A; van Kranenburg et al. 1997). An interesting (see also below) feature like chain length determination putatively depends on the activity of EpsA and EpsB, which share overall homology and specific consensus motifs with the family of ExoP-like proteins. ExoP is supposed to be involved in chain length determination of succinoglycan produced by *R. meliloti* (Becker et al. 1995). The homologous counterparts of the EpsK and EpsI play a role in *Salmonella* or *Shigella* O-antigen production as a lipid-linked repeating unit flippase-exporter and polymerase, respectively (Morona et al. 1994; MacPherson et al. 1995; Liu et al. 1996). The function of the B40-GTF encoding genes was determined *in vitro* using heterologous (*E. coli*) and homologous (non-EPS producing *L. lactis*) inducible expression of these genes. Induced cells were permeabilized and incubated with UDP-[^{14}C]-glucose or UDP-[^{14}C]-galactose, followed by

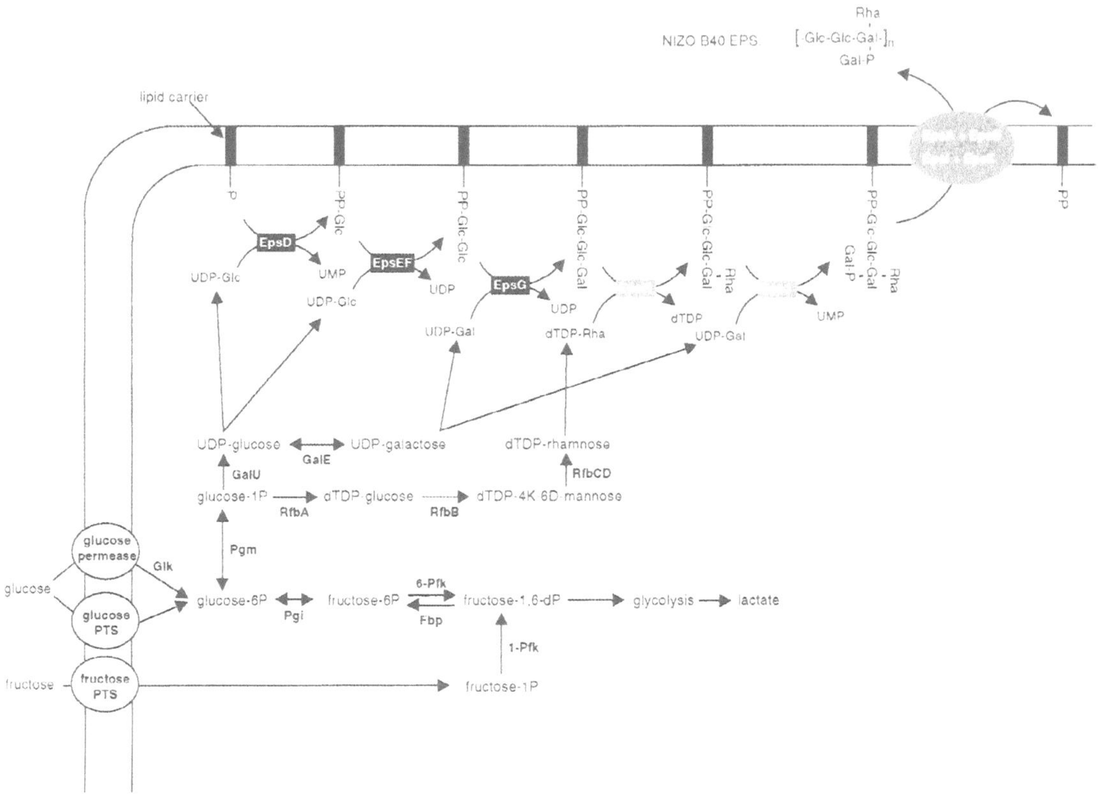

Figure 2. Schematic representation of pathways involved in NIZO B40 EPS biosynthesis. Both the general metabolism focussed on nucleotide-sugar biosynthesis from glucose or fructose as an external carbon source (lower left) and the *eps* genecluster encoded steps of repeating unit synthesis, export and polymerisation (upper part) are indicated. Enzymes involved in conversions in the general metabolism are indicated. The role of specific *eps* gene products has either been experimentally determined (black boxes) or is predicted by homologies (grey boxes). Glc, glucose; Gal, galactose; Rha, rhamnose. UDP-Glc, UDP-Gal and dTDP-Rha are nucleotide sugars.

identification of lipid-linked [^{14}C]-saccharides by thin layer chromatography and autoradiography. These experiments have shown that the biosynthesis of B40-polysaccharide backbone is initiated by the linkage of a glucose from UDP-glucose to the lipid carrier by priming-glucosyltransferase EpsD (van Kranenburg et al. 1997, 1999a). Subsequently, the addition of a second glucose from UDP-glucose to the lipid-linked glucose involves the combined activity of EpsE and EpsF, and finally, the addition of the third backbone-sugar moiety, galactose, from UDP-galactose to this carrier bound cellobiose involves the activity of EpsG (van Kranenburg et al. 1999a; Figure 1A, 2). The fourth step in repeating unit biosynthesis is expected to be the linkage of rhamnose from dTDP-rhamnose or galactosyl-phosphate from UDP-galactose to the galactose moiety of the lipid linked trisaccharide. The high unstability of dTDP-rhamnose prevents its use as substrate in the *in vitro* GTF assay, which could explain the lack of direct experimental data to support the suggested involvement of EpsH and EpsJ as rhamnosyltransferase and phosphogalactosyltransferase, respectively, in these later steps of B40 repeating unit biosynthesis (van Kranenburg et al. 1999a; Figure 2).

The diversity of EPSs produced by different *L. lactis* strains is illustrated by the variation of the chemical (sugar-) composition of the EPS produced and by the genetic organisation of the specific gene clusters involved (van Kranenburg et al. 1999b). Partial DNA sequence data of the *eps* gene clusters from *L. lactis* strains NIZO B891 and NIZO B35 revealed that both genetic organisation and sequence of the first genes, *epsRXABC*, are highly conserved (van Kranenburg et al. 1997, 1999b; Figure 1B). The organisation of the other known B891-*eps* genes (*epsDEF*) appears to be conserved relative to NIZO B40. The B35 *eps* cluster, however, appears to be interrupted by an *IS982* element, which possibly indicates IS-element mediated rearrangement or horizontal gene transfer of parts of polysaccharide gene clusters. The *in vitro* functional analysis of the B891 and B35 priming GTF encoding genes (*epsD*-homologues) revealed that they encode a glucosyl- and galactosyl-priming transferase, respectively (van Kranenburg et al. 1999b; Figure 1B). Moreover, functional analysis of the B891 *epsDEF* genes (Figure 2B) led to a carrier-linked lactose, indicating that B891 EpsEF are involved in coupling of a galactose moiety from UDP-galactose to the carrier-bound glucose (van Kranenburg et al. 1999b).

A first and essential step towards EPS engineering was accomplished by creating a non-polar disruption mutation in the NIZO B40 *epsD* gene, which resulted in loss of EPS production. Interestingly, the *epsD* mutation could subsequently be complemented by introduction of a second plasmid encoding either the homologous NIZO B40 *epsD* gene, or the NIZO B891 *epsD* gene, both encoding a lactococcal priming glucosyltransferase (van Kranenburg et al. 1999b). Moreover, heterologous complementation of the *epsD* mutation was achieved by expression of the *S. pneumoniae* serotype 14 priming glucosyltransferase encoded by the *cps14E* gene, which is normally involved in CPS biosynthesis (Kolkman et al. 1996, 1997). These results are the first demonstration of functional complementation of a GTF mutation in a Gram-positive host by *in trans* expression of a gene encoding a similar activity. Moreover, they show that although transcriptional coupling of the genes involved is disturbed, the functional coupling of the enzymes encoded can apparently still occur. In contrast, expression of a priming galactosyltransferase derived from either *L. lactis* NIZO B35 (*epsD*) or from *S. thermophilus* Sfi6 (*epsE*; Stingele et al. 1996, 1998) did not result in complementation of EPS production (van Kranenburg et al. 1999b), suggesting an exclusive specificity for its natural substrate of one or more components of the B40-EPS biosynthesis machinery.

Recently, other examples that hint at the possibilities for production of heterologous or new EPS structures in *L. lactis* have been reported. Expression of the genes responsible for type 3 CPS production in *S. pneumoniae* resulted in high level production of the type 3 polysaccharide by *L. lactis* (see below; Gilbert et al. 1998). *L. lactis* cells in which the entire *S. thermophilus* Sfi6 *eps* gene cluster is expressed appeared to produce low but significant amounts of EPS, of which the structure differed from the EPS produced by the original *S. thermophilus* strain both in backbone composition (Gal instead of GalNAC) and sidechain substitution (lactococcal polymer lacks Gal sidechain). In contrast to the B40-epsD complementation studies (see above), this latter example illustrates that GTFs do not necessarily exhibit exclusive specificity for donor and acceptor sugar moieties and that the polymerase and export machinery in its turn might not be exclusively specific for one repeating unit (Stingele 1998).

These findings offer good perspective for future EPS engineering, although several bottlenecks caused by substrate specificity of the biosynthesis machinery will have to be overcome. Increasing knowledge on EPS and CPS structures, combined with the availability of individual biosynthetic components from *eps* and *cps* gene clusters will allow evaluation of the possibilities and limitations of genetic engineering strategies directed towards production of 'tailor made' EPS molecules in *L. lactis* or other LAB. Moreover, the role of specific *eps* gene products in repeating unit export, polymerisation and determination of polymer chain-length should be studied in more detail in order to manipulate these features. Especially chain-length engineering could generate EPSs of higher molecular weight, leading to better rheological properties (see below) or on the other hand could lead to the production of tailor made short-chain oligosaccharides.

Metabolic pathway engineering of EPS production

The industrial application of EPS produced by LAB is strongly hampered by the low production level, which is highly variable but ranges from 50 to 500 mg per liter. In the above section we have discussed the final steps of EPS biosynthesis that are mediated by the proteins encoded by the *eps* gene cluster. Besides these specific *eps* genes, EPS production also requires a number of so-called housekeeping genes that are involved in the biosynthesis of the EPS building blocks, the nucleotide sugars. Therefore, the physiology of nucleotide sugar biosynthesis in relation to EPS production levels should be evaluated to eventually use metabolic engineering as a means to increase EPS production by LAB. This approach should be supported by a kinetic model that describes the contribution of the individual enzymes to level of a complex end-product like EPS.

A possible key-step in nucleotide-sugar biosynthesis is the interconversion of glucose-6-phosphate, an intermediate in sugar breakdown, and glucose-1-phosphate, the central precursor in nucleotide-sugar biosynthesis (Sjöberg & Hahn-Hägerdahl 1989), which is performed by the phosphoglucomutase PGM (Figure 2). *L. lactis* contains two distinct forms of PGM, one specific for β-glucose-1P (i.e. β-PGM) and the other apparently specific for α-glucose-1P (i.e. α-PGM) (Qian et al. 1994). Since the phosphorolysis of glucose, lactose and other carbon sources usually yields α-glucose-1P, α-PGM could be a key-enzyme in sugar-nucleotide biosynthesis. However, to our knowledge, only the gene encoding β-PGM has been cloned from *L. lactis* (Qian et al. 1997).

The central intermediate, glucose-1P, in its turn has to be converted to UDP-glucose, UDP-galactose and dTDP-rhamnose in order to fulfil the precursor-requirements for EPS biosynthesis by *L. lactis* NIZO B40 (Figure 2). Presently, all the genes that encode the enzymes that are putatively involved in the biosynthesis of these sugar-nucleotides from glucose-1P (*galU*, *galE*, and *rfbACBD*; Figure 2) have been cloned from *L. lactis* strain MG1363 (Grossiord et al. 1998a; Boels et al. 1998; Boels et al., unpublished results), thereby generating the tools to target specific endogenous enzymatic activities that could be bottlenecks in EPS biosynthesis, and thereby evaluate the potential of metabolic engineering strategies to improve EPS production levels.

The role of the Leloir enzyme galactose epimerase (GalE; Figure 2) was evaluated by analysis of NIZO B40 EPS biosynthesis in a *galE* mutant strain. EPS biosynthesis was abolished in *galE* mutant cells grown on media with glucose as a sole carbon source, but could be recovered by addition of galactose to the medium. These results indicate a crucial role for *galE* in the synthesis of UDP-galactose from glucose and thus in EPS biosynthesis by *L. lactis* (Boels et al. 1998). Moreover, the *galE* strain was also affected in cell division when grown on glucose alone, which probably reflects the critical role of UDP-galactose as a precursor in cell-wall biosynthesis (Grossiord et al. 1998b).

Gilbert et al. (1998) described the expression of the *cps3D* and *cps3S* genes from *S. pneumoniae* type 3 in *L. lactis*, leading to low heterologous production of type 3 polysaccharide. Interestingly, its production level could be strongly increased (to 15 μg per 10^8 cells) by co-expression of the *cps3U* gene that encodes a glucose-1-phosphate uridylyltransferase (GalU; Figure 2) analogue, indicating that the level of available UDP-glucose in lactococcal cells is limited by the endogenous GalU enzyme level. Functional overexpression of the lactococcal *galU* gene resulted in strongly increased levels of UDP-glucose and UDP-galactose in *L. lactis* (GalE maintains an almost stable ratio between the two nucleotide sugars; Figure 2), which confirms the hypothesis that the GalU enzyme activity level determines the intracellular levels of UDP-glucose and –galactose in wild-type cells (Boels et al., unpublished results). The effects of increased GalU levels on EPS production are currently being evaluated.

Rhamnose is an important residue in the O-antigen of lipopolysaccharides in many Gram-negative bacteria and the role of *rfb*-like genes in biosynthesis of the precursor dTDP-rhamnose is well established (Reeves 1993). Also in Gram-positive bacteria *rfb*-homologues appear to play an essential role in the production of rhamnose containing polysaccharides. The *cps19fL-O* genes are part of the *cps* gene cluster of *S. pneumoniae* type 19F and exhibit very strong homology with the *Shigella flexneri rfbA-D* genes and functionally complement a mutation deleting the respective *S. flexneri* homologues (Morona et al. 1997). An *S. pneumoniae cps19fL* mutant exhibited a so-called rough phenotype and had lost the capacity to produce CPS, confirming an essential role of the *rfb*-like genes in CPS-19F production (Morona et al. 1997). In *S. mutans*, the genes encoding the homologues of the *S. flexneri* RfbA, RfbB and RfbD proteins have been identified (designated *rmlA*, *rmlB* and *rmlC*, respectively) and functionally characterised. The gene encoding the RfbC homologue of *S. mutans* has not been identified sofar, but apparently does not map close to the *rml* locus (containing *rmlA-C*) on the *S. mutans* chromosomal map. In *S. mutans*, *rml* mutations led to a changed cell-wall composition (lacking rhamnose) and did no longer produce the serotype-specific polysaccharide antigen (Tsukioka et al. 1997). Functional analysis of the lactococcal *rfb* gene cluster (Boels et al. 1998) is especially interesting in terms of evaluation of EPS-engineering possibilities in *L. lactis*. Rhamnose is not a 'backbone-sugar' in NIZO B40 EPS, which is in contrast with the *S. pneumoniae* and *S. mutans* polysaccharides described above. Therefore, introduction of a plasmid carrying the NIZO B40-*eps* gene cluster in a *rfb* mutant strain might allow evaluation of side-chain specificity of the repeating-unit biosynthesis, export and polymerisation machinery. That this approach could be successful in creating a novel EPS with an altered structure, is supported by the experiments described by Stingele (1998; see above).

Important indications of blockage points or rate limiting steps for EPS production by LAB could be obtained from studying the physiological conditions that influence the EPS production level in LAB. Evaluation of NIZO B40 EPS production levels under various physiological conditions showed that fructose grown cells produced much less EPS compared to glucose (or several other carbon sources) grown cells. This observation was also described for the level of EPS production by *Lactobacillus delbrueckii* subsp. *bulgaricus* (Grobben et al. 1996). This fructose specific effect is possibly explained by the low activity level of fructose 1,6-bisphosphatase (Fbp; Figure 2) in *L. lactis* (Looijesteijn et al. unpublished results). This enzyme is required for conversion of fructose 1,6-dP to fructose-6-P, an essential step in generation of the central sugar nucleotide precursor glucose-1-P from glucose (via Pgi and Pgm; Figure 2). Preliminary results indicate that EPS production levels in fructose grown cells can be increased by functional overexpression of the heterologous *fbp* gene from *B. subtilis*, which confirms the assumed limitation of Fbp in *L. lactis*. Interestingly, also the growth rate of *L. lactis* on media containing fructose as sole carbon source was significantly increased by Fbp overexpression (Looijesteijn et al., unpublished results), suggesting that the sugar nucleotide requirements for efficient growth are also limited by the low Fbp activity in this organism. In analogy, the observed reduced growth rate of *L. lactis galE* mutant cells in media containing glucose as a sole carbon source, could be explained by limiting UDP-galactose levels (see also above).

Alternatively, the maximal EPS production level could in some cases be determined by the activity level of the factors encoded by the *eps* gene cluster, rather than by the level of sugar nucleotides. This possibility is supported by the finding that NIZO B40 EPS production can be increased by overexpression of *epsD* (van Kranenburg et al. 1999b). However, this hypothesis might not hold true for the biosynthesis of EPSs other than NIZO B40 EPS. The latter is supported by the findings that the level of *S. pneumoniae* type 3 CPS production in *L. lactis* can be dramatically increased by the expression of a pneumococcal GalU analogue (Gilbert et al. 1998).

Physical properties of EPS in watery solutions and products

A crucial factor in design of EPSs that would be desirable for a certain application is knowledge of the physical properties of different EPSs in solution and its interactions with other components of the product. Since in most cases, the desired contribution of EPS to the properties of a product is their thickening effect, it is important to understand the parameters that determine the viscosity of EPS solutions. Most EPSs can be described as random coil polymers that do not have a fixed shape but a randomly fluctuating tertiary structure. The relative viscosity η_r (with respect to the solvent) of EPS solutions is a function only of the intrinsic viscosity ($[\eta]$; a specific volume of the dispersed spherical particles in m^3/kg) and the concentration (c; g/l).

Recently, the following equation was derived, describing the master curve for random coil polysaccharide solutions (Tuinier et al. 1999a):

$$\eta_r = 1 + [\eta]c + 1/25\{[\eta]c\}^{3.5}$$

The differences between polysaccharides can thus be explained by differences in the intrinsic viscosity $[\eta]$. This quantity is proportional to the specific volume of the polysaccharide in solution and can be expressed as a function of the molar mass (M) and radius of gyration (R_g; a measure for the size of a polymer in solution). For random coil polysaccharides R_g depends on M ($R_g \sim M^{0.5-0.6}$) and for more expanded polysaccharides R_g is larger at a given M. These observations mean that, in order to obtain a higher (intrinsic) viscosity, either the molar mass (chain length) or the polysaccharide stiffness should be increased. Although the relation between polysaccharide chemistry and chain stiffness have not yet been established completely, some trends have been described (Rees 1977); (i) $\beta(1\rightarrow4)$ linkage of monosaccharides leads to a higher chain stiffness compared to $\beta(1\rightarrow2)$ or $\beta(1\rightarrow3)$ linkages (Lapasin & Pricl 1995), (ii) α linkages usually lead to more flexible chains than β linkages as is the case for $(1\rightarrow4)$ linkages (Lapasin & Pricl 1995) and (iii) branches and side groups influence the chain stiffness.

Static light scattering measurements of size-fractionated *L. lactis* NIZO-B40 EPS in aqueous solution at an ionic strength of 0.1 M generated a number-averaged molar mass of $1.47*10^6$ g/mol and radius of gyration of 86 $\pm$ 2 nm. Moreover, the relation of R_g and M for B40 EPS ($R_g \sim M^{0.57}$) is in good

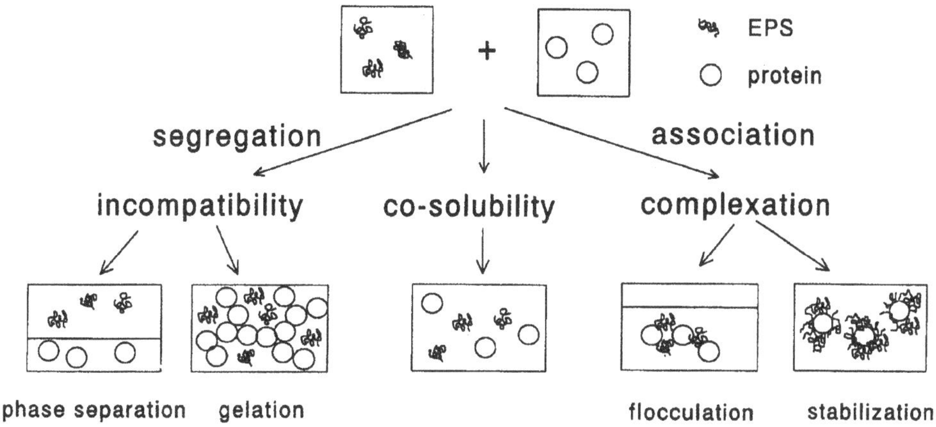

Figure 3. Schematic representation of the main trends in the behavior of protein/polysaccharide mixtures.

agreement with a random coil structure of this polysaccharide in solution (Tuinier et al. 1999b). Similar measurements with *L. lactis* NIZO B891 EPS showed that although the molar mass of this polysaccharide ($2.38*10^6$ g/mol) is larger than that measured for B40 EPS, the radius of gyration of B891 EPS is only slightly larger (B891: Rg = 91± 2; $R_g \sim M^{0.51}$) (Tuinier 1999c). These observations suggest higher stiffness of the B40 EPS 'backbone' compared to B891 EPS. The β-(1-4) sugar linkages in the B40 EPS backbone support a relatively high stiffness of this polymer, which is also reflected in its radius of gyration and good rheological properties. However, since the primary structure of B891 EPS has not yet been fully determined one can only speculate about the reasons for the apparent higher flexibility of this polymer.

The developments described above, will in time allow a good prediction of the viscosity of a certain EPS molecule in solution. However, the contribution of EPSs to the properties of food products not only depends on the properties of the EPS itself but also on the interactions of the EPS with the various components in food products (e.g. proteins). When protein and polysaccharide solutions are mixed, the behaviour of the mixture depends on whether the two biopolymers are segregative (the biopolymers repel, are incompatible) or associative (the biopolymers attract, are compatible) (Tolstoguzov 1991). The main trends of mixing polysaccharides and proteins are (highly) schematically illustrated in Figure 3. For very dilute solutions the system is stable and proteins and polysaccharides are co-soluble. Upon increasing the concentration of the biopolymers the system may become unstable, depending on the type of interaction. As a rule proteins and polysaccharides tend to segregate (Tolstoguzov 1991), leading to a reduction of the polysaccharide concentration near the protein particle. This phenomenon is called depletion (Fleer et al. 1993) and exceeding a certain polymer concentration leads to a phase separation into protein-enriched and polysaccharide-enriched phases (Figure 3). These observations are illustrated by mixtures of B40 EPS with casein micelles, where a segregative interaction between the biopolymers takes place that leads to phase separation (Tuinier et al. 1999d). However, when the polysaccharides are very small compared to the protein particles and the volume fraction of particles is sufficiently high, aggregation of particles can lead to a space filling network, in a process called gelation (Syrbe et al. 1998).

A mixture of polysaccharides and proteins can also induce aggregation when associative interactions are operational (Syrbe et al. 1998) and the polysaccharides tend to adsorb onto the protein surfaces. If the amount of polymer is not sufficient to cover the protein surfaces, a polysaccharide may adsorb onto more than one protein surface, leading to (bridging) flocculation. In contrast, when the proteins are fully covered the polysaccharide-coated proteins are stabilised, which could be important if the proteins are initially unstable and the polysaccharide can thus act as stabiliser.

In fermented products EPS is produced during the growth phase of the EPS-producing LAB. In a product like yoghurt, EPSs play a significant role in the rheology of stirred yoghurt (van Marle 1998). Usually about 100 mg/l EPS is formed during yoghurt fermentation and the viscosity of stirred yoghurt depends very strongly on the type of EPS produced. Also here, R_g and M appear to be the most important characteristics in determination of viscosity (Faber et al. 1998; Zoon et al., unpublished results).

Concluding remarks

Overall, one can conclude that the knowledge of EPS production in *Lactococcus lactis* (and other LAB) has developed impressively over the last years. The specific genetics and molecular biology involved in EPS production (*eps* genes) have been (partially) resolved for several EPS producing lactococci, and have reached a very exciting stage, where evaluation of possibilities and limitations of true genetic EPS-engineering can be envisaged. Moreover, the studies aiming for EPS production improvement have led to the identification and characterisation of interesting housekeeping genes. The encoded enzymes will be essential for design of metabolic engineering strategies to increase the EPS production level by LAB. However, until now, this approach has been hampered by the lack of predictive kinetic metabolic models that allow the identification and manipulation of the rate limiting steps of EPS biosynthesis by these organisms. The development of such models will add a clear rational to the metabolic engineering approaches.

Finally, the models developed to predict the behaviour and, consequently the effect of EPS addition to (food) products have already led to the identification of several important molecular properties (e.g. molar mass or chain length, polymer stiffness or monomer linkage-type in the EPS backbone) that determine their effects on a product property like viscosity. These models require further validation and fine-tuning to predict the exact properties of a specific EPS with for example a modification of the chemical composition or sugar-linkage-type in its repeating unit. Such predictive models will in the future allow the design of virtual EPSs that provide the desired product properties. A major challenge in the future for EPS- and metabolic-engineering will be the construction of modified lactococcal strains (or other LAB) that produce these 'designer' EPSs at a sufficiently high level.

Acknowledgements

Part of the work described here was supported by the European Community Biotechnology program (contracts BIOT-CT94-3055 and BIOT-CT96-0498). We thank Roland Siezen, Dick van den Berg and Oscar Kuipers for critically reading the manuscript.

References

Boels IC, Kleerebezem M, Hugenholtz J & de Vos WM (1998) Metabolic engineering of exopolysaccharide production in *Lactococcus lactis*. Proceedings 5th ASM on the genetics and molecular biology of streptococci, enterococci and lactococci, p. 66

Becker A, Niehaus K & Pühler A (1995) Low molecular-weight succinoglycan is predominantly produced by *Rhizobium meliloti* strains carrying a mutated ExoP protein characterized by a periplasmic N-terminal domain and a missing C-terminal domain. Mol. Microbiol. 16: 191–203

Faber EJ, Zoon P, Kamerling JP & Vliegenthart JFG (1998) The exopolysaccharide produced by *Streptococcus thermophilus* Rs and Sts have the same repeating unit but differ in viscosity of their milk cultures. Carboh. Res. 310: 269–276

Fleer GJ, Cohen-Stuart MA, Scheutjens JMHM, Vincent B & Cosgrove T (1993) Polymers At Interfaces. Harper & Row

Gibson GR & Roberfroid MB (1995) Dietary modulation of the human colonic micobiota: introducing the concept of prebiotics. J. Nutr. 124: 1401–1412

Gilbert C, Robinson K, LePage RWF & Wells JM (1998) Heterologous biosynthesis of pneumococcal type 3 capsule in *Lactococcus lactis* and immunogenicity studies. Proceedings 5th ASM on the genetics and molecular biology of streptococci, enterococci and lactococci, p. 67

Grobben GJ, Smith MR, Sikkema J & de Bont JAM (1996) Influence of fructose and glucose on the production of exopolysaccharides and the activities of enzymes involved in the sugar metabolism and the synthesis of sugar nucleotides in *Lactobacillus delbrueckii* subsp. *bulgaricus* NCFB 2772. Appl. Microbiol. Biotechnol. 46: 279–284

Grossiord B, Vaughan EE, Luesink EJ & de Vos WM (1998a) Genetics of galactose-utilization via the Leloir pathway in lactic acid bacteria. le Lait 78: 77–84

Grossiord B (1998b) Metabolisme du galactose par la voie de Leloir l'operon gal de *Lactococcus lactis*. Thesis, l'École Nationale Supérieure Agronomique de Montpellier, Montpellier, France

Hosono A, Lee J, Ametani A, Natsume M, Hirayama M, Adachi T & Kaminogawa S (1997) Characterization of a water-soluble polysaccharide fraction with immunopotentiating activity from *Bifidobacterium adolescentis* M101-4. Biosci. Biotech. Biochem. 61: 312–316

Kitazawa H, Yamaguchi T, Miura M, Saito T & Itoh H (1993) B-cell mitogen produced by slime-forming, encapsulated *Lactococcus lactis* subspecies *cremoris* isolated from ropy sour milk, viili. J. Dairy. Sci. 76: 1514–1519

Kolkman MAB, Morrison DA, van der Zeijst BAM & Nuijten PJM (1996) The capsule polysaccharide synthesis locus of*S. pneumoniae* serotype 14: identification of the glycosyl transferase gene *cps14E*. J. Bacteriol. 178: 3736–3741

Kolkman MAB, van der Zeijst BAM & Nuijten PJM (1997) Functional analysis of glycosyltransferase encoded by capsular polysaccharide synthesis locus of *S. pneumoniae* serotype 14. J. Biol. Chem. 272: 19502–19508

Lamothe GT, Stingele F, Neeser JR & Mollet B (1998) Genetic analysis of exopolysaccharide (EPS) production in *Lactobacillus delbrueckii* subsp. *bulgaricus*. Proceedings 5th ASM on the

genetics and molecular biology of streptococci, enterococci and lactococci, p. 67
Lapasin R & Pricl S (1995) Rheology of Industrial Polysaccharides: Theory and Applications. Blackie Academic Press
Liu D, Cole RA & Reeves PR (1996) An O-antigen processing function for Wzx (RfbX): a promising candidate for O-unit flippase. J. Bacteriol. 178: 2102–2107
MacPherson DF, Manning PA & Morona R (1995) Genetic analysis of the *rfbX* gene of *Shigella flexneri*. Gene 155: 9–17
Morona R, Mavris M, Fallarino A & Manning PA (1994) Characterization of the *rfc* region of *Shigella flexneri*. J. Bacteriol. 176: 733–747
Morona JK, Morona R & Paton JC (1997) Characterization of the locus encoding the *Streptococcus pneumoniae* type 19F capsular polysaccharide biosynthesis pathway. Mol. Microbiol. 23: 751–763
Nakajima H, Suzuki Y, Kaizu H & Hirota T (1992a) Cholesterol-lowering activity of ropy fermented milk. J. Food Sci. 57: 1327–1329
Nakajima H, Hirota T, Toba T, Itoh T & Adachi S (1992b) Structure of the extracellular polysaccharide from slime-forming *Lactococcus lactis* subspecies *cremoris* SBT0495. Carbohydr. Res. 224: 245–253
Qian N, Stanley GA, Hahn- Hägerdahl B & Rådström P (1994) Purification and characterization of two phosphoglucomutases from *Lactococcus lactis* subsp. *lactis* and their regulation in maltose- and glucose utilizing cells. J. Bacteriol. 176: 5304–5311.
Qian N, Stanley GA, Bunte A & Rådström P (1997) Product formation and phosphoglucomutase activities in *Lactococcus lactis*: cloning and characterization of a novel phosphoglucomutase gene. Microbiology 143: 855–866
Rees DA (1977) Polysaccharide Shapes, Outline Studies in Biology. Chapman & Hall, London
Reeves P (1993) Evolution of *Salmonella* O-antigen variation by interspecific gene transfer on a large scale. Trends Genet. 9: 17–22
Sjöberg A & Hahn-Hägerdahl B (1989) ß-glucose-1-phosphate, a possible mediator for polysaccharide formation in maltose-assimilating *Lactococcus lactis*. Appl. Environ. Microbiol. 55: 1549–1554
Stingele F, Neeser JR & Mollet B (1996) Identification and characterization of the *eps* (exopolysaccharide) gene cluster from *Streptococcus thermophilus* Sfi6. J. Bacteriol. 178: 1680–1690
Stingele F (1998) Exopolysaccharide production and engineering in dairy *Streptococcus* and *Lactococcus*. Proceedings 5th ASM on the genetics and molecular biology of streptococci, enterococci and lactococci, p. 31–32
Sutherland IW (1993) Microbial polysaccharides, In: Whistler RL & Miller JN Eds. Industrial Gums; polysaccharides and their derivatives, 3rd edition, (pp 69–85). Academic Press, San Diego
Sutherland IW (1998) Novel and established applications of microbial polysaccharides. TIBTECH 16: 41–46.
Syrbe A, Bauer WJ & Klostermeyer H (1998) Polymer science concepts in dairy systems. An overview of milk protein and food hydrocolloid interaction. Int. Dairy J. 8: 179–193.
Tolstoguzov VB (1991) Functional properties of food proteins and role of protein-polysaccharide interaction. Food Hydrocolloids4: 429–468.
Tsukioka Y, Yamashita Y, Oho T, Nakano Y & Koga T (1997) Biological function of the dTDP-rhamnose synthetic pathway in *Streptococcus mutans*. J. Bacteriol. 179: 1126–1134
Tuinier R, Zoon P, Cohen-Stuart MA, Fleer GJ & de Kruif CG (1999a) Concentration and shear-rate dependence of the viscosity of an exocellular polysaccharide. (Submitted for publication)
Tuinier R, Zoon P, Olieman C, Cohen-Stuart MA, Fleer GJ & de Kruif CG (1999b) Isolation and physical characterization of an exocellular polysaccharide. Biopolymers 49: 1–9
Tuinier R (1999c) An exocellular polysaccharide and its interactions with proteins. Thesis, Wageningen Agricultural University, Wageningen, The Netherlands
Tuinier R & de Kruif CG (1999d) Phase behaviour of casein micelles/exocellular polysaccharide mixtures; experiment and theory. J. Chem. Phys. (In press)
van Kranenburg R, Marugg JD, van Swam II, Willem NJ, & de Vos WM (1997) Molecular characterization of plasmid-encoded *eps* gene cluster essential for exopolysaccharide biosynthesis in *Lactococcus lactis*. Mol. Microbiol. 24: 387–397
van Kranenburg R & de Vos WM (1998) Characterization of multiple regions involved in replication and mobilization of plasmid pNZ4000 coding for exopolysaccharide production in *Lactococcus lactis*. J. Bacteriol. 180: 5285–5290
van Kranenburg R, van Swam II, Marugg JD, Kleerebezem M & de Vos WM (1999a) Exopolysaccharide biosynthesis in *Lactococcus lactis* NIZO B40: functional analysis of the glycosyltransferase genes involved in synthesis of the polysaccharide backbone. J. Bacteriol. 181: 338–340
van Kranenburg R, Vos HR, van Swam II, Kleerebezem M & de Vos WM (1999b) Functional analysis of glycosyltransferase genes from *Lactococcus* and other Gram-positive cocci: complementation, expression and diversity. (Submitted for publication)
van Marle ME (1998) Steady shear viscosity of stirred yoghurts with varying ropiness. Thesis, University of Twente, Enschede, The Netherlands
Vedamuthu ER & Neville JM (1986) Involvement of a plasmid in production of ropiness (mucoidness) in milk cultures by *Streptococcus cremoris* MS. Appl. Environ. Microbiol. 51: 677–682
Von Wright A & Tynkkynen S (1987) Construction of *Streptococcus lactis* ssp. *lactis* strains with a single plasmid associated with mucoid phenotype. Appl. Environ. Microbiol. 53: 1385–1386

Antonie van Leeuwenhoek 76: 367–376, 1999.

Anchoring of proteins to lactic acid bacteria

Kees Leenhouts, Girbe Buist & Jan Kok*
*Department of Genetics, Groningen Biomolecular Sciences and Biotechnology Institute, University of Groningen, Haren, The Netherlands. (*Author for correspondence; E-mail: kokj@biol.rug.nl)*

Key words: *Lactococcus lactis*, *Lactobacillus* subsp., *Staphylococcus carnosus*, *Staphylococcus xylosus*, *Streptococcus gordonii*, surface display

Abstract

The anchoring of proteins to the cell surface of lactic acid bacteria (LAB) using genetic techniques is an exciting and emerging research area that holds great promise for a wide variety of biotechnological applications. This paper reviews five different types of anchoring domains that have been explored for their efficiency in attaching hybrid proteins to the cell membrane or cell wall of LAB. The most exploited anchoring regions are those with the LPXTG box that bind the proteins in a covalent way to the cell wall. In recent years, two new modes of cell wall protein anchoring have been studied and these may provide new approaches in surface display. The important progress that is being made with cell surface display of chimaeric proteins in the areas of vaccine development and enzyme- or whole-cell immobilisation is highlighted.

Introduction

There are at least three reasons to change the outside appearance of a bacterium by expressing foreign proteins at its surface. First, it should help to understand the fundamental mechanisms of protein targeting to the cell envelope. Second, it may enable to control the interactions between the bacterium and its environment. Third, it opens the way to a number of potentially important biotechnological applications. This paper focuses mainly on the third reason.

In the last decade there has been considerable progress towards the development of systems to anchor and display heterologous (poly)peptides at the surface of bacteria, such that they are detectable on the outside of the intact cells. The progress is most advanced in Gram-negative bacteria, notably *Escherichia coli*, for which several efficient fusion protein display systems have been described (Hofnung 1991; Georgiou et al. 1997; Ståhl & Uhlén 1997a). Most of these systems take advantage of the anchoring capacity of outer membrane proteins, employing transmembrane spanning domains (LamB, OmpA, PhoE) or domains that contain a lipomodification signal (Lpp, TraT). Others use the surface exposure capacity of fimbriae, pili or flagellae (FliC, FimH, PapA; Georgiou et al. 1997).

Gram-positive bacteria have also been taken into account for bacterial surface display purposes. The lack of an outer membrane and the presence of multiple peptidoglycan layers in the cell wall of these bacteria resulted in the use of a number of different targeting strategies that link proteins to the membrane or cell wall (Fischetti et al. 1993; Ståhl & Uhlén 1997a). The non-pathogenic staphylococci, *Staphylococcus xylosus* and *carnosus*, and the oral commensal *Streptococcus gordonii* were among the first Gram-positives for which anchored heterologous cell surface proteins were described (Hansson et al. 1992; Pozzi et al. 1992a,b). Recently, other lactic acid bacteria (LAB) like *Lactococcus lactis* and *Lactobacillus* subsp., are being examined for their capacity to target and attach heterologous proteins to the membrane and wall components (Piard et al. 1997b; Pouwels et al. 1998; Poquet et al. 1998; Steidler et al. 1998). The anchoring domains that have been used in these LAB, including *Staph. xylosus*, *Staph. carnosus* and *Strep. gordonii*, to attach heterologous (poly)peptides will be discussed in this short review. The vector systems, expression and translocation signals to produce and transport the hybrid surface proteins are also essential elements in the surface display strategies, but for these specific items the reader is referred to a recent comprehensive overview (Pozzi & Wells 1997a). The biotechnological applications that form the driving force behind research of surface linked proteins will be briefly addressed. A wide variety of applications are envisaged for LAB carrying surface exposed proteins, among which the immobilisation of enzymes at the bacterial surface, the immobilisation of the production strain at ligand-coated surfaces, generating whole-cell bioadsorbents for environmental purposes, diagnostic tools and display of entire peptide libraries. Hitherto, the most common motivation to attach proteins to the outer compartment of LAB has been the development of live-bacterial-vaccine delivery systems (Fischetti et al. 1996; Medaglini et al. 1998; Pouwels et al. 1998; Pozzi & Wells 1997a; Ståhl et al. 1997b).

Modes of anchoring

The general approach taken to position proteins at the cell surface consists of genetically linking a heterologous polypeptide in such a way to a protein known to be located at the cell surface that the chimaeric protein will be exposed, at least partly, to the outer environment. This approach requires rather detailed knowledge of the topology of the protein containing the anchoring signal, in order to determine an appropriate site for insertion or fusion. Such information may be obtained from protein homology comparisons to make an educated guess or from experimentation. Both strategies have been followed for LAB using cell surface anchoring domains of LAB- or non-LAB proteins.

The application of five different types of anchoring domains has been described or is currently under investigation. Figure 1 summarises these different modes of anchoring and gives also a proposal for nomenclature. Type 1 and type 2 anchors (A1 and A2, respectively) link the hybrid protein to the membrane. At present, the most commonly applied method of anchoring in LAB uses the type 3 anchor (A3), in which the displayed protein is covalently linked to cell-wall components. Attachment domains of type 4 (A4) and type 5 (A5) interact in a yet unknown manner with the cell wall and only very recently received attention. All types of anchors and their use in hybrid proteins are listed in Table 1 and will be discussed below. A1 and A2 domains reside in the N-terminal part of the hybrid proteins. The cell wall anchoring regions A3 to A5 are unique for Gram-positive bacteria and, in the cases described here, require the presence of the attachment domain in the C-terminus of the target protein. The hybrid protein gene is then preceded in the genetic construct by a secretion signal sequence that enables translocation of the expressed product across the cytoplasmic membrane.

Transmembrane anchors. In general, transmembrane spanning domains (TMSs) of any membrane protein may be used as an A1 domain. Protein topology studies that formed the basis for using one or more TMSs as an anchor were performed e.g. for the holin LytP of the lactococcal bacteriophage r1t. Alkaline phophatase (PhoA) fusion and insertions studies of the human immunodeficiency virus (HIV) gp41E epitope (ELDKWAS) revealed suitable sites for anchoring (Leenhouts et al., unpublished). The strategy to insert amino acid sequences in an exterior loop between TMSs may limit the insert size in order not to disturb the membrane protein topology (Hofnung 1991). In addition, this type of anchoring may not result in true surface exposure of the inserted domain since at least 100 amino acids are needed in the extended loop to cross the cell wall (Fischetti et al. 1990). For these reasons a fusion approach is often preferred in which the target protein is simply linked at its N-terminus to one or more TMDs of a cytoplasmic membrane protein. PhoA and β-galactosidase (LacZ) were fused to different N-terminal parts of the *L. lactis* bacteriocin-transport-accessory protein LcnD. Fusions located C-terminally to amino acid residue 44 of LcnD resulted in extracellularly anchored enzymes (Franke et al. 1996). A similar hybrid protein containing the N-terminal 80 amino acids of LcnD and at the C-terminus a 99 amino acid B-cell epitope of the human cytomegalovirus (hCMV) pp65 matrix protein was successfully anchored in this way. The pp65 moiety was accessible to proteinase K from the outside of protoplasts (Franke et al. 1998).

Poquet et al. (1998) identified in a random approach with an export specific reporter enzyme, seven lactococcal gene fragments encoding TMSs that function as membrane anchors. The nuclease reporter protein requires an extracellular location to be active and, therefore, important information on the topology of the fused membrane proteins is obtained.

For none of the above mentioned hybrid proteins has its accessibility from the outside of intact cells been described.

Lipoprotein anchors. A2 or lipoprotein anchoring domains are characterised by covalent binding to the

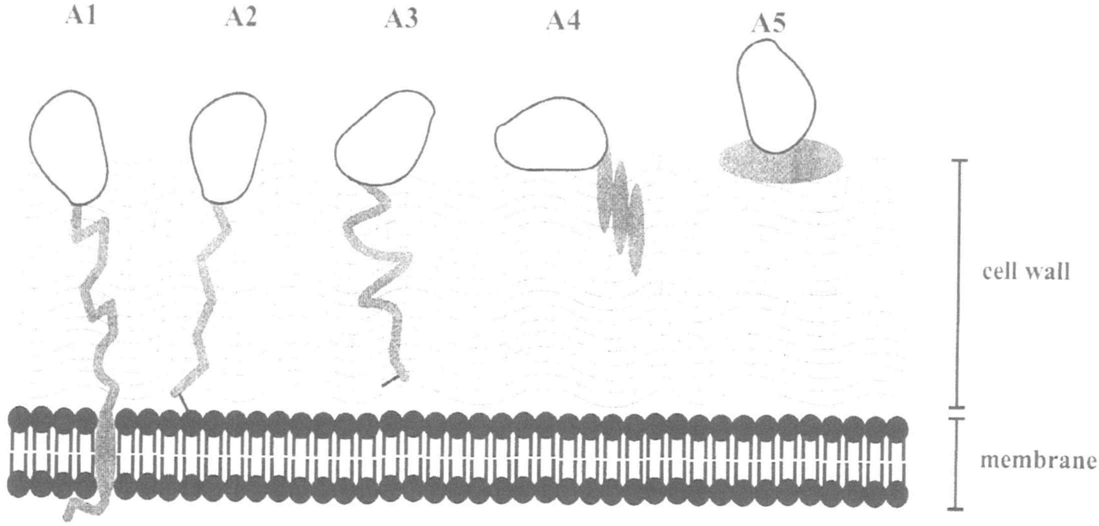

Figure 1. Modes of anchoring of chimaeric proteins to the cell surface of LAB. The light grey areas represent a heterologous (poly)peptide. The grey regions are the different types of anchoring domains. A1: transmembrane anchor region. A2: lipoprotein anchor domain. A3: LPXTG-type cell-wall anchoring domain. A4: AcmA-repeats cell-wall binding domain. A5: surface-layer-protein attachment region. The black lines extending from the A2 and A3 domains represent the covalent bond between the anchors and the lipid bilayer and the peptide crossbridge that connect the peptidoglycan layers, respectively. Components of the cell surface such as (lipo)teichoic acids and oligosaccharides are not shown.

lipid bilayer through N-acyl diglyceride modification of a cysteine residue located immediately C-terminal to the signal sequence cleavage site (Pugsley 1993). Haandrikman et al. (1991) have shown in their study on the proteinase maturation lipoprotein PrtM that this type of modification mechanism operates in *L. lactis*. An A2 signal has been identified on the basis of homology in the *L. lactis* oligopeptide binding protein OppA (Tynkkynen et al., 1993). This putative OppA anchoring signal was used to position the merozoite stage surface antigen MSA2 of *Plasmodium falciparum* on the outside of *L. lactis* (Leenhouts et al. unpublished). Data on the surface display of this OppA::MSA2 fusion protein and of other anchored MSA2 hybrids (Table 1 and below) should be available soon.

In the same random procedure as described above for the transmembrane anchors Poquet et al. (1998) identified four new lactococcal lipoprotein-anchoring signals. In one case, Nlp1, degradation of the fusion protein by proteinase K treatment on intact cells demonstrated its surface exposure.

LPXTG-motif anchor. The type 3 anchor signals (A3) contain highly conserved sequences and start at the C-terminus with a short tail of positively charged residues that remain in the cytoplasm. Upstream of the cytoplasmic domain a stretch of approximately 30 hydrophobic amino acids (optimal length is species dependent; Schneewind et al. 1993) is preceded by the well-conserved pentapeptide LPXTG (Fischetti et al. 1990). The charged tail and hydrophobic domain are thought to function as a temporary stop to position the LPXTG motif for proteolytic cleavage. Correct positioning results in cleavage between the threonine and glycine residues followed by amide-linkage of the threonine residue to the peptide crossbridge in the peptidoglycan of the cell wall, by the action of a postulated sortase (Navarre & Schneewind 1994; Schneewind et al. 1995). The amino acid composition of the peptide crossbridge, which varies among the different LAB species, is flexible with respect to the sorting reaction (Strauss et al. 1998; Ton-That et al. 1998). The sorting signal (LPXTG box, hydrophobic region and charged tail) is preceded by a wall-associated region of about 50 to 125 residues and is characterised by a high percentage of proline/glycine and/or threonine/serine residues (Fischetti et al. 1990). The mechanism by which these A3 domains are cell-surface targeted and subsequently covalently anchored to the cell wall has been elucidated in detail for immunoglobulin binding protein A

Table 1. Anchoring domains that have been used to immobilise hybrid proteins in or on the cell envelope of LAB

Host bacterium	Anchor domain[type]	Origin	Displayed protein	Origin	References
Staph. carnosus and *xylosus*	ProtA[A3]	*Staph. aureus*	CTB	*V. cholerae*	Liljeqvist, 1997
			ABP	*Strep.* subsp.	Andréoni, 1997
			ScFv anti-hIgE	mouse	Gunneriusson, 1996
			gpG (101aa epitope)	hRSV	Nguyen, 1993[a], 1995; Robert, 1996
			M3 (Pf155/RESA)	*P. falciparum*	Samuelson,1995; Hansson,1992
	FnBPB[A3]	*Staph. aureus*	lipase	*Staph. hyicus*	Strauss, 1996[b]
			β-lactamase	*E. coli*	Strauss, 1996[b]
Strep. gordonii	M6[A3]	*Strep. pyogenes*	M6	*Strep. pyogenes*	Pozzi, 1992b; Oggioni, 1996
			E7	HPV	Pozzi, 1992a; Oggioni, 1995; Medaglini, 1997
			gp120 (T-cell epitope)	HIV	Pozzi, 1994
			gp120/E7 (V3 epitope)	HIV/HPV	Di Fabio, 1998; Medaglini, 1998
			Ag5.2	hornet venom	Medaglini, 1995
			HA	measles virus	Medaglini, 1998; Pozzi, 1997
			LTB	*E. coli*	Ricci, 1996; Pozzi, 1997b
			gp120/LTB (V3 epitope)	HIV/*E. coli*	Medaglini, 1998; Pozzi, 1997b
L. lactis	ProtA[A3]	*Staph. aureus*	streptavidin	*Sm. avidinii*	Steidler, 1998
	M6[A3]	*Strep. pyogenes*	nuclease	*Staph. aureus*	Piard, 1997b
	holin[A1]	phage r1t	gp41E (Katinger epitope)	HIV	Leenhouts, unp.
	LcnD[A1]	*L. lactis*	pp65 (99aa epitope)	hCMV	Franke, 1998
	Tmp1-7[A1]	*L. lactis*	nuclease	*Staph. aureus*	Poquet, 1998
	Nlp1-4[A2]	*L. lactis*	nuclease	*Staph. aureus*	Poquet, 1998
	OppA[A2]	*L. lactis*	MSA2	*P. falciparum*	Leenhouts, unp.
	PrtP[A3]	*L. lactis*	TTFC	*C. tetanii*	Norton, 1995; Norton, 1996
			MSA2	*P. falciparum*	Leenhouts, unp.
	AcmA[A4]	*L. lactis*	β-lactamase	*E. coli*	Buist, 1997
			α-amylase	*B. licheniformis*	Buist, 1997
			MSA2	*P. falciparum*	Leenhouts, unp.
Lactobacillus subsp.	M6[A3]	*Strep. pyogenes*	LTB	*E. coli*	Rush, 1997
			gp41E (Katinger epitope)	HIV	Mercenier, 1996
	AcmA[A4]	*L. lactis*	β-lactamase	*E. coli*	Steen, unp.
	PrtP[A3]	*Lb. paracasei*	GusA	*E. coli*	Pouwels, 1996; 1998
			HA (Hackett epitope)	Infuenza virus	Pouwels, 1996
			TTFC	*C. tetanii*	Maassen, 1999
			VP7 and 8	Rotavirus	Leer, 1996; Pouwels, 1998
			Urease A and B	*H. pylori*	Leer, 1996; Pouwels, 1998
	Slp[A5]	*Lb. brevis*	VP1 (11 aa epitope)	Enterovirus	Palva et al., unp.

[A1]: transmembrane anchor; [A2]: lipoprotein membrane anchor; [A3]: LPXTG-type cell-wall anchor; [A4]: AcmA repeats cell-wall anchor; [A5]: surface-layer-protein anchor. [a]: only in *Staph. xylosus*. [b]: only in *Staph. carnosus*.

(ProtA) of *Staphylococcus aureus*. Its description is beyond the scope of this short review and the reader is referred to other reviews and the original literature (Schneewind et al. 1992, 1993, 1995; Ton-That et al. 1997; Navarre et al. 1998, 1999).

The anchoring domain of ProtA has been employed extensively in *Staph. xylosus* and *carnosus* to surface display various antigens for vaccine development purposes (Ståhl et al. 1997b). Electron microscopy, immunofluorescence microscopy and fluorescence-activated cell sorting (FACScan) techniques have been used to determine the proper display of the proteins. Many of the hybrid proteins in these studies contain the albumin binding protein fragment (ABP) from streptococcal protein G, allowing easy isolation of the hybrid proteins and rapid colorimetric analysis of successful display. In a comparative study, mice were immunised with live recombinant bacteria with surface expressed ABP as the model immunogen (Robert et al. 1996; Andréoni et al. 1997). Higher serum antibody titers were obtained with recombinant *Staph. carnosus* than with the corresponding *Staph. xylosus* strain. This result was attributed to the higher number of hybrid proteins present on the surface of the *Staph. carnosus* strain which was calculated to be as much as 10^4 per cell and 3×10^3 for the *Staph. xylosus* strain. Nguyen et al. (1995) engineered the hydrophobicity of a non-secretable portion of the G protein of the respiratory syncytial virus (RSV) by replacing or deleting hydrophobic phenylalanine residues. This resulted in a properly surface displayed protein and showed for the first time that hydrophobicity engineering can be used to optimise surface display. Such a strategy may have important implications in the area of vaccine development if surface display is required of proteins that are normally not secreted. Hydrophobicity engineering may influence the native structure of the protein, which is acceptable as long as protective immunity can be elicited. Correct folding of the displayed proteins is a requirement if they have to exert certain (enzymatic) activities. It was shown that the ProtA anchoring system is efficient in expressing a functional single chain antibody fragment (scFv), the mouse anti-human IgE scFv, on the surface of *Staphylococcus* (Gunneriusson et al. 1996). These recombinant *Staph. carnosus* cells were able to recognise human IgE and this result paves the way for applications such as using recombinant LAB as whole cell diagnostic devices or the surface display of scFv libraries. Cholera toxin fragment B (CTB) of *Vibrio cholerae* was also functionally displayed on *Staph. carnosus* (Liljeqvist et al. 1997). The functionality of the non-toxic CTB requires pentamerization as was demonstrated by binding of the recombinant bacteria to monosialoganglioside GM1, which is present on all epithelial cells of mucosal surfaces.

In another application of the ProtA surface anchor it was fused to the streptavidin monomer of *Streptomyces avidinii* and expressed in *L. lactis* (Steidler et al. 1998). Lysostaphin treatment of producer cells released the fusion protein, indicating that it was linked to the peptidoglycan layer. Immobilisation of the recombinant strain on a biotinylated alkaline phosphatase-coated polystyrene support suggested that the streptavidin moiety of the hybrid proteins is accessible from the outside. These results open possibilities to immobilise LAB on solid surfaces for production purposes.

The C-termini of numerous Gram-positive bacterial surface proteins are highly homologous to the ProtA anchoring signal (Fischetti et al. 1996). Another *Staph. aureus* A3 region, that of the fibronectin binding protein B (FnBPB), was effective in immobilisation of the normally soluble enzymes lipase of *Staphylococcus hyicus* and β-lactamase of *E. coli* on the cell surface of *Staph. carnosus* (Strauss & Götz 1996). A spacer region which exceeded a critical length of approximately 90 amino acids between the LPXTG box and the C-terminus of the enzymes was required to allow efficient folding of the enzymes in their catalytically active form.

An A3 cell wall binding region that has found application in various LAB, like the one of ProtA, is the anchor domain of the fibrillar M6 protein of *Streptococcus pyogenes*. The M6 sorting signal appears to be functional in *Strep. gordonii, L. lactis, Lb. fermentum, Lb. sake,* and *Streptococcus thermophilus* (Pozzi et al. 1992a; Piard et al. 1997a). The M6 anchoring domain was extensively exploited in *Strep. gordonii*. Reported were the cell surface display of the subunit B of heat labile toxin (LTB) of *E. coli*, which has a similar function as CTB, the allergen Ag5.2 of white-faced hornet venom and a variety of relevant immunodominant antigens of human virusses, including the E7 protein of papillomavirus (HPV), parts of gp120 of HIV and the measles virus hemoagglutinin. The hybrid proteins were detected on the cell surface by immunofluorescence microscopy and elicited relevant immune responses in parenteral and local immunisations (Pozzi et al. 1997b). An important step in the development of LAB carriers for vaccine purposes was the successful colonisation of the vaginal mucosa of

Cynomolgus monkeys with recombinant *Strep. gordonii* expressing the HPV E7 protein and part of HIV gp120 on the cell surface, which resulted in antigen-specific vaginal IgA, serum IgG and T-cell responses (Di Fabrio et al. 1998; Medaglini et al. 1998).

A heterologous enzyme, a *Staph. aureus* nuclease, was immobilised on the surface of *L. lactis* using 139 C-terminal residues of M6. The presence of active nuclease in the cell wall fraction suggested correct surface display of the hybrid protein (Piard et al. 1997b). A similar M6 anchoring domain was exploited to surface expose *E. coli* LTB on *Lb. casei, Lb. paracasei, Lb. acidophilus, Lb. plantarum* and *Lb. fermentum* (Rush et al. 1997). The presence of the LTB immunogen on the surface of this set of *Lactobacillus* species allows to investigate the mucosal immune responses to the antigen expressed in colonising and non-colonising strains. Mercenier et al. (1996) expressed M6 hybrid proteins containing parts of HIV gp41 or gp120 on the cell surface of *Lb. paracasei*, as was demonstrated by immunofluorescence techniques. Local administration in mice of recombinant colonising *Lb. paracasei* strains carrying the M6-anchored HIV parts induced significant systemic and mucosal responses.

LAB A3 domains that are being examined for surface display purposes are those of the proteinases (PrtP) of *L. lactis* and *Lb. paracasei* (Kok et al. 1988; Vos et al. 1989; Holck & Næs 1992). The tetanus toxin fragment C (TTFC) of *Clostridium tetanii* was fused to the lactococcal PrtP anchor but, although the hybrid protein was produced in *L. lactis*, it appeared that TTFC could not be detected on the surface of intact cells by immunogold or immunofluorescent labelling techniques. Lysozyme/lysostaphin treatment of the producer demonstrated that all TTFC was present in the cell membrane and not in the cell wall. Nevertheless, membrane located TTFC proved to be approximately 13 to 20-fold more immunogenic than intracellular soluble TTFC. Both forms were capable of evoking protective immune responses in mice subcutaneously immunised with the recombinant strains (Norton et al. 1995; Norton et al. 1996). Whether the inefficient targeting to the cell wall is caused by the lactococcal PrtP anchoring domain or is a TTFC-specific effect, are questions that need to be addressed. At present, a fusion with MSA2 is under investigation (Leenhouts et al., unpublished). The PrtP sorting signal of *Lb. paracasei* seems to work efficiently in a number of different *Lactobacillus* species to surface display various antigens, including TTFC of *C. tetanii*, rotavirus proteins VP7 and 8, urease A and B of *Helicobacter pylori*, and an influenza virus hemoagglutinin epitope fused to β-glucoronidase (GusA) of *E. coli* as a carrier protein (Leer et al. 1996; Pouwels et al. 1996, 1998). FACScan analyses indicated that *Lb. casei* producing surface-anchored TTFC exposed approximately 4×10^3 antigenic molecules per cell to the environment. High levels of serum IgG specific for TTFC were induced following parenteral immunisation of mice with this recombinant strain (Maassen et al. 1999), but low antibody levels were observed after oral administration, which was attributed to the poor viability of the *Lb. casei* strain in the gut of mice (Pouwels et al. 1998).

AcmA-repeats anchor. Nearly all bacterial cell wall hydrolases have a modular design, in which an active site domain degrades the peptidoglycan and a cell wall binding domain immobilises the enzyme in or on the peptidoglycan layer. The cell wall binding domain is often comprised of repeated amino acid sequences. The C-terminal region of the lactococcal cell wall hydrolase AcmA contains three repeated sequences of 44 amino acids separated by stretches of 21 to 31 amino acids rich in serine, threonine and asparagine residues (Buist et al. 1995). This repeat region is defined here as an A4 domain. There are no reports providing insight in the mechanism of binding of these repeats or details on the interaction of the A4 attachment region with the cell wall. Repeats homologous to the ones in AcmA are present in many cell wall- or membrane-associated proteins in Gram-positive and Gram-negative bacteria, but do not necessarily reside in the C-terminus of the protein (Buist 1997). Proteins of Gram-positives with an A4-like domain are usually detected in the wall fraction as well as in the supernatant. This observation suggests an interaction of the A4 attachment region with the cell wall that is less strong than that of the LPXTG type (A3) and is, most likely, of a non-covalent nature.

Although the cell wall binding domain of AcmA contains 3 repeats, 1 repeat proved to be sufficient for cell wall binding (Buist 1997). Interestingly, AcmA present in the supernatant of an *L. lactis* producer was able to bind from the outside to an *L. lactis* strain devoid of AcmA and to a range of Gram-positive bacteria, including *Lactobacillus*, *Clostridium*, *Listeria* and *Bacillus* (Buist and Steen, unpublished results). The A4 domain of AcmA could also be used to immobilise the normally soluble enzymes α-amylase of *E. coli* and β-lactamase of *Bacillus licheniformes* in an active form on the cell surface of *L. lactis*. Like AcmA, these chimaeric enzymes could bind to *L. lactis* cells when added from the outside (Buist 1997). An attractive possibility of this type of binding is to couple hybrid proteins to non-recombinant Gram-positive bacteria. The possibility to use the AcmA A4 attachment region to display antigens, e.g. like MSA2, on the cell surface of *L. lactis* is currently under investigation (Leenhouts et al., unpublished).

Surface-layer-protein anchor. Some LAB strains contain, like various other Gram-positive bacteria, a layer of proteins exterior to the cell wall. These so-called surface layer (S-layer) proteins form porous lattices of identical subunits completely covering the cell surface and may constitute up to 20% of the total cell protein content. These properties make S-layer proteins (SLPs) an attractive target for protein anchoring studies. At present, the genes of *Lb. brevis*, *Lb. acidophilus*, *Lb. crispatus* and *Lb. helveticus* SLPs have been genetically characterised (Vidgrén et al. 1992; Boot et al. 1993; Toba et al. 1995; Callegari et al. 1998). Comparison of the SLP amino acid sequences of the latter three bacteria revealed a conserved C-terminal one-third of the proteins (>80% identity), that is thought to interact with the cell wall (Pouwels et al. 1998). Expression of SLPs in heterologous hosts (strains of *Lb. casei* and *L. lactis* which lack S-layer proteins) resulted in secretion of the S-layer proteins in the medium (Callegari et al. 1998; Martinez et al., unpublished results). This finding suggests that *Lb. casei* and *L. lactis* lack (a) cell wall component(s) required for proper attachment and this may implicate that cell surface anchoring of SLP hybrid proteins is limited to the host from which the *slp* gene was isolated.

Palva et al. (unpublished) used the *Lb. brevis* SLP to produce chimaeric proteins in *Lb. brevis* with an 11 amino acid insert of an enteroviral capsid protein VP1. Whole cell-ELISA techniques were used to identify two permissive hydrophilic sites in the C-terminal region of this SLP that allow surface exposure of the epitope. These results may suggest that the *Lb. brevis* SLP anchoring domain resides in another part of the protein than it does in the other three SLPs.

Conclusions and perspectives

A number of different anchoring domains have been explored in LAB for the cell surface anchoring of heterologous (poly)peptides and, although still rather premature, a few interesting applications arise at the horizon. From a practical point of view, LAB may have certain advantages over Gram-negative and other Gram-positive bacteria that make them more suitable for bacterial surface-display purposes. LAB have the status of being generally recognised as safe (GRAS) making them certainly more useful in food and medical applications than some other bacterial species. It is exactly in these areas that research on surface display in LAB has concentrated so far. Important progress has been made in the immobilisation of enzymes, the immobilisation of LAB for production purposes and in vaccine development. Although the latter research area has been the major driving force in this field, it is still a matter of debate whether cell-surface display of antigens for vaccine purposes is essential for eliciting effective immune responses. Experimental data for *Staph. xylosus* and *carnosus* indicated that surface exposure is essential (Nguyen et al. 1995). On the other hand, immunisations with *L. lactis* expressing intracellular soluble, membrane-located or cell wall-associated immunogens revealed that all forms were effective in evoking protective immune responses, the particulate form of the antigen being more immunogenic (Norton et al. 1996; Wells et al. 1996; Robinson et al. 1997). Although surface exposure of antigens may, thus, be advantageous in terms of immunogenicity, it may also have drawbacks as, e.g., higher susceptibility to proteolytic degradation in the gastrointestinal tract (GIT) in oral vaccinations. Antigens anchored to the cytoplasmic face of the cell membrane may have the same elevated level of immunogenicity as proteins anchored at the exterior and may be better protected against proteolytic activity in the GIT. This approach has been taken in Gram-negative bacteria (Eko et al. 1999) but, to the best of our knowledge, it has not been investigated in Gram-positives.

The expression of an scFv on the surface of *Staph. carnosus* opens the possibility to express entire (artificial) antibody libraries in LAB. Although construction of such libraries is limited to strains that are highly transformable, strategies like that could help to identify antibody fragments that are reactive to surface components of mucosal cells, such as M-cells that are present in immunoreactive sites. Co-expression of these antibody fragments on the surface of LAB carrying an immunogen may target the bacteria to the desired area and enhance the immune response (Ståhl & Uhlén 1997a). Another interesting application of recombinant antibody- or artificial binding-protein libraries that warrants further exploration is the selection of ligands that bind specifically to e.g. pollutants, such as heavy metals (Sousa et al. 1996). LAB carrying such surface-located ligands could then be exploited in environmental applications like whole-cell bioadsorbants or biofilters.

The cell wall sorting signals that contain the LPXTG box are clearly the most exploited anchoring signals in LAB. The systems developed in *Staph. carnosus* and *Strep. gordonii* for cell-surface display using this type of anchor belong to the most advanced in LAB. New techniques, such as FACScan, were used for the first time for *Staph. carnosus* and, more recently also for *Lb. casei*, to quantify the number of surface exposed protein molecules on intact cells. A broader use of these techniques will be required to reliably assess each anchoring system in the various bacteria.

In addition to the successful use of heterologous anchor domains from pathogenic bacteria, an important development is the identification and application of LAB anchors. For several purposes the presence of heterologous DNA fragments should be minimised, especially for those in which the GRAS status of the production organism plays a critical role.

The AcmA-repeats and the S-layer proteins represent new types of anchoring devices in LAB and offer alternative approaches for cell surface attachment. Both need further characterisation as they are potentially very versatile. For instance, the AcmA repeat domain binds to the cell wall of a wide range of Gram-positive bacteria, also when it is added from the outside. This property allows to anchor chimaeric proteins, e.g. antigens or active enzymes, to the surface of non-recombinant bacteria, which would prevent the release of recombinant DNA in the environment. Hybrid S-layer proteins may have the limitation that surface anchoring is restricted to the bacterium from which the gene was isolated, but they do have the attractive ability to form crystalline structures *in vitro*. The ability to form lattices was retained for *Lb. acidophilus* and *Lb. crispatus* SLPs that were produced in *E. coli* (Smit et al., unpublished). This property may be exploited in yet other approaches using the isolated protein for, e.g., ultrafiltration membranes, diagnostics, vaccine development and nanotechnology (Sleytr et al. 1997).

As LAB research on membrane proteins, cell wall proteins and surface display advances, the putative biotechnological applications will become increasingly realistic and still new ones may emerge from this exciting and rapidly developing research area.

Acknowledgements

The authors thank Peter Pouwels and Airi Palva for communicating data prior to publication. Part of the work on AcmA was carried out in a research project with Unilever Research Laboratorium, The Netherlands.

References

Andréoni C, Goetsch L, Libon C, Samuelson P, Nguyen TN, Robert A, Uhlén M, Binz H & Ståhl S (1997) Flow cytometric quantification of surface-displayed recombinant receptors on staphylococci. BioTechniques 23: 696–704

Boot HJ, Kolen CPAM, Van Noort JM & Pouwels PH (1993) S-layer protein of *Lactobacillus acidophilus* ATCC 4356: purification, expression in *Escherichia coli*, and nucleotide sequence of the corresponding gene. J. Bacteriol. 175: 6089–6096

Buist G (1997) AcmA of *Lactococcus lactis*, a cell-binding major autolysin. Ph.D. thesis, University of Groningen, Haren, The Netherlands

Buist G, Kok J, Leenhouts KJ, Dabrowska M, Venema G & Haandrikman AJ (1995) Molecular cloning and nucleotide sequence of the gene encoding the major peptidoglycan hydrolase of *Lactococcus lactis*, a muramidase needed for cell separation. J. Bacteriol. 177: 1554–1563

Callegari ML, Riboli B, Sanders JW, Cocconcelli PS, Kok J, Venema G & Morelli L (1998) The S-layer gene of *Lactobacillus helveticus* CNRZ 892: cloning, sequence and heterologous expression. Microbiol. 144: 719–726

Di Fabio S, Medaglini D, Rush CM, Corrias F, Panzini GL, Pace M, Verani P, Pozzi G & Titti F (1998) Vaginal immunization of *Cynomolgus* monkeys with *Streptococcus gordonii* expressing HIV-1 and HPV16 antigens. Vaccine 16: 485–492

Eko FO, Witte A, Huter V, Kuen B, Fürst-Ladani S, Haslberger A., Katinger A, Hensel A, Szostak MP, Resch S. Mader S. Raza P, Brand E, Marchart J, Jechlinger W, Haidinger W & Lubitz W (1999) New strategies for combination vaccines based on the extended recombinant bacterial ghost system. Vaccine 17: 1643–1649

Fischetti VA, Medaglini D & Pozzi G (1996) Gram-positive commensal bacteria for mucosal vaccine delivery. Curr. Opin. Biotechnol. 7: 659–666

Fischetti VA, Medaglini D, Oggioni M & Pozzi G (1993) Expression of foreign proteins on Gram-positive commensal bacteria for mucosal vaccine delivery. Curr. Opin. Biotechnol. 4: 603–610

Fischetti VA, Pancholi V & Schneewind O (1990) Conservation of a hexapeptide sequence in the anchor region of surface proteins from Gram-positive cocci. Mol. Microbiol. 4: 1603–1605

Franke CM (1998) Topology of a type I secretion system for bacteriocins of *Lactococcus lactis*. Ph.D. thesis, University of Groningen, Haren, The Netherlands

Franke CM, Leenhouts KJ, Haandrikman AJ, Kok J, Venema G & Venema K (1996) Topology of LcnD, a protein implicated in the transport of bacteriocins from *Lactococcus lactis*. J. Bacteriol. 178: 1766–1769

Georgiou G, Stathopoulos C, Daugherty PS, Nayak AR, Iverson BL & Curtiss III R (1997) Display of heterologous proteins on the surface of microorganisms: from the screening of combinatorial libraries to live recombinant vaccines. Nature Biotechnol. 15: 29–34

Gunneriusson E, Samuelson P, Uhlén M, Nygren PA & Ståhl S (1996) Surface display of a functional single chain Fv antibody on staphylococci. J. Bacteriol. 178: 1341–1346

Haandrikman AJ, Kok J & Venema G (1991) Lactococcal proteinase maturation protein PrtM is a lipoprotein. J. Bacteriol. 173: 4517–4525

Hansson M, Ståhl S, Nguyen TN, Bächi T, Robert A, Binz H, Sjölander A & Uhlén M (1992) Expression of recombinant proteins on the surface of the coagulase-negative bacterium *Staphylococcus xylosus*. J. Bacteriol. 174: 4239–4245

Hofnung M (1991) Expression of foreign polypeptides at the *Escherichia coli* cell surface. Methods Cell Biol. 34: 77–105

Holck A & Næs H (1992) Cloning, sequencing and expression of the gene encoding the cell-envelope-associated proteinase from *Lactobacillus paracasei* subsp. *Paracasei* NCDO151. J. Gen. Microbiol. 138: 1353–1364

Kok J, Leenhouts KJ, Haandrikman AJ, Ledeboer AM & Venema G (1988) Nucleotide sequence of the cell wall-associated proteinase gene of *Streptococcus cremoris* Wg2. Appl. Environ. Microbiol. 54: 231–238

Leer RJ, Antonissen C, Bergmans A, Jore J, Boersma WJA & Pouwels PH (1996) A new series of *Lactobacillus* expression vectors. Abstracts of the Fifth Symposium of Lactic Acid Bacteria: Genetics, Metabolism and Applications. Veldhoven, The Netherlands, Abstract no. E8

Liljeqvist S, Samuelsen P, Hansson M, Nguyen TN, Binz H & Ståhl S (1997) Surface display of the cholera toxin subunit on *Staphylococcus xylosus* and *Staphylococcus carnosus*. Appl. Environ. Microbiol. 63: 2481–2488

Maassen CBM, Laman JD, Heijne Den Bak-Glashouwer MJ, Tielen FJ, Van Holten-Neelen JCPA, Hoogteijling L, Antonissen C, Leer RJ, Pouwels PH, Boersma WJA & Shaw DM (1999) Instruments for oral disease-intervention strategies: recombinant *Lactobacillus casei* expressing tetanus toxin fragment C for vaccination or myelin proteins for oral tolerance induction in multiple sclerosis. Vaccine 17: 2117–2128

Medaglini D, Oggioni MR & Pozzi G (1998) Vaginal immunization with recombinant gram-positive bacteria. Am. J. Repr. Immunol. 39: 199–208

Medaglini D, Pozzi G, King TP & Fischetti VA (1995) Mucosal and systemic immune responses to a recombinant protein expressed on the surface of the oral commensal bacterium *Streptococcus gordonii* after oral colonization. Proc. Natl. Acad. Sci. USA 92: 6868–6872

Medaglini D, Rush CM, Sestini P & Pozzi G (1997) Commensal bacteria as vectors for mucosal vaccines against sexually transmitted diseases: vaginal colonization with recombinant streptococci induces local and systemic antibodies in mice. Vaccine 15: 1330–1337

Mercenier A, Dutot P, Kleinpeter P, Aguirre M, Paris P, Reymund J & Slos P (1996) Development of lactic acid bacteria as live vectors for oral or local vaccines. Adv. Food Sci. 18: 73–77

Navarre WW & Schneewind O (1994) Proteolytic cleavage and cell wall anchoring at the LPXTG motif of surface proteins in Gram-positive bacteria. Mol. Microbiol. 14: 115–121

Navarre WW & Schneewind O (1999) Surface proteins of Gram-positive bacteria and mechanisms of their targeting to the cell wall envelope. Microbiol. Mol. Biol. Rev. 63: 174–229

Navarre WW, Ton-That H, Faull KF & Schneewind O (1998) Anchor structure of staphylococcal surface proteins. II. COOH-terminal structure of muramidase and amidase-solubilized surface protein. J. Biol. Chem. 273: 29135–29142

Nguyen TN, Gourdon MH, Hansson M, Robert A, Samuelson P, Libon C, Andréoni C, Nygren PA, Binz H, Uhlén M, & Ståhl S (1995) Hydrophobicity engineering to facilitate surface display of heterologous gene products on *Staphylococcus xylosus*. J. Biotechnol. 42: 207–219

Nguyen TN, Hansson M, Ståhl S, Bächi T, Robert A, Domzig W, Binz H & Uhlén M (1993) Cell-surface display of heterologous epitopes on *Staphylococcus xylosus* as a potential delivery system for oral vaccination. Gene 128: 89–94

Norton PM, Brown HWG, Wells JM, Macpherson AM, Wilson PW & Le Page RWF (1996) Factors affecting the immunogenicity of tetanus toxin fragment C expressed in *Lactococcus lactis*. FEMS Immunol. Med. Microbiol. 14: 167–177

Norton PM, Le Page RWF & Wells JM (1995) Progress in the development of *Lactococcus lactis* as a recombinant mucosal vaccine delivery system. Folia Microbiol. 40: 225–230

Oggioni MR & Pozzi G (1996) A host-vector system for heterologous gene expression in *Streptococcus gordonii*. Gene 169: 85–90

Oggioni MR, Manganelli R, Contorni M, Tommasimo M & Pozzi G (1995) Immunization of mice by oral colonization with live recombinant commensal streptococci. Vaccine 13: 775–779

Piard JC, Hautefort I, Fischetti VA, Ehrlich SD, Fons M & Gruss A (1997a) Cell wall anchoring of the *Streptococcus pyogenes* M6 protein in various lactic acid bacteria. J. Bacteriol. 179: 3068–3072

Piard JC, Jimenez-Diaz R, Fischetti VA, Ehrlich SD & Gruss A (1997b) The M6 protein of *Streptococcus pyogenes* and its potential as a tool to anchor biologically active molecules at the surface of lactic acid bacteria. In: Horaud et al. (Eds) Streptococci and the Host, (pp. 545–550). Plenum Press, New York, USA

Poquet I, Ehrlich SD & Gruss A (1998) An export-specific reporter designed for gram-positive bacteria: application to *Lactococcus lactis*. J. Bacteriol. 180: 1904–1912

Pouwels PH, Leer RJ & Boersma WJA (1996) The potential of *Lactobacillus* as a carrier for oral immunization: Development and preliminary characterization of vector systems for targeted delivery of antigens. J. Biotechnol. 44: 183–192

Pouwels PH, Leer RJ, Shaw M, Heijne den Bak-Glashouwer MJ, Tielen FD, Smit E, Martinez B, Jore J & Conway PL (1998) Lactic acid bacteria as antigen delivery vehicles for oral immunization purposes. Int. J. Food Microbiol. 41: 155–167

Pozzi G & Wells JM (1997a) Gram-Positive Bacteria. Vaccine Vehicles for Mucosal Immunization. Landes Biosciences Georgetown, TX, USA

Pozzi G, Contorni M, Oggioni MR, Manganelli R, Tommasino M, Cavalieri F & Fischetti VA (1992a) Delivery and expression of a heterologous antigen on the surface of streptococci. Infect. Immun. 60: 1902–1907

Pozzi G, Oggioni MR & Medaglini D (1997b) Recombinant *Streptococcus gordonii* as live vehicle for vaccine antigens. In: Pozzi G & Wells JM (Eds) Gram-positives Bacteria. Vaccine Vehicles for Mucosal Immunization (pp. 35–60). Landes Bioscience Georgetown, TX, USA

Pozzi G, Oggioni MR, Manganelli R & Fischetti VA (1992b) Expression of M6 protein gene of *Streptococcus pyogenes* in *Streptococcus gordonii* after chromosomal integration and transcription. Res. Microbiol. 143: 449–457

Pozzi G, Oggioni MR, Manganelli R, Medaglini D, Fischetti VA, Fenoglio D, Valle MT, Kunkl A & Manca F (1994) Human

T-helper cell recognition of an immunodominant epitope of HIV-1 gp120 expressed on the surface of *Streptococcus gordonii*. Vaccine 12: 1071–1077

Pugsley AP (1993) The complete general secretory pathway in Gram-negative bacteria. Microbiol. Rev. 57: 50–108

Ricci S, Rush CM, Medaglini D & Pozzi G (1996) Expression of the *Escherichia coli* heat labile toxin subunit B in *Streptococcus gordonii*. Abstr. XII European Meeting on Bacterial Gene Transfer and Expression. Siena, Italy. p. 88

Robert A, Samuelson P, Andréoni C, Bächi T, Uhlén M, Binz H, Nguyen TN & Ståhl S (1996) Surface display on staphylococci: a comparative study. FEBS Lett. 390: 327–333

Robinson K, Chamberlain LM, Schofield KM, Wells JM & Le Page RWF (1997) Oral vaccination of mice against tetanus with recombinant *Lactococcus lactis*. Nature Biotechnol. 15: 653–657

Rush CM, Mercenier A & Pozzi G (1997) Expression of vaccine antigens in *Lactobacillus*. In: Pozzi G & Wells JM (Eds) Gram-positive Bacteria. Vaccine Vehicles for Mucosal Immunization (pp. 107–144). Landes Bioscience Georgetown, TX, USA

Samuelson P, Hansson M, Ahlborg N, Andréoni C, Götz F, Bächi T, Nguyen TN, Binz H, Uhlén M & Ståhl S (1995) Cell surface display of recombinant proteins on *Staphylococcus carnosus*. J. Bacteriol. 177: 1470–1476

Schneewind O, Fowler A & Faull KF (1995) Structure of the cell wall anchor of surface proteins in *Staphylococcus aureus*. Science 268: 103–106

Schneewind O, Mihaylova-Petkov D & Model P (1993) Cell wall sorting signals in surface proteins of Gram-positive bacteria. EMBO J. 12: 4803–4811

Schneewind O, Model P & Fischetti VA (1992) Sorting of protein A to the staphylococcal cell wall. Cell 70: 267–281

Sleytr UB, Bayley H, Sára M, Breitwieser A, Küpcü S, Mader C, Weigert S, Unger FM & et al. (1997) Applications of S-layers. FEMS Microbiol. Rev. 20: 151–175

Sousa C, Cebolla A & De Lorenzo V (1996) Enhanced metalloadsorption of bacterial cells displaying poly-His peptides. Nature Biotechnol. 14: 1017–1020

Ståhl S & Uhlén M (1997a) Bacterial surface display: trends and progress. Trends Biotechnol. 15: 185–192

Ståhl S, Samuelson P, Hansson M, Andréoni C, Goetsch L, Libon C, Liljeqvist S, Gunneriusson E, Binz H, Nguyen TN & Uhlén M (1997b) Development of non-pathogenic staphylococci as vaccine delivery vehicles. In: Pozzi G & Wells JM (Eds) Gram-positive Bacteria. Vaccine Vehicles for Mucosal Immunization (pp. 61–81). Landes Bioscience Georgetown, TX, USA

Steidler L, Viaene J, Fiers W & Remaut E (1998) Functional display of a heterologous protein on the surface of *Lactococcus lactis* by means of the cell wall anchor of *Staphylococcus aureus* protein A. Appl. Environ. Microbiol. 64: 342–345

Strauss A & Götz F (1996) *In vivo* immobilization of enzymatically active polypeptides on the cell surface of *Staphylococcus carnosus*. Mol. Microbiol. 21: 491–500

Strauss A, Thumm G & Götz F (1998) Influence of Lif, the lysostaphin immunity factor, on acceptors of surface proteins and cell wall sorting efficiency in *Staphylococcus carnosus*. J. Bacteriol. 180: 4960–4962

Toba T, Virkola R, Westerlund B, Björkman Y, Sillanpää J, Vartio T, Kalkkinen N & Korhonen TK (1995) A collagen-binding S-layer protein in *Lactobacillus crispatus*. Appl. Environ. Microbiol. 61: 2467–2471

Ton-That H, Faull KF & Schneewind O (1997) Anchor structure of staphylococcal surface proteins. A branched peptide that links the carboxyl terminus of proteins to the cell wall. J. Biol. Chem. 272: 22285–22292

Ton-That H, Labischinski H, Berger-Bächi & Schneewind O (1998) Anchor structure of staphylococcal surface proteins. III. Role of the FemA, FemB, and FemX factors in anchoring surface proteins on the bacterial cell wall. J. Biol. Chem. 273: 29143–29149

Tynkkynen S, Buist G, Kunji E, Kok J, Poolman B, Venema G & Haandrikman AJ (1993) Genetic and biochemical characterization of the oligopeptide transport system of *Lactococcus lactis*. J. Bacteriol. 175: 7523–7532

Vidgrén G, Palva I, Pakkanen R, Lounatmaa K & Palva A (1992) S-layer protein gene of *Lactobacillus brevis*: cloning by polymerase chain reaction and determination of the nucleotide sequence. J. Bacteriol. 174: 7419–7427

Vos P, Simons G, Siezen RJ & De Vos WM (1989) Primary structure and organization of the gene for a prokaryotic cell envelope-located serine proteinase. J. Biol. Chem. 264: 13579–13585

Wells JM, Robinson K, Chamberlain LM, Schofield KM & Le Page RWF (1996) Lactic acid bacteria as vaccine delivery vehicles. Antonie van Leeuwenhoek 70: 317–330

Antonie van Leeuwenhoek **76:** 377–382, 1999.

Applications of phage resistance in lactic acid bacteria

Sylvain Moineau
Department of Biochemistry and Groupe de Recherche en Ecologie Buccale (GREB), Faculté de Médecine Dentaire, Université Laval, Québec, Canada, G1K 7P4; E-mail: Sylvain.Moineau@bcm.ulaval.ca.

Introduction

For the past twenty years, our understanding of phage-host interactions in *Lactococcus lactis* has led to the discovery of innate anti-phage systems. The advent of biotechnology allowed the characterization of these natural plasmid-encoded defense mechanisms and their transfer into industrial strains. Phage resistant starter cultures have been successfully used for a decade in industrial large-scale fermentations and many of the anti-phage systems have been patented. Over the same period, new phages have continued to emerge and the biodiversity of the phage population has never been so apparent. Modern strategies to prevent phage infections are now designed to minimize the appearance of phage mutants. Over the last five years, the dairy industry has also faced increasing phage problems with *Streptococcus thermophilus*. Although the diversity of thermophilic phages appears limited, the development of commercial phage-resistant strains may prove very challenging. Since the 5th LAB symposium, many commendable review articles have addressed microbiological and molecular aspects of bacteriophage and phage resistance in lactic acid bacteria (Daly et al. 1996; Schouler 1996; Djordjevic & Klaenhammer 1997; Allison & Klaenhammer 1998; Brüssow et al. 1998; Josephsen & Neve 1998). Applications of patented anti-phage strategies and their effectiveness in commercial strains were rarely covered and will be addressed.

Bacteriophage detection and identification

Lactic phages are naturally found in raw milk; they resist pasteurization, have a short latent period, and a relatively large burst size (Daly et al. 1996). As they can spread rapidly within a plant, constant monitoring is essential. Phages are always present in dairy environments and, therefore, our aim is to control rather than eradicate them. Phage control relies on a variety of practical approaches ranging from adapted factory design, improved sanitation, adequate ventilation, processing changes, starter medium, culture rotation, and the use of phage-resistant starters.

When the acidification process is retarded or arrested during milk fermentation, the first usual step is to examine for the presence of lytic phages. Standard microbiological methods are available to detect lactic phages (reviewed by Everson 1991). In recent years, some culture suppliers have extented their analysis to identify the phage species. Lactococcal phages are classified into twelve genetically distinct species (Jarvis et al. 1991). Three of them (936, P335 and c2) are the main causes of trouble since members of these species are regularly isolated worldwide. The three species share some important features, including a double-stranded DNA genome and a long non-contractile tail (*Siphoviridae* family, Figure 1). The 936 and P335 phages possess a small-isometric head, whereas the c2 phages have a prolate head. Monoclonal antibodies, DNA probes, or PCR methods are available to rapidly identify the three common species. These detection techniques should also provide ways to identify new emerging species. The results of the identification efforts are applied to adjust starter rotations and to develop phage-resistant strains (see below). Lastly, it is worth noting that antimicrobials or technical problems may also be responsible for the delays in acid production (reviewed by Champagne & Moineau 1998).

Natural anti-phage systems in Lactococcus

Developing a starter culture is a relatively long operation that proceeds via careful strain selection based on microbiological, physiological, and technical criteria. The quest for constant fermentations (within a production schedule) and uniform products favours the

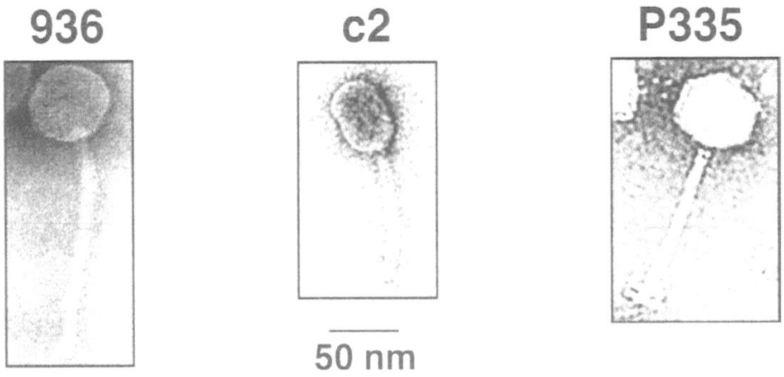

Figure 1. The three main lactococcal phage species.

Table 1. List of patented natural phage resistance plasmids or mechanisms in *Lactococcus lactis*.

Plasmids (systems)	Efficiency			Authors	Year*	Patent numbers
	936	c2	P335			
pTN1060	10^{-9}	10^{-3}	NA	Klaenhammer & Sanozky	1989	US4883756
pTR2030 (AbiA, *Lla*I)	10^{-9}	10^{-1}	10^{-9}	Klaenhammer et al.	1990	US4931396, US5139950
pCI528 (Ads)	10^{-9}	10^{-9}	NA	Daly et al.	1991	EP0246909, IE66528,
pCI750 (AbiG)	10^{-9}	10^{-2}	NA			US5019506
pCI829 (AbiA)	10^{-9}	10^{-2}	NA			
pFW094 (*Lla*AI)	10^{-5}	10^{-4}	NA	Josephsen et al.	1996	WO9625503
pJW563 (*Lla*BI)	10^{-3}	10^{-3}	NA			
pHW393 (*Lla*DII)	10^{-4}	10^{-2}	10^{-6}			
pIL415 (AbiB)	10^{-6}	NA	NA	Chopin & Cluzel	1997	US5629182
Chromosome (AbiH)	10^{-9}	10^{-3}	NA	Prevots et al.	1997	AU6421796, CA2183606, EP0764724, NZ299209
pPF144 (AbiO)	10^{-9}	10^{-2}	NA	Prevots et al.	1997	AU6888094, CA2129812, EP0643134, JP7170987, NZ264191, US5658770
Chromosome (AbiN)	10^{-5}	10^{-1}	NA	Prevots et al.	1998	AU6210696, CA2183196, EP0764723, FR2737896, JP9107979, NZ299182, US5712150.
pSRQ700 (*Lla*DCHI)	10^{-6}	10^{-4}	10^{-6}	Moineau et al.	1998	AU4497296, CA2208107, EP0805861, US5824523, WO9621017.
pSRQ800 (AbiK)	10^{-6}	10^{-1}	10^{-6}	Moineau et al.	1998	AU7705696, CA2237494, EP0868514, US5814499, WO9720917.
pSRQ900 (AbiQ)	10^{-9}	10^{-9}	1.0	Moineau et al.	1999	Pending

NA, not available; * year of publication.

repeated use of the best performing strains. History has proven that phages will eventually appear and attack these preferred strains, thereby preventing their long-term utilization. Various techniques are available to protect the sensitive strains by transforming them into phage-resistant derivatives. The development of bacteriophage-insensitive mutant strains (BIMs) was formerly the method of choice. BIM's are a spontaneous phage-resistant derivative that survives a long-term exposure to lytic phages, most likely as a result of a change in the phage receptor. Although this technique has the convenience of simplicity and speed, this natural evolutionary process is now rarely used for commercial *Lactococcus* strains due to substantial drawbacks including narrow phage-specificity, reversion of the resistance phenotype, and altered cell growth.

Decades of progress in the genetics of *Lactococcus* has established a solid basis for constructing new phage-resistant strains. Some *L. lactis* strains were found to naturally possess plasmids coding for anti-phage systems. More than 40 different lactococcal anti-phage plasmids have been reported in the literature (Daly et al. 1996). These mechanisms are divided into four groups, based on their mode of action: adsorption inhibition, blocking of DNA ejection, restriction/modification (R/M), and abortive infection (Abi) (Allison & Klaenhammer 1998; Josephsen & Neve 1998). The resistance is often broad and effective against most phages of the same species. Some systems are even potent, to a certain degree, against more than one phage species. Anti-phage barriers hold important commercial value and, therefore many systems were patented worldwide (reviewed in Table 1). This discovery had a significant impact, as resistance strategies became based on phage species rather than on phage isolates (BIMs).

Effectiveness of anti-phage systems

Unfortunately, not all systems can withstand phage attacks in the dairy battlefield. It became quickly obvious that the long term potential of a phage resistance mechanism had to be evaluated thoroughly. First, the efficacy of the system (EOP=phage titer on the resistant host contaning the anti-phage system divided by the phage titer on the sensitive host) must be determined against the three main phage species (936, c2, and P335). Still today, the EOP values for the three species are available for only a few systems (Table 1). A strong phage resistance mechanism will give EOPs in the range of 10^{-7} to 10^{-9}, with 10^{-9} being the maximum strength technically possible to measure. An EOP in the range of 10^{-4} to 10^{-6} is characterized as a medium potency system. Finally, a weak anti-phage barrier has an EOP of 10^{-1} to 10^{-3}. Obviously, strong systems (10^{-7} to 10^{-9}) are premium for starter culture companies. Medium systems can be effective, if supplemented in a good rotation strategy. Natural weak systems (10^{-1} to 10^{-3}) will not provide sufficient protection against phages. Another way of measuring the efficacy of an anti-phage system is via a milk-based assay known as the modified Heap-Lawrence method (Durmaz & Klaenhammer 1995). Essentially, the improved starter culture will go through (in the presence of phages) a time-temperature curve that mimics a fermentation process. If the anti-phage system is able to protect the strain against a high initial concentration of phages ($>10^6$ PFU/ml), it will likely be beneficial in the field. Interestingly, some anti-phage systems are less efficient at higher temperatures (>37 °C) (Allison & Klaenhammer 1998). Thus, the type of dairy fermentation for which the starter is intended will also influence the evaluation of the resistance mechanism.

Anti-phage systems do not necessarily need to confer resistance against all three species. Some *L. lactis* strains are only sensitive to phages that belong to one species. Other strains are sensitive to members of two species. For example, 936/c2 sensitive strains are often encountered, whereas 936/P335 and c2/P335 strains are relatively rare. Thus far in our experience, no *L. lactis* strains have been found sensitive to members of the three common phage species. Although the availability of such a strain would be welcomed in fundamental research (to study resistance mechanism in a single host background), a permanent removal of that strain from starter products would be the practical option.

After establishing the species sensitivity pattern of a strain using classified phages, an anti-phage mechanism is selected. For example, an anti-936-phage system would be introduced into a strain sensitive to phages of the 936 species. The plasmid-encoded anti-phage system can be introduced into a strain in many ways, but the two main delivery approaches are conjugation and transformation by electroporation. The former is widely accepted, whereas the use of the latter is legally limited in many countries. Multiple anti-phage systems can also be introduced within a strain to increase its overall resistance. It is also an effective way to find a use for the weaker systems. The compounded anti-phage effects may prevent the emergence of phage mutants, as they will need to overcome diverse biological hurdles (Sing & Klaenhammer 1993).

Transformation of LAB

Although the introduction of a phage resistance mechanism seems simplistic and straightforward, scientists with experience in using these technologies to improve commercial *Lactococcus* strains, have a somewhat different opinion. First and foremost, many industrial *L. lactis* strains are highly refractory to the acquisition of new DNA. Procedures to efficiently transform laboratory strains are useless for these strains. The challenge of transforming them has yet to be addressed. The next issue is the selection of the transformants. Some large anti-phage plasmids also encode additional features (lactose, bacteriocin, and proteinase) that enable their selection using multiple markers, but these phenotypes can be problematic in practical applications. For smaller plasmids, the selection is often solely based on phage resistance. Scientists will vary parameters to prevent the emergence of BIMs during the transformation procedure. Another reality rarely studied is plasmid compatibility. Introduction of new plasmids can substantially modify the overall plasmid content of a commercial strain. The modified strain may lose plasmids with key functional attributes like lactose-fermenting ability and proteinase activity. Stability of anti-phage plasmids is another critical issue for which data is lacking. Stable plasmids are more desirable, but often unstable plasmids find their way into direct vat set cultures. Insufficient attention has been given to the effect of adding multiple plasmids on cell growth and acid production. Larger plasmids are likely to increase the metabolic load and elicit a detrimental effect on the phenotypes that are commercially relevant. Thus, the development of a phage-resistant strain remains even today, a daunting task. Fortunately, success stories are found in many University laboratories and culture companies.

Rotation: an old tradition with modern twists

Cheesemakers have been rotating starter cultures for years to avoid phage attacks. The advantage of the rotation is based on the availability of a sufficient number of phage-unrelated strains. The difficulty in identifying genuine phage-unrelated strains, coupled with non-uniform fermentation due to strain variations, are among the problems associated with rotation systems. The use of many strains also favours the presence of a heterologous phage population. Klaenhammer's group patented a novel approach that addresses these concerns (Table 2). The strategy is based on the rotation of isogenic *Lactococcus* strains, generated by the introduction of different plasmid-encoded anti-phage systems into the same parent strain. Presenting diverse, anti-phage barriers within isogenic strains should, in theory, limit the emergence of new phages. The use of isogenic strains should also limit the diversity of the phage population within a plant, while ensuring a constant quality of the product. A sufficient number of strong anti-phage systems are needed to apply such an attractive strategy. As another approach to control phage, it has been proposed to rotate strains based on their phage-species sensitivity rather than on phage isolates (Moineau et al. 1996). Rotation in the production of cheese varieties is another means of limiting phage problems (Moineau 1997). A cheese diversification scheme supports the use of various starters and, consequently, the dilution of specific phages within a plant.

Genetically modified anti-phage systems

Antibacterial or antiviral systems are always offset by the emergence of resistant strains. Natural phage-resistant systems are not immune to this evolutionary process. After considerable use of the constructed strains, modified or new phages, capable of overcoming introduced anti-phage systems have been isolated (reviewed by Allison & Klaenhammer 1998; Josephsen & Neve 1998). Interestingly, some of these insensitive mutants were not always fit enough to delay milk fermentation. Researchers then turned their attention to the phage genome as a potential source of novel resistance traits (Table 2). These new anti-phage technologies do rely on recombinant DNA strategies and appear to have a much narrower spectrum than the natural systems. Given the current social and regulatory climate, it is very unlikely that they will find immediate industrial applications. As new phage resistance plasmids are still isolated worldwide, one could argue that we have yet to dry up the source of natural anti-phage systems. Nonetheless, phage-based strategies have the benefit of providing significant insights into the fundamentals of the phage lytic cycle. These studies should still be encouraged, as they are likely to provide innovative tactics for phage control. Another potential of genetic engineering is to increase the effectiveness of weak anti-phage systems.

Table 2. List of patented phage resistance strategies.

Strategies	Authors	Year	Patent numbers
L. lactis			
PER (phage origin of replication)	Hill & Klaenhammer	1996	EP0474464, US5538864
Phage defense rotation strategy	Klaenhammer et al.	1997	AU8356291, AU664537, CA2050533, EP0474463, US5593885
Phage-triggered cell suicide	Klaenhammer et al.	1998	US5792625
S. thermophilus			
Pediococcus Lac+	Broadbent et al.	1997	US5677166
Phage DNA fragment	Mollet et al.	1998	EP0748871, JP9000274, US5766904
L. lactis R/M system	Moineau et al.	1998	AU4497296, CA2208107, EP0805861, US5824523, WO9621017

The case of Streptococcus thermophilus

Due to the extensive use of thermophilic cultures, dairy plants are now facing phage problems with *S. thermophilus*. These phages can be detected using similar standard microbiological methods (Everson 1991), but plaques tend to be smaller in size. A number of studies (reviewed by Brüssow et al. 1998) described the general characteristics of *S. thermophilus* phages. All investigated *S. thermophilus* phages have an isometric head, a long non-contractile tail and they share DNA homology, albeit to different degrees. This shared DNA homology led Mercenier et al. (1994) to hypothesize that all *S. thermophilus* phages were derived from a common ancestor. Le Marrec et al. (1997) proposed two groups of *S. thermophilus* phages based on the number of major structural proteins (MSP) and their mode of DNA packaging. The first group comprises phages with two MSP and cohesive genomic extremities (cos-type). The second group includes phages with three MSP and a DNA packaging scheme which proceeds via the headful mechanism (pac-type). The biological significance of this classification is unknown, but will most likely be considered in the development of phage-resistant strains and rotation programs. Understanding the mechanisms creating this biodiversity is, from a fundamental perspective, of considerable interest.

Since many *S. thermophilus* phages have a narrow host range, the traditional starter rotation system and BIMs (when possible) are the methods currently used by the industry. Very few natural phage resistance mechanisms have been published for *S. thermophilus*. The relative absence of plasmids is the likely explanation for the scarcity of known anti-phage barriers. Five chromosomally encoded R/M systems have been identified in *S. thermophilus* (Moineau 1997). None of the genes have been cloned or sequenced, let alone used for improvement of commercial strains. The rarity of plasmids and the chromosomal location of the anti-phage hurdles will make it more difficult to develop food-grade strategies for phage control without relying on recombinant DNA techniques. A possible alternative is to introduce anti-phage mechanisms from *Lactococcus* into *S. thermophilus*. The expression of a *L. lactis* R/M system in *S. thermophilus* was demonstrated recently (Moineau et al. 1997; Table 1). This genetic modification conferred strong resistance against the two major phage groups attacking *S. thermophilus* (EOPs from 10^{-5} to 10^{-8}). Another more practical method to control bacteriophage may well be to replace *S. thermophilus* strains in starter blends with other lactic acid bacteria. *Pediococcus* strains, genetically modified to ferment lactose, were proposed as replacement cocci for *S. thermophilus* (Broadbent et al. 1997). Research groups are now looking into this novel battle, fueled by the success with *Lactococcus*.

Concluding remarks

The field of bacteriophage resistance is arguably the LAB research area that had the most significant impact on the dairy sector. The investments have been justified, fundamental research promises have been kept and the application milestones have been attained with undeniable success. Over the years, the phage diversity has allowed researchers to maintain a range of strong basic and applied studies. This phage-host battle will certainly carry through to the next millenium. Hopefully, the ultimate winner will continue to be the dairy industry.

Acknowledgments

I would like to acknowledge Frédéric Bissonnette, Julie Bouchard, Isabelle Boucher, Eric Dion, Eric Emond, Steve Labrie, Robert St-Laurent, Denise Tremblay and Nathalie Turgeon for their critical review of the manuscript. I am also very grateful to the following organizations for funding our research program on bacteriophages of lactic acid bacteria: Natural Sciences and Engineering Research Council of Canada (NSERC), Fonds pour la Formation de Chercheurs et Aide à la Recherche (FCAR), Conseil des Recherches en Pêche et Agroalimentaire du Québec (CORPAQ), Novalait and Quest International.

References

Allison GE & Klaenhammer TR (1998) Phage resistance mechanisms in Lactic Acid Bacteria. Int. Dairy J. 8: 207–226

Broadbent JR, Oberg CJ & Caldwell S (1997) Compositions and methods for phage resistance in dairy fermentations. Patent US5677166

Brüssow H, Bruttin A, Desière F, Lucchini S & Foley S (1998) Molecular ecology and evolution of *Streptococcus thermophilus* bacteriophages – a review. Virus genes. 16: 95–109

Champagne CP & Moineau S (1998) Les Bactériophages. In: Champagne CP (Ed) Production de ferments lactiques dans l'industrie laitière. (pp. 89–115). Edisem & Fondation des Gouverneurs, St-Hyacinthe, PQ, Canada

Chopin MC & Cluzel PJ (1997) DNA fragments coding for a bacteriophage-resistant mechanism. Patent US5629182

Daly C, Fitzgerald GF, Coffey A, Costello VA, Murphy MC & Baumgartner A (1991) Plasmid conferring phage insensitivity on bacteria. Patents EP0246909, IE66528, US5019506

Daly C, Fitzgerald GF & Davis R (1996) Biotechnology of lactic acid bacteria with special reference to bacteriophage resistance. Antonie van Leeuwenhoek. 70: 99–110

Djordjevic GM & Klaehammer TR (1997) Genes and gene expression in *Lactococcus lactis*. Int. Dairy J. 7: 489–508

Durmaz E & Klaenhammer TR (1995) A starter culture rotation strategy incorporating paired restriction/modification and abortive infection bacteriophage defenses in a single *Lactococcus lactis* strain. Appl. Environ. Microbiol. 61: 1266–1273

Everson TC (1991) Control of phage in dairy plant. Bull. IDF. 263: 24–28

Hill CJ & Klaenhammer TR (1996) Bacteriophage resistant recombinant bacteria. Patents EP0474464, US5538864

Jarvis AW, Fitzgerald GF, Mata M, Mercenier A, Neve H, Powell IA, Ronda C, Saxelin M & Teuber M (1991) Species and type phages of lactococcal bacteriophages. Intervirology. 32: 2–9

Josephsen J & Neve H (1998) Bacteriophages and lactic acid bacteria. In: Salminen S & von Wright S & A (Eds) Lactic Acid Bacteria, Microbiology and Functional Aspects. (pp. 385–436). Marcel Dekker Inc., New York, US

Josephsen J, Nyengaard NR, Vogensen FK & Madsen A (1996) Plasmid-derived type II restriction-modification systems from *Lactococcus lactis*. Patent WO9625503

Klaenhammer TR, Conkling MA, O'Sullivan D, Djordjevic G, Walker SA & Taylor CG (1998) Bacteriophage-triggered cell suicide systems and fermentation methods. Patent US5792625

Klaenhammer TR & Sanozky-Dawes RB (1989) pTN1060, a conjugal plasmid and derivatives thereof that confer phage resistance to group N streptococci. Patent US4883756

Klaenhammer TR, Sanozky RB & Steenson LR (1990) pTR2030, a conjugal plasmid and derivatives thereof that confer phage resistance to group N streptococci. Patents US4931396, US5139950

Klaenhammer TR, Sing WD & Hill CJ (1997) Phage defense rotation strategy. Patents AU8356291, AU664537, CA2050533, EP0474463, US5593885

Le Marrec C, van Sinderen D, Walsh L, Stanley E, Vlegels E, Moineau S, Heinze P, Fitzgerald GF & Fayard B (1997) Two groups of bacteriophages infecting *Streptococcus thermophilus* can be distinguished on the basis of mode of packaging and genetic determinants for major structural proteins. Appl. Environ. Microbiol. 63: 3246–3253

Mercenier A, Pouwels PH & Chassy BM (1994) Genetic engineering of lactobacilli, leuconostocs, and *Streptococcus thermophilus*. In: Gasson MJ & DeVos WM (Eds) Genetics and Biotechnology of Lactic Acid Bacteria. (pp. 253-293). Blackie Academic & Professional, Glasgow, UK

Moineau S (1997) Bacteriophages and phage resistance in *Streptococcus thermophilus*: An update. Proc. 34th Marschall Italian & Specialty Cheeses. (pp. 9–17). Madison, WI

Moineau S, Borkaev M, Holler BJ, Walker SA, Kondo JK, Vedamuthu ER & Vandenbergh PA (1996) Isolation and characterization of lactococcal bacteriophages from cultured buttermilk plants in the United States. J. Dairy Sci. 79: 2104–2111

Moineau S, Holler BJ, Vandenbergh PA, Vedamuthu ER & Kondo JK (1998) DNA encoding phage abortive infection protein from *Lactococcus lactis* and method of use thereof. Patents AU7705696, CA2237494, EP0868514, US5814499, WO9720917

Moineau S, Walker SA, Vedamuthu ER & Vandenbergh PA (1998) Isolated DNA encoding enzyme for phage resistance. Patents AU4497296, CA2208107, EP0805861, US5824523, WO9621017

Mollet B, Pridmore D & Zwahlen MC (1998) Phage-resistant *Streptococcus*. Patents EP0748871, JP9000274, US5766904

Prevots F, Remy E & Ritzenthaler P (1997) Nucleic acid sequence and plasmids comprising at least one phage resistance mechanism, bacteria in which they are present, and their use. Patents AU6888094, CA2129812, EP0643134, JP7170987, NZ264191, US5658770

Prevots F, Tolou S & Daloyau M (1997) Nucleic acid treatment and plasmids comprising at least a phage resistance mechanism, bacteria containing them and their utilization. Patents AU6421796, CA2183606, EP0764724, NZ299209

Prevots F, Tolou S & Daloyau M (1998) Nucleic acid sequence and plasmids comprising at least one phage resistance mechanism, bacteria containing them and their use. Patents AU6210696, CA2183196, EP0764723, FR2737896, JP9107979, NZ299182, US5712150

Schouler C (1996) Genomic organization of lactic acid bacteria phages. Lait. 76: 81–89

Sing WD & Klaenhammer TR (1993) A strategy for rotation of different bacteriophage defenses in a lactococcal single-strain starter culture system. Appl. Environ. Microbiol. 59: 365–372

Antonie van Leeuwenhoek **76:** 383–389, 1999.

Immunomodulatory function of lactic acid bacteria

Hisako Yasui, Kan Shida, Takeshi Matsuzaki & Teruo Yokokura
Yakult Central Institute for Microbiological Research, 1796 Yaho, Kunitachi-shi, Tokyo, 186-8650 Japan

Key words: Bifidobacterium breve, Lactobacillus casei, immunomodulation, IgG, IgA, IgE, prevention against diseases, rotavirus infection, influenza virus infection, tumor

Abstract

Using mice, we found that oral administration of *Bifidobacterium breve* YIT4064 (*B. breve*) activated the humoral immune system, augmented anti-rotavirus IgA production or anti-influenza virus (IFV) IgG production and protected against rotavirus infection or influenza infection, respectively. Furthermore, when the *B. breve* was given to infants from an infant home, there was a significant reduction of the frequency of rotavirus shedding in stool samples during the administration of the bacteria. It was also found, again using mice, that oral administration of *Lactobacillus casei* strain Shirota (LcS) stimulated type 1 helper T (Th1) cells, activated the cellular immune system and inhibited incidence of tumors and IgE production. These results demonstrated that these two strains of lactic acid bacteria modulated the different immune systems each in its own way and prevented against various diseases.

Introduction

At the beginning of the present century, the Russian scientist Metchnikoff proposed that the consumption of fermented milk was useful in maintaining good health (Metchnikoff 1908). Since that time, the works of numerous researchers have substantiated the nutritional advantages of fermented milk as a source of calcium, as well as the effects of its various ingredients and the probiotic action of lactic acid bacteria to improve digestion and absorption, reduce diarrhea and constipation, and normalize intestinal flora (Gilliland 1990; Goldin 1998). Fermented milk has also been reported to provide other beneficial effects such as the lowering of serum cholesterol levels (Agerbaek et al. 1995). Recently, it has been reported that fermented milk and the lactic acid bacteria in it act on the immune mechanism *in vivo* and increase the body's resistance to infection, cancer and allergy (Trapp et al. 1993; Salminen et al. 1998).

When pathogenic bacteria and viruses invade the body from the exterior, the body has the capability to recognize these as foreign substances and to act to remove them. This capability is generally termed the "immune system", which includes the humoral and the cellular immune systems. Lactic acid bacteria and its components can act on the immune system within the body, and in many cases enhance the immune response.

We found two strains of lactic acid bacteria which modulated the different immune systems each in its own way. One strain (*B. breve*) enhanced the humoral immunity and protected against virus infection. The other strain (LcS) enhanced the cellular immunity and inhibited the incidence of tumors and the production of IgE that mediates allergic reaction. Centering primarily on the results of these studies, we would like to discuss the activity of lactic acid bacteria as a immunological response-modifier.

Function of *Bifidobacterium breve* YIT4064 (*B. breve*)

Bifidobacterium is a primary element of intestinal flora, is a strict anaerobe gram-positive bacterium and produces mainly lactic acid and acetic acid (Mitsuoka & Kaneuchi 1977). *Bifidobacterium* synthesizes B-group vitamins and enhances digestion and absorption and prevents against pathogenic microorganism infection (Poupard et al. 1973; Yamazaki et al. 1982). *Bifidobacterium* was also reported to show immunopotentiating effects. In this section, we report about augmentation of the humoral immune response and protective activity against virus-infection of one strain of *Bifidobacterium* (*B. breve*).

Augmentation of local immune response

Augmentative effect on IgA production

It is known that Peyer's patches (PP), assembled lymphoid tissues, are in the small intestine and a lot of solitary lymphoid tissues are in the large intestine. M cells of these lymphoid tissues in both intestines process antigens in the gut and initial cellular events of the immune response occur in these lymphoid tissues.

When a loop of rabbit small intestine was inoculated with heat killed *B. breve* and observed under an electron microscope after 5 h, it was confirmed that the administered bacteria had been taken up by M cells in PP and transferred into PP (Takahashi et al. 1991). Using the PP cell culture method, we selected the *B. breve* strains capable of inducing large quantities of IgA from 120 strains of bifidobacteria isolated from human feces (Yasui et al. 1992). This strain also elevated the production of interleukin(IL)-4 and IL-5, and increased the number of IgA^+ cells and IgA-producing cells. Based on these findings, it was assumed that this strain activated type 2 helper T (Th2) cells in PP, promoted to differentiate toward IgA-producing cells due to cytokines such as IL-4 and IL-5 and thus enhanced IgA production (Yasui & Ohwaki 1991). Also, when this strain was added to PP cells together with antigens such as rotavirus, polio virus, or influenza virus, antigen-specific IgA were produced in greater quantity than ones from the antigens alone (Yasui et al. 1994; Yasui et al., unpublished data). This suggests that this strain has an adjuvant (immune response-enhancing) effect, and clearly augments antigen-specific IgA production.

Protective effect against rotavirus infection

Rotavirus is a major viral cause of diarrhea in nursing infants during the winter season, and is responsible for one million deaths a year in developing countries. Secretory IgA is known to play an important role in the prevention of this form of diarrhea. We used a mouse model to clearly indicate whether oral administration of *B. breve* prevents against rotavirus infection (Yasui et al. 1995). One group was given a diet containing the strain (the *B. breve* group), and the other group was given a diet with no such additives (the control group). After female mice in both groups were mated, they were immunized orally with rotavirus, and after delivery the antibody titer in the dam's milk was determined. All of the pups in both groups were orally inoculated with rotavirus and were observed for diarrhea. Anti-rotavirus IgA level in the milk in the *B. breve* group was significantly higher than that in the control group (Figure 1) and incidence of diarrhea in pups in the *B. breve* group was significantly lower than that in the control group (Figure 2). These find-

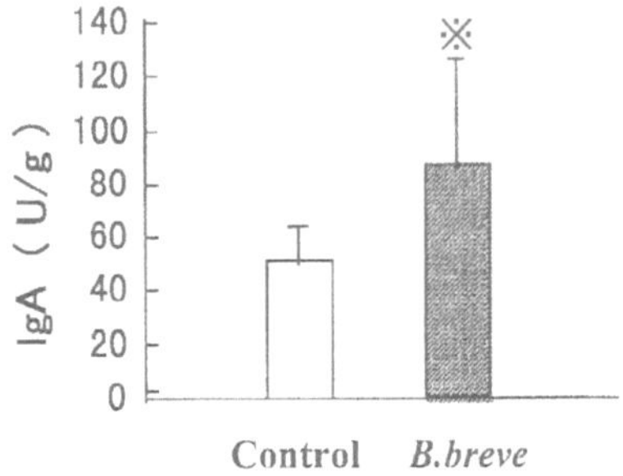

Figure 1. Antibody titers in milk from dams immunized orally with rotavirus and fed *B. breve*. Dams were immunized orally with a single 10^6 PFU dose of rotavirus (SA-11) 9–12 days before delivery. Dams were fed a diet containing (wt/wt) 0.05% *B. breve* for 12 weeks before and 14 days after delivery. Anti-rotavirus IgA titers in milk received by pups were measured by ELISA 14 days after birth. *P<0.02 vs. the control group. Student's *t* test.

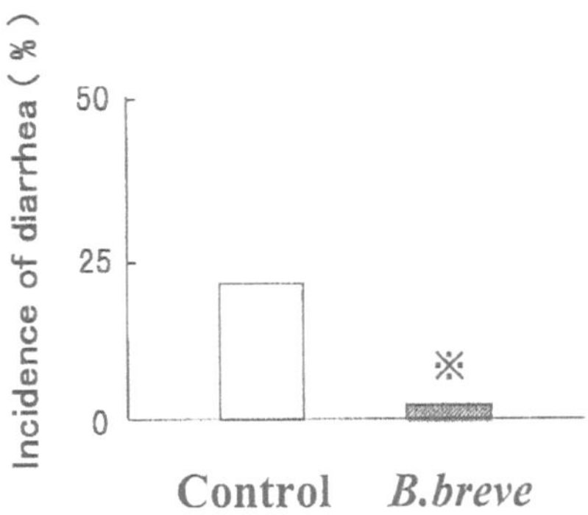

Figure 2. Passive protection against rotavirus-induced diarrhea in pups born to and nursed by mouse dams orally administered *B. breve*. Dams were immunized orally with a single 10^6 PFU dose of rotavirus (SA-11) 9–12 days before delivery. Dams were fed a diet containing (wt/wt) 0.05% *B. breve* strain for 12 weeks before and 14 days after delivery. 5 days after birth, pups in each litter were orally inoculated with 2×10^6 PFU rotavirus and observed for diarrhea 1–4 days after inoculation. *P<0.0081 vs. the control group. Fisher's exact test.

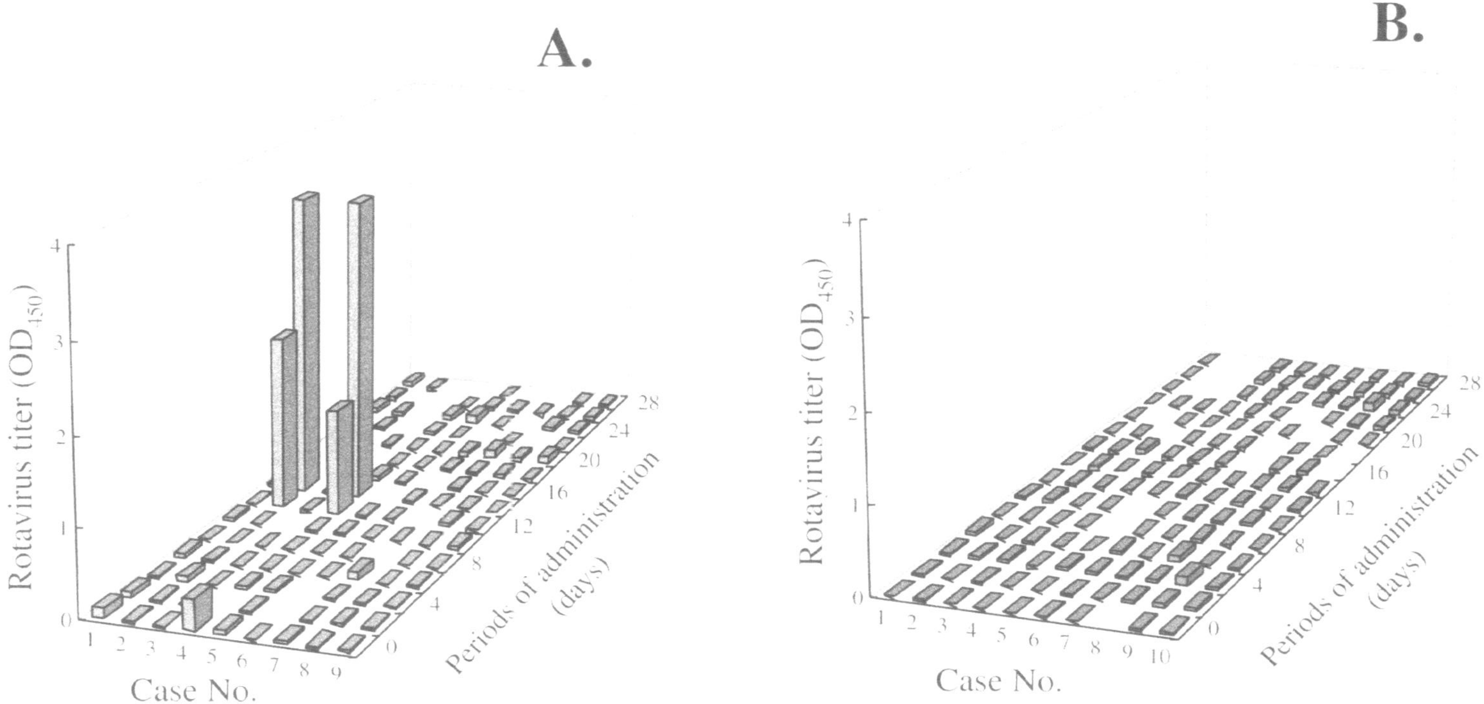

Figure 3. Prevention of rotavirus shedding in stool samples by oral administration of *B. breve*. Nine infants from an infant home were not administered *B. breve* (A; the control group) and ten infants were administered 50 mg of *B. breve* every day for 28 days (B; the *B. breve* group). Stool samples were collected every two days during test periods of administration and rotavirus titer was measured.

Table 1. Number of mice with serum antibody enhanced by oral administration of *B. breve*[1]

Group	No. of mice with >400U of anti-IFV/tested	% with >400U of anti-IFV IgG
Control	0/10	0
B. breve	5/10*	50

[1]Mice were immunized orally with two doses of $10^{4.9}$ EID_{50} of IFV(PR8) at 6 and 2 weeks before nasal inoculation ($10^{3.5}EID_{50}$ IFV(PR8)). Mice were not fed (the control group) or fed (the *B. breve* group) a diet containing (wt/wt) 0.05% *B. breve* for 17 weeks throughout these experiments. The anti-IFV IgG titer in serum was measured just before nasal inoculation ($10^{3.5}EID_{50}$ IFV) and the number of mice with anti-IFV IgG levels of >400U was determined.
*P<0.02 vs. the number of antibody-enhanced mice in the control group.
χ^2test.

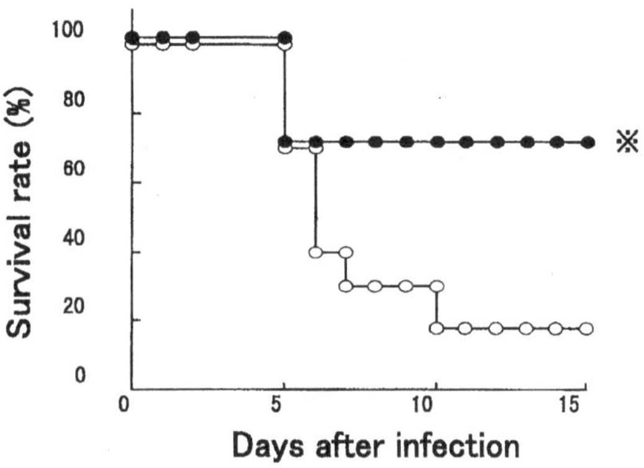

Figure 4. Protection against mortality due to influenza infection in mice administered *B. breve* orally. Mice were immunized with two doses of $10^{4.9}EID_{50}$ of IFV(PR8) at 6 and 2 weeks before nasal inoculation ($10^{3.5}$ EID_{50} IFV(PR8)). Mice were not fed (○; the control group) or fed (●; the *B. breve* group) a diet containing (wt/wt) 0.05% *B. breve* for 17 weeks throughout these experiments. After nasal inoculation of IFV, the mice were observed for survival for 15 days. *P<0.05 vs. the control group. Fisher's exact test.

ings demonstrate that oral administration of this *B. breve* augments antigen-specific IgA production in the mammary gland, and that pups nursed by the milk are protected against rotavirus-induced diarrhea. It was assumed that oral administration of *B. breve* stimulated IgA precursor cells in gut-associated lymhoid tissues (GALT), especially PP, and the cells were disseminated to mammary glands and the other mucosal tissues with maturation toward IgA plasma cells.

Furthermore, the effect of the same strain (heat killed *B. breve*) was evaluated in ten infants from an infant home who received 50 mg of the bacterium every day for 28 days (the *B. breve* group). Nine infants did not receive this (the control group). Stool samples were collected every two days and rotavirus titers of them were measured. Though rotavirus shedding in the control group was detected from 2 (a total of 5 stool samples) of 9 infant (a total of 112 stool samples), it was not detected in any infants (a total of 133 stool samples) in the *B. breve* group during the administration period. The frequency of rotavirus shedding in the *B. breve* group was significantly lower than that in the control group (Figure 3) and anti-rotavirus IgA levels of stool samples in the *B. breve* group showed a tendency to be higher than those in control group (Araki et al. 1999).

Augmentation of systemic immune response

The augmentative effect on IgG production and the protective effect against influenza infection

Influenza is an acute viral respiratory infectious disease that results in high morbidity and significant mortality in humans. The degree of protection against influenza infection was found to be correlated with the level of IgG in serum or mucosal IgA in the respiratory tract. We investigated whether oral administration of *B. breve* augmented anti-influenza IgG in serum and the antibody-enhanced mice were protected against influenza infection in the lower respiratory tract and saved from death. One group was given a diet containing the strain (the *B. breve* group) and another group was given a diet with no such additives (the control group). Both groups were orally immunized with influenza virus (IFV) and just before nasal inoculation with IFV, anti-IFV IgG titer in the serum was determined. After inoculation, mice were observed for mortality. The number of mice with enhanced anti-IFV IgG in the *B. breve* group was significantly greater than that in the control group (Table 1). Further, the survival rate against influenza infection in the *B. breve* group was significantly higher than that in the control group (Figure 4). These findings demonstrate that oral administration of *B. breve* augments anti-IFV IgG production in serum and protects against influenza infection (Yasui et al. 1999).

The function of *Lactobacillus casei* strain Shirota (LcS)

LcS is a nonpathogenic gram-positive bacterium that is isolated from human feces. It was cloned in 1930 and has been used for the production of fermented milk since that time. In this section, we report that LcS activates the cellular immune response and inhibits the incidence of tumors and IgE production.

Anti-tumor effect

It was reported that the growths of both chemical-induced and secondary implantable tumor cells were inhibited in rodents fed LcS. 3-methylcholanthrene(MC) was injected intradermally into all mice and mice were observed for incidence of sarcoma at the site of the injection for 150 days after the injection. The incidence of tumors (59.0%) in the mice fed a diet containing LcS (the LcS group) was significantly lower than that (91.8%) of the control mice fed a diet with no such additives (the control group). Splenocytes from the mice at 120 days after MC treatment were cultured *in vitro* in the presence of concanavalin A. The result showed that proliferative response and IL-2 production of T cells from the LcS group were higher than those from the control group (Yokokura 1994). The effect of orally administered LcS on the growth of secondary tumors was investigated after surgical removal of primary tumors. As a result of this experiment, tumors grew gradually in the control group because tumor immunity was not sufficiently established, while tumor growth in the LcS group was significantly inhibited compared with that in the control (Kato et al. 1994). Further, It was demonstrated that oral administration of LcS into surgical operated patients suppressed the recurrence of superficial bladder cancer in a double-blind clinical trial (Aso et al. 1995). These results suggest that LcS decreases the incidence of tumors through activating the cellular immune response in rodents and humans.

Inhibitory effect on IgE production

Because allergen-specific IgE is directly involved in mediating many allergic reactions, an effective substance for inhibiting IgE production may be useful for preventing allergic diseases. Cytokines derived from T helper (Th) cells are major modulators of IgE synthesis. IL-4 induces IgE isotype class switching and plays a crucial role in IgE production. IL-5 enhances IL-4-dependent IgE production. All of these cytokines are produced by Th2 cells. On the other hand, interferon (IFN)-γ produced by Th1 cells suppresses IgE production both by interfering with the IL-4-derived isotype switching and by inhibiting the proliferation of Th2 cells.

Therefore, the activable effect of heat-killed LcS on Th1 cells and the inhibitory effect of it on IgE

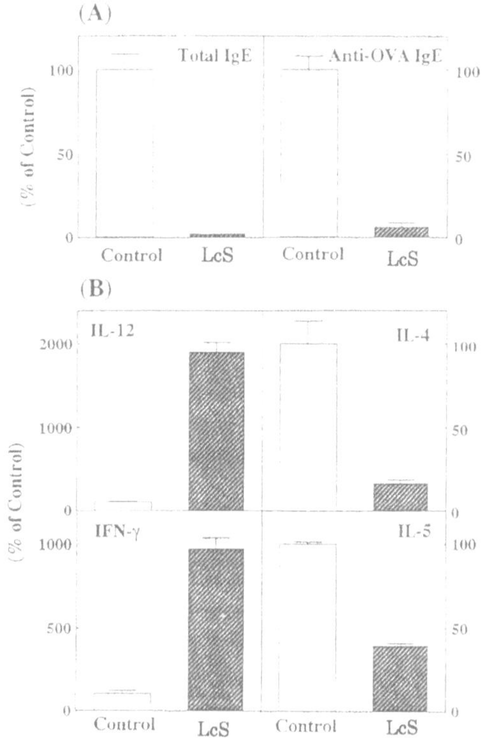

Figure 5. Effect of LcS on OVA-induced IgE, IL-12, IFN-γ, IL-4 and IL-5 production by splenocytes. Splenocytes from OVA-primed mice were cultured with OVA in the absence (control) or presence of LcS (1 μm/ml) (LcS). The levels of total IgE and anti-OVA IgE (A) on day 14 and cytokines (B) on day 7 in supernatants were determined by ELISA. Results are means ±SD of three independent cultures expressed as relative values to control experiments.

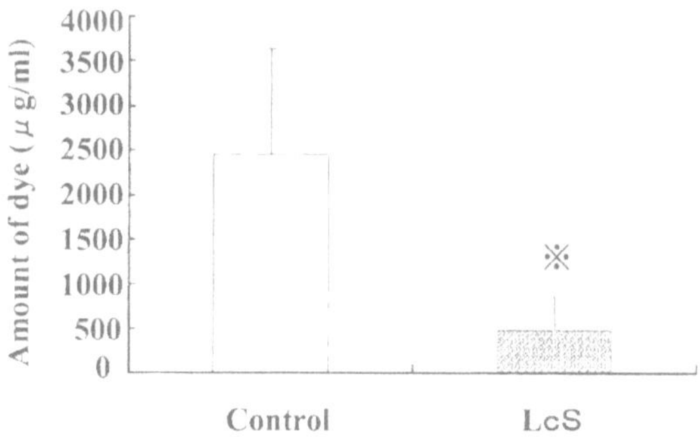

Figure 6. Effect of orally fed LcS on OVA-specific IgE production. Mice were intraperitoneally injected on day 0 and 14 with 20 μg of OVA and 2 mg of $Al(OH)_3$ at a total volume of 0.2 ml. The mice were then fed a diet containing (wt/wt) 0.05% LcS for 21 days. Control mice were injected with OVA only. Sera were collected on day 21 and the level of OVA-specific IgE was determined by passive cutaneous anaphylaxis reaction. The bar represents mean values (±SD) for 5 mice. *P<0.01 vs. the control group. Dunnett's multiple comparison test.

production were examined. Splenocytes obtained from ovalbumin (OVA)-primed mice were restimulated *in vitro* with the same antigen in the presence of LcS. LcS induced IL-12 and IFN-γ production, but inhibited IL-4 and IL-5 secretion, and markedly suppressed anti-OVA IgE secretion (Figure 5) (Shida et al. 1998). Further, LcS was orally administered to mice that had been preinjected intraperitoneally with OVA. The oral feeding of LcS was effective in inhibiting IgE production in serum (Figure 6), and the *in vitro* production of IgE by the splenocytes from mice fed LcS (the LcS group) in response to restimulation with OVA was inhibited in contrast to that of splenocytos from mice fed a diet with no such additives (the control group). In the LcS group, the production by the splenocytes of Th1 cell-associated cytokines, such as IFN-γ and IL-2, was higher than that of the splenocytes from the control group. In contrast, the production of Th2 cell-associated cytokines, such as IL-4, IL-5, IL-6 and IL-10, by splenocytes in the LcS group was lower than that by the cells from the control group. Furthermore, the IL-12 production of the splenocytes from the LcS group was also higher than that of the control group (Matsuzaki et al. 1998). These results suggest that LcS activates macrophage, induces Th1 cell-associated cytokines and inhibits IgE production in *in vitro* and *in vivo* experiments.

Concluding remarks

We have described that one strain of *Bifidobacterium* (*B. breve*) activates the humoral immune system and can protect against rotavirus infection and influenza infection and that one strain of *Lactobacillus* (LcS) stimulates the cellular immune system and inhibits incidence of tumors and production of IgE. Though *B. breve* enhanced IL-4 and IL-5 productions by P.P. cells, it did not induce IgE production. IL-4 and IL-5 play a crucial role in IgA production, too. *B. breve* may promote to differentiate specifically toward IgA-producing cells and not toward IgE-producing cells. We are studying the mechanism now.

Recently, it has been reported that oral administrations of some strains of *Bifidobacterium* modulate the immune system or protect against rotavirus infection. Fukushima et al. (1998) has reported that oral administration of live *Bifidobacterium lactis* enhances total IgA and anti-poliovirus IgA levels in stools of healthy children. Duffy et al. (1994) showed that oral administration of live *Bifidobacteroum bifidum* into mouse pups inhibited adhesion of rotavirus onto intestinal epitherial cells and protected against rotavirus infection. Further, Saavedra et al. (1994) reported that feeding live *Bifidobacterium bifidum* and *Streptococcus thermophilus* to infants in hospitals prevented diarrhea and reduced rotavirus shedding in stools. They hypothesized that the acids produced by these bacteria prevented against rotavirus infection. Using heat killed *B. breve*, we found that protection against rotavirus infection was due to anti-rotavirus IgA induced by the bacterium.

It has been already reported that oral administrations of some strains of *Lactobacillus* modulate immune systems and protect against various diseases. Perdigon et al. (1998) reported that yogurt prepared using *Lactobacillus delbrueckii* subsp. *bulgaricus* and *Streptococcus salivarius* subsp. *thermophilus* inhibited tumor growth through its immunomodulatory activity. It was reported that oral administration of *Lactobacillus* sp. enhanced the antibody production to IFV vaccine applied subcutaneously (Pouwels et al. 1996) and oral administration of *Lactobacillus casei* subsp. enhanced anti-salmonella IgA level and inhibited the growth of Salmonella (Perdigon et al. 1995). It was also reported that oral administration of *Lactobacillus* GG decreased atopic dermatitis (Majamaa et al. 1995).

Results presented in both our data and the data of other authors have shown that some strains of *Bifidobacterium* and *Lactobacillus* can be probiotics, having immunomodulatory activities and being able to protect against various diseases. Further, as *Bifidobacterium breve* YIT4064 has a strong adjuvant activity, it may be used as a live antigens-delivery-vehicle for oral vaccines to various pathogens.

Acknowledgments

We are grateful to Miss Junko Kiyoshima for assistance with the preparation of this manuscript.

References

Agerbaek M, Gerdes LU & Richelsen B (1995) Hypocholesterolaemic effect of a new fermented milk product in healthy middle-aged men. Eur. J. Clin. Nutr. 49: 346–352

Araki T, Shinozaki T, Irie Y & Miyazawa Y (1999) Trial of oral administration of *Bifidobacterium breve* for the prevention of rotavirus infection. J. Jpn. Assoc. Infect. Dis. 73; 305–310

Aso Y, Akaza H, Kotake T, Tsukamoto T, Imai K, Naito S & the BLP Study Group (1995) Preventive effect of *Lactobacillus casei* preparation on the recurrence of superficial bladder cancer in a double-blind trial. Eur. Urol. 27: 104–109

Duffy LD, Zielezny MA, Riepenhoff-Talty M, Dryja D, Sayahtaheri-Altaie S, Griffiths E, Ruffin D, Barrett H, Rossman J & Ogra PL (1994) Effectiveness of *Bifidobacterium bifidum* in mediating the clinical course of murine rotavirus diarrhea. Pediatr. Res. 35: 690–695

Fukushima Y, Kawata Y, Hara H, Terada A & Mitsuoka T (1998) Effect of a probiotic formula on intestinal immunoglobulin A production in healthy children. Int. J. Food Microbiol. 42: 39–44

Gilliland SE (1990) Health and nutritional benefits from lactic acid bacteria. FEMS Microbiol. Rev. 87: 175–188

Goldin BR (1998) Health benefits of probiotics. Br. J. Nutr. 80: S203–S207

Kato I, Endo K & Yokokura T (1994) Effects of oral administration of *Lactobacillus casei* on anti-tumour responses induced by tumour resection in mice. Int. . Immunopharmacol. 16: 29–36

Majamaa H, Isolauri E, Saxelin M & Vesikari T (1995) Lactic acid bacteria in the treatment of acute rotavirus gastroenteritis. Pediatr. Gastroenterol. Nutr. 20: 333–338

Matsuzaki T, Yamazaki R, Hashimoto S, & Yokokura T (1998) The effect of oral feeding of *Lactobacillus casei* strain Shirota on immunoglobulin E production in mice. J. Dairy Sci. 81: 48–53

Metchnikoff E (1908) The prolongation of life. G.P. Putnam, 4

Mitsuoka T & Kaneuchi C (1977) Ecology of bifidobacteria. Am. J. Clin. Nutr. 30: 1799–1810

Perdigon G, Alvarez S, Rachid M, Aguero G & Gobbato N (1995) Symposium: Probiotic bacteria for humans: clinical systems for evaluation of effectiveness. J. Dairy Sci. 78: 1597–1606

Perdigon G, Valdez JC & Rachid M (1998) Antitumor activity of yogurt: study of possible immune mechanisms. J. Dairy Res. 65: 129–138

Poupard JA, Husain I & Norrits RF (1973) Biology and bifidobacteria. Bacteriol. Rev. 37: 136–165

Pouwels PH, Leer RJ & Boersma JA (1996) The potential of *Lactobacillus* as a carrier for oral immunization: development and preliminary characterization of vector systems for targeted delivery of antigens. J. Biotechnol. 44: 183–192

Saavedra M, Bauman NA, Oung I, Perman A & Yolken RH (1994) Feeding of *Bifidobacterium bifidum* and *Streptococcus thermophilus* to infants in hospital for prevention of diarrhoea and shedding of rotavirus. 344: 1046–1049

Salminen S, Deighton MA, Benno Y. & Gorbach SL (1998) Lactic acid bacteria in health and disease. Lactic acid bacteria. Marcel Dekker, Inc. 211–253

Shida K, Makino K, Morishita A, Takamizawa K, Hachimura S, Ametani A, Sato T, Kumagai Y, Habu S & Kaminogawa S (1998) *Lactobacillus casei* inhibits antigen-induced IgE secretion though regulation of cytokine production in murine splenocyte cultures. Int. Arch. Allergy Immunol. 11: 278–287

Takahashi M, Iwata S, Yamazaki N & Fujiwara H (1991) Phagocytosis of the lactic acid bacteria by M cells in the rabbit Peyer's patches. J. Clin. Electron. Microsopy. 24: 5–6

Trapp CL, Chang CC, Halpern GM, Keen CL & Gershwin ME (1993) The influence of chronic yogurt consumption on populations of young and elderly adults. Int. J. Immunotherapy IX (1): 53–63

Yamazaki S, Kamimura H, Momose H, Kawashima T & Ueda K (1982) Protective effect of *Bifidobacterium* monoassociation against lethal activity in *Escherichia coli.* Bifidobacteria Microflora. 1: 55–59

Yasui H & Ohwaki M (1991) IgA production and intestinal microflora: augmentation of IgA production by *Bifidobacterium breve.* Jpn. Clin. Microscopy. 24: 426–430

Yasui H, Nagaoka N, Mike A, Hayakawa K & Ohwaki M (1992) Detection of *Bifidobacterium* strains that induce large quantities of IgA. Microb. Ecol. Health Dis. 5: 155–162

Yasui H, Nagaoka N & Hayakawa K (1994) Augmentation of anti-influenza virus hemagglutinin antibody production by Peyer's patch cells with *Bifidobacterium breve* YIT4064. Clin. Diagn. Lab. Immunol. 1: 244–246

Yasui H, Kiyoshima J & Ushijima H (1995) Passive protection against rotavirus-induced diarrhea of mouse pups born to and nursed by dams fed *Bifidobacterium breve* YIT4064. J. Infect Dis. 172: 403–409

Yasui H, Kiyosima J, Hori T & Shida K (1999) Protection against influenza virus infection of mice fed *Bifidobacterium breve* YIT4064. Clin. Diagn. Lab. Immunol. 6: 186–192

Yokokura T (1994) Anti-tumor and immuno-stimulating activities of *Lactobacillus casei.* Jpn. J. Dairy and Food Science. 43: 141–150

Antonie van Leeuwenhoek **76:** 391–394, 1999.

The role of lactic acid bacteria in colon cancer prevention: mechanistic considerations

Kazuhiro Hirayama* & Joseph Rafter
Department of Medical Nutrition, Karolinska Institute, NOVUM, S-141 86 Huddinge, Sweden
*(*Author for correspondence)*

Abstract

Colorectal cancer is one of the most important causes of cancer morbidity and mortality in Western countries. While a myriad of healthful effects have been attributed to the probiotic lactic acid bacteria, perhaps the most controversial remains that of anticancer activity. It should be pointed out already at this point that there is no direct experimental evidence for cancer suppression in humans as a result of consumption of lactic cultures in fermented or unfermented dairy products. However, there is a wealth of indirect evidence, based largely on laboratory studies, in the literature. The precise mechanisms by which lactic acid bacteria may inhibit colon cancer are presently unknown. However, such mechanisms might include: enhancing the host's immune response; binding and degrading potential carcinogens; quantitative and/or qualitative alterations in the intestinal microflora incriminated in producing putative carcinogen(s) and promoters (e.g. bile acid-degrading bacteria); producing antitumorigenic or antimutagenic compounds in the colon; alteration of the metabolic activities of intestinal microflora; alteration of physicochemical conditions in the colon; effects on physiology of the host. These potential mechanisms are discussed in the present paper.

Enhancing the host's immune response

One explanation for tumour suppression by lactic acid bacteria may be mediated through an immune response of the host. Sekine et al. (1985) suggested that *B. infantis* stimulates the host-mediated response, leading to tumour suppression or regression. In addition, there are studies to suggest that lactic acid bacteria play an important role and function in the host's immunoprotective system by increasing specific and non-specific mechanisms to have an anti-tumour effect (Kato et al. 1983; De Simone et al. 1993; Schiffrin et al. 1995). *Lactobacillus casei* strain Shirota (LcS) has been shown to have potent anti-tumor and anti-metastatic effects on transplantable tumor cells and to suppress chemically-induced carcinogenesis in rodents. Also, intrapleural administration of LcS into tumor-bearing mice has been shown to induce the production of several cytokines, such as IFN-γ, IL-1β and TNF-α, in the thoracic cavity of mice, resulting in the inhibition of tumor growth and increased survival (Matsuzaki 1998). These findings suggest that treatment with LcS has the potential to ameliorate or prevent tumourigenesis through modulation of the host's immune system, specifically cellular immune responses. An additional study has indicated that oral administration of BLP, a preparation of viable *Lactobacillus casei* YIT 9018, potentiated systemic immune responses that modified T-cell functions in tumor bearing mice (Kato et al. 1994). It has also been demonstrated that *B. longum* and *B. animalis* promote the induction of inflammatory cytokines (IL-6, TNF-α) in mouse peritoneal cells (Sekine et al. 1994).

Binding and degrading potential carcinogens

As mentioned above, bacterial cells in addition to certain plant cell walls may be an important factor in determining the ratio of bound to free (bioavailable) toxins in the intestine. Morotomi & Mutai (1986) investigated the ability of 22 strains of intestinal bacteria to bind the mutagenic pyrolyzates and compared their ability to that of some dietary fibers. 3-amino-1,4-dimethyl-5*H*-pyrido[4,3-*b*]indole (Trp-P-1) and 3-amino-1-methyl-5*H*-pyrido[4,3-*b*]indole (Trp-P-2) were effectively bound to all gram-positive and some gram-negative bacterial cells, corn bran, apple pulp and soy bean fiber. When the mechanism of binding of Trp-P-2 to *Lactobacillus casei* YIT 9018 and corn bran was investigated, it was shown to be pH dependent, occurred instantaneously and was inhibited by the addition of metal salts, indicating a cation-exchange mechanism. The mutagenicity of Trp-P-2 for *Salmonella typhimurium* TA98 in the presence of S9 mix was inhibited by the addition of *Lactobacillus casei* YIT 9018 to the reaction mixture, indicating that bound Trp-P-2 did not cause mutation under the assay conditions.

Although little is known about the fate of bound mutagens in the human gastrointestinal system, Zhang & Ohta (1993) showed that freeze-dried cells of lactic acid bacteria, intestinal bacteria and yeast significantly reduced the absorption of Trp-P-1 from the small intestine in rats and that this was accompanied by decreased levels of this food mutagen in blood. Lactobacilli have also been shown to degrade nitrosamines (Rowland & Grasso 1975).

Quantitative and/or qualitative alterations in the intestinal microflora

Consumption of fermented milk containing *L. acidophilus* has been shown to reduce significantly the counts of faecal putrefactive bacteria such as coliforms and increase the levels of lactobacilli in the intestine (Ayebo et al. 1980; Shahani & Ayebo 1980) suggesting that supplemental *L. acidophilus* has a beneficial effect on the intestinal microecology by suppressing the putrefactive organisms that are possibly involved in the production of tumour promoters and putative precarcinogens. However, the mechanisms underlying these effects are still to a large extent not understood.

Production of antitumorigenic or antimutagenic compounds in the colon

Milk fermented by *B. infantis*, *B. bifidum*, *B. animalis*, *L. acidophilus* and *L. paracasei* inhibited the growth of the MCF7 breast cancer cell line and the anti-proliferative effect was not related to the presence of bacteria (Biffi et al. 1997). These findings suggest the presence of an *ex novo* soluble compound produced by lactic acid bacteria during milk fermentation or the microbial transformation of some milk components in a biologically active form. Arimochi et al. (1997) showed an inhibitory effect of *L. acidophilus* on aberrant crypt foci (ACF) formation in the colon of rats, induced by azoxymethane, and enhanced removal of O^6-methylguanine from the colon mucosal DNA and that these effects came from culture supernatants, not from bacterial cells. ACF are putative preneoplastic lesions from which adenomas and carcinomas may develop. In addition, it has been demonstrated that dietary administration of lyophilized cultures of *Bifidobacterium longum* strongly suppressed azoxymethane-induced colonic tumor development and that this effect was associated with a decrease in colonic mucosal cell proliferation and colonic mucosal and tumor ornithine decarboxylase and ras-p21 activities (Reddy 1998).

Alteration of the metabolic activities of intestinal microflora

Many foreign compounds are detoxified by glucuronide formation in the liver before entering the intestine via the bile. The bacterial enzyme β-glucuronidase has the ability to hydrolyse many glucuronides due to its wide substrate specificity, and thus may liberate carcinogenic aglycones in the intestinal lumen. Several other bacterial enzymes have also been suggested to be implicated in the carcinogenic process, releasing carcinogens in the intestinal tract. Interestingly, it was these earlier observations that feeding lactic acid bacteria supplements in the diets of rodents significantly decreased the activities of some of the above faecal enzyme activities which focused attention on these bacteria as possible anticancer agents (Goldin & Gorbach 1984a; Kulkarni & Reddy 1994; Abdelali et al. 1995; Rowland et al. 1998). Lactic acid bacteria also reduced the specific activities of faecal enzymes in human volunteer studies (Ayebo et al. 1980; Goldin et al. 1980; Ling et al. 1994; Spanhaak et al. 1998). Goldin & Gorbach (1984b) studied the effect of feeding *L. acidophilus* strains NCFM and N-2 on the activity of three bacterial enzymes – β-glucuronidase, nitroreductase and azoreductase – in 21 healthy volunteers. Both strains had similar effects and caused a significant decline in the specific activity of the three enzymes in all subjects after 10 days of feeding. A reversal of the effect was observed within 10–30 days of stopping *Lactobacillus* feeding, suggesting that continuous consumption of these bacteria was necessary to maintain the effect. The authors suggested that the observed reduction of these enzymes may explain the reduced colon cancer incidence in rats fed viable *L. acidophilus* (Goldin & Gorbach 1980). Thus in summary, the animal and human studies do indicate that feeding certain lactic cultures can result in a decrease of faecal enzymes that may be involved in formation of carcinogens. It is important to mention here that the reports published to date do not always find reductions in the same enzymes, although findings with β-glucuronidase and nitroreductase are most consistently positive. However, we still do not know how or whether a reduction in these enzyme activities affects cancer rates in humans. Indeed, the origin of the carcinogens causing this disease in humans is still to a large extent unknown.

Alteration of physicochemical conditions in the colon

Modler et al. (1990) have suggested that large bowel cancer could be influenced directly by reducing intestinal pH, thereby preventing the growth of putrefactive bacteria. In rats given inulin-containing diets with or without *B. longum*, an increase in caecal weight and β-glucosidase and a decrease in caecal pH were observed (Rowland et al. 1998), though some other studies did not detect a significant change in intestinal pH (Bartram et al. 1994; Abdelali et al. 1995).

Dietary fat has been considered a risk factor for colon cancer, and it has been suggested that this phenomenon may be mediated by increased levels of bile acids (mainly secondary bile acids, produced by the action of bacterial 7α-dehydroxylase on primary bile acids) in the colon (Weisburger & Wynder 1987). One hypothesis regarding colon carcinogenesis involves a cytotoxic effect on the colonic epithelium exerted by bile acids in the aqueous phase of faeces (soluble bile acids), followed by an increased proliferation of cells in the intestine (Bruce 1987). It has been demonstrated that a 6 week administration of *L. acidophilus* fermented milk supplements to colon cancer patients resulted in lower concentrations of soluble bile acids in faeces (Lidbeck et al. 1991). Although the decrease in the concentration of bile acids in this fraction of faeces was not significant (perhaps due to a low number of patients or a limited supplementation period), it was of interest that decreased levels of soluble bile acids were observed in the colon cancer patients receiving *L. acidophilus* fermented milk supplements. In another study, patients with colonic adenomas participated in a 3-month study, where *L. acidophilus* was administered together with *B. bifidum* (Venitt 1988). During this period, the faecal pH was reduced significantly and patients having a higher proliferative activity in the upper colonic crypts than that calculated for subjects at low risk for colon cancer, showed a significant decrease after therapy with the lactic acid bacteria. In view of the results in the above mentioned study (Lidbeck et al. 1991) it is interesting to speculate that this latter effect was in part due to decreased levels of bile acids in the aqueous phase of faeces.

Effects on physiology of the host

Lactobacilli are one of the dominant species in the small intestine, and these micro-organisms presumably affect metabolic reactions occurring in this part of the gastrointestinal tract. The ileal mucosa (Venitt 1988) as well as the colonic mucosa (Fang & Strobel 1978) have the capacity to absorb mutagenic compounds from the intestinal lumen whereafter the compounds are passed into the bloodstream, either unchanged or as metabolites. In addition, lactic acid bacteria have been shown to increase colonic NADPH-cytochrome P-450 reductase activity (Pool-Zobel et al. 1996) and glutathione S-transferase levels (Challa et al. 1997) and to reduce hepatic uridine diphosphoglucuronyl transferase activity (Abdelali et al. 1995), enzymes which are involved in the metabolism of carcinogens in rats.

Conclusion

Thus, as is obvious from the above, there are several possible mechanisms which might explain how lactic acid bacteria might protect against tumor development in the colon. However, questions such as: what is the major contributing mechanism for a particular bacterial strain or how are the different mechanisms linked are questions to which we presently do not have the answers. It is possible that different strains target different mechanisms. All of the mechanisms have various degrees of support, mainly originating from *in vitro* and animal experiments and some of them even have some support from human clinical studies. Thus, more work needs to be done to identify the specific strains and strain characteristics responsible for specific antitumour effects and the mechanisms by which these effects are mediated. However, even with these reservations in mind and mindful of the limited number of human studies available, the use of lactic cultures for human cancer suppression is interesting, holds promise and certainly deserves more scrutiny. The latter should involve carefully designed human clinical trials to corroborate the wealth of experimental studies.

Acknowledgements

This work was supported by grants from the Swedish Dairies Association and the Swedish Cancer Society.

References

Abdelali H, Cassand P, Soussotte V, Daubeze M, Bouley C & Narbonne JF (1995) Effect of dairy products on initiation of precursor lesions of colon cancer in rats. Nutr. Cancer 24: 121–132

Arimochi H, Kinouchi T, Kataoka K, Kuwahara T & Ohnishi Y (1997) Effect of intestinal bacteria on formation of azoxymethane-induced aberrant crypt foci in the rat colon. Biochem. Biophys. Res. Commun. 238: 753–757

Ayebo AD, Angelo IA & Shahani KM (1980) Effect of ingesting *Lactobacillus acidophilus* milk upon fecal flora and enzyme activity in humans. Milch Wissenschaft 35: 730–733

Bartram HP, Scheppach W, Gerlach S, Ruckdeschel G, Kelber E & Kasper H (1994) Does yogurt enriched with *Bifidobacterium longum* affect colonic microbiology and fecal metabolites in healthy subjects. Am. J. Clin. Nutr. 59: 428–432

Biffi A, Coradini D, Larsen R, Riva L & Di Fronzo G (1997) Antiproliferative effect of fermented milk on the growth of a human breast cancer cell line. Nutr. Cancer 28: 93–99

Bruce WR (1987) Recent hypotheses for the origin of colon cancer. Cancer Res. 47: 4237–4242

Challa A, Rao DR, Chawan CB & Shackelford L (1997) *Bifidobacterium longum* and lactulose suppress azoxymethane-induced colonic aberrant crypt foci in rats. Carcinogenesis 18: 517–521

De Simone C, Vesely R, Bianchi Salvadori B & Jirillo E (1993) The role of probiotics in modulation of the immune system in man and in animals. Int. J. Immunother. 9: 23–28

Fang W-F & Strobel HW (1978) Activation of carcinogens and mutagens by rat colon mucosa. Cancer Res. 38: 2939–2944

Goldin BR & Gorbach SL (1980) Effect of *Lactobacillus acidophilus* dietary supplements on 1,2-dimethylhydrazine dihydrochloride-induced intestinal cancer in rats. J. Natl. Cancer Inst. 64: 263–265

Goldin BR, Swenson L, Dwyer J, Sexton M & Gorbach SL (1980) Effect of diet and *Lactobacillus acidophilus* supplements on human fecal bacterial enzymes. J. Natl. Cancer Inst. 64: 255–261

Goldin BR & Gorbach SL (1984a) Alterations of the intestinal microflora by diet, oral antibiotics and *Lactobacillus*: decreased production of free amines from aromatic nitro compounds, azo dyes and glucuronides. J. Natl. Cancer Inst. 73: 689–695

Goldin BR & Gorbach SL (1984b) The effect of milk and lactobacillus feeding on human intestinal bacterial enzyme activity. Am. J. Clin. Nutr. 39: 756–761

Kato I, Yokokura T & Mutai M (1983) Macrophage activation by *Lactobacillus casei* in mice. Microbiol. Immunol. 27: 611–618

Kato I, Endo K & Yokokura T (1994) Effects of oral adminstration of *Lactobacillus casei* on antitumor responses induced by tumor resection in mice. Int. J. Immunopharmacol. 16: 29–36

Kulkarni N & Reddy BS (1994) Inhibitory effect of *Bifidobacterium longum* cultures on the azoxymethane-induced aberrant crypt foci formation and fecal bacterial β-glucuronidase. Proc. Soc. Exp. Biol. Med. 207: 278–283

Lidbeck A, Geltner-Allinger U, Orrhage KM, Ottava L, Brismar B, Gustafsson J-Å, Rafter JJ & Nord CE (1991) Impact of *Lactobacillus acidophilus* supplements on the faecal microflora and soluble faecal bile acids in colon cancer patients. Microbial Ecol. Health Dis. 4: 81–88

Ling WH, Korpela R, Mykkanen H, Salminen S & Hanninen O (1994) *Lactobacillus* strain GG supplementation decreases colonic hydrolytic and reductive enzyme activities in healthy female adults. J. Nutr. 124: 18–23

Matsuzaki T (1998) Immunomodulation by treatment with *Lactobacillus casei* strain Shirota. Int. J. Food Microbiol. 41: 133–140

Modler GW, McKellar RC & Yaguchi M (1990) Bifidobacteria and bifidogenic factors. Can. Inst. Food Sci. Technol. J. 23: 29–41

Morotomi M & Mutai M (1986), In vitro binding of potent mutagenic pyrolysates to intestinal bacteria. J. Natl. Cancer Inst. 77: 195–201

Pool-Zobel BL, Neudecker C, Domizlaff I, Ji S, Schillinger U, Rumney C, Moretti M, Vilarini I, Scassellati-Sforzolini R & Rowland IR (1996) Lactobacillus- and Bifidobacterium-mediated antigenotoxicity in the colon of rats. Nutr. Cancer 26: 365–380

Reddy BS (1998) Prevention of colon cancer by pre- and probiotics: evidence from laboratory studies. Br. J. Nutr. 80: S219–223

Rowland IR & Grasso P (1975) Degradation of *N*-nitrosamines by intestinal bacteria. Appl. Microbiol. 29: 7–12

Rowland IR, Rumney CJ, Coutts JT & Lievense LC (1998) Effect of *Bifidobacterium longum* and inulin on gut bacterial metabolism and carcinogen-induced aberrant crypt foci in rats. Carcinogenesis 19: 281–285

Schiffrin, EJ, Rochat F, Link-Amster H, Aeschlimann JM & Donnet-Hughes A (1995) Immunomodulation of human blood cells following the ingestion of lactic acid bacteria. J. Dairy Sci. 78: 491–496

Sekine K, Toida T, Saito M, Kuboyama M, Kawashima T & Hashimoto Y (1985) A new morphologically characterized cell wall preparation (whole peptidoglycan) from *Bifidobacterium infantis* with a higher efficacy on the regression of an established tumor in mice. Cancer Res. 45: 1300–1307

Sekine K, Kawashima T & Hashimoto Y (1994) Comparison of the TNF-α levels induced by human-derived *Bifidobacterium longum* and rat-derived *Bifidobacterium animalis* in mouse peritoneal cells. Bifidobacteria Microflora 13: 79–89

Shahani KM & Ayebo AD (1980) Role of dietary lactobacilli in gastrointestinal microecology. Am. J. Clin. Nutr. 33: 2448–2457

Spanhaak S, Havenaar R & Schaafsma G (1998) The effect of consumption of milk fermented by *Lactobacillus casei* strain Shirota on the intestinal microflora and immune parmeters in humans. Eur. J. Clin. Nutr. 52: 899–907

Venitt S (1988) Mutagens in human faeces and cancer of the large bowel. In: Rowland IR (Ed.), Role of the gut flora in toxicity and cancer (pp. 399–460). Academic Press, London

Weisburger JH & Wynder EL (1987) Etiology of colorectal cancer with emphasis on mechanism of action and prevention. In: De Vita VT, Hellman S & Rosenberg SA (Eds.), Important advances in oncology (pp. 197–220). JB Lippincott, Philadelphia

Zhang XB & Ohta Y (1993) Microorganisms in the gastrointestinal tract of the rat prevent absorption of the mutagen-carcinogen 3-amino-1,4-dimethyl-5*H*-pyrido[4,3-*b*]indole. Can. J. Microbiol. 39: 841–845

Antonie van Leeuwenhoek 76: 395–401, 1999.

Lactic acid food fermentation in tropical climates

M.J.R. Nout[1,*] & P.K. Sarkar[2]
[1]*Food Hygiene and Microbiology Group, Department of Food Technology and Nutritional Sciences, Wageningen University, Bomenweg 2, 6703 HD Wageningen, The Netherlands.*
[2]*Microbiology Laboratory, Department of Botany, University of North Bengal, Siliguri 734 430, India.*
*(*Author for correspondence)*

Introduction

Lactic acid fermentation is widely applied in tropical climates as a low-cost method of enhancing food quality, safety and shelf-life (Cooke et al. 1987; Nout & Motarjemi 1997). In many developing countries, the manufacture, preservation and distribution of foods are still seriously restrained by the lack of required technological infrastructure. Whereas refrigeration, freezing, and canning require cold-chains and industrial-scale processing to be economically feasible, lactic acid fermentation processes can be successfully carried out using simple means and at a smaller scale.

As most of such fermentation processes are carried out under non-sterile conditions, the microbiological composition of the resulting products is of a mixed nature. Whereas lactic acid bacteria are often predominant, other micro-organisms such as yeasts are usually present as well.

Lactic acid fermentation occurs during the preparation of a wide variety of foods, made from raw materials of plant and animal origin. Some ingredients, their fermented products, mode of consumption and references to their microbiological composition are presented in Table 1. Two examples will be discussed, namely the fermentation of cassava, and of mixtures of cereals and legumes.

Cassava

Cassava (*Manihot esculenta* Crantz) is a tuberous rootcrop indigenous to South America, which is now grown in many tropical countries. Its global annual production is estimated at $1.5 \cdot 10^{11}$ kg. Its good yield of starchy tubers, even from unfertile soils or under conditions of drought, combined with its ease of cultivation, makes it a very important source of dietary energy for low-income farmers and consumers. One of the limitations of cassava is that once harvested, the tubers are highly perishable. In addition, potentially toxic cyanogenic glycosides (mainly linamarin) occur in variable concentrations in the fresh roots of bitter cassava cultivars. Immediate post-harvest processing is required to preserve and detoxify this important food source.

Agbéli mawè is the name used in the Republic of Bénin (West Africa) for a fermented dough made of cassava. A similar product is known as agbelima in Ghana. As a result of the fermentation process, the chemical and microbiological stability of the product is improved, the toxicity is eliminated and a characteristic flavour is formed. This product must be cooked prior to consumption. The fermentation process contributes to the desirable visco-elastic properties of the resulting paste, which is usually consumed with a gluey sauce.

The major unit operations for producing agbéli mawè are illustrated in Figure 1. Cassava tubers are washed, peeled and grated to pulp. The pulp is pressed to remove water to 50–55% moisture content, sieved to remove fibrous materials, and milled to a fine flour. This is kneaded into a dough that is kept during 1–3 days at ambient temperature (25–35 °C) allowing a natural fermentation to take place which is characterized by a succession of lactic acid bacteria.

Facultative heterofermentative *Lactobacillus plantarum* and obligate heterofermentative *Lb.brevis* and *Leuconostoc mesenteroides* occurred in all samples of Ghanaian agbelima (Amoa-Awua et al. 1996). The common occurrence of *Lb.plantarum* in fermenting cassava had also been reported earlier by Oyewole & Odunfa (1990). They studied the succession of lactic acid bacteria species (Figure 2) and observed that *Lb.plantarum* is predominant after 3 days of fermentation.

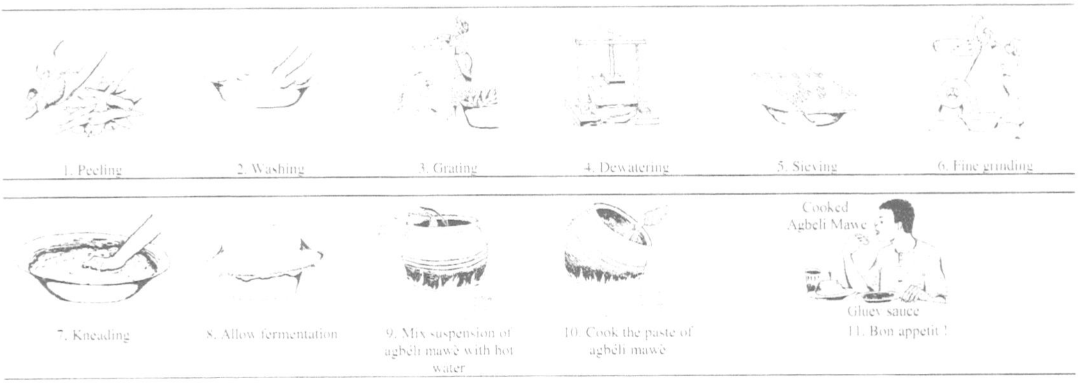

Figure 1. Processing of cassava, and preparation and consumption of agbéli mawè (adapted from Nago & Hounhouigan 1998, with permission).

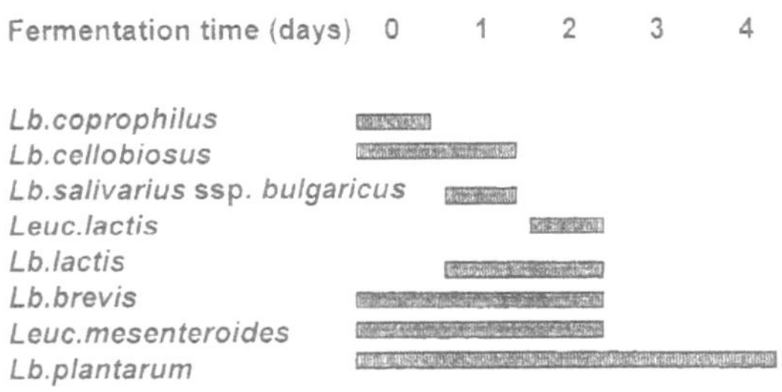

Figure 2. Succession of lactic acid bacteria in fermenting cassava pulp (adapted from Oyewole & Odunfa 1990).

Traditional cassava fermentation processes are fairly adequate for detoxification, for preservation and for creating a tasteful product with a desirable visco-elastic texture when cooked.

Several authors found that *Lb.plantarum* strains are able to degrade linamarin (Oyewole & Odunfa 1992, Giraud et al. 1993). The degradation of linamarin is a 2-stage process catalyzed by ß-glucosidase (linamarin → glucose + acetone cyanohydrin), followed by hydroxynitrile lyase (acetone cyanohydrin ↔ acetone + HCN). The latter equilibrium reaction is pH-dependent, towards cyanohydrin at lower pH. The data (Table 2) recalculated from Westby & Choo (1994) and Giraud et al. (1993) show that detoxification of cassava pulp by the endogenous enzymes of cassava, is most rapid under sterile conditions (neutral pH) whereas slow acidification (natural fermentation) and rapid acidification (added *Lb.plantarum* starter) allow degradation of linamarin by ß-glucosidase, but are increasingly detrimental to final detoxification of acetone cyanohydrin. In processes where cassava is not grated to pulp, bacterial cell wall degrading enzymes cause softening (Amoa-Awua & Jakobsen 1995) and facilitate the linamarin degradation by endogenous ß-glucosidase. In all cases, lactic acidification should be moderate initially in order to adequately detoxify the cyanohydrin moiety.

Of more positive importance is the impact of lactic acid bacteria on the sensory qualities and preservation of cassava. Important sensory properties of fermented cassava are its flavour and the visco-elastic properties of the cooked paste, made from the fermented pulp. The aroma compounds of agbelima were mainly low molecular weight alcohols, 1-propanol, isoamylalcohol, 3-methyl-1-butanol, ethyl acetate and acetoin (Amoa-Awua et al. 1996).

Whereas the swelling capacity of the cassava starch is not affected by lactic acid fermentation, improved visco-elastic properties of the cooked product are achieved (Moorthy & Mathew 1998). Presently, the mechanism for this effect has not been convincingly demonstrated. It is hypothesized that amylolytic enzyme activity, chemical depolymerization of native starch, or extraction of amylose from native starch are involved. Amylolytic strains of *Lb.plantarum* (Giraud et al. 1993, Olasupo et al. 1996) and *Lb.manihotivorans* (Morlon-Guyot et al. 1998) were reported. A study of the effect of natural lactic fermentation on physicochemical characteristics of cassava starch (Mestres & Rouau 1997) revealed that fermentation and mild drying did not affect the crystallinity of native starch granules or their gelatinization properties. However, Numfor et al. (1995) observed significant decrease of hardness, gumminess and adhesiveness (stickiness) of fermented cooked cassava. Probably, leaching and enzymatic degrada-

Table 1. Examples of lactic acid fermented foods of tropical climates

Ingredients	Fermented product (Geography)	Mode of consumption	References
Plant origin: *Cereals:*			
maize	kenkey (Ghana)	stiff, cooked dough	Nche et al. 1996
maize	pozol (Mexico)	sour beverage	Nuraida et al. 1995
maize, sorghum	uji (Kenya)		Mbugua &Njenga 1992
maize	busaa (Kenya)	sour-alcoholic	Nout 1980
millet	jnard (India)	beverage (beer)	Tamang et al. 1996
Starchy roots and stems:			
cassava	agbéli mawè (Bénin), agbelima (Ghana)	cooked paste from fermented pulp	Nago &Hounhouigan 1998 Amoa-Awua et al. 1996
cassava	gari (Nigeria)	fermented, toasted and dried pulp	Kemdirim et al. 1995
cassava	peujeum (Indonesia)	sour-sweet-alcoholic softened root	Arihantana & Buckle 1987
radish tap root	sinki (India, Nepal)	soup base, pickle	Tamang & Sarkar 1993
young bamboo shoots	mesu (India, Nepal)	pickle	Tamang & Sarkar 1996
Leafy vegetables:			
cabbage, garlic, peppers	kimchi (Korea)	pickles	Cheigh &Park 1994
rayo[1] / radish / cauliflower	gundruk (India, Nepal)	soup base	Yonzan &Tamang 1998
Pulses and beans:			
black gram and rice	idli (India, Sri Lanka)	steamed cake	Reddy et al. 1986
Bengal gram and rice / wheat	dhokla (India)	steamed cake	Ramakrishnan 1979
Animal origin: *Milk:*			
cow or buffalo	dahi (India)	yoghurt	Balasubramanyam & Varadaraj 1994
Meat:			
pork or beef	chorizo (Mexico)	dry sausage	Santos et al. 1997
Fish:			
fish, salt, roasted rice	pla-ra (Thailand)	fish paste	Tanasupawat et al. 1998

[1] *Brassica campestris* L. var. *cumifolia* Roxb.

Table 2. Changes in cyanogens in cassava pulp at 30°C.

Incubation (h)	Irradiated (sterile) (Westby & Choo 1994)				Natural Fermentation (Westby & Choo 1994)				*Lb.plantarum* A6 (high linamarase and acidifying activity) (Giraud et al. 1993)			
	Total	Linamarin	Cyanohydrin	HCN	Total	Linamarin	Cyanohydrin	HCN	Total	Linamarin	Cyanohydrin	HCN
0	100	88	-	12	100	88	-	12	100	77	8	15
6	53	6	29	18	76	-	18	58	77	-	31	46
12	47	-	26	21	71	-	24	47	77	-	54	23
24	41	-	24	17	53	-	29	24	77	-	62	15
48	35	-	21	14	47	-	29	18	77	-	62	15
72	26	-	18	7	41	-	24	17	77	-	63	14
96	23	-	12	11	41	-	24	17	74	-	65	9

Values represent % of levels of total cyanogens at start, i.e. 170 and 325 mg $HCN.kg^{-1}$ dry matter , for Westby & Choo (1994) and Giraud et al. (1993), respectively.

Table 3. Bacterial flora of raw ingredients and fermented product of idli (From Ramakrishnan 1979)

Bacteria	% of total counts		
	Rice	Black gram dal	Fermented batter of rice and dal
Leuconostoc mesenteroides	33	69	43
Lactobacillus fermentum	17	-	14
Lactobacillus delbrueckii	8	7	14
Enterococcus faecalis	8	7	14
Pediococcus acidilactici	17	-	-
Bacillus sp.	17	15	-

Table 4. Microbial and physico-chemical changes during idli batter fermentation (from Soni & Sandhu 1991)

Parameters	Before fermentation	After fermentation
Bacteria (Log cfu.g^{-1})	6.3	10.9
Yeasts (Log cfu.g^{-1})	2.6	5.9
pH	6.6	4.9
Volume (ml)	200	410
Amylases (IU.ml^{-1})	16.0	22.0
Reducing sugars (mg.g^{-1})	3.6	6.0
Free amino acids (mg.g^{-1})	8.0	13.0
Crude protein (% w/w)	12.0	12.4
Thiamine (mg.kg^{-1})	2.0	7.5
Riboflavin (mg.kg^{-1})	2.5	7.5

tion of amylose from native starch play an important and desirable role. Amylolytic lactic acid bacteria can possibly enhance this effect.

The production of organic acids and other antimicrobial components by lactic acid bacteria is well documented (Adams & Nicolaides 1997). Bacteriocin producing *Enterococcus faecium* (Olasupo et al. 1994b) and *Lactobacillus* spp. (Olasupo et al. 1994a) have been isolated from various African fermented foods. Olasupo et al. (1997) reported that bacteriocin producing *Lb.casei* strain 012 had a significant inhibitory effect on an enterotoxigenic strain of *E.coli* in an experimental fermentation of maize dough. The reported data do not permit a conclusion about the mechanism of this inhibition. All the same, there is ample evidence for the inhibitory and sometimes bactericidal impact of lactic acid bacteria on susceptible bacterial pathogens, provided that lactic acid bacteria can dominate fermentations at an early stage. However, other pathogens e.g. enteric viruses and parasites appear to be less sensitive and may survive in lactic fermented foods in the absence of subsequent heat treatment (Adams & Nicolaides 1997). Thus the antimicrobial properties of lactic acid bacteria can improve the microbiological quality of the food, but only in combination with good hygienic practices.

Cereal-legume mixtures

From a nutritional point of view, it is advantageous to consume mixtures of cereals and legumes because this results in an improved balance of carbohydrates and proteins as well as a better balance of dietary essential amino acids (Nout 1993). Acidified and leavened foods like Indian idli, dosa and dhokla constitute an interesting group of products made from cereal-legume mixtures.

Idli is a popular food, prepared and consumed throughout India, particularly in southern parts, and in many places of Sri Lanka. It is relished mainly as a breakfast dish along with chutney (a batter containing fresh ground coconut and spices) and sambar, a thin spiced soup prepared from red gram dal (dehulled legume seeds), vegetables, tamarind juice, salt and herbs. An early history of the use of this traditional food dates back to 1100 A.D. (Ramakrishnan 1979). The ever-increasing popularity of idli is for its unique organoleptic quality (a typical sour flavour and spongy texture), nutritional quality and improved digestibility. For this reason, it is fed to infants as a weaning food, and as a main dish in diets in many South Indian hospitals.

The main ingredients used in the traditional preparation of idli are white polished rice (*Oryza sativa* L.) and black gram (*Phaseolus mungo* L.) dal. The proportion of these two ingredients may vary. In general, it consists of three parts of rice and one part dal. Rice and dal are washed several times with tap water, soaked separately in water at room temperature for 5-10 h before grinding in a stone mortar or other grinders. While rice is coarsely ground, the dal is ground to a fine smooth paste. The rice and dal slurries are mixed and stirred well manually to form a thick batter. Salt is added to taste. The batter, put in a closed container, is kept at a warm place to ferment overnight or longer. The fermentation period must allow a definite leaven-

Table 5. Examples of effects of lactic acid fermentation on anti-nutritional factors in cereals and legumes, and impact on *in-vitro* and *in-vivo* availability of nutrients.

Food	Parameter	Not Fermented	Fermented	Starter	Reference
Degradation of anti-nutritional factors:					
maize:sorghum:ground-nut flour 30:30:35	Phytate (IP-6) (%dm)	0.87	0.41	*Lb.plantarum*	Nout (unpublished)
	IP-5	0.10	0.20		
millet flour	Polyphenols (%dm)	0.69	0.60	*Lb.brevis*	Khetarpaul & Chauhan 1991b
In-vitro availability:					
low-tannin sorghum	Fe solubility		40 - 50% increased	natural lactic fermentation	Svanberg et al. 1993; Lorri & Svanberg 1993
	Protein digestibility (%)	61 - 79	70 - 82		
high-tannin sorghum	Fe solubility		10 - 20% increased	natural lactic fermentation	Svanberg et al. 1993; Lorri & Svanberg 1993
	Protein digestibility (%)	33 - 40	45 - 53		
In-vivo availability:					
millet	Protein Efficiency Ratio (PER)	1.41	1.80 - 1.91	natural lactic fermentation	Khetarpaul & Chauhan 1991a
sorghum	PER	2.06	2.34	*Lb.plantarum*	Nout 1993

ing of the batter and development of a pleasant acid flavour. The fermented batter is poured in small cups or in a special idli pan having cups (8–10 cm dia), and steamed until the starch is gelatinized and the idli cakes are soft and spongy. The fermented batter is consumed the same day and there is no effort to preserve it.

Mukherjee et al. (1965) concluded that *Leuc. mesenteroides* is the micro-organism essential for leavening of the batter and also responsible, along with *Ent. faecalis*, for acid production in idli. Soni & Sandhu (1989) reported the involvement of *Lb. fermentum* and *Pediococcus cerevisiae* in addition to the aforementioned bacteria in idli batter fermentation.

Idli is a natural fermented food; no inoculum is added generally for fermentation. This is because the essential micro-organisms have been found to be naturally present in the ingredients (Table 3). When the product is made daily, it is often the practice of adding a bit of freshly fermented batter ('back-slop') to the newly ground one. In addition to lactic acid bacteria, the slightly acid environment favours the growth and activity of yeasts (Venkatasubbaiah et al. 1985). The principal yeasts in the fermenting idli batter are *Saccharomyces cerevisiae*, *Debaromyces hansenii*, *Hansenula anomala*, *Torulopsis candida* and *Trichosporon beigelii*.

The major functions of the fermentation include the leavening of the batter (Table 4) and the improvement of taste and nutritional value of idli. *Leuc.mesenteroides* is the main responsible for the production of CO_2 which results in about 2–3 times increase in the original volume of batter (Soni & Sandhu 1989).

Certain oligosaccharides such as raffinose, stachyose and verbascose are prevalent in legumes and are associated with abdominal distention and flatulence. Lactic acid fermentation reduces these flatulence factors.

Cereals and legumes have restricted bio-availability of nutrients as a result of the presence of anti-nutritional factors (ANF) such as phytates and polyphenols (tannins). Lactic acid bacteria are able to reduce their levels, and it has been demonstrated that this results in improved bio-availability of minerals and starch, as well as increased protein efficiency ratio (PER) (Table 5).

Not all ANF can be degraded by lactic acid bacteria. For example, lectins and protease inhibitors occurring in legumes can only be inactivated adequately by heat treatments. Thus, the steaming of fermented idli batter complements the impact of fermentation.

Research needs

Taking into account the increasing demand particularly by urban populations, for safe and affordable food of high quality, there are certainly prospects for industrialization of traditional fermentation processes. Commercial availability of ready-to-use fermented products saves much labour and time in the household.

Scaling-up and specialization require better control of fermentation and quality. This includes the choice of optimum raw materials, fermentation conditions and starters used. Concerning the development of starter cultures for use in larger-scale industrial processes, selection for, or genetic modification to introduce a number of properties could be valuable. Mention was made already of the partially understood effects on rheological properties. Optimization of flavour and texture would greatly enhance consumer appeal. For assurance of safety and maximum nutritional value, starter properties must include improved inhibition of acid-tolerant pathogens and the ability to degrade a number of potentially toxic or anti-nutritive food components such as lectins, protease inhibitors, allergens and mycotoxins. In addition the ability to degrade plant cell walls would contribute to detoxification and increased digestibility of foods of plant origin.

References

Adams MR & Nicolaides L (1997) Review of the sensitivity of different foodborne pathogens to fermentation. Food Control 8: 227–240

Amoa-Awua WKA & Jakobsen M (1995) The role of *Bacillus* species in the fermentation of cassava. J. Appl. Bacteriol. 79: 250–256

Amoa-Awua WKA, Appoh FE & Jakobsen M (1996) Lactic acid fermentation of cassava dough into agbelima. Int. J. Food Microbiol. 31: 87–98

Arihantana MB & Buckle KA (1987) Cassava detoxication during tape fermentation with traditional inoculum. Int. J. Food Sci. Technol. 22: 41–48

Balasubramanyam BV & Varadaraj MC (1994) Dahi as a potential source of lactic acid bacteria active against foodborne pathogenic and spoilage bacteria. J. Food Sci. Technol. - Mysore 31: 241–243

Cheigh HS & Park KY (1994) Biochemical, microbiological, and nutritional aspects of Kimchi (Korean fermented vegetable products). Crit. Rev. Food Sci. Nutr. 34: 175–203

Cooke RD, Twiddy DR & Reilly PJA (1987) Lactic acid fermentation as a low-cost means of food preservation in tropical countries. FEMS Microbiol. Rev. 46: 369–379

Giraud E, Gosselin L & Raimbault M (1993) Production of a *Lactobacillus plantarum* starter with linamarase and amylase activities for cassava fermentation. J. Sci. Food Agric. 62: 77–82

Kemdirim OC, Chukwu OA & Achinewhu SC (1995) Effect of traditional processing of cassava on the cyanide content of gari and cassava flour. Plant Foods for Human Nutr. 48: 335–339

Khetarpaul N & Chauhan BM (1991a) Biological utilisation of pearl millet flour fermented with yeasts and lactobacilli. Plant Foods for Human Nutr. 41: 309–319

Khetarpaul N & Chauhan BM (1991b) Sequential fermentation of pearl millet by yeasts and lactobacilli – effect on the antinutrients and *in vitro* digestibility. Plant Foods for Human Nutr. 41: 321–327

Lorri W & Svanberg U (1993) Lactic-fermented cereal gruels with improved *in vitro* protein digestibility. Int. J. Food Sci. Nutr. 44: 29–36

Mbugua SK & Njenga J (1992) The antimicrobial activity of fermented uji. Ecol. Food Nutr. 28: 191–198

Mestres C & Rouau X (1997) Influence of natural fermentation and drying conditions on the physicochemical characteristics of cassava starch. J. Sci. Food Agric. 74: 147–155

Moorthy SN & Mathew G (1998) Cassava fermentation and associated changes in physicochemical and functional properties. Crit. Rev. Food Sci. Nutr. 38: 73–121

Morlon-Guyot J, Guyot J-P, Pot B, Jacobe de Haut I & Raimbault M (1998) *Lactobacillus manihotivorans* sp. nov., a new starch-hydrolyzing lactic acid bacterium isolated from cassava sour starch fermentation. Int. J. Syst. Bacteriol. 48: 1101–1109

Mukherjee SK, Albury MN, Pederson CS, van Veen AG & Steinkraus KH (1965) Role of *Leuconostoc mesenteroides* in leavening the batter of idli, a fermented food of India. Appl. Microbiol. 13: 227–231

Nago MC & Hounhouigan DJ (1998) La transformation alimentaire traditionelle des racines et tubercules au Benin. Les publications du CERNA (2), 01 BP 526, Cotonou, Benin

Nche PF, Odamtten GT, Nout MJR & Rombouts FM (1996) Soaking determines the quality of aflata for kenkey production. J. Cereal Sci. 24: 291–297

Nout MJR (1980) Microbiological aspects of the traditional manufacture of Busaa, a Kenyan opaque maize beer. Chem. Mikrobiol. Technol. Lebensm. 6: 137–142

Nout MJR (1993) Processed weaning foods for tropical climates. Int. J. Food Sci. Nutr. 43: 213–221

Nout MJR & Motarjemi Y (1997) Assessment of fermentation as a household technology for improving food safety: a joint FAO/WHO workshop. Food Control 8: 221–226

Numfor FA, Walter WM Jr & Schwartz SJ (1995) Physicochemical changes in cassava starch and flour associated with fermentation: effect on textural properties. Starch/Stärke 47: 86–91

Nuraida L, Wacher MC & Owens JD (1995) Microbiology of pozol, a Mexican fermented maize dough. World J. Microbiol. Biotechnol. 11: 567–571

Olasupo NA, Olukoya DK & Odunfa SA (1994a) Plasmid profiles of bacteriocin-producing Lactobacillus isolates from African fermented foods. Folia Microbiol. 39: 181–186

Olasupo NA, Schillinger U, Franz CMAP & Holzapfel WH (1994b) Bacteriocin production by Enterococcus faecium NA01 from 'wara' - A fermented skimmed cow milk product from west Africa. Lett. Appl. Microbiol. 19: 438–441

Olasupo NA, Olukoya DK & Odunfa SA (1996) Studies on local strains of amylolytic Lactobacillus from Nigerian fermented foods. Nahrung 40: 45–46

Olasupo NA, Olukoya DK & Odunfa SA (1997) Assessment of a bacteriocin-producing Lactobacillus strain in the control of spoilage of a cereal-based African fermented food. Folia Microbiol. 42: 31–34

Oyewole OB & Odunfa SA (1990) Characterization and distribution of lactic acid bacteria in cassava fermentation during fufu production. J. Appl. Bacteriol. 68: 145–152

Oyewole OB & Odunfa SA (1992) Extracellular enzyme activities during cassava fermentation for 'Fufu' production. World J. Microbiol. Biotechnol. 8: 71–72

Ramakrishnan CV (1979) Terminal Report of American PL480-project. Study of Indian fermented foods from legumes and production of similar fermented foods from U.S. soybean. Biochem. Dep., M.S. University of Baroda, Baroda, India

Reddy NR, Pierson MD & Salunkhe DK (1986) Idli. In: Reddy NR, Pierson MD & Salunkhe DK (Eds) Legume-based Fermented Foods (pp. 145–160). CRC Press, Boca Raton, Florida, USA

Santos EM, Gonzalez Fernandez C & Rovira IJYJ (1997) Identification and characterization of lactic acid bacteria isolated from traditional chorizo made in castilla leon. Food Sci. Technol. Int. 3: 21–29

Soni SK & Sandhu DK (1989) Fermentation of idli: Effects of changes in raw materials and physico-chemical conditions. J. Cereal Sci. 10: 227–238

Soni SK & Sandhu DK (1991) Role of yeast domination in Indian idli batter fermentation. World J. Microbiol. Biotechnol. 7: 505–507

Svanberg U, Lorri W & Sandberg AS (1993) Lactic fermentation of non-tannin and high-tannin cereals – effects on *in vitro* estimation of iron availability and phytate hydrolysis. J. Food Sci. 58: 408–412

Tamang JP & Sarkar PK (1993) Sinki: a traditional lactic acid fermented radish tap root product. J. Gen. Appl. Microbiol. 39: 395–408

Tamang JP & Sarkar PK (1996) Microbiology of mesu, a traditional fermented bamboo shoot product. Int. J. Food Microbiol. 29: 49–58

Tamang JP, Thapa S, Tamang N & Rai B (1996) Indigenous fermented food beverages of Darjeeling hills and Sikkim: process and product characterization. J. Hill Res. 9: 401–411

Tanasupawat S, Okada S & Komagata K (1998) Lactic acid bacteria found in fermented fish in Thailand. J. Gen. Appl. Microbiol. 44: 193–200

Venkatasubbaiah P, Dwarakanath CT & Sreenivasamurthy V (1985) Involvement of yeast flora in idli batter fermentation. J. Food Sci. Technol. 22: 88–90

Westby A & Choo BK (1994) Cyanogen reduction during lactic fermentation of cassava. Acta Horticulturae 375: 209–216

Yonzan H & Tamang JP (1998) Consumption pattern of traditional fermented foods in the Sikkim Himalaya. J. Hill Res. 11: 112–115

Antonie van Leeuwenhoek **76:** 403–411, 1999.

Non-dairy lactic fermentations: the cereal world*

Rudi F. Vogel*, Ruth Knorr, Martin R.A. Müller, Ulrich Steudel, Michael G. Gänzle & Matthias A. Ehrmann
Lehrstuhl für Technische Mikrobiologie, Technische Universität München, 85350 Freising, Germany
*(*Author for correspondence)*

Key words: cereal fermentation, genetics, *Lactobacillus sanfranciscensis*, metabolism, sourdough

Abstract

Sourdough is the foremost cereal fermentation performed in a variety of technologies with almost any cereal. The lactobacilli studied most intensely include *Lactobacillus sanfranciscensis*, *L. reuteri* and *L. pontis* isolated from traditional and modern rye and wheat fermentations. Molecular biology tools are available for their rapid identification and monitoring throughout a process. The currently available insight on their metabolism can be used to explain their prevalence in this environment and their interactions. Key genes of the sugar degradation pathway were cloned and characterised from *L. sanfranciscensis*. In addition some strains were found to have special properties including the production of antagonistic compounds or the adhesion to human intestinal cells.

Introduction

The use of defined strains of lactic acid bacteria and mixtures thereof is state of the art in most dairy fermentations. Some strains especially of *Lactococcus lactis* have been subjected to thorough metabolic and genetic analyses as well as for the construction of recombinant starter organisms. In non-dairy fields of food fermentations the development of starter organisms is increasingly advanced for meat fermentations, whereas other man-made fermentation habitats have only been discovered for a more comprehensive analysis in the last few years. Among these, cereal fermentations generally differ from most other fermentations because the fermentation product is not consumed as such but subject to further processing to beer, bread, crackers and the like. Cereals are subjected to fermentations in almost all regions of the world where cereals are consumed as staple food and therefore the variety of raw materials and the conditions of the respective fermentations are quite large. Nevertheless, there are general ecological principles which are decisive for and which are in turn mirror-imaged by the microbial flora found. As a general rule, lactic acid bacteria are the predominant organisms and in many cases yeasts are present in significant numbers (for review, see Hammes & Vogel 1997).

The major focus of recent studies on cereal fermentations was the characterisation of sourdough fermentations based on wheat or rye flour as raw materials. Basically, tree types of sourdough fermentations can be distinguished (i) type I doughs sustained by repeated inoculation at ambient temperature (20–30 °C) according to traditional procedures, (ii) type II doughs using elevated temperatures, longer fermentation times (greater 2d) , and higher water content, mostly used in industrialised processes to produce bakery pre-products, and (iii) type III processes initiated by defined starter cultures. The latter preparations often contain lactic acid bacteria from other environments because of their easy handling properties. The lactobacilli from 'condition-induced' types I and II fermentations have been characterised in the past years with respect to their impact on product quality, the response to changes in fermentation conditions, the metabolism, and their interactions with yeasts, and several comprehensive reviews are available on

* We dedicate this presentation to Walter P. Hammes, Universität Hohenheim, Germany on occasion of his 60th birthday.

Table 1. Oligonucleotide probes derived from 16S rDNA sequences for a specific identification of lactic acid bacteria in cerceal fermentations

	Probe 5′→3′	References
L. sanfranciscensis	TAA GAA TCA ATT GGG CGA C	Vogel et al. (1994)
L. amylovorus	TCT TTA AAA CAC ATC CAC	Ehrmann (1994)
L. brevis	TGT TGA AAT CAA GTG CAA G	Vogel et al. (1994)
L. reuteri	GAT CCA TCG TCA ATC AGG	Vogel et al. (1994)
L. fermentum	GCG ACC AAA ATC AAT CAG	Vogel et al. (1994)
L. pontis/panis	GGT AAT CCA TCG TCA AAT C	Vogel et al. (1994)
L. delbrueckii	AAG GAT AGC ATG TCT GCA	Hertel et al. (1993)
Lc. lactis	ATG CTA GAA TAG GGA ATG AT	Betzl et al. (1990)
L. acidophilus	GAA TCT GTT GGT TCA GCT CGC	Pot et al. (1993)

the subject (Hammes & Vogel 1997; Hammes et al. 1996; Gobbetti, 1998; Röcken & Voysey 1995). In this communication, we present recent advances in the development of tools for the rapid identification and monitoring of the fermentation flora, the characterisation of genes of key metabolic enzymes, and strategies for the conditioning of the recalcitrant organisms for facilitated use in starter preparations.

Molecular taxonomy

The taxonomic classification of lactic acid bacteria from cereal fermentations is not reliable when only classical methods are used such as the determination of sugar fermentation or cell wall composition. Therefore, classification strategies were developed using consecutive application of molecular methods which are useful beyond cereal fermentations. For rapid classification a set of probes was developed recognising specific sequences in the rRNA genes of sourdough lactobacilli (Vogel et al. 1994). Table 1 provides an overview of currently available probes. Nevertheless, the use of probes at the species level is restricted as their specificity may be lost during the discovery of new species sharing the same part of an rRNA sequence as found for *L. pontis* and *L. panis* (Wiese et al. 1996). Alternatively, RAPD allows elucidation of strain biodiversity below the species level. RAPD typing using a single primer has proven useful to reduce the number of isolates to be further characterised by initially detecting genomic similarities (Vogel et al. 1996). The resulting electrophoretic patterns can be clustered and compared to a database. With a given standard procedure these patterns are nicely reproducible and often allow one step classification of a known species. By using this procedure, 120 strains of *L. sanfranciscensis* were assorted to 7 different clusters partly sharing metabolic properties, e.g. arginine metabolism or origin of isolation (data not shown). The coherence of a cluster obtained by RAPD can be confirmed by the use of several different primers and comparison of the cluster patters (Zapparoli et al. 1998). For the confirmation of the phylogenetic position, selected strains from each cluster and strains which are not reacting with a probe or whose patterns can not be assorted to known patterns can be further characterised by (partial) 16S rRNA sequencing. Genomic comparison can also be accomplished by pulse field gel electrophoresis (Zapparoli et al. 1998) or determination of the occurrence and copy number of insertion sequences (IS). Although knowledge about IS activities within lactic acid bacteria stems mostly from investigation with *Lactococcus* (Romero & Klaenhammer 1993), IS isolated from a *Lactobacillus* background are increasingly available. Figure 1 demonstrates the use of IS*153* isolated from *L. sanfranciscensis* TMW1.53 for typing purposes of *L. sanfranciscensis* strains. The overall similarity as well as structural features identify IS*153* as a member of the IS3 family (Schwartz et al. 1988) of bacterial insertion sequences. Analyses of integration sites currently does not suggest any target specificity of insertion.

Table 2 gives an overview about lactic acid bacteria whose presence in types I and II fermentations in significant numbers was proven by the use of molecular methods, namely M13 typing and 16S rRNA sequencing. In quite a few cases strains were discovered which do not share complete 16S rRNA homology to

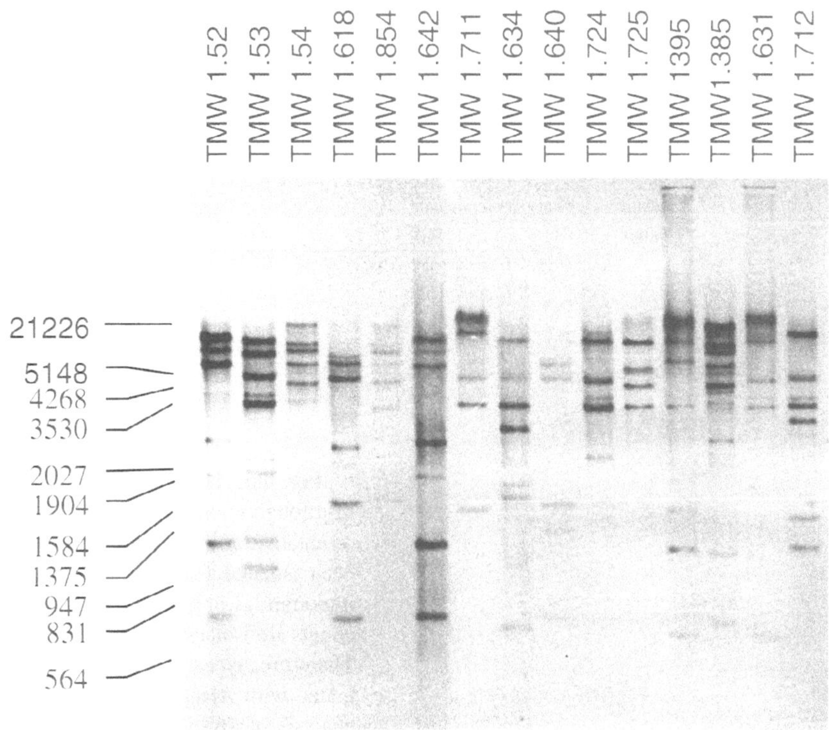

Figure 1. Detection of sequences homologous to IS*153* in *Lactobacillus sanfranciscensis* strains. DNA was digested with *Eco*RV. Hybridization was performed with a biotin labelled PCR fragment (1000 base pairs.)

a known species-type strain. Whereas some of these may justify the description of a new species, others of these 'genomovars' will not always justify this due to the lack of other significant differences. These are listed as e.g. *L. pontis*-like. Whereas traditional type I processes are dominated by the heterofermentative *L. sanfranciscensis* and *L. pontis*, type II sourdough fermentations also contain homofermentative *L. amylovorus*. Other predominant strains in the latter fermentations are closely related to or even identical with species found in the animal and human intestinal tract, i.e. *L. reuteri*, *L. oris*, *L. acidophilus*. Sour wort fermentations used as a barley type II pre-fermentation in beer production typically contain a similar flora as rye and wheat fermentations of this type. These fermentations are dominated by the recently described new species *L. amylolyticus* which is closely related to *L. amylovorus* (Bohak et al. 1998). African and Turkish cereal fermentations using sorghum, maize or rice flours as raw materials have been reported to be dominated by strains of the species of *L. reuteri*, *L. fermentum* and *Leuconostoc* spp (Hancioglu & Karapinar 1997; Olsen et al. 1995; Hamad et al. 1997). The fermentation conditions with respect to time and temperature resemble those of type II fermentations.

Monitoring the microflora

Cereals can not be pasteurised prior to fermentation and therefore the alteration of the process is the only means to induce ecological and metabolic changes. Tools to follow the development of the fermentation flora in dependence on the process parameters are a prerequisite for the standardisation of industrial processes and rapid methods are required to monitor the microflora, mainly lactobacilli, conducting the process (Vogel & Ehrmann 1996). As a rule the number of relevant organisms to be followed in such a process is limited to 2–5, each of which belonging to a different 'genomovar' or even species. Therefore, statistical M13 typing can be a helpful but laborious tool. For sourdough lactobacilli it has however been found that amplification of a specific part of the 16S rRNA gene (bp 7–342 according to the *E. coli* numbering system,

Table 2. Lactobacilli conducting type I and type II sourdough fermentations

Type I		Type II		
Wheat	Rye	Wheat	Rye	Barley
		L. panis		
		L. pontis		
L. sanfranciscensis		*L. reuteri*		
L. pontis		*L. fermentum*		*L. amylolyticus*
L. panis				
		L. delbrueckii		
		L. acidophilus	*L. amylovorus*	
		Lc. lactis		
		L. brevis		

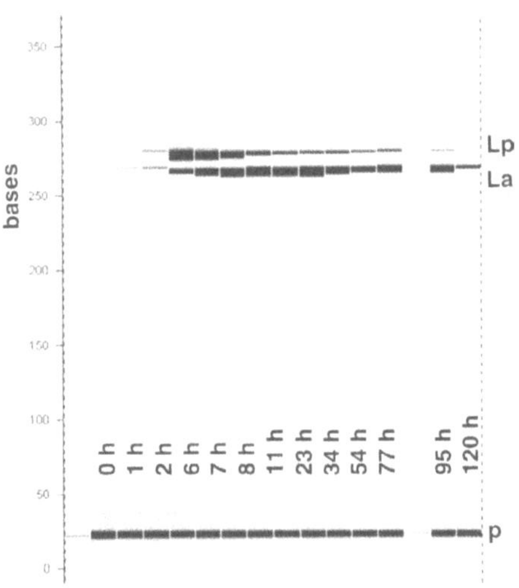

Figure 2. Monitoring a 5-day type II fermentation by ss16S rDNAFLP. La: *Lactobacillus amylovous* amplificate (269 bases); Lp: *Lactobacillus pontis* amplificate (273 bases); p: primers.

Brosius et al. 1981) results in amplicons of different sizes. When PCR amplification is performed with a DNA preparation directly obtained from sourdough, the relative intensity of the resulting amplificates corresponds to the proportion of one strain of the total flora. Figure 2 shows the monitoring of *L. pontis* and *L. amylovorus* in a type II fermentation by determination of this single strand 16S rDNA fragment length polymorphism (ss 16S rDNAFLP). Upon initial proof with statistical M13 typing this method can be used to easily monitor the microflora within a given range of process parameters.

For the characterisation and monitoring of the sourdough microflora it is crucial to prove that a representative collection of all relevant organisms has been isolated from the process. This is even more important as only about 10% of the microscopic cell counts are isolated as CFU even on the richest media. Therefore, types I and II fermentations were investigated with ARDRA, i.e. representative cloning and analysis of 16S rRNA (Weidner et al. 1996; Ulrich & Müller 1998). The clones were screened by RAPD and single(T)-lane sequencing, and no previously unknown patterns were found. This clearly proves that all organisms present in relevant numbers were isolated from these processes.

Metabolism

The metabolism of sourdough lactobacilli has been characterised with respect to their utilisation of maltose (Stolz et al. 1993) electron acceptors (Stolz et al. 1995a,b), arginine metabolism (Hammes et al. 1996) and proteolytic activity (Gobbetti et al. 1996). Following these investigations the competitiveness of heterofermentative lactobacilli in sourdoughs can be explained mainly by their combined use of maltose and electron acceptors (Figure 3). Maltose is continuously delivered by flour amylases or by *L. amylovorus* present in type II doughs and split by maltose phosphorylase without the expenditure of ATP. Additional energy may be generated by the activity of acetate kinase in the presence of electron acceptors which allows the recycling of NAD^+ without the need of ethanol formation. Electron acceptors used by *L. sanfranciscensis* include fructose which is reduced to mannitol (Stolz et al. 1995a,b). Furthermore, NAD^+ can be recycled in the NADH-oxidase reaction. Fructose

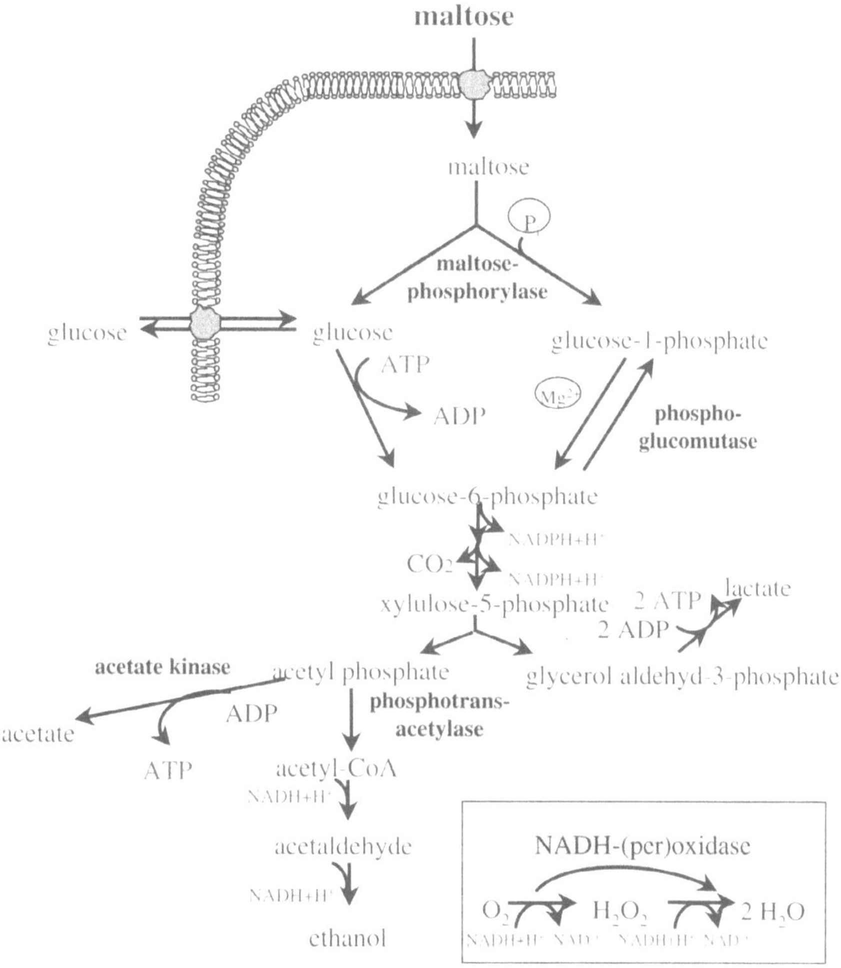

Figure 3. Key reactions in the maltose metabolism of *Lactobacillus sanfranciscensis*. As an example for NAD recycling using electron acceptors the reaction of oxygen is given (for further explanations see text).

is present in the flour in gluco-fructosanes which are degraded by maltose negative *Candida milleri* strains. This yeast usually shares type I cereal environments with the lactobacilli (Yarrow 1978; Mäntynen et al. 1999). While the growth requirements of lactobacilli and yeasts with respect to pH, temperature and organic acids (Gänzle et al. 1998) as well as metabolic interactions between these organisms (Gobbetti 1998) may contribute to their stable coexistence in sourdough, a symbiotic role of these yeasts could further be this supply of fructose (Saunders et al. 1972) which is used as electron acceptor by the lactobacilli to increase their energy yield. The practical relevance of this interaction is the change in the lactate/acetate ratio affecting the baking and sensorial properties of sourdough bread (Röcken et al. 1992, Martinez-Anaya et al. 1994). In the abundance of maltose and under stress conditions *L. sanfranciscensis* and *L. reuteri* release glucose into the medium which originates from the maltose phosphorylase reaction (Neubauer et al. 1994). This glucose may affect the ecological system as it may be metabolised by its producers, by other lactobacilli and by the yeasts and possibly initiate glucose repression in competitors for maltose. The maltose phosphorylase reaction is not repressed by glucose (Stolz et al. 1996). For a review see Hammes et al. (1996).

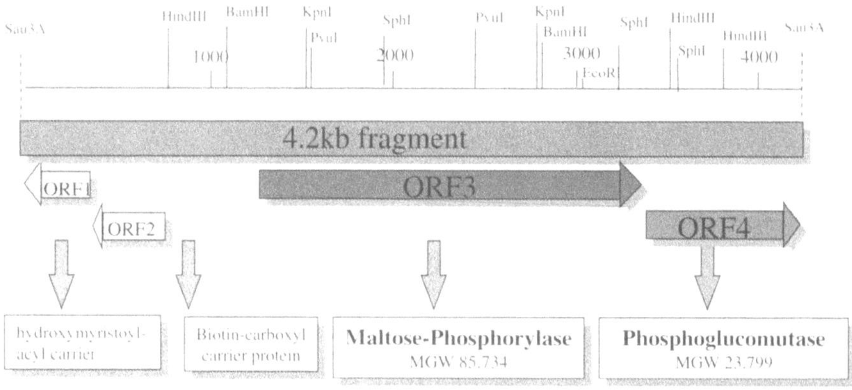

Figure 4. Map of pMap1.0 (4.2 kbp) harbouring key genes of the maltose metabolism of *Lactobacillus sanfranciscensis* DSM20451^T.

Gene cloning

Gene cloning in and from sourdough lactobacilli is at an early stage. By cloning of an α-amylase from *Bacillus subtilis* into *L. sanfranciscensis* it has been shown, that it accepts replicons stemming from other gram positive bacteria and may use their promoter signals (Gobbetti et al. 1995). Very recently the genes of several metabolic key enzymes were cloned and characterised (Ehrmann & Vogel 1998). The maltose degradation operon from *Lactobacillus sanfranciscensis* DSM20451^T contains genes coding for maltose phosphorylase *mapA* and phosphoglucomutase *pgmA*. They were cloned and expressed in *Escherichia coli*. *MapA*, encodes a 754 amino acids polypeptide representing maltose phosphorylase, MapA, with a calculated molecular weight of 85.7 kDa. Following comparative sequence analysis, *mapA* is of a new type distinct from other α-glucosidases. *PgmA* is separated by a single nucleotide from the preceding *mapA* gene indicating translational coupling. Figure 4 shows a map of this operon.

In the heterolactic sugar degradation acetylphosphate, the substrate of acetate kinase and phosphotransacetylase, is a key metabolic intermediate (Figure 3). It represents the branching point of the carbon flux which will lead to the formation of either ethanol or acetate. Therefore, the genes for acetate kinase *ack* and phosphotransacetylase *pta* of *L. sanfranciscensis* DSM20451^T were cloned and characterised. They show 60 and 61% homology with the respective genes of *B. subtilis* and encodes polypeptides with calculated molecular weights of 44,5 and 35,5 kD. *Ack* and *pta* are not located adjacent to each other as observed with other bacteria where the encoded enzymes catalyse the same reactions. After cloning in *E. coli*, expression as biotinylated fusion proteins and cleavage of the fusion protein, the purified enzymes showed acetate kinase and phosphotransacetylase activity. Whether or not the energetically favourable pathway of acetate can be used depends on a recycling of NAD^+ as an electron acceptor. In *L. sanfranciscensis* this can be achieved by the NADH-oxidase reaction. The respective *nox* gene of *L. sanfranciscensis* DSM20451^T encodes a polypeptide with a calculated molecular weight of 49,7 kD. It shares 64% homology to the *Enterococcus faecalis nox* gene and could also be heterologously expressed as a biotinylated fusion protein.

A major problem for the deliberate use of organisms in sourdoughs is their limited resistance to stress conditions occurring upon preparations, e.g. drying. The understanding of the role of specific stress proteins that stabilise organisms at various stress conditions may therefore facilitate the preparation of starter cultures. Therefore, the reaction to stress was determined of *L. pontis* in 2D-electrophoreses. It was demonstrated by western analysis with the respective antibodies that the *Dan*K and *gro*EL proteins are produced at increased levels upon heat shock, stimulating resistance to further stress applications. The *dna*K genes of *L. pontis* and *L. sanfranciscensis* are currently under investigation.

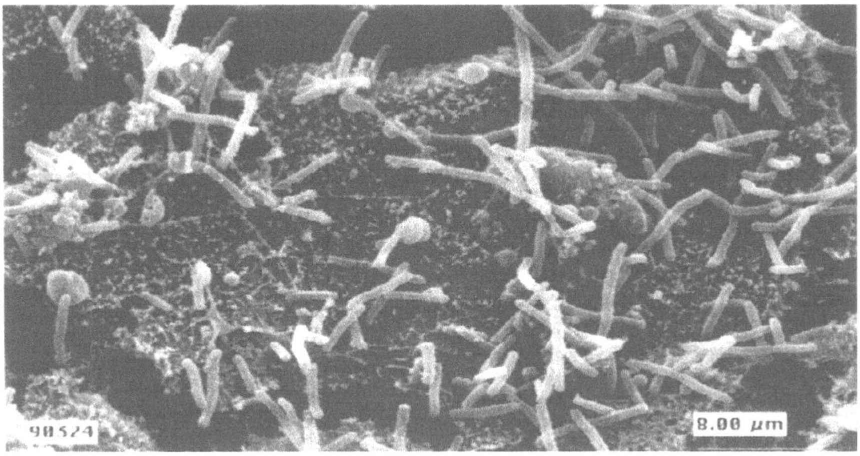

Figure 5. Adhesion of *Lactobacillus amylovorus* TMW1.102 to a human Caco-2 cell monolayer.

Special features

While the metabolism of the sourdough lactobacilli may be considered special as such, there are some properties which deserve mentioning. Several bacteriocins produced by *L. reuteri* have been characterised and strains of *L. sanfranciscensis* have been reported to produce inhibitory activity that was attributed to proteinaceous compounds (Corsetti et al. 1996). Growth of *L. sanfranciscensis* in wheat flour hydrolysate results in formation of acetic and caproic acids inhibitory to moulds (Corsetti et al. 1998). Some strains of *L. reuteri* isolated from sourdoughs produce antagonistic compounds not fitting the general bacteriocin definition. Reuterin (2-hydroxy-propanal) is a product glycerol metabolism of *L. reuteri* inhibitory against a wide range of Gram-positive as well as Gram-negative organisms, and *L. reuteri* LTH2584 produces reutericyclin, a novel antibiotic with a molecular weight of 349 (Gänzle, 1998). This compound is active against a broad range of Gram-positive bacteria including spoilage organisms and pathogens as *Staphylococcus aureus*, *Enterococcus faecalis*, *Listeria monocytogenes* and *Bacillus cereus* as well as vegetative cells and spores of rope forming bacilli. Gram-negative bacteria are not inhibited by reutericyclin due to the permeability barrier of their outer membrane. The optimum pH and temperature for reutericyclin production match the conditions prevailing in sourdough fermentation. Preliminary evidence suggests that it is produced *in situ* in active concentrations.

Strains of this *L. reuteri* and *L. amylovorus* deserve attention as they have the capability to adhere to human intestinal cells (Müller et al. 1998). Figure 5 shows the adhesion of *L. amylovorus* TMW1.102 to human Caco-2 cells. The combination of adhesion and formation of antagonistic compounds is especially tempting for the development of probiotic strains of cereal origin, as *L. reuteri* is a normal part of the human intestinal flora.

Conclusion

The cereal world is a fascinating environment harbouring lactobacilli with a lot of interesting features far beyond sourdough fermentations. The analytical phase of investigations is now entering a constructive one. Still the major handicaps in the deliberate use of these recalcitrant organisms is their difficult handling in technical preparation processes and the lack of an effective cloning system.

Acknowledgements

Parts of the presented original work was supported by the Commission of the European Communities' FAIR programme contract no. CT 961126. Ruth Knorr and Martin Müller are furthermore grateful for stipendia from the 'Technische Universität München' and the 'Stiftung für Industrieforschung e.V.', respectively.

References

Betzl D, Ludwig W & Schleifer KH (1990) Identification of lactococci and enterococci by colony hybridization with 23S rRNA-targeted oligonucleotides, Appl. Environ. Microbiol., 56: 2927–2929

Bohak I, Back W, Richter L, Ehrmann M, Ludwig W & Schleifer KH (1998) *Lactobacillus amylolyticus* sp. nov. isolated from beer malt and beer wort. System. Appl. Microbiol. 21: 360–364

Brosius J, Dull TJ, Sleeter DD & Noller HF (1981) Gene organization and primary structure of a ribosomal RNA operon from *Escherichia coli*. J. Mol. Biol. 148: 107–127

Corsetti A, Gobbetti M & Smacchi E (1996) Antibacterial activity of sourdough lactic acid bacteria: isolation of a bacteriocin-like inhibitory substance from *Lactobacillus sanfrancisco* C57. Food Microbiol. 13: 447–456

Corsetti A, Gobbetti M, Rossi J & Damiani P (1998) Antimould activity of sourdough lactic acid bacteria: identification of a mixture of organic acids produced by *Lactobacillus sanfrancisco* CB1. Appl. Microbiol. Biotechnol. 50: 253–256

Ehrmann A & Vogel RF (1998) Maltose metabolism *of Lactobacillus sanfranciscensis*: cloning and heterologous expession of the key enzymes, maltose phosphorylase and phosphoglucomutase. FEMS Microbiol. Lett 169: 81–86

Gänzle MG, Ehmann M & Hammes WP (1998) Modelling of growth of *Lactobacillus sanfranciscensis* and *Candida milleri* in response to process parameters of the sourdough fermentation. Appl. Environ. Microbiol. 64: 2616–2623

Gänzle M (1998) Useful properties of lactobacilli for application as protective cultures in food. PhD thesis, Universität Hohenheim, Germany

Gobbetti M (1998) Interactions of lactic acid bacteria and yeasta in sourdoughs. Trends Food Sci. Technol. 9: 267–274

Gobetti M, Corsetti A & Rossi J (1996) *Lactobacillus sanfrancisco*, a key sourdough lactic acid bacterium: Physiology, genetic and biotechnology. Adv. Food Sci. 18: 167–175

Gobetti M, Corsetti A, Morelli L & Elli M (1995) Expression of α-amylase gene from *Bacillus stearothermophilus* in *Lactobacillus sanfrancisco*. Biotechnol. Lett. 18: 969–974

Hamad SH, Dieng MC, Ehrmann MA & Vogel RF (1997) Characterization of the bacteial flora of sudanese sorghum flour and sorghum sourdough. J. Appl. Microbiol. 83: 764–770

Hammes WP, Stolz P & Gänzle M (1996) Metabolism of lactobacilli in traditional sourdoughs. Adv. Food Sci. 18: 176–184

Hammes WP & Vogel RF (1997) Mikrobiologie von Sauerteig. in: Müller G, Holzapfel W & Weber H (Eds) Mikrobiologie der Lebensmittel: Lebensmittel pfanzlicher Herkunft. Behr's Verlag, Hamburg

Hancioglu Ö & Karapinar M (1997) Microflora of Boza, a traditional fermented Turkish beverage. Int. J. Food Microbiol. 35: 271–274

Hertel C, Ludwig W, Pot B, Kersters K & Schleifer KH (1993) Differentiation of lactobacilli occurring in fermented milk products by using oligonucleotide probes and electrophoretic protein profiles, System. Appl. Microbiol. 16: 453–467

Mäntynen VH, Korhola M, Gudmundsson H, Turkainen H, Alfredsson GA, Salovaara H & Lindström K (1999) A polyphasic study on the taxonomic position of industrial sour dough yeasts. System. Appl. Microbiol. 22: 87–96

Martinez-Anaya MA, Llin Z, Macias MP & Collar C (1994) Regulation of acetic acid production by homo- and heterofermentative lactobacilli in whole wheat sour-doughs. Z. Lebensm. Unters. Forsch. 199: 186–190

Mora D, Fortina MG, Parini P & Manachini PL (1997) Identification of *Pediococcus acidilactici* and *Pediococcus pentosaceus* based on 16S rRNA and *ldh*D gene-targeted multiplex PCR analysis. FEMS Microbiol. Lett. 231–236

Müller MRA, Rouvet M, Brassart D, Böcker G, Ehrmann MA & Vogel RF (1998) Adhesion of *Lactobacillus* strains from cereal fermentations to human intestinal cells. Int. Dairy J. 8: 584

Neubauer H, Glaasker E, Hammes WP, Poolmann B & Konings W (1994) Mechanisms of maltose uptake and glucose excretion on *Lactobacillus sanfrancisco*. J. Bacteriol. 176: 3007–3012

Olsen A, Halm M & Jakobsen K (1995) The antimicrobial activity of lactic acid bacteria from fermented maize (kenkey) and their interactions during fermentation. J. Appl. Bacteriol. 79: 506–512

Pot B, Hertel C, Ludwig W, Descgeemaeker P, Kersters K & Schleifer KH (1993) Identification and classification of *Lactobacillus acidophilus, L. gasseri* and *Lactobacillus johnsonii* strains by SDS-PAGE and rRNA-targeted oligonucleotide probe hybridization, J. Gen. Microbiol. 139: 513–517

Röcken W, Rick M & Reinkemeier MZ (1992) Controlled production of acetic acid in wheat sour doughs. Lebensm. Unters. Forsch. 195: 259–263

Röcken W & Voysey PA (1995) Sour-dough fermentation in bread making. J. Appl. Bacteriol. 79: 38S–48S

Romero DA & Klaenhammer TR (1993) Transposable elements in lactococci: A review. J. Dairy Sci. 76: 1–19

Saunders RM, Ng H & Kline L (1972) The sugars of flour and their involvement in the San Francisco sour dough French bread process. II. Isolation and characterization of undescribed bacteria responsible for the souring activity. Appl. Microbiol. 21: 459–465

Schwartz E, Kroeger M & Rak B (1988) Distribution, nucleotide sequence and phylogenetic relationships of a new *Escherichia coli* insertion element. Nucleic Acid Res. 16: 6789–6802

Stolz P, Böcker G, Hammes WP & Vogel RF (1995a) Utilization of electron acceptors by lactobacilli isolated from sourdough. I. *Lactobacillus sanfrancisco*. Z. Lebensm. Unters. Forsch. 201: 91–96

Stolz P, Böcker G, Vogel RF & Hammes WP (1993) Utilisation of maltose and glucose by lactobacilli isolated from sourdough. FEMS Microbiol. Lett. 109: 237–242

Stolz P, Hammes WP & Vogel RF (1996) Maltose-phosphorylase and hexokinase activity in lactobacilli from traditionally prepared sourdoughs. Adv. Food Sci. 18: 1–6

Stolz P, Vogel RF & Hammes WP (1995b) Utilization of electron acceptors by lactobacilli isolated from sourdough. II. *Lactobacillus pontis, L. reuteri, L. amylovorus,* and *L. fermentum*. Z. Lebensm. Unters. Forsch. 201: 402–410

Ulrich A & Müller T (1998) Heterogeneity of plant-associated streptococci as characterized by phenotypic features and restriction analysis of PCR-amplified 16S rDNA, In: J. Appl. Microbiol. 84: 293–303

Vogel RF & Ehrmann M (1996) Genetics of lactobacilli in food fermentations. In: El-Gewely MR (Ed) Biotechnology Annual Review II pp. 123–150

Vogel RF, Böcker G, Stolz P, Ehrmann M, Fanta D, Ludwig W, Pot B, Kersters K, Schleifer KH & Hammes WP (1994) Identification of lactobacilli from sourdough and description of *Lactobacillus pontis* sp. nov. Int. J. System. Bacteriol. 44: 223–229

Vogel RF, Müller M, Stolz P & Ehrmann M (1996) Ecology in sourdoughs produced by traditional and modern technologies. Adv. Food. Sci. 18: 152–159

Weidner S, Arnold W & Pühler A (1996) Diversity of uncultured microorganisms associated with the seagrass *Halophila stipu-*

lacea estimated by restriction fragment length polymorphism analysis of PCR-amplified 16S rRNA genes, in: Appl. Environ. Microbiol. 62: 766–771

Wiese B, Strohmar W, Rainey FA & Diekmann H (1996) *Lactobacillus panis* sp. nov., from sourdough with a long fermentation period. Int. J. System. Bacteriol. 46: 449–453

Yarrow D (1978) *Candida milleri* sp. nov. Int. J. System. Bacteriol. 28: 608–610

Zapparoli G, Torriani S & Dellaglio F (1998) Differentiation *of Lactobacillus sanfranciscensis* strains by randomly amplified polymorphic DNA and pulsed field gel electrophoresis. FEMS Microbiol. Lett. 166: 325–332

MIX
Papier aus verantwortungsvollen Quellen
Paper from responsible sources
FSC® C105338

If you have any concerns about our products,
you can contact us on
ProductSafety@springernature.com

In case Publisher is established outside the EU,
the EU authorized representative is:
Springer Nature Customer Service Center GmbH
Europaplatz 3, 69115 Heidelberg, Germany

Printed by Libri Plureos GmbH
in Hamburg, Germany